Joachim Weiss

Handbook of Ion Chromatography

Further Titles of Interest:

G. Gauglitz, T. Vo-Dinh (Eds.)

Handbook of Spectroscopy

2 Volumes
2003, ISBN 3-527-29782-0

H. Günzler, A. Williams (Eds.)

Handbook of Analytical Techniques

2 Volumes
2001, ISBN 3-527-30165-8

J. S. Fritz, D. T. Gjerde

Ion Chromatography

2000, ISBN 3-527-29914-9

Joachim Weiss

Handbook of Ion Chromatography

Third, completely revised and updated edition

Volume 1

Translated by Tatjana Weiss

WILEY-VCH

WILEY-VCH Verlag GmbH & Co. KGaA

Dr. Joachim Weiss
Dionex GmbH
Am Wörtzgarten 10
65510 Idstein
Germany

■ All books published by Wiley-VCH are carefully produced. Nevertheless, author and publisher do not warrant the information contained in these books, including this book, to be free of errors. Readers are advised to keep in mind that statements, data, illustrations, procedural details or other items may inadvertently be inaccurate.

Library of Congress Card No. applied for.

British Library Cataloguing-in-Publication Data:
A catalogue record for this book is available from the British Library.

Bibliographic information published by Die Deutsche Bibliothek
Die Deutsche Bibliothek lists this publication in the Deutsche Nationalbibliografie; detailed bibliographic data is available in the Internet at <http://dnb.ddb.de>.

© 2004 WILEY-VCH Verlag GmbH & Co. KGaA, Weinheim

Printed on acid-free paper.

All rights reserved (including those of translation into other languages). No part of this book may be reproduced in any form − by photoprinting, microfilm, or any other means − nor transmitted or translated into a machine language without written permission from the publishers. Registered names, trademarks, etc. used in this book, even when not specifically marked as such, are not to be considered unprotected by law.

Composition   pagina media gmbh, Hemsbach
Printing   betz-druck gmbh, Darmstadt
Bookbinding   J. Schäffer GmbH i.G., Industrie- und Verlagsbuchbinderei, Grünstadt

Printed in the Federal Republic of Germany.
ISBN   3-527-28701-9

Foreword

Over the past few years, ion chromatography has developed into a significant chromatographic technique for ion analysis within the field of separation science. In addition to publications on the basic principles, it is above all the applied research and the broad applicability of ion chromatography that has made this technique practically indispensable for the analytical chemist. In view of the ever-growing number of publications in this field, the numerous international conferences, as well as the diversity of applications, qualitative, quantitative, and quality-assured data acquisition become extremely important. It is thus of great importance to have a fundamental book on ion chromatography, such as this one by Dr. Joachim Weiss.

Today, in addition to the classical inorganic ions, the field of ion chromatography and the economical significance of this method cover such organic ionic compounds as organic acids, carbohydrates, and glycoproteins, to name but a few.

The previous books by Dr. Weiss in the field of ion chromatography are already regarded as classics. This new edition again accommodates all possible needs in terms of the science and industrial applications involved in this technique.

The volume describes in great detail the theoretical principles, considerations regarding the separation mechanisms, new stationary phases for anion- and cation-exchange chromatography, ion-exclusion chromatography, and ion-pair chromatography along with new detection methods. Even the statistical data acquisition together with the related fundamental principles, which is indispensable nowadays for a modern certified analytical laboratory, is outlined, as are special applications important, for example, to the semiconductor industry.

I trust that the present book, succeeding the previous successful works by Dr. Joachim Weiss, will once again gain recognition as an important reference work in science, the laboratory, and industry.

I wish the author and the publishing house much success with this auspicious work.

Innsbruck, January 2004 *Dr. Günther Bonn*
Professor for Analytical Chemistry
University of Innsbruck (Austria)

Table of Contents

Foreword *V*

Preface *XIII*

Volume 1

1	**Introduction** *1*	
1.1	Historical Perspective *1*	
1.2	Types of Ion Chromatography *3*	
1.3	The Ion Chromatographic System *5*	
1.4	Advantages of Ion Chromatography *7*	
1.5	Selection of Separation and Detection Systems *9*	
2	**Theory of Chromatography** *13*	
2.1	Chromatographic Terms *13*	
2.1.1	Asymmetry Factor A_s *14*	
2.2	Parameters for Assessing the Quality of a Separation *15*	
2.2.1	Resolution *15*	
2.2.2	Selectivity *16*	
2.2.3	Capacity Factor *17*	
2.3	Column Efficiency *17*	
2.4	The Concept of Theoretical Plates (Van-Deemter Theory) *19*	
2.5	Van-Deemter Curves in Ion Chromatography *24*	
3	**Anion Exchange Chromatography (HPIC)** *27*	
3.1	General Remarks *27*	
3.2	The Ion-Exchange Process *27*	
3.3	Thermodynamic Aspects *29*	
3.4	Stationary Phases *34*	
3.4.1	Polymer-Based Anion Exchangers *35*	
3.4.1.1	Styrene/Divinylbenzene Copolymers *35*	
3.4.1.2	Polymethacrylate and Polyvinyl Resins *47*	

3.4.2	Latex-Agglomerated Anion Exchangers	54
3.4.2.1	Review of Different Latex-Agglomerated Anion Exchangers	57
3.4.3	Silica-Based Anion Exchangers	86
3.4.4	Other Materials for Anion Separations	86
3.4.4.1	Crown Ether and Cryptand Phases	86
3.4.4.2	Alumina Phases	95
3.5	Eluants for Anion Exchange Chromatography	98
3.6	Suppressor Systems in Anion Exchange Chromatography	103
3.6.1	Suppressor Columns	104
3.6.2	Hollow Fiber Suppressors	108
3.6.3	Micromembrane Suppressors	111
3.6.4	Self-Regenerating Suppressors	115
3.6.5	Suppressors with Monolithic Suppression Beds	120
3.7	Anion Exchange Chromatography of Inorganic Anions	121
3.7.1	Overview	121
3.7.2	General Parameters Affecting Retention	123
3.7.3	Experimental Parameters Affecting Retention When Applying Suppressor Systems	123
3.7.3.1	Choice of Eluant	124
3.7.3.2	Eluant Concentration and pH Value	137
3.7.3.3	Influence of Organic Solvents	144
3.7.4	Experimental Parameters Affecting Retention When Applying Direct Conductivity Detection	145
3.7.4.1	Choice of Eluant	145
3.7.4.2	System Peaks	153
3.7.4.3	Eluant Concentration and pH Value	156
3.7.5	Polarizable Anions	159
3.8	Anion Exchange Chromatography of Organic Anions	168
3.8.1	Organic Acids	168
3.8.2	Polyvalent Anions	184
3.9	Gradient Elution Techniques in Anion Exchange Chromatography of Inorganic and Organic Anions	191
3.9.1	Theoretical Aspects	192
3.9.2	Choice of Eluants	194
3.9.3	Possibilities for Optimizing Concentration Gradients	201
3.9.4	Isoconductive Techniques	204
3.10	Carbohydrates	205
3.10.1	Sugar Alcohols	209
3.10.2	Monosaccharides	212
3.10.3	Oligosaccharides	226
3.10.4	Polysaccharides	231
3.10.5	Carbohydrates Derived from Glycoproteins	236

3.10.5.1	Compositional and Structural Analysis 240
3.10.5.2	Chosen Examples 254
3.11	Amino Acids 257
3.12	Proteins 267
3.13	Nucleic Acids 275

4	**Cation Exchange Chromatography (HPIC)** 279
4.1	Stationary Phases 279
4.1.1	Polymer-Based Cation Exchangers 280
4.1.1.1	Styrene/Divinylbenzene Copolymers 280
4.1.1.2	Ethylvinylbenzene/Divinylbenzene Copolymers 282
4.1.1.3	Polymethacrylate and Polyvinyl Resins 296
4.1.2	Latexed Cation Exchangers 298
4.1.3	Silica-Based Cation Exchangers 303
4.2	Eluants in Cation Exchange Chromatography 308
4.3	Suppressor Systems in Cation Exchange Chromatography 310
4.3.1	Suppressor Columns 310
4.3.2	Hollow Fiber Suppressors 311
4.3.3	Micromembrane Suppressors 312
4.3.4	Self-Regenerating Suppressors 313
4.3.5	Suppressors with Monolithic Suppression Beds 318
4.4	Cation Exchange Chromatography of Alkali Metals, Alkaline-Earth Metals, and Amines 318
4.5	Transition Metal Analysis 325
4.5.1	Basic Theory 326
4.5.2	Transition Metal Analysis with Non-Suppressed Conductivity Detection 330
4.5.3	Transition Metal Analysis with Spectrophotometric Detection 331
4.6	Analysis of Polyamines 344
4.7	Gradient Techniques in Cation Exchange Chromatography of Inorganic and Organic Cations 349

5	**Ion-Exclusion chromatography (HPICE)** 359
5.1	The Ion-Exclusion Process 359
5.2	Stationary Phases 361
5.3	Eluants in Ion-Exclusion Chromatography 365
5.4	Suppressor Systems in Ion-Exclusion Chromatography 366
5.5	Analysis of Inorganic Acids 368
5.6	Analysis of Organic Acids 372
5.7	HPICE/HPIC-Coupling 375
5.8	Analysis of Alcohols and Aldehydes 379

5.9	Amino Acid Analysis	382
5.9.1	Separation of Amino Acids	384
5.9.2	Post-Column Derivatizations of Amino Acids	389
5.9.3	Sample Preparation	391

6 Ion-Pair Chromatography (MPIC) 393

6.1	Survey of Existing Retention Models	394
6.2	Suppressor Systems in Ion-Pair Chromatography	399
6.3	Experimental Parameters that Affect Retention	400
6.3.1	Type and Concentration of Lipophilic Counter Ions in the Mobile Phase	400
6.3.2	Type and Concentration of the Organic Modifier	403
6.3.3	Inorganic Additives	406
6.3.4	pH Effects and Temperature Influence	408
6.4	Analysis of Surface-Inactive Ions	409
6.5	Analysis of Surface-Active Ions	424
6.6	Applications of the Ion-Suppression Technique	440
6.7	Applications of Multi-Dimensional Ion Chromatography on Multimode Phases	444

7 Detection Methods in Ion Chromatography 461

7.1	Electrochemical Detection Methods	461
7.1.1	Conductivity Detection	461
7.1.1.1	Theoretical Principles	462
7.1.1.2	Application Modes of Conductivity Detection	469
7.1.2	Amperometric Detection	474
7.1.2.1	Fundamental Principles of Voltammetry	475
7.1.2.2	Amperometry	478
7.2	Spectrometric Detection Methods	498
7.2.1	UV/Vis Detection	498
7.2.1.1	Direct UV/Vis Detection	498
7.2.1.2	UV/Vis Detection in Combination with Derivatization Techniques	499
7.2.1.3	Indirect UV Detection	510
7.2.2	Fluorescence Detection	514
7.3	Other Detection Methods	521
7.4	Hyphenated Techniques	522
7.4.1	IC-ICP Coupling	523
7.4.2	IC-MS Coupling	530

Index   I 1

Volume 2

8 Quantitative Analysis 549
8.1 General 549
8.2 Analytical Chemical Information Parameters 550
8.3 Determination of Peak Areas 551
8.3.1 Manual Determination of Peak Areas and Peak Heights 552
8.3.2 Electronic Peak Area Determination 554
8.4 Statistical Quantities 557
8.4.1 Mean Value 558
8.4.2 Standard Deviation 558
8.4.3 Scatter and Confidence Interval 559
8.5 Calibration of an Analytical Method (Basic Calibration) 560
8.5.1 Acquisition of the Calibration Function 561
8.5.1.1 Method Characteristic Parameters of a Linear Calibration Function 561
8.5.1.2 Method Parameters of a Calibration Function of 2nd Degree 564
8.5.2 Testing of the Basic Calibration 565
8.5.3 Testing the Precision 566
8.5.3.1 Homogeneity of Variances 566
8.5.3.2 Outlier Tests 567
8.5.4 Calibration Methods 569
8.5.4.1 Area Normalization 569
8.5.4.2 Internal Standard 570
8.5.4.3 External Standard 571
8.5.4.4 Standard Addition 572
8.6 Detection Criteria, Limit of Detection, Limit of Determination 574
8.6.1 Determination of Detection Criteria, Limit of Detection 574
8.7 The System of Quality Control Cards 578
8.7.1 Types of Quality Control Cards and their Applications 579

9 Applications 587
9.1 Ion Chromatography in Environmental Analysis 588
9.2 Ion Chromatography in Power Plant Chemistry 626
9.2.1 Analysis of Conditioned Waters 634
9.2.2 Cooling Water Analysis 640
9.2.3 Flue Gas Scrubber Solutions 644
9.2.4 Analysis of Chemicals 648
9.3 Ion Chromatography in the Semiconductor Industry 651
9.3.1 High-Purity Water Analysis 652
9.3.2 Surface Contaminations 654
9.3.3 Solvents 658

9.3.4	Acids, Bases, and Etching Agents	664
9.3.5	Other Applications	673
9.4	Ion Chromatography in the Electroplating Industry	678
9.4.1	Analysis of Inorganic Anions	679
9.4.2	Analysis of Metal Complexes	686
9.4.3	Analysis of Organic Acids	688
9.4.4.	Analysis of Inorganic Cations	689
9.4.5	Analysis of Organic Additives	690
9.5	Ion Chromatography in the Detergent and Household Product Industry	697
9.5.1	Detergents	697
9.5.2	Household Products	704
9.6	Ion Chromatography in the Food and Beverage Industry	711
9.6.1	Beverages	713
9.6.2	Dairy Products	733
9.6.3	Meat Processing	739
9.6.4	Baby Food	742
9.6.5	Groceries and Luxuries	746
9.6.6	Sweeteners	752
9.7	Ion Chromatography in the Pharmaceutical Industry	756
9.7.1	Fermentation	772
9.8	Ion Chromatography in Clinical Chemistry	781
9.9	Oligosaccharide Analysis of Membrane Coupled Glycoproteins	797
9.10	Other Applications	803
9.11	Sample Preparation and Matrix Problems	823

Index 871

Preface to the Third Edition

Precisely ten years have passed since the publication of the second edition of this book in 1994. Over this decade the method of ion chromatography has rapidly developed and been further established. One can name the many new separator columns with partly extraordinary selectivities and separation power in this connection. Nowadays, new grafted polymers, compatible with large volume injections, enable ion analysis down to the sub-µg/L range without time-consuming pre-concentration. Of particular importance is the development of continuous and contamination-free eluant generation by means of electrolysis, which considerably facilitates the use of gradient elution techniques in ion chromatography. Furthermore, only de-ionized water is used as a carrier with ion chromatography systems configured for the use of such technology. Along with this development, hydroxide eluants, which are particularly suitable for concentration gradients in anion exchange chromatography, are increasingly replacing the classical carbonate/bicarbonate buffers predominantly used so far. In contrast to carbonate/bicarbonate buffers, which will still be used for relatively simple applications, higher sensitivities are now achieved with hydroxide eluants. This trend is supported by an exciting development of hydroxide-selective stationary phases. In line with classical liquid chromatography, hyphenation with atomic spectrometry and molecular spectrometry, such as ICP and ESI-MS, is becoming increasingly important in ion chromatography, too. The section *Hyphenated Techniques* underlines this importance. Because carbohydrates, proteins, and oligonucleotides are also analyzed by ion-exchange chromatography, but only carbohydrates were included in the second edition, the sections *Proteins* and *Oligonucleotides* have been added in this new edition. In combination with integrated amperometry as a direct detection method, ion-exchange chromatography also revolutionized amino acid analysis. Thus, in many application areas ion chromatography has become almost indispensable for the analysis of inorganic and organic anions and cations.

Since the publication of the second edition, all these developments have made it necessary to rewrite major parts, such that this third edition can be confidently regarded as a new text. Almost every chapter has been renewed or significantly revised. For better clarity, the previous chapter *Ion-Exchange Chromatography* is now split into the chapters *Anion Exchange Chromatography* and *Cation Exchange Chromatography*. The remaining structure of this book proved to be of value, and has thus remained unchanged. The sections *Carbohydrates Derived from Glycoproteins* (Section 3.10.5), *Proteins* (Section 3.12), *Nucleic Acids* (Section 3.13), and *Oligosaccharide Analysis of Membrane-Coupled Glycoproteins* (Section 9.9) have been completely rewritten by an expert, Dr. Dietrich Hauffe, to whom I would like to express my sincere gratitude. The chapter Quantitative Analysis was also rewritten and expanded with information on validation parameters and quality control cards. In addition, the chapters on detection and applications were significantly expanded with new material and with numerous practical examples in the form of chromatograms, while applications of ion chromatography in the petrochemical industry and in the pulp and paper industry were also added.

The objective for this third edition is the same as for the previous two editions: The author addresses analytical chemists, who wish to familiarize themselves with this method, as well as practitioners who employ these techniques on a day-to-day basis and are looking for a reference book that can help to facilitate method development and provide an overview on existing applications.

At this point, I would like to express my sincere gratitude to many of my colleagues in all parts of the world, who contributed their experience and knowledge to the preparation of this third edition. I am particularly grateful to Dr. Detlef Jensen (Germany) for his willingness to always discuss the various aspects of ion chromatography and for his valuable suggestions, and to Jennifer Kindred (Sunnyvale, USA) for her incredible effort, patience, and diligence in editing the translated manuscript. I would be grateful for any criticisms or suggestions that could serve to improve future editions of this book.

Finally, many thanks to my wife and children who for quite some time, with an amazing amount of understanding and tolerance, did not see much of their husband or dad, who spent many evenings, weekends, and public holidays at the computer.

Idstein, February 2004   *Joachim Weiss*

1
Introduction

1.1
Historical Perspective

"Chromatography" is the general term for a variety of physico-chemical separation techniques, all of which have in common the distribution of a component between a mobile phase and a stationary phase. The various chromatographic techniques are subdivided according to the physical state of these two phases.

The discovery of chromatography is attributed to Tswett [1, 2], who in 1903 was the first to separate leaf pigments on a polar solid phase and to interpret this process. In the following years, chromatographic applications were limited to the distribution between a solid stationary and a liquid mobile phase (**L**iquid **S**olid **C**hromatography, LSC). In 1938, Izmailov and Schraiber [3] laid the foundation for **T**hin **L**ayer **C**hromatography (TLC). Stahl [4, 5] refined this method in 1958 and developed it into the technique known today. In their noteworthy paper of 1941, Martin and Synge [6] proposed the concept of theoretical plates, which was adapted from the theory of distillation processes, as a formal measurement of the efficiency of the chromatographic process. This approach not only revolutionized the understanding of liquid chromatography, but also set the stage for the development of both gas chromatography (GC) and paper chromatography.

In 1952, Martin and James [7] published their first paper on gas chromatography, initiating the rapid development of this analytical technique.

High **P**erformance **L**iquid **C**hromatography (HPLC) was derived from the classical column chromatography and, besides gas chromatography, is one of the most important tools of analytical chemistry today. The technique of HPLC flourished after it became possible to produce columns with packing materials made of very small beads (≈ 10 µm) and to operate them under high pressure. The development of HPLC and the theoretical understanding of the separation processes rest on the basic works of Horvath [8], Knox [9], Scott [10], Snyder [11], Guiochon [12], Möckel [13], and others.

Ion **C**hromatography (IC) was introduced in 1975 by Small, Stevens, and Bauman [14] as a new analytical method. Within a short period of time, ion chromatography evolved from a new detection scheme for a few selected inorganic anions and cations to a versatile analytical technique for ionic species in general. For a sensitive detection of ions via their electrical conductance, the separator column effluent was passed through a "suppressor" column. This suppressor column chemically reduces the eluant background conductance, while at the same time increasing the electrical conductance of the analyte ions.

In 1979, Fritz et al. [15] described an alternative separation and detection scheme for inorganic anions, in which the separator column is directly coupled to the conductivity cell. As a prerequisite for this chromatographic setup, low capacity ion-exchange resins must be employed, so that low ionic strength eluants can be used. In addition, the eluant ions should exhibit low equivalent conductances, thus enabling sensitive detection of the sample components.

At the end of the 1970s, ion chromatographic techniques were used to analyze organic ions for the first time. The requirement for a quantitative analysis of organic acids brought about an ion chromatographic method based on the ion-exclusion process that was first described by Wheaton and Bauman [16] in 1953.

The 1980s witnessed the development of high efficiency separator columns with particle diameters between 5 µm and 8 µm, which resulted in a significant reduction of analysis time. In addition, separation methods based on the ion-pair process were introduced as an alternative to ion-exchange chromatography, because they allow the separation and determination of both anions and cations.

Since the beginning of the 1990s column development has aimed to provide stationary phases with special selectivities. In inorganic anion analysis, stationary phases were developed that allow the separation of fluoride from the system void and the analysis of the most important mineral acids as well as oxyhalides such as chlorite, chlorate, and bromate in the same run [17]. Moreover, high-capacity anion exchangers are under development that will enable analysis of, for example, trace anionic impurities in concentrated acids and salinary samples. Problem solutions of this kind are especially important for the semiconductor industry, sea water analysis, and clinical chemistry. In inorganic cation analysis, simultaneous analysis of alkali- and alkaline-earth metals is of vital importance, and can only be realized within an acceptable time frame of 15 minutes by using weak acid cation exchangers [18]. Of increasing importance is the analysis of aliphatic amines, which can be carried out on similar stationary phases by adding organic solvents to the acid eluant.

The scope of ion chromatography was considerably enlarged by newly designed electrochemical and spectrophotometric detectors. A milestone of this development was the introduction of a pulsed amperometric detector in 1983, allowing a very sensitive detection of carbohydrates, amino acids, and divalent sulfur compounds [19, 20].

A growing number of applications utilizing post-column derivatization in combination with photometric detection opened the field of polyphosphate, polyphosphonate, and transition metal analysis for ion chromatography, thus providing a powerful extension to conventional titrimetric and atomic spectrometry methods.

These developments made ion chromatography an integral part of both modern inorganic and organic analysis.

Even though ion chromatography is still the preferred analytical method for inorganic and organic ions, meanwhile, ion analyses are also carried out with capillary electrophoresis (CE) [21], which offers certain advantages when analyzing samples with extremely complex matrices. In terms of detection, only spectrometric methods such as UV/Vis and fluorescence detection are commercially available. Because inorganic anions and cations as well as aliphatic carboxylic acids cannot be detected very sensitively or cannot be detected at all, applications of CE are rather limited as compared to IC, with the universal conductivity detection being employed in most cases.

Dasgupta et al. [22] as well as Avdalovic et al. [23] independently succeeded to miniaturize a conductivity cell and a suppressor device down to the scale required for CE. Since the sensitivity of conductivity detection does not suffer from miniaturization, detection limits achieved for totally dissociated anions and low molecular weight organics compete well with those of ion chromatography techniques. Thus, capillary electrophoresis with suppressed conductivity detection can be regarded as a complementary technique for analyzing small ions in simple and complex matrices.

1.2
Types of Ion Chromatography

This book only discusses separation methods which can be summarized under the general term *Ion Chromatography*. Modern ion chromatography as an element of liquid chromatography is based on three different separation mechanisms, which also provide the basis for the nomenclature in use.

Ion-Exchange Chromatography (HPIC)
(**H**igh **P**erformance **I**on **C**hromatography)
This separation method is based on ion-exchange processes occurring between the mobile phase and ion-exchange groups bonded to the support material. In highly polarizable ions, additional non-ionic adsorption processes contribute to the separation mechanism. The stationary phase consists of polystyrene, ethylvinylbenzene, or methacrylate resins co-polymerized with divinylbenzene and modified with ion-exchange groups. Ion-exchange chromatography is used for the separation of both inorganic and organic anions and cations. Separation of anions is accomplished with quaternary ammonium groups attached to the polymer, whereas sulfonate-, carboxyl-, or phosphonate groups are used as ion-exchange sites for the separation of cations. Chapters 3 and 4 deal with this type of separation method in greater detail.

Ion-Exclusion Chromatography (HPICE)
(High Performance Ion Chromatography Exclusion)

The separation mechanism in ion-exclusion chromatography is governed by Donnan exclusion, steric exclusion, sorption processes and, depending on the type of separator column, by hydrogen bonding. A high-capacity, totally sulfonated cation exchange material based on polystyrene/divinylbenzene is employed as the stationary phase. In case hydrogen bonding should determine selectivity, significant amounts of methacrylate are added to the styrene polymer. Ion-exclusion chromatography is particularly useful for the separation of weak inorganic and organic acids from completely dissociated acids which elute as one peak within the void volume of the column. In combination with suitable detection systems, this separation method is also useful for determining amino acids, aldehydes, and alcohols. A detailed description of this separation method is given in Chapter 5.

Ion-Pair Chromatography (MPIC)
(Mobile Phase Ion Chromatography)

The dominating separation mechanism in ion-pair chromatography is adsorption. The stationary phase consists of a neutral porous divinylbenzene resin of low polarity and high specific surface area. Alternatively, chemically bonded octadecyl silica phases with even lower polarity can be used. The selectivity of the separator column is determined by the mobile phase. Besides an organic modifier, an ion-pair reagent is added to the eluant (water, aqueous buffer solution, etc.) depending on the chemical nature of the analytes. Ion-pair chromatography is particularly suited for the separation of surface-active anions and cations, sulfur compounds, amines, and transition metal complexes. A detailed description of this separation method is given in Chapter 6.

Alternative Methods

In addition to the three classical separation methods mentioned above, reversed-phase liquid chromatography (RPLC) can also be used for the separation of highly polar and ionic species. Long-chain fatty acids, for example, are separated on a chemically bonded octadecyl phase after protonation in the mobile phase with a suitable aqueous buffer solution. This separation mode is known as ion suppression [24].

Chemically bonded aminopropyl phases have also been successfully employed for the separation of inorganic ions. Leuenberger et al. [25] described the separation of nitrate and bromide in foods on such a phase using a phosphate buffer solution as the eluant. Separations of this kind are limited in terms of their applicability, because they can only be applied to UV-absorbing species.

Moreover, applications of multidimensional ion chromatography utilizing multimode phases are very interesting, too. In those separations, ion-exchange and reversed-phase interactions equally contribute to the retention mechanism of ionic and polar species [26]. These alternative techniques are also described in Chapter 6.

1.3
The Ion Chromatographic System

The basic components of an ion chromatograph are shown schematically in Fig. 1-1. It resembles the setup of conventional HPLC systems.

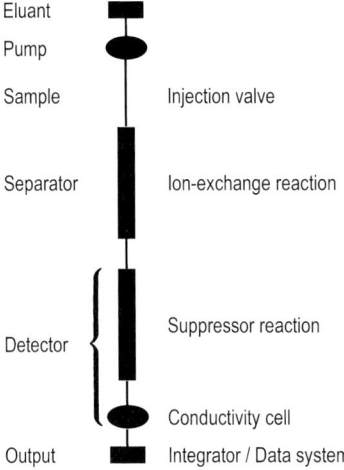

Figure 1-1. Basic components of an ion chromatograph.

A pump delivers the mobile phase through the chromatographic system. In general, either single-piston or dual-piston pumps are employed. A pulse-free flow of the eluant is necessary for employing sensitive UV/Vis and amperometric detectors. Therefore, pulse dampers are used with single-piston pumps and a sophisticated electronic circuitry with dual-piston pumps.

The sample is injected into the system via a loop injector, as schematically shown in Fig. 1-2. A three-way valve is required, with two ports being connected to the sample loop. The sample loading is carried out at atmospheric pressure. After switching the injection valve, the sample is transported to the separator column by the mobile phase. Typical injection volumes are between 5 µL and 100 µL.

The most important part of the chromatographic system is the separator column. The choice of a suitable stationary phase (see Section 1.5) and the chromatographic conditions determine the quality of the analysis. The column tubes are manufactured from inert material such as Tefzec, epoxy resins, or PEEK

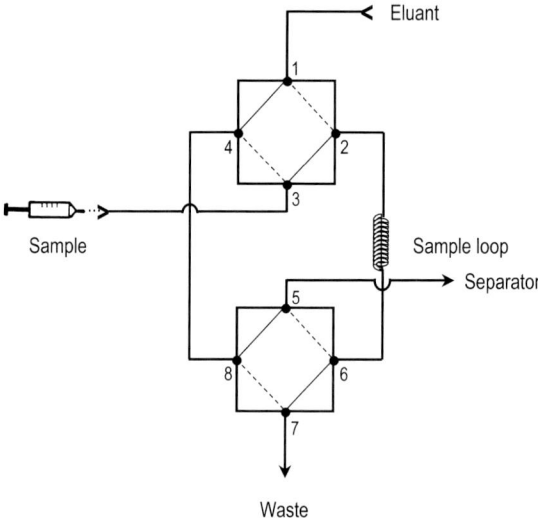

Figure 1-2. Schematic representation of a loop injector.

(polyether ether ketone). In general, separation is achieved at room temperature. Only in very few cases – for example for the analysis of long-chain fatty acids – an elevated temperature is required to improve analyte solubility. An elevated column temperature is also recommended for the analysis of polyamines in order to improve peak efficiencies.

The analytes are detected and quantified by a detection system. The performance of any detector is evaluated according to the following criteria:
- Sensitivity
- Linearity
- Resolution (detector cell volume)
- Noise (detection limit)

The most commonly employed detector in ion chromatography is the conductivity detector, which is used with or without a suppressor system. The main function of the suppressor system as part of the detection unit is to *chemically* reduce the high background conductivity of the electrolytes in the eluant, and to convert the sample ions into a more conductive form. In addition to conductivity detectors, UV/Vis, amperometric, and fluorescence detectors are used, all of which are described in detail in Chapter 7.

The chromatographic signals can be displayed on a recorder. Quantitative results are obtained by evaluating peak areas or peak heights, both of which are proportional to the analyte concentration over a wide range. This was traditionally performed using digital integrators which are connected directly to the analog signal output of the detector. Due to low computer prices and lack of GLP/GLAP conformity, digital integrators are hardly used anymore. Modern detectors feature an additional parallel interface (e.g., RS-232-C), that enables the connection to a personal computer or a host computer with a suitable chromatography software. Computers also take over control functions, thus allowing a fully automated operation of the chromatographic system.

Because corrosive eluants such as diluted acids and bases are often used in ion chromatography, all parts of the chromatographic system being exposed to these liquids should be made of inert, metal-free materials. Conventional HPLC systems with tubings and pump heads made of stainless steel are only partially suited for ion chromatography, because even stainless steel is eventually corroded by aggressive eluants. Considerable contamination problems would result, because metal ions exhibit a high affinity towards the stationary phase of ion exchangers, leading to a significant loss of separation efficiency. Moreover, metal parts in the chromatographic fluid path would make the analysis of orthophosphate, complexing agents, and transition metals more difficult.

1.4 Advantages of Ion Chromatography

The determination of ionic species in solution is a classical analytical problem with a variety of solutions. Whereas in the field of cation analysis both fast and sensitive analytical methods (AAS, ICP, polarography, and others) have been available for a long time, the lack of corresponding, highly sensitive methods for anion analysis is noteworthy. Conventional wet-chemical methods such as titration, photometry, gravimetry, turbidimetry, and colorimetry are all labor-intensive, time-consuming, and occasionally troublesome. In contrast, ion chromatography offers the following advantages:

- Speed
- Sensitivity
- Selectivity
- Simultaneous detection
- Stability of the separator columns

Speed

The time necessary to perform an analysis becomes an increasingly important aspect, because enhanced manufacturing costs for high quality products and additional environmental efforts have lead to a significant increase in the number of samples to be analyzed.

With the introduction of high efficiency separator columns for ion-exchange, ion-exclusion, and ion-pair chromatography in recent years, the average analysis time could be reduced to about 10 minutes. Today, a baseline-resolved separation of the seven most important inorganic anions [27] requires only three minutes.

Therefore, quantitative results are obtained in a fraction of the time previously required for traditional wet-chemical methods, thus increasing the sample throughput.

Sensitivity

The introduction of microprocessor technology, in combination with modern high efficiency stationary phases, makes it a routine task to detect ions in the medium and lower µg/L concentration range without pre-concentration. The detection limit for simple inorganic anions and cations is about 10 µg/L based on an injection volume of 50 µL. The total amount of injected sample lies in the lower ng range. Even ultrapure water, required for the operation of power plants or for the production of semiconductors, may be analyzed for its anion and cation content after pre-concentration with respective concentrator columns. With these pre-concentration techniques, the detection limit could be lowered to the ng/L range. However, it should be emphasized that the instrumentation for measuring such incredibly low amounts is rather sophisticated. In addition, high demands have to be met in the creation of suitable environmental conditions. The limiting factor for further lowering the detection limits is the contamination by ubiquitous chloride and sodium ions.

High sensitivities down to the pmol range are also achieved in carbohydrate and amino acid analysis by using integrated pulsed amperometric detection.

Selectivity

The selectivity of ion chromatographic methods for analyzing inorganic and organic anions and cations is ensured by the selection of suitable separation and detection systems. Regarding conductivity detection, the suppression technique is of vital importance, because the respective counter ions of the analyte ions as a potential source of interferences are exchanged against hydronium and hydroxide ions, respectively. A high degree of selectivity is achieved by using solute-specific detectors such as a UV/Vis detector to analyze nitrite in the presence of high amounts of chloride. New developments in the field of post-column derivatization show that specific compound classes such as transition metals, alkaline-earth metals, polyvalent anions, silicate, etc. can be detected with high selectivity. Such examples explain why sample preparation for ion chromatographic analyses usually involves only a simple dilution and filtration of the sample. This high degree of selectivity facilitates the identification of unknown sample components.

Simultaneous Detection

A major advantage of ion chromatography – especially in contrast to other instrumental techniques such as photometry and AAS – is its ability to simultaneously detect multiple sample components. Anion and cation profiles may be obtained within a short time; such profiles provide information about the sample composition and help to avoid time-consuming tests. However, the ability of ion chromatographic techniques for simultaneous quantitation is limited by extreme concentration differences between various sample components. For example, the major and minor components in a wastewater matrix may only be detected simultaneously if the concentration ratio is <1000:1. Otherwise, the sample must be diluted and analyzed in a separate chromatographic run.

Stability of the Separator Columns

The stability of separator columns very much depends on the type of the packing material being used. In contrast to silica-based separator columns commonly used in conventional HPLC, resin materials such as polystyrene/divinylbenzene copolymers prevail as support material in ion chromatography. The high pH stability of these resins allows the use of strong acids and bases as eluants, which is a prerequisite for the wide-spread applicability of this method. Strong acids and bases, on the other hand, can also be used for rinsing procedures. Meanwhile, most organic polymers are compatible with organic solvents such as methanol and acetonitrile, which can be used for the removal of organic contaminants (see also Chapter 9). Hence, polymer-based stationary phases exhibit a low sensitivity towards complex matrices such as wastewater, foods, or body fluids, so that a simple dilution of the sample with de-ionized water prior to filtration is often the only sample preparation procedure.

1.5 Selection of Separation and Detection Systems

As previously mentioned, a wealth of different separation techniques is summarized under the term "ion chromatography". Therefore, what follows is a survey of criteria for selecting stationary phases and detection modes being suitable for solving specific separation problems.

The analyst usually has some information regarding the nature of the ion to be analyzed (inorganic or organic), its surface activity, its valency, and its acidity or basicity, respectively. With this information and on the basis of the selection criteria outlined schematically in Table 1-1, it should not be difficult for the analytical chemist to select a suitable stationary phase and detection mode. In many cases, several procedures are feasible for solving a specific separation problem. In these cases, the choice of the analytical procedure is determined by the type of matrix, the simplicity of the procedure, and, increasingly, by financial aspects. Two examples illustrate this:

Various sulfur-containing species in the scrubber solution of a flue-gas desulfurization plant (see also Section 9.2) are to be analyzed. According to Table 1-1, non-polarizable ions such as sulfite, sulfate, and amidosulfonic acid with pK values below 7, are separated isocratically by HPIC using a conventional anion exchanger and are detected via electrical conductivity. A suppressor system may be used to increase the sensitivity and specificity of the procedure. Often, scrubber solutions also contain thiocyanate and thiosulfate in small concentrations. However, due to their polarizability, these anions exhibit a high affinity towards the stationary phase of conventional anion exchangers. Three different approaches are feasible for the analysis of such anions. A conventional anion exchanger may be used with a high ionic strength mobile phase. Depending on the analyte concentration, difficulties with the sensitivity of the subsequent conductivity detection may arise. Alternatively, a special methacrylate-based anion exchanger with hydrophilic functional groups may be employed. Polarizable anions are not adsorbed as strongly on this kind of stationary phases and, therefore, elute together with non-polarizable anions. Taking into account that other sulfur-containing species such as dithionate may also have to be analyzed, a gradient elution technique has to be employed, which allows *all* compounds mentioned above to be separated in a single run utilizing a high efficiency separator column and conductivity detection. However, the required concentration gradient makes the use of a suppressor system inevitable. Concentration gradients on anion exchangers reach the limit when extremely polarizable anions such as nitrilotrisulfonic acid have to be analyzed. In this case, ion-pair chromatography (MPIC) is the better separation mode, because organic solvents added to the mobile phase determine analyte retention.

A second example is the determination of organic acids in soluble coffee. According to Table 1-1, aliphatic carboxylic acids are separated by HPICE on a totally sulfonated cation exchange resin with subsequent conductivity detection. While this procedure is characterized by a high selectivity for aliphatic monocarboxylic acids with a small number of carbon atoms, sufficient separation cannot be obtained for the aliphatic open-chain and cyclic hydroxy acids that are also present in coffee. Only after introducing a new stationary phase with specific selectivity for hydroxycarboxylic acids did it become possible to separate the most important representatives of this class of compounds in such a matrix. Ion-exclusion chromatography is not suited for the separation of aromatic carboxylic acids, which are present in coffee in large numbers. Examples are ferulic acid, caffeic acid, and the class of chlorogenic acids. Due to π-π-interactions with the aromatic rings of the organic polymers used as support material for the stationary phase, aromatic acids are strongly retained and, thus, cannot be analyzed. A good separation is achieved by reversed-phase chromatography using chemically bonded octadecyl phases with high chromatographic efficiencies. These compounds are then detected by measuring their light absorption at 254 nm.

Further details on the selection of separation and detection modes are given in Chapters 3 to 6.

1.5 Selection of Separation and Detection Systems

Table 1-1. Schematic representation of selection criteria for separation and detection modes.

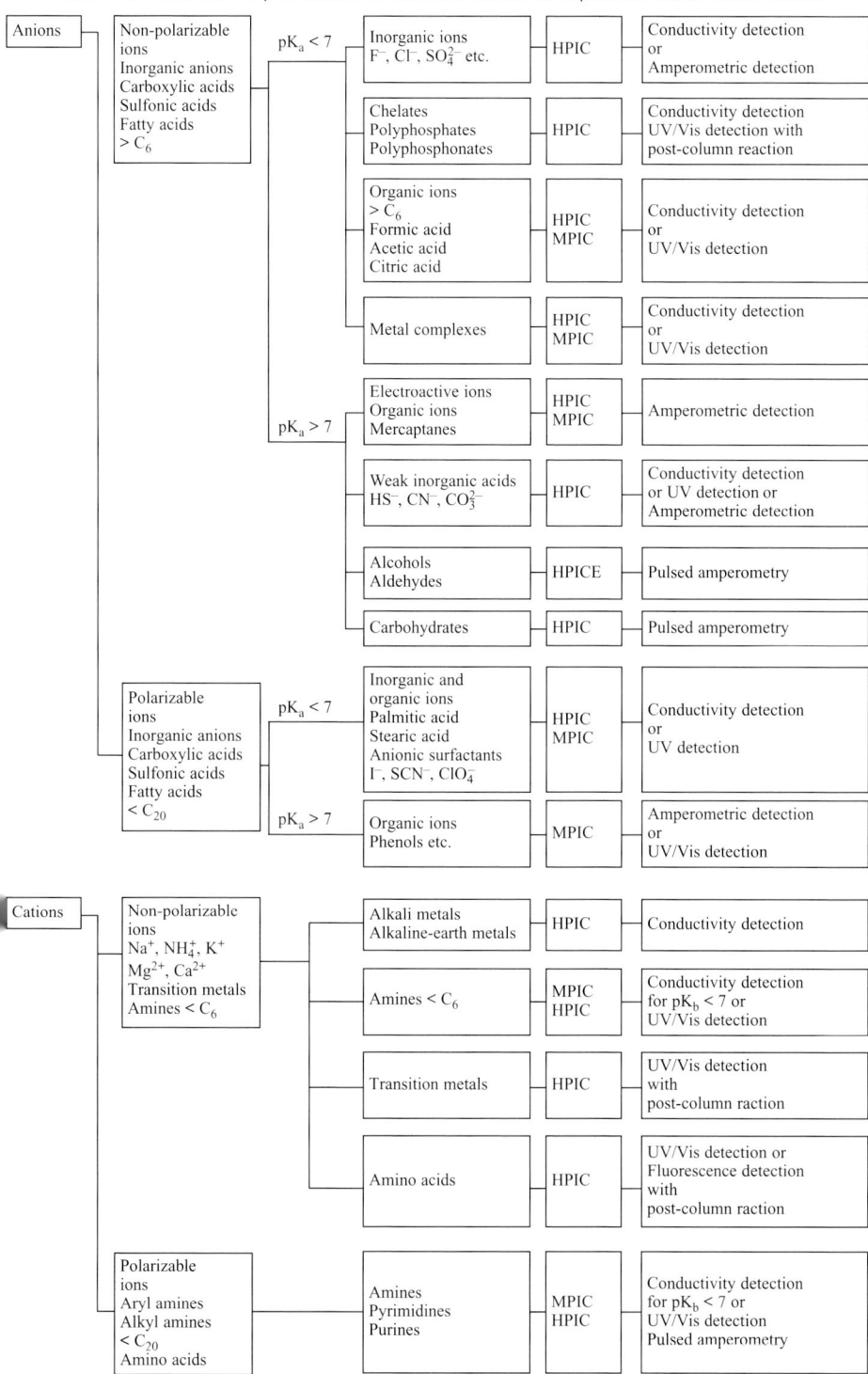

2
Theory of Chromatography

2.1
Chromatographic Terms

Chromatographic signals are usually registered in form of a chromatogram. A typical chromatogram is schematically depicted in Fig. 2-1:

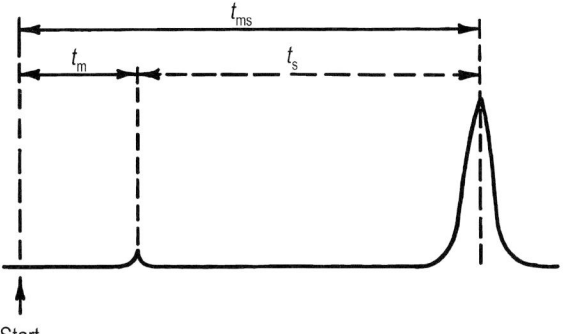

Fig. 2-1. General illustration of a chromatogram.

Two different components are separated in a chromatographic column only if they spend different times in or at the stationary phase. The time in which the components do not travel along the column is called the solute retention time, t_s. The column dead time, t_m, is defined as the time necessary for a non-retained component to pass through the column. The gross retention time, t_{ms}, is calculated from the solute retention time and the column dead time:

$$t_{ms} = t_m + t_s \tag{1}$$

The chromatographic terms for the characterization of a separator column can be seen in Fig. 2-1.

In a first approximation, the shape of a chromatographic peak is described by a Gaussian curve (Fig. 2-2).

The peak height at any given position x can be derived from Eq. (2):

$$y = \frac{1}{\sigma \cdot \sqrt{2\pi}} \cdot e^{-\frac{(x-y)^2}{2\sigma^2}} \cdot Y_0 \tag{2}$$

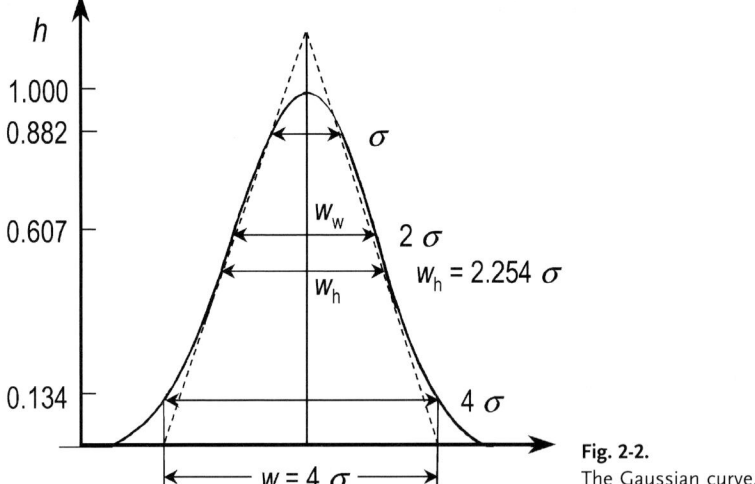

Fig. 2-2. The Gaussian curve.

σ	Standard deviation (half the peak width at the point of inflexion)
$y = h$	Peak height at maximum
x	Any particular point within the peak
μ	Position of peak maximum
w	Peak width at the baseline as determined by the intersection points of the tangents drawn to the peak above its points of inflection

2.1.1
Asymmetry Factor A_s

The signals (called the peaks) due to elution of a species from a chromatographic column are rarely perfectly Gaussian. Normally, the peaks are asymmetrical to some extent, this is expressed by Eq. (3) (see also Fig. 2-3):

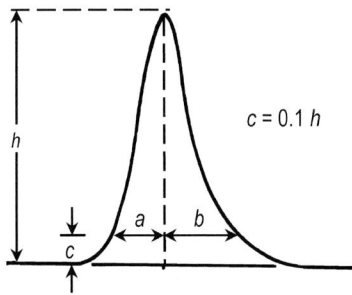

Fig. 2-3. Definition of the asymmetry factor.

$$A_s = \frac{b}{a} \tag{3}$$

At A_s-values higher than 1, this asymmetry is called "tailing". The peak shape is characterized by a rapid increase of the chromatographic signal followed by a comparatively slow decrease. Adsorption processes are mainly responsible for such tailing effects. At A_s-values lower than 1, the asymmetry is called "leading" or "fronting". This effect is characterized by a slow increase of the signal followed by a fast decrease. The leading effect occurs if the stationary phase does not have a sufficient number of suitable adsorption sites and, hence, some of the sample molecules (or sample ions) pass the peak center. For practical applications, separator columns are considered to be good when the asymmetry factor is between 0.9 and 1.2.

2.2
Parameters for Assessing the Quality of a Separation

2.2.1
Resolution

The objective of a chromatographic analysis is to separate the components of a mixture into separate bands. The resolution R of two neighboring peaks is defined as the quotient of the difference of the two peak maxima (expressed as the difference between the gross retention times) and the arithmetic mean of their respective peak widths, w, at the peak base.

$$R = \frac{t_{ms_1} - t_{ms_2}}{\frac{w_1 + w_2}{2}} \quad (4)$$

$$= \frac{2\Delta t_{ms}}{w_1 + w_2}$$

t_{ms_1}, t_{ms_2} Gross retention times for signals 1 and 2, respectively
w_1, w_2 Peak widths at baseline, as determined by the intersection points of the tangents drawn to the peak above its points of inflection

As shown in Fig. 2-4, these parameters can be obtained directly from the chromatogram. If the peaks exhibit a Gaussian peak shape, a resolution of $R = 2.0$ (corresponding to an 8σ-separation) is sufficient for quantitative analysis. Thus, the two peaks are completely resolved to baseline, because peak width at the base is given by Eq. (5):

$$w = 4\,\sigma \quad (5)$$

Higher values of R would result in excessively prolonged analysis times. At a resolution of $R = 0.5$, two sample components can still be recognized as separate peaks.

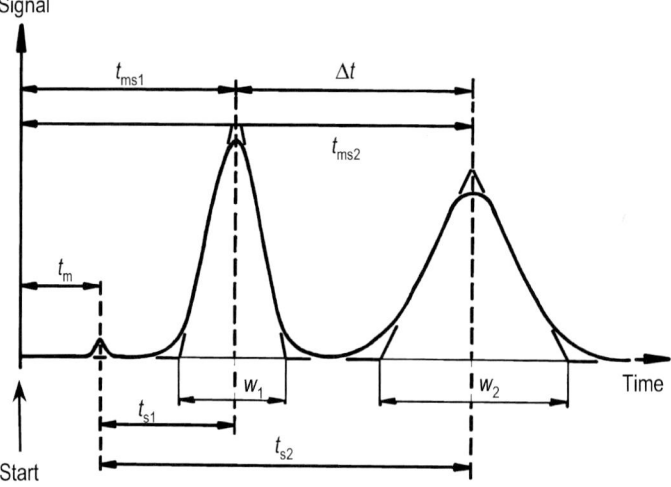

Fig. 2-4. Parameter for assessing resolution and selectivity.

2.2.2
Selectivity

The decisive parameter for the separation of two components is their relative retention, which is called selectivity α. The selectivity is defined as the ratio of the solute retention times of two different signals, as shown in Eq. (6):

$$\alpha = \frac{t_{s_2}}{t_{s_1}}$$

$$= \frac{t_{ms_2} - t_m}{t_{ms_1} - t_m}$$

(6)

According to Fig. 2-4, these parameters may also be obtained from the chromatogram. The selectivity is determined by the properties of the stationary phase. In HPLC, the selectivity is also affected by the mobile phase composition. If $\alpha = 1$, there are no thermodynamic differences between the two sample components under the given chromatographic conditions; therefore, no separation is possible. At equilibrium, which is established in reasonably close approximation in chromatography, the selectivity α is a thermodynamic quantity. At a constant temperature, selectivity α depends on the specific properties of the sample components to be separated and on the properties of the mobile and stationary phases being used.

2.2.3
Capacity Factor

The capacity factor, k, is the product of the phase ratio Φ between stationary and mobile phases in the separator column and the Nernst distribution coefficient, K, as shown in Eq. (7):

$$\begin{aligned} k &= K \cdot \frac{V_s}{V_m} \\ &= \frac{C_s}{C_m} \cdot \frac{V_s}{V_m} \\ &= \frac{t_{ms} - t_m}{t_m} \\ &= \frac{t_s}{t_m} \end{aligned} \quad (7)$$

K Nernst distribution coefficient
V_s Volume of the stationary phase
V_m Volume of the mobile phase
C_s Solute concentration in the stationary phase
C_m Solute concentration in the mobile phase

The capacity factor is independent of the equipment being used, and is a measure of the column's ability to retain a sample component. Small values of k imply that the respective component elutes near the void volume; thus the separation will be poor. High values of k, on the other hand, are tantamount to longer analysis times, peak broadening, and a decrease in sensitivity.

2.3
Column Efficiency

A fundamental disadvantage of chromatography is the broadening of the sample component zone during its passage through the separation system. Peak broadening is caused by diffusion processes and flow processes. Peak broadening can be measured by the plate number, N, or the plate height, H.

The height of a theoretical plate, in which the distribution equilibrium of sample molecules between stationary and mobile phases is established, is related to the plate number via the length of the separator column, as shown in Eq. (8):

$$H = \frac{L}{N} \quad (8)$$

N Plate number
L Length of the separator column

Based on chromatographic data, the theoretical plate height, H, which is defined as the ratio of the peak variance and the column length, L, can be calculated via Eq. (9):

$$H = \frac{\sigma^2}{L}$$

$$= \frac{1}{16} \cdot \frac{w^2}{L} \qquad (9)$$

$$= \frac{1}{8 \ln 2} \left(\frac{b_{0.5}}{L}\right)^2$$

$b_{0.5}$ Peak width at half the peak height
w Peak width at the baseline between the tangents drawn to the peak above the points of inflection

The term "8 ln 2" arises from the approximation of a peak as a Gaussian curve. Using Eq. (8), the number of theoretical plates is shown in Eq. (10):

$$N = \frac{L^2}{\sigma^2}$$

$$= \frac{t_{ms}^2}{\sigma_t^2} \qquad (10)$$

Two sample components may only be separated from each other if their k values are different. The effective plate number, N_{eff}, or the effective plate height, H_{eff}, is used to describe the separation efficiency of a column, as shown in Eqs. (11) and (12):

$$N_{eff} = 5.54 \cdot \left(\frac{t_s}{b_{0.5}}\right)^2 \qquad (11)$$

$$H_{eff} = \frac{L}{N_{eff}} \qquad (12)$$

The resolution of a column can be coupled to multiple parameters including: (i) efficiency of the separator column, (ii) selectivity, and (iii) capacity factor in a single equation. These parameters are defined by:

$$R = \underbrace{\frac{1}{4}\sqrt{N}}_{a} \cdot \underbrace{\left(\frac{\alpha - 1}{\alpha}\right)}_{b} \cdot \underbrace{\left(\frac{k}{k+1}\right)}_{c} \qquad (13)$$

a Term for the column efficiency
b Term for the column selectivity
c Term for the capacity factor

Eq. (13) may be reordered, as shown in Eq. (14), so that the plate number being required to afford the desired resolution may be calculated for any given values of k and α:

$$N = 16 R^2 \cdot \left(\frac{k+1}{k}\right) \cdot \left(\frac{\alpha}{\alpha-1}\right)^2 \tag{14}$$

2.4
The Concept of Theoretical Plates (Van-Deemter Theory)

Based on the work of Martin and Synge [1], van Deemter, Zuiderweg, and Klinkenberg [2] introduced the concept of the theoretical plate height, H, as a measure for relative peak broadening in correlation with the terminology used in distillation technology. In general form, the plate height, H, is given by Eq. (15):

$$H = A + \frac{B}{v} + C \cdot v \tag{15}$$

(Van-Deemter equation)

v Linear flow velocity of the mobile phase in cm/s

The individual terms of the sum vary depending on the mobile phase velocity v.
Accordingly, term A is independent of the flow velocity and characterizes the peak dispersion caused by the Eddy diffusion. This effect considers the different pathways for solute molecules in the column packing. The longitudinal diffusion is described by the term B/v. The term $C \cdot v$ comprises the lateral diffusion and the resistance to mass-transfer between mobile and stationary phase. These effects depend linearly on the flow velocity.

With the peak width expressed in terms of length units σ_1, it follows by taking Eq. (8):

$$N = \left(\frac{L}{\sigma_1}\right)^2 \tag{16}$$

or

$$H = \frac{\sigma_1}{L} \tag{17}$$

Various portions σ_1 contribute to the broadening of a peak. The sum of their variances σ_i^2 gives the total band spreading:

$$\sigma_1^2 = \sum_{i=1}^{n} \sigma_i^2 \tag{18}$$

The portion of the peak broadening which arises due to the irregularity of the column packing is called Eddy diffusion. It is approximated by Eq. (19):

$$\sigma_E^2 = 2\lambda \cdot d_p \tag{19}$$

λ Coefficient describing the quality of the packing
d_p Particle diameter

All molecules present in the mobile phase at time t_m may diffuse in and against the flow direction. The contribution of the longitudinal diffusion in the mobile phase is described by Eq. (20):

$$\sigma_L^2 = \frac{2\gamma_m \cdot D_m}{v} \tag{20}$$

γ_m Obstruction factor, which takes into account the obstruction of the free longitudinal diffusion due to collisions with particles of the column packing
D_m Diffusion coefficient in the mobile phase
v Linear velocity of the mobile phase in cm/s

Lateral diffusion and resistance to mass-transfer are the predominating effects for the total peak broadening and, thus, mainly determine the efficiency of the separator column. Any one sample molecule which interacts with the column packing, diffuses back and forth between the stationary and the mobile phase. It is retained at the stationary phase and, therefore, trails the center of the peak, which passes through the separator column. This effect is illustrated in Fig. 2-5 [3]. In the mobile phase, on the other hand, the sample molecule travels with the eluant. The mass-transfer effect causes a peak broadening, because sample molecules pass through the column ahead of, as well as behind, the peak center. Due to the eluant flow through the column, the equilibrium between the solute concentration in the mobile phase and in the adjacent stationary phase is not attained. Both phases contribute to the resistance to mass-transfer. Therefore, two terms must be considered for the peak broadening. The peak broadening by the stationary phase is given by:

$$\sigma_s^2 = \frac{f \cdot k_e \cdot d_f^2}{D_s} \cdot v \tag{21}$$

f Form factor for the stationary phase
d_f Thickness of the stationary phase
D_s Diffusion coefficient of the solute in the stationary phase

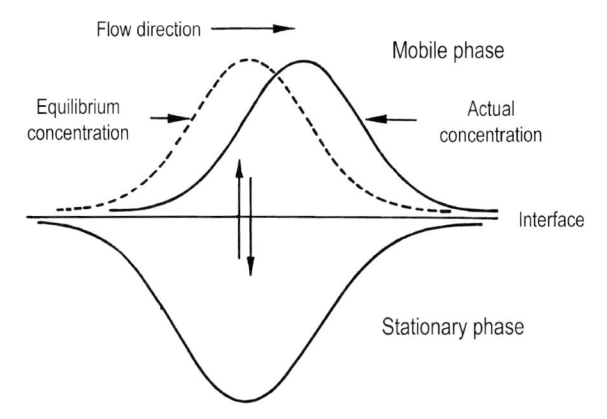

Fig. 2-5. Illustration of the mass-transfer effect.

Instead of the capacity factor, k, the capacity ratio, k_e, which is calculated by Eq. (22), is used in the equation above:

$$k_e = \frac{V_R - V_e}{V_e} \tag{22}$$

V_R Retention volume of the solute
V_e Exclusion volume of the column

The surface-functionalized column packings used in ion chromatography minimize the contribution to the total peak broadening caused by the stationary phase, because solute molecules are not able to penetrate into the packing material.

The dependence of the peak broadening on the mobile phase is given by Eq. (23):

$$\sigma_m^2 = \frac{\omega \cdot d_p^2}{D_m} \cdot v \tag{23}$$

The column coefficient, ω, is a measure for the regularity of the packing. The contribution σ_m^2 to the total peak broadening is significant, but may be substantially influenced by decreasing the particle diameter, d_p. The use of eluants with low viscosity also leads to a reduction in the peak broadening, σ_m^2, because the diffusion coefficient of the solutes in the mobile phase increases accordingly.

The total plate height, H, may be expressed by combining Equations (19) to (23):

$$H = 2\lambda \cdot d_p + \frac{2\gamma_m \cdot D_m}{v} + \frac{f \cdot k_e \cdot d_f^2}{D_s} \cdot v + \frac{\omega \cdot d_p^2}{D_m} \cdot v \tag{24}$$

The first experimental results published by Keulemans and Kwantes [4] confirmed the applicability of the equation to gas chromatography, but it soon became apparent that the equation introduced by van Deemter et al. is only of limited validity for liquid chromatography.

In 1961, Giddings [5] proposed a HETP equation, which may be considered as a special case of the Van-Deemter equation:

$$H = \frac{A}{1 + E/v} + \frac{B}{v} + C \cdot v \tag{25}$$

Giddings' main criticism of the Van-Deemter equation was that a finite contribution to the peak broadening by the Eddy diffusion term is predicted even for zero flow velocity. However, at the flow velocities encountered in practical applications, the equation proposed by Giddings reduces to the Van-Deemter equation, because all other terms remain the same.

The equation introduced in 1967 by Huber and Hulsman [6] accounts for this phenomenon.

$$H = \frac{A}{1 + E/v^{1/2}} + \frac{B}{v} + C \cdot v + D \cdot v^{1/2} \tag{26}$$

They introduced a coupling term, which causes the Eddy diffusion term to vanish if the flow velocity approaches zero. In contrast to van Deemter and Giddings, the resistance to mass-transfer in the mobile phase is described by an additional term $D \cdot v^{1/2}$. However, this factor resembles the coupling term proposed by Giddings in its physical interpretation and in its dependence on the flow velocity.

In the early 1970s, Knox et al. [7, 8] suggested another HETP equation based on their extensive data. Their equation differed significantly from the equation discussed thus far. This equation was derived by curve fitting on the authors' extensive data:

$$H = A \cdot v^{1/3} + \frac{B}{v} + C \cdot v \tag{27}$$

Finally, Horvath and Lin [9, 10] developed an equation very similar to the one introduced by Huber and Hulsman:

$$H = \frac{A}{1 + E/v^{1/3}} + \frac{B}{v} + C \cdot v + D \cdot v^{2/3} \tag{28}$$

The only difference between the two equations is the description of the resistance-to-mass-transfer effect, which Horvath et al. interpret to depend on the square of the cubic root of the flow velocity instead of a quadratic root dependence.

Although the various HETP equations differ significantly from each other, Scott et al. [11] showed that, on the basis of their extensive experimental data, the dependence of the theoretical plate height H on the linear flow velocity may be satisfactorily described by the Van-Deemter equation in the range between 0.02 cm/s and 1 cm/s. In Scott's tests, porous silicas with four different particle sizes were employed as stationary phases, on which nine solute compounds were separated using six different eluant mixtures.

Although the experimental data for H and v may be depicted by any hyperbolic function, not all of them provide a meaningful physical insight into the dispersion process. According to Scott et al. [11], the Van-Deemter equation in the form

$$H = 2\lambda \cdot d_p + \frac{2\gamma_m \cdot D_m}{v} + \frac{(a + b \cdot k_e + c \cdot k_e^2) \cdot d_p^2 \cdot v}{24(1 + k_e)^2 \cdot D_m} \tag{29}$$

is applicable to liquid chromatography at normal operating conditions. The coefficients λ and γ are numbers that describe the quality of the packing; in case of well-packed columns, they are between 0.5 and 0.8. The coefficients a, b, and c were calculated by the authors to be 0.37, 4.69, and 4.04, respectively.

The coupling term introduced by Giddings seems to be particularly significant only for surface-functionalized packings. This is due to the low porosity of these materials, which reduces the volume of mobile phase being retained in the packing. Hence, the term describing the resistance-to-mass-transfer is smaller, so that the mass-transfer effect between the particles gains significance.

Instead of the total plate height H and the linear flow velocity v, often the reduced plate height, h,

$$h = \frac{H}{d_p} \tag{30}$$

and the reduced flow velocity, u, are used.

$$u = \frac{v \cdot d_p}{D_m} \tag{31}$$

The graphical representation of $\ln h$ as a function of $\ln u$ (Fig. 2-6) is known in the literature as the Knox plot. The dependence of the curve's position on the retention of the compound is disadvantageous. Minima in this kind of illustration are only obtained for compounds having no retention ($k = 0$).

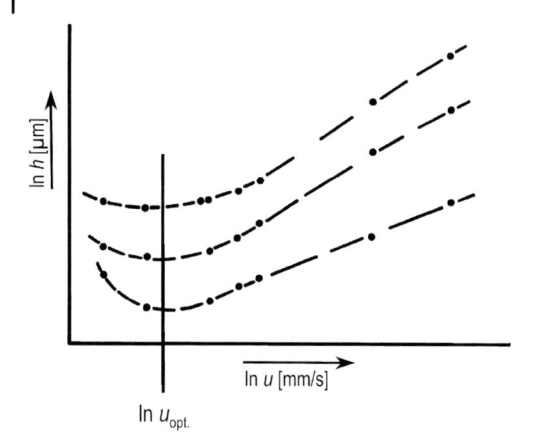

Fig. 2-6. General illustration of a Knox-plot.

2.5 Van-Deemter Curves in Ion Chromatography

The terms for the various contributions to the peak broadening combined in Eq. (24) give the impression that they are independent from each other. In practice, however, an interdependence exists between these terms. This leads to a much smaller decrease in separation efficiency than predicted by the simplifying Van-Deemter theory.

By plotting the theoretical plate height calculated via Eqs. (8) and (10) as a function of the flow rate, u, the dependence obtained for an anion exchange column (IonPac AS4) is shown in Fig. 2-7.

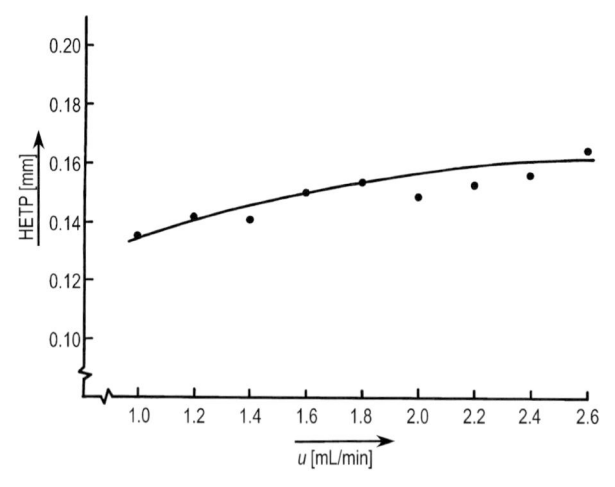

Fig. 2-7. HETP versus linear velocity u for an IonPac AS4 anion separator using $t_{ms}(SO_4^{2-})$.

From Fig. 2-7 it is clear that the plate height is almost invariant at higher flow rates. Similar dependencies have also been observed by Majors [12] for silica-based HPLC columns with smaller particle size. Such dependencies show that higher flow rates may lead to drastically reduced analysis times without any significant loss in chromatographic efficiency.

Cation separator columns exhibit a more pronounced dependence of the plate height on the flow rate (Fig. 2-8).

In this case, a compromise between separation efficiency and required analysis time has to be made. Flow rates between 2.0 mL/min and 2.3 mL/min have proved to be most suitable for practical applications.

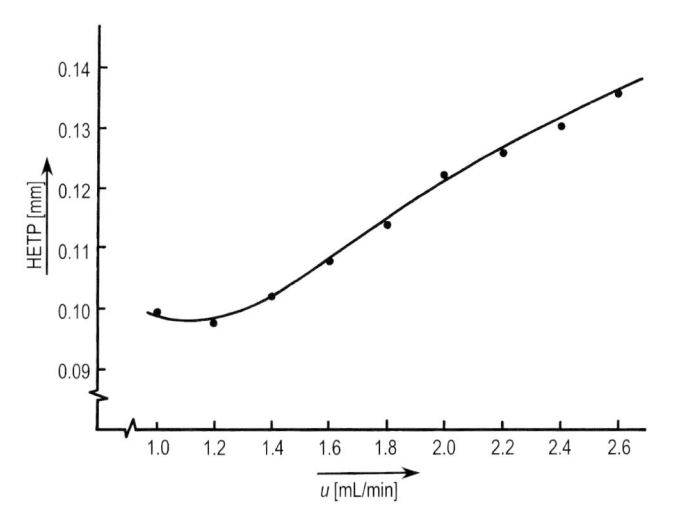

Fig. 2-8. HETP versus linear velocity u for an IonPac CS1 cation separator using t_{ms} (K$^+$).

3
Anion Exchange Chromatography (HPIC)

3.1
General Remarks

Ion-exchange is one of the oldest separation processes described in literature [1, 2]. The classical column chromatography on macroporous ion-exchange resins was a precursor of modern ion-exchange chromatography. The major differences between both processes are the method of sample introduction and the type of separation and detection systems being used. In classical ion-exchange chromatography, a column (typically 10 cm to 50 cm long) was filled with an anion or cation exchange resin having a particle size between 60 and 200 mesh (0.075 to 0.25 mm). After the sample was applied to the top of the column, it migrated down the column driven by gravitational force, and became more or less separated. Individual fractions of the eluant were collected using a fraction collector, and subsequently analyzed in a separate work-step. Due to the high ion-exchange capacity of the columns, high electrolyte concentrations were necessary to ensure the elution of the sample ions from the column. In many cases, several liters of eluant had to be worked up.

The enormous improvement in the performance of modern ion-exchange chromatography is attributed to the pioneering work of Small et al. [3]. Their major achievement was the development of low-capacity ion-exchange resins of *high chromatographic efficiencies*, which could be prepared reproducibly. The required injection volume was reduced to 10–100 µL, which resulted in an enhanced resolution with very narrow peaks. Another important improvement was that of automated detection, which allowed continuous monitoring of the signal. The introduction of conductivity detection for ionic species added a new dimension to ion-exchange chromatography.

3.2
The Ion-Exchange Process

The resins employed in ion-exchange chromatography carry functional groups with a fixed charge. The respective counter ions are located in the vicinity of these functional groups, thus rendering the whole entity electrically neutral. In anion exchange chromatography, quaternary ammonium bases are generally

Handbook of Ion Chromatography, Third, Completely Revised and Enlarged Edition. Joachim Weiss
Copyright © 2004 WILEY-VCH Verlag GmbH & Co. KGaA, Weinheim
ISBN: 3-527-28701-9

used as ion-exchange groups; sulfonate groups are used in strong acid cation exchange chromatography. Weak acid cation exchangers are usually functionalized with carboxyl- or phosphonate groups, or with a mixture of both.

When the counter ion of the ion-exchange site is replaced by a solute ion, the latter is temporarily retained by the fixed charge. The various sample ions remain for a different period of time within the column due to their different affinities towards the stationary phase and, thus, separation is brought about.

For example, if a solution containing bicarbonate anions is passed through an anion exchange column, the quaternary ammonium groups attached to the resin are exclusively in their bicarbonate form. If a sample with the anions A^- and B^- is injected onto the column, these anions are exchanged for bicarbonate ions according to the reversible equilibrium process given by Eqs. (32) and (33):

$$\text{Resin-NR}_3^+ \text{HCO}_3^- + A^- \underset{}{\overset{K_1}{\rightleftharpoons}} \text{Resin-NR}_3^+ A^- + \text{HCO}_3^- \tag{32}$$

$$\text{Resin-NR}_3^+ \text{HCO}_3^- + B^- \underset{}{\overset{K_2}{\rightleftharpoons}} \text{Resin-NR}_3^+ B^- + \text{HCO}_3^- \tag{33}$$

The separation of the anions is determined by their different affinities towards the stationary phase. The constant determining the equilibrium process is the selectivity coefficient, K, which is defined as follows:

$$K = \frac{[X^-]_s \cdot [\text{HCO}_3^-]_m}{[\text{HCO}_3^-]_s \cdot [X^-]_m} \tag{34}$$

$[X^-]_{m,s}$ Concentration of the sample ion in the mobile (m) or the stationary phase (s)
$[\text{HCO}_3^-]_{m,s}$ Bicarbonate concentration in the mobile (m) or the stationary phase (s)

The selectivity coefficient can be determined experimentally by adding a certain amount of resin material to a solution with known concentrations of X^- and HCO_3^-. The resulting concentration of the exchanged ions is determined in the mobile and stationary phase, respectively, after equilibrium is achieved. To precisely calculate the selectivity coefficient, the activities a_i have to be used instead of the concentrations c_i. As a prerequisite, the determination of the activity coefficient f_i according to Eq. (35) is required, which is difficult to perform in the matrix of an ion-exchange resin.

$$a_i = f_i \cdot c_i \tag{35}$$

Since in ion chromatography the concentration of ions in the solutions to be analyzed is small, they may be equated with the activity coefficient in first approximation. For the sake of simplicity, we only use the concentrations c_i within the scope of this discussion.

The efficiency of an ion as eluant for ion chromatography may be estimated on the basis of the selectivity coefficient. Ions with high selectivity coefficients are preferentially used as eluant ions, because they exhibit a high elution power even in diluted solutions. If the sample ions elute too quickly with a given eluant, the ionic strength must be lowered or another eluant ion with a smaller selectivity coefficient must be used. In general, the selectivity coefficients of eluant and sample ions should be comparable when choosing an eluant.

Another measure for the affinity of a solute ion towards the stationary phase of an ion-exchange resin is the weight distribution coefficient, D_g, which is defined for an exchanged ion X^- as follows:

$$D_g = \frac{[X^-]_s}{[X^-]_m} \qquad (36)$$

Instead of the weight distribution coefficient, in most cases the capacity factor, k, is used, which is defined by Eq. (37):

$$k = D \cdot \frac{m_{Resin}}{V_{Solution}} \qquad (37)$$

It follows from Eq. (37) that the capacity factor can be derived from the weight distribution coefficient. However, the determination of k from chromatographic data [see Eq. (7) in Section 2.2] is much more convenient.

3.3
Thermodynamic Aspects

In addition to pure ion-exchange processes, non-ionic interactions with the stationary phase are observed with certain ionic species. The most important non-ionic interaction is adsorption.

If the substrate material for ion-exchange resins is made of an organic polymer with an aromatic backbone, in ions with aromatic and olefinic carbon skeleton the ion-exchange process is superposed by sorption interactions, which can be attributed to π-π-interactions with the aromatic resin backbone. Adsorption-type interactions are not only observed in aromatic or olefinic solutes, but generally in all polarizable inorganic and organic ions. In some cases, even the separation of simple inorganic anions such as bromide and nitrate is entirely attained by various non-ionic sorption properties. This effect may be demonstrated by a simple experiment. It is possible to block the adsorption sites on the surface of the stationary phase by adding p-cyanophenol, which is especially effective in this respect, to the carbonate/bicarbonate eluant. As shown in Fig. 3.1, bromide and nitrate co-elute under these conditions, although the ion-exchange groups of the stationary phase remain unaffected.

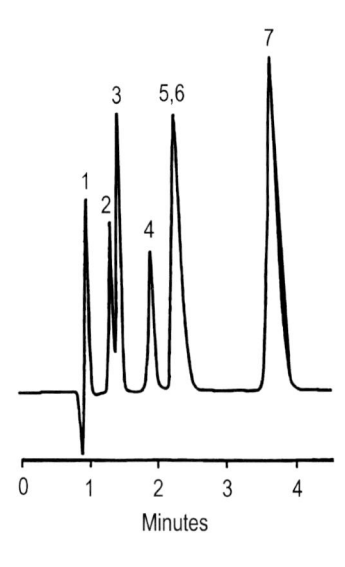

Fig. 3-1. Representation of the sorption effects in the separation of bromide and nitrate. − Separator: IonPac AS4; eluant: 4.3 mmol/L $NaHCO_3$ + 3.4 mmol/L Na_2CO_3 + 100 mg/L p-cyanophenol; flow rate: 2 mL/min; detection: suppressed conductivity; injection volume: 50 µL; solute concentrations: 3 mg/L fluoride (1), 4 mg/L chloride (2), 10 mg/L nitrite (3), 10 mg/L phosphate (4), 10 mg/L bromide (5), 20 mg/L nitrate (6), and 25 mg/L sulfate (7).

Such sorption phenomena may also be characterized thermodynamically. It should be mentioned here that in ion chromatography the quantities k and K can be defined as the sum of ionic and non-ionic interactions. However, the following discussion refers only to non-ionic interactions. It is based on the general representation of the retention value of compound i in terms of the capacity factor, k, according to Eq. (38).

$$k_i = \frac{t_{ms}^i - t_m^i}{t_m^i} \tag{38}$$

The capacity factor is directly connected to the distribution coefficient, K, and thus to the thermodynamic quantities $\Delta H_{m \to s}$ and $\Delta S_{m \to s}$, which determine the chromatographic sorption process. (The indices m→s characterize the mass-transfer from the mobile phase to the stationary phase.)

$$k_i = K_i \cdot \Phi$$
$$= K_i \cdot \frac{v_{stat}}{v_{mob}} \tag{39}$$

Φ Phase volume ratio
v_{stat} Volume of the stationary phase
v_{mob} Volume of the mobile phase

3.3 Thermodynamic Aspects

Thermodynamically [4-6], the retention of a solute, A, can be expressed as a function of its free sorption energy, $\Delta G_{m \to s}(A)$, and the phase volume ratio, ϕ:

$$\ln k = -\frac{\Delta G_{m \to s}(A)}{RT} + \ln \Phi \tag{40}$$

If A is a member of a homologous or quasi-homologous series of compounds, the result is with $\Delta\Delta G_{m \to s}$ corresponding to the change of free sorption energy with each incremental step of the homologous series.

$$\Delta G_{m \to s} = \Delta G^*_{m \to s} + \Delta\Delta G_{m \to s} \cdot n \tag{41}$$

The term $\Delta G^*_{m \to s}$ takes into account both the non-linearity of $\Delta G_{m \to s}$ with n for small members of the series and the contribution of the functional group. It then follows for $\ln k$:

$$\ln k = -\frac{\Delta G^*_{m \to s}}{RT} + \ln \Phi - \frac{\Delta\Delta G_{m \to s}}{RT} \cdot n \tag{42}$$

By plotting $\ln k$ as function of n, a linear dependence is derived for each homologous series with the general form

$$\ln k = a + b \cdot n \tag{43}$$

By substituting into Eq. (42) it follows for a und b:

$$a = -\frac{\Delta G^*_{m \to s}}{RT} + \ln \Phi \tag{44}$$

$$b = -\frac{\Delta\Delta G_{m \to s}}{RT} \tag{45}$$

Since

$$\Delta G = \Delta H - T \cdot \Delta S \tag{46}$$

it follows:

$$a = -\frac{\Delta H^*_{m \to s}}{RT} + \frac{\Delta S^*_{m \to s}}{R} + \ln \Phi \tag{47}$$

$$b = -\frac{\Delta\Delta H_{m \to s}}{RT} + \frac{\Delta\Delta S_{m \to s}}{R} \tag{48}$$

Therefore, the following expression is derived for the retention:

$$\ln k = -\frac{\Delta H^*_{m \to s}}{RT} + \frac{\Delta S^*_{m \to s}}{R} - \left(\frac{\Delta \Delta H_{m \to s}}{RT} - \frac{\Delta \Delta S_{m \to s}}{R} \right) \cdot n + \ln \Phi \qquad (49)$$

Both the $\ln k$ values and the sorption enthalpies, $\Delta H_{m \to s}$, may be determined experimentally from the temperature dependence of retention. To calculate the sorption entropy, the phase volume ratio must be known. However, the thermodynamic data may be regarded simply as formal quantities, because the capacity factors correlate directly with $\Delta G_{m \to s}$ via the distribution coefficient according to Eq. (39), and because sorption exhibits both distributive and adsorptive character.

The retention model developed by Eon and Guiochon [7, 8] to describe the adsorption effects at both gas-liquid and liquid-solid interfaces, which was later modified by Möckel et al. [6] to account for the retention at chemically bonded reversed-phase materials in HPLC, is not applicable to ion chromatography. But if the dependence of the capacity factors of various inorganic anions on the column temperature is studied, certain parallels with HPLC are observed. The linear dependences shown in Fig. 3-2 are obtained for the ions bromide and nitrate when the $\ln k$ values are plotted versus the reciprocal temperature (van't Hoff plot).

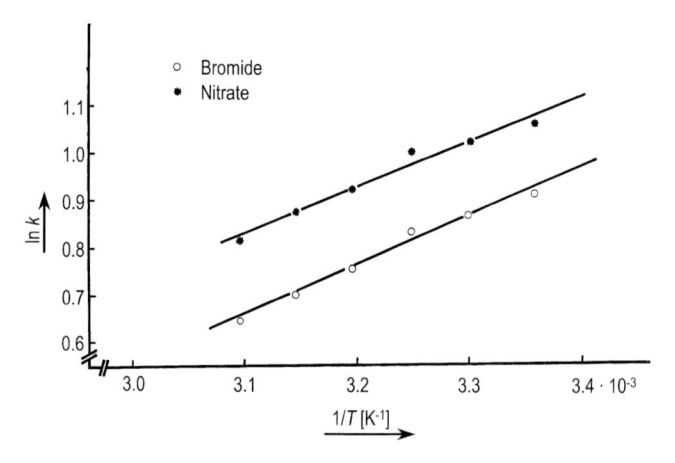

Fig. 3-2. Variation of the retention of bromide and nitrate with temperature. – Separator: IonPac AS3; eluant: 2.8 mmol/L $NaHCO_3$ + 2.2 mmol/L Na_2CO_3; flow rate: 2.3 mL/min; detection: suppressed conductivity; injection: 50 µL anion standard.

However, in the case of fluoride, chloride, nitrite, orthophosphate, and sulfate, the $\ln k$ values were found to be constant within experimental error limits in the temperature range investigated. Upon linear regression of the values in Table 3-1, the following relations are derived for bromide and nitrate:

Bromide

$$\ln k = 1036.03\ T^{-1} - 2.56$$
$$r = 0.994 \tag{50}$$

Nitrate

$$\ln k = 968.38\ T^{-1} - 2.17$$
$$r = 0.983 \tag{51}$$

The heats of sorption $\Delta H_{m \to s}$ are derived from the slope b via

$$b = -\frac{\Delta H_{m \to s}}{R} \tag{52}$$

With $R = 8.313$ J/(mol K) the following sorption enthalpies are calculated:

Bromide

$$\Delta H_{m \to s} = -8.6\ \text{kJ/(mol K)}$$

Nitrate

$$\Delta H_{m \to s} = -8.1\ \text{kJ/(mol K)}$$

Table 3-1. ln k values for bromide and nitrate at different temperatures.

| T | T^{-1} | ln k | |
[K]	[10^3 K^{-1}]	Bromide	Nitrate
298	3.356	0.909	1.060
303	3.300	0.862	1.019
308	3.247	0.829	1.004
313	3.195	0.752	0.918
318	3.145	0.699	0.871
323	3.096	0.645	0.813

The values thus derived are in the same order of magnitude as the sorption enthalpies determined by Möckel et al. [6] for various non-ionic inorganic and organic solutes. It is not possible to calculate the sorption entropy, $\Delta S_{m \to s}$, which is normally derived from the intercept, a, by using Eq. (53), because the phase volume ratio is unknown.

$$a = \frac{\Delta S_{m \to s}}{R} + \ln \Phi \tag{53}$$

Investigations by Möckel et al. [6] revealed that the latter may be approximately estimated in reversed-phase chromatography. For that, the capacity factors of individual members of a homologous series are determined using two different eluant compositions, and are then plotted versus n. Since the free sorption energy becomes zero for solutes with $K = 1$, the phase volume ratio may be directly taken via Eq. (54) from the intersection of the two straight lines.

$$\ln k \approx \ln \Phi \tag{54}$$

However, suitable eluant systems are still required to analogously calculate the phase volume ratio in ion chromatography.

The sorption enthalpy $\Delta H_{m \to s}$ for fluoride, chloride, nitrite, orthophosphate, and sulfate is small or zero, respectively, because no relation between $\ln k$ and the reciprocal temperature could be established. The change in the molar sorption energy is therefore entirely due to an entropy increase. Ongoing investigations will reveal the contributions to the molar sorption entropy from changes in the mixing entropy, from the orientation of water caused by the shrinkage or the swelling of hydration shells, or from the configuration entropy of the matrix.

3.4
Stationary Phases

In contrast to silica-based column packings used in classical HPLC, organic polymers are employed as the predominant support materials used in ion chromatography. These materials show a much higher stability in extreme pH conditions. While silica-based HPLC columns can only be used in a pH range between 2 and 8, ion exchangers based on organic polymers are also stable in the alkaline pH region. Nonetheless, few silica-based ion exchangers have been developed that exhibit a much higher chromatographic efficiency in comparison to organic polymers. However, the stationary phases used in anion exchange chromatography differ not only in the type of support material used; they can also be classified according to their different pore sizes and ion-exchange capacities.

Ion-Exchange Capacity
The ion-exchange capacity of a resin is defined as the number of ion-exchange sites per weight equivalent of the column packing. It is typically expressed in terms of milli-equivalent per gram resin (mequiv/g)*).

Generally:
The retention times for the analyte ions increase with the increasing ion-exchange capacity of the resin. This effect can be partially compensated for by using eluants of higher ionic strength.

---

*) The correct unit is mmol/g referring to monovalent ion-exchange groups. Here, we still employ the expression commonly used in the literature for the unit of the ion-exchange capacity.

Independent of the type of the support materials, anion exchangers with comparatively high exchange capacities (in the range of 1 mequiv/g) play only a minor role. They are mentioned here only for the sake of completeness. Such materials are quite limited in applicability, because inorganic anions have to be eluted at high ionic strength. This renders conductivity detection very difficult, if not impossible. Therefore, high-capacity anion exchangers are only used in combination with amperometric [9] or photometric detection after derivatization of the column effluent [10]. For example, nitrite and nitrate can be separated on pellicular ion exchangers based on silica [11], organic polymers [12], or cellulose [13]. They are then detected by direct UV detection. Recent developments in the field of indirect detection methods [14, 15] show that refractive index detection may provide an alternative to conductivity detection.

Much more important are ion exchangers with low ion-exchange capacities. The various types are described in detail below.

3.4.1
Polymer-Based Anion Exchangers

Styrene/divinylbenzene copolymers, ethylvinylbenzene/divinylbenzene copolymers, polymethacrylate, and polyvinyl resins are the most important of all the organic compounds that were tested for their suitability as substrate materials in the manufacturing process for polymer-based anion exchangers.

3.4.1.1 Styrene/Divinylbenzene Copolymers

Styrene/divinylbenzene copolymers are the most widely used substrate materials. Because they are stable in the pH range between 0 and 14, eluants with extreme pH values may be used. This allows the conversion of compounds such as carbohydrates, which are not ionic at a neutral pH, into the anionic form, making them available for ion chromatographic analysis (see also Section 3.10).

The copolymerization of styrene with divinylbenzene (DVB) is necessary to obtain the required mechanical stability of the resin. Upon adding divinylbenzene to styrene, the two functional groups of DVB cross-link two polystyrene chains with each other. Part of the resulting skeleton is depicted schematically in Fig. 3-3. The percentage of DVB in the resin is indicated as "percent cross-linking". The degree of cross-linking determines the porosity of the resin, which is another characteristic used in the classification of PS/DVB resins.

In general, a distinction is made between microporous and macroreticular resins. The former are by far more important for anion exchange chromatography.

Microporous (gel-type) substrates are prepared by bead polymerization. In this process, the two monomeric compounds, styrene and divinylbenzene, are suspended as small droplets in water by rapid, even stirring. The polymerization is initiated by appropriate catalysts and leads to uniform particles, the microporous structure and the size distribution depending on the stirring speed. In general, a decrease of the average particle size is observed with increased stirring speed.

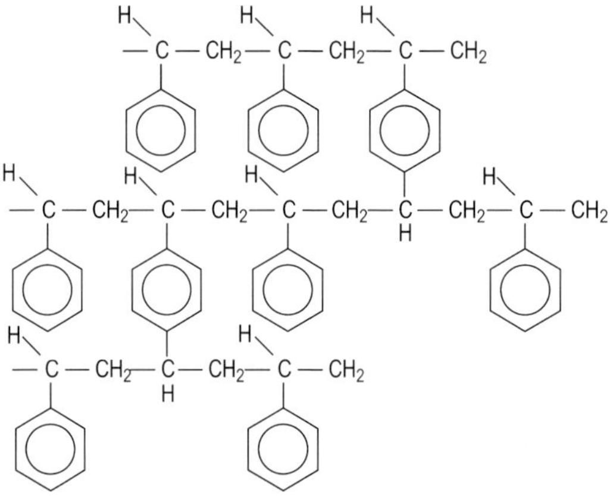

Fig. 3-3. Schematic illustration of the polystyrene/divinylbenzene skeleton.

The polarity of the particles is substantially augmented upon functionalization of the resin to the desired anion exchange material. Polar solvents such as water or acetonitrile solvate the resin according to the relative number of functional groups. The associated swelling becomes less pronounced as the degree of cross-linking increases. Unpolar solvents lead to resin dehydration and, thus, to its shrinkage. The column packing collapses as a result of the particle shrinkage, creating a void volume at the top end of the column, which is accompanied by a drastic loss of separation efficiency. An eluant change can also cause swelling and shrinkage, because the hydration of the ion-exchange resin depends on the counter ion in the mobile phase, i.e., the ionic form of the resin. Ion-exchange materials with a low degree of cross-linking are particularly prone to strong swelling and shrinkage if the ionic form of the resin changes significantly (e.g., transition from inorganic sodium hydroxide to organic potassium hydrogen phthalate). The optimum degree of cross-linking for a microporous PS/DVB resin can be determined experimentally. Normally, this value lies between 2% and 8%, which is a compromise between the mechanical stability and the chromatographic efficiency of the stationary phase. Higher values have been found to be disadvantageous, because ion-exclusion effects are increasingly observed with ions having large radii. High amounts of DVB also limit the mass-transfer, leading to a low column efficiency. On the other hand, resins that are cross-linked less than 2% are not mechanically stable enough. However, this is only true for directly functionalized resins. If the resin just serves as support material, as in latexed anion exchangers (see Section 3.4.2), a higher degree of cross-linking is required to achieve solvent compatibility.

Macroporous (macroreticular) [16, 17] substrates are also prepared by bead polymerization in polar solvents. To form these structures, a chemically inert solvent is added to the monomer suspension, in which the monomer is soluble, but the resulting polymer is not. When the polymerization process is complete, the solvent inclusions formed during the polymerization are washed out. Particles prepared by this method are spherical and have a large specific surface between 25 m^2/g and 800 m^2/g depending on the type of polymer. In other methods, finely ground inorganic salts such as calcium carbonate are used instead of chemically inert solvents. These salts are extracted from the polymer after completing the polymerization. Both methods result in the formation of copolymers having an average pore size in the range of 20 to 500 Å. Therefore, these particles should be called mesoporous.

Macroporous resins exhibit a remarkable mechanical stability due to their high degree of cross-linking. They are preferably employed in anion exchange chromatography under conditions where the mobile phase contains significant amounts of organic solvents. Swelling and shrinkage, which could result from changing the polarity of the solvent, are also suppressed due to the high degree of cross-linking.

Functionalization of the Resin

With the exception of latex-based anion exchangers (see Section 3.4.2), where totally porous latex particles act as ion-exchange material, organic polymers are functionalized directly at their surface. Surface-functionalized, so-called "pellicular" substrates, show a much higher chromatographic efficiency than fully functionalized resins.

The organic polymers mentioned above are functionalized in a two-step process:

1. Addition of a chloromethyl group to the aromatic skeleton of the resin

$$\text{Resin}-\text{C}_6\text{H}_5 \xrightarrow[\text{ZnCl}_2]{\text{ClCH}_2\text{OCH}_3} \text{Resin}-\text{C}_6\text{H}_4-\text{CH}_2\text{Cl} \tag{55}$$

2. Amination of the choromethylated resin with a tertiary amine:

$$\text{Resin}-\text{C}_6\text{H}_4-\text{CH}_2\text{Cl} \xrightarrow{\text{NR}_3} \text{Resin}-\text{C}_6\text{H}_4-\text{CH}_2\overset{+}{\text{N}}\text{R}_3 \quad \text{Cl}^- \tag{56}$$

In the past, the chloromethyl group was added to the aromatic skeleton of the resin via reaction with chloromethyl methyl ether. The polarity of the C-X bond in the primary halide was enhanced for the electrophilic attack by adding zinc chloride as Lewis acid to the reaction mixture. Since chloromethyl methyl ether is suspected to be carcinogenic and is no longer commercially available, Fritz et al. [18] devised a method in which chloromethyl methyl ether is prepared *in situ* and reacted with the resin.

However, the use of chloromethyl methyl ether can be avoided, if the polymer is reacted with formaldehyde and hydrochloric acid in the presence of a Lewis acid:

$$\text{Resin}-C_6H_5 + HCHO + HCl \xrightarrow{ZnCl_2} \text{Resin}-C_6H_4-CH_2Cl + H_2O \tag{57}$$

The conjugated acid of formaldehyde, i.e., the hydroxymethylene cation, acts as an electrophile:

$$H-CHO + H^+ \longrightarrow \left\{ H-\overset{+}{C}H-OH \longleftrightarrow H-\overset{+}{C}H-OH \right\}$$

Hydroxymethylene Cation

In a second reaction step, the alcohol formed via electrophilic substitution is transformed by hydrochloric acid into the corresponding halogenide:

$$\text{Resin}-C_6H_5 + \overset{+}{C}H_2-OH \longrightarrow \text{Resin}-[C_6H_6-CH_2OH]^+ \tag{58}$$

$$\downarrow -H^+$$

$$\text{Resin}-C_6H_4-CH_2Cl \xleftarrow{HCl} \text{Resin}-C_6H_4-CH_2OH$$

The resin is subsequently aminated with a tertiary amine. This method allows good control of the ion-exchange capacity of the final anion exchange resin. Depending on the reaction pathway, ion-exchange capacities between 0.001 and 0.09 mequiv/g are possible.

Overview of Surface-aminated Styrene/Divinylbenzene Copolymers

In connection with the 1978 introduction of a chromatographic system for the analysis of inorganic anions, in which the separator column is coupled directly to the conductivity cell, Fritz et al. [16, 17] developed a series of low-capacity, mesoporous, macroreticular anion exchangers, featuring highly cross-linked PS/DVB particles as substrate material (XAD-1, Rohm & Haas, Philadelphia, USA). The substrate was chloromethylated and aminated with triethyl amine. Investigations into the dependence of the retention of inorganic anions on the ion-exchange capacity of the resin [19] revealed that the distribution coefficients decrease as the ion-exchange capacity decreases. On the other hand, the selectivity coefficients, which are decisive for the separation, remained constant within the investigated capacity range of 0.04 to 1.46 mequiv/g.

Surface-functionalized XAD-1 resins are not in use anymore, because stationary phases of this kind exhibit a low chromatographic efficiency (about 1000 to 2000 theoretical plates per meter), that is not sufficient for most applications. Such resins can only be used, if high resolution is not required [20].

Surface-functionalized PS/DVB polymers are also commercially available. For example, Hamilton (Reno, USA) introduced an anion exchange resin under the trade name PRP-X100. This resin features spherical PS/DVB particles, which are surface-aminated with trimethyl amine [21, 22]. As can be seen from Fig. 3-4, seven inorganic anions, which are commonly referred to as "standard anions", can be separated within a short time. Sodium p-hydroxybenzoate was used as the eluant. Similar stationary phases are also available from SYKAM (Gilching, Germany) under the trade name LCA A01. The different elution order of anions on LCA A01 as compared with the PRP-X100 is due to the use of a carbonate/bicarbonate eluant (Fig. 3-5). In addition, Sykam offers a stationary phase under the trade name LCA A04, on which bromide and nitrate elute behind sulfate. The resulting chromatogram is comparable to the AS2 column from Dionex (Sunnyvale, USA). (See Fig. 3-25 in Section 3.4.2).

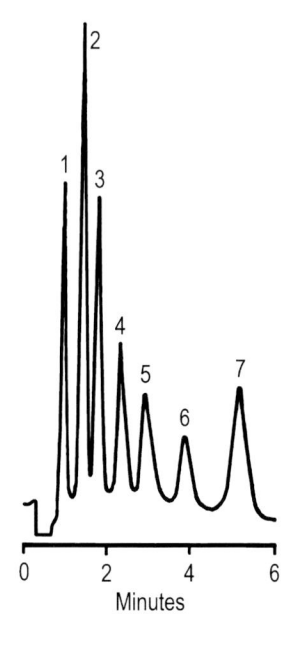

Fig. 3-4. Separation of various inorganic anions on PRP-X100. – Eluant: 4 mmol/L sodium p-hydroxybenzoate (pH 8.6); flow rate: 3 mL/min; detection: direct conductivity; injection volume: 100 µL; solute concentrations: 20 mg/L each of fluoride (1), chloride (2), nitrite (3), bromide (4), nitrate (5), orthophosphate (6), and sulfate (7); (taken from [22]).

All three phases show the same problem regarding the retention behavior of fluoride, which elutes very close to the void volume under standard chromatographic conditions. Therefore, fluoride determinations in complex matrices are not possible under these conditions; quantitation of fluoride in less complex matrices can only be considered at concentrations >0.5 mg/L (see also Section 3.7.2).

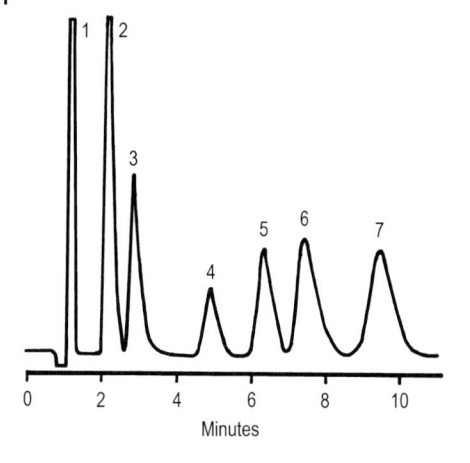

Fig. 3-5. Separation of various inorganic anions on LCA A01. — Eluant: 3 mmol/L $NaHCO_3$ + 2 mmol/L Na_2CO_3 in water/butanol (99.5:0.5 v/v); flow rate: 2.5 mL/min; detection: suppressed conductivity; injection volume: 50 µL; solute concentrations: 1 mg/L each of fluoride (1), chloride (2), nitrite (3), orthophosphate (4), bromide (5), nitrate (6), and sulfate (7).

In contrast to the above-mentioned separators, the ExcelPak ICS-A23 from Yokogawa was designed to separate fluoride from the system void using carbonate-based eluants in combination with a suppressor system. As can be seen from the respective chromatogram in Fig. 3-6, the seven standard inorganic anions are separated with high resolution within ten minutes. Although the manufacturer recommends an eluant of 3 mmol/L Na_2CO_3, similar results are obtained with an eluant mixture of 1.7 mmol/L $NaHCO_3$ and 1.8 mmol/L Na_2CO_3, which has a certain chemical buffer capacity. This is important, because sample pH usually differs from eluant pH. Despite the excellent resolution between the seven standard anions, other important anions such as bromate and chlorate, resulting from drinking water disinfection, co-elute with chloride and nitrate, respectively. As with previous generations of latexed anion exchangers (see Section 3.4.2), the low degree of substrate cross-linking prohibits the use of organic solvents as eluant components; for the same reason, organic solvents cannot be used for cleaning separator columns.

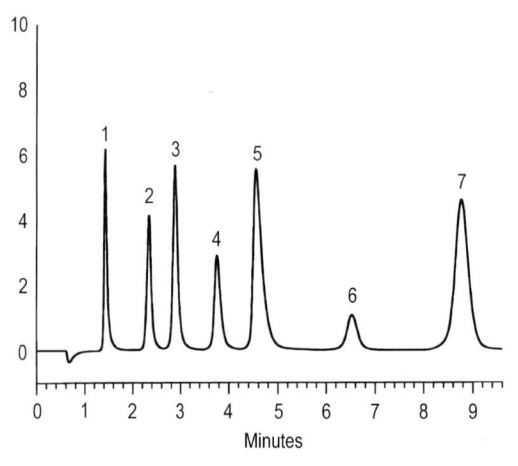

Fig. 3-6. Separation of various inorganic anions on ExcelPak ICS-A23. — Column temperature: 40 °C; eluant: 1.7 mmol/L $NaHCO_3$ + 1.8 mmol/L Na_2CO_3; flow rate: 1.5 mL/min; detection: suppressed conductivity; injection volume: 25 µL; solute concentrations: 3 mg/L fluoride (1), 4 mg/L chloride (2), 10 mg/L nitrite (3), 10 mg/L bromide (4), 20 mg/L nitrate (5), 10 mg/L orthophosphate (6), and 20 mg/L sulfate (7).

While the ExcelPak ICS-A23 is only available in Japan, separators from Sarasep (Santa Clara, USA) with similar selectivities that are designed for non-suppressed and suppressed conductivity detection are available in most countries. Separators such as AN1, AN2 and AN300 are highly cross-linked and surface-modified with an alkyldimethylethanolamine. Despite the high degree of cross-linking these packing materials are not solvent compatible. This might be due to pure adsorptive interactions between the methacrylate-based ion-exchange polymer and the support material. Rinsing these columns with pure organic solvents leads to an irreversible loss of selectivity. In addition, mechanical stability is limited, so that these columns can only be operated up to a system pressure of 10 MPa. Flow rates of 2 mL/min are only possible with the Sarasep AN300. For this reason, the AN300 has a large inner column diameter of 7.5 mm, which leads to a reduced column backpressure. Figure 3-7 indicates that the chromatographic resolution between fluoride and the system void and the co-elution of oxyhalides with some of the standard anions are similar to the Yokogawa column. Therefore, that column can also only be used for simple water analysis applications. Chlorite and chloride are not resolved under standard conditions, and nitrite and bromate as well as chlorate and sulfate are incompletely separated. In contrast to common latexed anion exchangers (see Section 3.4.2), the separation of fluoride and acetate is improved under standard conditions. However, acetate and formate, which can be separated on latexed anion exchangers, co-elute on the Sarasep columns.

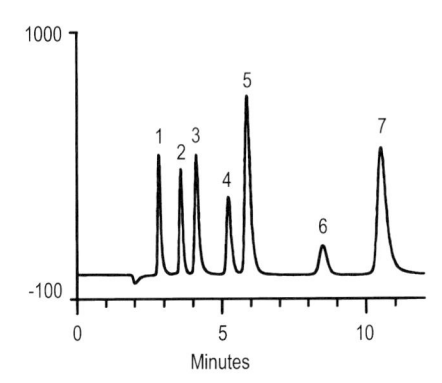

Fig. 3-7. Separation of various inorganic anions on Sarasep AN300. – For chromatographic conditions, see Fig. 3-6.

The Star Ion A300 IC Anion from Phenomenex (Torrance, USA) is a universal anion exchanger with relatively short analysis times for standard anions when using a carbonate/bicarbonate eluant and suppressed conductivity detection. As with all columns of this type, the separation between fluoride and the system void is not as good as with acrylate-based materials. The selectivity of this column for standard anions is very similar to the IonPac AS4A-SC latexed anion exchanger (see Section 3.4.2). It only differs in the separation of fluoride and acetate, which is almost baseline resolved on Star Ion A300 IC Anion. Star Ion

A300 IC Anion is also available in a 100 mm × 10 mm i. d. format (Star Ion A300 HC), which has a higher ion-exchange capacity due to the larger inner diameter of the column. According to the manufacturer, this column is particularly well-suited for trace bromate analysis in drinking water.

The characteristic structural and technical properties of these columns are summarized in Table 3-2.

Table 3-2. Structural and technical properties of surface-aminated styrene/divinylbenzene copolymers.

Separator	PRP-X100	PRP-X110	LCA A01	ExcelPak ICS-A23	AN1	AN300	Star Ion A300 IC Anion
Dimensions (Length × I.D.) [mm]	100 × 4.1 150 × 4.1 250 × 4.1	50 × (4.6, 4.1, 2.1, 1.0) 100 × (4.6, 4.1, 2.1, 1.0) 150 × (4.6, 4.1, 2.1, 1.0) 250 × (4.6, 4.1, 2.1, 1.0)	200 × 4.0	75 × 4.6	250 × 4.6	100 × 7.5	100 × 4.6 100 × 10
Manufacturer	Hamilton	Hamilton	Sykam	Yokogawa	Sarasep	Sarasep	Phenomenex
pH range	1 – 13	0 – 14	1 – 14	2 – 12	2 – 12	2 – 12	1 – 12
Maximum pressure [MPa]	35	not specified	25	13	10	7	7
Maximum flow rate [mL/min]	8	not specified	3	1.5	1.5	2	2
Solvent compatibility [%]	100	100	10	<5	<5	<5	<5
Capacity [mequiv/g]	0.2	not specified	0.04	0.05	0.05	not specified	not specified
Particle diameter [μm]	10	7	12	5	10	7	not specified
Type of column packing	Spherical PS/DVB aminated with triethylamine	PS/DVB with quaternary ammonium functions	PS/DVB with quaternary ammonium functions	PS/DVB with quaternary ammonium functions	PS/DVB aminated with alkyldimethylethanolamine	PS/DVB aminated with alkyldimethylethanolamine	PS/DVB with quaternary ammonium functions

Summary of Surface-Aminated Ethylvinylbenzene/Divinylbenzene Resins

A surface-functionalized polymer based on ethylvinylbenzene/divinylbenzene (EVB/DVB) is available under the trade name IonPac AS14 (Dionex, Sunnyvale, USA). This universal anion exchanger for the analysis of inorganic anions is an advanced successor of the IonPac AS12A [23] (see Section. 3.4.2). The IonPac AS14 improves the resolution between fluoride and the negative water dip, as well as the resolution among fluoride, acetate, and formate without increasing the total analysis time for the seven standard anions. This was achieved by grafting an anion exchange polymer to the surface of a highly cross-linked support material having a particle diameter of 9 μm [24]. Such an anion exchange particle is schematically depicted in Fig. 3-8.

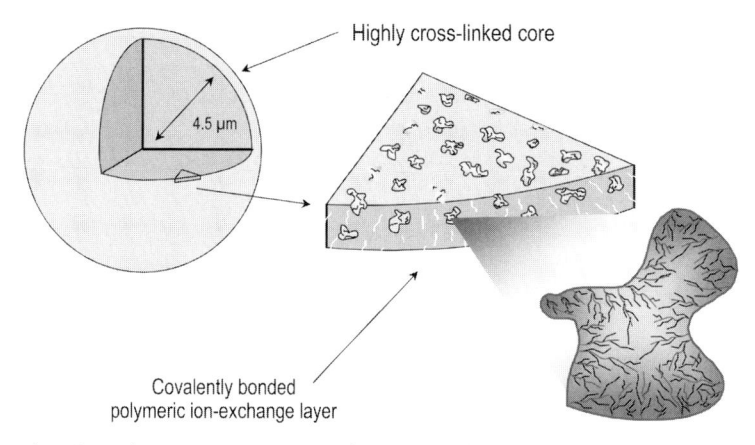

Fig. 3-8. Schematic representation of an anion exchange particle in separator columns such as the IonPac AS14.

In polymer chemistry, *grafting* describes a process in which two or more separately synthesized polymers are attached covalently. Polymers that are not covalently bonded are described as *mixtures of polymers*. The term grafting was adopted from horticulture, where it is a method commonly used to refine fruit trees. In both cases, the goal is to produce a material with properties that cannot be obtained with the individual components or with mixtures of them. As with polymerization techniques, there are a variety of grafting technologies. One method, for instance, is to use a base polymer carrying reactive functional end groups to which a second monomer can be bonded.

While the grafting technology is commonly used in polymer chemistry, its application for the preparation of stationary phases in chromatography is less common. Nevertheless, it is not a new application; grafting was first used in the preparation of stationary phases over 20 years ago. Schomburg et al. successfully used this technology to introduce weak acid cation exchangers based on polymer-coated silica to effect the simultaneous separation of alkali- and alkaline-earth metals (see Section 4.1.3), a feat that revolutionized cation exchange chromatography. The support material used for the IonPac AS14 consists of ethylvinylbenzene cross-linked with 55% divinylbenzene, so it is completely solvent compatible. Three different monomers are attached to this support material via *co-grafting*. Each of these monomers contributes its specific properties to the selectivity of the stationary phase, which can be modulated by varying the percentages of the monomers. The extremely small and controlled thickness of the anion exchange polymer layer results in rapid mass-transfer characteristics and, consequently, in a high chromatographic efficiency. For the elution of inorganic anions in combination with suppressed conductivity detection, the IonPac AS14 is used, mostly with carbonate/bicarbonate-based eluants. The respective chromatogram in Fig. 3-9 shows that the separation between fluoride and the system void is clearly improved in comparison with IonPac AS12A (see Section 3.4.2). In addition, no chromatographic interference of fluoride with short-chain fatty acids such as acetate and formate is observed, so that this column is suitable for the isocratic analysis of fluoride even in complex matrices. Moreover, an almost baseline-resolved separation between fluoride, glycolate, acetate, and formate is obtained when applying a tetraborate gradient. Interestingly, the selectivity of this stationary phase for oxyhalides differs from the selectivity achieved with latexed anion exchangers because of the resin structure. In contrast to IonPac AS12A, chlorate is more strongly retained than bromide and nitrate and, consequently, elutes between nitrate and orthophosphate when a slightly modified carbonate/bicarbonate eluant (2.7 mmol/L and 1.0 mmol/L) is applied. Samples with large concentration differences between the individual analytes can easily be analyzed with IonPac AS14 because of its larger ion-exchange capacity of 65 µequiv/column as compared with IonPac AS12A. The resulting analysis time for sulfate can be reduced to about 8 minutes by increasing the ionic strength and/or the flow rate of the eluant [25].

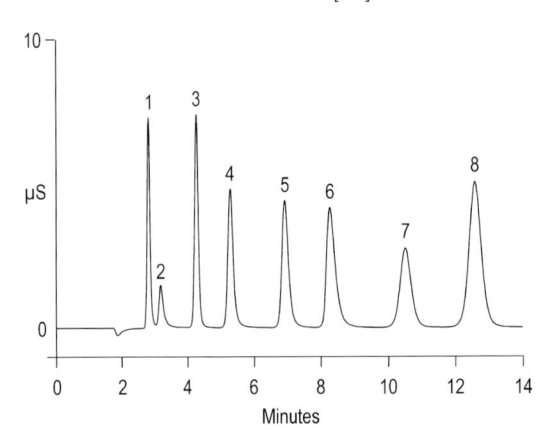

Fig. 3-9. Separation of various inorganic anions on IonPac AS14. – Eluant: 3.5 mmol/L Na_2CO_3 + 1 mmol/L $NaHCO_3$; flow rate: 1.2 mL/min; detection: suppressed conductivity; injection volume: 10 µL; solute concentrations: 5 mg/L fluoride (1), 20 mg/L acetate (2), 10 mg/L chloride (3), 15 mg/L nitrite (4), 25 mg/L each of bromide (5) and nitrate (6), 40 mg/L orthophosphate (7), and 30 mg/L sulfate (8).

It should be pointed out, however, that stationary phases currently prepared by the mentioned grafting process show a disturbed Gaussian peak form as compared to modern latexed anion exchangers. As can be seen from Fig. 3-9, peaks are curved at the peak base on both sides of the peak, which can be attributed to simultaneous tailing and fronting effects. This, in turn, implies that mass-transfer in the stationary phase is somewhat limited with this kind of separator column. Recent research indicates that, besides surface effects, a certain solubility of the monomer used for the grafting process in the support polymer is responsible for this phenomenon. As the reactive vinyl groups of the support polymer are not only found on its surface but also in the inner part, there is a certain probability that some of the grafted chains have their origins in this inner part. This region, however, is highly cross-linked, so that ion-exchange functional groups found in that area are weakly hydrated, leading to the observed peak tailing. Depending on their position, the ion-exchange functional groups on the grafted chains are also differently hydrated. Because the surface of the support material is hydrophobic, ion-exchange functions close to the surface are more hydrophobic than those further away from the surface, towards the end of the monomer chain. The latter ones are more strongly hydrated and hence responsible for fronting effects. A different explanation is based on the thermodynamic properties of the ion-exchange functionality. An ideal chromatographic behavior resulting in symmetrical peaks requires a uniform environment of the ion-exchange functionality. Because this cannot be realized at present with covalently bonded ion-exchange polymers, current research efforts are aimed to find ways for improving the uniformity of the ion-exchange functionality.

Significant improvements are obtained when using a support material with a smaller particle diameter, as it is realized with the 5-μm IonPac AS14A column. This column's dimensions are 150 mm × 3 mm i. d., which reduces the pressure drop along this separator column. Nevertheless, a relatively high ion-exchange capacity of 40 μequiv/column is obtained. Operating this column with an eluant consisting of 8 mmol/L sodium carbonate and 1 mmol/L sodium bicarbonate not only reduces the run time, but also improves sensitivity for the standard inorganic anions (Fig. 3-10).

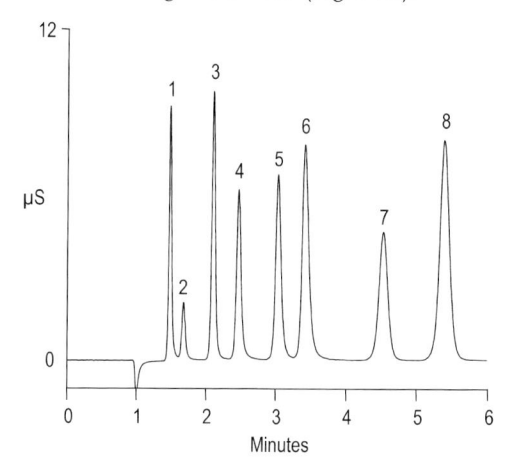

Fig. 3-10. Fast separation of inorganic anions on IonPac AS14A. – Column temperature: 30 °C; eluant: 8 mmol/L Na_2CO_3 + 1 mmol/L $NaHCO_3$; flow rate: 0.8 mL/min; detection: suppressed conductivity; injection volume: 5 μL; solute concentrations: 5 mg/L fluoride (1), 20 mg/L acetate (2), 10 mg/L chloride (3), 15 mg/L nitrite (4), 25 mg/L each of bromide (5) and nitrate (6), 40 mg/L orthophosphate (7), and 30 mg/L sulfate (8).

Recently, Dionex introduced another surface-functionalized EVB/DVB resin featuring a covalently attached anion exchange polymer under the trade name IonPac AS15. This material is suitable for gradient elution of inorganic and organic anions. As an improvement of the IonPac AS11-HC (see Section 3.4.2), the AS15 column is specifically designed for the separation of fluoride, glycolate, acetate, and formate using a hydroxide concentration gradient. The support material of IonPac AS15 is identical to that of IonPac AS14. Figure 3-11 shows the respective chromatogram of inorganic anions and the above-mentioned organic acids on IonPac AS15. As can be seen from this chromatogram, the resolution between various organic acids is significantly improved relative to IonPac AS11-HC (see Fig. 3-42 in Section 3.4.2). The pronounced hydrophobicity of the ion-exchange functional groups of the AS15 column leads to a special selectivity

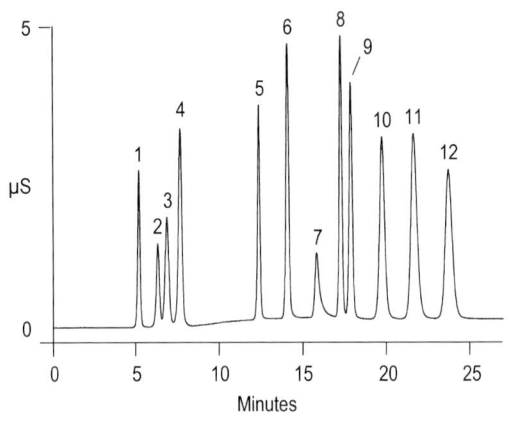

Fig. 3-11. Gradient elution of inorganic and organic anions on IonPac AS15. — Eluant: (A) water, (B) 0.1 mol/L NaOH; gradient: 8% B isocratically for 7 min, then to 45% B in 8 min; column temperature: 30 °C; flow rate: 1.6 mL/min; detection: suppressed conductivity; injection volume: 10 µL; solute concentrations: 2 mg/L fluoride (1), 10 mg/L each of glycolate (2), acetate (3), and formate (4), 5 mg/L each of chloride (5) and nitrite (6), 50 mg/L carbonate (7), 10 mg/L each of sulfate (8), oxalate (9), bromide (10), nitrate (11), and 30 mg/L orthophosphate (12).

for later eluting species: bromide and nitrate are eluting between sulfate and orthophosphate because of adsorptive interactions. The column's high ion-exchange capacity of 225 µequiv allows the complete separation of standard anions under isocratic conditions (Fig. 3-12), although total analysis time of 25 minutes for orthophosphate is significantly higher as compared to the AS11-HC column. However, the total analysis time can be reduced to approximately 15 minutes by using 5-µm particles and smaller column dimensions (150 mm × 3 mm i. d.) which has been realized with the IonPac AS15A. Best chromatographic separations under isocratic and gradient conditions as well as highly symmetric peaks are obtained if the column is operated at slightly elevated temperatures (30 °C).

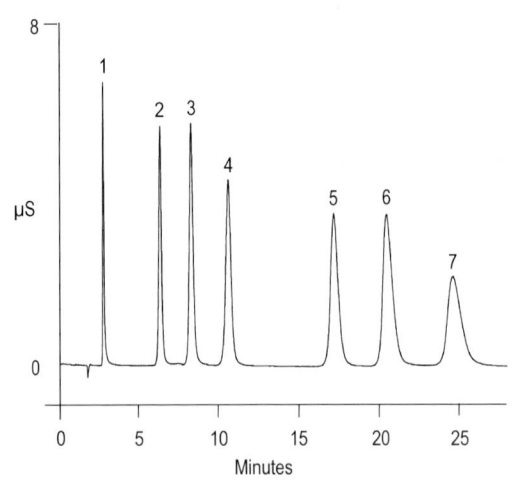

Fig. 3-12. Isocratic separation of inorganic anions on IonPac AS15. — Eluant: 38 mmol/L NaOH; column temperature: 30 °C; flow rate: 1.2 mL/min; detection: suppressed conductivity; injection volume: 25 µL; solute concentrations: 2 mg/L fluoride (1), 5 mg/L chloride (2), 10 mg/L each of nitrite (3) and sulfate (4), 20 mg/L each of bromide (5) and nitrate (6), and 30 mg/L orthophosphate (7).

Table 3-3 summarizes the characteristic structural and technical properties of surface-aminated EVB/DVB-polymers.

Table 3-3. Structural and technical properties of ethylvinylbenzene/divinylbenzene copolymers.

Column	IonPac AS14	IonPac AS14A	IonPac AS15	IonPac AS15A
Dimension (Length × I.D.) [mm]	250 × 2 250 × 4	150 × 3 250 × 4	250 × 2 250 × 4	150 × 3
Manufacturer	Dionex	Dionex	Dionex	Dionex
pH range	2 – 11	0 – 14	0 – 14	0 – 14
Maximum pressure [MPa]	27	27	27	27
Maximum flow rate [mL/min]	2	2	2	2
Solvent compatibility [%]	100	100	100	100
Capacity [mequiv/column]	0.016 0.065	0.04 0.12	0.056 0.225	0.097
Particle diameter [µm]	9	5 7	9	5
Type of column packing	EVB/DVB with quaternary ammonium functions	EVB/DVB with quaternary ammonium functions	EVB/DVB with quaternary ammonium functions	EVB/DVB with quaternary ammonium functions

3.4.1.2 Polymethacrylate and Polyvinyl Resins

In 1983, Toyo Soda (Japan) introduced TSK Gel IC-PW, a high efficiency anion exchange resin. It is a hydrophilic, porous methacrylate polymer with an ion-exchange capacity of 0.03 mequiv/g that is surface-aminated with methyldiethylamine. The chromatographic efficiency of this 50 mm × 4.6 mm i. d. column is surprisingly high. For the nitrate peak, 20,000 theoretical plates per meter have been calculated. However, polymethacrylate resins are only stable at pH values between 1 and 12. Figure 3-13 shows the respective anion chromatogram obtained, with potassium gluconate in a borate buffer solution as the eluant.

The signal in the void volume of the chromatogram in Fig. 3-13 is characteristic for all ion chromatographic separations when applying non-suppressed conductivity detection. The signal represents the sample cations that are not retained by the stationary phase. They move as a discrete band with the mobile phase through the column and eventually pass the conductivity cell together with the eluant anions. If the cation conductivity is higher than the eluant conductivity, a positive signal results. If the cation concentration is lower than the eluant concentration, a negative signal is observed. In both cases, interferences with early eluting analyte ions are likely to be observed. Potassium hydroxide can also be used for the separation of monovalent anions [27]. The TSK Gel IC-PW is also traded by Waters Millipore under the name "IC Pak A".

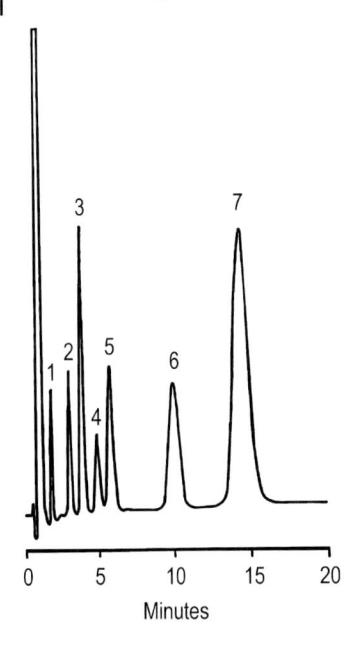

Fig. 3-13. Separation of various inorganic anions on TSK Gel IC-PW. — Eluant: 1.3 mmol/L $Na_2B_4O_7$ + 5.8 mmol/L H_3BO_3 + 1.4 mmol/L potassium gluconate (pH 8.5) + 120 mL/L acetonitrile; flow rate: 1.2 mL/min; detection: direct conductivity; injection volume: 100 µL; solute concentrations: 5 to 40 mg/L; (taken from [26]).

Much shorter analysis times can be achieved by using the acrylate-based stationary phase LCA A03 from Sykam and the IC-A1 being manufactured by Shimadzu. The selectivities of these phases are clearly shown in Figs. 3-14 and 3-15, respectively.

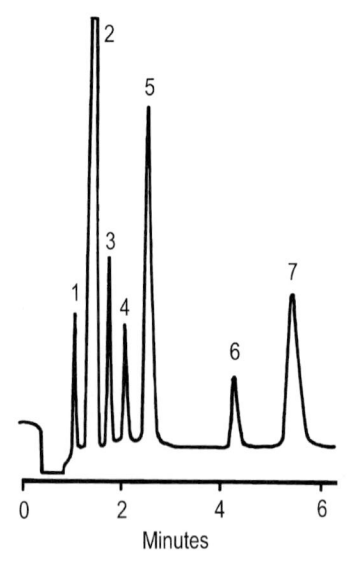

Fig. 3-14. Separation of various inorganic anions on LCA A03. — Eluant: 4 mmol/L sodium p-hydroxybenzoate (pH 8.5); flow rate: 2 mL/min; detection: direct conductivity; injection volume: 50 µL; peaks: fluoride (1), chloride (2), nitrite (3), bromide (4), nitrate (5), orthophosphate (6), and sulfate (7).

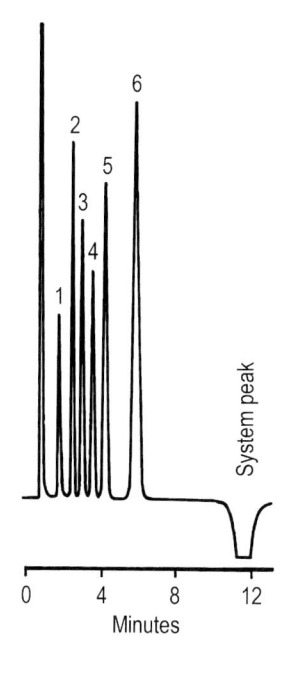

Fig. 3-15. Separation of various inorganic anions on Shimpack IC-A1. — Column temperature: 40 °C; eluant: 2.5 mmol/L phthalic acid + 2.4 mmol/L Tris; flow rate: 1.5 mL/min; detection: direct conductivity; injection volume: 20 µL; solute concentrations: 5 mg/L fluoride (1), 10 mg/L chloride (2), 15 mg/L nitrite (3), 20 mg/L bromide (4), 30 mg/L nitrate (5), and 40 mg/L sulfate (6).

Polyspher® IC AN-1 from Merck AG (Darmstadt, Germany), consisting of a hydrophilic polymethacrylate packing with a particle diameter of 12 µm, exhibits a similar selectivity. Depending on the type of application, this column can be operated with eluants based on phthalic acid, p-hydroxybenzoic acid, tartaric acid, oxalic acid, or boric acid. Figure 3-16 shows a typical chromatogram obtained with a phthalic acid eluant. Under these chromatographic conditions (pH 4) orthophosphate elutes as a monovalent anion ahead of chloride.

With the Metrosep Anion Dual 1, Metrohm (Herisau, Switzerland) also offers a stationary phase based on a hydroxyethylmethacrylate (HEMA). According to the manufacturer, organic sample components are barely retained on this support material due to its hydrophilicity and, thus, interferences with early eluting analyte anions are avoided. The column is available as a glass cartridge system and is operated with a relatively low flow rate of 0.5 mL/min due to the small inner column diameter of 3 mm. As with Polyspher® IC AN-1, Metrosep Anion Dual 1 is suitable for the analysis of simple inorganic anions (Fig. 3-17) as well as for oxyhalides, short-chain fatty acids, and oxalic acid. The selectivity of the Metrosep Anion Dual 2 is designed for carbonate/bicarbonate eluants employing chemically suppressed conductivity detection. A remarkable feature of this separator is the outstanding separation of fluoride from the system void; the separation between acetate and formate is also notable. As can be seen from the respective chromatogram in Fig. 3-18, this high resolution is obtained at the cost of a relatively long total analysis time of about 20 minutes until sulfate elutes.

3 Anion Exchange Chromatography (HPIC)

According to the manufacturer, fluoride determination can be compromised by glycolate and lactate in the sample; likewise, tartrate, malate, and malonate can interfere with sulfate.

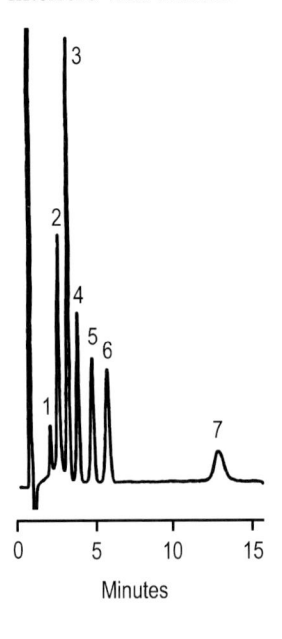

Fig. 3-16. Separation of various inorganic anions on Polyspher® IC AN-1. — Column temperature: 35 °C; eluant: 1.5 mmol/L phthalic acid + 1.38 mol/L Tris + 0.3 mol/L H_3BO_3, pH 4.0; flow rate: 1.3 mL/min; detection: direct conductivity; injection volume: 20 µL; solute concentrations: 10 mg/L each of orthophosphate (1), fluoride (2), chloride (3), nitrite (4), bromide (5), nitrate (6), and sulfate (7).

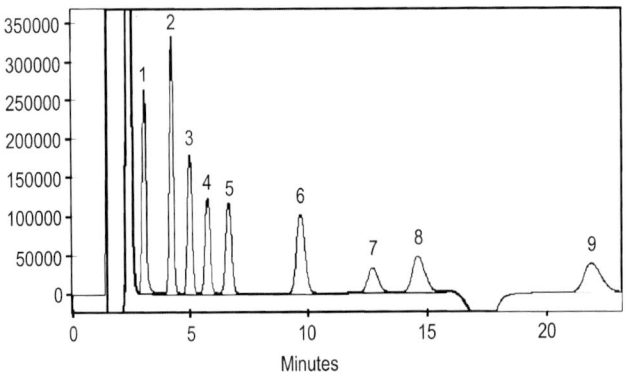

Fig. 3-17. Separation of various inorganic anions on Metrosep Anion Dual 1. — Eluant: 9 mmol/L phthalic acid — acetonitrile (98:2 v/v), pH 4.2 with Tris; flow rate: 0.5 mL/min; detection: direct conductivity; injection volume: 50 µL; solute concentrations: 5 mg/L each of fluoride (1), chloride (2), nitrite (3), bromide (4), nitrate (5), sulfate (6), iodide (7), thiosulfate (8), and thiocyanate (9).

Similar observations are made with the macroporous anion exchanger based on hydroxyethylmethacrylate from Alltech (Deerfield, USA) known under the trade name "Universal Anion" [28]. This stationary phase consists of a 2-hydroxyethylmethacrylate polymer that is cross-linked with ethylenedimethacrylate and

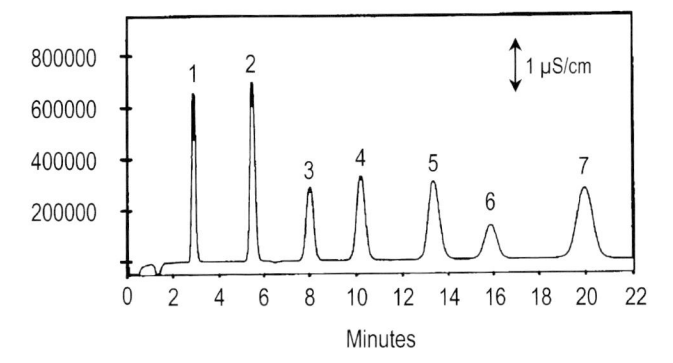

Fig. 3-18. Separation of various inorganic anions on Metrosep Anion Dual 2. – Eluant: 1.3 mmol/L Na_2CO_3 + 2 mmol/L $NaHCO_3$; flow rate: 0.8 mL/min; detection: suppressed conductivity; injection volume: 20 µL; solute concentrations: 2 mg/L fluoride (1), 5 mg/L each of chloride (2) and nitrite (3), 10 mg/L each of bromide (4), nitrate (5), orthophosphate (6), and sulfate (7).

functionalized with trimethylamine [29]. According to the manufacturer this 10-µm anion exchanger, the capacity of which is listed with 0.1 mequiv/g, shows good ion-exchange kinetics with various eluants such as p-hydroxybenzoic acid, phthalic acid, borate/gluconate, hydroxide/benzoate, and carbonate/bicarbonate. As with all methacrylate polymers these eluants can be used in the pH range between 2 and 12. Mechanical and chemical stability of this material is attributed to the high degree of cross-linking. Free hydroxide groups on the HEMA matrix result in a marked hydrophilicity of this material, which has a positive influence on the peak form of polarizable anions. Good separations of standard anions are obtained with p-hydroxybenzoic acid in the pH range between 8.0 und 8.6. Under these conditions all seven anions are separated on the Alltech Universal Anion column within 20 minutes (Fig. 3-19).

Mitsubishi Kasei (Japan) offers a methacrylate-based 5-µm anion exchanger exclusively for non-suppressed conductivity applications under the trade name MCI Gel SCA04. An eluant mixture consisting of vanillic acid and N-methyldiethanolamine has been developed especially for this column. This eluant is patented in Japan. According to the manufacturer, a relatively low background conductivity is obtained when the eluant pH is adjusted to 6.2. Under these chromatographic conditions the system peak is not observed. This pH value has to be very precisely adjusted, because the elution order between fluoride and orthophosphate reverses at lower pH values. In addition, a system peak between nitrate and the late eluting sulfate is then observed. On the other hand, at pH values around 7 co-elution of orthophosphate and chloride occurs. Figure 3-20 shows the respective chromatogram of the seven standard anions under optimized chromatographic conditions.

The characteristic structural-technical properties of these columns are summarized in Table 3-4.

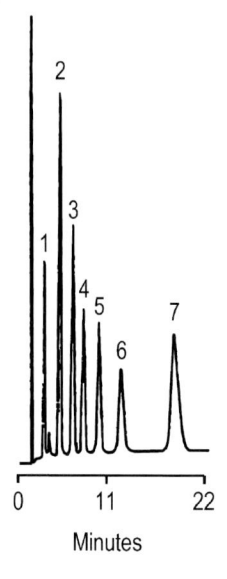

Fig. 3-19. Separation of various inorganic anions on Alltech Universal Anion. — Separator: 150 mm × 4.6 mm i. d. Universal Anion; eluant: 5 mmol/L p-hydroxybenzoic acid, pH 7.9 with LiOH; flow rate: 1 mL/min; detection: direct conductivity; injection volume: 100 µL; solute concentrations: 10 mg/L fluoride (1), 20 mg/L chloride (2), 20 mg/L nitrite (3), 20 mg/L bromide (4), 20 mg/L nitrate (5), 30 mg/L orthophosphate (6), and 30 mg/L sulfate (7); (taken from [28]).

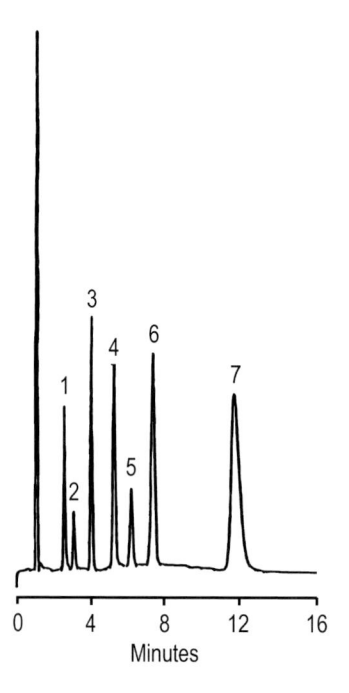

Fig. 3-20. Separation of various inorganic anions on MCI Gel SCA04. — Column temperature: 40°C; eluant: 3 mmol/L vanillic acid + 2.8 mmol/L N-methyl-diethanolamine, pH 6.2; flow rate: 1.2 mL/min; detection: direct conductivity; injection volume: 10 µL; solute concentrations: 5 mg/L fluoride (1), 40 mg/L orthophosphate (2), 10 mg/L chloride (3), 15 mg/L nitrite (4), 10 mg/L bromide (5), 30 mg/L nitrate (6), and 40 mg/L sulfate (7).

Polyvinyl-based anion exchange resins have been available since 1984 from Interaction Chemicals (Mountain View, CA, USA) under the trade names "ION-100" and "ION-110" [30]. According to the manufacturer, these macroreticular resins are stable at pH values between 0 and 14, thus allowing the use of a

3.4 Stationary Phases

Table 3-4. Structural and technical properties of surface-aminated polymethacrylate resins.

Separator	TSK-Gel[1] IC-PW	LCA A03	Shimpack IC-A1	Metrosep Anion Dual 1	Metrosep Anion Dual 2	Polyspher IC AN-1	Universal Anion	MCI SCA04
Dimensions (Length × I.D.) [mm]	50 × 4.6	100 × 4.0	100 × 4.6	150 × 3.0	75 × 4.6	100 × 4.6	150 × 4.6	150 × 4.6
Manufacturer	Toyo Soda	Sykam	Shimadzu	Metrohm	Metrohm	Merck	Alltech	Mitsubishi Kasei
pH range	1 – 12	2 – 12	2 – 11	2 – 12	1 – 12	2 – 10	2 – 12	2 – 12
Max. pressure [MPa]	7	20	5	9	7	10	17	not specified
Max. flow rate [mL/min]	2	5	2	0.7	1.2	2	2	2
Solvent compatibility [%]	20	10	10	10	20	20	5	5
Capacity [mequiv/g]	0.03	0.1	not specified	0.009	0.034	not specified	0.1	0.03
Particle diameter [µm]	10	10	10	10	6	12	10	5
Type of packing material	Polymethacrylate aminated with methyldiethylamine	Polymethacrylate with quaternary ammonium functions	Polymethacrylate with quaternary ammonium functions	Hydroxyethylmethacrylate with quaternary ammonium functions	Polymethacrylate with quaternary ammonium functions	Polymethacrylate with quaternary ammonium functions	Polymethacrylate aminated with trimethylamine functions	Polymethacrylate with quaternary ammonium functions

[1] Identical to Waters IC Pak A

variety of eluants. In comparison with TSK Gel IC-PW, this stationary phase exhibits a much lower chromatographic efficiency [31]. For example, refer to the separation of simple inorganic anions on ION-110 with a salicylate eluant as shown in Fig. 3-21.

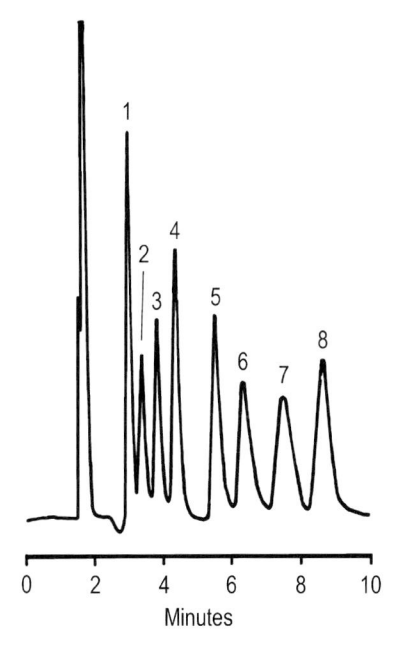

Fig. 3-21. Separation of various inorganic anions on ION-110. — Eluant: 3 mmol/L salicylate (pH 8.2); flow rate: 0.75 mL/min; injection volume and solute concentrations: no reference; peaks: (1) fluoride, (2) bicarbonate, (3) chloride, (4) nitrite, (5) bromide, (6) nitrate, (7) orthophosphate, and (8) sulfate.

Recently, Metrohm (Herisau, Switzerland) introduced two new polyvinyl-based anion exchangers: Metrosep Anion Supp 4 and 5. While the Metrosep Anion Supp 4 has been developed for routine applications such as the analysis of standard anions in all kinds of water, the Metrosep Anion Supp 5 is suited for separating more complex samples, because it exhibits a much higher chromatographic efficiency due to a smaller particle size. The Metrosep Anion Supp 5 is offered in three different column lengths; each size has a different ion-exchange capacity. As can be seen from the chromatogram of standard anions in Fig. 3-22, fluoride is very well separated from the system void when using the 250-mm column. The exceptional resolution between fluoride and short-chain monocarboxylic acids such as acetate, glycolate and lactate, however, results in a relatively long analysis time of approximately 25 minutes for sulfate.

The characteristic structural and technical properties of theses columns are summarized in Table 3-5.

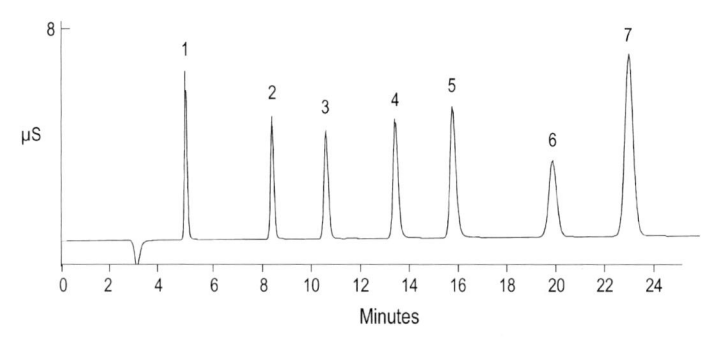

Fig. 3-22. Separation of various inorganic anions on Metrosep Anion Supp 5. − Eluant: 3.2 mmol/L Na_2CO_3 + 1 mmol/L $NaHCO_3$; flow rate: 0.7 mL/min; detection: suppressed conductivity; injection volume: 20 µL; solute concentrations: 2 mg/L fluoride (1), 3 mg/L chloride (2), 5 mg/L nitrite (3), 10 mg/L each of bromide (4) and nitrate (5), 15 mg/L each of orthophosphate (6) and sulfate (7).

3.4.2
Latex-Agglomerated Anion Exchangers

A special type of pellicular anion exchangers was first introduced in 1975 by Small et al. [3] in their introductory paper on ion chromatography. These stationary phases, which are called latex-based anion exchangers, have been further developed by Dionex (Sunnyvale, USA). The structure of these stationary phases is schematically depicted in Fig. 3-23.

Table 3-5. Structural and technical properties of surface-aminated polyvinyl resins.

Column	ION-100 (ION-110)	Metrosep Anion Supp 4	Metrosep Anion Supp 5[1]
Dimensions (Length × I.D.) [mm]	100 × 3.0 (250 × 3.0)	250 × 4.0	100 × 4.0 150 × 4.0 250 × 4.0
Manufacturer	Interaction Chemicals	Metrohm	Metrohm
pH range	0 – 14	3 – 12	3 – 12
Max. pressure [MPa]	10	12	7 11 15
Max. flow rate [mL/min]	1 – 2	2	0.8
Solvent compatibility [%]	10	100	100
Capacity [mequiv/g]	0.1	0.046	0.038 0.057 0.094
Particle diameter [µm]	10	9	5
Type of packing material	Polyvinyl resin with quaternary ammonium functions	Polyvinyl resin with quaternary ammonium functions	Polyvinyl resin with quaternary ammonium functions

[1] Identical to Shodex SI-50 4E

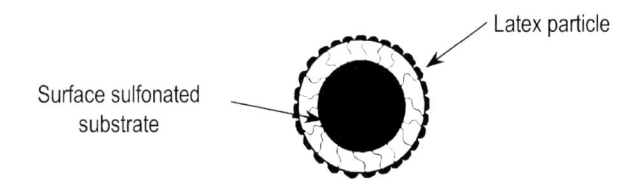

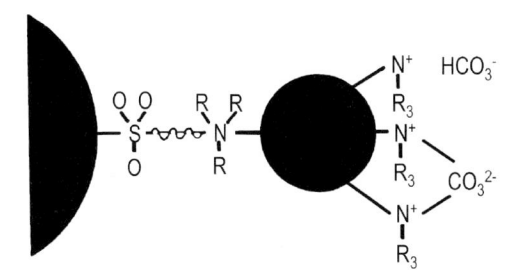

Fig. 3-23. Structure of a latex-based anion exchange resin.

Latex-based anion exchangers are comprised of a surface-sulfonated polystyrene/divinylbenzene substrate with particle diameters between 5 µm and 25 µm and fully aminated, high-capacity porous polymer beads made of polyvinylbenzyl chloride or polymethacrylate, which are called latex particles. The latter have a much smaller diameter (about 0.1 µm) and are agglomerated to

the surface by electrostatic and van-der-Waals interactions. A scanning electron micrograph of this material is shown in Fig. 3-24. The stationary phase features three chemically distinct regions:
- An inert and mechanically stable substrate
- A thin coating of sulfonic acid groups which covers the substrate
- An outer layer of latex beads that carry the quaternary ammonium groups as anion exchange functions

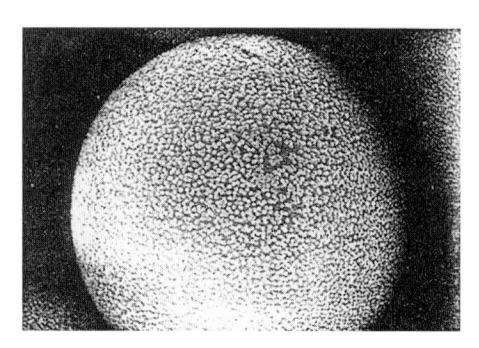

Fig. 3-24. Scanning electron micrograph of a latex-agglomerated anion exchange particle.

Although the latex polymer exhibits a very high ion-exchange capacity due to its complete amination, the small size of the beads finally results in a low anion exchange capacity of about 0.03 mequiv/g. The pellicular structure of these anion exchangers is responsible for their high chromatographic efficiencies. Two parameters play a significant role in this respect:
- Degree of surface sulfonation
- Size of the latex beads

The surface sulfonation of the substrate prevents the diffusion of inorganic species into the inner part of the stationary phase via Donnan exclusion (see Section 5.1). Therefore, the diffusion process is dominated by the functional groups bonded to the latex beads. The size of these beads determines the length of the diffusion paths, and thus the rate of the diffusion process [32].

This complex system offers several advantages compared with other column packings such as silica-based anion exchangers (see Section 3.4.3) and directly aminated resins developed by Fritz et al. [33]. The advantages include:
- The inner substrate provides mechanical stability and a moderate backpressure
- The small size of the latex beads ensures fast ion-exchange processes and thus, a high chromatographic efficiency of the separator column
- Swelling and shrinkage are considerably reduced due to the surface functionalization

Latex-agglomerated anion exchangers are chemically very stable. Even 4 mol/L sodium hydroxide is unable to cleave the ionic bond between the substrate particle and the latex beads. The selectivity is altered by changing the chemical nature of the quaternary ammonium base. Since the latex material is synthesized in a separate step, it is possible to optimize the selectivity of a separator column for a specific analytical problem, either by varying the functional groups bonded to the latex beads or by changing the degree of cross-linking. For the analyst, economical eluants and especially the customized selectivities are of great interest. Since the functionality and the substrate of latex-agglomerated anion exchangers are two separate entities, the Dionex family of such anion exchangers is an adequate example for a discussion of the various parameters affecting selectivity.

3.4.2.1 Review of Different Latex-Agglomerated Anion Exchangers

At present, 16 different anion exchangers (IonPac AS1 to AS17) with diverse selectivities are available from Dionex. The structural and technical characteristics of these separator columns are summarized in Table 3-6. Additionally, special columns for the fast separation of simple inorganic anions as well as for amino acids, peptides, proteins, oligonucleotides, and carbohydrates are also available.

The different selectivities are mainly governed by the following parameters:
- Degree of the latex cross-linking
- Type of functional groups attached to the latex bead

The ion-exchange capacity, on the other hand, is determined by:
- Substrate particle size
- Size of the latex beads
- Degree of latex coverage on the substrate surface

While the ion-exchange capacities of these separator columns are directly proportional to both the size of the latex beads and the degree of latex coverage on the substrate surface, they are inversely proportional to the particle size of the substrate.

For historical reasons, Fig. 3-25 shows the separation of standard anions on the polystyrene/divinylbenzene-based IonPac AS1. Introduced in 1975, this separation was obtained with a carbonate/bicarbonate eluant mixture. The elution order of anions such as chloride and sulfate, which are separated exclusively via anion exchange processes, cannot be altered by varying chromatographic conditions. Exceptions are polyvalent ions, because their valency and thus their retention are a function of the pH value. On the other hand, the elution order for anions such as bromide and nitrate, for which the separation is mainly based on a different adsorption behavior (see Section 3.3), is governed by the hydrophobicity of the ion-exchange functional groups. If this hydrophobicity is increased, bromide and nitrate are more strongly retained. This effect is of particular importance for the analysis of anions in nitric acid digests, which cannot be carried out on conventional anion exchangers with hydrophilic functional groups due to the large excess of nitrate anions. Developed for this application, the

Table 3-6. Structural and technical properties of latex-agglomerated anion exchangers.

Separator	Particle diameter [µm]	Degree of cross-linking [%]	Size of latex particle [nm]	Applicability
IonPac AS1	25	5	150	Universal anion exchanger
IonPac AS2	25	5	150	Anion exchanger with high affinity to bromide and nitrate
IonPac AS3	25	2	100	Anion exchanger for highly contaminated samples
IonPac AS4	15	3.5	75	High-performance separator for less contaminated samples
IonPac AS4A-SC	13	0.5	160	Universal high-performance separator
IonPac AS5	15	0.5	180	Anion exchanger for polarizable anions
CarboPac PA1 (IonPac AS6)	10	5	350	Special separator for sulfide and cyanide
IonPac AS7	10	5	350	Special separator for polyvalent anions
IonPac AS9-SC	13	20	110	Universal high-performance separator, specially suited for oxyhalides
IonPac AS9-HC	9	15	90	Universal high-capacity, high-performance separator, specially suited for oxyhalides
IonPac AS10	8.5	5	65	High-capacity separator with high affinity to bromide and nitrate
IonPac AS11	13	6	85	High-performance separator for gradient elution
IonPac AS11-HC	9	6	70	High-capacity, high-performance separator for gradient elution
IonPac AS12A	9	0.2	140	High-performance separator, specially suited for fluoride and oxyhalides
IonPac AS16	9	1	200	High-performance separator for polarizable anions
IonPac AS17	10.5	6	75	High-performance separator for gradient elution of standard anions

IonPac AS2 differs from the IonPac AS1 only in the type of functional groups, which are considerably more hydrophobic. To achieve a baseline-resolved separation between chloride and orthophosphate, and between sulfate and bromide the eluant has to be slightly modified. The standard chromatogram of the AS2 column shown in Fig. 3-26 has been obtained using a carbonate/hydroxide eluant.

The elution profile of the separator column AS3 (Fig. 3-27) with considerably shorter analysis times is obtained by reducing both the degree of cross-linking and the size of the latex beads. The comparison between the AS1 and AS3 columns demonstrates impressively how the analysis time for a baseline-resolved

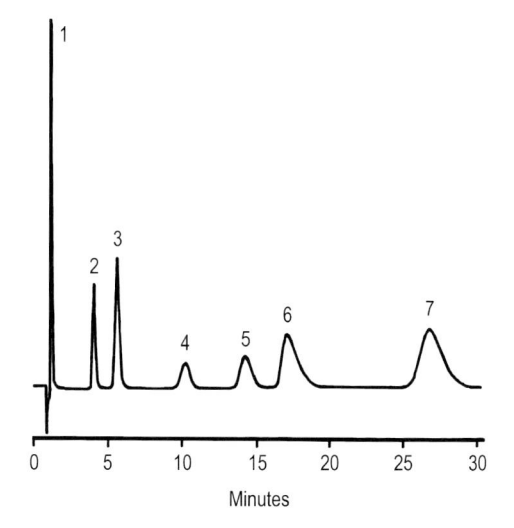

Fig. 3-25. Separation of various inorganic anions on IonPac AS1. – Eluant: 2.8 mmol/L $NaHCO_3$ + 2.2 mmol/L Na_2CO_3; flow rate: 2.3 mL/min; detection: suppressed conductivity; injection volume: 50 µL; solute concentrations: 3 mg/L fluoride (1), 4 mg/L chloride (2), 10 mg/L nitrite (3), 10 mg/L orthophosphate (4), 10 mg/L bromide (5), 20 mg/L nitrate (6), and 25 mg/L sulfate (7).

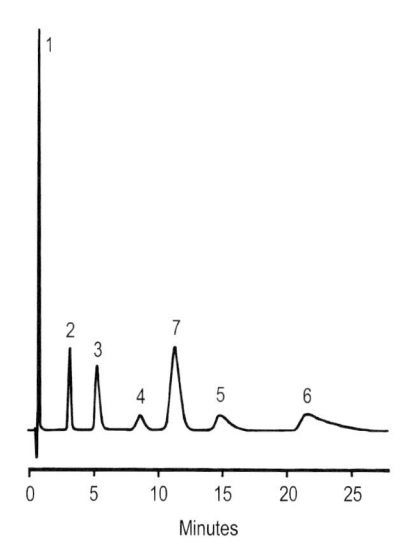

Fig. 3-26. Separation of various inorganic anions on IonPac AS2. – Eluant: 3 mmol/L Na_2CO_3 + 2 mmol/L NaOH; other chromatographic conditions: see Fig. 3-25.

separation of the seven standard anions has been reduced by about 60% just by improving the latex technology. Yet the particle diameter of 25 µm was held constant. In terms of chromatographic efficiency, ion-exchange capacity, and resistance to column fouling, the IonPac AS3 was the final product in the development of latex-agglomerated anion exchangers of the *first* generation. It is particularly suited for the analysis of wastewater and other heavily contaminated samples featuring large concentration differences between the individual sample components.

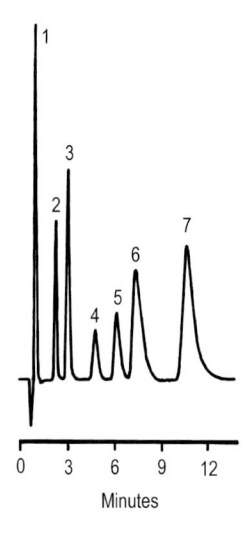

Fig. 3-27. Separation of various inorganic anions on IonPac AS3. – Chromatographic conditions: see Fig. 3-25.

The objective for the *second* generation of latex-agglomerated anion exchangers was a further reduction of the retention times by decreasing the substrate particle size. The concurrent increase of the chromatographic efficiency results in a significantly higher column backpressure. This development was initiated in 1982 with the introduction of the IonPac AS4, the selectivity of which is identical to the IonPac AS1. Although the chemical nature of the functional groups and the physical properties of the latex beads, such as their size and their degree of cross-linking, were altered only slightly, a reduction in the retention times by another 50% in comparison to the AS3 column was accomplished by decreasing the substrate particle size to 15 µm. Disadvantageous is the low ion-exchange capacity and the less pronounced ruggedness of this high-performance separator column. Today, this column is only used in combination with IonPac AS4A-SC for the analysis of inorganic and organic anions in beverages (see Section 9.6).

This disadvantage was overcome four years later when a universal high-performance separator column – the IonPac AS4A – was introduced. This column combines the positive properties of both IonPac AS3 and AS4 separators. While the particle size of the substrate is identical to the AS4 column, the loading capacity could be brought up to the capacity of the AS3 column by significantly reducing the degree of cross-linking. The tailing of the bromide and nitrate peaks, which is observed with the AS4 column could be reduced by altering the chemical nature of the functional groups. A prerequisite for this is the stronger hydrophilicity of the functional groups of the AS4A separator. In comparison to the AS4 column, bromide and nitrate are less strongly retained by means of adsorption. In other ions, the modification in the functionality of the latex beads does not result in any noticeable change of the retention behavior. The only exception to this is orthophosphate, which would elute much later with the standard carbonate/bicarbonate eluant mixture and thus interferes with the nitrate peak. A slight change in the concentration ratio of the two eluant components leads to the standard elution profile of the AS4A column (Fig. 3-28).

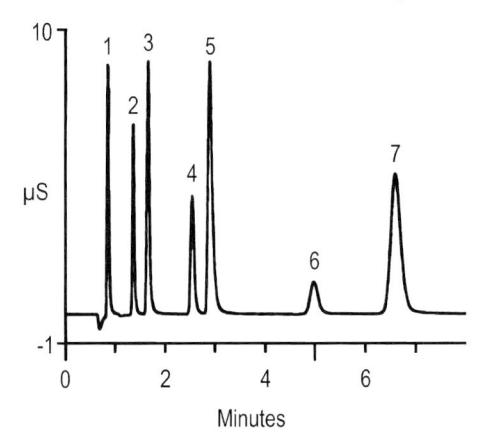

Fig. 3-28. Elution profile of the IonPac AS4A (-SC) separator column for standard inorganic anions. — Eluant: 1.7 mmol/L $NaHCO_3$ + 1.8 mmol/L Na_2CO_3; flow rate: 2 mL/min; detection: suppressed conductivity; injection volume: 50 µL; solute concentrations: 3 mg/L fluoride (1), 4 mg/L chloride (2), 10 mg/L each of nitrite (3) and bromide (4), 20 mg/L nitrate (5), 10 mg/L orthophosphate (6), and 25 mg/L sulfate (7).

A limiting factor for pellicular anion exchangers based on polystyrene/divinylbenzene is their incompatibility with conventional HPLC solvents such as methanol and acetonitrile, a result of the low degree of cross-linking in the support material. Since ion chromatographic applications in the field of inorganic anions are usually carried out in a purely aqueous medium, separator columns with a low degree of cross-linking are perfectly suitable. However, in the field of organic ion analysis there is a demand for polymer-based, solvent-compatible anion exchangers. Solvent compatibility can be achieved by using highly cross-linked support materials. Especially suitable are microporous ethylvinylbenzene/divinylbenzene (EVB/DVB) copolymers. With 55% cross-linking, 100% solvent compatibility with common HPLC solvents such as methanol and acetonitrile is achieved [34]. After sulfonating the surface in the usual way, latex beads can be agglomerated. The first column of this kind has been commercialized under the name IonPac AS4A-SC. Since sulfonation of the support material and the physico-chemical properties remained unchanged, the standard chromatogram of the AS4A-SC is identical to that of the AS4A column. The introduction of polymer-based, solvent-compatible anion exchangers opened a new dimension in anion exchange chromatography. Organic solvents can now be used for removing organic contaminants as well as for modifying ion-exchange selectivities (see Section 3.7.2).

The IonPac AS5, a special 15-µm anion exchanger, was developed for the analysis of strongly polarizable ions such as iodide, thiocyanate, and thiosulfate. Since the kind of support material and its particle diameter are not any different from the IonPac AS4, the resulting retention times are comparable. The different selectivity of the AS5 column is based on the hydrophilic properties of the functional groups bonded to the latex beads; the hydrophilicity prohibits a separation of bromide and nitrate. However, this hydrophilicity is a prerequisite for the separation of polarizable anions. Due to their large radii, such anions are strongly retained by adsorptive forces at the stationary phase of anion exchangers with latex particles made of polyvinylbenzyl chloride. Thus, polarizable anions give rise to a strong tailing effect. Taking the separation of iodide and thiocyanate as an example, Fig. 3-29 shows that this tailing can be significantly reduced by adding p-cyanophenol to the eluant, which is especially suited in this respect. Today, the IonPac AS5 is outdated and has been replaced by the IonPac AS16 (see Fig. 3-47).

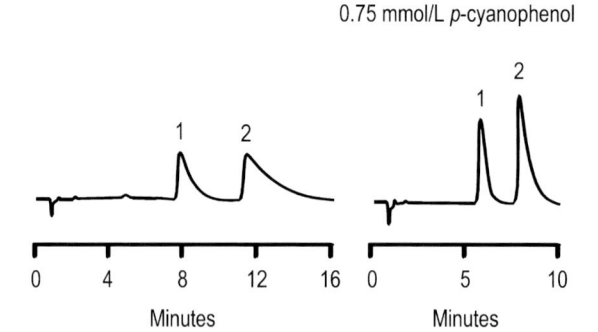

Fig. 3-29. Influence of p-cyanophenol on the tailing properties of iodide and thiocyanate separated on IonPac AS5. – Chromatographic conditions: see Fig. 3-1.

Alternatively, these anions may be separated at a stationary phase that was originally developed for the chromatographic analysis of transition metals and has a defined anion and cation exchange capacity (see Section 4.5.3). This separator column (IonPac CS5) contains a surface-sulfonated PS/DVB substrate with a particle diameter of 13 µm and a degree of cross-linking of about 2%. Like the AS5 column, the CS5 column has also latex beads of about 150 nm. However, the functionality of the AS5 column was replaced by another one with a similar hydrophilicity. Iodide and thiocyanate are separated at this stationary phase using a carbonate/bicarbonate mixture without any organic modifiers. The corresponding chromatogram differs slightly in analysis time and peak shape from the chromatogram in Fig. 3-29. In terms of its loading capacity, the CS5 column is inferior to the AS5 column because of its higher degree of cross-linking.

The *simultaneous* separation of polarizable and non-polarizable anions will be discussed below (see IonPac AS9-SC).

The CarboPac PA1 (IonPac AS6), originally developed for carbohydrate analysis, is based on a 10-µm support material to which relatively large latex beads are bonded. The comparatively high ion-exchange capacity is necessary, because high ionic strength eluants are required for carbohydrate analysis. In order to separate carbohydrates by anion exchange, they must first be converted into the ionic form. Because the pK values of carbohydrates and sugar alcohols are between 12 and 14, 0.1 to 0.15 mol/L NaOH is employed as eluant.

The CarboPac PA1 column is also suitable for the separation of sulfide and cyanide (see Section 3.7.2). Due to its high ion-exchange capacity the strongly alkaline absorber solution (c = 1 mol/L) described in the DIN 38405 (Part 13) method [35] can be injected directly. Conventional anion exchangers such as the IonPac AS4A-SC would be completely overloaded. With the CarboPac PA1 column, this problem can be solved by using the NaOH absorber solution in combination with sodium acetate as the eluant.

Monocarboxylic acids, which elute in the void volume of the AS4A-SC column, are strongly retained when a CarboPac PA1 column is conditioned with a carbonate/bicarbonate buffer (see Section 3.8.1).

The AS7 separator column was developed especially for the analysis of polyvalent anions. In the past, polyvalent anions such as aminopolycarboxylic and aminopolyphosphonic acids could not be analyzed by ion chromatography, because their retention times increased drastically with increasing charge number. Moreover, strong adsorptive forces, caused by the organic skeleton of these compounds, resulted in significant tailing effects. However, if strongly acidic mobile phases of a high elution strength are utilized (see Section 3.8.2), polyvalent anions elute as symmetrical peaks.

Both CarboPac PA1 and IonPac AS7 columns have in common a relatively low loading capacity resulting from the high degree of cross-linking in the substrate material. The functionality of the latex beads corresponds to that of a conventional anion exchanger, for example, the AS3 column.

The second generation of separator columns, with particle diameters between 10 μm and 15 μm, enable high chromatographic efficiencies and short analysis times. It was assumed that further efforts in this direction were not appropriate, as they would result in a further decrease of the loading capacity. Even samples with low electrolyte content would have to be diluted in order to avoid an overloading of the separator column. The time required for the sample preparation would in no way negate the advantages of a shorter analysis time. However, once again, this train of thought is only valid, if the reduction in analysis time is achieved by the reduction of the particle diameter and by the miniaturization of the separator column. Experiments with 5-μm substrate particles and column dimensions of 35 mm × 3 mm i. d. were indeed carried out. The experimental results confirmed that such an approach is pointless, especially in view of optimizing the chromatographic system. Only under laboratory conditions could the analysis time for the seven most important inorganic anions be reduced to less than two minutes. In addition, a chromatographic efficiency of about 50,000 theoretical plates per meter – related to sulfate – was achieved. However, because of handling and reliability problems, the utilization of this system in routine analysis is inconceivable.

Reducing the analysis times by increasing the flow rate and/or the eluant concentration, is also not feasible due to an insufficient sensitivity and ion-exchange capacity.

Only the use of various *acrylate-based* latex polymers aided the successful development of a separator column with analysis times shorter than three minutes. In 1987, the Fast-Sep Anion column was introduced [36]. The substrate particle size and the elution order of the seven most important inorganic anions do not differ from those of the AS4A-SC column, which for a long time represented the state-of-the-art in chromatographic anion analysis. However, the amination was carried out using a different amine. Figure 3-30 displays the elution profile of the Fast-Sep column, which is again obtained by using a carbonate/bicarbonate mixture as the mobile phase.

Fig. 3-30. Elution profile of a Fast-Sep anion exchange column. — Eluant: 0.15 mmol/L NaHCO$_3$ + 2 mmol/L Na$_2$CO$_3$; flow rate: 2 mL/min; detection: suppressed conductivity; injection volume: 20 µL; solute concentrations: 1.5 mg/L fluoride (1), 2.5 mg/L chloride (2), 7.5 mg/L each of nitrite (3) and bromide (4), 10 mg/L each of nitrate (5) and sulfate (6), and 15 mg/L orthophosphate (7).

To avoid overloading the column, the injection volume should be limited to 20 µL. Upon installation in an existing chromatographic system, a short tubing should be employed to obtain optimum resolution between the individual signals. However, such a column is only suitable for analyzing slightly contaminated samples (e.g., potable waters and surface waters) in which the primary components such as chloride, nitrate, bromide, and sulfate can be separated to baseline. For samples with complex matrices, the separation efficiency of this column is usually insufficient to separate major and minor components. Compared to the conventional exchanger AS4A-SC, the Fast-Sep column has the advantage of a higher sensitivity, especially for later-eluting compounds such as orthophosphate and sulfate that have much shorter retention times and, thus, elute as sharper signals. Although the ion-exchange capacity of the Fast-Sep column is only half that of the AS4A column, the values for the loading capacity — at least for monovalent anions — are comparable. The loading capacity is defined as the solute concentration at which the peak efficiency falls below 90% of its original value. As seen in Table 3-7, both columns differ in their loading capacity for sulfate, which is twice as high for a conventional anion exchanger. This phenomenon clearly establishes the higher affinity of divalent anions towards styrene/divinylbenzene-based latex polymers.

Table 3-7. Comparison of the loading capacity between the Fast-Sep column and the conventional IonPac AS4A-SC anion exchanger.

Solute ion	Fast-Sep Anion [ppm][1]	[ng]	IonPac AS4A [ppm][2]	[ng]
Chloride	30	600	12	600
Sulfate	35	700	30	1500

[1] Injection volume: 20 µL   [2] Injection volume: 50 µL

Compared with conventional anion exchangers such as the AS4A-SC column, the acrylate-based latex polymer of the Fast-Sep column exhibits a lower pH stability. However, long-term experiments with the Fast-Sep column showed no deterioration as long as the pH value of the mobile phase was kept between 2 and 11. The pH value of the sample to be analyzed should not exceed 13.

By agglomerating acrylate-based latex beads on polystyrene/divinylbenzene substrates, a separation of chlorate and nitrate can also be achieved; this was previously not possible using conventional anion exchangers. These two anions exhibit the same interactions with both polyvinylbenzene-based latexed anion exchangers and directly aminated substrates and, thus, co-elute at these stationary phases. In this case, ion-pair chromatography was used for this separation, which allows baseline separation of chlorate and nitrate due to their different sizes. This selectivity problem was solved with the IonPac AS9 anion exchanger, which was developed specifically for the separation of chlorate and nitrate. With 35 µequiv/g, the ion-exchange capacity of the AS9 column is slightly higher than that of the AS4A-SC. The latex beads are functionalized with a medium hydrophobic tertiary amine. Figure 3-31 shows a standard chromatogram obtained using a carbonate/bicarbonate mixture as an eluant. To obtain optimum resolution among bromide, chlorate, and nitrate, it is advisable to use an eluant flow rate of 1 mL/min, even though this increases the analysis time for sulfate to about 20 minutes. With this stationary phase, chlorite and bromate are also much better resolved from chloride, which allows the determination of comparatively low concentrations of chlorite and bromate in the presence of high amounts of chloride. This analytical problem becomes increasingly important in the analysis of potable water that is disinfected with chlorine dioxide or ozone (see Section 9.1). As can also be seen in Fig. 3-31, a much better separation between sulfite and sulfate is obtained with IonPac AS9 as compared with IonPac AS4A. Even large concentration differences between these two anions do not prohibit their determination. Therefore, this stationary phase is especially suited for the analysis of flue gas scrubber solutions in desulfurization plants,

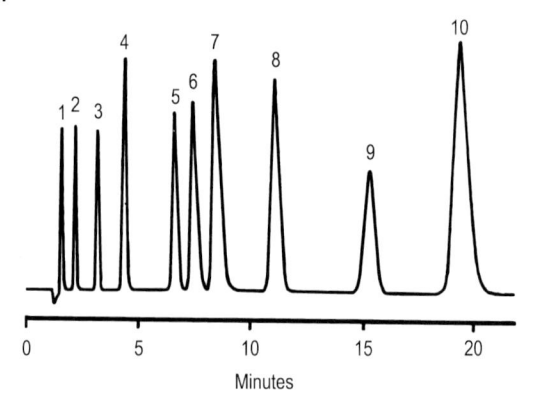

Fig. 3-31. Separation of standard inorganic anions on IonPac AS9 (-SC). — Eluant: 1.7 mmol/L NaHCO$_3$ + 1.8 mmol/L Na$_2$CO$_3$; flow rate: 1 mL/min; detection: suppressed conductivity; injection volume: 50 µL; solute concentrations: 1 mg/L fluoride (1), 5 mg/L chlorite (2), 1.5 mg/L chloride (3), 6 mg/L nitrite (4), 10 mg/L bromide (5), 15 mg/L each of chlorate (6) and nitrate (7), 20 mg/L each of orthophosphate (8) and sulfite (9), and 25 mg/L sulfate (10).

where similar analytical problems occur. Compared with conventional anion exchangers such as the AS4A-SC column, acrylate-based latex polymers are less pH stable.

A remarkable property of the AS9 column with its acrylate-based latex beads is the relatively short retention times of polarizable anions such as iodide, thiocyanate and thiosulfate, for which the IonPac AS5 has been used in the past (see Fig. 3-29). On IonPac AS9 (-SC), these anions, together with mineral acids, can be separated within 20 minutes using a sodium carbonate eluant (Fig. 3-32). Although retention times are very different for non-polarizable and polarizable anions, the latter ones do not exhibit tailing. Therefore, the AS5 column is currently only used for analyzing polarizable anions if short retention times are required.

Acrylate-based latex beads can also be agglomerated on a 55% cross-linked ethylvinylbenzene/divinylbenzene (EVB/DVB)-based microporous support material, which represents the solvent-compatible version of this stationary phase (IonPac AS9-SC). Both separators exhibit the same selectivity, so that a particular method can be transferred from one column to the other.

Using a macroporous support with a pore size of 200 nm instead of a microporous one, the acrylate-based latex beads can also be agglomerated electrostatically inside the pores after sulfonation. Based on this technology, a solvent-compatible, pellicular packing material with a relatively high ion-exchange capacity of about 190 µequiv/column (250 mm × 4 mm i. d.) is obtained. This development was commercialized under the trade name IonPac AS9-HC. Figure 3-33 schematically shows the structure of such a macroporous support material.

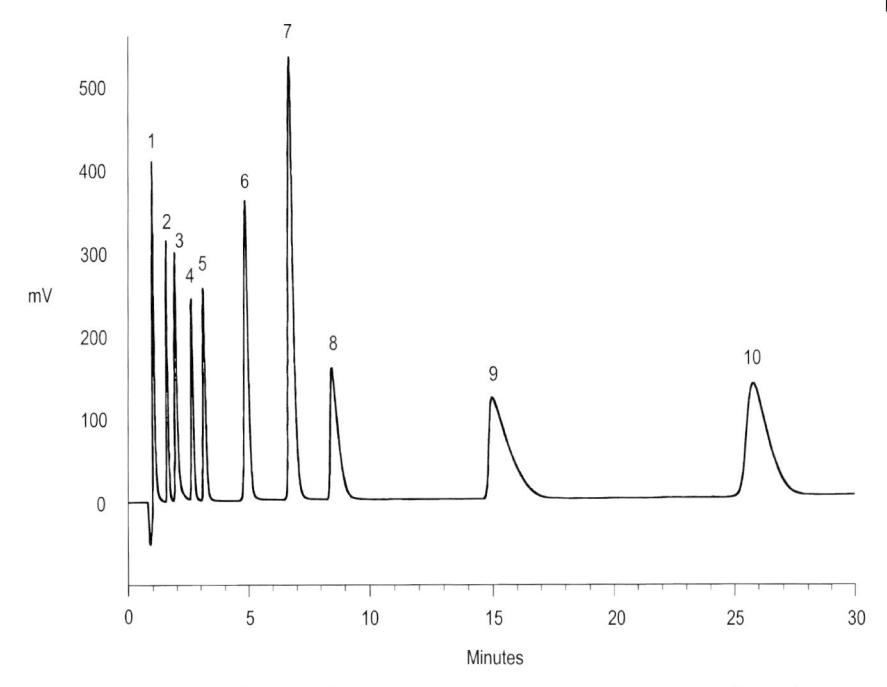

Fig. 3-32. Separation of polarizable and non-polarizable anions on IonPac AS9 (-SC). — Eluant: 3 mmol/L Na_2CO_3; flow rate: 2 mL/min; detection: suppressed conductivity; injection volume: 25 µL; solute concentrations: 2 mg/L fluoride (1), 2 mg/L chloride (2), 5 mg/L nitrite (3), 5 mg/L bromide (4), 5 mg/L nitrate (5), 20 mg/L orthophosphate (6), 10 mg/L sulfate (7), 20 mg/L iodide (8), 20 mg/L thiocyanate (9), and 30 mg/L thiosulfate (10).

Both IonPac AS9-HC and IonPac AS9-SC are suitable for the analysis of standard anions and oxyhalides in potable water and groundwater. Due to its higher capacity, the IonPac AS9-HC is operated with a 9 mmol/L carbonate eluant. A representative standard chromatogram is shown in Fig. 3-34.

In comparison with the AS9-SC column, the AS9-HC column offers improved resolution between bromate/chloride, chloride/nitrite, and chlorate/nitrate pairs. The resolution between fluoride and the system void is also increased significantly. The IonPac AS9-HC column is especially suited for trace analysis of bromate in potable water (see Section 9.1), because it is compatible with large volume injections.

Following the 1987 introduction of the gradient elution of anions in combination with a subsequent conductivity detection, the *third* generation of latex-based anion exchangers with 5-µm substrate particles was developed. The small latex beads agglomerated on the substrate surface are barely visible in the scanning electron micrograph shown in Fig. 3-35. As already discussed in connection with the Fast-Sep column, the objective of the third generation improvements was not a further reduction of the analysis time. Rather, the decrease of the

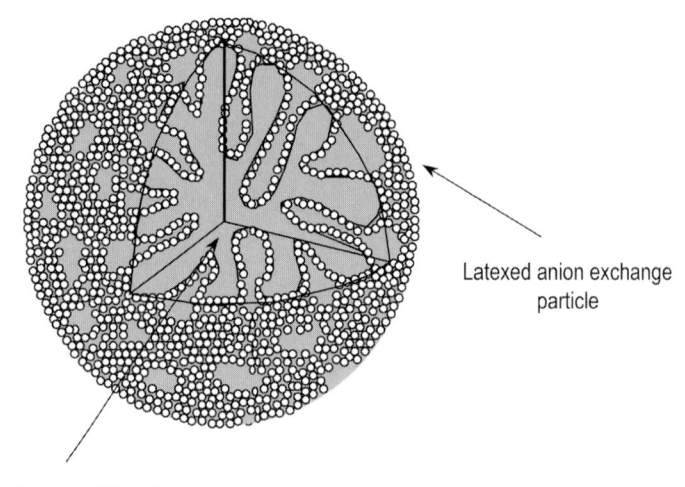

Fig. 3-33. Schematic representation of the structure of a pellicular, macroporous support material.

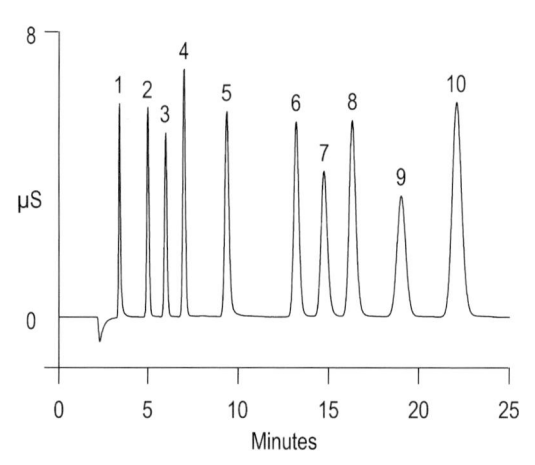

Fig. 3-34. Separation of standard anions and oxyhalides on IonPac AS9-HC. — Eluant: 9 mmol/L Na_2CO_3; flow rate: 1 mL/min; detection: suppressed conductivity; injection volume: 25 µL; solute concentrations: 3 mg/L fluoride (1), 10 mg/L chlorite (2), 20 mg/L bromate (3), 6 mg/L chloride (4), 15 mg/L nitrite (5), 25 mg/L each of bromide (6), chlorate (7) and nitrate (8), 40 mg/L orthophosphate (9), and 30 mg/L sulfate (10).

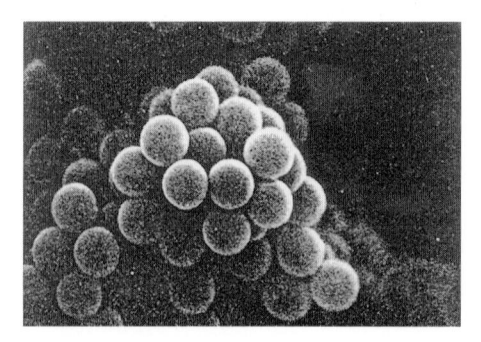

Fig. 3-35. Raster electron micrograph of a 5-µm anion exchange particle.

particle diameter to 5 µm was aimed at a further increase of the chromatographic efficiency and, thus, of the separation power. However, this development offers no advantage for the isocratic analysis of inorganic anions, because longer retention times at lower resolution between bromide and nitrate are obtained with the eluant mixture developed for the IonPac AS4A. This is not surprising, because the 5-µm exchanger can only be operated with a flow rate of 1 mL/min due to the considerably higher back pressure. The lower resolution between bromide and nitrate is caused by the functionality of the latex material, which is identical to the AS5 column. The degree of cross-linking and the size of the latex beads were modified, so that the loading capacity of the AS5A column corresponds to that of the AS4 column. Therefore, the elution profile is very similar to that of the AS5 column. The comparatively better resolution between bromide and nitrate can be attributed to the higher chromatographic efficiency of the AS5A column.

A high separation efficiency is obtained with the AS5A column when it is employed for the gradient elution of anions (see Section 3.9). The objective of the gradient elution technique is to analyze anions having a wide range of retention characteristics − e.g., monovalent, divalent, and trivalent anions − within the same chromatographic run via a gradual increase of the ionic strength of the mobile phase. Considering that a suppressor system is essential for the subsequent conductivity detection, a successful gradient elution is only feasible using a sodium hydroxide eluant. Because sodium hydroxide as a monovalent eluant ion exhibits only a low elution power, the functionality and dimension of the latex material were adjusted respectively. The chromatogram in Fig. 3-36 impressively demonstrates the high separation efficiency of this column when applying the gradient technique with sodium hydroxide as the mobile phase.

The gradient elution of different inorganic and organic anions in Fig. 3-36 illustrates the enormous power of anion exchange chromatography with latex-based anion exchangers of the *third* generation, but it no longer represents the state-of-the-art in the resulting analysis times. The relatively small particle diameter of the latex material (60 nm) results in a low ion-exchange capacity, which easily leads to column overloading when analyzing samples with large concentration differences between the analytes. Furthermore, a long equilibration time after ending the gradient run leads to total analysis time of 45 minutes, which is also obviously problematic.

For this reason, every effort was made to develop a separator column with a high affinity towards hydroxide ions that would allow the elution of polyvalent anions with a much lower NaOH concentration. Other development goals included:

- Increase of ion-exchange capacity by using larger latex particles
- Higher selectivity
- Moderate column backpressure by increasing the substrate particle diameter
- Provide solvent compatibility by utilizing a highly cross-linked EVB/DVB substrate

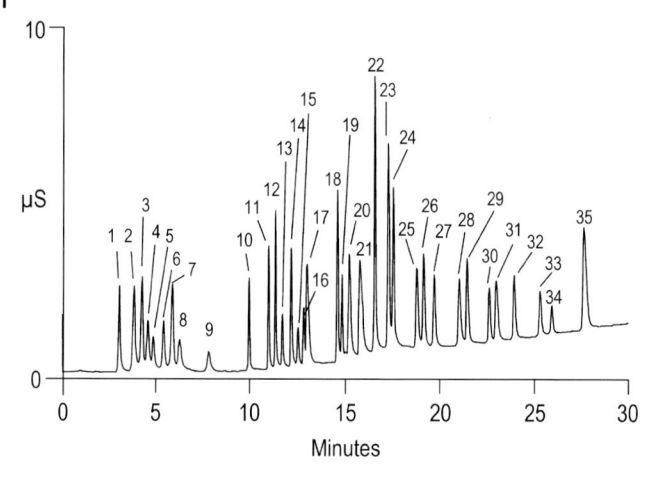

Fig. 3-36. Gradient elution of inorganic and organic anions on IonPac AS5A. – Eluant: (A) 0.75 mmol/L NaOH, (B) 0.1 mol/L NaOH; gradient: 100% A isocratically for 5 min, then linearly to 30% B in 15 min, then linearly to 86% B in 15 min; flow rate: 1 mL/min; detection: suppressed conductivity; injection volume: 50 µL; solute concentrations: 1.5 mg/L fluoride (1), 10 mg/L each of acetate (2), α-hydroxybutyrate (3), butyrate (4), glycolate (5), gluconate (6), and α-hydroxyvalerate (7), 5 mg/L formate (8), 10 mg/L each of valerate (9), pyruvate (10), monochloroacetate (11), and bromate (12), 3 mg/L chloride (13), 10 mg/L galacturonate (14), 5 mg/L nitrite (15), 10 mg/L each of glucoronate (16), dichloroacetate (17), trifluoroacetate (18), phosphite (19), selenite (20), bromide (21), nitrate (22), sulfate (23), oxalate (24), selenate (25), a-ketoglutarate (26), fumarate (27), phthalate (28), oxaloacetate (29), phosphate (30), arsenate (31), chromate (32), citrate (33), isocitrate (34), cis-aconitate (35), and trans-aconitate (36).

These development goals were realized with the IonPac AS11. The diameter of the latex beads has been increased to 85 nm, yielding an ion-exchange capacity of 45 µequiv/column. As with all solvent-compatible columns, a highly cross-linked EVB/DVB polymer serves as substrate material. The AS11 particle diameter of 13 µm is large enough to ensure a flow rate of 2 mL/min at moderate backpressure.

The affinity of hydroxide ions to the stationary phase is determined by the degree of hydration of the quaternary ammonium functional groups. This, in turn, depends on the number of hydroxide groups and their distance from the ion-exchange functional group. With increasing number of hydroxide groups and with decreasing distance of these hydroxide groups to the ion-exchange functionality, the region around the ion-exchange functional groups will be more strongly hydrated and, thus, be more selective for hydroxide ions. Because hydroxide ions are strongly hydrated themselves, they predominantly enter this region and are, therefore, more strongly retained. In other words: With increasing degree of hydration of the quaternary ammonium functional groups, the affinity of hydroxide ions to the stationary phase increases and, thus, the elution power increases. A practical way to achieve this is the synthesis of quaternary ion-exchange functional groups with alkyl chains carrying hydroxide groups or the synthesis of a monomer with hydroxide groups in the direct vicinity of the ion-exchange functional groups. In the development of the IonPac AS11, both pathways were used. The difference in hydroxide selectivity between IonPac AS4A-SC and AS11 is demonstrated by comparing the two chromatograms; one was obtained with a pure carbonate and the other one with a pure hydroxide eluant (see Fig. 3-37).

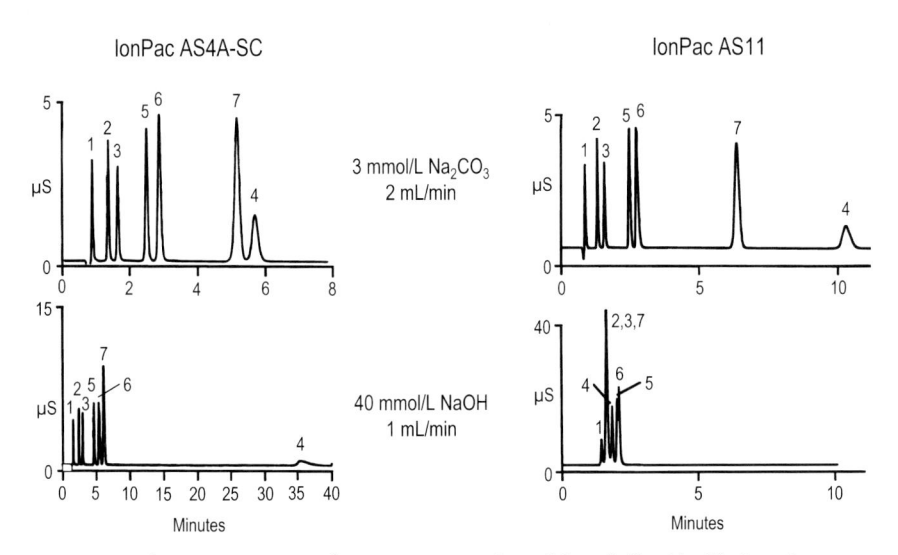

Fig. 3-37. Selectivity comparison for hydroxide ions between IonPac AS4A-SC and AS11. − Detection: suppressed conductivity; injection volume: 25 µL; solute concentrations: 1.5 mg/L fluoride (1), 3 mg/L chloride (2), 5 mg/L nitrite (3), 15 mg/L orthophosphate (4), 10 mg/L bromide (5), 10 mg/L nitrate (6), and 15 mg/L sulfate (7).

As can be seen from both chromatograms on the left side, the less hydrated carbonate is the better eluant for the IonPac AS4A-SC, which has an ion-exchange functionality carrying only *one* hydroxide group. Although orthophosphate can be eluted with 0.04 mol/L NaOH, the resulting retention time of 35 minutes is rather high. In contrast, there are three hydroxide groups in the region around the quaternary ammonium functionality of the AS11 column, yielding a much higher degree of hydration. Therefore, sodium hydroxide is a very effective eluant for this separator column, which is demonstrated by the extremely short retention times for the seven most important inorganic anions. Even the trivalent orthophosphate elutes before the monovalent bromide and nitrate with 0.02 mol/L NaOH. As can be seen from Fig. 3-38, the resolution between mono- and divalent anions can even be increased by employing a shallow NaOH gradient without a significant increase of the total analysis time. Chloride and nitrite as well as bromide and nitrate are better resolved under these conditions.

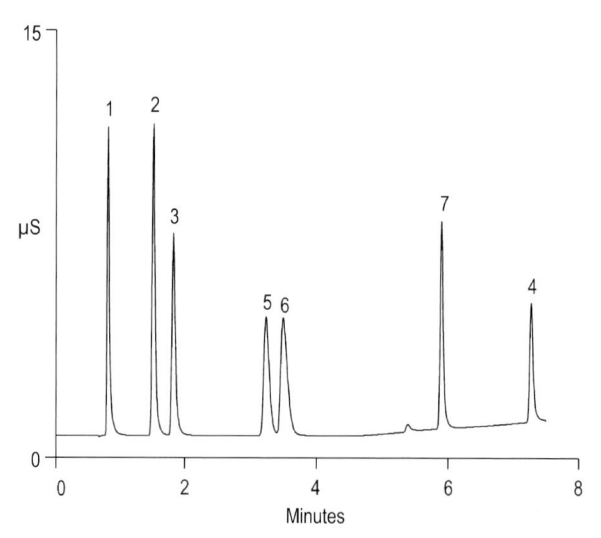

Fig. 3-38. Fast gradient separation of mineral acids on IonPac AS11. — Eluant: (A) water, (B) 0.1 mol/L NaOH; gradient: 5% B isocratically for 3 min, then linearly to 45% B in 8 min; flow rate: 2 mL/min; detection: suppressed conductivity; injection: 25 µL anion standard with 2.5 mg/L fluoride (1), 5 mg/L each of chloride (2) and nitrite (3), 10 mg/L orthophosphate (4), 10 mg/L bromide (5), 10 mg/L nitrate (6), and 5 mg/L sulfate (7).

In comparison to Fig. 3-36, Fig. 3-39 shows the significant time saving when separating various inorganic and organic anions on IonPac AS11, although the final NaOH concentration is only 38 mmol/L.

Due to its solvent compatibility, the selectivity of the IonPac AS11 can be altered by adding organic solvents. This effect is demonstrated with the gradient elution of inorganic and organic anions in Fig. 3-40. When eluting with purely aqueous NaOH, a separation of, for example, succinate/malate, malonate/tartrate, and fumarate/sulfate pairs is not possible. A baseline-resolved separation of these anions can only be achieved by adding 16% (v/v) methanol to the mobile phase. This leads to a shorter retention time of the more surface-active anion, respectively. It is remarkable that in some cases the retention times of the less surface-active anions actually increase.

Organic solvents not only influence the selectivity for surface-active anions. The retention of polarizable anions, which are retained by ion-exchange and adsorptive interactions, can be reduced significantly by organic solvents. Figure

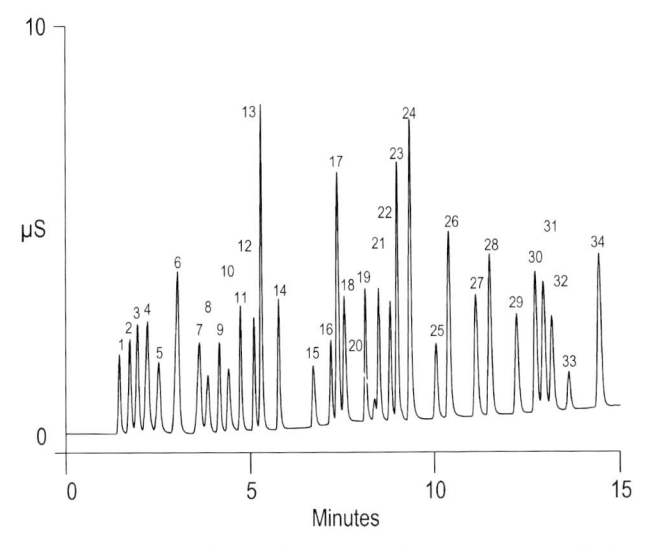

Fig. 3-39. Gradient elution of inorganic and organic anions on IonPac AS11. — Eluant: (A) water, (B) 5 mmol/L NaOH, (C) 0.1 mol/L NaOH; gradient: 2 min 90% A + 10% B isocratically, then linearly to 100% B in 3 min, then linearly to 65% B + 35% C in 10 min; flow rate: 2 mL/min; detection: suppressed conductivity; injection volume: 25 µL; solute concentrations: 5 mg/L each of isopropylethylphosphonate (1) and quinate (2), 1 mg/L fluoride (3), 5 mg/L each of acetate (4), propionate (5), formate (6), methylsulfonate (7), pyruvate (8), chlorite (9), valerate (10), monochloroacetate (11), and bromate (12), 2 mg/L chloride (13), 5 mg/L each of nitrite (14) and trifluoroacetate (15), 3 mg/L each of bromide (16), nitrate (17), chlorate (18), and selenite (19), carbonate (20), 5 mg/L each of malonate (21), maleate (22), sulfate (23), and oxalate (24), 10 mg/L each of ketomalonate (25), tungstate (26), phthalate (27), orthophosphate (28), chromate (29), citrate (30), tricarballylate (31), isocitrate (32), *cis*-aconitate (33), and *trans*-aconitate (34).

3-41 shows an isocratic separation of iodide, thiocyanate, and thiosulfate together with the five most important mineral acids, which can be achieved with a NaOH/MeOH eluant mixture. The high methanol content in the mobile phase (40% v/v) allows the separation of these anions in less than 10 minutes.

Like the IonPac AS9-HC, the AS11 is also available in a high-capacity version, the IonPac AS11-HC. Its support material is identical to that of the IonPac AS9-HC (see Fig. 3-32). The relatively high capacity of the AS11-HC (290 µequiv for a 250 mm × 4 mm i. d. column) not only allows the injection of more concentrated samples, which would cause column overloading and peak broadening on IonPac AS11, but also the isocratic separation of the seven standard anions with 30 mmol/L NaOH. As can be seen from the correlative chromatogram in Fig. 3-42, such a separation can be obtained in less than 10 minutes. Bromide and nitrate are more strongly retained than sulfate under these conditions; this

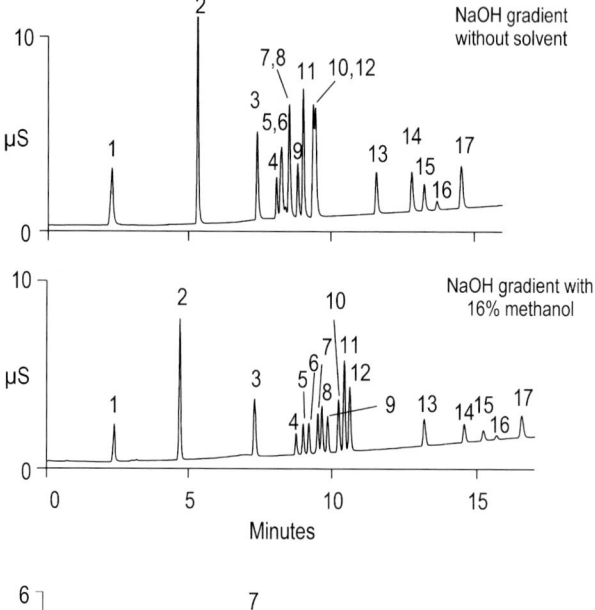

Fig. 3-40. Influence of methanol on the selectivity of the IonPac AS11. — Chromatographic conditions: see Fig. 3-39; solute concentrations: 5 mg/L each of acetate (1), chloride (2), nitrate (3), glutarate (4), succinate (5), malate (6), malonate (7), tartrate (8), maleate (9), fumarate (10), sulfate (11), oxalate (12), orthophosphate (13), citrate (14), isocitrate (15), cis-aconitate (16), and trans-aconitate (17).

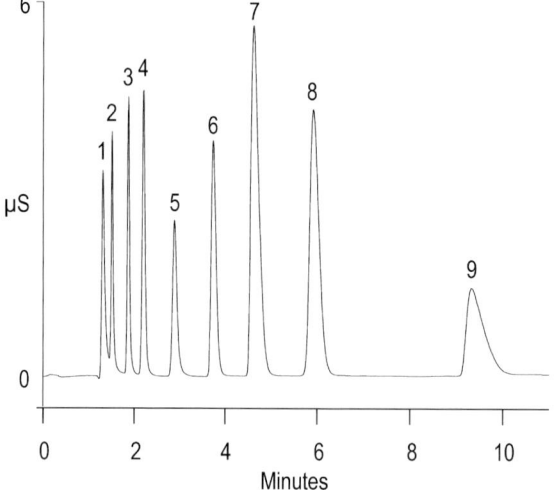

Fig. 3-41. Separation of polarizable anions on IonPac AS11. — Eluant: 45 mmol/L NaOH — methanol (60:40 v/v); flow rate: 1 mL/min; detection: suppressed conductivity; injection volume: 25 µL; solute concentrations: 2 mg/L fluoride (1), 2 mg/L chloride (2), 5 mg/L nitrate (3), 5 mg/L sulfate (4), 10 mg/L orthophosphate (5), 20 mg/L each of iodide (6), thiocyanate (7), and thiosulfate (8), and 30 mg/L perchlorate (9).

can be attributed to the pronounced hydrophobicity of the ion-exchange functionality. In general, the IonPac AS11-HC is operated under gradient conditions utilizing NaOH gradients. Figure 3-43 shows the separation of a number of inorganic and organic anions.

As compared to the chromatogram in Fig. 3-39, monovalent carboxylic acids such as quinic, lactic, formic, acetic, propionic, and butyric acid are very well resolved on IonPac AS11-HC. However, the separation is strongly influenced by the column temperature. When decreasing the column temperature of 30 °C in Fig. 3-43 by 7 °C (room temperature), resolution between formic acid and butyric

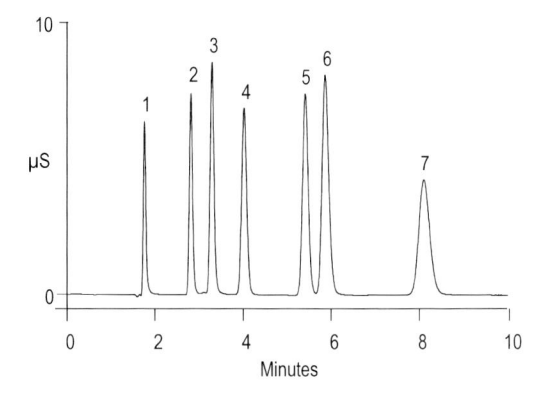

Fig. 3-42. Isocratic separation of inorganic anions on IonPac AS11-HC. − Eluant: 30 mmol/L NaOH; flow rate: 1.5 mL/min; detection: suppressed conductivity; injection volume: 10 µL; solute concentrations: 2 mg/L fluoride (1), 5 mg/L chloride (2), 10 mg/L each of nitrite (3) and sulfate (4), 20 mg/L each of bromide (5) and nitrate (6), and 30 mg/L orthophosphate (7).

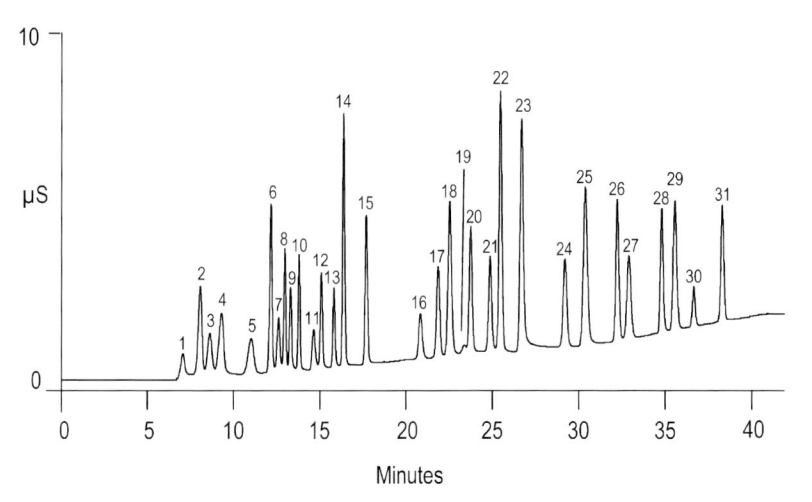

Fig. 3-43. Gradient elution of inorganic and organic anions on IonPac AS11-HC. − Eluant: (A) water, (B) 0.1 mol/L NaOH; gradient: 1% B linearly to 60% B in 38 min; column temperature: 30 °C; flow rate: 1.5 mL/min; detection: suppressed conductivity; injection volume: 10 µL; solute concentrations: 10 mg/L quinate (1), 3 mg/L fluoride (2), 10 mg/L each of lactate (3), acetate (4), propionate (5), formate (6), butyrate (7), methanesulfonate (8), pyruvate (9), chlorite (10), valerate (11), monochloroacetate (12), and bromate (13), 5 mg/L chloride (14), 10 mg/L each of nitrite (15), trifluoroacetate (16), bromide (17), and nitrate (18), 20 mg/L carbonate (19), 15 mg/L each of malonate (20), maleate (21), sulfate (22), and oxalate (23), 20 mg/L each of ketomalonate (24), tungstate (25), orthophosphate (26), phthalate (27), citrate (28), chromate (29), cis-aconitate (30), and trans-aconitate (31).

acid decreases, while nitrate and malonate nearly co-elute. Like with the AS11 column, the selectivity of the AS11-HC can also be altered by adding organic solvents to the mobile phase, which results in a better resolution between the divalent organic acids, but compromises the resolution of short-chain fatty acids. Optimum resolution is achieved with a combined NaOH and methanol gradient, the latter one being decreased towards the end of the run (Fig. 3-44).

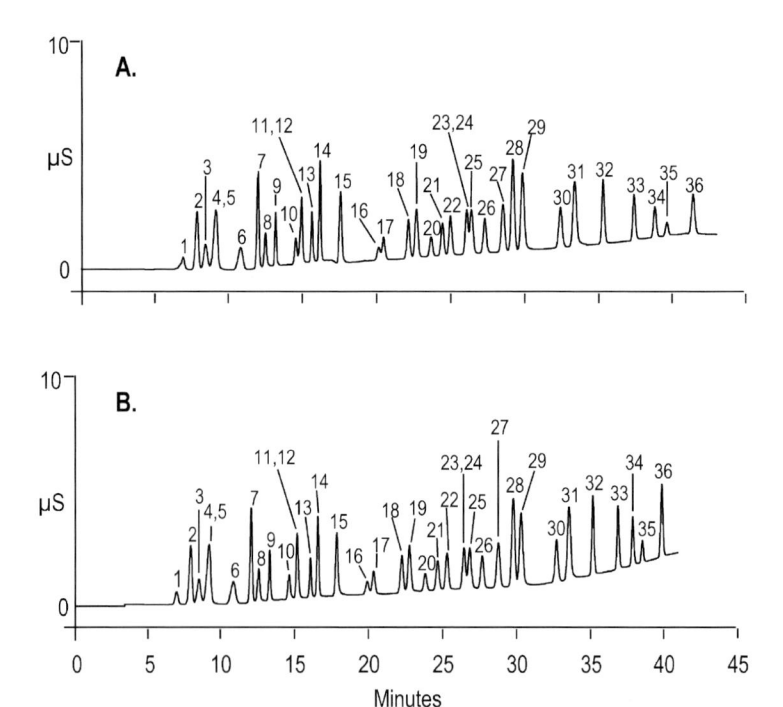

Fig. 3-44. Influence of methanol on gradient elution of inorganic and organic anions on IonPac AS11-HC. – Eluant: (A) water, (B) 0.1 mol/L NaOH, (C) methanol; gradient: 1% B for 8 min isocratically, then linearly to 30% B and 20% C in 10 min, then linearly to 60% B and 20% C in 10 min, then from 60% B and 20% C to 60% B and 0% C in 10 min; column temperature: 30 °C; flow rate: 1.5 mL/min; detection: suppressed conductivity; injection volume: 10 µL; solute concentrations: 10 mg/L quinate (1), 3 mg/L fluoride (2), 10 mg/L each of lactate (3), acetate (4), glycolate (5), propionate (6), formate (7), butyrate (8), pyruvate (9), valerate (10), galacturonate (11), monochloroacetate (12), and bromate (13), 5 mg/L chloride (14), 10 mg/L each of nitrite (15), sorbate (16), trifluoroacetate (17), bromide (18), nitrate (19), and glutarate (20), 15 mg/L each of succinate (21) and malate (22), 20 mg/L carbonate (23), 15 mg/L each of malonate (24), tartrate (25), maleate (26), fumarate (27), sulfate (28), and oxalate (29), 20 mg/L each of ketomalonate (30), tungstate (31), orthophosphate (32), citrate (33), iso-citrate (34), cis-aconitate (35), and trans-aconitate (36).

Macroporous support materials with pore sizes of 200 nm, which have been mentioned above in the discussion of the IonPac AS9-HC, were originally developed for the isocratic analysis of fluoride and short-chain fatty acids together with other inorganic anions. In the past, the reliable determination of fluoride represented one of the biggest challenges for conventional anion exchange chromatography. Introduced in 1992, the IonPac AS10 was the first latexed anion exchanger allowing the separation of fluoride in such a matrix. The AS10 column is based on a 8.5-µm EVB/DVB support material with 65-nm latex beads made of polyvinylbenzyl chloride and functionalized with a tertiary amine being agglomerated to its surface. The resulting ion-exchange capacity is about 170 µequiv/column. The respective standard chromatogram showing a separation of inorganic and organic anions is depicted in Fig. 3-45.

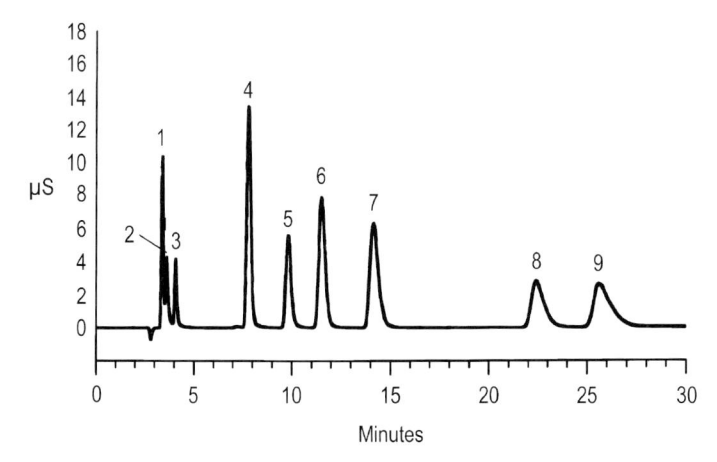

Fig. 3-45. Separation of inorganic and organic anions on IonPac AS10. – Eluant: 0.1 mol/L NaOH; flow rate: 1 mL/min; detection: suppressed conductivity; injection volume: 25 µL; solute concentrations: 3 mg/L fluoride (1), 10 mg/L acetate (2), 5 mg/L formate (3), 10 mg/L chloride (4), 10 mg/L nitrite (5), 15 mg/L sulfate (6), 20 mg/L each of oxalate (7), bromide (8), and nitrate (9).

Because of the high ion-exchange capacity of this column a high ionic strength eluant is required. Sodium hydroxide is especially suitable, because even high NaOH concentrations (<0.15 mol/L) are converted to water by modern suppressor systems resulting in background conductivities of about 1 µS/cm. As can be seen from Fig. 3-45, fluoride is well separated from the system void despite the high eluant concentration. Acetate, which co-elutes with fluoride on conventional latexed anion exchangers, is not completely resolved from fluoride under these chromatographic conditions, but resolution is usually sufficient for quantitation. A baseline-resolved separation can be achieved by reducing the NaOH concentration or by employing gradient elution. As with IonPac AS2, the ion-exchange functional groups of the AS10 column are very hydrophobic, so that bromide and nitrate are strongly retained. This is intentional, because the IonPac AS10 is predominantly used for trace anion analysis in ultra-pure chemicals and in high-ionic strength samples. Due to its high-capacity and superior chromatographic efficiency, the AS10 technologically replaced the AS2 column for virtually every application.

The IonPac AS10 was the first latexed anion exchanger to separate fluoride from the system void by combining high ion-exchange capacity and a relatively weak eluant. However, the long retention times for bromide and nitrate of more than 40 minutes under standard conditions are unacceptable for conventional water analysis applications. Therefore, a special latexed anion exchanger was developed which allowed the simultaneous analysis of fluoride and other mineral acids [23]. A major problem is the high degree of hydration of fluoride, which can only be separated from the system void on a stationary phase of high water content. To achieve this the quaternary ion-exchange functional groups must be very hydrophilic. However, fluoride is so strongly hydrated that a significant retention can only be realized with latex materials having a very low degree of cross-linking (<1%). This requirement is in total contradiction to the resulting low ion-exchange capacity, which is inversely proportional to the water content of the stationary phase of latex materials. Considering that the water content in the latex polymer acts as a diluent, the concentration of ion-exchange functional groups decreases with increasing water content. The only way of compensation is the utilization of a macroporous support material like the one developed for the IonPac AS10.

The resulting anion exchanger has been commercialized under the trade name IonPac AS12A. For separating fluoride from the system void, a latex polymer based on vinylbenzyl chloride with a degree of cross-linking of 0.15% has been synthesized. For solvent compatibility the latex beads were agglomerated on a highly cross-linked (55%) 9-μm macroporous EVB/DVB polymer. The average pore size of 200 nm is identical to that of the AS10 column; the specific surface area is 15 m^2/g. Surface sulfonation of the support material has been optimized in terms of completeness to exclude to a large extent adsorptive effects accompanying the separation of bromide and nitrate. The ion-exchange capacity of this column is twice as high as that of the IonPac AS4A-SC, which is somewhat remarkable considering the high water content of the latex polymer. As with all other conventional latexed anion exchangers, the standard eluant is a carbonate/bicarbonate mixture, which allows a rapid and efficient elution of inorganic anions under isocratic conditions. As can be seen from the respective standard chromatogram in Fig. 3-46, fluoride is well separated from the system void and can be analyzed together with other mineral acids and oxyhalides in less than 15 minutes. Flow rate and eluant composition have been optimized for maximum resolution between all analytes. The extremely high resolution between chloride and nitrite, which at present cannot be achieved with any other latexed anion exchanger, is somewhat surprising. Therefore, both anions can be separated and quantified with conductivity detection even when present in largely different concentration levels. This application was previously only possible with low wavelength UV detection. In comparison to the IonPac AS9-SC, resolution between fluoride, chlorite, bromate, and chloride is significantly better. Even acetic acid, which co-elutes with fluoride on an IonPac AS9-SC, can be

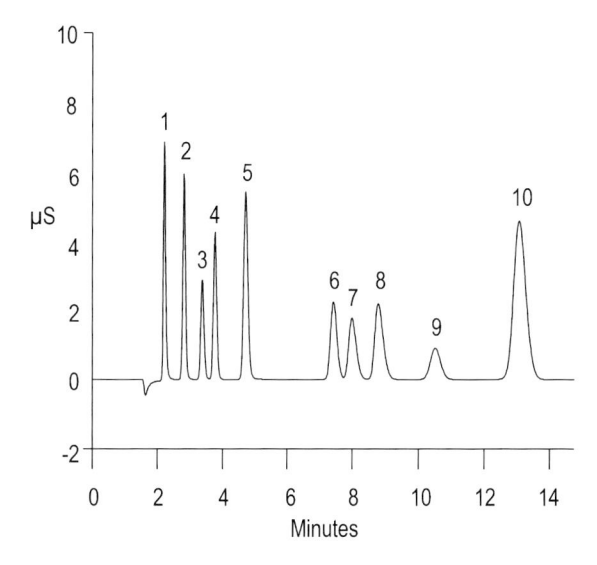

Fig. 3-46. Separation of various inorganic anions on IonPac AS12A. — Eluant: 0.3 mmol/L NaHCO$_3$ + 2.7 mmol/L Na$_2$CO$_3$; flow rate: 1.5 mL/min; detection: suppressed conductivity; injection volume: 25 µL; solute concentrations: 3 mg/L fluoride (1), 10 mg/L chlorite (2), 10 mg/L bromate (3), 4 mg/L chloride (4), 10 mg/L each of nitrite (5), bromide (6), chlorate (7), nitrate (8), and orthophosphate (9), and 20 mg/L sulfate (10).

separated from fluoride on IonPac AS12A under standard conditions. However, baseline resolution cannot be achieved. The only disadvantage of the AS12A column as compared to the acrylate-based AS9-SC is the retention behavior of polarizable anions such as iodide, thiocyanate, and thiosulfate, which are strongly retained on an IonPac AS12A and, therefore, cannot be analyzed together with mineral acids.

When it was introduced, the IonPac AS12A was clearly the most modern latexed anion exchanger for the separation of fluoride, oxyhalides, and mineral acids. The latest development in the field of universal anion exchangers for water analysis is the IonPac AS14 discussed in Section 3.4.1, which is based on a support material with a covalently bonded ion-exchange polymer (see Fig. 3-33 in Section 3.4.1).

The IonPac AS16, which was developed for the analysis of polarizable anions such as iodide, thiocyanate, thiosulfate, and perchlorate, has replaced the IonPac AS5 (see Fig. 3-29). The substrate material is also based on a 55% cross-linked macroporous EVB/DVB polymer with a particle diameter of 9 µm. To minimize sorption interactions between the polarizable anions and the stationary phase, the latex particles carry extremely hydrophilic ion-exchange groups. The latex cross-linking is 1%, the particle diameter is 80 nm. The IonPac AS16 has been optimized for isocratic analysis of polarizable anions with a hydroxide eluant. Its high capacity of 170 µequiv/column (250 mm × 4 mm i. d.) allows large volume injections up to 2 mL, so that polarizable anions can also be analyzed in the low µg/L level without pre-concentration techniques. Moreover, hydroxide eluants in combination with a membrane-based suppressor system and the low background conductivity resulting there from, simplify trace analysis of polarizable anions. As an example, Fig. 3-47 shows the separation of such anions together with fluoride, chloride, and sulfate. For obtaining optimum peak shapes, a column temperature of 30 °C is important. At this temperature, even perchlorate anions, which exhibit an enormous affinity towards conventional anion exchangers, elute as completely symmetrical peaks. While the absolute value of the column temperature is not critical, it should be kept constant throughout the analysis to obtain reproducible retention times. When employing a gradient elution technique (Fig. 3-48), a variety of other inorganic and organic anions can be separated together with polarizable anions. The only exceptions are bromide and nitrate, which cannot be separated because of the marked hydrophilicity of the resin. An important field of application for such a stationary phase is the analysis of polyphosphates (see Section 3.7.4). With a pure aqueous NaOH eluant polyphosphates up to P_{20} can be eluted from this column.

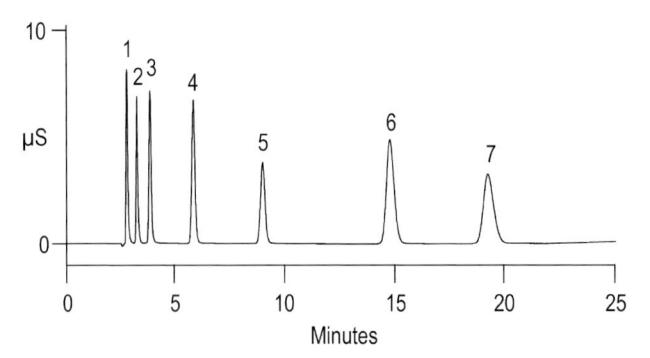

Fig. 3-47. Separation of polarizable inorganic anions on IonPac AS16. – Column temperature: 30 °C; eluant: 35 mmol/L NaOH; flow rate: 1 mL/min; detection: suppressed conductivity; injection volume: 10 µL; solute concentrations: 2 mg/L fluoride (1), 3 mg/L chloride (2), 5 mg/L sulfate (3), 10 mg/L thiosulfate (4), 20 mg/L iodide (5), 20 mg/L thiocyanate (6), and 30 mg/L perchlorate (7).

The IonPac AS17 represents the latest development in the field of hydroxide-selective stationary phases. This column was designed for the fast gradient elution of standard inorganic anions in potable water, wastewater, and soil extracts. Thus, it is an analogue to the IonPac AS14 (see Fig. 3-9 in Section 3.4.1), which uses mostly carbonate/bicarbonate eluants under isocratic conditions for the same analyses. The IonPac AS17 is based on a microporous EVB/DVB polymer with a particle diameter of 10.5 µm and a degree of cross-linking of 55%. Its latex particles have a degree of cross-linking of 6% and a diameter of 75 nm and carry strongly hydrophilic ion-exchange groups. A special characteristic of the IonPac AS17 is the high resolution between fluoride and the system void, allowing fluoride quantitation at very low concentrations. Short-chain fatty acids such as formic, acetic, and propionic acid are also separated to baseline and

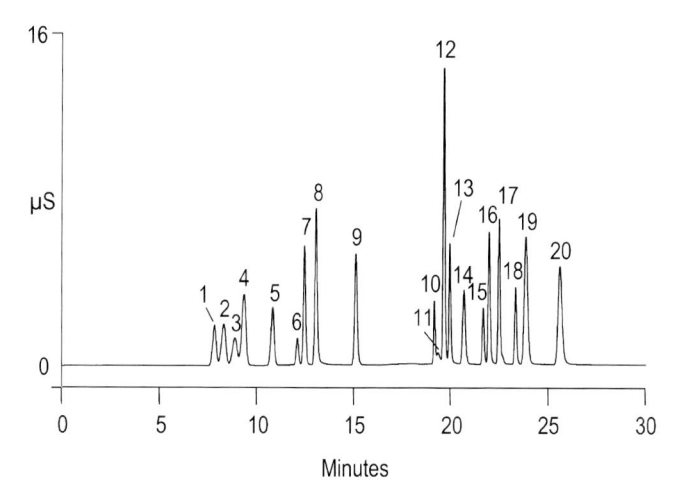

Fig. 3-48. Gradient elution of standard inorganic anions and polarizable anions on IonPac AS16. – Column temperature: 30 °C; eluant: electrolytically generated KOH; gradient: 1.5 mmol/L KOH for 8.3 min isocratically, then to 10 mmol/L in 6 min, then to 55 mmol/L in 10 min; flow rate: 1.5 mL/min; detection: suppressed conductivity; solute concentrations: 2 mg/L fluoride (1), 10 mg/L each of acetate (2), propionate (3), formate (4), chlorite (5), and bromate (6), 5 mg/L chloride (7), 10 mg/L each of nitrite (8), nitrate (9), and selenite (10), carbonate impurity (11), 10 mg/L sulfate (12), 10 mg/L selenate (13), 20 mg/L iodide (14), 10 mg/L thiosulfate (15), 20 mg/L each of chromate (16), orthophosphate (17), arsenate (18), and thiocyanate (19), and 30 mg/L perchlorate (20).

elute behind fluoride. In the format 250 mm × 4 mm i. d., the ion-exchange capacity of the IonPac AS17 column is 30 µequiv, which is comparable to that of the IonPac AS4A-SC. Standard anions are separated on this column in about eight minutes with an electrolytically generated hydroxide eluent under gradient conditions (Fig. 3-49). Including the time for re-equilibration, the total analysis time from injection to injection is approximately ten minutes.

3.4.3
Silica-Based Anion Exchangers

Parallel to the development of organic polymers as anion exchange substrates, a number of silica-based anion exchangers were introduced over several years [37, 38]. Once again, the development of low-capacity ion-exchange materials was a high priority; the goal was to be able to dispense with the suppressor system by using eluants with low background conductivities.

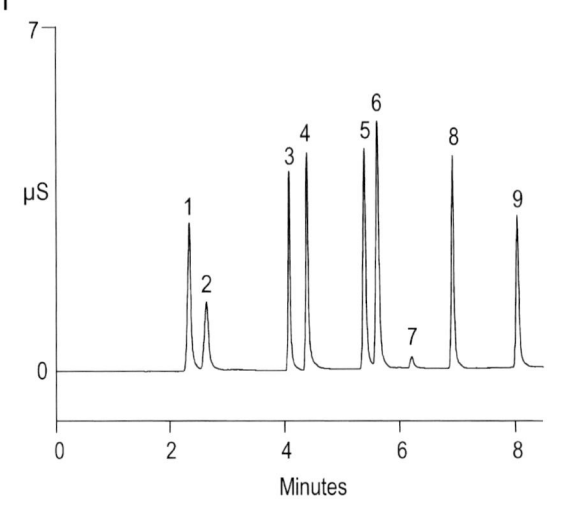

Fig. 3-49. Fast gradient separation of standard inorganic anions on IonPac AS17. — Column temperature: 30 °C; eluant: KOH (EG40); gradient: 0.1 mmol/L for 1.5 min isocratically, then from 1 mmol/L to 20 mmol/L in 3.5 min, then to 40 mmol/L in 2 min; flow rate: 2 mL/min; detection: suppressed conductivity; injection volume: 10 µL; solute concentrations: 2 mg/L fluoride (1), 5 mg/L acetate (2), 3 mg/L chloride (3), 5 mg/L nitrite (4), 10 mg/L bromide (5), 10 mg/L nitrate (6), carbonate (7), 5 mg/L sulfate (8), and 10 mg/L orthophosphate (9).

In contrast to organic polymers, silica-based substrates have the advantages of higher chromatographic efficiency and greater mechanical stability. In general, no swelling or shrinkage problems are encountered, even if the ionic form of the ion exchanger is changed or an organic solvent is added to the eluant. Column temperatures up to 80 °C also have no adverse effect on the stationary phase. Although the chromatographic efficiency of these stationary phases with theoretical plate numbers up to 15,000 to 20,000 is very high [39], they can only be used within a narrow pH range (pH 2 to pH 7). This reduces both the number of possible eluants and the type of samples that can be analyzed using silica-based ion exchangers. The behavior of transition metal ions in ion chromatography systems without a suppressor system is quite remarkable. Jenke and Pagenkopf [40] showed that metal cations such as Cu^{2+}, Pb^{2+}, and Zn^{2+} can adversely affect the stability and efficiency of silica-based ion exchangers. The presence of free silanol groups leads to an interaction with these metal ions, which are retained at the stationary phase. A co-elution of lead, zinc, and copper ions with inorganic anions such as chloride, bromide, and nitrate is observed. Thus, the resulting data are misleading.

In general, silica substrates are grouped according to their particle size. Totally porous substrates with small particle sizes in the range of 3 µm to 10 µm, which are called microparticulate beads, are preferred. Ion-exchange capacities between 0.1 and 0.3 mequiv/g are obtained by bonding quaternary ammonium functions via reaction of the free silanol groups with an appropriate chlorosilane. Pellicular materials, on the contrary, consist of much larger particles with a diameter between 25 µm and 40 µm. Their surface is covered with a polymer such as lauryl methacrylate (Zipax SAX) which carries the actual ion-exchange functionality. The thickness of the polymer film is kept deliberately low at 1 µm to 3 µm to minimize the contribution, σ_s^2, of the peak broadening caused by the stationary phase [see Eq. (22) in Section 2.4]. The ion-exchange capacity of pellicular materials is about 0.01 mequiv/g.

Silica-based anion exchangers are offered by manufacturers such as Wescan (Santa Clara, USA), Separations Group (Hesperia, USA), Toyo Soda (Tokyo, Japan), and Macherey & Nagel (Düren, Germany). Table 3-8 summarizes the characteristic structural and technical properties of the most important column packings available today.

Table 3-8. Structural and technical properties of various silica-based anion exchangers.

Column	Vydac 302 IC 4.6	Vydac 300 IC 405	Wescan 269-001	Nucleosil 10 Anion	TSK Gel IC-SW
Dimensions (Length × ID) [mm]	50 × 4.6	250 × 4.6	250 × 4.6	250 × 4.0	50 × 4.6
Manufacturer	Separations Group	Separations Group	Wescan	Macherey & Nagel	Toyo Soda
Capacity [mequiv/g]	ca. 0.1	ca. 0.1	0.08	0.06	0.4
Particle size [µm]	10	15	13	10	5
Type of column packing	spherical particles with quaternary ammonium functions			spherical particles aminated with trimethyl- amine	spherical particles aminated with methyldi- methylamine

Figure 3-50 shows a typical example of an inorganic anion separation on a silica-based ion exchanger, Vydac 302 IC 4.6. The stationary phase recently introduced by the same manufacturer under the trade name Vydac 300 IC 405 is a novelty. Although the substrate of the stationary phase also consists of spherical silica particles, the pH stability of this material was enhanced by a polymer coating [41]. According to the manufacturer's specifications, eluants in the pH range between 2 and 10 may be used. The standard anion chromatogram shown in Fig. 3-51 is obtained using o-phthalic acid as the eluant. While the selectivity of this stationary phase is identical to that of Vydac 302 IC 4.6, the retention times could be reduced by 50% if the column length is shortened to 50 mm.

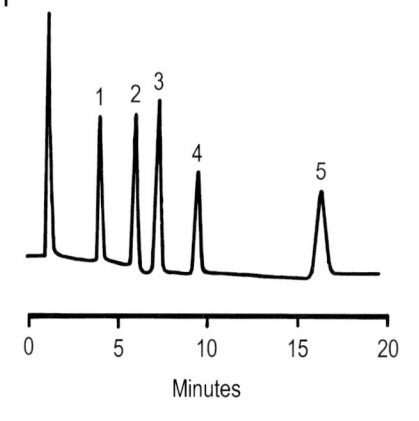

Fig. 3-50. Separation of various inorganic anions on a Vydac 302 IC 4.6 silica-based anion exchanger. — Eluant: 2 mmol/L o-phthalic acid (pH 5.0); flow rate: 2 mL/min; detection: direct conductivity; injection volume: 10 µL; solute concentrations: 100 mg/L each of chloride (1), nitrite (2), bromide (3), nitrate (4), and sulfate (5); (taken from [31]).

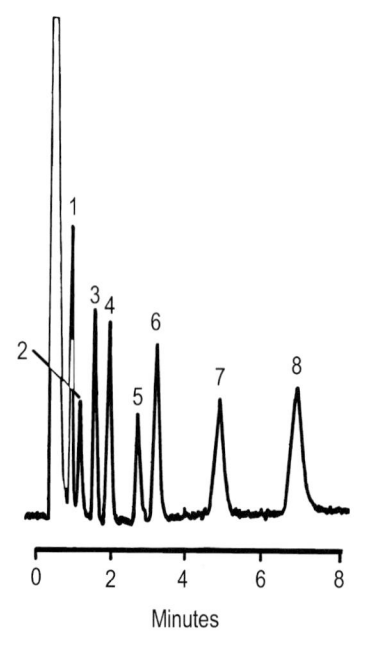

Fig. 3-51. Separation of various inorganic anions on a Vydac 300 IC 405 silica-based anion exchanger. — Eluant: 1.5 mmol/L o-phthalic acid (pH 8.9); flow rate: no specification; detection: direct conductivity; solute concentrations: 5 mg/L fluoride (1), carbonate (2), 1 mg/L chloride (3), 1.5 mg/L nitrite (4), 3 mg/L bromide (5), 2.5 mg/L nitrate (6), 3 mg/L each of orthophosphate (7) and sulfate (8).

The silica-based ion exchanger manufactured by Wescan is based on a macroporous substrate with a pore width of 300 Å. A specific feature of this column is the large retention difference between monovalent and divalent anions as shown in Fig. 3-52. The negative signal appearing at about 20 minutes is the "system peak" (see Section 3.7.3), which is inevitable when employing phthalates as eluants.

The chromatogram shown in Fig. 3-53 exemplifies how strongly the selectivity of a separation system depends on the eluant. A silica ion exchanger, TSK GEL IC-SW from Toyo Soda Company, was employed as a stationary phase. The separation of chlorate and nitrate that is achieved using 1 mmol/L tartaric acid eluant is remarkable. Such a separation is not possible on a polymer-based latex-agglomerated anion exchanger such as IonPac AS4A-SC at alkaline pH, but can be done on modern latexed anion exchangers such as IonPac AS9 (-SC) and AS12A. However, it should be pointed out that under these chromatographic conditions only monovalent anions can be eluted from the column with acceptable analysis times.

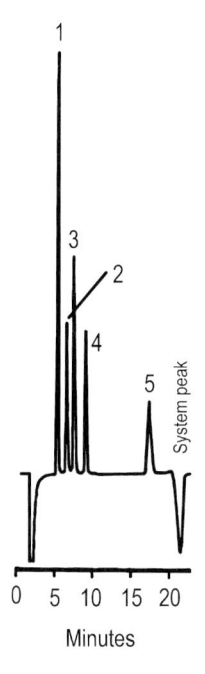

Fig. 3-52. Separation of various inorganic anions on a Wescan 269-001 silica-based anion exchanger. — Eluant: 4 mmol/L potassium hydrogen phthalate; flow rate: 1.5 mL/min; detection: direct conductivity; injection volume: no specification; solute concentrations: 10 mg/L each of chloride (1), nitrite (2), bromide (3), nitrate (4), and sulfate (5); (taken from [42]).

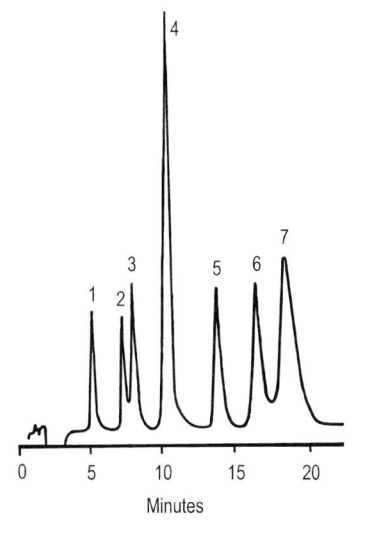

Fig. 3-53. Separation of various inorganic anions on a TSK Gel IC-SW silica-based anion exchanger. — Eluant: 1 mmol/L tartaric acid (pH 3.2); flow rate: 1.5 mL/min; detection: direct conductivity; injection volume: 100 µL; solute concentrations: 10 mg/L each of iodate (1), nitrite (2), bromate (3), chloride (4), chlorate (5), bromide (6), and nitrate (7).

As a comparison, Fig. 3-54 shows the chromatogram of inorganic anions using a Nucleosil 10 Anion silica-based anion exchanger. Orthophosphate elutes as a monovalent ion due to the comparatively low pH value of the mobile phase.

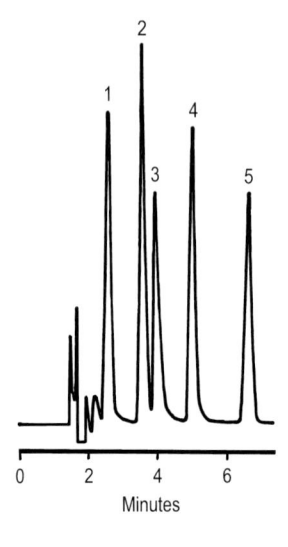

Fig. 3-54. Separation of various inorganic anions on a Nucleosil 10 Anion silica-based anion exchanger. — Eluant: 25 mmol/L sodium salicylate (pH 4.0 with salicylic acid); flow rate: 1.5 mL/min; detection: refractive index; injection volume: 10 µL; solute concentrations: 1 to 2 g/L of orthophosphate (1), chloride (2), nitrite (3), nitrate (4), and sulfate (5).

3.4.4
Other Materials for Anion Separations

3.4.4.1 Crown Ether and Cryptand Phases

Inorganic and organic anions can be separated on resins other than strongly basic anion exchangers based on polymeric or silica substrates. In the mid 1970s, Blasius et al. [43] described the separation of ionic species on cross-linked polymers modified with cyclic polyethers. Cyclic polyethers are neutral macrocyclic ligands with more than nine ring atoms, at least three of which act as donor atoms [44]. The possibility of complexation of inorganic salts, as well as of ionic and non-ionic organic compounds, can be explained by the dipole effect of the C-O bond in the polyether ring. Other hetero atoms such as nitrogen or sulfur may also act as donor atoms. The ring system most thoroughly investigated and most commonly employed is 18-crown-6 (1,4,7,10,13,16-hexaoxacyclooctadecane), a planar cyclic polyether with 18 ring atoms:

18-Crown-6

The selective complexation of a specific cation, which is always associated with an anion because of electroneutrality, depends on the number and position of the hetero atoms. Complexation with organic molecules occurs very often via hydrogen bonding. Figure 3-55 shows the molecular structure of a RbNCS ion-pair that is coordinated by dibenzo-18-crown-6.

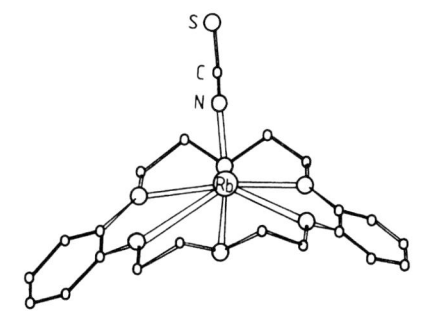

Fig. 3-55. Molecular structure of a RbNCS ion-pair coordinated by 18-crown-6.

Crown ethers can be introduced into the ion chromatographic system via the mobile and/or the stationary phase. Sousa et al. [45] used them for the first time as mobile phase additives describing a separation of amino acid isomers separated with dinaphthal-18-crown-6. But even simple inorganic anions can be separated with this method [46]. However, the number of applicable crown ethers is limited because of their solubilities. Moreover, crown ethers are relatively expensive, so that their application at higher concentrations is not justified.

The introduction of crown ethers via the stationary phase can be carried out in three different ways:
- Adsorption of the crown ether on a solid support material
- Polymerization of the crown ether to form a stationary phase
- Covalent bonding of a crown ether onto a polymeric or silica-based support material.

Pioneering work in the field of adsorbing crown ethers onto chemically bonded reversed phases and PS/DVB polymers have been carried out by Kimura et al. [47]. They employed dodecyl-substituted crown ethers such as 12-crown-4, 15-crown-5, and 18-crown-6, all of which have a very hydrophobic side chain which enables these molecules to adsorb onto the stationary phase. The capacity of the separator column can be controlled by the amount of adsorbed crown ether. Stationary phases of this type are stable over a longer period of time only when used with purely aqueous eluants. If the mobile phase contains more than 40% (v/v) organic solvents, the crown ether layer is rapidly washed off.

Polymeric crown ether phases have been thoroughly investigated for their applications in chromatography by Blasius et al. [48-53]. They are characterized by chemical and thermal stability, are compatible with a variety of organic solvents, and are of high capacity. Thus, Blasius et al. separated anions with pure methanol as the mobile phase. Because the stability constants of the cation crown ether complexes depend on the type of solvent, the solvent content in the mobile phase can be used as an additional parameter for optimizing the chromatographic separation.

Polymeric crown ether resins can be prepared in two different ways:
- Synthesis of solid polymer particles
- Coating of a solid support material with a crown ether polymer

The first stationary phases were prepared by packing columns with polymeric particles where benzo- or dibenzocrown ethers were cross-linked with formaldehyde in formic acid. The single crown ether molecules are bridged by methylene groups [49]. Dibenzo-18-crown-6 is a preferred starting material. Igawa et al. [54] also developed a polyamide crown ether resin to separate inorganic anions.

Monomer of a polyamide crown ether resin

Polymeric crown ether resins are mechanically unstable and are, therefore, operated with low flow rates between 0.05 and 0.1 mL/min resulting in long analysis times. Typical for all polymeric crown ether phases is the relatively low chromatographic efficiency, which does not meet today's requirements. However, this problem can be overcome by immobilizing crown ether polymers on the surface of a solid support. Modified and non-modified silica are predominantly used as support materials. Igawa et al., for instance, coated silica particles with the above-mentioned polyamide crown ether resin and obtained significantly better separations then with the uncoated resin.

The immobilization of crown ethers on these support materials may take place in different ways. If silica beads are impregnated with a solution of dibenzo-18-crown-6 in formic acid, a substrate that is stable to hydrolysis is obtained when cross-linking with formaldehyde (see Fig. 3-56, upper left).

Polymeric crown ethers may also be bonded chemically to the silica surface. For that purpose, the support material is modified with 3-(N-methacryloylamino)-propyl groups, and subsequently co-polymerized with (4-methacryloylamino)-benzo-15-crown-5 [55]. Stationary phases, in which the crown ether molecules are chemically bonded to silica via Si-O-C linkages, are obtained with silica treated with thionyl chloride via reaction with 4-hydroxymethylbenzo-18-crown-6 (see Fig. 3-56, upper right). Since Si-O-C-bonds are prone to hydrolysis, water-free methanol is used as eluant for such stationary phases. Much more stable are silicas modified via Si-C-bonds (see Fig. 3-56, bottom left and right) that are prepared by using 4-vinylbenzo-18-crown-6 or 4-butene-18-crown-6 and dimethylchlorosilane in the presence of $H_2[PtCl_6]$. The remaining free silanol groups are reacted with trimethylchlorosilane to prevent interactions with the solutes. The literature refers to this process as "endcapping".

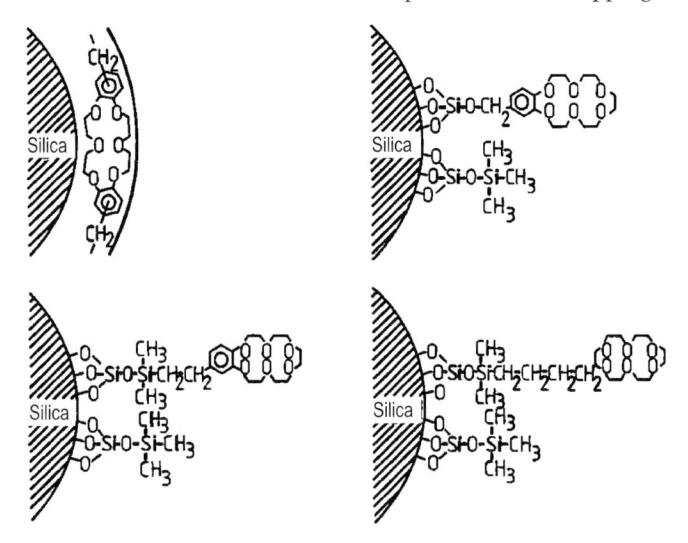

Fig. 3-56. Schematic structures of various silica-based crown ether phases.

Experimental Retention Determining Parameters
While ionic strength is the most important retention-determining parameter in conventional anion exchange chromatography, anion retention on crown ether phases is determined by both ionic strength and column capacity, the latter one being constant for strongly basic anion exchangers. The capacity of crown ether phases, in turn, depends on the amount of bonded ligands and on the number of formed complexes between cations and crown ethers. The latter complexes are influenced by the type and concentration of cation and the organic solvent content in the mobile phase. Independent of the type of crown ether being used, anions associated to a common cation elute in the order $SCN^- > I^- > NO_3^- > Br^- > Cl^- > F^- > SO_4^{2-}$. This elution order is very different from that in conventional anion exchange, where divalent anions, in particular sulfate, are more strongly retained. Anion retention directly depends on the formation constant of the eluant cation with the respective crown ether. Therefore, cations with a larger formation constant increase retention of the associated anions. In this respect, a much better separation of anions as a mixture of their potassium salts as compared to sodium salts is observed with stationary phases based on 18-crown-6.

Of marked interest is the use of organic solvents with crown ether phases [56]. In anion exchange chromatography, organic solvents as mobile phase additives have only a limited effect on selectivity; however, the influence of organic solvents on the binding process of cations on crown ethers is well known. In comparison with pure aqueous systems, formation constants in organic solvents such as methanol and acetonitrile are larger by several orders of magnitude [57, 58]. Increasing solvent content in the mobile phase causes increases in the capacity of a crown ether phase and, thus, in the retention of anions. This effect has been shown by Lamb et al. [56]. However, the stability of crown ether phases is problematic in presence of organic solvents. If the crown ether is adsorbed onto the support by hydrophobic interactions, it can potentially be washed off by the solvent in the mobile phase. Good stabilities are achieved with tetradecyl-substituted crown ethers based on 18-crown-6, which — according to Lamb et al. [56] — loose their capacity only with solvent contents larger than 20% (v/v).

As an example, Fig. 3-57 shows a separation of various inorganic anions on a surface-coated silica. As can be seen in this chromatogram, the chromatographic efficiency of this stationary phase is very poor as compared to modern anion exchangers. However, having the option to elute anions with pure de-ionized water is advantageous, because in this case it means that the suppressor system (normally necessary for a sensitive conductivity detection) is dispensable. This scenario also applies for the separation of the sodium salts of various chloro-containing compounds on silica modified with 4-hydroxymethylbenzo-18-crown-6 shown in Fig. 3-58. Pure methanol was used as a mobile phase for this separation.

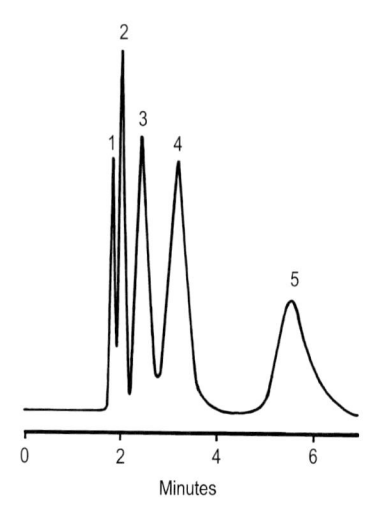

Fig. 3-57. Separation of the sodium salts of various anions on silica coated with crown ether polymers. — Stationary phase: dibenzo-18-crown-6; eluant: water; flow rate: 1 mL/min; detection: direct conductivity; solute concentrations: 0.7 mg/L Na_2SO_4, 0.1 mg/L NaCl; 1 mg/L NaI, 4 mg/L NaSCN, and 8 mg/L $NaHCO_3$; (taken from [50]).

More efficient separations are obtained with the above-mentioned tetradecyl-substituted materials based on 18-crown-6. Ion-exchange interactions are dominating the separation mechanism in this case [46]. By using two of those stationary phases in series, 14 different inorganic and organic anions can be separated utilizing potassium hydroxide and acetonitrile as mobile phase (Fig. 3-59). The use of two stationary phases in series is necessary to ensure adequate capacity for the sufficient separation of all analytes. In comparison with conventional anion exchangers, crown ether phases exhibit less selectivity for divalent anions;

this can be attributed to the lower density of ion-exchange groups on the stationary phase surface. Therefore, interactions with monovalent anions are preferred, resulting in a different elution order. Sulfate, for instance, elutes ahead of nitrate!

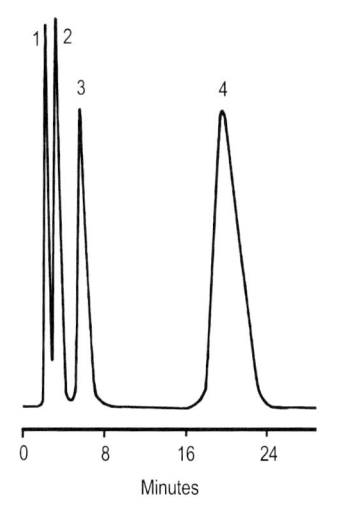

Fig. 3-58. Separation of the sodium salts of various chloro-containing species on silica modified with 4-hydroxymethylbenzo-18-crown-6. — Eluant: methanol; flow rate: 1.6 mL/min; detection: direct conductivity; solute concentrations: 23 mg/L $NaClO_2$, 6 mg/L NaCl, 37 mg/L $NaClO_3$, and 326 mg/L $NaClO_4$; (taken from [50]).

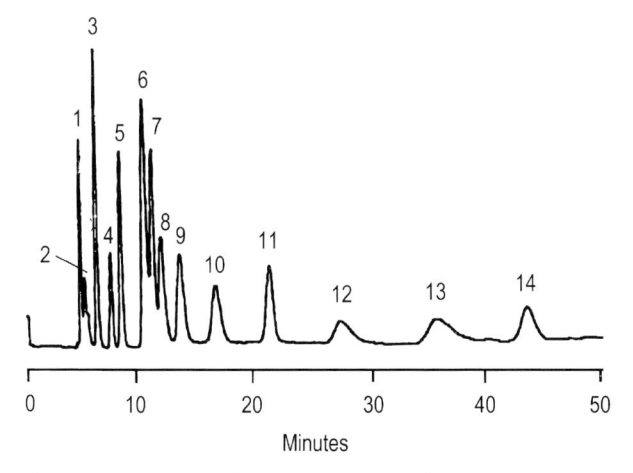

Fig. 3-59. Separation of various inorganic and organic anions on a DVB polymer (IonPac NS1, Dionex) coated with tetradecyl-18-crown-6. — Eluant: 0.05 mol/L KOH — acetonitrile (80:20 v/v); flow rate: 1 mL/min; detection: suppressed conductivity; injection volume: 20 µL; solute concentrations: 1.5 mg/L fluoride (1), 10 mg/L acetate (2), 2.5 mg/L chloride (3), 10 mg/L each of nitrite (4), bromide (5), sulfate (6), nitrate (7), oxalate (8), chromate (9), phthalate (10), iodide (11), orthophosphate (12), citrate (13), and thiocyanate (14); (taken from [56]).

Since the type of eluant cation determines column capacity, capacity gradients can be employed by changing the type of cation during the chromatographic run. This can be done by changing from potassium hydroxide to sodium hydroxide during the run, which leads to significantly shorter analysis times. Taking the set of analytes from Fig. 3-59 as an example, the chromatogram in Fig. 3-60 illustrates how this change shortens retention by decreasing capacity. Because the hydroxide concentration remains constant, baseline disturbances are not observed. Re-equilibration at the end of the run is also reduced to a minimum. Today, this type of gradient is often employed for efficient separations of a number of anions on crown ether phases.

Capacity gradients can also be carried out on CRYPTAND columns. Cryptands provide a three-dimensional cavity for metal cation entrapment. The basic structure of cryptands is illustrated by the ligand 2.2.2:

Cryptand 2.2.2

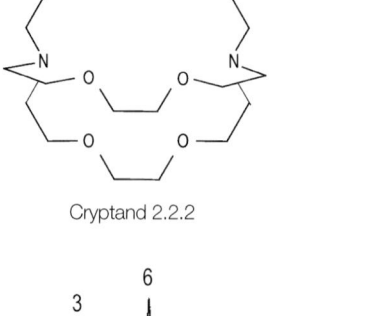

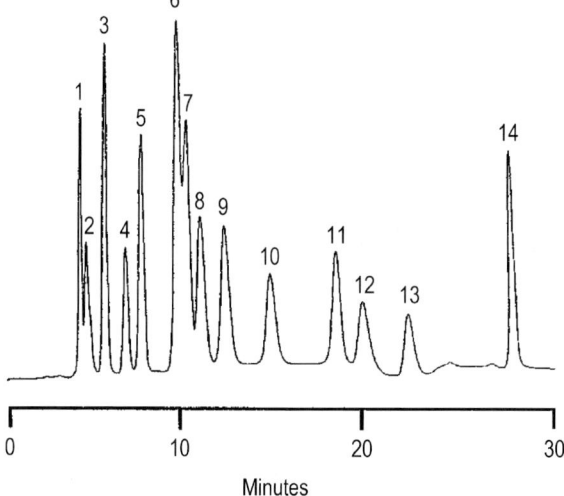

Fig. 3-60. Separation of inorganic and organic anions on a DVB polymer (IonPac NS1, Dionex) coated with tetradecyl-18-crown-6 utilizing a capacity gradient. – Eluant: (A) 0.05 mol/L KOH – acetonitrile (80:20 v/v), (B) 0.05 mol/L NaOH – acetonitrile (80:20 v/v); gradient: 100% A isocratically for 5 min, then linearly in 15 min to 100% B; other chromatographic conditions: see Fig. 3-59; (taken from [56]).

The selectivity of macrocycles such as crown ethers and cryptands is often determined by the ability of the cation to fit into the central cavity of the macrocycle.

$$\text{Binding Constant } K = [\text{complex}] / [L][K^+] \tag{59}$$

The drastic change of cryptand selectivities vis-a-vis the size of the alkali metal cations is shown in Table 3-9. As cryptands are usually neutral molecules, anion separator columns are resulting when cryptand-cation complexes are formed serving as anion exchange sites (Fig. 3-61). Anions must be associated with the positively charged complexes to maintain electrical neutrality. Since cryptands are generally hydrophobic and are associated with the hydrophobic environment of the column, polarizable anions are more strongly retained [49].

Table 3-9. Binding constants for cryptand 2.2.2 in water and methanol.

Cation	log K (in water)	log K (in methanol)
Li^+	~1	~2
Cs^+	<2	4.4
Na^+	3.9	7.9
Rb^+	4.35	8.9
Ca^{2+}	4.4	8.16
NH_4^+	4.5	–
K^+	5.4	10.4
Sr^{2+}	8	11.75
Ba^{2+}	9.5	12.9
Cu^{2+}	6.8	–
Ag^+	9.6	12.22

Already in 1994, Lamb et al. [59] developed a series of cryptand-based anion separators by loading decyl-2.2.2 (D222) onto polystyrene-based resins [46]. The preparation of the stationary phase was carried out by slurrying D222 with the resin in a methanol-water (60:40 v/v) solution and then evaporating off the methanol. The resin thus prepared was then packed into chromatographic columns. However, there are some problems with adsorbed macrocycles:
- Adsorbed macrocycles slowly bleed off the stationary phase resulting in a loss of capacity
- The use of organic solvents is prohibited due to rapid loss of capacity
- Ruggedness
- Poor efficiencies

Those problems were overcome by Woodruff et al. [60] who succeeded in synthesizing a cryptand 2.2.2 monomer. The respective separator column was introduced in 2002 under the trade name IonPac Cryptand A1 (Dionex Corp. Sunnyvale, USA). The 150 mm × 3 mm i. d. column with an ion-exchange capacity of 110 µequiv contains cryptand 2.2.2 moieties covalently attached to a 5-µm PS/DVB support material.

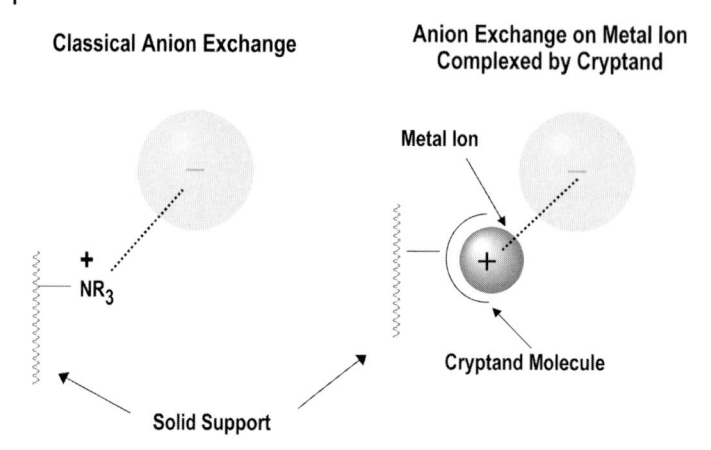

Fig. 3-61. Comparison between classical anion exchange and anion exchange on a metal ion complexed by a cryptand.

Cryptand 2.2.2 shows a higher affinity for metal ions than crown ethers such as 18-crown-6. For this reason, columns incorporating cryptands have higher capacities due to the ability of the cryptands to form more stable cation complexes [61]. The effect of eluant cation on the separation of anions can be illustrated by comparing separations achieved with different alkali metal cations. Excellent separations of anions are obtained with alkali metal hydroxide eluants. As can be seen from the three chromatograms in Fig. 3-62 which were obtained on a 250 mm × 4 mm i. d. prototype column, potassium, sodium, and lithium hydroxide eluants each produce a unique column capacity range. While a potassium hydroxide eluant is recommended for high capacity applications, sufficient resolution between standard anions is achieved with sodium hydroxide. By scaling down substrate particle diameter to 5 µm and column dimensions to 150 mm × 3 mm i. d., eluant concentration can be decreased significantly. However, under isocratic conditions, multivalent anions such as sulfate and orthophosphate are strongly retained (Fig. 3-63, top chromatogram). A baseline-resolved separation of all anions is achieved in a much shorter time by applying a capacity gradient, which is carried out by changing the cation component of the hydroxide eluant shortly after injecting the sample (Fig. 3-63, bottom chromatogram).

The biggest advantage of capacity gradients on a cryptand column is the possibility to elute non-polarizable and polarizable anions in the same run; this is extremely difficult, if not impossible, with both latex-agglomerated anion exchangers and aminated grafted polymers. As impressively demonstrated in Figure 3-64, even the highly polarizable perchlorate is rapidly eluting without sacrificing resolution among all the non-polarizable anions.

The cryptand A1 column should be operated at a controlled temperature to ensure reproducible retention times. The cryptand A1 is designed to operate at 35 °C. Higher temperatures will enable shorter run times.

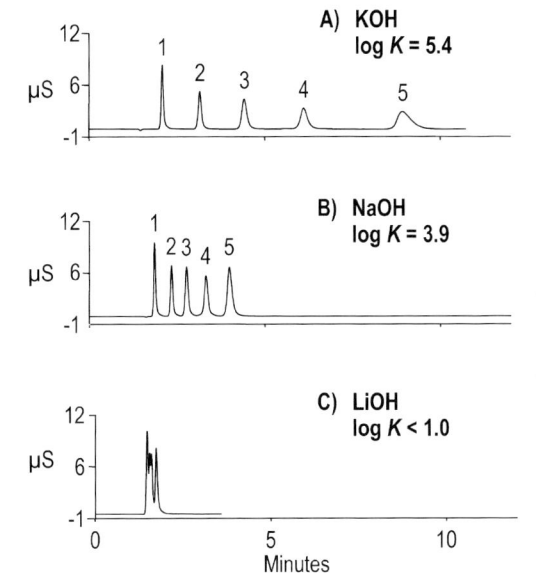

Fig. 3-62. Comparison of standard anion separations on a cryptand column with different alkali metal hydroxide eluants. — Column: Cryptand Prototype (250 mm × 4 mm i. d.); column temperature: 35 °C; eluants: 70 mmol/L alkali metal hydroxide; flow rate: 1 mL/min; detection: suppressed conductivity; solute concentrations: 2 mg/L fluoride (1), 3 mg/L chloride (2), 5 mg/L sulfate (3), 10 mg/L nitrite (4), and 10 mg/L bromide (5).

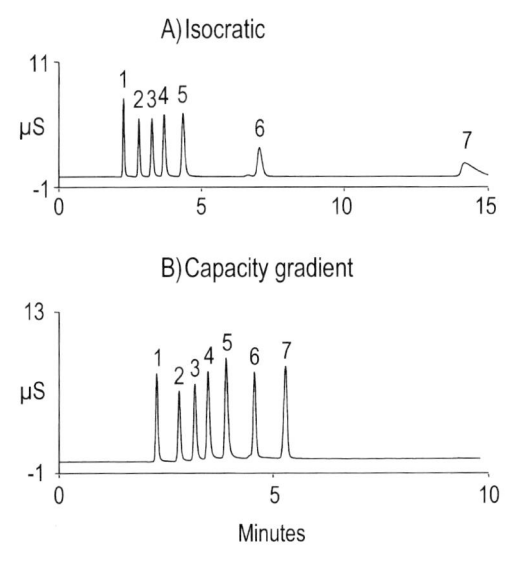

Fig. 3-63. Comparison of isocratic and capacity gradient separation of standard anions on IonPac Cryptand A1. — Column temperature 35 °C; eluants: (A) 10 mmol/L NaOH, (B) 10 mmol/L NaOH, step at 0.1 min to 10 mmol/L LiOH; flow rate: 0.5 mL/min; detection: suppressed conductivity; solute concentrations: 2 mg/L fluoride (1), 3 mg/L chloride (2), 5 mg/L nitrite (3), 10 mg/L bromide (4), 10 mg/L nitrate (5), 5 mg/L sulfate (6), and 15 mg/L orthophosphate (7).

3.4.4.2 Alumina Phases

In addition to silica, $(SiO_2)_x$, alumina, $(Al_2O_3)_x$, is among the most common adsorbents in liquid-solid chromatography. Although highly efficient separator columns were developed with spherical beads of small diameter, alumina is of only minor importance in HPLC after the introduction of silica-based, chemically bonded reversed phases.

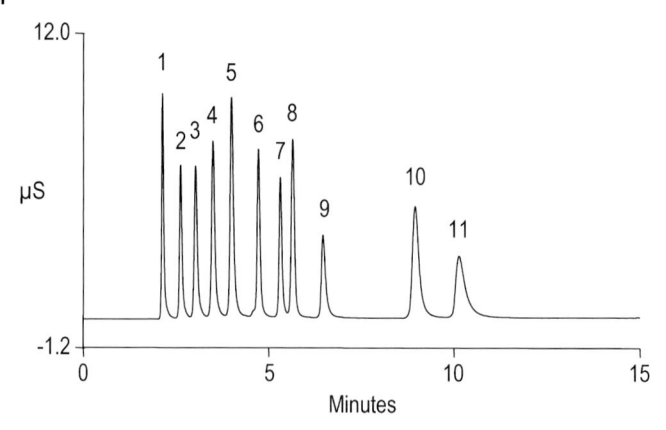

Fig. 3-64. Capacity gradient elution of non-polarizable and polarizable anions on IonPac Cryptand A1. — Column temperature: 35 °C; eluants: (A) 10 mmol/L NaOH, (B) 10 mmol/L NaOH, step at 0.1 min to 10 mmol/L LiOH; flow rate: 0.5 mL/min; detection: suppressed conductivity; solute concentrations: 2 mg/L fluoride (1), 3 mg/L chloride (2), 5 mg/L nitrite (3), 10 mg/L bromide (4), 10 mg/L nitrate (5), 5 mg/L sulfate (6), 5 mg/L thiosulfate (7), 5 mg/L orthophosphate (8), 10 mg/L iodide (9), 10 mg/L thiocyanate (10), and 15 mg/L perchlorate (11).

Like many other metal oxides, alumina has typical ion-exchange properties [62]. It is also mechanically and thermally stable, and exhibits only slight swelling and shrinkage in aqueous media. Nonetheless, alumina was seldom used in the past as a substrate for ion exchangers because of the low ion-exchange capacity and the inadequate stability against strong acids and bases. However, with the introduction of sensitive HPLC detectors, high ion-exchange and loading capacities were no longer necessary, creating an intensified interest in alumina phases for the separation of ionic species [63, 64].

The first detailed studies of the retention behavior of inorganic anions and cations on alumina were carried out by Schwab et al. [65]. They confirmed that predominantly ion-exchange processes are responsible for the retention of ionic species on alumina. The ion-exchange model considers the development of an electrical double layer at the alumina surface which is formed by the dissociation of Al-OH groups present at the surface and abstraction of H^+ and OH^- ions. This process, which is described in a simplified way by the dissociation equilibria (60) and (61), leads to the formation of positive and negative charges at which according to Eqs. (62) and (63), anion or cation exchange occurs with the solute ions X^- and X^+, respectively.

$$\diagdown\!\!\!\!\!\!\diagup\!\!\text{Al}-\text{OH} \rightleftharpoons \diagdown\!\!\!\!\!\!\diagup\!\!\overset{+}{\text{Al}}-\text{OH}^- \qquad (60)$$

$$\diagdown\!\!\!\!\!\!\diagup\!\!\text{Al}-\text{OH} \rightleftharpoons \diagdown\!\!\!\!\!\!\diagup\!\!\overset{+}{\text{Al}}-\text{O}^- \ \text{H}^+ \qquad (61)$$

$$\diagdown\!\!\!\!\!\!\diagup\!\!\overset{+}{\text{Al}}-\text{OH}^- + \text{X}^- \rightleftharpoons \diagdown\!\!\!\!\!\!\diagup\!\!\overset{+}{\text{Al}}-\text{X}^- + \text{OH}^- \qquad (62)$$

$$\diagdown\!\!\!\!\!\!\diagup\!\!\overset{+}{\text{Al}}-\text{O}^- \text{H}^+ + \text{X}^+ \rightleftharpoons \diagdown\!\!\!\!\!\!\diagup\!\!\overset{+}{\text{Al}}-\text{O}^- \text{X}^+ + \text{H}^+ \qquad (63)$$

The presence of hydroxide groups at the alumina surface was unequivocally confirmed by IR spectroscopy [66], whereas the number and nature of these groups are determined by the thermal preconditioning of the material. The amphoteric character of alumina and its transformation into an anion or cation exchanger (via hydration and subsequent treatment with acid or base) are illustrated in the following reaction scheme:

$$Al_2O_3 + H_2O \longrightarrow \begin{array}{c} HH \\ \diagdown\!\!\bar{O}\!\!\diagup \\ -Al-\bar{O}-Al-\bar{O}- \end{array}$$

$$\downarrow$$

$$H_2O + \begin{array}{c} Na \\ | \\ |\bar{O}| H \\ -Al-\bar{O}-Al-\bar{O}- \end{array} \xleftarrow{\ NaOH\ } \begin{array}{c} H \\ | \\ |\bar{O}| H \\ -Al-\bar{O}-Al-\bar{O}- \end{array} \xrightarrow{\ HCl\ } \begin{array}{c} Cl H \\ | \\ -Al-\bar{O}-Al-\bar{O}- \end{array}$$

Accordingly, the anion exchange process occurs between solute anions and OH^- ions attached to the alumina surface, while cation exchange occurs between solute cations and H^+ ions, which are released by dissociation of Al-OH groups.

The retention behavior of inorganic anions on an alumina phase was investigated in detail by Schmitt and Pietrzyk [67]. The most important experimental parameters determining retention are the pH value and the ionic strength of the mobile phase. Because alumina is weakly basic, anion exchange can only occur at pH values below the isoelectric point, which depends on the type of eluant. In general, an increase of the anion exchange capacity and, thus, the retention, is observed with decreasing pH value of the mobile phase.

Most interesting are the selectivity differences between an alumina phase and a strongly basic anion exchanger with quaternary ammonium groups. In comparison with a conventional anion exchanger, halide ions, for example, are eluted on an alumina phase in reverse order: $I^- < Br^- < Cl^- < F^-$. This order corresponds to the formation constants of aluminum halide complexes [46], which suggests an interaction between aluminum and halide ions within the Al_2O_3 structure.

The influence of the ionic strength on retention does not differ from conventional anion exchange chromatography; retention decreases with increasing ionic strength of the mobile phase. However, due to the different selectivity of an alumina phase, other eluant ions must be chosen. Fluoride belongs to the strongest eluant ions, but its use is not recommended because the interaction with alumina is too strong. Hydroxide ions also exhibit a strong elution power, because they convert the alumina phase into a cation exchanger at certain concentrations. Sodium perchlorate was found to be a suitable eluant for the determination of simple inorganic anions. A typical chromatogram, displayed in Figure 3-65, reveals that modern alumina phases with spherical 5-μm particles are highly efficient. However, it should be pointed out that the eluants usually employed for the separation of inorganic anions are not compatible with suppressed conductivity detection.

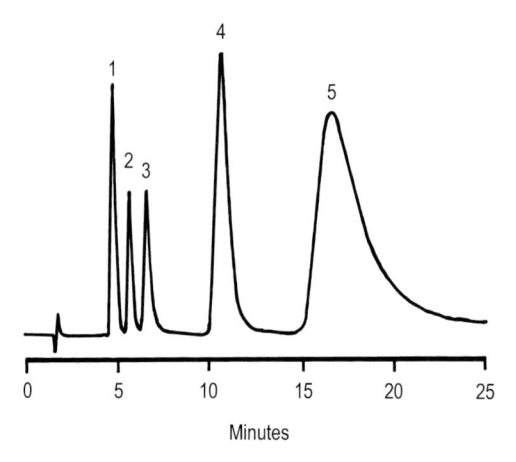

Fig. 3-65. Separation of various anions on an alumina phase. — Column: Spherisorb A5Y; eluant: 1 mmol/L sodium acetate buffer (pH 5.6) + 0.05 mol/L $NaClO_4$; flow rate: 1 mL/min; detection: UV (214 nm); injection volume: 2 μL; solute concentrations: 1 g/L each of iodide (1), nitrate (2), nitrite (3), bromate (4), and benzoate (5); (taken from [67]).

3.5
Eluants for Anion Exchange Chromatography

The kind of eluant that is used for anion exchange chromatography depends mainly on the detection system being employed. Since the detection of inorganic and organic anions in many cases is performed with conductivity detection, the eluants used are classified into two groups:
- Eluants for conductivity detection with *chemical* suppression of the background conductivity
- Eluants for conductivity detection with *electronic* compensation for the background conductivity.

3.5 Eluants for Anion Exchange Chromatography

Independently, the affinities of eluant ions and solute ions must be comparable; i.e., as a rule of thumb, divalent solute ions may only be eluted with divalent eluant ions.

Eluants of the *first* kind (for conductivity detection with chemical suppression) include the salts of weak inorganic acids, which exhibit a low background conductivity after chemical modification within a suppressor system (see Section 3.6). In their introductory 1975 paper, Small et al. [3] used sodium phenolate for the separation of non-polarizable inorganic and organic anions. Although good separations were achieved with this eluant, even minute traces of oxygen in the mobile phase lead to oxidation of this compound and, thus, to a gradual poisoning of the stationary phase by the oxidation products. Today, sodium phenolate is no longer used as an eluant.

The versatile mixture of sodium carbonate and sodium bicarbonate, on the other hand, finds widespread application, because the elution power and the selectivity resulting there from are determined solely by the concentration ratio of these two compounds. A great variety of inorganic and organic anions can be separated with this eluant combination. As the product of the suppressor reaction, the carbonic acid is only weakly dissociated, so that the background conductivity is very low.

As an alternative to carbonate/bicarbonate systems, amino acids (α-aminocarboxylic acids) may be used as an eluant [69, 70]. Their dissociation behavior is depicted in Fig. 3-66. At alkaline pH, amino acids exist in the anionic form due to the dissociation of the carboxyl group and, thus, may act as an eluant ion. The product of the suppressor reaction is the zwitterionic form with a correspondingly low background conductivity. This depends on the isoelectric point, pI, of the amino acid.

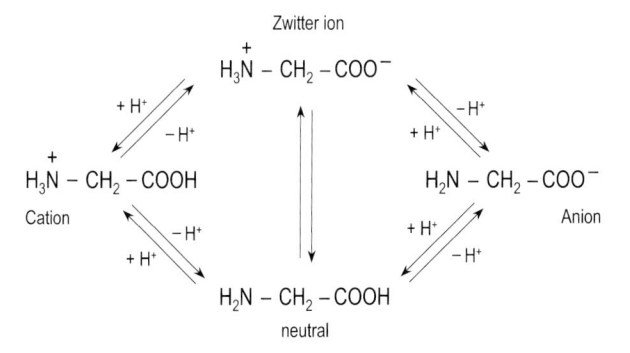

$$\text{pI} = \frac{\text{p}K_1 + \text{p}K_2}{2} \tag{64}$$

Fig. 3-66. Dissociation behavior of amino acids on the example of glycine.

The residual dissociation of the zwitter ion and the background conductivity of the eluant is even lower than for the carbonate/bicarbonate system, if amino acids are selected with an isoelectric point near the neutral pH value. Similar low background conductivities are obtained with N-substituted aminoalkylsulfonic acids, which were introduced as eluants by Irgum [70], and which can also be used for gradient elution due to their sufficient purity [71].

The application of pure sodium hydroxide eluants was regarded to be disadvantageous in the past, although water as the suppressor product produces virtually no background conductivity. However, since hydroxide ions exhibit only a small affinity towards the stationary phase, it was necessary to work with relatively high concentrations to elute anions with more than one negative charge. This has an adverse effect on the background conductivity because suppressor systems used in the past (packed bed suppressors, hollow fiber suppressors) possessed only a limited ion-exchange capacity. Only since the introduction of high-capacity micromembrane suppressors it became possible to use sodium hydroxide in concentrations up to 0.15 mol/L. Hence, this eluant is of special significance, because it is possible to elute even polyvalent solute ions at these high concentrations. Sodium hydroxide is therefore perfectly suited for gradient elution of anions in combination with conductivity detection.

Tetraborate ions exhibit a similar low affinity towards the stationary phase allowing for the separation of fluoride and short-chain aliphatic carboxylic acids. To a limited extent, sodium tetraborate is also suited for gradient elution, because boric acid as the suppression product is only weakly dissociated.

Aromatic amino acids (e.g., tyrosine) which favorably reduce the retention of polarizable anions at alkaline pH, have a comparatively high elution power. *p*-Cyanophenol acts in a similar way. When it is added to a carbonate/bicarbonate mixture, a wealth of polarizable anions can be separated at a suitable stationary phase (IonPac AS4A and AS5) and detected via their electrical conductivity. Today, tyrosine as well as *p*-cyanophenol are only rarely used as eluants, because both acrylate-based and cryptand anion exchangers allow the separation of polarizable and non-polarizable anions in the same run.

Table 3-10 presents an overview of the various eluants that are compatible with suppressor systems, including a rating to their elution power. With the eluants listed in Table 3-10 and with the aid of organic additives described above, nearly all anions detectable via conductivity may be analyzed using one of the many available anion exchangers. This multitude of stationary phases with their different selectivities enables the vast number of potential eluants to be reduced to a few versatile systems. Therefore, the statement often seen in the literature − that suppressor systems limit the choice of eluants − is not valid.

Eluants of the *second* kind (for conductivity detection with electronic compensation) should have a low background conductivity to enable sensitive conductivity detection of anions being analyzed. The selection of eluants suited for conductivity detection with electronic compensation of the background conduc-

3.5 Eluants for Anion Exchange Chromatography

Table 3-10. Eluants commonly used for suppressed conductivity detection.

Eluant	Eluant ion	Suppressor product	Elution strength
$Na_2B_4O_7$	$B_4O_7^{2-}$	H_3BO_3	very weak
NaOH	OH^-	H_2O	weak
$NaHCO_3$	HCO_3^-	$[CO_2 + H_2O]$	weak
$NaHCO_3/Na_2CO_3$	HCO_3^-/CO_3^{2-}	$[CO_2 + H_2O]$	medium strong
$H_2NCH(R)COOH/NaOH$	$H_2NCH(R)COO^-$	$H_3N^+CH(R)COO^-$	medium strong
$RNHCH(R')SO_3H/NaOH$	$RNHCH(R')SO_3^-$	$RNH_2^+CH(R')SO_3^-$	medium strong
Na_2CO_3	CO_3^{2-}	$[CO_2 + H_2O]$	strong

tivity has been the subject of numerous investigations [16, 17, 19, 72]. Benzoates, phthalates, and o-sulfobenzoates are used most frequently, because they exhibit both a sufficient affinity towards the stationary phase of surface-aminated ion exchangers and a relatively low conductivity. When aromatic carboxylic acids are used as eluants, the pH value of the mobile phase must be adjusted between pH 4 and 7, because it affects the degree of dissociation of the organic acid and, thus, determines the retention behavior of the species to be analyzed. In addition, an almost neutralized eluant is necessary to minimize the concentration of oxonium ions with their high equivalent conductivity. Table 3-11 comprises the resulting background conductivities of the eluants used by Fritz et al. in their introductory paper [19]. The background conductivities of these eluants are much higher than those of carbonate/bicarbonate-based eluants after passing the suppressor system ($\sim$15 to 20 µS/cm). The high background conductivities with values between 60 and 160 µS/cm reduce the detection limit and the linear range of the detector. Investigations by Gjerde and Fritz [72] show significantly lower background conductivities when adjusting the pH value of the mobile phase to neutral.

Table 3-11. Background conductivities of the eluants used by Fritz et al. in their introductory paper [19].

Eluant	Concentration [mol/L]	pH value	Conductivity [µS/cm]
Sodium benzoate	$6.5 \cdot 10^{-4}$	4.6	65.9
Potassium hydrogen phthalate	$5.0 \cdot 10^{-4}$	4.4	74.3
	$5.0 \cdot 10^{-4}$	6.2	112
	$6.5 \cdot 10^{-4}$	4.4	90.5
	$6.5 \cdot 10^{-4}$	6.2	158
Ammonium-o-sulfobenzoate	$5.0 \cdot 10^{-4}$	5.8	132

Problems with adjusting the pH value are encountered when commercially available pH electrodes are used. In general, they contain a Ag/AgCl reference electrode with saturated KCl solution as supporting electrolyte, as well as porous frits as salt bridges. Therefore, eluants can be contaminated with chloride, which would result in false data for a subsequent chloride determination [73]. This problem may be circumvented by spatially separating the reference electrode and eluant from each other.

Using aromatic carboxylic acids as eluants, a separation of the most important inorganic anions such as fluoride, chloride, bromide, iodide, nitrite, nitrate, orthophosphate, sulfate, thiocyanate, and thiosulfate is possible. However, at best, seven anions may be separated within a single run under suitable chromatographic conditions. Sodium and potassium benzoate are primarily used for the separation of monovalent anions. The affinity of benzoate towards the stationary phase is comparable with that of sodium bicarbonate. The corresponding phthalic acid salts, exhibiting a higher elution strength, have to be used for the elution of monovalent *and* divalent species. An increase of the benzoate concentration does not work, because the resulting background conductivity becomes too high for a sensitive conductivity detection. Strongly polarizable and polyvalent anions might be eluted when employing the salts of trimesic acid, which exhibits a strong affinity towards the stationary phase due to its three carboxyl groups.

Potassium hydroxide is a suitable eluant for the determination of inorganic anions with pK values above 7. Due to the high pH value of the mobile phase, even weak acids are completely dissociated and, thus, may be detected via their conductivity. However, the conductivity of the analytes is lower than that of the mobile phase, so that negative signals are observed for these species. This method is called indirect conductivity detection.

The classification of eluants into the two categories mentioned above makes sense and is necessary only in the framework of conductivity detection with its different application requirements. Eluant choice is much easier for applications using spectrophotometric or amperometric detectors. In photometric detection, both the photometric properties of the eluant ions and their chemical properties have to be taken into account; nevertheless, a large number of eluants are available. The alkali salts of phosphoric acid, sulfuric acid, and perchloric acid have proved to be successful, because they all feature good UV transmittance. In the field of amperometric detection, the choice of eluants is even much higher. The electrolyte concentration in the mobile phase must be about 50 to 100 times higher than the analyte ion concentration. The mobile phase acts as a supporting electrolyte which, by the reduction of the mobile phase resistance, R_L, ensures that the voltage drop, $i \cdot R_L$, is kept low. Chlorides, chlorates, and perchlorates of alkali metals and alkali hydroxides and carbonates are suited for use as support electrolytes.

3.6
Suppressor Systems in Anion Exchange Chromatography

A "suppressor system" is used for the sensitive detection of ions via their electrical conductivity. Its function is to chemically reduce the background conductivity of the electrolyte used as eluant before it enters the conductivity cell. Therefore, the suppressor system may be regarded as a part of the detection system.

In its simplest form, a suppressor system consists of a column containing a strongly acidic cation exchange resin in the hydrogen form. The function of this "suppressor column", originally used by Small [3], is illustrated by the analysis of chloride and bromide anions. After these anions are separated with an eluant such as sodium bicarbonate on one of the anion exchangers described above, the column effluent is passed through the suppressor column prior to entering the conductivity cell, where the following reactions occur:

- Strongly conducting sodium bicarbonate is converted to weakly dissociated carbonic acid by exchanging sodium ions of the eluant with hydronium ions of the cation exchanger:

$$\text{Resin} - SO_3H + NaHCO_3 \longrightarrow \text{Resin} - SO_3Na + [CO_2 + H_2O] \qquad (65)$$

- Similarly, sodium chloride and sodium bromide are converted into their corresponding acids:

$$\text{Resin} - SO_3H + NaCl \longrightarrow \text{Resin} - SO_3Na + HCl \qquad (66)$$

$$\text{Resin} - SO_3H + NaBr \longrightarrow \text{Resin} - SO_3Na + HBr \qquad (67)$$

As the result of the suppressor reaction, strongly-conducting mineral acids in the presence of weakly dissociated carbonic acid enter the conductivity cell and, consequently, are easily detected.

Higher sensitivity is the primary advantage of the suppressor technique as compared to direct conductivity detection. In addition, the specificity of the method is also increased, because chemical modification of eluant and sample in the suppressor system turns the conductivity detector from a bulk-property detector into a solute-specific detector [74]. Thus, exchanging eluant and sample cations with hydronium ions means that only analyte anions are detected by the conductivity detector and appear in the resulting chromatogram.

In the early 1980s, conventional suppressor columns became technologically outdated and were replaced by membrane-based suppressor systems. Nevertheless, miniaturized forms of the original packed bed suppressors are still available. The characteristic properties of the various suppressor devices are discussed below.

3.6.1
Suppressor Columns

As already discussed, a packed bed suppressor suitable for anion exchange chromatography contains a strongly acidic cation exchange resin in the hydrogen form. The properties of this resin are extremely important to the quality of the analysis. In addition to the conversion of eluant and sample into their corresponding acids, other phenomena ensue in packed bed suppressors. Since a regenerated cation exchanger exists in the hydrogen form, it resembles an ICE column. Therefore, weak and strong electrolytes are subject to the Donnan effect (see Section 5.1). While strong acids are excluded from the stationary phase and travel with the mobile phase along the column, weak acids such as nitrite and acetate, that can partly be non-dissociated, may pass through the Donnan layer and may be adsorbed on the surface of the stationary phase. This results in a small increase in the retention times for weak acids, which depends on the pK_a value of the respective species, the total volume of the packed bed suppressor, and the degree of its use. The resulting change in peak height adversely affects routine analysis. Thus, gel-type polystyrene/divinylbenzene-based cation exchangers with 8% cross-linking, particle diameters between 20 µm and 40 µm, and a low specific surface area are generally used. Although, in principle, cation exchangers with a high specific surface area are suited for suppressor columns because of their fast ion-exchange kinetics, they have to be considered disadvantageous regarding the adsorption processes described above.

A major drawback of packed bed suppressors is the requirement of periodic regeneration. Depending on the degree of use, the retention time of the negative water dip will change. The latter arises due to the fact that the conductivity of water, as solvent for the sample, is smaller than the eluant background conductivity after passing through the suppressor column. Pure de-ionized water injected into the ion chromatograph travels with the mobile phase through the column. The negative dip is observed at a time that is almost identical to the system void. With the water dip changing its position, analytes eluting close to the system void, such as fluoride and chloride are difficult to quantitate.

The particle size of the resin and the void volume of the suppressor column affect the quality of the separation, because both parameters determine the peak broadening. The total volume of the suppressor column should be as small as possible to prevent mixing of the already separated signals. For the resulting suppression capacity, however, the total volume of the suppressor column should be as large as possible. These two requirements are incompatible, so the dimensioning of packed bed suppressors is always a compromise between the suppression capacity and the peak broadening caused by the void volume. Therefore, it is advisable to use suppressor columns only in combination with separator columns packed with resins that have a particle diameter of more than 15 µm. Under these conditions, the contribution to the total peak variance is small.

The regeneration of conventional packed bed suppressors is usually carried out for about 15 minutes using 0.5 mol/L sulfuric acid with a flow rate of 4 mL/min. Subsequently, the system is flushed with de-ionized water for several minutes, and then it is conditioned with eluant again.

In 1996, Metrohm (Herisau, Switzerland) introduced a periodically regenerated suppressor module (MSM) consisting of *three* suppressor cartridges connected to a rotor valve. The cartridges are filled with a strong acid cation exchange resin in the hydrogen form [75]. While one of the three cartridges in the position "suppression" is used for the current chromatographic run, the other two are regenerated with diluted sulfuric acid (position "regeneration") and rinsed with de-ionized water (position "rinsing"), respectively. After every chromatographic run, the rotor valve switches to the next position, so that a freshly regenerated and rinsed cartridge is inline with the separator column for the next chromatographic run. Regenerant and de-ionized water are delivered via a 4-channel peristaltic pump.

This setup represents a certain improvement over conventional packed bed suppressors, because the void volume of the cartridges is very small. However, as a consequence, the suppression capacity of the MSM is also very small, so that this kind of packed bed suppressor can only be used for low ionic strength eluants (see Fig. 3-18 in Section 3.4.1). High-capacity anion exchangers as well as gradient elution techniques cannot be utilized with this suppressor module. The basic problem of packed bed suppressors, i. e. periodic regeneration of cartridges, is not addressed by the Metrohm Suppressor Module. On the other hand, the MSM does deliver low noise, as do all chemically regenerated packed bed suppressor systems.

A similar device was introduced several years ago by Alltech (Deerfield, USA) under the trade name ERIS. The ERIS suppressor consists of *two* packed bed cartridges filled with a strong acid cation exchange resin in the hydrogen form, that are connected to a rotor valve. In contrast to the Metrohm device, the cation exchange cartridges of the ERIS suppressor are regenerated electrolytically, i. e. the regenerant ions are generated from the aqueous mobile phase inside the cartridge by means of electrolysis. Therefore, only two cartridges are necessary to operate the system: one is inline with the separator column during the chromatographic run, the other cartridge being regenerated. The only advantage of the ERIS suppressor over the Metrohm Suppressor Module is the fact that chemical regenerants are not required. Apart from this, it suffers from the same drawbacks as described above for the MSM. Both types of suppressor devices are only available for anion analysis.

In 2001, the ERIS suppressor was replaced by the DS-Plus^{TM} suppressor, the first packed bed suppressor that is continuously regenerated. As shown in Fig. 3-67, the eluant enters the resin-based suppressor and is split into three paths. Two paths are directed to the electrodes where electrolysis takes place, thereby providing the regenerant ions. The third path carries the analyte ions through the suppression bed. Hydronium ions generated at the anode replace sodium ions in the eluant thus suppressing background conductivity. All three effluents are then directed to a degassing chamber to remove electrolysis gases and CO_2, which is formed when carbonate-based eluants are used for chromatography. The three flow paths are balanced by using significant lengths of tubing. Because the eluant flow is split into different paths, all gas must be removed from each path to maintain the split ratios between the flow paths. If there is compressible gas in these flow paths, the flows will not be balanced correctly and the device will not work correctly. The internal dead volume of the DS-Plus packed bed suppressor itself is approximately 200 µL. In combination with the excessive tubing required for flow splitting, these conditions lead to a relatively low chromatographic efficiency.

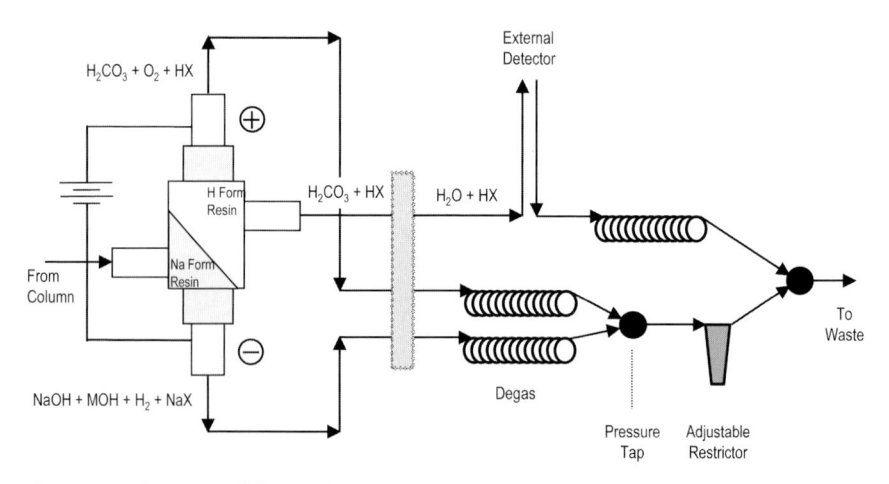

Fig. 3-67. Schematics of the DS-Plus suppressor.

The beneficial side effects of degassing are the lower background when running carbonate eluants and a reduced water dip. The DS-Plus suppressor can also suppress carbonate gradients, reducing the baseline shift that normally occurs as the carbonate concentration is increased. In comparison with hydroxide concentration gradients on hydroxide-selective stationary phases (see Section 3.9), the applicability of carbonate gradients – composition and concentration gradients – is somewhat limited. This is due to the relatively strong eluting power of carbonate-based eluants, which makes it difficult, if not impossible, to separate early eluting low-molecular weight organic acids from fluoride and dicarboxylic acids from nitrate and sulfate. The following two examples illustrate this limitation. The first example shown in Fig. 3-68 was obtained with a combined composition and concentration gradient, i. e. the eluting ion is changed from bicarbonate to carbonate during the run, and the total carbonate concentration is increased. A small concentration of an organic modifier (*p*-cyanophenol) is added during the run to improve the symmetry of late eluting peaks and to elute polarizable anions such as perchlorate and citrate within 20 minutes. Although mono-, di- and tri-valent anions can be separated in this way, resolution between orthophosphate, sulfate, and oxalate is very limited.

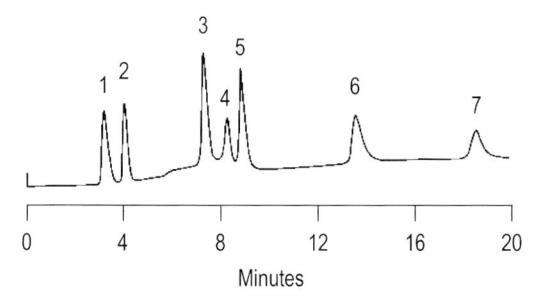

Fig. 3-68. Separation of mono-, di- and tri-valent anions with a combined carbonate-based composition and concentration gradient. – Column: Alltech Novosep A-1 (100 mm × 4.6 mm i. d.); eluant: 1.7 mmol/L NaHCO$_3$ linearly in 5 min to 2.8 mmol/L NaHCO$_3$ + 2.2 mmol/L Na$_2$CO$_3$ + 0.22 mmol/L p-cyanophenol, then to 5 mmol/L NaHCO$_3$ + 5 mmol/L Na$_2$CO$_3$ + 0.5 mmol/L p-cyanophenol in 5 min, then to 10 mmol/L Na$_2$CO$_3$ + 1 mmol/L p-cyanophenol in 5 min; flow rate: 1.5 mL/min; detection: suppressed conductivity; solute concentrations: 5 mg/L acetate (1), 1 mg/L chloride (2), 5 mg/L orthophosphate (3), 1 mg/L sulfate (4), 5 mg/L each of oxalate (5), perchlorate (6), and citrate (7).

Even with a shallower carbonate gradient as shown in Fig. 3-69, fluoride, acetate, and formate are not resolved to baseline. This is not surprising, because the dilute sodium bicarbonate eluant still exhibits a much stronger eluting power in comparison to a dilute sodium hydroxide eluant. Moreover, separation between orthophosphate, selenite, arsenate, and sulfate is not sufficient at all, although a longer column has been used. Due to the absence of an organic modifier and a lower flow rate, a total analysis time of 30 minutes results, which is no longer acceptable for this kind of analysis.

The DS-Plus suppressor produces very little noise despite the electrochemical regeneration, which leads to significantly higher noise with membrane-based suppressors. Using a standard carbonate/bicarbonate eluant (1.8 mmol/L/ 1.7 mmol/L), the resulting noise level is approximately 0.5 nS/cm. Due to the removal of carbonic acid by eluant degassing, the background conductivity is about 2 µS/cm; for comparison this value is 14 µS/cm with standard suppressors.

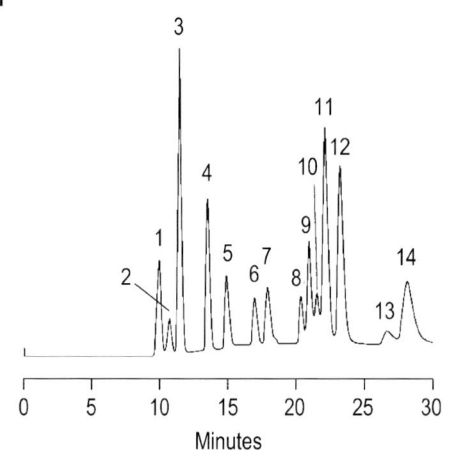

Fig. 3-69. Separation of early and late eluting anions with a combined carbonate-based composition and concentration gradient. — Column: Alltech Novosep A-1 (150 mm × 4.6 mm i. d.); eluant: 1 mmol/L $NaHCO_3$ isocratically for 5 min, then linearly to 2.8 mmol/L $NaHCO_3$ + 2.2 mmol/L Na_2CO_3 in 10 min; flow rate: 1 mL/min; detection: suppressed conductivity; solute concentrations: 1 mg/L fluoride (1), 10 mg/L acetate (2), 10 mg/L formate (3), 2 mg/L each of chloride (4), nitrite (5), bromide (6), and nitrate (7), 3 mg/L orthophosphate (8), 10 mg/L selenite (9), 10 mg/L arsenate (10), 3 mg/L sulfate (11), 10 mg/L each of selenate (12), tungstate (13), and molybdate (14).

3.6.2
Hollow Fiber Suppressors

Conventional suppressor columns are not in use anymore. They became obsolete with the introduction of modern membrane technologies in 1982. The disadvantages of suppressor columns mentioned above could be eliminated with the development of a hollow fiber suppressor. It consists of a semipermeable membrane, which is wrapped around a cylindrically shaped body. This tubular coil is housed in a jacketed vessel. While the column effluent is passed through the interior of the fiber that is packed with inert beads, a dilute sulfuric acid solution flows countercurrent to the eluant, in contact with the exterior of the fiber. The flow rate can easily be controlled by the difference in height between the regenerant reservoir and the suppressor housing. When working in extremely sensitive detector ranges (<1 µS/cm full scale), it is recommended to pump the regenerant solution through the fiber suppressor, because the constantly changing hydrostatic pressure leads to baseline drifts. A much more stable baseline is obtained by pneumatic delivery of the regenerant. For that purpose, the sulfuric acid reservoir is pressurized, resulting in a pulse-free flow.

In contrast to conventional suppressor columns, hollow fiber suppressors are continuously regenerated, and thus do not require an additional delivery system. The reactions occurring at the membrane wall are shown in Fig. 3-70. Because fiber suppressors suited for anion exchange chromatography act as cation exchangers, the eluant cations are exchanged with hydronium ions in the regenerant solution. The driving force for the diffusion of hydronium ions through the membrane is provided by their subsequent reaction with bicarbonate to form carbonic acid. In order to maintain the ionic balance, the cations diffuse into the regenerant solution. As illustrated in Fig. 3-70, three distinct regions in the fiber can be defined:

- An expended region where the eluant enters the fiber suppressor (cation exchanger in the sodium form)
- A region of "dynamic equilibrium" in which the actual suppression reactions occur
- A regenerated region where the eluant exits the fiber suppressor (cation exchanger in the hydrogen form).

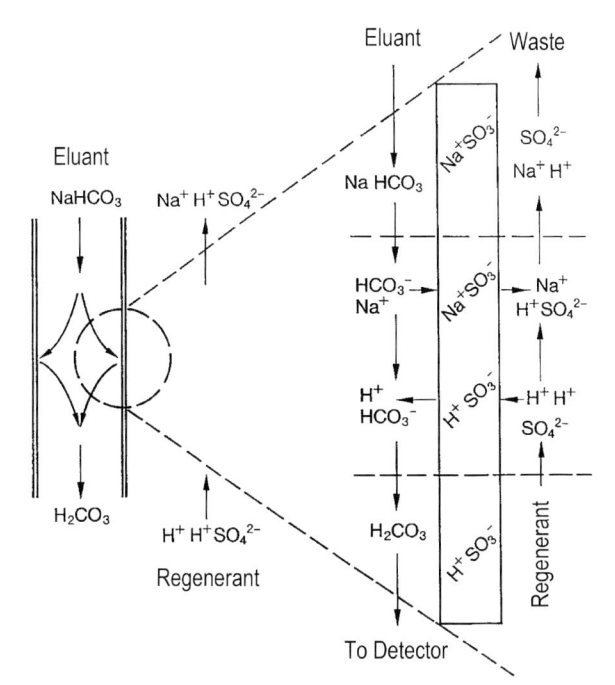

Fig. 3-70. Schematic of a hollow fiber suppressor for anion exchange chromatography.

Once established, the dynamic equilibrium is maintained as long as pertinent operating conditions are not altered.

Since the percentage of ion-exchange groups in the sodium form remains constant due to continuous regeneration, peak heights of weak acids also remain constant. Thus, standardization of the method is possible. In addition, quantitation of fluoride and chloride is simplified, because the water dip precedes the fluoride peak and remains constant in retention time. In comparison to suppressor columns, hollow fiber suppressors have a void volume that is smaller by an order of magnitude. This drastically reduces mixing and band broadening effects and enhances sensitivity. Most remarkable is the 300% sensitivity increase for nitrite.

Looking at the schematic representation of the flow profile within the fiber shown in Fig. 3-71, it becomes apparent that a further sensitivity increase can be accomplished by changing the parabolic flow profile. When the fiber is packed with inert Nafion beads, the translational diffusion of ions is favored over longitudinal diffusion. This, in turn, improves the mass-transfer through the membrane, which leads to a further increase in sensitivity, particularly pronounced in the case of orthophosphate as the salt of a weak acid.

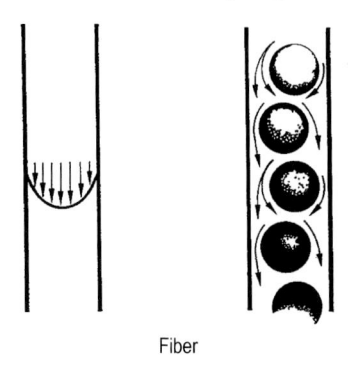

Fig. 3-71. Schematic representation of the flow profile in the interior of a fiber.

10 mmol/L sulfuric acid is recommended for the continuous regeneration of a hollow fiber suppressor. The flow rate should be adjusted between 2 and 3 mL/min. Under these conditions, a background conductivity of 18 to 25 µS/cm results for the bicarbonate/carbonate eluant (2.8 mmol/L; 2.2 mmol/L) depending on the purity of the water used for eluant preparation. If the ionic strength is lowered, the regenerant concentration should still be maintained, resulting in a decrease of the background conductivity. If more concentrated eluants are required, the suppression capacity of the hollow fiber suppressor might not be sufficient. However, the regenerant concentration can only be increased to a certain extent because, at higher concentrations, sulfate ions can overcome Donnan exclusion forces and diffuse into the interior of the fiber. Table 3-12 lists the background conductivities resulting from different concentrations of the regenerant solution. When using more concentrated carbonate-based eluants, outgassing of carbon dioxide may occur within the conductivity cell. This effect may be prevented by applying an adequate backpressure at the outlet capillary of the detector cell. However, the backpressure must not exceed about 13 bar (200 psi) due to the limited pressure stability of the membrane suppressor. If the use of more concentrated carbonate eluants cannot be avoided, a micromembrane suppressor with a much higher suppression capacity has to be employed.

Table 3-12. Dependence of background conductivity on the regenerant concentration.

$[H_2SO_4]$ [mol/L]	Background conductivity [µS/cm]
0.0125	21.7
0.025	28.2
0.125	288

3.6.3
Micromembrane Suppressors

Although suppressor systems are very effective in terms of a sensitivity increase, their application limits the choice of eluants and their concentrations. Therefore, the objective for developing a new generation of suppressors was to create a system that combines the positive properties of packed bed and hollow fiber suppressors. This objective was achieved with the introduction of the micromembrane suppressor in 1985 [76], which, like the hollow fiber suppressor, operates by using the principle of continuous regeneration but features a much higher suppression capacity. Its design allows the suppression capacity to be enhanced by more than one order of magnitude, so that, in terms of capacity, a micromembrane suppressor is comparable to a conventional suppressor column. Significant progress was made to further decrease the void volume of the system, which is only 50 µL for a micromembrane suppressor, a volume that has little effect on chromatographic separation. (For comparison: the void volume of a conventional suppressor column is about 2000 µL; that of a hollow fiber suppressor is about 200 µL.) Further improvements have been made regarding system ease-of-use. For example, the membrane in a hollow fiber suppressor ages and has to be replaced twice a year, but a micromembrane suppressor is almost maintenance-free when the operating instructions are followed.

Figure 3-72 is a schematic illustration of the sandwich structure of a micromembrane suppressor. It consists of a flat, two-part enclosure in which strongly sulfonated ion-exchange screens and thin ion-exchange membranes are sandwiched together in alternating order. The two parts of the enclosure keep them together. The ion-exchange screen functions as an eluant or regenerant channel.

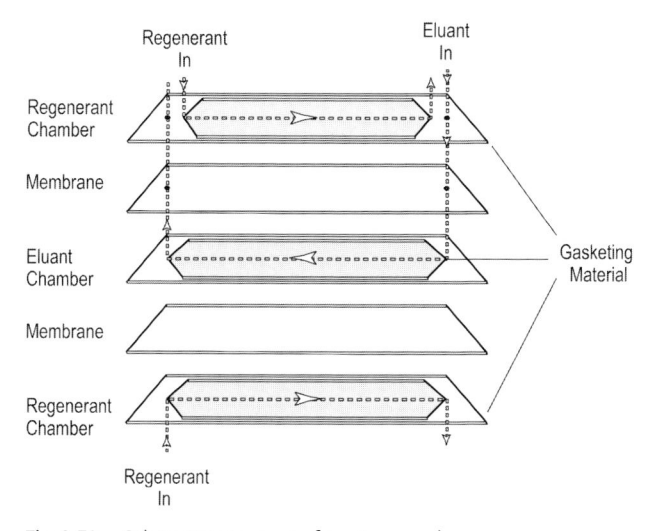

Fig. 3-72. Schematic structure of a micromembrane suppressor.

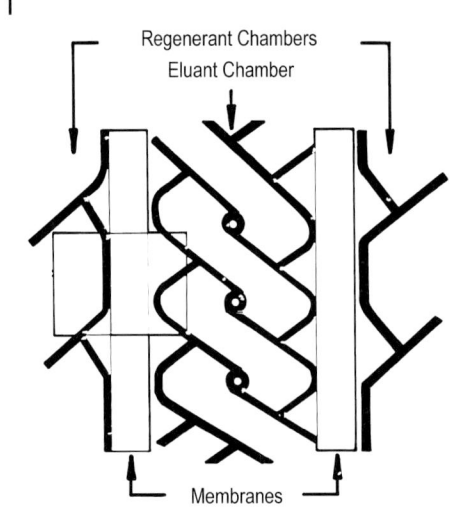

Fig. 3-73. Enlarged view of an eluant chamber in a micromembrane suppressor.

Gasketing material is attached at the side, so that the screen is porous only in the central region. In analogy to a hollow fiber suppressor, the eluant is passed through the eluant chamber located in the middle while regenerant flows countercurrent through the two regenerant chambers. It is clear from the enlarged view of the eluant chamber in Fig. 3-73 that, due to the screen structure, eluant cations are transported much more efficiently to the adjacent membrane wall than in a packed hollow fiber membrane suppressor.

The comparatively higher suppression capacity of a micromembrane suppressor is not only caused by the higher diffusion efficiency of eluant cations to the membrane wall. The screen's ion-exchange characteristics have an even greater impact; they are directly proportional to the suppression capacity. In continuously regenerated suppressors, one talks about the dynamic cation exchange capacity. Therefore, the suppression capacity is defined for both the hollow fiber suppressor and the micromembrane suppressor according to Eq. (68) as the product of eluant concentration and eluant flow rate:

$$\text{Suppressor capacity} \, [\mu\text{equiv/min}] = \text{Eluant concentration} \, [\mu\text{equiv/mL}] \cdot \text{Flow rate} \, [\text{mL/min}] \tag{68}$$

Because a micromembrane suppressor allows the use of a sodium hydroxide eluant at a maximum concentration of 0.1 mol/L (100 µequiv/mL) and with a flow rate of 2 mL/min, the suppression capacity is about 200 µequiv/min.

The high capacity of a micromembrane suppressor and the ability to provide continuous suppression allow the number of possible eluants to be significantly enlarged. In general, any weak acid eluant can be used as long as it exists in an anionic form above pH 8 and in a neutral form between pH 5 and 8. Above all, this includes amino acids already mentioned in Section 3.5 which, for example, allow the elution of all halide ions within a short time.

The high capacity of a micromembrane suppressor also allows the use of high ionic strength eluants, thereby reducing analysis times significantly. The low void volume of the suppressor, which barely affects separation efficiency, also contributes to the sensitivity increase.

The greatest achievement for ion chromatography that was made possible by the development of micromembrane suppressors was the introduction of gradient elution techniques with subsequent conductivity detection. As in conventional HPLC and in anion exchange chromatography, a distinction is made between composition gradients and concentration gradients. Simple composition gradients, such as the stepwise or continuous replacement of bicarbonate ions by carbonate ions, were successfully employed in the past. Although the background conductivity increases slightly when the eluant ion changes due to increased formation of carbonic acid as the suppression product, the capacity of conventional hollow fiber suppressors is sufficient for this kind of gradient technique. However, because the ionic strength in the mobile phase remains constant in the case of composition gradients, ions with strongly different retention characteristics (e.g., linear polyphosphates of the type $M_{n+2}P_nO_{3n+1}$) cannot be analyzed within a single run.

This becomes possible by applying a concentration gradient in which the ionic strength in the mobile phase is increased during the gradient run. When sodium hydroxide (which is suitable for suppression) is used as an eluant, the increase in ionic strength often covers more than two orders of magnitude because of the low elution power of hydroxide ions. Without the application of a high-capacity suppressor system, the resulting baseline drift would be so steep that analyte signals could not be evaluated, even after baseline subtraction by a computer. However, this baseline increase is limited to a few µS/cm when using the micromembrane suppressor. Further details about the handling and applicability of gradient techniques are discussed separately in Section 3.9.

The regeneration of a micromembrane suppressor differs from that of a hollow fiber suppressor in that the required regenerant flow rate cannot be obtained via gravity feed. Regeneration can be carried out in three different modes: the conventional Pressurized Bottle mode, the AutoRegen mode, or the new Displacement Chemical Regeneration mode. The conventional Pressurized Bottle mode uses a pressurized reservoir to deliver the chemical regenerant to the micromembrane suppressor (Fig. 3-74). The pressure is set at 5-10 psi, which delivers the regenerant to the suppressor at approximately 5-10 mL/min. The spent regenerant is then diverted to waste. While a sulfuric acid concentration of 10 mmol/L suffices for isocratic operation, a two-fold regenerant concentration is recommended for gradient techniques. The flow rate should be adjusted to ensure a sufficiently low background conductivity when the maximum eluant concentration is reached. Maintaining these conditions, one can then switch to the initial eluant concentration.

3 Anion Exchange Chromatography (HPIC)

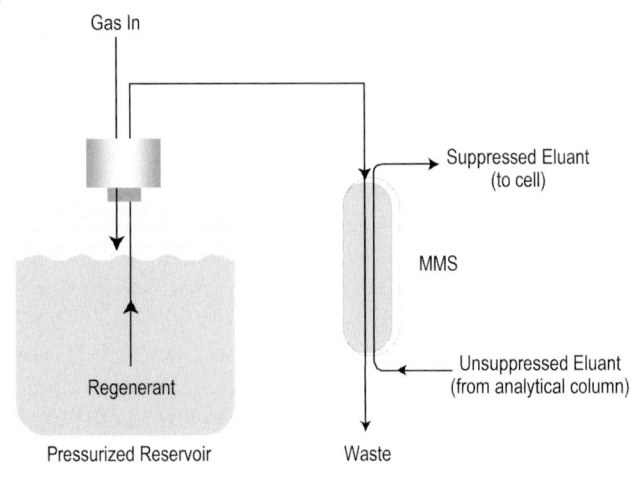

Fig. 3-74. Flow diagram of a micromembrane suppressor operated in the Pressurized Bottle mode.

If the ion chromatograph is to be operated online for several days without interruption, the regeneration of the suppressor system must not be interrupted. Thus, the regenerant cannot be delivered pneumatically from a reservoir, because even a relatively large volume must be opened and refilled after some time. Such interruptions can only be avoided by using a closed loop for the regeneration process. The spent regenerant is pumped through an ion-exchange cartridge which contains a cation exchanger in the hydrogen form to maintain a constant hydronium ion concentration. In that way, sodium sulfate formed in the suppressor is converted back into the corresponding acid for reuse. The regenerant flow rate is approximately 10 mL/min. As shown in the flow diagram in Fig. 3-75, a small pump is used instead of pneumatic delivery. The lifetime of such an ion-exchange cartridge having a fixed capacity of 500,000 µequiv depends on eluant concentration and flow rate. It may be calculated according to Eq. (69):

$$\text{Cartridge lifetime} = \frac{\text{Cartridge capacity}}{\text{Eluant normality} \cdot \text{Flow rate}}$$

$$[\min] = \frac{[\mu\text{equiv}]}{[\mu\text{equiv/mL}] \cdot [\text{mL/min}]} \tag{69}$$

When the chromatographic conditions for the analysis of inorganic anions are taken as a basis*), the lifetime of the ion-exchange cartridge is about 780 h. Even if this cartridge is already changed as suggested when it is expended to 90%, the suppressor system can be operated for about 30 days without any interruption.

*) Eluant: 1.7 mmol/L $NaHCO_3$ + 1.8 mmol/L Na_2CO_3; flow rate: 2 mL/min.

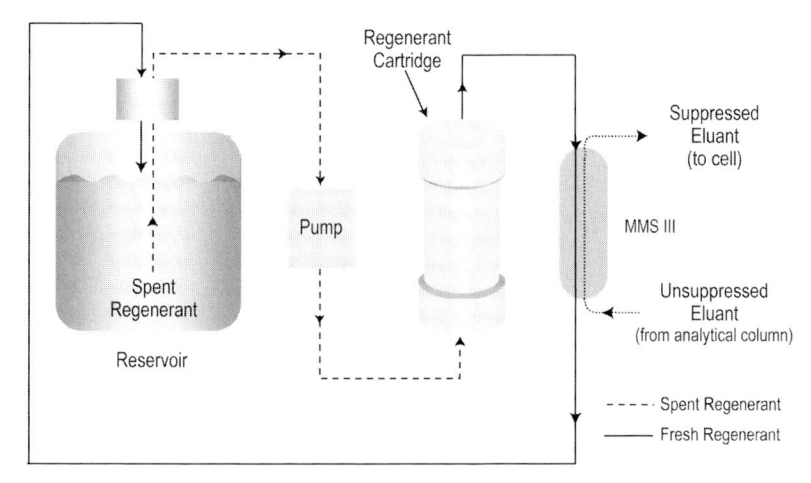

Fig. 3-75. Flow diagram of a micromembrane suppressor operated in the AutoRegen mode for continuous operation.

The Displacement Chemical Regeneration mode (DCR™) is a new mode of operation for chemical suppressors in which the regenerant is displaced with the conductivity cell effluent, delivering the regenerant to the suppressor at a flow rate equal to the eluant flow rate (Fig. 3-76). In this mode, the regenerant bottle is completely filled with regenerant upon startup. As the cell effluent is pumped into the regenerant bottle, the regenerant is forced out into the suppressor regenerant chambers. No additional pump or pressure is required. Eluant and regenerant bottles have to be of equivalent volumes, and new regenerant is prepared together with a new eluant. The low regenerant flow rate minimizes waste and allows unattended operation, offering an economical option to the AutoRegen of Pressurized Bottle mode.

3.6.4
Self-Regenerating Suppressors

Ever since the introduction of membrane-based suppressor systems, attempts have been made to support the ion transport through the ion-exchange membranes with an electric field, thus utilizing the principle of electrodialysis. The basic idea behind this is to enhance suppression by applying an electric field which will impel the ions involved in the suppression reaction to penetrate the ion-exchange membranes.

Tian et al. [77] implemented this idea by constructing a suppressor, in which the resin-packed eluant chamber was separated from the electrodes by cation exchange membranes. Usually, 0.10 mol/L sulfuric acid is used as a static regenerant in this device. According to the authors, the applied electric field supports the removal of sodium ions from the eluant chamber. However, they did not

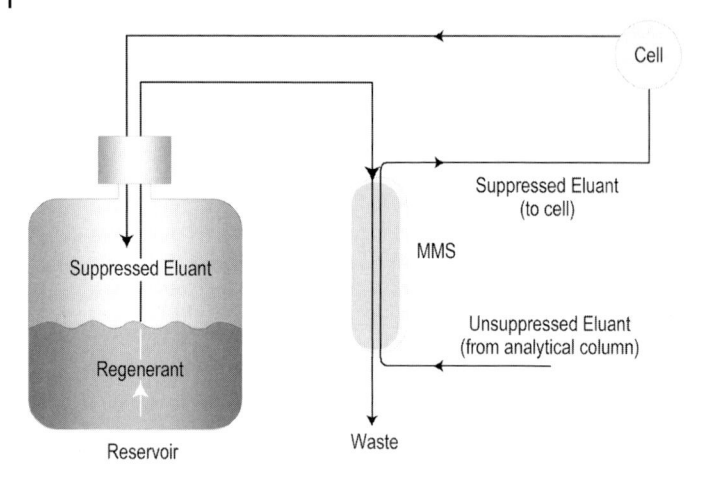

Fig. 3-76. Flow diagram of a micromembrane suppressor operated in the Displacement Chemical Regeneration (DCR™) mode.

mention whether this suppressor also works without applying an electric field. With such a high sulfuric acid concentration on one side of the cation exchange membrane, it is more than likely that diffusion of hydronium ions due to the concentration gradient similar to micromembrane suppressors will occur. With the setup used by Tian et al. it is also unclear what happens to the sodium ions, which enrich in a static regenerant. Thus, the suppressor is expended when the sodium ion concentration in the regenerant reaches that of the mobile phase.

Similar suppressor systems were described by Jansen et al. [78] and Ban et al. [79], which were operated contrary to that developed by Tian et al. with an acidic regenerant in a continuous way. Under typical operating conditions, the resulting current is 120 mA when applying a potential of 200 V. According to the authors, in both cases the applied electric field supports ion migration through the ion-exchange membranes. However, the charge at the electrodes cannot be generated without applying a current between the electrodes. Although ions can take over the transport of charge between the electrodes, the respective redox chemistry has to then take place at the electrode surface in order to close the cycle. Therefore, one has to assume electrochemistry, not electrodialysis, to be the driving force for the suppressor reaction in the device developed by Jansen et al. and Ban et al. In fact, the presence of an acid in the regenerant chamber of an electrochemical suppressor presents certain disadvantages [80].

Electrochemical suppressors can also be operated without an external chemical regenerant. Hydronium or hydroxide ions, required for the suppressor reaction, can be generated from water by means of electrolysis. The respective reactions are given by the Equations 70 and 71:

$$3H_2O \rightarrow 2H_3O^+ + {}^1/_2O_2 \text{ (g)} + 2e^- \qquad (70)$$

$$2e^- + 2H_2O \rightarrow 2OH^- + H_2 \text{ (g)} \qquad (71)$$

Reactions of this type are common on metal surfaces with a small overpotential such as platinum [81]. Hydronium or hydroxide ions formed in those reactions in combination with suitable ion-exchange membranes can be utilized for suppression [82]. The first suppressor based on this principle was developed by Strong and Dasgupta [80]. It housed a spiral-type double membrane having a suppression capacity high enough for sodium hydroxide eluants with a maximum concentration of 0.2 mol/L. The water required for electrolysis was delivered with a peristaltic pump.

In summer 1992, the first commercial electrochemical suppressor system was introduced: the self-regenerating suppressor (SRS) [83, 84]. Its schematic differs from a micromembrane suppressor (see Fig. 3-72) only in the two platinum electrodes for the electrolysis of water, which are placed at the top and at the bottom between the regenerant chamber and the enclosure. The SRS comes complete with a power supply for applying an electric field for the electrolytic reactions. One advantage of this suppressor is that the water required for electrolysis can be delivered through the regenerant channels at a relatively low flow rate. Moreover, the de-ionized effluent of the detector cell after suppression can be used as a water source, so that an external water supply is no longer necessary.

Neutralization reactions occurring in a self-regenerating suppressor do not differ much from those in conventional membrane suppressors. Taking an SRS for anion analysis as an example (Fig. 3-77), hydronium ions (after their formation at the anode) permeate the cation exchange membrane and neutralize the sodium hydroxide eluant. The sodium counter ions are attracted by the cathode, permeate through the membrane into the cathodic regenerant chamber and combine therein with hydroxide ions to maintain electroneutrality. Both hydrogen and oxygen gas formed at the cathode and anode, respectively, are sent to waste together with the liquid reaction products. The amount of hydrogen gas formed during electrolysis (about 21 mL/h at $I = 50$ mA) is relatively small and poses no safety problem.

In contrast to conventional chemical suppressors, the suppressor reaction in an SRS is directed by the electrodes. In a micromembrane suppressor, for instance, the chemical regenerant flows countercurrent through both regenerant chambers. Therefore, regenerant ions required for neutralization are provided to the eluant chamber through both membranes. Hydronium ions in an ASRS are exclusively formed at the anode, so that only the ion-exchange membrane in the anodic regenerant chamber is permeable for hydronium ions. Conversely,

3 Anion Exchange Chromatography (HPIC)

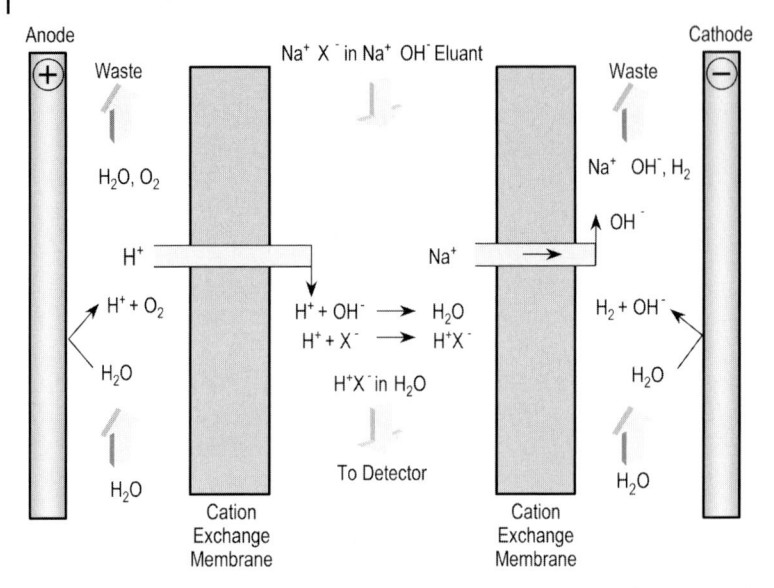

Fig. 3-77. Neutralization reactions in a self-regenerating suppressor for anion analysis.

the exchange of counter ions only occurs at the membrane in the cathodic regenerant chamber, where they associate with hydroxide ions formed therein. Thus, the anodic membrane is partly in the hydrogen form, whereas the cathodic membrane is partly in the sodium form.

Close contact between the electrodes and the membrane material is important for a suppressor in which the regenerant ions are formed via electrolysis of water. Since the charge transport in the interior of the suppressor is supported by ion migration via the ion-exchange functions at the membranes, the key for faultless functioning is the electrical resistance between the two electrodes; the resistance should be as low as possible.

Depending on the application, a self-regenerating suppressor can be operated in three different modes.

(1) AutoSuppression in the Recycle Mode. In this mode, the effluent of the conductivity cell is used as the source for the required de-ionized water. As can be seen in the respective flow diagram in Fig. 3-78, the column effluent is directed through the suppressor, where it is neutralized by exchanging counter ions for hydronium ions. The suppressor effluent, practically de-ionized (except for a few analyte ions), is then directed through the conductivity cell and back into the suppressor to be used as regenerant.

Recycling of the cell effluent offers a number of advantages. In addition to the simplicity of the plumbing, the waste volume is greatly reduced, because an external chemical regenerant such as dilute sulfuric acid is not needed. Therefore, operating costs are much lower, and the system is practically maintenance-free. AutoSuppression is the most widely used operating mode for an ASRS and can be applied in all applications with pure *aqueous* eluants below 40 °C.

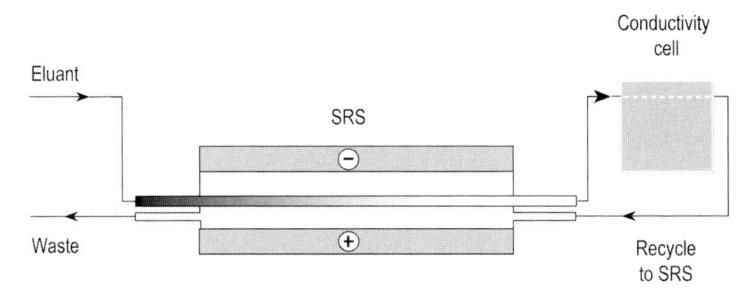

Fig. 3-78. Eluant flow in the AutoSuppression Recycle Mode with a self-regenerating suppressor for anion analysis.

(2) AutoSuppression with external water supply. In this operation mode, the deionized water needed for electrolysis is provided externally. Usually, a pressurized reservoir is used; i.e., water is delivered pneumatically to the regenerant chambers of the suppressor. Gases formed during the electrolysis process are diverted to waste together with the aqueous effluent containing the cations and other suppressor products. The conductivity cell effluent is also diverted to waste.

The external water mode is predominantly used when organic solvents are required for the separation. In the AutoSuppression mode, these solvents would oxidize and form ionic products, thus increasing noise and background conductivity. Moreover, this operation mode is chosen when low noise and high sensitivity are required. Suppression capacity can also be increased by applying the external water mode, since water can be delivered to the regenerant chambers at a higher flow rate, which results in a more efficient removal of eluant counter ions. This is especially important when using microbore columns, which are usually operated at flow rates between 0.25 and 0.5 mL/min. Those flow rates are too low for AutoSuppression in the recycle mode.

(3) Chemical regeneration. Since self-regenerating suppressors do not differ much from micromembrane suppressors, they can be regenerated conventionally with diluted sulfuric acid. There are no specific applications for this, although chemical regeneration is superior to AutoSuppression with external water supply regarding sensitivity, since the noise is always somewhat higher because of the electrolysis process.

3.6.5
Suppressors with Monolithic Suppression Beds

In 2001, a third type of suppressor – the Atlas Electrolytic Suppressor (AES) – was introduced. The Atlas suppressor is a continuously regenerated suppressor operated in the recycle mode and designed for optimal performance when using carbonate-based eluants. Suppressing up to 25 mmol/L sodium at 1 mL/min, it can be used in combination with all carbonate-based anion exchangers, including high-capacity columns such as IonPac AS9-HC (see Section 3.4.2).

In contrast to the membrane-based micromembrane and self-regenerated suppressor devices, the suppression bed of an Atlas suppressor is made of an ion-exchange monolith. As illustrated in the respective cross-section in Fig. 3-79, the electrodes are placed on both sides of the monolithic suppression bed that is approximately 1 cm long. The length of this monolithic bed is a compromise between the suppression capacity and the dead volume. To ensure maximum penetration with the eluant, the monolith is cut in slices (MonoDiscs), which are separated from each other by flow distribution discs. The small boreholes in those discs alternate between top and bottom, so that the eluant is forced to flow through every ion-exchange MonoDisc. This configuration facilitates efficient exchange of eluant ions for regenerant ions generated at the anode.

The new design of the Atlas suppressor results in much faster daily startup after an overnight shutdown, improving the throughput of routine sample analysis. A stable baseline is observed after less than 30 minutes, which is much faster than any SRS device. The Atlas suppressor also improves the analysis of standard anions by significantly lowering the noise, which is only slightly higher

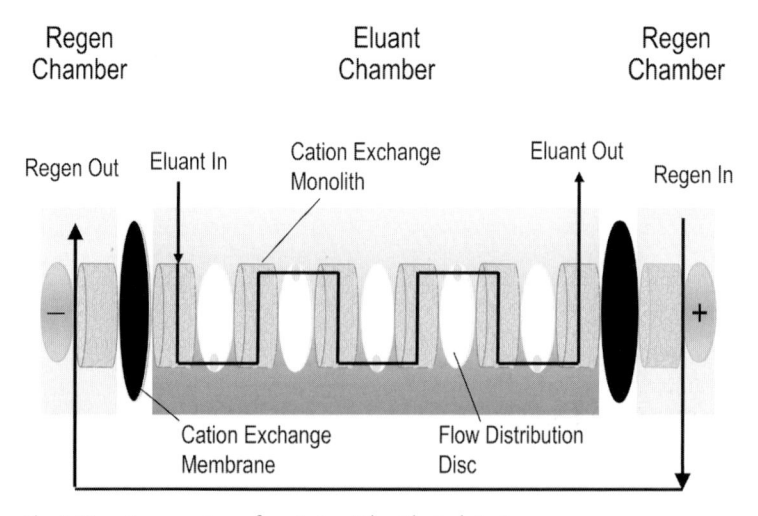

Fig. 3-79. Cross-section of an Anion Atlas Electrolytic Suppressor.

than that observed for a micromembrane suppressor in the chemical suppression mode. Like self-regenerated suppressors, the Atlas suppressor is not compatible with organic solvents. This is not a real drawback, as organic solvents are not required for routine carbonate-based separations.

3.7
Anion Exchange Chromatography of Inorganic Anions

3.7.1
Overview

A variety of inorganic anions may be separated using the stationary phases described in Section 3.4. These include:

Halides:
F^-, Cl^-, Br^-, and I^-

Oxygen-containing halides:
OCl^-, ClO_2^-, ClO_3^-, ClO_4^-, BrO_3^-, and IO_3^-

Oxygen-containing phosphorus compounds:
PO_2^{3-}, PO_3^{3-}, PO_4^{3-}, $P_2O_7^{4-}$, $P_3O_{10}^{5-}$, $P_4O_{13}^{6-}$, and PO_3F^{2-}

Sulfur compounds:
S^{2-}, SO_3^{2-}, SO_4^{2-}, $S_2O_3^{2-}$, and SCN^-

Nitrogen compounds:
CN^-, OCN^-, NO_2^-, NO_3^-, and N_3^-

Silicon compounds:
SiO_3^{2-} and SiF_6^{2-}

Boron compounds:
$B_4O_7^{2-}$ and BF_4^-

Oxy non-metal anions:
AsO_2^-, AsO_4^{3-}, SeO_3^{2-}, and SeO_4^{2-}

Oxy metal anions:
MoO_4^{2-}, WO_4^{2-}, and CrO_4^{2-}

When classifying these anions, it is important to remember that water, unlike other liquids, exhibits good solvent properties for salts because of its specific structure and the special interaction mechanism between the ion and the water molecule. When an ion is solvated by water, hydrogen bonds are broken (cavity effect) and the water structure is destroyed. The larger the ion, the higher the energy required for the formation of a cavity with molecular dimension. On the other hand, electrostatic ion-dipole interactions occur that lead to the formation of a new structure. Thus, the smaller the ionic radius and the higher the ionic charge, the stronger the effect. The values of the molar hydration enthalpies for halide ions given in Table 3-13 confirm that the hydration increases as the size of the ion decreases.

Table 3-13. Molar hydration enthalpies of halide ions [85].

Ion	$\Delta H_{Hydr.}$ [kcal/mol]
F^-	−98
Cl^-	−79
Br^-	−71
I^-	−62

Large ions such as iodide exhibit a very strong affinity towards the stationary phase of an anion exchanger. In these ions, the hydration enthalpy is partly counterbalanced by the cavity formation energy. Further, these anions are called *polarizable*; their chromatography is discussed separately.

The polarizability of an ion is directly related to the ionic radius in the hydrated state; it is one of the solute-specific properties that determines the affinity of an ion towards the stationary phase. In general, the retention time increases with increasing ionic radius in the hydrated state and, thus, with stronger polarizability. Accordingly, halide ions elute in the order of increasing retention times: fluoride < chloride < bromide < iodide. The retention time difference between bromide and iodide is already so large that the set of halide ions can only be analyzed in a single run by using special eluants or stationary phases.

In addition to the ionic radius in the hydrated state, the valency of an ion is another solute-specific property that affects retention. In general, retention increases with increasing valency. Thus, the monovalent nitrate elutes prior to the divalent sulfate. Exceptions are multivalent ions such as orthophosphate, where the retention depends on the eluant pH (the pH influences the dissociation equilibria). However, the size of an ion often influences the retention more strongly than the valency. Hence, the divalent sulfate elutes prior to the monovalent, but strongly polarizable, thiocyanate.

3.7.2
General Parameters Affecting Retention

Eluant Flow Rate

The van Deemter plot of a selected anion exchange column of the IonPac AS4 type shown in Fig. 2-7 (see Section 2.5) reveals that, as in HPLC on reversed phases, the height of a theoretical plate changes only at very low flow rates and thereafter approaches a plateau region. Hence, retention times can be reduced without a significant loss in separation efficiency by increasing the flow rate. An inverse proportionality exists between flow rate and resulting retention time. However, a flow rate increase is limited by the maximal operating pressure of the separator column. On the other hand, obtaining a better resolution via a flow rate reduction is only possible to a limited extent. Since the pH value and ionic strength and, thus, the order of elution, remain unaffected by a flow rate change, the control of the retention times via the flow rate can easily be accomplished.

Length of the Separator Column

The number of theoretical plates and, thus, the separation efficiency is determined by the length of the separator column. If two separator columns are used in series, the resulting enhancement of separation efficiency leads to a better resolution between ions with similar retention characteristics, with a corresponding increase in retention times. The separator column length also determines the ion-exchange capacity. An increase of the ion-exchange capacity via elongation of the separator column is recommended in all cases where the ion to be analyzed is present in an excess of another component.

Shortening of separator columns is generally not feasible, because the ageing process causes the separation efficiency to slowly diminish with time. An overly high resolution between two signals leads to unnecessarily long retention times. This can be prevented by choosing appropriate eluants and stationary phases. Under special circumstances, a drastic reduction in the retention time of an ion may be achieved by a separation on a shorter pre-column.

3.7.3
Experimental Parameters Affecting Retention when Applying Suppressor Systems

While the retention-determining parameters given in Section 3.7.2 have a fundamental character, further parameters related to the eluant depend on the detection system being used. This particularly applies to conductivity detection, which is possible directly or in combination with a suppressor system. These two modes of conductivity detection are fundamentally different and require eluants which not only differ in their type but also in their concentrations and pH values. Therefore, it is advisable to discuss separately the influence of these parameters on both modes of this very important detection system.

3.7.3.1 Choice of Eluant

The appropriate eluant to use when applying suppression techniques depends on the type of analyte; the possible eluant and eluant mixtures are listed in Table 3-10. In general, the eluant ion and the analyte ion should exhibit a similar affinity towards the stationary phase. For the majority of applications, the mixture of sodium carbonate and sodium bicarbonate has proved to be successful, because over a wide range the resulting selectivity is determined solely by the concentration ratio of both components. As can be seen from Fig. 3-80, the most important monovalent and divalent inorganic anions can be eluted with this eluant [86]. Under the same chromatographic conditions, it is possible to separate the nitrogen-containing compounds (Fig. 3-81). While suppressed conductivity detection is very sensitive for cyanate, this method is inappropriate for cyanide, because the corresponding acid, HCN, formed in the suppressor is not dissociated at neutral pH.

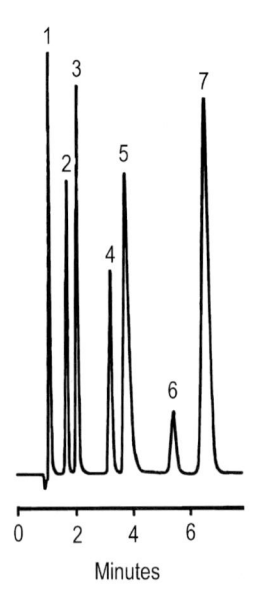

Fig. 3-80. Separation of inorganic anions according to DIN 38405, Part 19 [86]. – Separator column: IonPac AS4A; eluant: 1.7 mmol/L $NaHCO_3$ + 1.8 mmol/L Na_2CO_3; flow rate: 2 mL/min; detection: suppressed conductivity; injection volume: 50 µL; solute concentrations: 3 mg/L fluoride (1), 4 mg/L chloride (2), 10 mg/L nitrite (3), 10 mg/L bromide (4), 20 mg/L nitrate (5), 10 mg/L orthophosphate (6), and 25 mg/L sulfate (7).

Only a small modification of the concentration ratio between carbonate and bicarbonate, leading to a change in pH, is necessary to analyze oxy non-metal anions and common mineral acids in a single chromatographic run. However, an exception is arsenic(III) in the form of ortho- or metaarsenite. In analogy to cyanide, it cannot be detected via suppressed conductivity detection, because

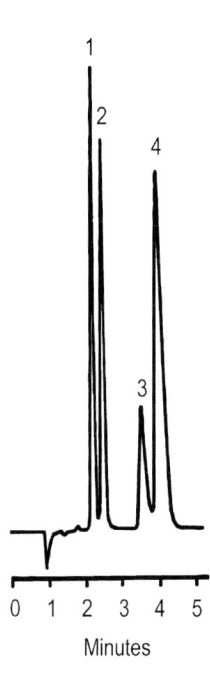

Fig. 3-81. Separation of various nitrogen-containing species. — Chromatographic conditions: see Fig. 3-80; solute concentrations: 10 mg/L nitrite (1), 10 mg/L cyanate (2), 20 mg/L azide (3), and 20 mg/L nitrate (4).

arsenic acid, H_3AsO_3 ($pK = 9.23$ [87]), formed in the suppressor is hardly dissociated. On the other hand, arsenite can easily be oxidized on a Pt working electrode. If the effluent of the separator column would be passed through an amperometric detection cell before entering the suppressor system, simultaneous detection of As(III) and As(V) is possible. In this way, the various detection modes would contribute to the selectivity of the analytical procedure. However, both species have to be present in a comparable concentration range in order to be above the minimum detection limit, in case the sample has to be diluted. Alternatively, element-specific detection via ICP-OES (see Section 7.4) can be applied for simultaneous analysis of arsenic(III) and arsenic(V). Therefore, the separation of non-metal anions that are detectable via conductivity, depicted in Fig. 3-82, provides an acceptable alternative to atomic absorption spectrometry, because the two most important oxidation states of arsenic and selenium can be distinguished by ion chromatography. However, regarding the detection limit, the hydride system in AAS is by far superior to suppressed conductivity detection in ion chromatography. Similar sensitivities can only be achieved by hyphenating ion chromatography with ESI-MS in the Selected Ion Monitoring (SIM) mode (see Section 7.4.2).

For the detection of mineral acids in the presence of an excessive amount of nitrate, the IonPac AS10 column is now used. On this column, bromide and nitrate elute after sulfate. The selectivity of this stationary phase is based on the hydrophobic properties of the ion-exchange groups bound to the latex beads (see

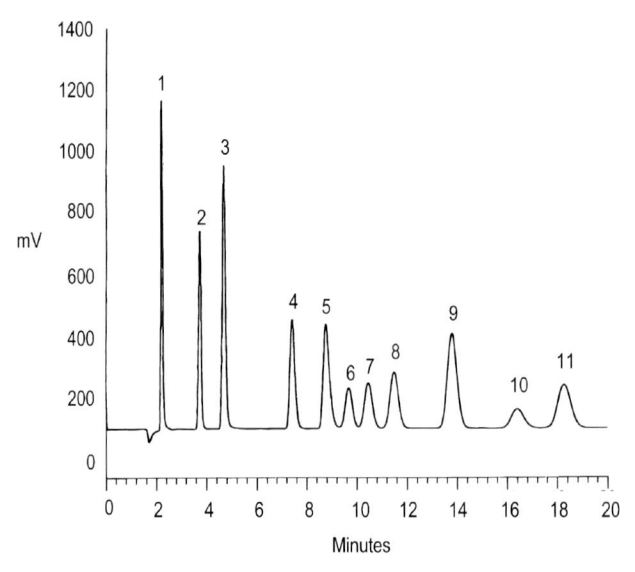

Fig. 3-82. Separation of mineral acids and oxy non-metal anions. — Separator column: IonPac AS12A; eluent: 0.8 mmol/L NaHCO$_3$ + 2.1 mmol/L Na$_2$CO$_3$; flow rate: 1.5 mL/min; detection: suppressed conductivity; injection volume: 25 µL; solute concentrations: 3 mg/L fluoride (1), 4 mg/L chloride (2), 10 mg/L each of nitrite (3), bromide (4), nitrate (5), selenite (6), orthophosphate (7), sulfite (8), sulfate (9), arsenate (10), and selenate (11).

Section 3.4.2). As shown in Fig. 3-83, small quantities of chloride, orthophosphate, and sulfate can be determined in the presence of high amounts of nitrate. Optimum separations are obtained with hydroxide eluants.

Pure sodium carbonate eluants were used in the past for the separation of several phosphorus species such as hypophosphite, orthophosphite, and orthophosphate; they were separated in one run on IonPac AS3 and detected via electrical conductivity. Today, an IonPac AS12A is used for this application, and the three phosphorus species are separated with a carbonate/bicarbonate eluant (Fig. 3-84).

Environmentally important anions such as sulfide and cyanide can be determined very sensitively via amperometric detection. For their separation on a conventional IonPac AS3 anion exchanger (Fig. 3-85), a mixture of sodium carbonate and sodium dihydrogen borate is used as an eluant. A small amount of ethylenediamine is added to the mobile phase to complex traces of transition metal ions, which could be present in the eluant [88, 89]. While the detection of these two anions is very easy, the interpretation of experimental results for the investigation of real-world samples is very difficult. These samples normally contain transition metal ions; in their presence, sulfide and cyanide do not exist or only partly exist as free ions. However, only free ions are detected under the chromatographic conditions listed in Fig. 3-85.

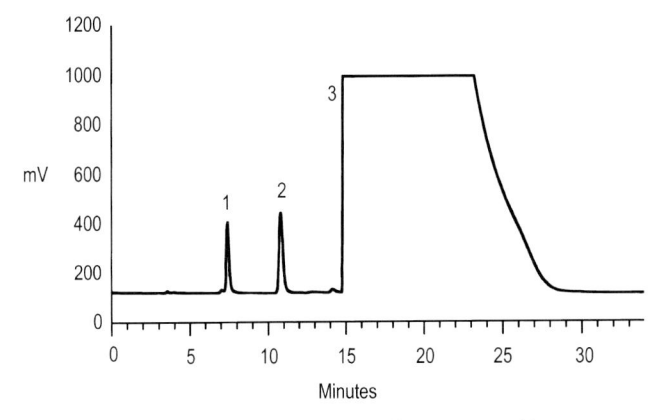

Fig. 3-83. Separation of mineral acids in the presence of large amounts of nitrate. — Separator column: IonPac AS10; eluant: 0.08 mol/L NaOH; flow rate: 1 mL/min; detection: suppressed conductivity; injection volume: 25 µL; solute concentrations: 5 mg/L chloride (1), 10 mg/L sulfate (2), and 5000 mg/L nitrate (3).

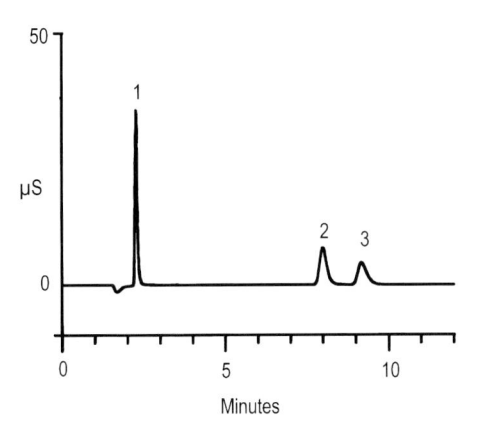

Fig. 3-84. Separation of hypophosphite, orthophosphite, and orthophosphate. — Separator column: IonPac AS12A; eluant: 0.7 mol/L $NaHCO_3$ + 2.7 mmol/L Na_2CO_3; flow rate: 1.5 mL/min; detection: suppressed conductivity; injection volume: 25 µL; solute concentrations: 10 mg/L each of hypophosphite (1), orthophosphite (2), and orthophosphate (3).

Disposal and drainage waters are among the few matrices in which free sulfide can be detected. Depending on the sample pH, transition metals present in these samples form sparingly soluble precipitates with sulfide. When the pH value of the sample is changed via dilution, the chemical equilibrium in the sample shifts, which could result in an increased amount of free sulfide. However, difficulties in sulfide determination arise not only because of complex matrices, but also due to the lack of a reference standard. Although sodium sulfide may be obtained in reagent-grade quality, the amount of crystal water cannot be determined unequivocally. Thus, the titer of a freshly prepared sodium sulfide solution must be determined via titration prior to each calibration.

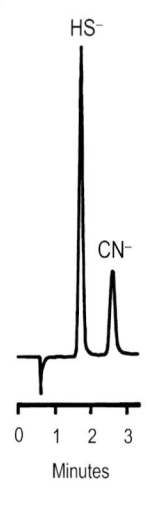

Fig. 3-85. Separation of free sulfide and cyanide. — Separator column: IonPac AS3; eluant: 1 mmol/L Na_2CO_3 + 10 mmol/L NaH_2BO_3 + 15 mmol/L ethylenediamine; flow rate: 2.3 mL/min; detection: amperometry on a Ag working electrode; injection volume: 50 µL; solute concentrations: 0.5 mg/L sulfide and 1 mg/L cyanide.

Conversely, the cyanide calibration poses no problem. For this purpose, reagent-grade sodium cyanide can be used. Free cyanide is found, for example, in untreated wastewater from electroplating processes. Because these samples also contain transition metal ions, part of the cyanide exists in complexed form, so the various metal cyano complexes differ significantly in their stability. Therefore, the cyanide signal produced amperometrically represents both the free cyanide initially present in the sample as well as the one that is released in the mobile phase via a shift of the complexation equilibrium according to Eq. (72).

$$M(CN)_x^{n-} \rightleftharpoons M(CN)_{x-1}^{(n-1)-} + CN^- \qquad (72)$$

The values summarized in Table 3-14 are obtained when investigating the content of free cyanide in aqueous solutions of various metal cyano complexes with concentrations below 1 mg/L, under the chromatographic conditions listed in Fig. 3-85. The cyanide dissociation from copper, nickel, zinc, or cadmium complexes is particularly pronounced. Thus, a much higher content of free cyanide is feigned in the respective samples. In contrast, iron and cobalt hexacyano complexes are kinetically extremely stable; therefore, the cyanide determination is not affected by their presence. According to Rocklin and Johnson [88], results for free cyanide in real-world samples are as doubtful regarding their accuracy as results obtained with wet-chemical methods. Investigations by Pohlandt [90] showed that the mobile phase is responsible for the decomposition of certain metal cyano complexes. The dissociation of these complexes occurs because of the ligand-exchange processes, which are caused by ethylenediamine as one of the eluant components. When the amine concentration in the mobile phase is reduced, a correspondingly lower dissociation of the nickel, copper, silver, and mercury complexes is observed. At an amine concentration of $4.4 \cdot 10^{-4}$ mol/L this dissociation does not occur at all. However, the reduction in the amine

Table 3-14. Percentage of free cyanide in an aqueous solution of various metal cyano complexes during separation on an anion exchanger.

Complex	Percentage of free cyanide
$Cd(CN)_4^{2-}$	100
$Zn(CN)_4^{2-}$	100
$Ni(CN)_4^{2-}$	86
$Cu(CN)_4^{3-}$	52
$Ag(CN)_2^{-}$	13
$Hg(CN)_4^{2-}$	8
$Au(CN)_2^{-}$	0
$Au(CN)_4^{-}$	0
$Fe(CN)_6^{3-}$	0
$Fe(CN)_6^{4-}$	0
$Co(CN)_6^{3-}$	0

concentration is accompanied by a significant peak broadening and tailing of the signal for free cyanide, which may be suppressed by the addition of small quantities of cyanide (approximately 50 µg/L) to the mobile phase. Using a modified eluant that consists of

$1.00 \cdot 10^{-3}$ mol/L sodium carbonate +
$1.00 \cdot 10^{-2}$ mol/L sodium dihydrogenborate +
$1.02 \cdot 10^{-6}$ mol/L sodium cyanide +
$4.40 \cdot 10^{-4}$ mol/L ethylenediamine,

hydrogen cyanide (HCN), cyanide ions (CN$^-$), and metal cyano complexes with formation constants of log $K_n < 20$ are detected ion chromatographically under the general term *free* cyanide. In contrast to titrimetric and spectrophotometric methods, no elaborate sample pretreatment is required. This does not apply to the determination of so-called *easily releasable cyanide*, where sample preparation is indispensable. For the determination of this parameter, which is only defined via this method, an air flow is passed through the sample solution, which is adjusted to pH 4. The gaseous hydrocyanic acid released under these conditions is then trapped in a strongly alkaline absorption solution [35]. Usually, 1 mol/L sodium hydroxide is used for this. Solutions that basic cannot be analyzed for their cyanide content using the chromatographic conditions given in Fig. 3-85. Because of its limited capacity, the anion exchanger would be totally overloaded by the high hydroxide ion concentration in the sample solution. Significant interferences close to the void volume would result. This problem can be circumvented when sodium hydroxide is used as an absorption solution and as an eluant. An anion exchanger with a relatively high capacity should be employed; this allows the injection of comparatively high NaOH concentrations. The high-capacity CarboPac PA1 is the most suitable anion exchanger for this application, because cyanide is strongly retained at this stationary phase, even in the presence of high hydroxide concentration in the mobile phase. However, the retention time can be significantly reduced by adding sodium acetate. Ethylenediamine is also added to the mobile phase, but does not interfere in this case, because the sample solution only contains cyanide ions due to sample preparation. Figure 3-86 shows the chromatogram of a sulfide and cyanide standard obtained with this method, which was directly injected into a 1 mol/L sodium hydroxide matrix. As can be seen in the chromatogram, the cyanide signal is not affected by the matrix. Although the sample preparation is the rate-determining step in this analysis, the ion chromatographic method offers the advantage of being less susceptible to interferences than the conventional wet-chemical determinations mentioned above. In addition, it can be easily automated and is applicable to the determination of total cyanide, whereby the cyanide in the sample is released in a different way but collected in the same absorption solution [35].

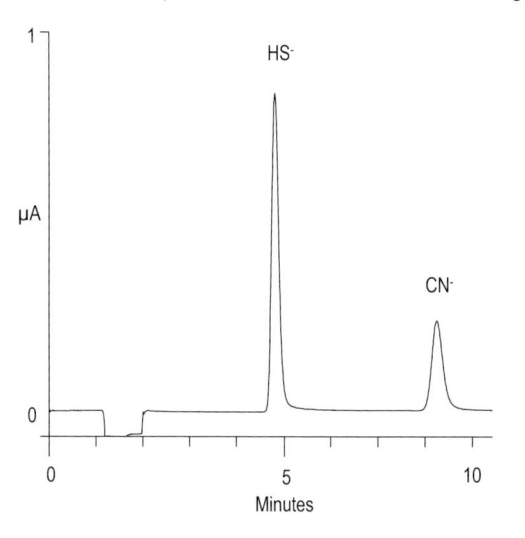

Fig. 3-86. Separation of sulfide and cyanide in a strongly alkaline solution. – Separator column: CarboPac PA1; eluant: 0.1 mol/L NaOH + 0.5 mol/L NaOAc + 0.0075 mol/L ethylenediamine; flow rate: 1 mL/min; other chromatographic conditions: see Fig. 3-85.

In addition to sulfide and cyanide, bromide and sulfite are also eluted under the chromatographic conditions given in Fig. 3-86. They can be detected at a slightly more positive oxidation potential. Of special importance is sulfite, the wet-chemical determination of which is exceptionally problematic. However, caution is also necessary in the case of ion chromatographic determination because, in the presence of oxygen, sulfite is subject to autoxidation, yielding sulfate. Therefore, it must be stabilized both in the sample – preferably directly at the time of sampling – and in the standard solution. Of all compounds that have been investigated in the past for their suitability as a stabilizing agent for sulfite, formaldehyde has proved to be particularly effective [91, 92]. Although the stabilizing effect is not fully understood, it is assumed to be a nucleophilic addition of the hydrogen sulfite anion with subsequent proton migration to form hydroxymethanesulfonic acid [93]:

$$\left[\text{H-C}^H_{\diagdown O} + HSO_3^- \rightleftharpoons \text{H-C}^H \overset{O^\ominus}{\underset{SO_3H}{|}} \longleftrightarrow \text{H-C}^H \overset{OH}{\underset{SO_3^\ominus}{|}} Na^+ \right] \quad (73)$$

The hydroxymethanesulfonic acid seems to hydrolyze completely in the alkaline pH region above pH 10, because commercially available hydroxymethanesulfonic acid and sulfite have identical retention times under the chromatographic conditions given in Fig. 3-80 and Fig. 3-82. But in acidic pH, a significantly shorter retention time typical of short-chain sulfonic acids is obtained for hydroxymethanesulfonic acid. Formaldehyde seems to have no effect on the chromatographic behavior of sulfite when basic eluants are used. The stabilizing effect is nevertheless very high. Thus, the effort to remove oxygen from the sample and the eluant is unnecessary when using formaldehyde as the stabilizing agent. Even when a stabilized sulfite solution is purged with air for about 15 minutes, no significant loss in the sulfite content is observed [92]. Only small quantities of formaldehyde are required to produce a stabilizing effect. Adding 0.5 mL of a 37% formaldehyde solution per liter sample solution has proved to be successful. Remarkably, higher amounts lead to a slight reduction in the sulfite retention time, an effect which is not understood at present. In stabilized conditions, sulfite standard solutions with a content of 1000 mg/L are stable for about 72 h. Standard solutions with lower sulfite content, however, should be prepared daily from a stock solution. For quantitative sulfite analyses, one should remember that even reagent-grade sodium sulfite contains a small amount of sulfate, which has to be taken into account in the calibration of both compounds. The chromatogram in Fig. 3-87 shows that the purity of the reference standard may be easily determined via ion chromatography.

As can be seen in the chromatogram in Fig. 3-80, the most important mineral acids can be eluted under isocratic conditions using a mixture of sodium carbonate and sodium bicarbonate. This does not apply to polarizable anions, which are more strongly retained at the stationary phase because of adsorption effects (see Section 3.3). However, if acrylate-based anion exchangers are not available, retention can be noticeably reduced by adding organic additives to the mobile phase. For example, if p-cyanophenol, which is especially suited for this purpose, is added to the mixture of sodium carbonate and sodium bicarbonate, certain polarizable anions such as monovalent iodide or tetrafluoroborate may be detected in the same run together with anions relevant for conventional water analysis. The resulting chromatogram is shown in Fig. 3-88. At first sight, such a separation seems to be of pure academic interest, but it is applied in the simultaneous determination of boron (as BF_4^-) and phosphorus (as PO_3^{3-} and PO_4^{3-}) in borophosphorosilicate glass films [94, 95], which is of great interest for the semiconductor industry. Characteristic of the separation shown in Fig.

3-88 is the pronounced tailing of the iodide peak, which illustrates the compromise between the peak shape and the analysis time in the simultaneous determination of anions with strongly different retention behavior. Moreover, a separation of bromide and nitrate is impossible under these chromatographic conditions, because this separation is based on the different adsorption behavior of both ions, which is not supported when *p*-cyanophenol is added to the eluant.

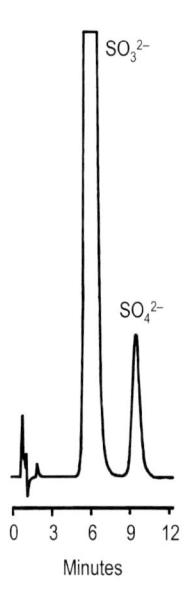

Fig. 3-87. Determination of sulfate in sodium sulfite. — Separator column: IonPac AS3; eluant: 2.8 mmol/L $NaHCO_3$ + 2.2 mmol/L Na_2CO_3; flow rate: 2.3 mL/min; detection: suppressed conductivity; injection: 50 µL; 500 mg/L sulfite.

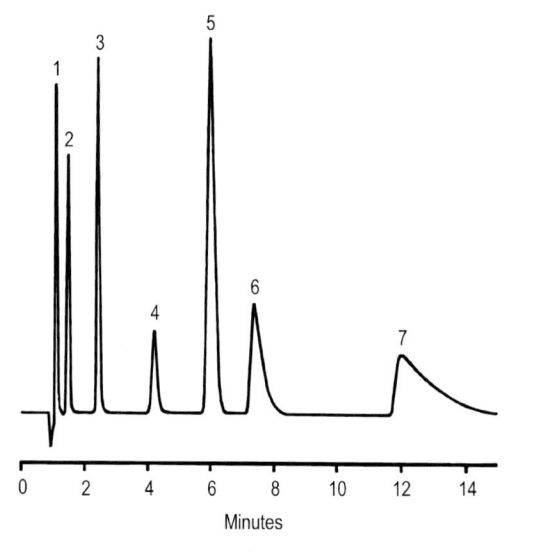

Fig. 3-88. Separation of polarizable and non-polarizable inorganic anions on a PS/DVB-based stationary phase by adding *p*-cyanophenol. — Separator column: IonPac AS4A; eluant: 1.7 mmol/L $NaHCO_3$ + 1.8 mmol/L Na_2CO_3 + 100 mg/L *p*-cyanophenol; flow rate: 2 mL/min; detection: suppressed conductivity; injection volume: 50 µL; solute concentrations: 2 mg/L fluoride (1), 4 mg/L chloride (2), 10 mg/L nitrate (3), 10 mg/L orthophosphate (4), 20 mg/L sulfate (5), 20 mg/L tetrafluoroborate (6), and 50 mg/L iodide (7).

In a similar way, the aromatic amino acid tyrosine has proved to be effective as an eluant for the simultaneous analysis of all halide anions utilizing PS/DVB-based stationary phases. Tyrosine also causes a reduction in the iodide retention at alkaline pH, while still allowing for the separation of bromide and nitrate, in contrast to *p*-cyanophenol-containing eluants. In the chromatogram obtained with tyrosine as an eluant (Fig. 3-89), a reversed retention is observed for orthophosphate and sulfate; this is caused by the comparatively high pH value of the mobile phase. Even with tyrosine as the eluant, the tailing of the iodide signal is not completely suppressed, but it is less pronounced than in Fig. 3-88 due to the shorter analysis time. As impressively demonstrated in Fig. 3-89, α-amino acids may be regarded as an alternative to eluants with other organic additives. However, not all of the potentially suitable compounds are applicable without restriction, because some of the amino acids are not available at high purity. In comparison, Fig. 3-90 shows the separation of halide ions on the acrylate-based IonPac AS9-SC, which can be obtained with a pure sodium carbonate eluant in approximately the same time. Under these conditions, mineral acids elute in the usual order.

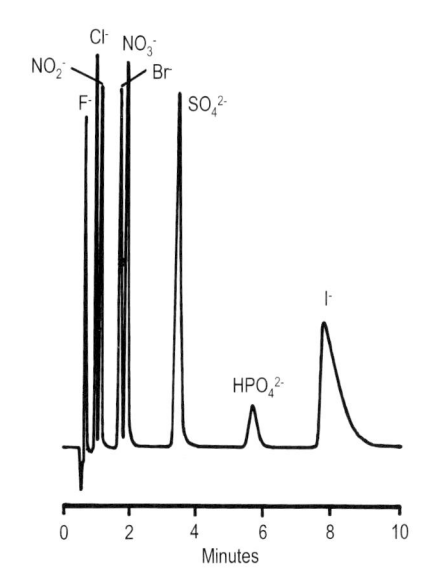

Fig. 3-89. Separation of halide ions with a tyrosine eluant on a PS/DVB-based anion exchanger. — Separator column: IonPac AS4A (-SC); eluant: 1 mmol/L tyrosine + 3 mmol/L NaOH; flow rate: 2 mL/min; detection: suppressed conductivity; injection volume: 50 µL.

A common problem in the isocratic analysis of inorganic anions is the verification of fluoride that elutes near the void volume. Under the chromatographic conditions in Fig. 3-80, monocarboxylic acids are only partly separated, or not at all separated, from the fluoride peak. Special caution is necessary in environmental samples (wastewaters, ground waters, digests, etc.). Fluoride can often be detected in such samples, but they may also contain a variety of organic acids. Because many of these organic acids co-elute with fluoride, interpreting the signals near the void volume is extremely difficult. Another potential reason for the

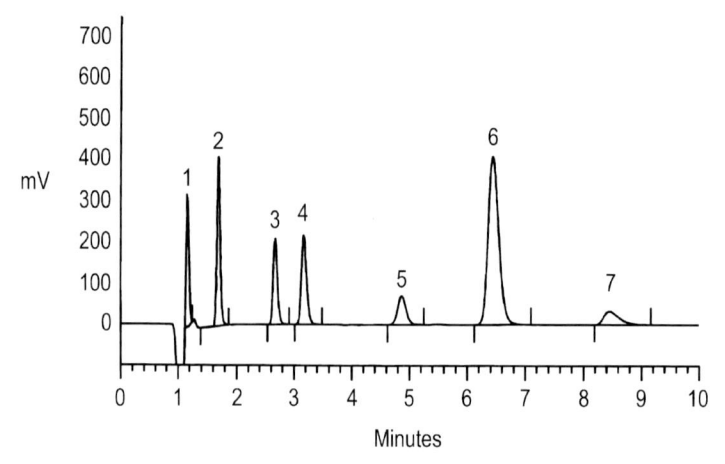

Fig. 3-90. Separation of halide ions on an acrylate-based anion exchanger. — Separator column: IonPac AS9-SC; eluant: 3 mmol/L Na_2CO_3; flow rate: 2 mL/min; detection: suppressed conductivity; injection volume: 100 µL; solute concentrations: 0.3 mg/L fluoride (1), 0.4 mg/L chloride (2), 1 mg/L each of bromide (3), nitrate (4), and orthophosphate (5), 2 mg/L sulfate (6), and 1 mg/L iodide (7); (taken from [96]).

appearance of supposed fluoride signals is the large amount of bicarbonate ions present in mineral waters. If their concentration is higher than the total carbonate concentration in the mobile phase, a positive signal is obtained within the column void volume instead of the negative water dip. This positive peak is hardly distinguishable from a fluoride peak, hence, the fluoride determination is impossible if the negative water dip is missing. Diluting the sample with deionized water does not solve this problem, even if no sign of interference can be discerned from the appearance of the resulting negative water dip. This becomes evident when the peak height is determined as a function of the bicarbonate concentration in the solution. Therefore, increasing amounts of bicarbonate were added to aqueous 2-mg/L fluoride standard solutions and the resulting fluoride peak heights were measured. The respective results are summarized in Table 3-15. The semi-logarithmic plot of peak height versus bicarbonate concentration in Fig. 3-91 clearly reveals that small amounts of bicarbonate in the sample lead to a significant decrease in the fluoride peak height. (The same result is obtained by adding carbonate to fluoride solutions of defined concentration.) In addition, the amount of reduction in peak height depends on the fluoride concentration of the sample. To illustrate this effect, four different fluoride standards were prepared in a concentration range between 0.5 and 2.0 mg/L, and 100 mg/L bicarbonate was added. The resulting peak heights were then compared with the values obtained with pure aqueous standards. The results in Table 3-16 show that the reduction of the fluoride peak height that is caused by the addition of bicarbonate is markedly augmented by increases in fluoride concentration. All these experiments substantiate that the determination of fluoride in real-world samples is problematic, if the fluoride is determined *in the same chromatographic run* together with other mineral acids.

Table 3-15. Fluoride peak height as a function of the carbonate content in solution.

$[CO_3^{2-}]$ [mg/L]	Peak height [cm]
0	8.50
5	8.50
10	8.45
25	8.35
50	8.20
100	7.90
150	7.50

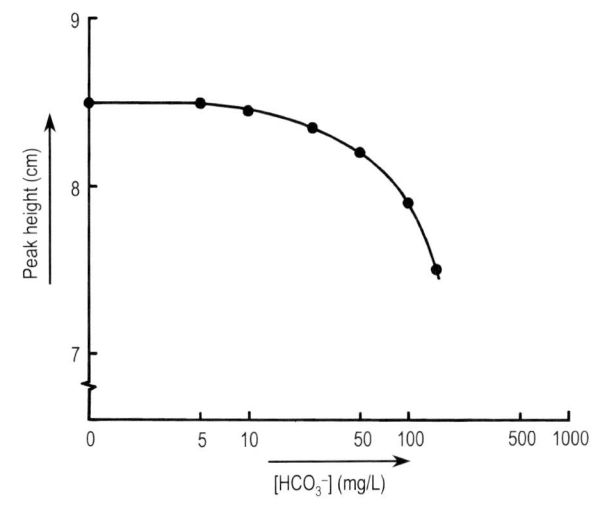

Fig. 3-91. Semi-logarithmic representation of the fluoride peak height versus the bicarbonate concentration being added.

An exact quantification of fluoride is possible if the advantage of simultaneous analysis is abandoned, and if the chromatographic conditions are changed so that fluoride is separated from the carbonate and bicarbonate travelling with the mobile phase. An increase in fluoride retention is achieved by using an eluant of lower elution strength. A dilute solution of sodium tetraborate is particularly suitable in this respect. As can be seen from the chromatogram in Fig. 3-92, baseline-resolved separation between fluoride, acetate, and formate is obtained with this eluant. However, this method has only limited applicability for routine analysis. Anions such as chloride, nitrate, and sulfate are more strongly retained under these chromatographic conditions, which may interfere with subsequent analyses. Therefore, the separator column must be rinsed occasionally with a 0.1 mol/L carbonate solution to remove the strongly retained anions. The relatively long time (one to two hours) required for the subsequent reconditioning of the separator column with the tetraborate eluant is problematic.

Table 3-16. Percent reduction of fluoride peak height upon addition of bicarbonate as a function of the fluoride content in solution.

Fluoride content [mg/L]	Percent reduction of peak height upon addition of 100 mg/L bicarbonate
0.5	3.9
1.0	3.9
1.5	4.8
2.0	9.8

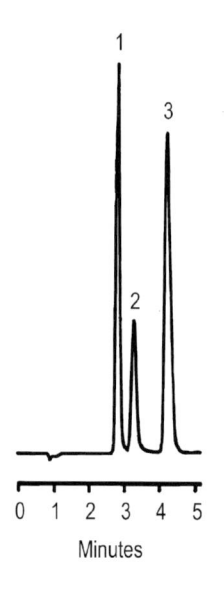

Fig. 3-92. Separation of fluoride, acetate, and formate. — Separator column: IonPac AS4A; eluant: 2 mmol/L $Na_2B_4O_7$; flow rate: 2 mL/min; detection: suppressed conductivity; injection volume: 50 µL; solute concentrations: 3 mg/L fluoride (1), 20 mg/L acetate (2), and 10 mg/L formate (3).

A much higher resolution between fluoride and chloride is obtained by using a CarboPac PA1 type stationary phase. Compared with a conventional anion exchanger such as IonPac AS4A (-SC), this separator column, which was initially developed for the analysis of carbohydrates, exhibits a significantly higher capacity. As a result, under the chromatographic conditions in Fig. 3-80 chloride retention increases to more than 20 minutes, resulting in an extremely high resolution between fluoride and chloride. Even the separations between fluoride and iodate, and between bromate and chloride, are possible under isocratic conditions; this is generally considered difficult. A corresponding chromatogram is depicted in Fig. 3-93. Because the particle size of the CarboPac PA1 is 10 µm, the standard flow rate for AS4A-SC column of 2 mL/min was halved, while the concentration of the carbonate/bicarbonate mixture was doubled for better chromatographic efficiency.

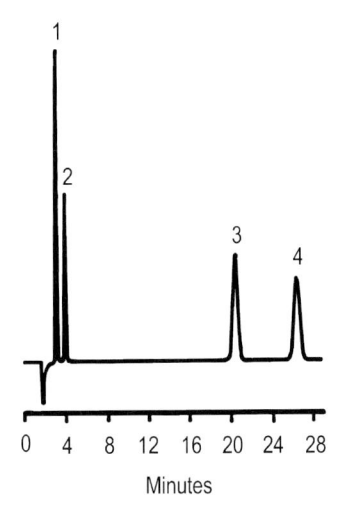

Fig. 3-93. Separation of fluoride, iodate, bromate, and chloride. – Separator column: CarboPac PA1; eluant: 3.4 mmol/L NaHCO$_3$ + 3.6 mmol/L Na$_2$CO$_3$; flow rate: 1 mL/min; detection: suppressed conductivity; injection volume: 50 µL; solute concentrations: 2 mg/L fluoride (1), 10 mg/L iodate (2), 20 mg/L bromate (3), and 5 mg/L chloride (4).

The above examples reveal that anions with similar as well as with strongly different chromatographic behavior can be analyzed simultaneously by choosing the appropriate eluant and detection system. However, the applicability of ion chromatographic techniques is limited by retention and concentration differences between two analytes, as is generally the case for chromatographic methods. While anions such as chloride and sulfate, eluting far from each other, may still be separated and determined in a ratio of 100,000:1, the ratio for anions such as chloride and nitrite, which elute very close to each other, is only 1000:1, especially when conductivity detection is applied. In those cases, the selectivity of the method may only be enhanced by combining different detection systems. Details will be discussed in Section 7.5.

3.7.3.2 Eluant Concentration and pH Value

In addition to the type of eluant, its concentration is one of the most important parameters affecting retention. The pH value of the mobile phase is closely connected to the eluant concentration. When a carbonate/bicarbonate mixture is selected as an eluant, parameters such as ionic strength and pH value cannot be considered separately, because the selectivity changes as the concentration ratio of the two components is varied.

In general, the retention of monovalent and divalent ions shifts forward as the eluant concentration increases. On the other hand, a change in pH primarily affects the retention behavior of multivalent ions, because the valency of such ions depends on the pH value of the mobile phase. This effect is illustrated using orthophosphoric acid as an example. Dissociation occurs in three steps [97]:

$$H_3PO_4 \rightleftharpoons H^+ + H_2PO_4^- \qquad pK_1 = 2.16 \qquad (74)$$

$$H_2PO_4^- \rightleftharpoons H^+ + HPO_4^{2-} \qquad pK_2 = 7.21 \qquad (75)$$

$$HPO_4^{2-} \rightleftharpoons H^+ + PO_4^{3-} \qquad pK_3 = 12.33 \qquad (76)$$

An ion chromatographic separation of the three orthophosphate species is impossible because of these pK equilibria. When using the relatively weak sodium hydroxide eluent, the pH value increases as the concentration increases. With a 1 mmol/L NaOH (pH 11) eluent, orthophosphate and sulfate do not elute at all, primarily because the concentration of the monovalent eluent ions is too low for divalent analyte ions to elute. Moreover, at pH 11 about 10% of the orthophosphate exist as PO_4^{3-} ions, which have a much longer retention time than HPO_4^{2-} ions. When the sodium hydroxide concentration is increased, the monovalent and divalent ions are shifted forward in retention; however, orthophosphate elutes later, because the pH value and, thus, the percentage of trivalent orthophosphate ions increases with increasing sodium hydroxide concentration.

A more complex situation is the carbonate/bicarbonate buffer mixture. Due to the dissociation equilibrium,

$$HCO_3^- \rightleftharpoons H^+ + CO_3^{2-} \qquad pK_2 = 10.31 \qquad (77)$$

the ratio of carbonate and bicarbonate changes at a constant total carbonate concentration, as the pH value of the solution is lowered or raised by adding boric acid or sodium hydroxide, respectively. To investigate these phenomena in more detail — starting from standard conditions (2.8 mmol/L $NaHCO_3$ + 2.2 mmol/L Na_2CO_3) — the carbonate concentration was varied between 0.5 and $2.22 \cdot 10^{-3}$ mol/L at a constant bicarbonate concentration of 2.8 mmol/L. Conversely, the carbonate concentration was kept constant while the bicarbonate concentration was varied accordingly. The capacity factors of the seven standard anions were determined under these chromatographic conditions. Plotting ln k versus the pH values that are calculated from the carbonate/bicarbonate concentrations according to Eq. (78) gives the correlations depicted in Fig. 3-94. The related data are summarized in Table 3-17.

$$pH = pK_2 + \log \frac{[CO_3^{2-}]}{[HCO_3^-]} \qquad (78)$$

Clearly, the retention times of the anions investigated decrease with an increasing amount of carbonate at a constant bicarbonate content (pH 8.35 to 10.23). However, when bicarbonate is added at constant carbonate concentrations (pH

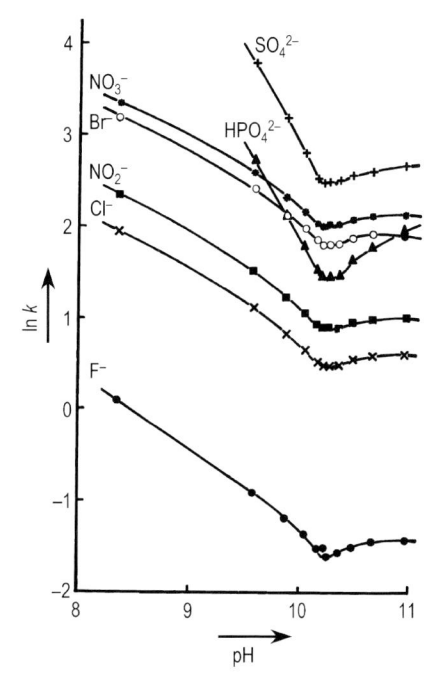

Fig. 3-94. Dependence of the retention of various inorganic anions as a function of eluant pH after adjustment via different carbonate/bicarbonate concentrations. — Separator column: IonPac AS4; flow rate: 2 mL/min; detection: suppressed conductivity; injection: 50 µL anion standard.

range 10.98 to 10.28), the retention times hardly decrease. This effect is understandable as bicarbonate exhibits only a small elution power. The decisive factor for the retention behavior of multivalent ions is the pH value resulting from the concentration ratio of both eluant components. As can be seen from Fig. 3-94, bromide and nitrate are superimposed by orthophosphate within a very narrow pH range between 9.6 and 10.0. Interferences also occur at pH > 10.8. In both cases, this may be attributed to changes in the dissociation equilibria of orthophosphoric acid.

Figure 3-95 shows the change in the retention behavior of inorganic anions when the ionic strength is kept constant and the pH value is varied by adding boric acid or sodium hydroxide. The relevant data are summarized in Table 3-18. As expected, the retention is shifted forward with increasing pH value. However, the addition of boric acid leads to increased retention times by shifting the dissociation equilibrium of carbonate to the bicarbonate site. Interferences in the determination of bromide and nitrate by orthophosphate are observed between pH 9.0 and 9.5.

Figures 3-94 and 3-95 clearly reveal that the influence of the pH value on the retention behavior of inorganic anions can only be investigated indirectly, because changes in the pH value also affect the dissociation equilibrium between carbonate and bicarbonate. Keeping the carbonate/bicarbonate ratio constant allows one to study the changes in retention behavior that result when ionic strength is varied, and to draw conclusions about the pH influence. Therefore,

Table 3-17. ln k values of inorganic anions at different mobile phase pH values. (see Fig. 3-94 for chromatographic conditions)

c Eluant	pH	ln k						
		F^-	Cl^-	NO_2^-	HPO_4^{2-}	Br^-	NO_3^-	SO_4^{2-}
2.8 mmol/L HCO_3^-	8.35	0.11	1.96	2.35	– –	3.19	3.35	– –
2.8 mmol/L HCO_3^- + 0.5 mmol/L CO_3^{2-}	9.58	−0.89	1.13	1.52	2.72	2.42	2.60	3.79
2.8 mmol/L HCO_3^- + 1.0 mmol/L CO_3^{2-}	9.88	−1.17	0.84	1.23	2.13	2.11	2.33	3.19
2.8 mmol/L HCO_3^- + 1.5 mmol/L CO_3^{2-}	10.06	−1.35	0.67	1.07	1.79	1.99	2.17	2.82
2.8 mmol/L HCO_3^- + 2.0 mmol/L CO_3^{2-}	10.18	−1.51	0.54	0.94	1.53	1.87	2.05	2.55
2,8 mmol/L HCO_3^- + 2.2 mmol/L CO_3^{2-}	10.23	−1.51	0.51	0.91	1.47	1.83	2.02	2.48
2.2 mmol/L CO_3^{2-} + 0.5 mmol/L HCO_3^-	10.98	−1.43	0.63	1.03	1.94	1.93	2.16	2.69
2.2 mmol/L CO_3^{2-} + 1.0 mmol/L HCO_3^-	10.68	−1.51	0.61	1.00	1.78	1.94	2.14	2.62
2.2 mmol/L CO_3^{2-} + 1.5 mmol/L HCO_3^-	10,50	−1.56	0.57	0.97	1.64	1.90	2.10	2.58
2.2 mmol/L CO_3^{2-} + 2.0 mmol/L HCO_3^-	10.38	−1.56	0.51	0.90	1.48	1.84	2.03	2.44
2.2 mmol/L CO_3^{2-} + 2.5 mmol/L HCO_3^-	10.28	−1.61	0.51	0.92	1.48	1.84	2.04	2.48

the total carbonate content was varied between $3 \cdot 10^{-3}$ and $6 \cdot 10^{-3}$ mol/L without changing the carbonate/bicarbonate ratio. Plotting the ln k values as a function of ionic strength (see Table 3-19) produces the dependences shown in Fig. 3-96. From the pH value of 10.28, established by the carbonate/bicarbonate concentration ratio, it follows that in addition to divalent sulfate ions, orthophosphate elutes as a divalent HPO_4^{2-} ion. As expected, ionic strength exerts a much stronger effect on the retention behavior of the two divalent anions than of the monovalent anions. When the dependences shown in Fig. 3-96 for the anions orthophosphate, bromide, and nitrate are extrapolated to low ionic strength, bromide and nitrate interfere with orthophosphate at an ionic strength between $2 \cdot 10^{-3}$ and $3 \cdot 10^{-3}$ mol/L. At much higher ionic strengths, a reversal in the elution order between nitrate and sulfate may be expected.

By comparing Figs. 3-95 and 3-96, one can conclude that the retention behavior of inorganic anions is only controlled by the ionic strength of the eluant. Changes in the pH value merely shift the carbonate/bicarbonate concentration ratio, thus only indirectly affecting the retention behavior of anions.

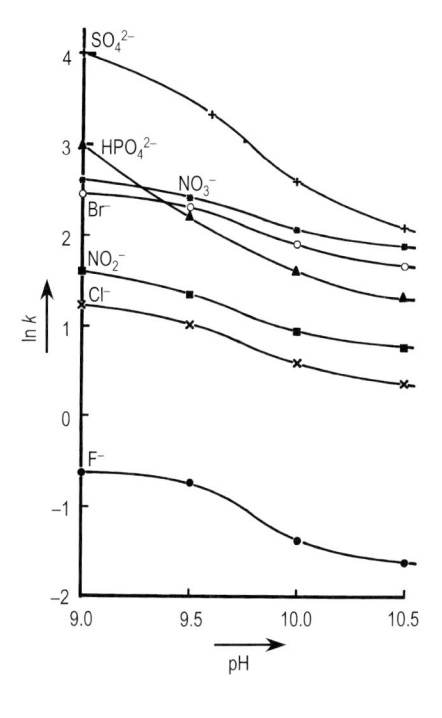

Fig. 3-95. Dependence of the retention of various inorganic anions on the pH value of the buffer mixture at constant ionic strength. — Separator column: IonPac AS4; eluant: 2.8 mmol/L NaHCO$_3$ + 2.2 mmol/L Na$_2$CO$_3$, pH adjusted with H$_3$BO$_3$/NaOH; flow rate: 2 mL/min; detection: suppressed conductivity; injection: 50 µL anion standard.

Table 3-18. ln k values of inorganic anions at different pH values but constant ionic strength. (see Fig. 3-95 for chromatographic conditions)

Species	ln k			
	pH 9.0	pH 9.5	pH 10.0	pH 10.5
F$^-$	−0.63	−0.73	−1.39	−1.66
Cl$^-$	1.22	1.00	0.57	0.34
NO$_2^-$	1.59	1.36	0.92	0.75
HPO$_4^{2-}$	2.98	2.19	1.59	1.30
Br$^-$	2.44	2.30	1.87	1.68
NO$_3^-$	2.61	2.40	2.06	1.87
SO$_4^{2-}$	4.00	3.33	2.59	2.07

3.7.3.3 Influence of Organic Solvents

As outlined in Sections 3.4.1 and 3.4.2, polymer-based anion exchangers with a low degree of cross-linking (<5 %) are not solvent compatible. Even in the presence of small amounts of organic solvents, polymer-based materials start swelling, which leads to a high backpressure. When conditioning a separator column packed with polymer-based materials with aqueous eluants after using organic solvents, void volumes at the column head result, which can be attributed to shrinking processes. Polymeric stationary phases can be made compatible with common HPLC solvents by increasing the degree of cross-linking to more than

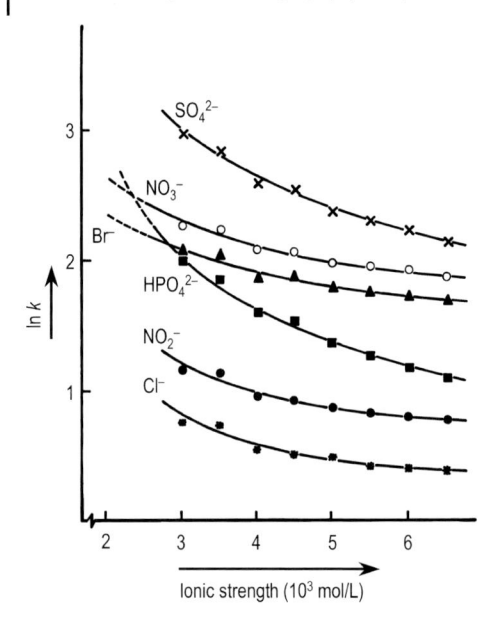

Fig. 3-96. Effect of ionic strength on the retention behavior of various inorganic anions at constant pH value. — Separator column: IonPac AS4; flow rate: 2 mL/min; detection: suppressed conductivity; injection: 50 μL anion standard.

Table 3-19. ln k values of inorganic anions at different ionic strengths but constant pH value. (see Fig. 3-96 for chromatographic conditions)

Ionic strength [10^3 mol/L]	ln k						
	F^-	Cl^-	NO_2^-	HPO_4^{2-}	Br^-	NO_3^-	SO_4^{2-}
3.02	−1.26	0.76	1.15	2.00	2.07	2.27	2.97
3.52	−1.11	0.74	1.13	1.86	2.05	2.23	2.84
4.02	−1.50	0.55	0.96	1.60	1.89	2.08	2.59
4.52	−1.55	0.51	0.92	1.55	1.88	2.07	2.55
5.02	−1.35	0.49	0.88	1.37	1.80	1.98	2.38
5.52	−1.61	0.42	0.83	1.28	1.77	1.96	2.31
6.02	−1.50	0.40	0.80	1.18	1.73	1.93	2.23
6.52	−1.61	0.38	0.77	1.10	1.70	1.88	2.14

50% [98]. Swelling in organic solvents is so insignificant with such stationary phases that organic solvents are not only used for cleaning the column, but also for altering selectivity [99].

The ion-exchange process in aqueous systems has already been described in Sections 3.2 and 3.3. When adding organic solvents to the mobile phase, the complexity of this process increases, because several competing parameters determine the retention characteristics.

In general, ions in solution are surrounded by a solvation shell, which consists of solvent molecules grouping around the ions in a relative order. Thus, they are in a lower entropic state than those in the bulk mobile phase. In aqueous solution, ions are solvated by water. Interactions between ions and water as a solvent include the destruction of the water structure by breaking hydrogen bonds (cavity effect). The larger the ion, the higher the energy required for the formation of a cavity of molecular dimension. On the other hand, electrostatic ion-dipole interactions occur that lead to the formation of a new structure. Thus, the smaller the ionic radius and the higher the ionic charge, the stronger the cavity effect. The structure of the hydration shell is imagined to be composed as follows: Direct ion-dipole interaction leads to *primary* hydration, to which, depending on the type of ion, a limited number of water molecules contribute via maximum approximation between ion and water molecule. In this sphere, water molecules are relatively rigid under the influence of electrostatic interaction, which is equivalent to an entropy decrease of water. Small and multivalent ions are especially strongly hydrated. Beyond the sphere of primary hydration, the water structure in the vicinity falls under the purview of the electric field of the ion (*secondary* hydration). In this zone, partial destruction of the water structure occurs via breaking hydrogen bonds, which lead to higher mobility of the water molecules as compared with an undisturbed liquid structure. If the aqueous solution contains an additional organic solvent, the hydration shell breaks open, so that solvent molecules can penetrate the solvation shell. The degree of penetration of the solvation shell by solvent molecules depends on the solvent's ability to form hydrogen bonds with the respective ions; penetration also depends on the polarizability of the ions. As mentioned above, the type of solvation shell determines the mobility of an ion in solution. If the solvation shell of an ion changes, the ion mobility changes as well. As a prerequisite for the interaction of an ion with an ion-exchange group, the ion has to cast — at least partially — its solvation shell to be able to get close to the ion-exchange group. This approximation is all the better, the easier the respective ion can cast its solvation shell. Thus, the binding strength between ion and ion-exchange group depends on the degree of approximation between the two binding partners. In an analogous manner, the solvation shell of the ion-exchange group also has to renew its orientation when an ion comes close. Basically, the ability of an ion to partially cast its solvation shell plays an important role in the ion-exchange process. Therefore, the affinity of an ion towards the ion-exchange group depends on the solvation degree of the ion and on its ability to repel surrounding solvent molecules. Because eluant ions are solvated as well and exhibit a certain affinity towards the ion-exchange groups, selectivity can be interpreted as a competition between the relative affinities of analyte and eluant ions and the ion-exchange group.

Selectivity is influenced by the type and concentration of the organic solvent. The type of organic solvent is an important factor in optimizing separation, because the solvation power and hydrophobicity of the solvent influence the retention mechanism. Investigations of the influence of various organic solvents on the retention behavior of inorganic anions were carried out by Stillian et al. [98] utilizing a solvent-compatible anion exchanger with a hydroxide eluant under gradient conditions. In comparison with pure aqueous eluants, they observed significantly higher retention times when using methanol-containing eluants; this results in an increased resolution between anions that have similar retention behaviors (see also Fig. 3-40 in Section 3.4.2). The authors interpret this phenomenon as such: there is a limited tendency of the strongly hydrated hydroxide ion (as compared with the stationary phase) to cast its hydration shell in presence of methanol. This leads to a decreased selectivity of the ion-exchange group for hydroxide ions. In contrary, shorter retention times are achieved with aprotic solvents such as acetonitrile. As shown in Fig. 3-97, retention of inorganic anions decreases with increasing content of acetonitrile in the mobile phase.

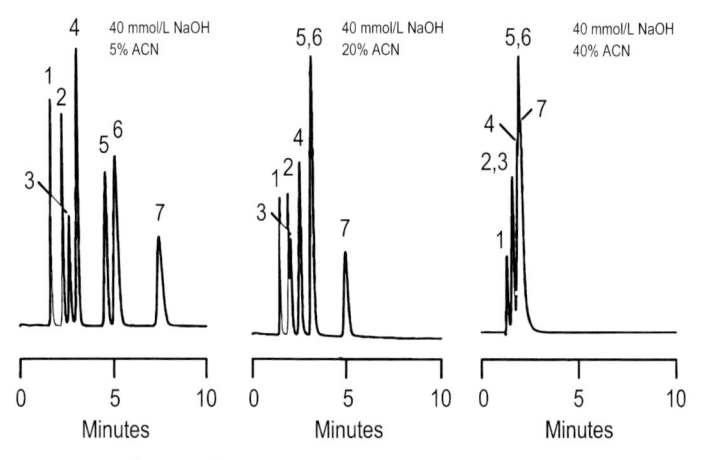

Fig. 3-97. Influence of acetonitrile on the retention of inorganic anions. – Separator column: OmniPac PAX-100; eluant: 0.04 mol/L NaOH – acetonitrile, (A) 95:5 v/v, (B) 80:20 v/v, (C) 60:40 v/v; flow rate: 1 mL/min; detection: suppressed conductivity; injection: anion standard with fluoride (1), chloride (2), nitrite (3), sulfate (4), bromide (5), nitrate (6), and orthophosphate (7); (taken from [99]).

Since the same effect is observed with a stepwise decrease of latex cross-linking, one can assume that the latex polymer swells more strongly in acetonitrile than in water, decreasing the *effective* degree of cross-linking. This, in turn, leads to a decrease in the number of ion-exchange groups per volume unit of the latex polymer and, thus, to a decrease in retention. To a certain extent, the dielectricity constant of acetonitrile, being smaller than the one for water, also affects the selectivity. However, this effect is much more pronounced with more hydrophobic solvents such as 2-propanol.

Besides methanol and acetonitrile, other water-miscible organic solvents such as ethanol and 2-propanol can also be utilized for optimizing separation. Longer-chain alcohols such as 2-propanol cause a much stronger swelling of the latex polymer. For a number of anions, Stillian et al. [98] observed a retention increase when adding small amounts of 2-propanol to an NaOH eluant. Thus, small concentrations of 2-propanol result in a similar selectivity change as observed with methanol. As expected, ethanol has an intermediate effect. When adding ethanol to the mobile phase, the retention decrease is comparable to that achieved with methanol. Because the dielectricity constant of ethanol is much smaller than that of methanol [100], an ethanolic solution is less polar than a methanolic one, which influences the solvation of the ions and thus the ion-exchange process.

3.7.4
Experimental Parameters Affecting Retention when Applying Direct Conductivity Detection

3.7.4.1 Choice of Eluant

As already discussed in Section 3.5, the choice of eluants suitable for applying direct conductivity detection has been the subject of numerous investigations [16, 17, 19, 72]. In general, organic acids with an aromatic backbone exhibit good elution properties, because they have a high affinity towards the stationary phase of an anion exchanger and, thus, may be used at comparatively low concentrations. An eluant with fairly low background conductivity is also desirable. Aromatic carboxylic acids, with their low equivalent conductances, are a good choice. The most important criterion for choosing an eluant in the application of direct conductivity detection is the resulting conductivity difference between the eluant and the analyte ions. Thus, the eluant pH value does not necessarily have to be close to the neutral point. Strongly alkaline eluants such as sodium hydroxide [27], with a high equivalent conductance, have been successfully employed. However, aqueous solutions of the sodium, potassium, and ammonium salts of various organic acids in the concentration range between 10^{-3} and 10^{-4} mol/L are also commonly used. The pH value of the mobile phase is adjusted between 4 and 7, because it determines the dissociation of the organic acid and, thus, the retention behavior of the analyte species. Although in many cases fully dissociated salts are employed as eluants, partially dissociated acids such as benzoic acid and partially neutralized acids such as potassium hydrogen phthalate are used as well. Mixtures of eluant ions with different valencies are also common. Often, the pH value of the mobile phase is adjusted with sodium borate, because borate ions act as eluant ions. A well-known example is the mixture of gluconate and borate, which is used in combination with polymethacrylate-based separator columns. An anionic complex with a low background conductivity is formed that is especially effective as eluant. As shown in Fig. 3-98, the seven most important inorganic anions [86] can be analyzed in a single run. Although the separator column being used is characterized by a high chromatographic efficiency, a relatively long analysis time is required for a baseline-resolved separation of all components. This is primarily due to the positive signal that appears in the void volume, which may not be separated from the subsequent fluoride peak without

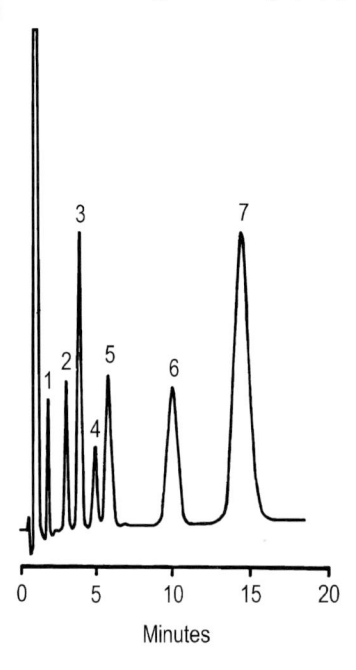

Fig. 3-98. Separation of inorganic anions using a borate/gluconate buffer as an eluant. – Separator column: TSK Gel IC-PW; eluant: 1.3 mmol/L $Na_2B_4O_7$ + 5.8 mmol/L H_3BO_3 + 1.4 mmol/L potassium gluconate/acetonitrile (88:12 v/v), pH 8.5; flow rate: 1.2 mL/min; detection: direct conductivity; injection: 100 µL anion standard (5 to 40 mg/L); peaks: fluoride (1), chloride (2), nitrite (3), bromide (4), nitrate (5), orthophosphate (6), and sulfate (7); (taken from [26]).

compromising on analysis time. The reason for the appearance of this signal, which is typical for all ion chromatographic systems with direct conductivity detection, was discussed in Section 3.4.1.

A selectivity comparable to the gluconate/borate mixture is obtained with potassium hydrogen phthalate, the eluant most often used for the application of both direct conductivity and indirect UV detection. Depending on the type of stationary phase being used, the pH value of the mobile phase is adjusted between 5.3 and 7.0. In this pH region, phthalate exists mainly as a divalent anion. Figure 3-99 displays an optimized separation of inorganic anions on Waters IC-PAK A. Potassium hydrogen phthalate is a slightly stronger eluant than the gluconate/borate mixture and, thus, is a good fit for the analysis of divalent anions. Polarizable anions such as iodide and thiocyanate may also be separated together with standard inorganic anions by using potassium hydrogen phthalate. Only orthophosphate cannot be determined under the chromatographic conditions given in Fig. 3-99.

The significant advantage of the carbonate/bicarbonate eluant mixture described in Section 3.5 for eluting monovalent and divalent anions in the same run can also be achieved with *p*-hydroxybenzoic acid when applying direct conductivity detection. When the pH value of the mobile phase is adjusted to 8.5, the carboxyl group is fully dissociated, while the hydroxide group with a pK value of 9.3 is only partially dissociated. Therefore, this eluant also contains a mixture of monovalent and divalent eluant ions, so fluoride and sulfate can be analyzed in a single run within 15 minutes. Figure 3-100 shows the results obtained with this eluant on a Wescan 269-029 polymeric stationary phase. The negative signal appearing after 25 minutes is called the *system peak*. The origin and significance of such signals will be discussed below in more detail.

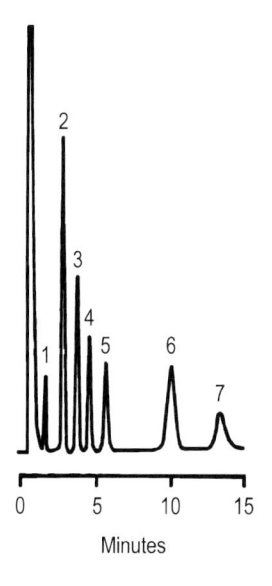

Fig. 3-99. Separation of inorganic anions using a potassium hydrogen phthalate eluant. — Separator column: Waters IC-PAK Anion; eluant: 1 mmol/L KHP, pH 7; flow rate: 2 mL/min; detection: direct conductivity; injection volume: 10 µL; solute concentrations: 100 mg/L each of fluoride (1), chloride (2), nitrite (3), bromide (4), nitrate (5), sulfate (6), and iodide (7).

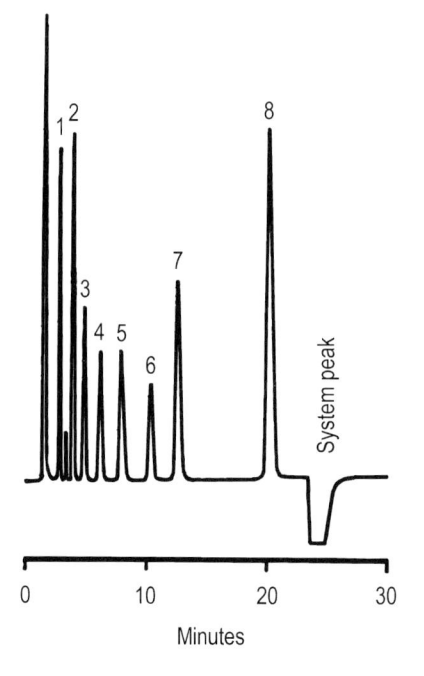

Fig. 3-100. Separation of inorganic anions with a *p*-hydroxybenzoic acid eluant. — Separator column: Wescan 269-029; eluant: 4 mmol/L PHBA, pH 8.7; flow rate: 1.5 mL/min; detection: direct conductivity; injection volume: 100 µL; solute concentrations: 5 mg/L each of fluoride (1), chloride (2), nitrite (3), bromide (4), and nitrate (5), 10 mg/L each of orthophosphate (6) and sulfate (7), and 20 mg/L chromate (8).

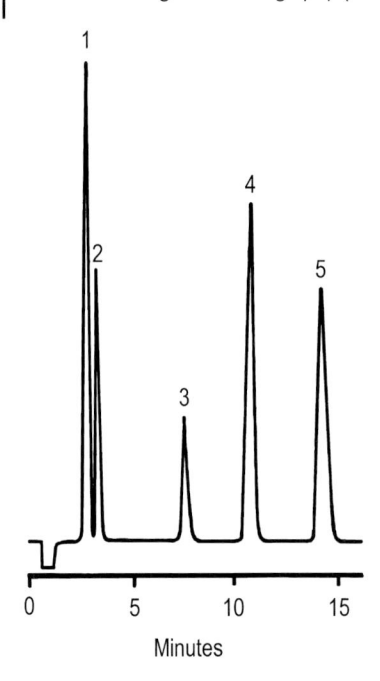

Fig. 3-101. Separation of inorganic anions with a potassium citrate eluant. — Separator column: TSK Gel IC-SW; eluant: 1 mmol/L potassium citrate, pH 5.2; flow rate: 1.2 mL/min; detection: direct conductivity; injection volume: 100 µL; solute concentrations: 5 mg/L chloride (1), 10 mg/L each of bromide (2), iodide (3), thiocyanate (4), and sulfate (5).

The potassium salt of citric acid has an elution strength similar to potassium hydrogen phthalate. When fully dissociated, citrate as a trivalent ion exhibits a strong affinity towards the stationary phase and, therefore, may be used as an eluant at relatively low concentration. However, the elution order of standard inorganic anions that is obtained with potassium citrate depends on the stationary phase being used. When, for example, a polystyrene/divinylbenzene-based resin is used, the elution order is the same as that obtained with potassium hydrogen phthalate. On the other hand, when using silica-based anion exchangers, Matsushita et al. [39] observed significantly shorter retention times for polarizable anions such as iodide and thiocyanate, which elute ahead of the divalent sulfate (Fig. 3-101). The higher affinity of these two anions towards a styrene polymer can be attributed to non-ionic interactions, which supersede the actual ion-exchange process.

As for the sensitivity of the conductivity detection, phthalate eluants have an advantage over citrate eluants, because the conductivity difference between eluant and analyte ions is much higher when using salts of aromatic acids as eluant. On the other hand, when the column effluent is passed through a suppressor system prior to entering the conductivity cell, a much higher sensitivity can be obtained with potassium citrate eluants. This is due to the incomplete dissociation of the corresponding acid formed in the suppressor (see also Section 3.5), which reduces the background conductivity and, thus, increases the conductivity difference to the analyte ions.

The above suggests that free carboxylic acids may act directly as eluants. The separation of inorganic anions on a Shimpack IC-A1 separator column shown in Fig. 3-102 reveals how legitimate this assumption is. The chromatogram impressively demonstrates the high chromatographic efficiency and relatively short retention times for the individual components obtained with a free phthalic acid eluant. However, the total analysis time increases significantly in this case, as the system peak does not appear until about 12 minutes. Furthermore, the high elution power of phthalate – either in the form of a salt (see also Fig. 3-99) or a free acid – renders impossible the analysis of orthophosphate in the same run.

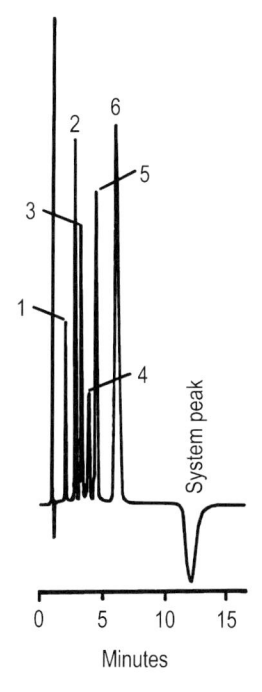

Fig. 3-102. Separation of inorganic anions with a phthalic acid eluant. – Separator column: Shimpack IC-A1; eluant: 2.5 mmol/L phthalic acid + 2.4 mmol/L tris(hydroxymethyl)aminomethane, pH 4.0; flow rate: 1.5 mL/min; detection: direct conductivity; injection volume: 20 µL; solute concentrations: 5 mg/L fluoride (1), 10 mg/L chloride (2), 15 mg/L nitrite (3), 10 mg/L bromide (4), 30 mg/L nitrate (5), and 40 mg/L sulfate (6).

In some cases, however, the analysis of inorganic anions in the acidic pH range is of particular interest. Above all, this holds for anions which are not stable in the alkaline pH range. This includes, for example, peroxomonosulfate, SO_5^{2-}, that, in contrast to peroxodisulfate, $S_2O_8^{2-}$, decays rapidly at pH values above 5. Another application is the determination of fully dissociated anions in presence of high amounts of weakly dissociated inorganic or organic acids. Because the latter are not dissociated at low pH and, therefore, travel with the eluant through the column, the separation from the more strongly retained anions is facilitated.

The concept of employing free acids as eluants dates back to the investigations of Fritz et al. [72]. In 1981, they demonstrated that free benzoic acid may be employed as an eluant instead of sodium or potassium salts, which had been used until then. This is understandable, taking into account that benzoic acid is partly dissociated in an aqueous solution. The degree of dissociation may be calculated from the dissociation constant K_a according to the Ostwald dilution law [85]:

$$K_a = \frac{\alpha^2}{1-\alpha} \cdot c_0 \qquad (79)$$

c_0 Total acid concentration

The degree of dissociation decreases with increasing electrolyte concentration. With the dissociation constant for benzoic acid, $K_a = 6.25 \cdot 10^{-5}$, the degree of benzoic acid dissociation in solution, with the concentration of $1.25 \cdot 10^{-3}$ mol/L used as an example, is calculated to be $\alpha = 0.2$. Therefore, such a solution contains $2.5 \cdot 10^{-4}$ mol/L of oxonium and benzoate ions. Comparing such a benzoic acid solution with a sodium benzoate solution of the concentration $c = 2.5 \cdot 10^{-4}$ mol/L, an almost identical elution behavior for inorganic anions is observed.

In general, with a benzoic acid eluant, higher sensitivities are obtained for the analyte anions as compared with either sodium or potassium benzoate eluants. This is illustrated in the following reaction scheme:

$$\begin{array}{c} \text{Resin} - A^- + H^+ + Bz^- \rightleftharpoons \text{Resin} - Bz^- + H^+ + A^- \\ \updownarrow \\ HBz \end{array} \qquad (80)$$

According to this scheme, analyte ions bound for a short time to the fixed stationary charge are exchanged for an equivalent amount of benzoate ions, which leads to a corresponding reduction in the benzoate concentration under the chromatographic signal. However, benzoate ions are delivered from undissociated benzoic acid, because the system is in a state of dynamic equilibrium. With the degree of dissociation of $\alpha = 0.2$ for the benzoic acid concentration of $1.25 \cdot 10^{-3}$ mol/L mentioned above, 80% of the oxonium and benzoate ions are derived from the molecular benzoic acid, and are subsequently transformed into the fully dissociated species H^+A^-. The much higher equivalent conductance of the oxonium ion as the counter ion for A^- is the actual reason for the sensitivity increase upon application of a free benzoic acid eluant.

However, a number of disadvantages counter this advantage. Weakly basic anions such as bicarbonate cannot be determined under these chromatographic conditions, since they are hardly dissociated or not dissociated at all at pH 3.6 in the mobile phase. Another limiting factor for practical application is the long time required for column conditioning because of the tendency of molecular acids to adsorb at the surface of the anion exchanger. Connected with this tendency is the formation of system peaks, whose appearance in the chromatogram may be interpreted as a disturbance of the adsorption equilibrium.

Eluants based on sodium benzoate or benzoic acid exhibit about the same elution power as bicarbonate and, thus, are used for the analysis of monovalent anions. When looking for less strongly adsorbing acids, Fritz, DuVal, and Barron [101] found other compounds that may also be employed as eluants. The properties of these compounds are listed in Table 3-20.

Table 3-20. Properties of various organic acids that may be utilized as eluants [101].

Compound	pK_1	pK_2	pK_3	Degree of dissociation	Elution power
Nicotinic acid	4.87			11	weak
Benzoic acid	4.19			22	strong
Succinic acid	4.16	5.61		25	weak
Citric acid	3.14	4.77	6.39	58	weak
Fumaric acid	3.03	4.44		62	weak
Salicylic acid	2.97	13.40		63	strong

Several factors have to be taken into account when choosing an eluant. The elution power, for instance, depends upon the acid's degree of dissociation, which determines the concentration of the acid anions acting as eluent ions. The affinity of the eluant towards the stationary phase is important to its selectivity. Eluant ions with a structure comparable to the matrix of the stationary phase, are characterized by a high affinity. At the same time, they also have a high selectivity due to intense interactions with the stationary phase. Although the organic acids listed in Table 3-20, depending on their degree of dissociation, elute inorganic analyte ions at different speeds, the investigations of Fritz et al. [102] into the relative retention of the solutes reveal only small differences between the various organic acids. However, in an individual case, small differences in the relative retention may be decisive for a successful separation.

There are a variety of organic acids which can be used as eluants. As an example, the chromatogram obtained with a nicotinic acid eluant is shown in Fig. 3-103. Nicotinic acid is most suitable for the separation of monovalent anions. In contrast to other monocarboxylic acids, it is characterized by a good solubility in water and a low equivalent conductance. In addition, its propensity to adsorb at the stationary phase is very low, which prevents the formation of the usual system peak in most cases. However, divalent anions are strongly retained under these chromatographic conditions. For a rapid elution of divalent anions, salicylic acid or trimesic acid eluants are more suitable.

While the eluants described so far are useful to separate a variety of inorganic (and organic) anions, they cannot be employed for the analysis of weak inorganic acids such as sulfide, cyanide, borate, and silicate, because these anions are fully dissociated only in a strongly alkaline solution. Therefore, the idea of using a strongly basic eluant for the separation of such anions seems inherently logical.

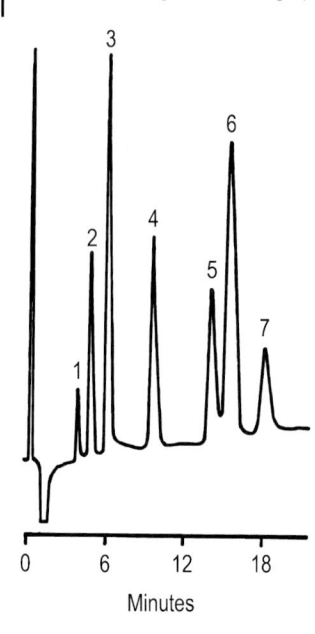

Fig. 3-103. Separation of monovalent anions using a nicotinic acid eluant. — Separator column: Wescan 269-031; eluant: 0.01 mol/L nicotinic acid; flow rate: 2.7 mL/min; detection: direct conductivity; injection volume: 100 µL; solute concentrations: 20 mg/L azide (1), 10 mg/L each of formate (2) and fluoride (3), 20 mg/L orthophosphate (4), 10 mg/L each of nitrite (5) and chloride (6), and 15 mg/L bromate (7); (taken from [103]).

Corresponding procedures utilizing sodium hydroxide eluants and indirect conductivity detection were introduced by Fritz et al. [104] a long time ago. Because a sodium hydroxide solution conducts much stronger than the analyte ions, the latter appear in the chromatogram as negative signals. Hence, the proportionality between peak area and concentration still exists. Suppressor systems cannot be used in this case because they would convert the anions mentioned above into the corresponding acids, which are not suitable for conductivity detection owing to their weak dissociation. Although sodium hydroxide fulfills all requirements for indirect conductivity detection, its elution power is in many cases not sufficient to elute anions of practical relevance. Retention times are significantly shortened by adding small amounts of sodium benzoate to the sodium hydroxide solution, which barely affects the conductivity difference between eluant and analyte ions. However, shorter retention times improve the peak shape, thus increasing the sensitivity of the method. Figure 3-104 shows such separation. Also in this case, the appearance of a system peak after about 28 minutes significantly increases total analysis time and is therefore disadvantageous. However, this is not the only factor limiting the applicability of this method to routine analysis. In real-world samples, environmentally relevant anions such as sulfide and cyanide occur only at very low concentrations in the presence of high amounts of chloride and sulfate. Since divalent anions are strongly retained under the chromatographic conditions given in Fig. 3-104, the separator column must be rinsed after only a few injections to avoid overloading the stationary phase with matrix components. For these reasons, stronger eluants in combination with more selective detection modes such as amperometry (see Fig. 3-85) offer significant advantages, at least for routine analysis of such anions. For the analysis of borate and silicate, ion-exclusion chromatography (see Chapter 5) is an alternative method. In ion-exclusion chromatography, the selectivity increase is achieved by a different separation mechanism whereby all strongly dissociated anions elute as a single peak within the void volume and, thus, do not interfere.

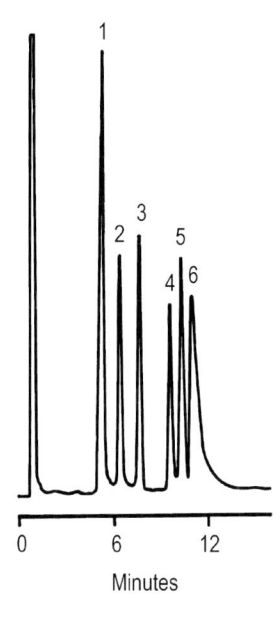

Fig. 3-104. Separation of anions derived from weak inorganic acids using a strongly basic eluant. — Separator column: Wescan 269-029; eluant: 4 mmol/L NaOH + 0.5 mmol/L sodium benzoate; flow rate: 1.5 mL/min; detection: indirect conductivity; injection volume: 100 µL; solute concentrations: 5 mg/L borate (as B) (1), 10 mg/L silicate (as SiO_2) (2), 10 mg/L each of formate (3) and sulfide (4), and 20 mg/L each of chloride (5) and cyanide (6); (system peak appears after 28 min); (taken from [103]).

3.7.4.2 System Peaks

The term *system peak* refers to signals that cannot be attributed to solutes. System peaks are characteristic for ion chromatographic systems that use weak organic acid eluants without a suppressor system. Despite numerous publications regarding this subject [105-109], system peaks were often the cause of misinterpretations. A significant amount of information about the thermodynamic and kinetic processes taking place within the separator column may be inferred from system peaks [110, 111], contributing to a better understanding of the chromatographic process.

The occurrence of system peaks will be discussed, taking an anion exchanger in equilibrium with a benzoic acid (BA) mobile phase as an example. This equilibrium is maintained as long as the chromatographic system is not disturbed. The equilibrium processes of interest are illustrated in Fig. 3-105. They comprise:

- Equilibrium between the solute ions in the mobile phase and those that are bound to the fixed stationary charge
- Dissociation equilibrium of benzoic acid in the mobile phase
- Equilibrium between the benzoic acid dissolved in the mobile phase and adsorbed on the hydrophobic surface of the stationary phase

If this equilibrium is disturbed by injecting a sample, a new equilibrium is established via relaxation; i.e., the kind of relaxation process depends on the pH value of the sample injected. If the sample pH is lower than the pH value of the mobile phase, benzoate ions in the mobile phase are protonated due to the sample injection. Thus, the concentration of molecular benzoic acid in the mobile phase increases as does the amount adsorbed onto the stationary phase. What is not adsorbed travels through the column and appears as a chromatographic signal: the *system peak*. A qualitatively similar chromatogram is obtained when a sample is injected into the system that contains the solute ions and the corresponding eluant component. However, only the position of the system peak is comparable, not its area and direction.

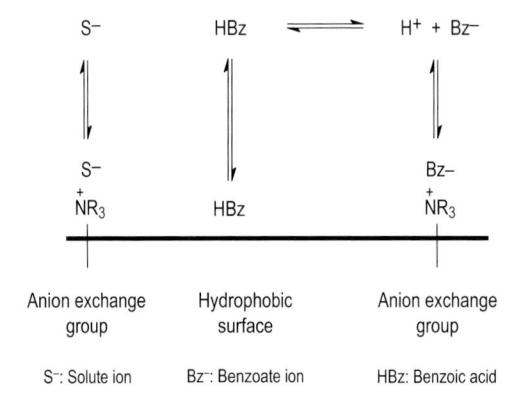

Fig. 3-105. Equilibrium processes between the stationary phase of an anion exchanger and benzoic acid as an example of a weak organic acid eluant.

S$^-$: Solute ion Bz$^-$: Benzoate ion HBz: Benzoic acid

Correspondingly, the injection of a sample with a pH higher than that of the mobile phase shifts the dissociation equilibrium of benzoic acid to the ionic side. To maintain the equilibrium, molecular benzoic acid is desorbed from the stationary phase surface. The resulting deficit in undissociated acid appears as a negative peak in the chromatogram with a characteristic elution time. The migration velocity of the system peak, v_i, through the column is smaller than that of the mobile phase. It is defined according to Eq. (81) as:

$$v_i = R_i\, u \qquad (81)$$

with R_i representing the equilibrium amount of the respective eluant component, i, in the mobile phase and u representing the mobile phase velocity. In the given chromatographic system the velocity of the system peak is constant, independent of the kind of sample injected. When the mobile phase contains more than one organic acid, several system peaks are observed.

This explanation is corroborated by the fact that no system peaks occur when fully dissociated eluant ions are used. On the other hand, position of the system peaks can be affected by adding organic solvents.

As mentioned above, a number of important conclusions can be drawn from the occurrence of system peaks. In this connection, the investigations of Levin and Grushka [110, 111] are worth mentioning. From the system peak area, the authors derived the amount of eluant component adsorbed at the stationary phase and calculated the capacity ratio, k. Remarkably, this approach allows the calculation of the capacity ratio without prior knowledge of the column dead time, t_d.

Levin and Grushka performed their studies using an ion-pair chromatographic system. A chemically bonded ODS phase was used as a stationary phase. The aqueous mobile phase contained an acetate buffer, copper ions, and sodium heptanesulfonate as an ion-pair reagent. Thus, three system peaks are obtained in addition to the negative water dip when pure water is injected. The attribution of each individual eluant component is facilitated, because the system peak area changes when different concentrations of the respective eluant component are injected as a sample, or when the mobile phase contains different concentrations of these components. The desorption of the eluant component that is initially adsorbed at the stationary phase upon injection of pure water occurs in a volume that approximately corresponds to the injection volume. The knowledge of the real volume is insignificant in this respect, because these quantities are put in relation to each other for the calculation of the capacity ratio. However, the transformation of area units into concentration units is of importance for calculating k. For this, Levin and Grushka prepared a calibration curve by injecting different concentrations of the eluant component at defined concentrations of this component in the mobile phase. Plotting the resulting peak area versus the injected concentration leads to a straight line with a positive intercept. When repeating these measurements with different concentrations of the eluant component in the mobile phase, straight lines are obtained, too; they only differ in their intercept. The intercept represents the system peak area at injection of pure water. The pertinent concentration of eluant component desorbed from the stationary phase upon the injection of water is determined via extrapolation of the calibration curve to the abscissa. It corresponds to the absolute value of the negative abscissa intercept. Provided that the desorption of the adsorbed component is complete, the concentration of the adsorbed component, w_s, is obtained. Since the concentration of the eluant component in the mobile phase, w_m, is known, k can be calculated according to Eq. (82):

$$k = \frac{w_s}{w_m} \qquad (82)$$

Since the capacity ratio is typically determined according to Eq. (83) from chromatographic data, inversely, the column dead time may be calculated when k is known.

$$k = \frac{t_{ms} - t_d}{t_d} \tag{83}$$

t_d Column dead time
t_m Gross retention time of the system peak

Transforming Eq. (83) to t_d yields

$$t_d = \frac{t_{ms}}{1+k} \tag{84}$$

The great difficulties connected with an exact measurement of the dead volume are circumvented when using this method for the determination of the dead time.

As useful as the information is that may be deduced from the system peaks, it is important for the practicing analyst to separate these signals from the analyte signals. Some options for optimizing the separation are described below.

3.7.4.3 Eluant Concentration and pH Value

The retention behavior of inorganic anions in ion chromatographic systems without a suppressor device was put on a theoretical base by Jenke and Pagenkopf [112]. They investigated several models for their abilities to mathematically describe the retention behavior. Another paper by Jenke and Pagenkopf [113] described options for optimizing the chromatographic conditions by varying the pH value while keeping the eluant concentration constant, and by varying the eluant concentration at constant pH value.

When varying the eluant concentration at constant pH value and plotting the logarithm of the net retention volume as a function of the logarithm of the eluant concentration, linear dependences are obtained for various inorganic anions. The slope of the straight line depends on both the type of eluant and analyte ions. Figure 3-106 illustrates this effect, taking both eluant and solute ions investigated by Small [114] as an example. The co-elution of some anions and the reversal of the elution order under specific chromatographic conditions is remarkable. The retention behavior is described mathematically as:

$$\log V_s \propto -m \cdot \log[E^{x-}] \tag{85}$$

V_s Net retention volume
$[E^{x-}]$ Concentration of eluant ions

3.7 Anion Exchange Chromatography of Inorganic Anions | 157

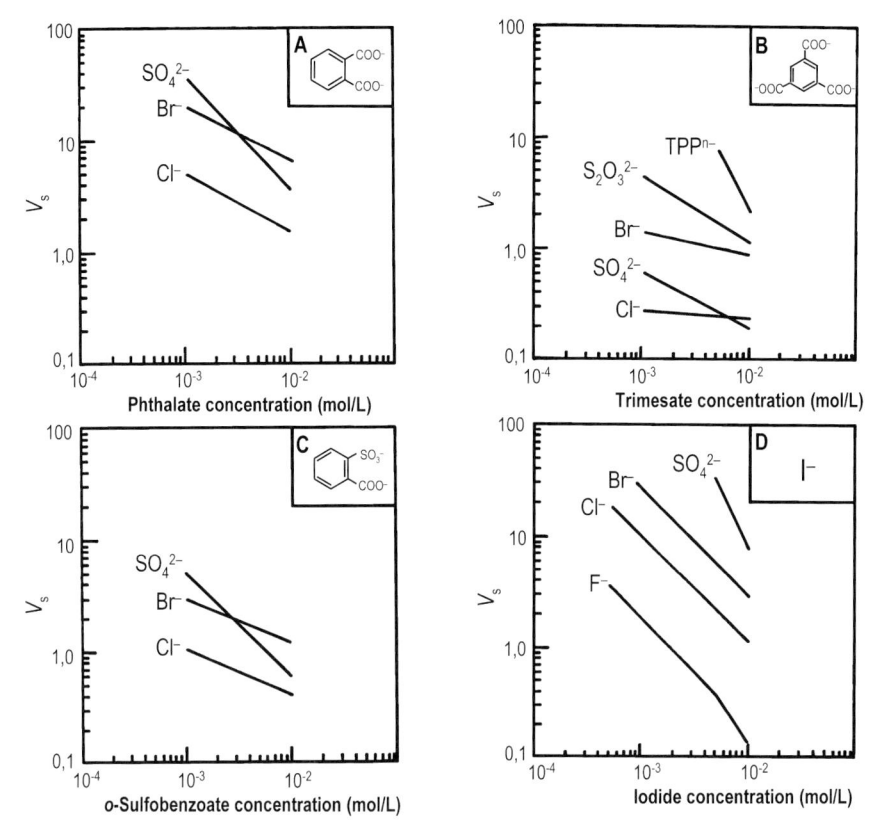

Fig. 3-106. Dependence of the net retention volume of various inorganic anions on the concentration of a) phthalate, b) trimesate, c) o-sulfobenzoate, and d) iodide. — Separator column: 250 mm × 2.8 mm i. d. SAR-40-0.6; (taken from [114]).

After transformation [115] one obtains:

$$\log V_s = \text{Constant} - \frac{y}{x} \cdot \log[\text{E}^{x-}] \tag{86}$$

x Charge of the eluant ion
y Charge of the analyte ion

According to Eq. (86), the slope of the straight line is proportional to the quotient of the charge numbers of analyte and eluant ions. When the calculated quotient, y/x, is compared to the measured values (Table 3-21), a good agreement is obtained for the eluants and solute ions studied by Small. Knowing the charge numbers of eluant and solute ions is sufficient to determine the retention volume of a compound at one eluant concentration, and to predict the retention of this species at other concentrations. Furthermore, the dependence of the slope of the straight line explains the observed reversal of the elution order. Based on these findings, co-elutions can be prevented by choosing appropriate chromatographic conditions.

Table 3-21. Comparison of theoretical and measured values for the quotient y/x (taken from [114]).

Eluant Ion, E^{x-}	Solute Ion, S^{y-}	$y/x_{theor.}$	$y/x_{meas.}$
I^-	Br^-	1.00	1.00
I^-	SO_4^{2-}	2.00	2.00
Trimesate^{3-}	Br^-	0.33	0.21
Trimesate^{3-}	SO_4^{2-}, $S_2O_3^{2-}$	0.66	0.60
Phthalate^{2-}	Br-	0.50	0.47
Phthalate^{2-}	SO_4^{2-}	1.00	0.98

Variation of the pH value at constant eluant concentration primarily affects the dissociation of the organic acid employed as an eluant. Therefore, in most cases polybasic acids such as phthalic acid ($pK_1 = 2.9$; $pK_2 = 5.5$) are used as an eluant to control the elution power via dissociation into monovalent or divalent anions. Figure 3-107 illustrates how the retention of a number of inorganic anions depends on pH when using a phthalate eluant ($c = 4$ mmol/L) on a commercially available silica-based anion exchanger. The position of the system peak allows the pH value of the mobile phase to be changed only within a limited range. At pH 2.7 – the pK value of phthalic acid itself – the system peak appears behind the water dip (Fig. 3-108). Under these conditions, the elution power of phthalic acid is very low, because only monovalent hydrogen phthalate anions exist as eluant ions at this pH. Thus, divalent analyte ions such as sulfate are strongly retained. When raising the pH value to 3.7, the mobile phase contains divalent phthalate ions in addition to hydrogen phthalate ions. Although monovalent and divalent ions may be analyzed in a single run, the system peak unfortunately lies between nitrate and sulfate. The retention times decrease as the pH value is further increased to 4.5, because the concentration of divalent eluant ions increases. The system peak elutes after the sulfate signal under these conditions. These three examples illustrate the importance of a correct pH adjustment for positioning the system peak in the chromatogram.

The pH value of the mobile phase not only affects the dissociation of the organic acid used as an eluant, but also the valency of some analyte ions. This is demonstrated in Fig. 3-109, where the pH dependence of the retention of inorganic anions using a p-hydroxybenzoic acid eluant is displayed. As expected, the retention decreases as pH increases. At higher pH, not only the carboxyl group ($pK_1 = 3.4$), but also the hydroxide group of p-hydroxybenzoic acid ($pK_2 = 9.4$) is dissociated. As is evident from Fig. 3-109, orthophosphate changes its valency at pH 7, as it is converted from the monovalent to the divalent form, which increases retention. Starting at pH 8, divalent eluant ions also exist in the mobile phase, which again decreases the retention of orthophosphate, which is

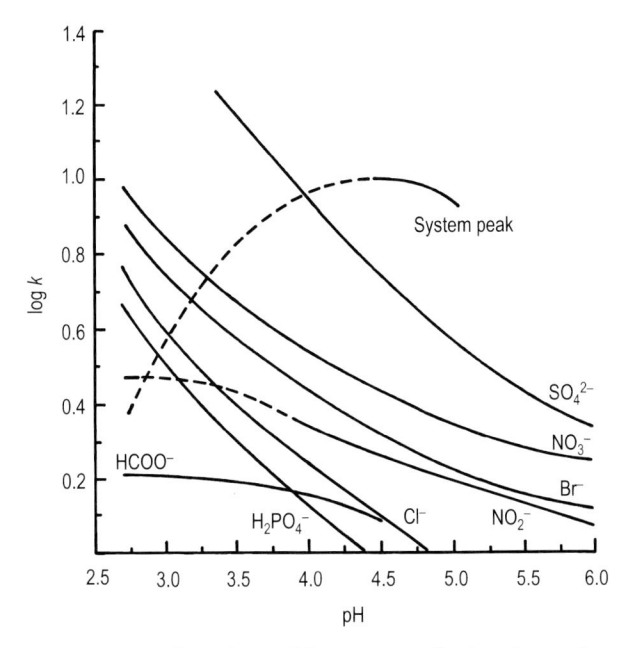

Fig. 3-107. pH dependence of the retention of various inorganic anions using a phthalate eluant (c = 4 mmol/L); (taken from [103]).

predominantly in the divalent form above pH 9. The optimum pH range for the separation of anions specified in DIN 38405 D19 [86] lies between 8.4 and 8.6. Under these chromatographic conditions, orthophosphate is eluted between nitrate and sulfate; thus, the elution order is comparable to that of Fig. 3-28, which was obtained with a carbonate/bicarbonate mixture.

3.7.5
Polarizable Anions

Polarizable inorganic anions comprise iodide, thiocyanate, and thiosulfate, as well as oxygen-containing metal anions such as tungstate, molybdate, and chromate. Due to their large radii such anions exhibit a very high affinity towards the stationary phase of anion exchangers. In the past, polarizable anions were separated on conventional anion exchangers such as IonPac AS1-AS4 with more concentrated carbonate eluants (c = 8 mmol/L Na_2CO_3). Because of the high carbonate concentration, hollow fiber suppressors could not be used to reduce background conductivity. The drawback of conventional packed bed suppressors, however, is that they require periodical regeneration. Moreover, strong tailing effects attributed to adsorption processes prevented determinations in the sub-mg/L range.

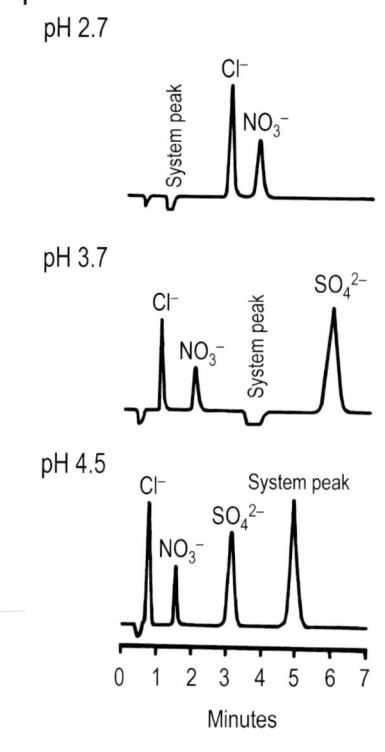

Fig. 3-108. Comparison of the retention of various inorganic anions using a phthalate eluant (c = 4 mmol/L) at three different pH values; (taken from [103]).

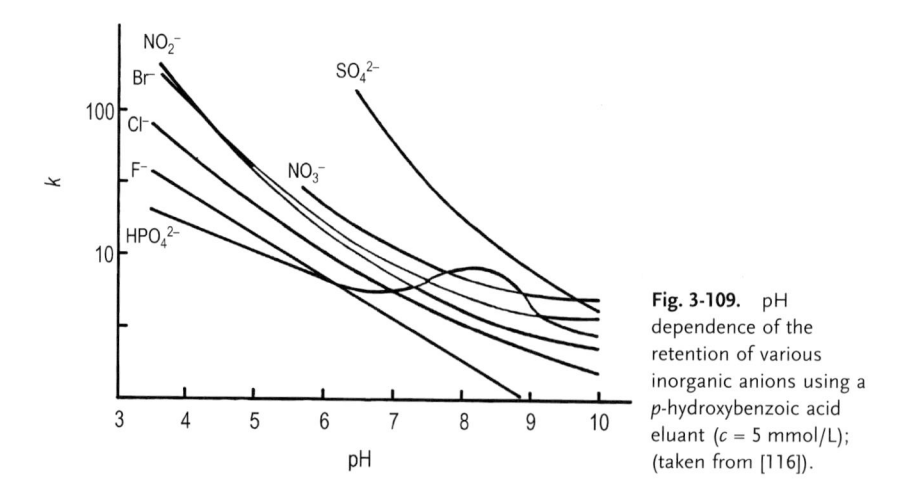

Fig. 3-109. pH dependence of the retention of various inorganic anions using a p-hydroxybenzoic acid eluant (c = 5 mmol/L); (taken from [116]).

The introduction of the IonPac AS5 separator column significantly facilitated the analysis of polarizable anions. The hydrophobicity of the functional groups bonded to the latex beads was lowered, so that polarizable anions can be eluted with a standard mixture of sodium bicarbonate and sodium carbonate. To minimize adsorption effects, some *p*-cyanophenol is added to this mixture. The influence of *p*-cyanophenol on the peak shape is illustrated in Fig. 3-29 (see Section 3.4.2). Due to the compatibility of this eluant with commercial membrane suppressors and the subsequent decrease in dead volume, peak broadening was significantly reduced.

Figures 3-110 and 3-111 show chromatograms of polarizable anions. Again, the concentration of carbonate/bicarbonate is the retention-determining parameter. If this concentration is raised, retention times decrease accordingly. As can be seen from the chromatograms, the retention behavior of individual components in both groups of compounds is comparable.

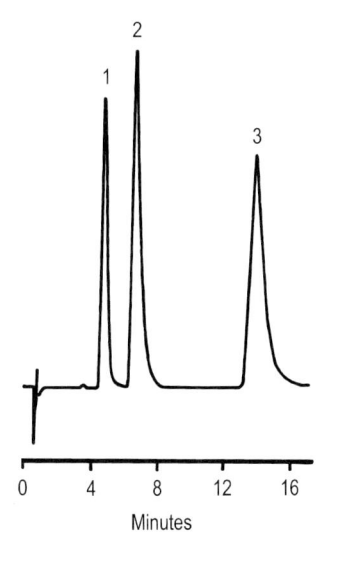

Fig. 3-110. Separation of iodide, thiocyanate, and thiosulfate on IonPac AS5. — Eluant: 3.5 mmol/L $NaHCO_3$ + 2.9 mmol/L Na_2CO_3 + 100 mg/L *p*-cyanophenol; flow rate: 2 mL/min; detection: suppressed conductivity; injection volume: 50 µL; solute concentrations: 20 mg/L each of iodide (1), thiocyanate (2), and thiosulfate (3).

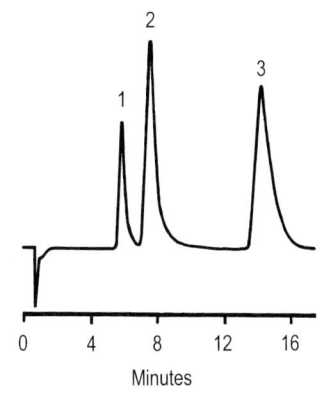

Fig. 3-111. Separation of oxygen-containing metal anions on IonPac AS5. — Chromatographic conditions: see Fig. 3-110; solute concentrations: 40 mg/L each of tungstate (1), molybdate (2), and chromate (3).

When polarizable anions are to be analyzed together with non-polarizable ones, the most suitable stationary phase available today is an acrylate-based anion exchanger such as IonPac AS9-SC. Pure sodium carbonate (c = 3 mmol/L) is used as an eluant. As can be seen from the chromatogram in Fig. 3-32 (see Section 3.4.2), the analysis time for such a separation is not much longer than that of polarizable anions on IonPac AS5. Thiosulfate and chromate cannot be separated under these chromatographic conditions, but they do not co-exist anyway for chemical reasons.

An important field of application for the AS5 column is the analysis of polarizable anions in presence of high amounts of sodium chloride. As impressively illustrated in Fig. 3-112, the separator column, due to its high capacity, can be loaded with a 1% NaCl solution without affecting the separation of polarizable anions that elute afterwards. The application area of this method is limited to concentration differences of no more than four orders of magnitude when electrical conductivity detection is used. Higher concentration differences are difficult to realize chromatographically, even when applying a more selective detection method such as amperometry on an Ag working electrode, since the processes occurring at the electrode surface are significantly disturbed by high amounts of chloride.

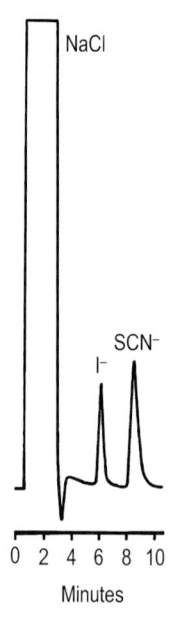

Fig. 3-112. Analysis of iodide and thiocyanate in the presence of high amounts of sodium chloride. − Separator column: IonPac AS5; eluant: 1.7 mmol/L NaHCO$_3$ + 1.8 mmol/L Na$_2$CO$_3$ + 100 mg/L p-cyanophenol; flow rate: 2 mL/min; detection: suppressed conductivity; injection volume: 50 µL; solute concentrations: 10 mg/L iodide and thiocyanate in 1% NaCl.

In combination with *direct* conductivity detection for the analysis of iodide, thiocyanate, and thiosulfate, potassium hydrogen phthalate is a suitable eluant. When silica-based anion exchangers are used, some methanol or 2-propanol is added to the mobile phase to lessen adsorption effects. Figure 3-113 shows such a separation; the elution of monovalent iodide prior to divalent sulfate is a characteristic feature. The relatively long analysis time required for a baseline-resolved separation of all components is notable, and may only be partly compensated for by increasing the flow rate. In comparison to the anion separation obtained on a polymeric phase (shown in Fig. 3-110), it is evident that the higher chromatographic efficiency of the silica-based ion exchanger is not exclusively decisive for optimizing the separation in terms of speed of analysis. Much more important is the selectivity of the stationary phase corresponding to the chemical nature of the analyte species, which is determined by the type of ion-exchange group. For the simultaneous analysis of polarizable and non-polarizable anions using direct conductivity detection, the stationary phase of choice – in analogy to suppressed conductivity detection – is also an acrylate-based anion exchanger. With the Metrosep Anion Dual 1 (see Fig. 3-17 in Section 3.4.1) and a phthalic acid eluant, such separation is achieved in about 20 minutes.

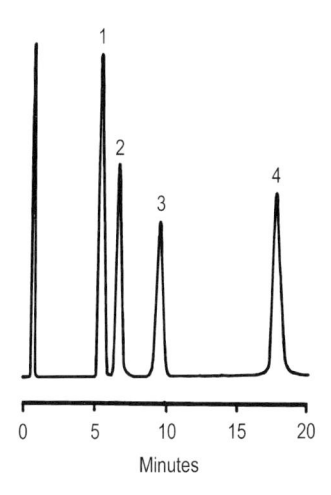

Fig. 3-113. Analysis of iodide, thiocyanate, and thiosulfate on a Vydac 300 IC 405 silica-based ion exchanger. – Eluant: 2 mmol/L KHP – methanol (90:10 v/v), pH 5.0; flow rate: 3 mL/min; detection: direct conductivity; injection volume: 100 µL; solute concentrations: 40 mg/L iodide (1), 20 mg/L sulfate (2), 20 mg/L thiocyanate (3), and 40 mg/L thiosulfate (4).

Anion exchangers such as IonPac AS5 are also suited for the analysis of complexed transition metals. Examples of chromatograms for the separation of metal-EDTA and metal-chloro complexes are displayed in Fig. 3-114 and Fig. 3-115, respectively.

Polyvalent inorganic anions also show a strong affinity towards the stationary phase of anion exchangers. In particular, this includes condensed phosphates such as pyrophosphate, tripolyphosphate, and higher polyphosphates which are of special interest for the detergent industry. Pyrophosphate and tripolyphosphate are separated on a special anion exchanger such as the IonPac AS7. The separation depicted in Fig. 3-116 was obtained using a relatively concentrated nitric acid eluant. The high acid concentration is necessary to restrain the degree of dissociation of the phosphates and, thus, to enable their elution. However, conductivity detection is not possible under these conditions, so detection of such compounds is carried out via post-column derivatization with ferric nitrate

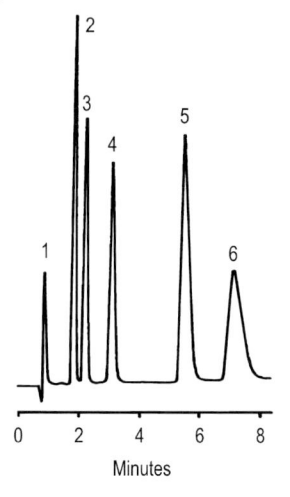

Fig. 3-114. Separation of various metal-EDTA complexes. — Separator column: IonPac AS5; eluant: 2 mmol/L NaHCO$_3$ + 2 mmol/L Na$_2$CO$_3$; flow rate: 2 mL/min; detection: suppressed conductivity; peaks: (1) acetate, (2) nitrate, (3) Pb EDTA^{2-}, (4) Ni EDTA^{2-}, (5) sulfate, (6) Cu EDTA^{2-}.

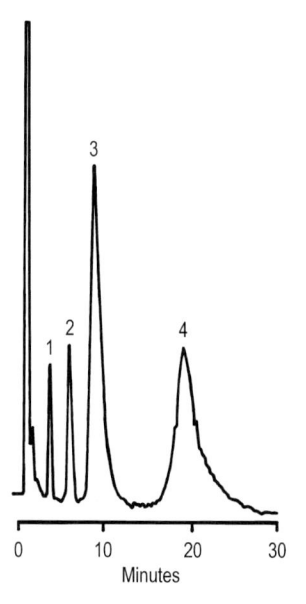

Fig. 3-115. Separation of various metal-chloro complexes. — Separator column: IonPac AS5; eluant: 0.2 mol/L NaClO$_4$ + 0.2 mol/L HCl; flow rate: 2 mL/min; detection: UV (215 nm); injection volume: 50 µL; solute concentrations: 0.2 mg/L Pt (1), 2 mg/L Pb (2), 5 mg/L Au (3), and 5 mg/L Fe (4).

and subsequent UV detection at 330 nm. Other inorganic species such as orthophosphate and sulfate can be determined in the same run, although their determination is much more sensitive in the alkaline pH range using conventional ion exchangers with suppressed conductivity detection.

The analysis of condensed phosphates together with standard inorganic anions using conductivity detection is also only possible in the alkaline pH range. Since the retention behavior of condensed phosphates, due to their polyvalent character, is strongly different from that of standard anions, the application of a gradient elution technique is unavoidable. Especially developed for gradient elution, the Ion Pac AS11 anion exchanger (see Fig. 3-39 in Section 3.4.2),

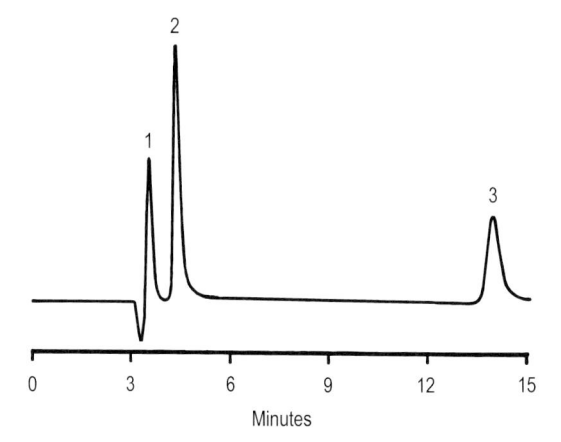

Fig. 3-116. Separation of orthophosphate, pyrophosphate, and tripolyphosphate. — Separator column: IonPac AS7; eluant: 0.07 mol/L HNO_3; flow rate: 0.5 mL/min; detection: photometry at 330 nm after post-column derivatization with ferric nitrate; injection volume: 50 µL; solute concentrations: 100 mg/L orthophosphate (1), 50 mg/L pyrophosphate (2), and 200 mg/L tripolyphosphate (3).

with its high selectivity for hydroxide ions, enables rapid elution of condensed phosphates with a hydroxide eluant. Figure 3-117 shows the separation of chloride and sulfate together with condensed phosphates up to P_4. Higher condensed phosphates can also be analyzed with this method, because modern suppressor systems are compatible with hydroxide concentrations up to 0.15 mol/L. The applicable hydroxide concentration can even be increased by utilizing narrow-bore columns with flow rates <0.5 mL/min, so that polyphosphates up to P_{10} can be analyzed. However, quantitation of those compounds via external calibration with reference substances is problematic, because condensed phosphates $>P_2$ are subject to hydrolysis, even in the crystalline state. Reference substances with the necessary purity are not commercially available. So thus far, this method can only be used to qualitatively analyze higher condensed phosphates. Investigations by Sekiguchi [117] revealed that plotting the slope of the respective calibration graphs of orthophosphate, pyrophosphate, and tripolyphosphate as a function of the number of P atoms yields a straight line (Fig. 3-118), so the slope of the condensed phosphate of interest can be calculated. High purity reference substances, therefore, are no longer needed for calibrating higher condensed phosphates.

Application of the phosphorus-specific detection (see Section 3.8.2) described by Vaeth et al. [118] allows the use of the AS7 column for the analysis of higher condensed phosphates. A mixture of potassium chloride and EDTA [119] is used as an eluant. The potassium chloride concentration determines retention; EDTA is only added for improving peak symmetry. However, post-column derivatization with ferric nitrate, as mentioned above, cannot be applied in this case because the Fe(III) ions of the derivatization reagent form iron-chloro complexes with the chloride ions of the mobile phase, which leads to a high background absorbance. Conductivity detection is also impossible due to the high salt concentration in the mobile phase. The detection system developed by Vaeth et al. comprises a post-column hydrolysis of the polyphosphates to orthophosphate

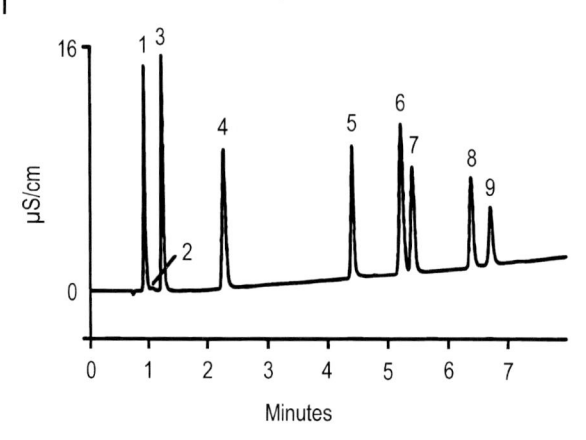

Fig. 3-117. Gradient elution of standard inorganic anions and condensed phosphates utilizing suppressed conductivity detection. − Separator column: IonPac AS11; eluent: (A) water, (B) 0.1 mol/L NaOH; gradient: 20% B linearly to 80% B in 10 min; flow rate: 2 mL/min; detection: suppressed conductivity; injection volume: 25 µL; solute concentrations: 3 mg/L chloride (1), carbonate (2), 5 mg/L sulfate (3), 10 mg/L each of orthophosphate (4), pyrophosphate (5), trimetaphosphate (6), tripolyphosphate (7), tetrapolyphosphate (8), and tetrametaphosphate (9).

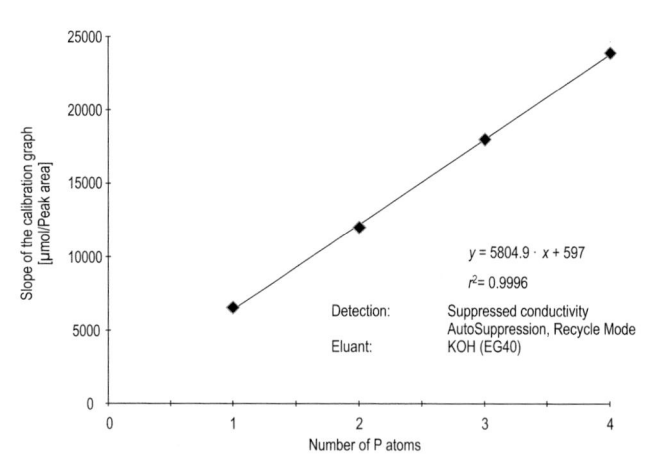

Fig. 3-118. Slopes of the calibration lines of orthophosphate, pyrophosphate, and tripolyphosphate as a function of the number of P atoms; (taken from [117]).

with nitric acid at 110 °C. In a second step, orthophosphate is reacted with a molybdate/vanadate reagent to yield the yellow-colored phosphorvanadomolybdic acid, which can be detected via UV/Vis detection at 410 nm. Hence, calibration can be carried out directly with an orthophosphate standard, because the

3.7 Anion Exchange Chromatography of Inorganic Anions

signal is proportional to the phosphorus content of the eluting compound. Another advantage of this method is the ability to analyze phosphorus species even in extremely complex matrices.

At sufficiently high KCl concentrations, polyphosphates with a molecular weight of up to 10,000 g/mol can be separated. Figure 3-119 shows the separation of a "tetrapolyphosphate" by gradient elution. Tetrapolyphosphate is the major component in this sample; it is also part of the distribution between P_1 and P_{20}. The chromatogram depicted in Fig. 3-120 is obtained under identical chromatographic conditions. It shows the separation of a polyphosphate with a chain length of P_{40}, performed according to the manufacturer's specification. The slight baseline rise during the chromatogram is explained by the elution of the high molecular weight fraction which is not separated under these conditions.

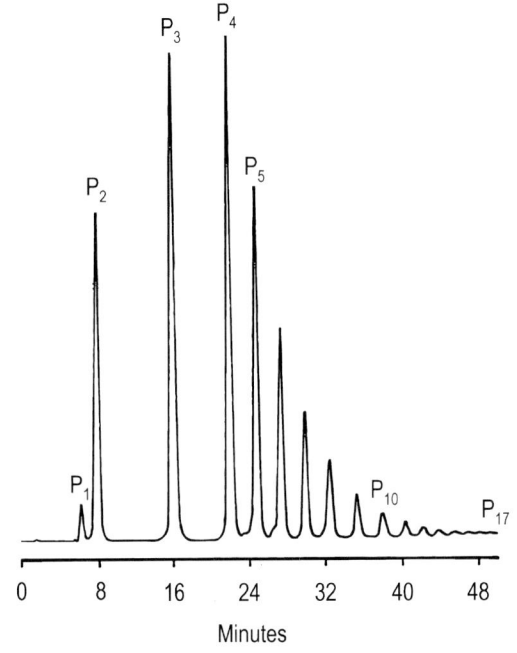

Fig. 3-119. Gradient elution of a tetrapolyphosphate. — Separator column: IonPac AS7; eluant: (A) 0.17 mol/L KCl + 3.2 mmol/L EDTA, pH 5.1, (B) 0.5 mol/L KCl + 3.2 mmol/L EDTA, pH 5.1; gradient: 100% A isocratically for 2 min, then to 30% B in 8 min, then to 100% B in 60 min; flow rate: 0.5 mL/min; detection: phosphorus-specific detection according to Vaeth et al. [118]; injection: 50 µL of a 0.1% solution.

In addition to polyvalent inorganic anions, numerous organic compounds can be analyzed under similar chromatographic conditions in combination with various post-column derivatization procedures. A detailed description of this type of analysis is given in Section 3.8.2.

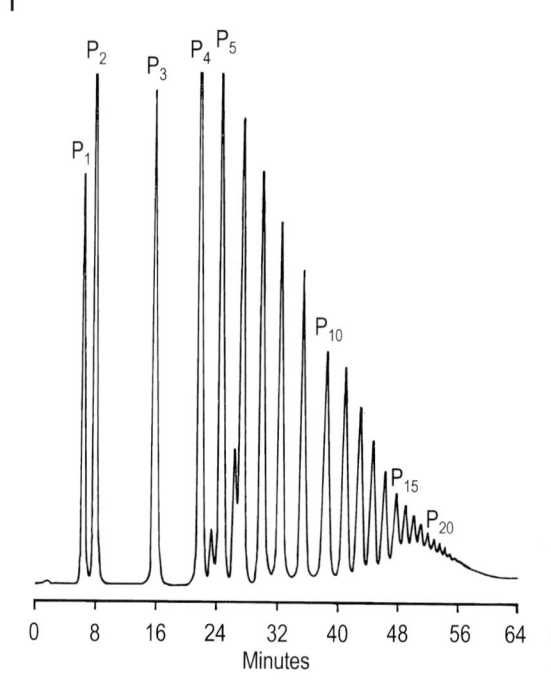

Fig. 3-120. Gradient elution of a polyphosphate. — Chromatographic conditions: see Fig. 3-119; injection: 50 µL of a 0.2% solution.

3.8
Anion Exchange Chromatography of Organic Anions

3.8.1
Organic Acids

Apart from inorganic anions, a number of organic anions can also be separated with conventional anion exchangers. The retention behavior of organic anions is, in part, very similar to that of mineral acids. Interferences are possible, because real-world samples often contain both inorganic and organic anions.

Short-chain aliphatic carboxylic acids, for instance, elute close to the void volume and render the separation and determination of fluoride more difficult. An eluant mixture of 1.7 mmol/L $NaHCO_3$ and 1.8 mmol/L Na_2CO_3 is suitable for separating mineral acids by applying suppressed conductivity detection. However, if this mixture is used, fluoride and formic acid are barely separated on a conventional anion exchanger such as IonPac AS4A (-SC). If the sample also contains acetic acid or a hydroxycarboxylic acid such as glycolic acid or lactic acid, a separation of fluoride is not possible under these chromatographic conditions. As already discussed in Section 3.7.2, this often leads to a misinterpretation of the supposed fluoride signal. To shed light on the kind of species that elutes close to the void volume, a dilute bicarbonate eluant solution is recommended. The low elution strength of bicarbonate makes it possible to separate formic and acetic acid from fluoride even on a conventional anion exchanger. (See Fig. 3-121 for the corresponding chromatogram.) However, this procedure is not suitable for the routine analysis of fluoride in such complex matrices, because even anions such as chloride exhibit extraordinarily long retention times. More strongly retained anions such as nitrate and sulfate interfere with subsequent analyses, so the separator column must be flushed frequently with a strong eluant to remove the anions. In individual cases, this may be circumvented by employing a step gradient. The advantages and disadvantages of using sodium tetraborate eluant to obtain a slightly higher resolution between fluoride and acetate were discussed in Section 3.7.2.

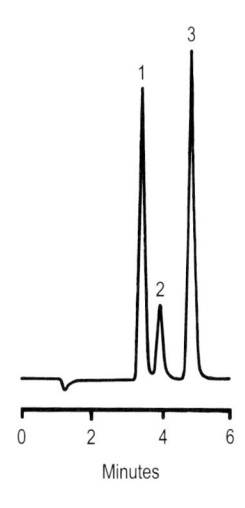

Fig. 3-121. Separation of fluoride, acetate, and formate. — Separator column: IonPac AS4A (-SC); eluant: 1 mmol/L $NaHCO_3$; flow rate: 1.5 mL/min; detection: suppressed conductivity; injection: 50 µL; solute concentrations: 3 mg/L fluoride (1), 20 mg/L acetate (2), and 10 mg/L formate (3).

For the analysis of this class of compounds, changing the stationary phase can be done instead of changing the eluant. A significantly better resolution in the retention area of fluoride is obtained with the IonPac AS12A (see Fig. 3-46 in Section 3.4.2), the ion-exchange groups of which are hydrophilic enough to enable the separation of the strongly hydrated fluoride from the water dip. Although the IonPac AS12A has been designed for simultaneous analysis of fluoride and other mineral acids, an unequivocal identification of fluoride — especially in samples with a complex matrix — is often impossible under isocratic conditions. An even higher retention of fluoride is obtained with the CarboPac PA1 (IonPac AS6), initially developed for carbohydrate analysis. If this separator column is conditioned with the carbonate/bicarbonate buffer mentioned above, even compounds exhibiting a low affinity to the stationary phase are strongly retained due to the comparatively high capacity of this stationary phase.

Thus, the retention difference between fluoride and chloride, which is less than 2 minutes under the same conditions on an AS4A column, may be expanded to more than 20 minutes with the CarboPac PA1. Figure 3-122 illustrates an example of the baseline-resolved separation of various monocarboxylic acids, which would elute as one peak close to the void volume when using an AS4A column. Monocarboxylic acids are relevant in many of today's analysis problems; this procedure provides an elegant solution. Amidosulfonic acid, for example, is one of the many components in flue gas scrubber solutions from desulfurization devices, which require frequent analysis.

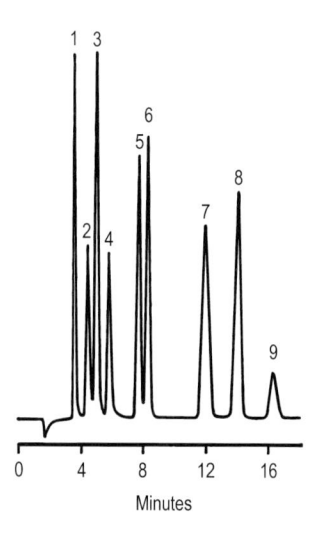

Fig. 3-122. Separation of various monocarboxylic acids on a CarboPac PA1. – Eluant: 1.7 mmol/L $NaHCO_3$ + 1.8 mmol/L Na_2CO_3; flow rate: 1 mL/min; detection: suppressed conductivity; injection: 50 µL; solute concentrations: 3 mg/L fluoride (1), 40 mg/L acetic acid (2), 20 mg/L glycolic acid (3), 10 mg/L α-hydroxyisocaproic acid (4), 20 mg/L each of formic acid (5), oxamic acid (6), methanesulfonic acid (7), amidosulfonic acid (8), and α-ketoisocaproic acid (9).

While acetic acid almost co-elutes with fluoride under chromatographic conditions suitable for analyzing mineral acids, haloacetic acids are more strongly retained due to their higher acid strength [120]. Figure 3-123 shows the separation of nine different haloacetic acids, which theoretically could occur at trace level in tap water that is disinfected with ozone or chlorine dioxide. The chromatogram in Fig. 3-123 was obtained utilizing gradient elution with a hydroxide/methanol eluant on the IonPac AS11 anion exchanger. Methanol was added to the mobile phase to improve the peak efficiency of the late-eluting dibromochloroacetic acid and tribromoacetic acid. The conventional method for determining haloacetic acids includes GC/ECD analysis after solvent extraction and derivatization with diazomethane [121]. Although detection limits in the lowest µg/L range can be achieved with this method, the necessary sample preparation is time-consuming and affects the relative standard deviation of the whole procedure in a negative way. It is significantly easier to determine haloacetic acids by anion exchange chromatography with suppressed conductivity detection, because sample preparation is unnecessary.

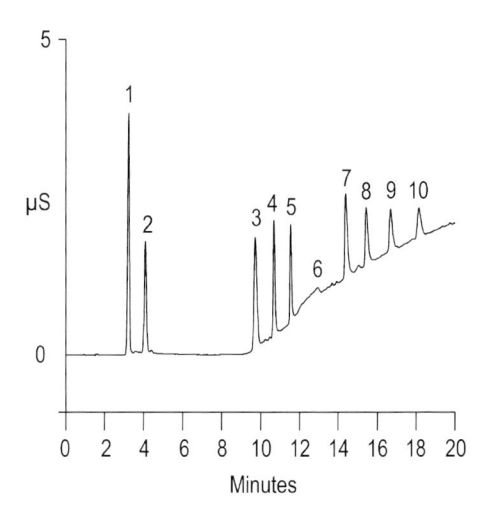

Fig. 3-123. Gradient elution of haloacetic acids. Separator column: IonPac AS11 (2-mm); eluent: (A) 0.1 mol/L NaOH, (B) 1 mmol/L NaOH, (C) methanol/water (80:20 v/v); gradient: 95% B + 5% C isocratically for 13 min, then linearly to 45% A + 15% B + 40% C in 16 min; flow rate: 0.5 mL/min; detection: suppressed conductivity; injection volume: 6.25 µL; solute concentrations: 10 mg/L each of monochloroacetic acid (1), monobromoacetic acid (2), dichloroacetic acid (3), monobromomonochloroacetic acid (4), dibromoacetic acid (5), carbonate (6), trichloroacetic acid (7), dichloromonobromoacetic acid (8), dibromomonochloroacetic acid (9), and tribromoacetic acid (10).

Compared to aliphatic monocarboxylic acids, aromatic monocarboxylic acids exhibit a slightly higher affinity towards the stationary phase. The simplest congener, benzoic acid, can be eluted together with nitrite on some stationary phases. Depending on the separator column, the interpretation of the nitrite peak may become a difficult task. Often, peak shape is the only distinguishing feature. While the nitrite peak is completely symmetrical, the benzoic acid peak shows a slight tailing, which is attributed to adsorption on the stationary phase due to the aromatic backbone of benzoic acid. A separation of benzoic acid and nitrite is possible, for example, on the IonPac AS4A (-SC) separator column under standard conditions, where benzoic acid elutes immediately after nitrite. An alternative to ion-exchange chromatography is ion-pair chromatography. Ion-pair chromatography employs a non-polar stationary phase at which both compounds show a markedly different retention behavior.

Under standard conditions (2.8 mmol/L $NaHCO_3$ + 2.2 mol/L Na_2CO_3), benzoate exhibits only a small retention on an IonPac AS4 separator column. Therefore, investigations were performed to determine to what extent other monocarboxylic acids may be separated using anion exchange chromatography. Figure 3-124 shows a separation of mandelic acid, hippuric acid, and methyl-substituted hippuric acid, which can only be obtained by employing a modified eluant and two separator columns in series. The separation between o-methylhippuric acid and p-methylhippuric acid is remarkably good. It is obvious from Table 3-22 that the elution order may be correlated with the pK values of the compounds.

The following elution order is generally observed: aliphatic monocarboxylic acids, followed by aromatic monocarboxylic acids, followed by aliphatic dicarboxylic acids. In Fig. 3-125, the retention behavior of aliphatic dicarboxylic acids is compared to that of inorganic anions. One can see from this chromatogram that the retention of compounds such as succinic acid, malonic acid, maleic acid, and tartaric acid is very similar to that of nitrate and sulfate. Sufficient separation of all these compounds is also, therefore, achieved only by using two AS4 columns in series. However, the separation of the two stereo isomers, maleic acid and fumaric acid, is much easier. It can be obtained under standard conditions and is shown in Fig. 3-126. In contrast to monocarboxylic acids, retention of aliphatic dicarboxylic acids increases with decreasing pK value. The corresponding data are summarized in Table 3-23.

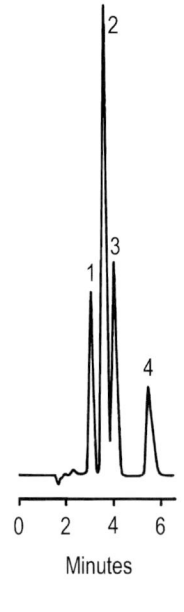

Fig. 3-124. Separation of various aromatic monocarboxylic acids. — Separator columns: 2 IonPac AS4; eluant: 0.5 mmol/L Na_2CO_3 + 0.5 mmol/L NaOH; flow rate: 2 L/min; detection: suppressed conductivity; injection: 50 µL; peaks: (1) o-methylhippuric acid, (2) mandelic acid, (3) hippuric acid, and (4) p-methylhippuric acid.

Table 3-22. Elution order and pK values [122] of some aromatic monocarboxylic acids.

Elution order	pK value
Mandelic acid	3.36
Hippuric acid	3.64
Benzoic acid	4.20

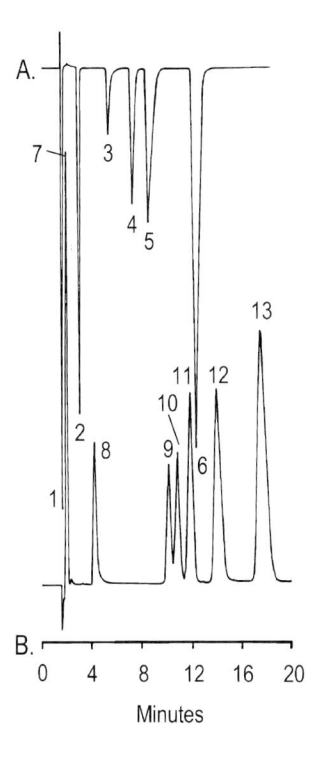

3.8 Anion Exchange Chromatography of Organic Anions

Fig. 3-125. Comparison between the retention behavior of inorganic anions and several carboxylic acids. — Separator columns: 2 IonPac AS4; eluant: 2.8 mmol/L $NaHCO_3$ + 2.2 mmol/L Na_2CO_3; flow rate: 1.6 mL/min; detection: suppressed conductivity; injection: 50 µL; solute concentrations: A) 1.5 mg/L fluoride (1), 2 mg/L chloride (2), 5 mg/L each of orthophosphate (3) and bromide (4), 10 mg/L nitrate (5), and 12.5 mg/L sulfate (6), B) 5 mg/L formic acid (7), 40 mg/L benzoic acid (8), 20 mg/L succinic acid (9), 10 mg/L malonic acid (10), 20 mg/L each of maleic acid (11), tartaric acid (12), and oxalic acid (13).

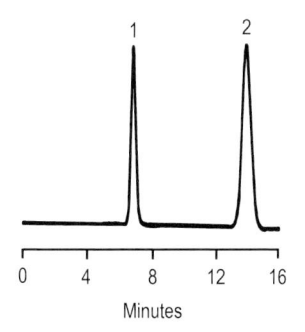

Fig. 3-126. Separation of maleic acid and fumaric acid. — Separator column: IonPac AS4A (-SC); eluant: 1.7 mmol/L $NaHCO_3$ + 1.8 mmol/L Na_2CO_3; flow rate: 2.0 mL/min; detection: UV (210 nm); injection: 50 µL; solute concentrations: 2 mg/L maleic acid (1) and 5 mg/L fumaric acid (2).

Table 3-23. Elution order and pK values [122] of some dicarboxylic acids

Elution order	pK value	
	pK_1	pK_2
Succinic acid	4.21	5.64
Malonic acid	2.88	5.68
Oxalic acid	1.27	4.29

This finding is explained by the charge-stabilizing effect exerted by the +I-effect of the methylene groups which decreases from succinic acid to oxalic acid:

+ I-effect

$$\text{}^{\ominus}\text{O-C(=O)} \longleftarrow CH_2-CH_2 \longrightarrow \text{C(=O)-O}^{\ominus}$$

$$\text{}^{\ominus}\text{O-C(=O)} \longleftarrow CH_2 \longrightarrow \text{C(=O)-O}^{\ominus}$$

$$\text{}^{\ominus}\text{O-C(=O)} — \text{C(=O)-O}^{\ominus}$$

The presence of additional hydroxide groups in dicarboxylic acids such as malic acid and tartaric acid increases the acidic character and, thus, the retention (Table 3-24). A corresponding chromatogram is shown in Fig. 3-127. However, a separation of malic acid (monohydroxysuccinic acid) and malonic acid is not possible under these conditions, while tartronic acid (hydroxymalonic acid) exhibits significantly higher retention.

Table 3-24. Comparison of elution order and pK values [122] between hydroxy-substituted and unsubstituted aliphatic dicarboxylic acids.

Elution order	pK value	
	pK_1	pK_2
Succinic acid	4.21	5.64
Malic acid	3.40	5.10
Tartaric acid	3.04	4.37

Aliphatic tricarboxylic acids such as citric acid exhibit a remarkably high affinity towards the stationary phase of an anion exchanger. Hence, low ionic strength bicarbonate/carbonate buffer solutions are not particularly suitable as eluants. However, when a sodium hydroxide solution at a comparatively high concentration ($c \sim 0.08$ mol/L) is used, citric acid may be eluted, and may even be separated from its structural isomer, isocitric acid. When the detection of these compounds is carried out via electrical conductivity, a suppressor system must be used to reduce background conductivity. For capacity reasons, only micromembrane (MMS) and self-regenerating suppressors (SRS) are applicable. Alternatively, a different stationary phase may be used for the separation of organic acids with high affinities. The IonPac AS5 separator column (see Section 3.7.4) has also proved to be suitable for this application problem. This is exemplified in Fig. 3-128 with 2,6-dihydroxy-isonicotinic acid (citrazinic acid), which can be eluted with a simple carbonate/bicarbonate buffer by adding some p-cyanophenol. A separation of this acid from thiocyanate, which exhibits a similar retention behavior, poses no problem under these conditions.

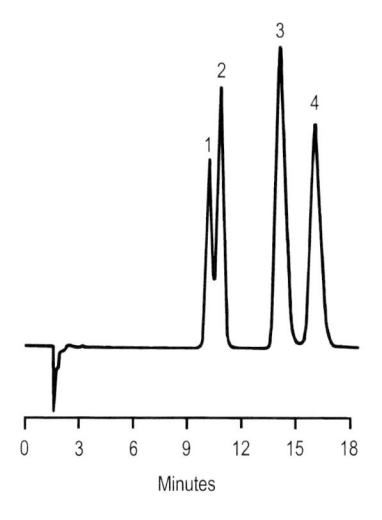

Fig. 3-127. Separation of aliphatic hydroxycarboxylic acids. – Chromatographic conditions: see Fig. 3-125; solute concentrations: 20 mg/L each of succinic acid (1), malic acid (2), tartaric acid (3), and tartronic acid (4).

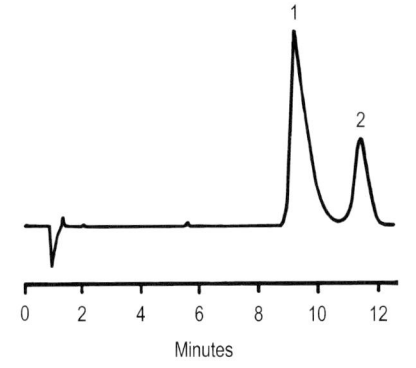

Fig. 3-128. Separation of citrazinic acid. – Separator column: IonPac AS5; eluant: 2.8 mmol/L $NaHCO_3$ + 2.2 mmol/L Na_2CO_3 + 100 mg/L p-cyanophenol; flow rate: 2 mL/min; detection: suppressed conductivity; injection volume: 50 µL; solute concentrations: 20 mg/L each of thiocyanate (1) and citrazinic acid (2).

An interesting application area for anion exchange chromatography is the analysis of herbicides based on phenoxycarboxylic acids and their derivatives. Generally, these compounds were separated at chemically modified silicas using a methanol/water mixture as an eluant to which acetic acid was added to suppress analyte dissociation. The list of herbicides that may be analyzed with anion exchange chromatography comprises:

- 2,4 D
  2,4-dichlorophenoxyacetic acid
- 2,4,5 T
  2,4,5-trichlorophenoxyacetic acid
- Silvex
  2-(2,4,5-trichlorophenoxy)propionic acid
- Mecoprop
  2-(2-methyl-4-chlorophenoxy)propionic acid

These compounds are also strongly retained due to their aromatic skeleton and the resulting π-π interactions with the stationary phase of the anion exchanger. Since they can be detected sensitively by measuring the UV absorption, strong eluants such as sodium nitrate may be employed. The resulting chromatogram is displayed in Fig. 3-129. For the analysis of herbicides lacking a chromophore, such as endothall, 7-oxabicyclo[2,2,1]-heptane-2,3-dicarboxylic acid, anion exchange chromatography with subsequent conductivity detection provides an alternative to RPLC.

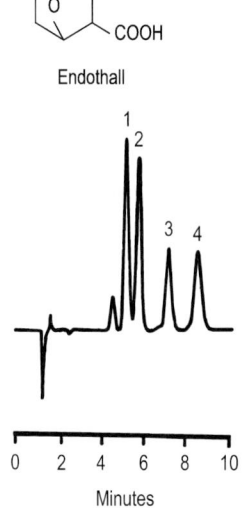

Fig. 3-129. Separation of phenoxycarboxylic acid-based herbicides. – Separator column: IonPac AS4A (-SC); eluant: 25 mmol/L NaNO$_3$ + 5 mmol/L NaOH; flow rate: 2 mL/min; detection: UV (280 nm); injection volume: 50 µL; peaks: (1) 2,4 D, (2) Mecoprop, (3) Silvex, and (4) 2,4,5 T.

Endothall can be analyzed together with standard inorganic anions under the chromatographic conditions given in Fig. 3-28. It elutes between nitrate and orthophosphate.

Conventional anion exchangers are also suited for the separation of inorganic and organic phosphates and sulfates. As a rule, organic species are eluted prior to inorganic ones. While long-chain alkyl phosphates, alkyl sulfates, and alkylsulfonates ($n_c > 4$) can only be separated using ion-pair chromatography, their short-chain homologues exhibit a significant retention. For the separation of monobutyl-, dibutyl-, and ortho-phosphate, a mixture of sodium carbonate and sodium hydroxide is suitable as an eluant (see Fig. 3-130). Increasing the pH value with sodium hydroxide is required to improve the separation between inorganic and organic phosphates.

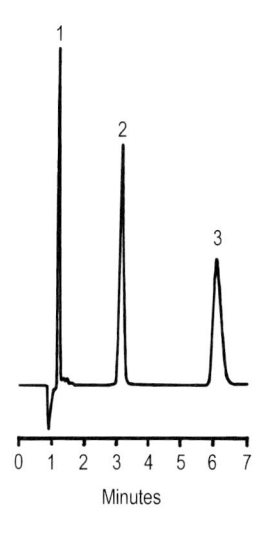

Fig. 3-130. Separation of various alkylphosphates and inorganic phosphates. – Separator column: IonPac AS4A (-SC); eluant: 3 mmol/L Na_2CO_3 + 1 mmol/L NaOH; flow rate: 2 mL/min; detection: suppressed conductivity; injection volume: 50 µL; solute concentrations: 60 mg/L dibutylphosphate (1, technical grade), monobutylphosphate (2, as impurity), and 20 mg/L orthophosphate (3).

In addition to alkyl phosphates, biologically relevant phosphates such as glycerophosphates can be separated on the high-performance separator column Ionpac AS4A (-SC). Glycerophosphates play a significant role in fat metabolism by promoting the esterification of fatty acids, which are then stored as glycerides. The separation of α- and β-glycerophosphates shown in Fig. 3-131 is another example of the high efficiency of modern anion exchangers.

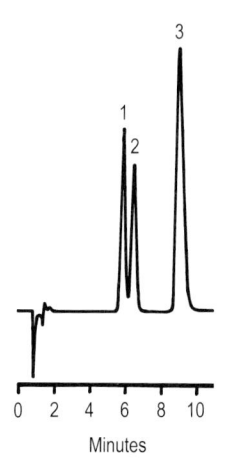

Fig. 3-131. Separation of α- and β-glycerophosphate. – Separator column: IonPac AS4A (-SC); eluant: 0.85 mmol/L $NaHCO_3$ + 0.9 mmol/L Na_2CO_3; flow rate: 2 mL/min; detection: suppressed conductivity; injection volume: 50 µL; solute concentrations: 20 mg/L each of α-glycerophosphate (1), β-glycerophosphate (2), and orthophosphate (3).

Essential components of ribonucleic acid and deoxyribonucleic acid (RNA and DNA) such as nucleosides and nucleotides can also be analyzed with anion exchange chromatography.

Nucleosides consist of a purine or pyrimidine base, to which a sugar component is attached via a βN-glycosidic linkage at the 1′-position. Depending on the type of sugar component, D-ribose or 2′-deoxy-D-ribose, one distinguishes between ribonucleosides and deoxyribonucleosides. Since nucleosides as weak acids exist in an anionic form at basic pH, they can be eluted with a low ionic strength bicarbonate/carbonate buffer. The determination of the UV absorption at 254 nm is suitable as a detection method based on the conjugated double bonds in the purine and pyrimidine units. For the separation of the most important nucleosides illustrated in Fig. 3-132 – the purine derivatives adenosine, guanosine, and inosine, as well as the pyrimidine derivatives cytidine, uridine, and thymidine – two separator columns were used in series in order to obtain the required separation efficiency in the shortest possible analysis time. The comparison between the retention behavior of 2′-deoxyguanosine and guanosine reveals that DNA units elute prior to RNA units that have the same carbon skeleton.

In addition to the sugar component, at the 5′-position of the ribose, nucleotides contain a mono-, di-, or tri-phosphate group.

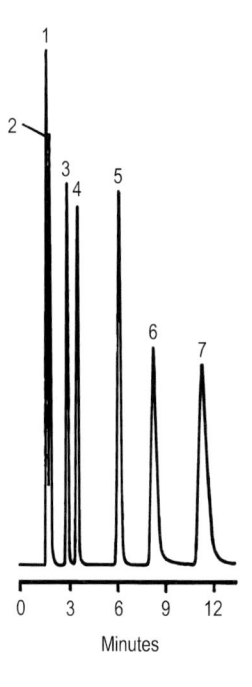

Fig. 3-132. Separation of nucleosides using anion exchange chromatography. – Separator columns: 2 IonPac AS4A (-SC); eluant: see Fig. 3-130; flow rate: 1.5 mL/min; detection: UV (254 nm); injection volume: 50 µL; solute concentrations: 10 mg/L cytidine (1), 5 mg/L adenosine (2), 10 mg/L each of thymidine (3) and uridine (4), 13 mg/L 2′-deoxyguanosine (5), 15 mg/L each of guanosine (6), and inosine (7).

```
   γ      β      α
   OH     OH     OH
   |      |      |
HO—P—O—P—O—P—O—H₂C   [Base]
   ||     ||     ||       O
   O      O      O      H   H
                        H     H
                        OH  OH
```

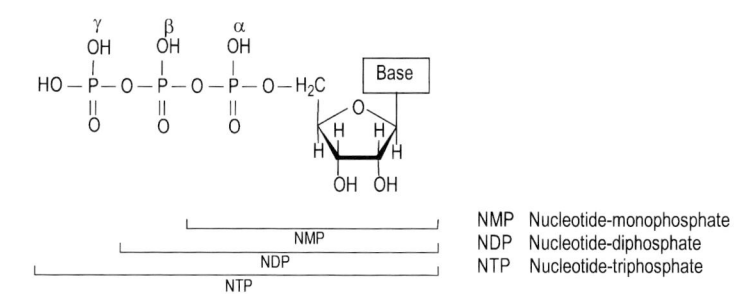

```
                    |_____|  NMP  Nucleotide-monophosphate
                    |     NMP        |  NDP  Nucleotide-diphosphate
              |_____NDP_____|  NTP  Nucleotide-triphosphate
        |_____NTP_____|
```

These represent the actual monomeric units of the nucleic acids where the phosphoric acid residues are always esterified at the 3'- and 5'-position with the hydroxide group of the pentose of the neighboring nucleotide. Generally, nucleotides are strong acids with a high affinity towards the stationary phase of anion exchangers. A phosphate buffer is typically used for the separation of nucleotides. The pH value of the mobile phase is the retention-determining parameter, because it determines both the charge at the purine and pyrimidine bases and the degree of dissociation of the phosphate groups. Figure 3-133 shows a chromatogram with the separation of various nucleotide monophosphates. The phosphate buffer concentration has to be increased for the elution of the di- and triphosphates. For the single run determination of the nucleotide monophosphates, -diphosphates, and -triphosphates, application of a gradient elution technique is inevitable. Such a chromatogram is shown in Fig. 3-134. The individual determination of ATP, ADP, and AMP with this method is of great practical significance in view of the dominant position of ATP (adenosine triphosphate) in the energy balance of cells.

Nucleosides and nucleotides can also be analyzed *simultaneously*. Lamb et al. [123], for example, used a neutral divinylbenzene polymer (IonPac NS1, Dionex Corporation), which they coated with an *n*-decyl-substituted bicyclic polyether such as 1,10-diaza-4,7,13,16,21,24-hexaoxabicyclo[8.8.8]hexacosane (D2.2.2). This coating results in a chromatographic matrix, on which both ionic and hydrophobic interactions can occur. So-called cryptands [124] (like their monocyclic analogues, bicyclic polyethers) form stable complexes with alkali metals. These metals are included in the cavity of the ring system; their presence results in weak covalent bonds between the cation and the two N atoms. Hence, when adding alkali metals to the mobile phase, these cations are dynamically bonded under the formation of positively charged ion-exchange centers. Thus, the stationary phase turns into a cation exchanger. Hydrophobic interactions are possible both on the support material itself and on the *n*-decyl residue of the polyether. While nucleosides can be eluted with pure de-ionized water on such a crown ether phase, the mobile phase must contain an alkali salt to elute nucleotides. Due to the high affinity of D2.2.2 to potassium, nucleotide retention is maximized with a potassium salt eluant. However, retention decreases with increasing salt concentration. The simultaneous separation of both classes of compounds can thus

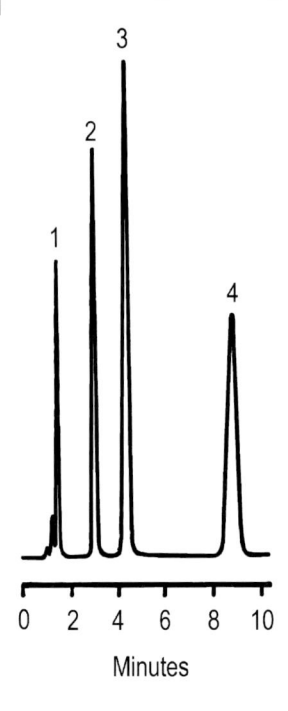

Fig. 3-133. Separation of nucleotide monophosphates. — Separator column: IonPac AS4A (-SC); eluant: 15 mmol/L NaH_2PO_4, pH 3.4 with H_3PO_4; flow rate: 1.5 mL/min; detection: UV (254 nm); injection volume: 50 µL; solute concentrations: 10 mg/L each of CMP (1), UMP (2), AMP (3), and GMP (4).

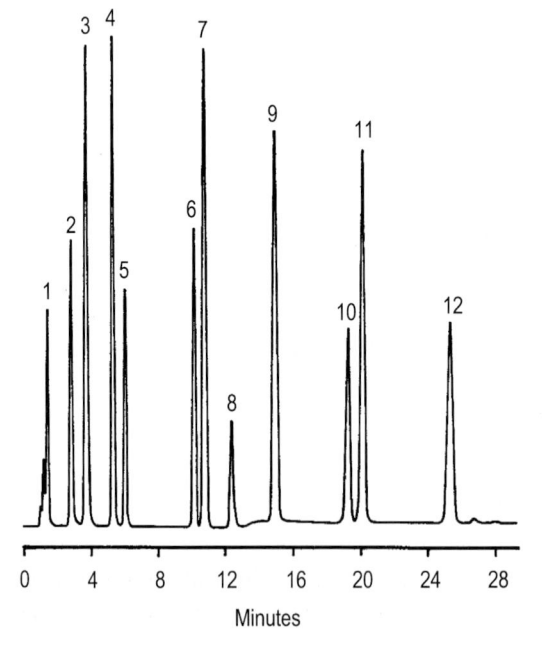

Fig. 3-134. Gradient elution of nucleotide monophosphates, diphosphates, and triphosphates. — Separator column: IonPac AS4A (-SC); eluant: (A) water, (B) 0.5 mol/L NaH_2PO_4, pH 3.4 with H_3PO_4; gradient: linear, 3% B to 100% B in 40 min; flow rate: 1.5 mL/min; detection: UV (254 nm); injection volume: 50 µL; solute concentrations: 10 mg/L each of CMP (1), UMP (2), AMP (3), and GMP (4), 20 mg/L each of CDP (5), UDP (6), ADP (7), GDP (8), CTP (9), UTP (10), ATP (11), and GTP (12).

be achieved by employing gradient elution with a stepwise increase of the salt concentration to elute strongly retained nucleotide diphosphates and -triphosphates as symmetrical peaks. Figure. 3-135 illustrates this with the separation of 15 different nucleosides and nucleotides with a KCl gradient. Retention can also be shortened by adding organic solvents, which assumes that hydrophobic interactions contribute to the retention mechanism. As expected, retention of nucleosides decreases more strongly than that of nucleotides when organic solvents are added; this can be explained by the different valencies.

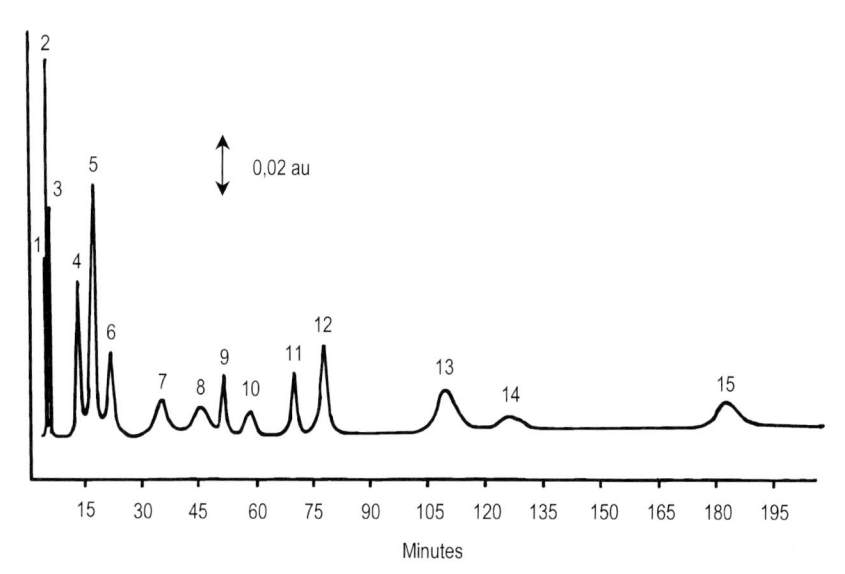

Fig. 3-135. Gradient elution of various nucleosides and nucleotides on a DVB polymer coated with D2.2.2. – Eluant: (A) water, (B) 0.1 mol/L KCl; gradient: 100% A isocratically for 5 min, then linearly to 5% B in 15 min, then linearly to 100% B in 130 min; flow rate: 1 mL/min; detection: UV (254 nm); injection volume: 20 µL; solute concentrations: 10 µmol/L each of cytidine (1), deoxycytidine (2), thymidine (3), cytidine-5'-monophosphate (4), deoxycytidine-5'-monophosphate (5), thymidine-5'-monophosphate (6), guanosine (7), adenosine/deoxyguanosine (8), uridine-5'-monophosphate (9), deoxyadenosine (10), inosine-5'-monophosphate (11), guanosine-5'-monophosphate (12), adenosine-5'-monophosphate (13), deoxyguanosine-5'-monophosphate (14), and deoxyadenosine-5'-monophosphate (15); (taken from [123]).

3.8.2
Polyvalent Anions

The degree of difficulty involved in the ion chromatographic separation and determination of polyvalent anions (e.g., aminopolycarboxylic acids and polyphosphonic acids) made this task virtually impossible in the past. The primary reason is that the characteristically high valencies of these compounds causes them to exhibit a very high affinity towards the stationary phase of anion exchangers.

The structure and identity of such compounds, which are of practical relevance as complexing agents, may be clearly elucidated by both one-dimensional and two-dimensional nuclear magnetic resonance spectroscopy of the isotopes H-1, C-13, and P-31. Sufficiently high concentrations make it possible to quantitatively analyze such polyvalent anions [125-129]. However, because of the low sensitivity, especially of the phosphorus nucleus, detection limitations are problematic for practical applications.

Preliminary attempts at a chromatographic separation and identification of acidic organophosphates using an autoanalyzer system were presented by Waldhoff and Sladek [130], but typical retention times were in the range of hours. Only with the introduction of a special anion exchanger, the IonPac AS7 (see Section 3.4.2), did it become possible to separate and determine a variety of polyvalent anions [131-133]. When 0.03 to 0.05 mol/L nitric acid is chosen as an eluant, the degree of analyte dissociation is pushed back via partial protonation, which is caused by the high acid strength of the eluant [134]. Thus, the effective charge of the analytes is reduced, and their elution is possible. In addition, the choice of nitric acid is based on the known high elution strength of the nitrate ion. The nitric acid concentration in the mobile phase is the only experimental parameter that determines retention.

Conductivity detection of the analytes is not possible owing to the high acid concentration in the mobile phase. Thus, complexing agents are detected via post-column derivatization with ferric nitrate in acidic solution, and subsequent measurement of the UV absorption in the wavelength region between 310 nm and 330 nm [135]. The reagent is added from a pressurized vessel to ensure a pulsation-free flow. The reaction with the reagent takes place in a reaction loop of appropriate dimension that is packed with chemically inert polymer beads. Alternatively, a knitted reaction coil can be used. In both cases, peak broadening effects are reduced and mixing is optimized.

The classes of compounds that may be analyzed with this procedure include complexing agents derived from the aminopolycarboxylic acids; their separation is shown in Fig. 3-136.

$$\text{HOOC}-\text{H}_2\text{C} \diagdown \text{N} - \left[-\text{CH}_2-\text{CH}_2-\text{N}(\text{CH}_2-\text{COOH})- \right]_n -\text{CH}_2-\text{COOH}$$
$$\text{HOOC}-\text{H}_2\text{C} \diagup$$

$n = 0$: NTA (Trilon A)
$n = 1$: EDTA (Trilon B)
$n = 2$: DTPA (Trilon C)

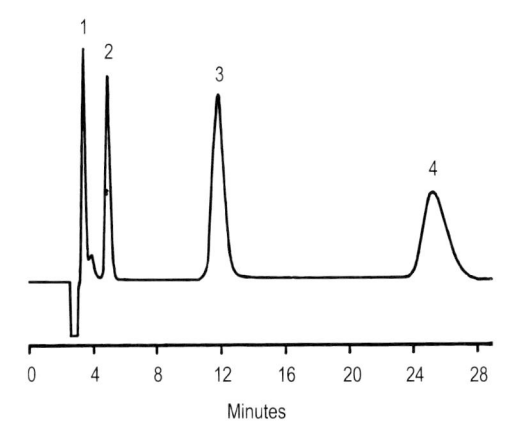

3.8 Anion Exchange Chromatography of Organic Anions | 183

Fig. 3-136. Separation of aminopolycarboxylic acids NTA, EDTA, HEDTA, and DTPA. – Separator column: IonPac AS7; eluant: 0.03 mol/L HNO_3; flow rate: 0.5 mL/min; detection: photometry at 310 nm after reaction with ferric nitrate; injection volume: 50 µL; solute concentrations: 20 mg/L each of NTA (1) and EDTA (2), 60 mg/L HEDTA (3), and 100 mg/L DTPA (4).

As can be seen from this chromatogram, retention of these compounds increases with the increasing number of (-CH_2-CH_2-N-CH_2-COOH)-units. The same chromatographic conditions allow the analysis of the hydroxyethyl-ethylenediaminetriacetic acid (HEDTA, Trilon D), a compound analogous to EDTA.

$$\begin{array}{c} HOOC-H_2C \\ \diagdown \\ HOOC-H_2C \end{array} N-CH_2-CH_2-N \begin{array}{c} \diagup CH_2-CH_2-OH \\ \\ \diagdown CH_2-COOH \end{array}$$

The degradation products from EDTA such as ethylenediaminetriacetic acid and ethylenediaminediacetic acid cannot be determined with this method, because they do not react with ferric nitrate. The analytical procedure for the determination of these compounds is discussed in Section 6.6.

Figure 3-137 shows the separation of some aminopolyphosphonic acids of the DEQUEST type that can also be eluted with nitric acid (c = 0.03 mol/L).

$$\begin{array}{c} H_2O_3P-H_2C \\ \diagdown \\ H_2O_3P-H_2C \end{array} N \left[-CH_2-CH_2-\underset{\underset{CH_2-PO_3H_2}{|}}{N}- \right]_n CH_2-PO_3H_2$$

n = 0: NTP (DEQUEST 2000)
n = 1: EDTP (DEQUEST 2041)
n = 2: DTPP (DEQUEST 2060)

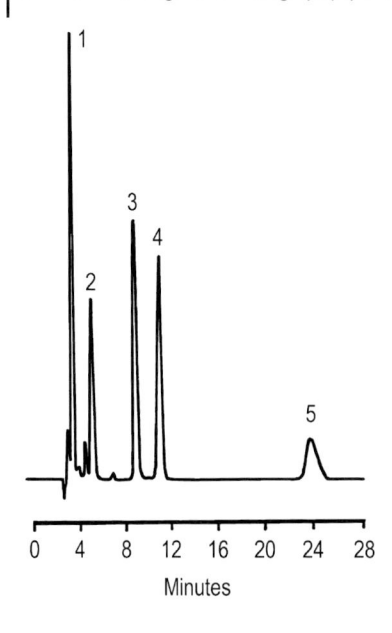

Fig. 3-137. Separation of aminopolyphosphonic acids of the DEQUEST type. – Separator column: IonPac AS7; chromatographic conditions: see Fig. 3-135; injection volume: 50 μL; solute concentrations: 50 mg/L each of DEQUEST 2010 (1), DEQUEST 2051 (2), DEQUEST 2000 (3), DEQUEST 2041 (4), and DEQUEST 2060 (5).

The analogous compound to EDTP, hexamethylenediamine-tetramethylenephosphonic acid (DEQUEST 2051) can be analyzed in the same chromatographic run.

$$\begin{array}{c} H_2O_3P-H_2C \\ \diagdown \\ N-(CH_2)_6-N \\ \diagup \\ H_2O_3P-H_2C \end{array} \begin{array}{c} \diagup CH_2-PO_3H_2 \\ \\ \diagdown CH_2-PO_3H_2 \end{array}$$

Aminopolyphosphonic acids have increasingly been included in detergent formulations. In the peroxide bleach, which is primarily used in Europe, these formulations contain sodium perborate, which hydrolyzes in water to form hydrogen peroxide. To obtain a good bleaching efficiency at temperatures below 60 °C, a bleaching activator such as tetraacetyl-ethylenediamine (TAED) is added. At pH values between 9 and 12 this compound forms peracetic acid in the presence of hydrogen peroxide. In the presence of these oxidants, aminopolyphosphonic acids are oxidized to stable N-oxides. First indications of the chromatographic analysis of such compounds are found in the paper by Vaeth et al. [118], which describes the separation of ethylenediamine-tetramethylenephosphonic acid (EDTP) from its respective di-N-oxide by anion exchange chromatography. The identity of the mono-N-oxide, postulated at that time as an intermediate in the oxidation reaction, has since been confirmed. This effect is illustrated below, taking diethylenetriamine-pentamethylenephosphonic acid (DTPP) as an example. When a 30% hydrogen peroxide solution is added to an aqueous DTPP solution and the mixture is then heated to about 90 °C for a certain time, DTPP is quantitatively converted into the respective di-N-oxide. Compared to the starting material, the di-N-oxide is much more strongly retained by the stationary phase. The resulting two chromatograms are opposite each other in Fig. 3-138 (far left and right sides). If the reaction solution is only heated to 60 °C for a short time, the mono-N-oxide is obtained. The middle chromatogram in Fig. 3-138 reveals that this intermediate can be separated from both the starting material and the di-N-oxide. Thus, ion chromatography can be used to investigate the kinetics of this oxidation.

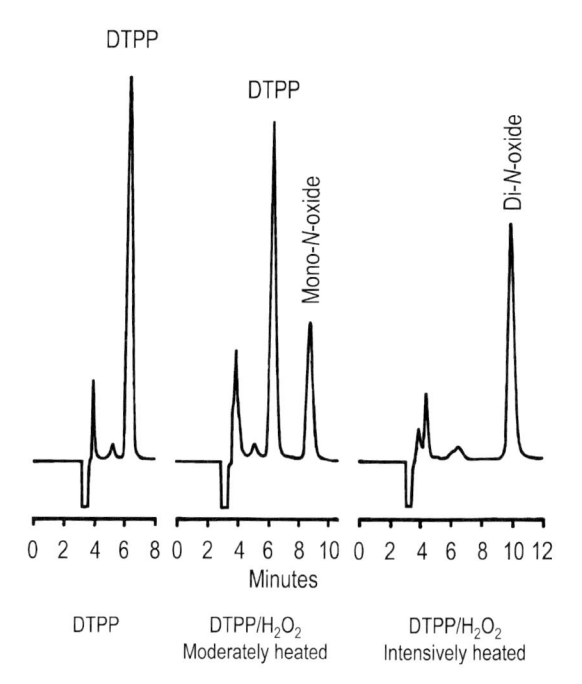

Fig. 3-138. Separation of diethylenetriamine-pentamethylenephosphonic acid (DEQUEST 2066) and its oxidation products. – Chromatographic conditions: see Fig. 3-136; solute concentration: 100 mg/L DEQUEST 2066.

In addition to aminopolyphosphonic acids, polyphosphonic acids without amino functions may also be analyzed by ion chromatography. Such compounds also serve as complexing agents for alkaline-earth metals and several auxiliary group elements. As for the monophosphonic acids, 2-phosphonobutane-1,2,4-tricarboxylic acid (PBTC) is of very high industrial importance. In diphosphonic acids, one distinguishes between geminal and vicinal structures. Geminally substituted phosphonic acids form stronger metal complexes.

A typical representative of geminal diphosphonic acids is 1-hydroxyethane-1,1-diphosphonic acid (HEDP, DEQUEST 2010), which is often employed in practical applications. As shown in Fig. 3-137, this compound may be analyzed in the same run with other aminopolyphosphonic acids.

Substituting a hydroxide group of the phosphonic acid with an alkyl or an aryl group leads to the compound class of polyphosphinic acids:

$$\begin{array}{c} HO \diagdown \nearrow O \\ P \\ R \diagup \diagdown OH \end{array} \longrightarrow \begin{array}{c} R \diagdown \nearrow O \\ P \\ R' \diagup \diagdown OH \end{array} \tag{87}$$

Generally, these compounds form less stable complexes with alkaline-earth metals than corresponding polyphosphonic acids. Weiß and Hägele [136] investigated the chromatographic behavior of a variety of aliphatic and olefinic polyphosphonic and polyphosphinic acids with 1 to 4 phosphorus atoms [125, 126]; these acids are used in medicine and in pharmaceutical chemistry.

#	Structure	Name
1	H–C(H)(P*)–C(H)(P*)–H	1,2-Ethanediphosphonic acid
2	H–C(H)(P°)–C(H)(P°)–H	Ethane-1,2-bis(P-methylphosphinic acid)
3	H–C(H)(P°)–C(P°)(P°)–H	Ethane-1,2,2-tris(P-methylphosphinic acid)
4	H–C(P°)(P°)–C(P°)(P°)–H	Ethane-1,1,2,2-tetrakis(P-methylphosphinic acid)
5	H–C(Ph)(P*)–C(H)(P*)–H	1-Phenylethane-1,2-diphosphonic acid
6	H–C(Ph)(P*)–C(P*)(P*)–H	1-Phenylethane-1,2,2-triphosphonic acid
7	H–C(Ph)(P°)–C(P°)(P°)–H	1-Phenylethane-1,2,2-tris(P-methylphosphinic acid)

8	Ph\C=C/H P*/ \H		1-Phenylethene-1-phosphonic acid
9	Ph\C=C/H H/ \P*		trans-1-Phenylethen-2-phosphonic acid

P*: -P(O)(OH)$_2$
P⁰: -CH$_3$P(O)(OH)

A comparison of the chromatographic behavior of the polyphosphinic acids **2**, **3**, and **4** (Fig. 3-139) reveals that the elution order depends strictly on the number of phosphinic acid groups. Thus, it follows the basic rule in ion chromatography, according to which retention increases with increasing valency of the respective anion. This observation also applies to the polyphosphonic acids **5** and **6**, whose separation is displayed in Fig. 3-140. The example of compounds **6** and **7** in Fig. 3-141 clearly shows that polyphosphinic acids always elute prior to polyphosphonic acids having the same carbon skeleton.

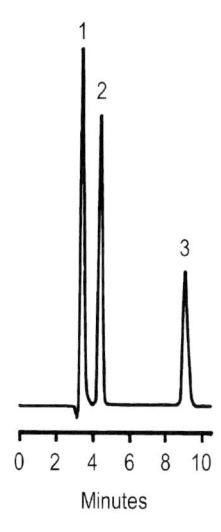

Fig. 3-139. Separation of various aliphatic polyphosphinic acids. – Chromatographic conditions: see Fig. 3-136; injection volume: 50 µL; solute concentrations; 20 mg/L each of **2**, **3**, and **4**.

As revealed by the comparison of the chromatographic behavior of **1** and **5** in Fig. 3-140, the introduction of a phenyl group into the molecule increases retention, assuming the number of phosphonic acid groups is kept constant. This finding is explained by π-π interactions between the aromatic ring systems of the stationary phase and the analytes.

In general, the chromatographic analysis of polyvalent anions shows an overlapping of ion-exchange and adsorption phenomena. Above all, the different adsorption behavior is responsible for the separation of structural isomers. This is evident, for example, from compounds **8** and **9** in Fig. 3-140, which are only

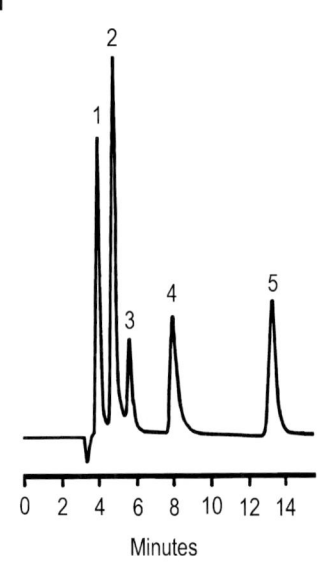

Fig. 3-140. Separation of aliphatic and aromatic polyphosphonic acids. — Separator column: IonPac AS7; eluant: 40 mmol/L HNO_3; other chromatographic conditions: see Fig. 3-136; injection volume: 50 µL; solute concentrations: 100 mg/L each of **1, 5, 8, 9,** and **6**.

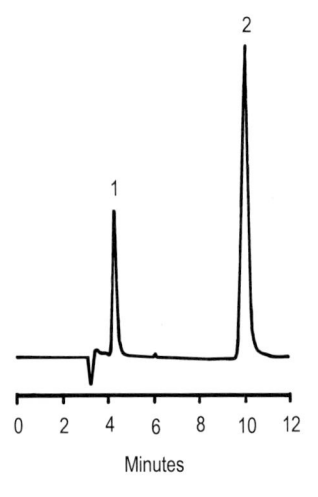

Fig. 3-141. Separation of an aromatic polyphosphinic and polyphosphonic acid of equal carbon backbone. — Chromatographic conditions: see Fig. 3-136; injection volume: 50 µL; solute concentrations: 200 mg/L each of **7** and **6**.

distinguished by the position of the phosphonic acid group. Since the dissociation behavior that governs the ion-exchange process is expected to be similar for both compounds, the separation of these compounds is probably based on a different adsorption behavior.

In addition to structural isomers, it is possible to separate stereo isomeric acids and, since just recently, rotational isomeric polyphosphonic and polyphosphinic acids as well. Most intriguing in this connection is the chromatogram in Fig. 3-142 of the sterically-crowded polyphosphonic acid **10**, which exists in two rotameric forms.

10
```
     tBu   P*
      |    |
  H – C –⟲– C – H      tert.-Butylethane-1,2,2-triphosphonic acid
      |    |
      P*   P*
```

P*: -P(O)(OH)$_2$

The two forms can be accurately identified and quantified using ^{31}P nuclear magnetic resonance spectroscopy. Figure 3-142 reveals that an ion chromatographic separation of both rotamers is possible, supposing a slow rotation around the C-C axis marked in the illustration above. It can be assumed with high probability that an equilibrium between both rotamers is established in the mobile phase which is characterized by the ratio of their peak areas. The rotation is interrupted only when the compound is retained for a short time at the fixed stationary charge. This is the only way to explain why both rotamer signals are well separated but merge into each other. Regarding the percentage distribution between both rotamers, comparable results are obtained with both analysis methods, i. e., NMR and ion chromatography. Remarkable results were obtained when investigating the temperature dependence of the rotamer's retention. As the column temperature decreases, a reduction in the retention difference between the signals is observed, although the opposite effect should be expected. Further studies concerning the dissociation behavior of these compounds will be required to understand this phenomenon.

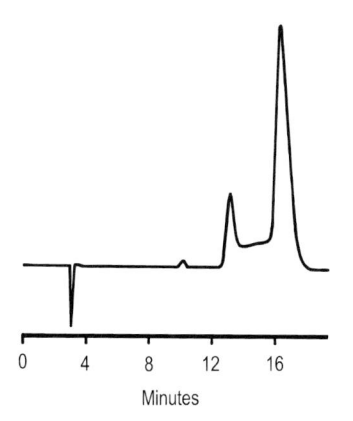

Fig. 3-142. Separation of the two rotamers of tert.-butylethane-1,2,2-triphosphonic acid. — Chromatographic conditions: see Fig. 3-136; column temperature: 25 °C; injection volume: 50 µL; solute concentration: 100 mg/L.

The successful separation of a diastereomeric phosphonic acid is also remarkable [137]. Figure 3-143 shows the separation of the two diastereomers of phosphonopropanetricarboxylic acid on IonPac AS7. In order to achieve a baseline-resolved separation, two separator columns were used in series. The percent area distribution between the two peaks can be confirmed by ^{31}P-NMR [138]. Thus, ion chromatography, due to its ease-of-use and fast analysis, is a welcome alternative to NMR for analyzing rotamers and diastereomers.

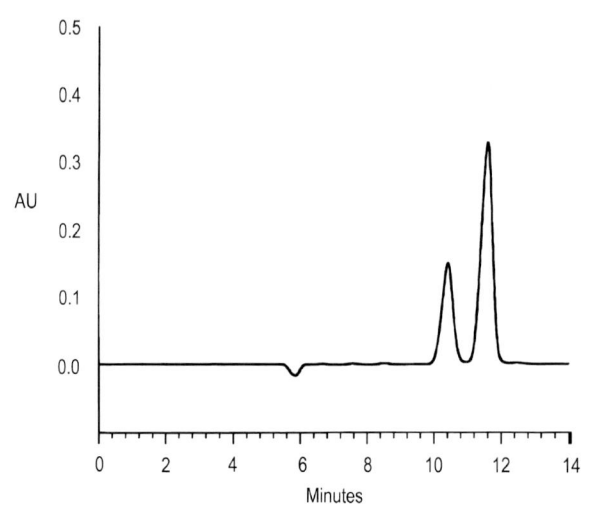

Fig. 3-143. Separation of the two diastereomers of phosphonopropanetricarboxylic acid. — Separator columns: 2 IonPac AS7; eluant: 30 mmol/L HNO$_3$; flow rate: 0.5 mL/min; detection: see Fig. 3-136; injection volume: 100 µL; solute concentration: 1000 mg/L.

Recently, another method for the separation and determination of polyphosphonic acids via anion exchange chromatography was described by Vaeth et al. [118]. The objective of this work was to develop a phosphorus-specific detection. The reaction of polyphosphonic acids with ferric nitrate is not specific and their determination is impeded by other complexing agents and by the potential presence of high amounts of inorganic anions such as chloride and sulfate in the sample matrix. In the method developed by Vaeth et al., polyphosphonic acids are hydrolyzed in the presence of ammonium peroxodisulfate at high temperature (110 °C) to orthophosphate that, as already described in Section 3.7.4, is converted in a second reaction step with a molybdate/vanadate reagent to the yellow phosphorvanadomolybdic acid. The light absorption of that product can be measured at a wavelength of 410 nm. Compared to the detection that uses ferric nitrate, this method offers the advantage that even in complex matrices only the phosphorus compound is detected. Again, a combination of potassium chloride and EDTA is utilized as an eluant; this eluant has already been qualified for the analysis of condensed phosphates. The retention behavior of polyphosphonic acids is affected by the pH value of the mobile phase. In general, an increase in the retention is observed with increasing pH value. The high selectivity of this special separation and detection system is exemplified by the separation of some selected polyphosphonic acids shown in Fig. 3-144.

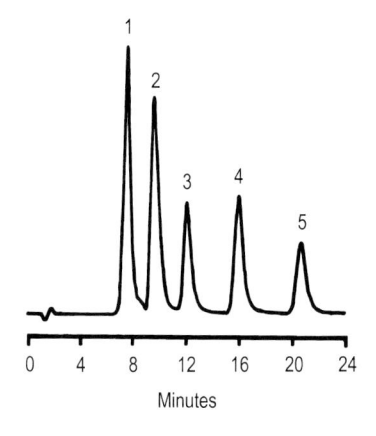

Fig. 3-144. Polyphosphonic acid analysis utilizing phosphorus-specific detection. — Separator column: IonPac AS7; eluant: 0.17 mol/L KCl + 3.2 mmol/L EDTA, pH 5.1; flow rate: 0.5 mL/min; detection: photometry at 410 nm after hydrolysis and derivatization with molybdate; injection volume: 50 µL; peaks: (1) 1-hydroxyethane-1,1-diphosphonic acid (HEDP), (2) aminotris-(methylenephosphonic acid) (ATMP), (3) ethylenediamine-tetramethylenephosphonic acid (EDTP), (4) 1,1-diphosphonopropane-2,3-dicarboxylic acid (DPD), and (5) 2-phosphonobutane-1,2,4-tricarboxylic acid (PBTC).

3.9 Gradient Elution Techniques in Anion Exchange Chromatography of Inorganic and Organic Anions

In conventional HPLC, the gradient elution of compounds is a well-established technique with a vast number of applications. For example, RPLC with photometric detection allows the separation of compounds with widely different retention behavior in a single run simply by changing the amount of organic solvent in the mobile phase. In general, a distinction between continuous and discontinuous gradients is made. The term *gradient elution* is reserved for separations obtained with continuous gradients, while *stepwise elution* is attributed to the class of discontinuous gradients. These chromatographic methods differ in practical application, but not substantially.

The first approach to an application of gradient elution is found in the paper by Mitchell, Gordon, and Haskins [139] published in 1949, although the authors were unaware of the fundamental advantages of this technique. Its unique potential to separate complex mixtures was recognized independently in several laboratories as late as the 1950s. Pioneers in this field were Hagdahl, Williams, and Tiselius [140] and Donaldson, Tulane, and Marshall [141]. Today, the technique of gradient elution is so mature that it has become impossible to completely cover all its applications. In this context, it is worth mentioning the review by Snyder [142], it focuses primarily on the theoretical aspects of gradient elution, in addition to some ideas concerning the apparatus.

So far, gradient elution techniques have found only limited application in the field of ion chromatography with conductivity detection, primarily because the most important inorganic anions such as fluoride, chloride, bromide, nitrite, nitrate, orthophosphate, and sulfate can be separated and determined under isocratic conditions. On the other hand, a variation of the eluant composition is inevitable if inorganic and organic anions or ions with different valencies are to be analyzed in a single run. The resulting change in background conductivity, which manifests itself in a strong baseline drift, is a serious problem for which no solution was offered in the first papers about this subject by Sundén et al. [143] and Tarter [144]. In addition, the application of gradient techniques in ion chromatography is hampered by a second problem, caused by anionic impurities in the eluants. These impurities may severely interfere with the analysis, because they are retained at the stationary phase at the beginning of the run. After the elution strength has been raised during the analysis, they may co-elute as peaks with the compounds of interest.

These problems have been circumvented both by the introduction of modern micromembrane suppressors and by the development of short clean-up columns for eliminating anionic impurities in the eluant [145]. A more elegant way to avoid anionic impurities in the mobile phase is to generate suitable eluants such as hydroxide *in situ* by means of electrolysis. The necessary hardware and the advantages of such a procedure will be discussed in detail below. As a result, gradient elution techniques in ion chromatography today are as common as in the field of conventional HPLC.

3.9.1
Theoretical Aspects

A mathematical description of the retention of ions under gradient elution conditions was introduced in 1957 by Schwab et al. [146]. It is based on parameters that are derived from the normal chromatographic elution process, for which the eluant composition is kept constant during the separation. Hence, the retention of an ion at isocratic elution may be described according to Eq. (88), taking into account the definitions for the capacity factor, k, and the selectivity coefficient, K, [see Eq. (34) and (38) in Section 3.2]:

$$k = \frac{V_{ms} - V_d}{V_d}$$

$$= \frac{V_s}{V_d} \cdot K^{1/x} \cdot Q^{y/x} \cdot [E]^{-y/x} \tag{88}$$

V_{ms}	Gross retention volume
V_d	Void volume
Q	Ion-exchange capacity of the resin
$[E]$	Eluant ion concentration
x	Charge number of eluant ion
y	Charge number of solute ion

With the exception of the eluant ion concentrations, all constants in Eq. (88) can be summarized to a general constant, Const_i:

$$k = \text{Const}_i \cdot [E]^{-y/x} \tag{89}$$

A linear relation is obtained by taking the logarithm on both sides of this equation

$$\log k = -y/x \cdot \log [E] + \log \text{Const}_i \tag{90}$$

which is described in a slightly different way in Eq. (86) in Section 3.7.3. According to Eq. (90), the slope of the straight line is proportional to the quotient of the charge number of solute and eluant ions.

In gradient elution, k changes as a function of the eluant ion concentration, so that it may not be equated with the ratio $(V_{ms}\text{-}V_d)/V_d$. For a gradient in which the eluant ion concentration increases linearly with time, starting from zero, the momentary eluant ion concentration is calculated by:

$$[E] = R \cdot V \tag{91}$$

V represents the eluant volume delivered since the beginning of the gradient run and R is the slope of the gradient ramp, which is defined as the ratio of the temporal change in the eluant ion concentration and the flow rate. The corresponding momentary capacity factor k_a for the peak maximum is obtained by inserting Eq. (91) into Eq. (89):

$$k_a = \text{Const}_i \cdot (RV)^{-y/x} \tag{92}$$

where k_a represents the capacity factor that would be valid if the eluant ion concentration is held constant from the moment of injecting a sample. It follows that in gradient elution, k_a must be integrated over this time period to calculate $(V_{ms}\text{-}V_d)/V_d$. Hence, k_a is replaced by dV/dx:

$$k_a = dV/dx \tag{93}$$

indicating that with each volume portion, dV, the peak maximum travels the portion dx through the column. Together with Eq. (92) one obtains:

$$dx = \frac{dV}{\text{Const}_i \cdot (RV)^{-y/x}} \tag{94}$$

Integrating both sides yields

$$\int_0^{V_{ms}-V_d} \text{Const}_i^{-1} \cdot (RV)^{y/x} \, dV = \int_0^{V_d} dx \tag{95}$$

$$\frac{V_{ms} - V_d}{V_d} = \left(\frac{x}{y+x}\right)^{-\frac{x}{y+x}} \cdot V_d^{-\frac{y}{y+x}} \cdot \text{Const}_i^{-\frac{x}{y+x}} \cdot R^{-\frac{y}{y+x}} \tag{96}$$

Again, the constants in Eq. (96) may be summarized in a general constant, Const_g:

$$\frac{V_{ms} - V_d}{V_d} = \text{Const}_g \cdot R^{-\frac{y}{y+x}} \tag{97}$$

Taking the logarithm on both sides it follows:

$$\log \frac{V_{ms} - V_d}{V_d} = -\frac{y}{y+x} \cdot \log R + \log \text{Const}_g \tag{98}$$

Equation (98) derived for gradient elution is very similar to Eq. (90) for isocratic elution. Plotting log $(V_{ms}\text{-}V_d)/V_d$ as a function of log R, straight lines with different slopes are obtained for each solute ion. However, Eq. (98) only applies to linear gradients with an initial eluant ion concentration of zero.

3.9.2
Choice of Eluant

Above all, the choice of eluant for gradient elution applications depends on the type of gradient. In ion-exchange chromatography two different techniques may be applied. While in a *concentration gradient* the eluant ion concentration is changed during the run, in a *composition gradient* the eluant composition is changed by replacing weakly-retained eluant ions with more strongly-retained ones. Occasionally, pH gradients are applied in which an increasing amount of a strong base is added to a defined concentration of a weak acid. Actually, pH gradients are also concentration gradients, because an increase in the concentration of dissociated acid is intended by increasing the pH value during the run. A new form of gradients in ion chromatography, a capacity gradient, has recently been introduced together with a cryptand anion exchanger. A capacity gradient is carried out by changing the cation component of the hydroxide eluant during the chromatographic run. Because in most cases the eluant strength is kept constant during the run, capacity gradients belong to the class of composition gradients.

Conventional composition gradients present a problem in practical applications. Taking into account the large retention differences between weakly- and strongly-retained solute ions such as the monovalent acetate and the trivalent citrate, the eluant ions employed must also differ significantly in their affinity towards the stationary phase. This means that for the elution of trivalent citrate, the weakly-retained eluant ions that dominate at the beginning of a run have to be completely replaced by strongly-retained eluant. The time required for this replacement depends on the ratio between ion-exchange capacity and eluant ion concentration. Furthermore, the reverse operation is necessary to repeat a run,

3.9 Gradient Elution Techniques in Anion Exchange Chromatography ...

which is only accomplished by flushing the column with a high concentration of the eluant used initially. However, this increases the equilibration time between two gradient runs. In contrast, such problems are not encountered when applying concentration gradients, because the ionic form of the resin remains unaffected.

As mentioned above, a change in eluant ion concentration causes a change in background conductivity. The latter is significant, because the eluant ion concentration has to be increased during the run by one to two orders of magnitude in order to elute anions with strongly different retention behavior. Thus, direct conductivity detection is impossible, because the expected baseline drift is much too strong. For a successful application of concentration gradients, therefore, the background conductivity has to be lowered significantly. This is only achieved chemically with micromembrane or self-regenerating suppressors (see Section 3.6), which reduce the background conductivity of eluants suitable for gradient elution to values of less than 1 µS/cm. This limits the background conductivity changes during a gradient run to a few nS/cm.

In principle, salts derived from inorganic or organic acids with pK values above 7 can be employed as an eluant. Hydroxide anions have proved to be particularly suitable, because they are converted into water independent of the initial concentration. The resulting background conductivity is barely affected by the hydroxide ion concentration in the mobile phase, as long as the suppression capacity of the suppressor is not exceeded. A typical example of a gradient elution using a sodium hydroxide eluant is displayed in Fig. 3-145. It shows a separation of chloride, sulfate, and various phosphorus compounds on an IonPac AS5 separator column. With a simple linear NaOH gradient, both the monovalent chloride and the hexavalent tetrapolyphosphate can be analyzed in a single run within 15 minutes.

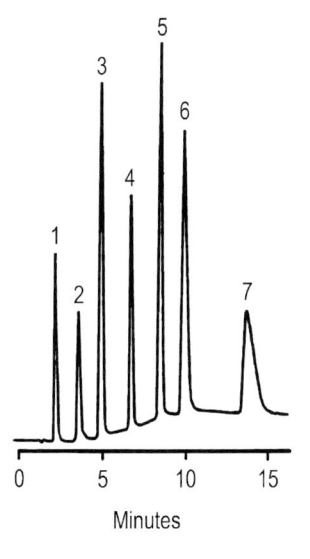

Fig. 3-145. Gradient elution of chloride, sulfate, and various phosphorus compounds. — Separator column: IonPac AS5; eluant: (A) water, (B) 0.1 mol/L NaOH; gradient: linear, 25% B to 100% B in 5 min; flow rate: 1 mL/min; detection: suppressed conductivity; injection volume: 50 µL; peaks: (1) chloride, (2) orthophosphite, (3) sulfate, (4) orthophosphate, (5) pyrophosphate, (6) tripolyphosphate, and (7) tetrapolyphosphate.

Even more impressive is the chromatogram in Fig. 3-39 (see Section 3.4.2), which was obtained using a modern AS11 latexed anion exchanger that was especially designed for gradient applications. It shows the separation of a standard solution containing 36 different inorganic and organic anions with a run time of about 15 minutes. Using a highly diluted sodium hydroxide solution, fluoride and various monocarboxylic acids that are only weakly retained are separated, while the final sodium hydroxide concentration suffices to elute trivalent anions such as orthophosphate and citrate. It is obvious that great progress has been made since the introduction of gradient elution; the complete analysis of this standard under isocratic conditions used to require three different operations [147].

When preparing hydroxide eluants, care must be taken that they are, to a high degree, carbonate-free! Because carbonate ions exhibit a much higher elution strength than hydroxide ions, even traces of carbonate in the eluant results in a lower resolution, especially at the beginning of a gradient run. The final dissociation of the carbonic acid formed in the suppressor leads to a strong baseline drift during the gradient run. (Incidentally, this is the reason that eluants based on carbonate/bicarbonate are totally unsuitable for a gradient technique, unless the suppressed eluant is directed through a vacuum degasser prior to entering the conductivity cell, so that part of the carbonic acid is removed.) Therefore, NaOH-based eluants should be prepared in a conventional way from a 50% concentrate in which the carbonate, originally adsorbed at the NaOH pellets, precipitates as a fine slurry. The de-ionized water used for eluant preparation should have an electrical conductance of 0.05 µS/cm and should be degassed with helium prior to adding the NaOH concentrate. After preparation, eluants are to be kept under inert gas. By observing these guidelines, the baseline drift for the concentration gradient given in Fig. 3-145 does not exceed 3 to 4 µS/cm. However, the anions can only be determined in the mg/L concentration range. As an additional precaution, a short anion trap column (ATC) is placed between the analytical pump and the injection valve, which contains a high-capacity anion exchange resin with a low chromatographic efficiency. The anion trap column retains anionic impurities from the eluant used at the beginning of an analysis. Those anionic impurities can be accumulated on the separator column at the beginning of a gradient run, and as the analysis progresses, may then elute as peaks. As the eluant strength increases during the gradient run the anionic impurities retained on the ATC are released again but do not appear as regular peaks in the chromatogram due to the low efficiency of the material.

A much more elegant way to prepare hydroxide eluants has recently been introduced by Liu et al. [148]. The respective device was commercialized under the trade name Eluant Generator™. The most important part of this module is a cartridge, in which potassium hydroxide is generated by means of electrolysis. As schematically depicted in Fig. 3-146, such a cartridge consists of an electrolyte reservoir, which is filled with KOH solution ($c = 4$ mol/L) and connected to an electrolysis chamber via a stack of cation exchange membranes.

3.9 Gradient Elution Techniques in Anion Exchange Chromatography ...

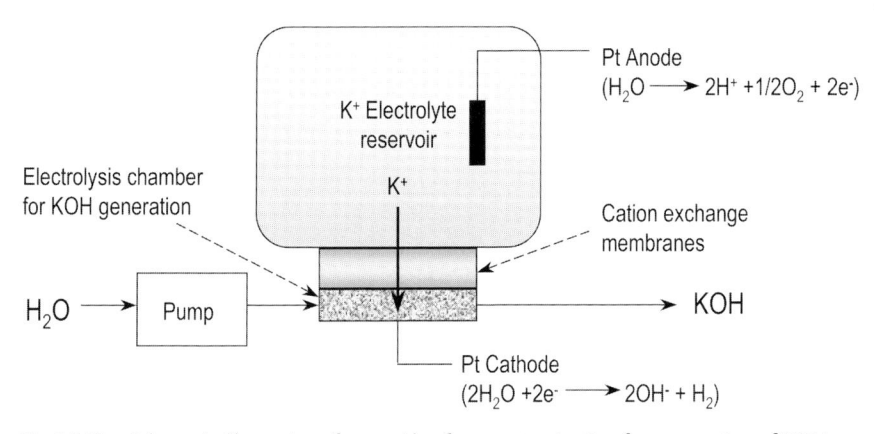

Fig. 3-146. Schematic illustration of a cartridge for a contamination-free generation of KOH.

While the electrolyte reservoir contains a perforated anode, the respective cathode is placed inside the electrolysis chamber. Both electrodes are made of platinum. For generating high-purity KOH, de-ionized water is pumped through the electrolysis chamber and an electric field is applied between the two electrodes. Under the influence of the electric field, hydroxide ions are generated at the cathode according to Eq. (99):

$$2\ H_2O + 2e^- \rightarrow 2\ OH^- + H_2 \qquad (99)$$

The hydroxide ions combine in the electrolysis chamber with the potassium ions migrating through the membranes to yield KOH. The concentration of KOH is directly proportional to the applied current and inversely proportional to the flow rate. The cation exchange membranes support the migration of the potassium ions from the electrolyte reservoir into the electrolysis chamber, while at the same time precluding the passage of co-ions from the electrolyte reservoir to the base generation chamber. Because the KOH being generated is directed to the injection valve via a capillary connector and then on to the separator column, the electrolysis chamber has to withstand the backpressure of the separator column. On its way to the injection valve, the KOH has to pass a high pressure degas unit for removing the hydrogen gas formed at the cathode. Thus, the second important function of the cation exchange membranes is to serve as a high pressure physical barrier that insulates the electrolyte reservoir from the acid generation chamber.

The greatest advantages of electrolytic KOH generation are that only de-ionized water is used as a carrier for the ion chromatograph and that KOH is generated free of any carbonate contaminants, because it is generated in a closed system. This is especially important for very low KOH concentrations ($c \ll 0.1$ mol/L), because time-consuming post-analytical rinsing steps to remove accumulated carbonate from the stationary phase are no longer necessary. Moreover, being able to adjust the KOH concentration via the applied current and the flow rate is also a real advantage. While the eluant flow rate is usually kept constant by programming the analytical pump, the applied current can be programmed electrically with minimal delay. Thus, gradient elution techniques are realized by electrical current gradients.

The maximum KOH concentration that can be generated with an Eluant Generator™ is 0.1 mol/L at a flow rate of 1 mL/min. The lifetime of a KOH cartridge naturally depends on the programmed KOH concentration, the flow rate, and the working time. Based on a flow rate of 1 mL/min and a KOH concentration of 0.05 mol/L the cartridge lifetime is about 1000 hours. Figure 3-147 shows the dependence of the resulting KOH concentration on the applied current, which is strictly linear up to the concentration of 0.1 mol/L. This is the prerequisite for a precise gradient formation via the applied current.

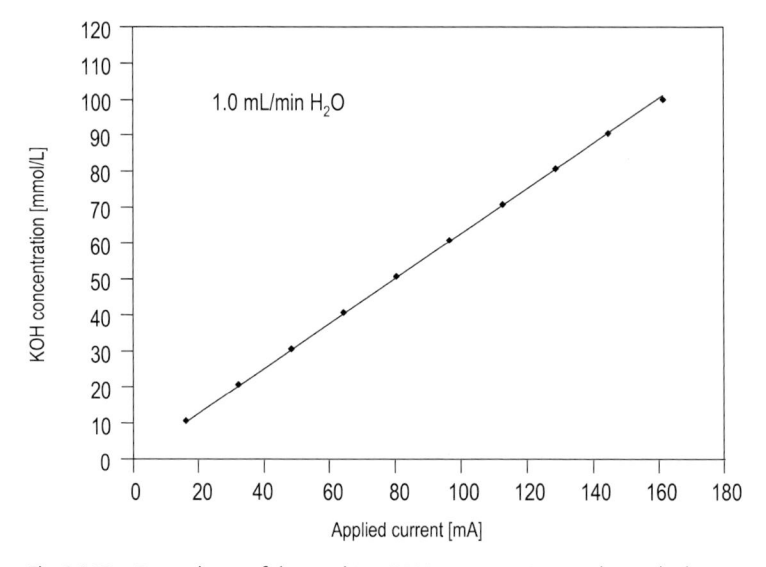

Fig. 3-147. Dependence of the resulting KOH concentration on the applied current.

The most impressive proof of performance of an Eluant Generator™ is depicted in Fig. 3-148, in which an isocratic separation of inorganic anions on IonPac AS11 is set against the respective gradient elution. Since the KOH eluant is not contaminated by carbonate when generated electrolytically, the baseline remains absolutely stable even under gradient conditions. Baseline drifts of several $\mu S/cm$, caused by carbonate impurities when preparing a hydroxide eluant in a conventional way, are not observed. Thus, the introduction of an Eluant Generator™ opens up a new era in ion chromatography, because baseline stability is so high that gradient and isocratic separations can no longer be distinguished. Further, the run time for the gradient separation in Fig. 3-148 is not much longer than that for the respective isocratic run.

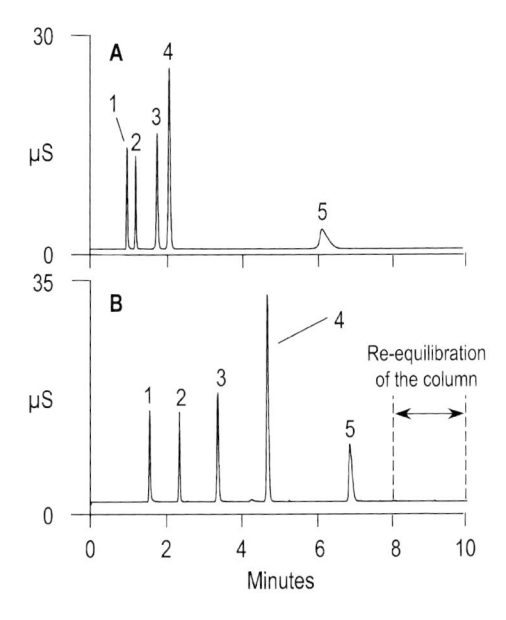

Fig. 3-148. Separation of inorganic anions with an electrolytically generated KOH eluant. — Separator column: IonPac AS11; eluant: (A) 15.5 mmol/L KOH, (B) 0.5 mmol/L KOH to 25 mmol/L KOH in 8 min; flow rate: 2 mL/min; detection: suppressed conductivity; injection volume: 25 µL; solute concentrations: 2 mg/L fluoride (1), 3 mg/L chloride (2), 10 mg/L nitrate (3), 15 mg/L sulfate (4), and 15 mg/L orthophosphate (5).

Since the Eluant Generator™ is software-controlled via a chromatography management system, gradient separations, programmed per mouse click, are as easy to perform as isocratic ones. The advantages of an Eluant Generator™ are especially evident in trace analysis. This is illustrated in Fig. 3-149 which shows a respective separation of inorganic anions in the lowest µg/L range. The low baseline drift of less than 80 nS/cm allows a sensitive setting of the conductivity detector, so that analyses in the sub-µg/L range are possible by direct injection. The largest peak in the chromatogram in Fig. 3-149 is carbonate, which in this case is a sample component and not an eluant impurity. By choosing appropriate gradient conditions, even a large excess of carbonate does not interfere with the analysis of other common inorganic and organic anions. A characteristic feature of the IonPac AS15 separator column is the high resolution between low molecular weight organic acids, including the ability to separate glycolate and acetate (see Fig. 3-11 in Section 3.4.1), which can only be achieved with a tetraborate gradient on other stationary phases (see below).

Salts of other weak acids may also be employed as eluants for gradient techniques. This includes sodium tetraborate, which has a lower elution strength than sodium or potassium hydroxide. Since the solubility of sodium tetraborate in water and, thus, the final concentration under gradient conditions is limited to 0.05 mol/L, tetraborate gradients are predominantly used for analyzing short-chain organic acids, which can be analyzed together with standard inorganic anions such as chloride, nitrate, and sulfate if this method is used. However, tetraborate is not an effective eluant for strongly-retained polyvalent anions due to their high affinities to the stationary phase. In this respect, the baseline-resolved separation of glycolate and acetate [149] on IonPac AS14, which is not possible with a hydroxide eluant, is quite remarkable (see Fig. 3-150).

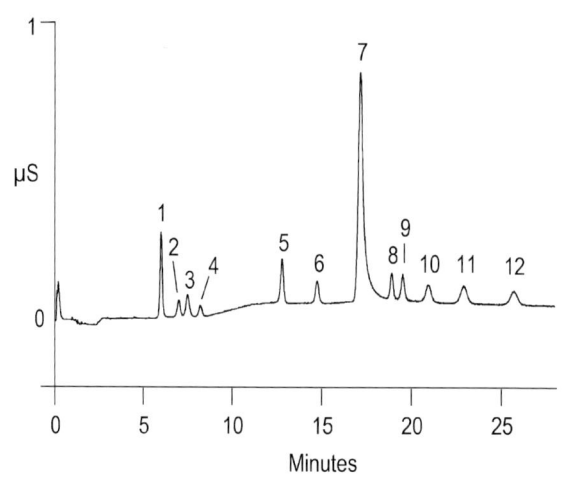

Fig. 3-149. Trace analysis of inorganic and organic anions with an electrolytically generated KOH eluant. − Separator column: IonPac AS15; column temperature: 30 °C; eluant: KOH (with Eluant Generator™); gradient: 9 mmol/L isocratically for 7 min, then to 46 mmol/L in 11 min; flow rate: 1.6 mL/min; detection: suppressed conductivity; injection volume: 2000 µL; solute concentrations: 2 µg/L fluoride (1), 4 µg/L glycolate (2), 4 µg/L acetate (3), 2 µg/L formate (4), 2 µg/L chloride (5), 2 µg/L nitrite (6), carbonate (7), 2 µg/L sulfate (8), 2 µg/L oxalate (9), 4 µg/L bromide (10), 4 µg/L nitrate (11), and 6 µg/L orthophosphate (12).

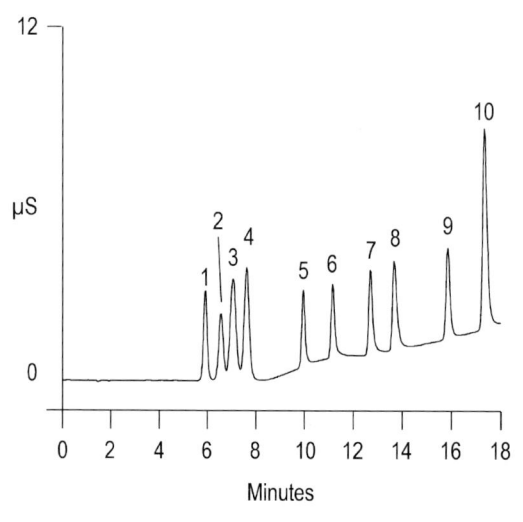

Fig. 3-150. Gradient elution of inorganic and organic anions with sodium tetraborate eluant. − Separator column: IonPac AS14; eluant: 2 mmol/L $Na_2B_4O_7$ isocratically for 6 min, then linearly to 17.5 mmol/L $Na_2B_4O_7$; flow rate: 1.5 mL/min; detection: suppressed conductivity; injection volume: 25 µL; solute concentrations: 5 mg/L fluoride (1), 10 mg/L glycolate (2), 20 mg/L acetate (3), 10 mg/L formate (4), 3 mg/L chloride (5), 10 mg/L nitrite (6), 10 mg/L bromide (7), 10 mg/L nitrate (8), 15 mg/L orthophosphate (9), and 15 mg/L sulfate (10).

Extensive investigations were also performed with p-cyanophenolate ions [145] which, compared to hydroxide ions, have a slightly stronger elution strength. A high selectivity for aliphatic dicarboxylic acids is obtained with p-cyanophenolate in combination with a CarboPac PA1 (IonPac AS6) anion exchanger. Nevertheless, this eluant did not gain acceptance, because it is not commercially available in the required purity.

A second class of available eluants comprises the family of zwitterionic compounds (see Section 3.5). At alkaline pH, they exist in an anionic form, so they can function as eluant ions. The product of the suppressor reaction is the zwitterionic form with a correspondingly low intrinsic conductivity. Promising experiments were carried out by Irgum [70] with N-substituted aminoalkylsulfonic acids, which can be employed for both composition and concentration gradients. The compounds taurine (2-aminoethanesulfonic acid) and CAPS [3-(N-cyclohexylamino)-1-propanesulfonic acid] used by Irgum are commercially available, but have to be thoroughly purified [150, 151] for gradient applications. For now, it is not known to what extent the eluant ion concentration, necessary for eluting strongly-retained anions, is limited by the slow transport of zwitterions through the membrane.

3.9.3
Possibilities for Optimizing Concentration Gradients

In theory, it is possible to derive the optimal conditions for a gradient elution with a sodium hydroxide eluant from the functional dependence of log $(V_{ms}-V_d)/V_d$ from log R. However, as mentioned before, this applies only to simple linear gradients with the initial eluant ion concentration of zero, which is rarely used for practical purposes. Much shorter analysis times are obtained when the gradient run starts at a higher eluant ion concentration than zero. Furthermore, gradient programs with different ramps, sometimes combined with isocratic periods, have to be developed to obtain optimal selectivity and speed of analysis. A mathematical description of the retention is impossible in all these cases, because the resulting equation for the calculation of the retention volume would be far too complex.

However, Eq. (98) can be employed to predict trends. Refer to Fig. 3-151 in which log $(V_{ms}-V_d)/V_d$ is plotted as a function of log R for various inorganic and organic anions. It appears from this representation that the elution order of anions with different valencies depends on the steepness of the gradient ramp. In case two anions with different valency co-elute, a separation of both components can be accomplished by employing a steeper ramp, whereas the ion with the higher valency elutes first.

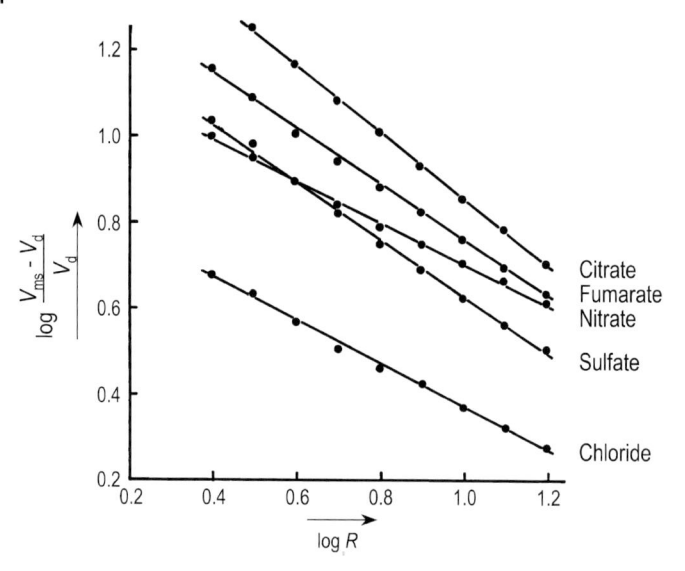

Fig. 3-151. Representation of log $(V_{ms}-V_d)/V_d$ as a function of log R for various inorganic and organic anions using an IonPac AS5A separator and a NaOH eluant. – Flow rate: 1 mL/min; solute concentrations: 3 mg/L chloride, 5 mg/L nitrate, 10 mg/L sulfate, 5 mg/L fumarate, and 50 mg/L citrate; (taken from [145]).

3.9.4
Isoconductive Techniques

A completely different approach for the gradient elution of anions was pursued by Jandik et al. with the development of "isoconductive eluants" [152]. Intended to be an alternative to the concept of chemical suppression, this technique employs two mobile phases which differ in their elution strength but not in the resulting background conductivity. This is achieved by substituting the cations of the eluant with those serving as counter ions to the eluant anions. These cations do not contribute to the separation process. The background conductivity remains unaffected, for example, when turning from a cation with high equivalent conductance in the weaker eluant to a cation with lower equivalent conductance in the comparatively stronger eluant, and compensating the resulting conductivity change by a slight increase in the eluant ion concentration. Thus, the gradient technique performed with isoconductive eluants represents a combination of both a concentration gradient and a composition gradient.

A quantitative statement regarding the increase of the elution strength when switching from one eluant to another may be derived from the equation proposed by Fritz et al. [102] for calculating the background conductivity of a mobile phase with fully dissociated eluant ions:

$$G = \frac{\lambda^+ + \lambda^-}{10^{-3}\,K} \cdot c \qquad (100)$$

- G Eluant background conductivity
- λ^-, λ^+ Equivalent conductances of eluant anions and cations
- K Cell constant
- c Eluant concentration

Consider two eluants of different elution strength containing the same eluant anions but different cations with equivalent conductances λ_1^+ and λ_2^+. Under isoconductive conditions, the background conductivity, G_1, of the weaker eluant equals that of the stronger eluant G_2. Given $G_1 = G_2$, Eq. (100) may be rearranged, so that the concentrations, c_1 and c_2, of both eluants are related to the corresponding equivalent conductances:

$$\frac{c_2}{c_1} = \frac{\lambda_1^+ + \lambda^-}{\lambda_2^+ + \lambda^-} \qquad (101)$$

Eq. (101) facilitates the search for a suitable combination of eluant and counter ions. The right combination would maximize the ratio c_2/c_1 as a measure for the increase in elution strength. Considerable improvements may only be obtained with eluant ions of low equivalent conductance. Particularly well-suited is the gluconate/borate mixture used by Jandik et al. in their pioneering paper [152]. The increase in gluconic acid concentration in the presence of a suitable concentration of boric acid was compensated by adjusting with potassium or lithium hydroxide. Figure 3-152 illustrates this by comparing the conductivity change that occurs in a pure concentration gradient to the conductivity change resulting from isoconductive conditions. The eluant compositions used for this comparison are summarized in Table 3-25.

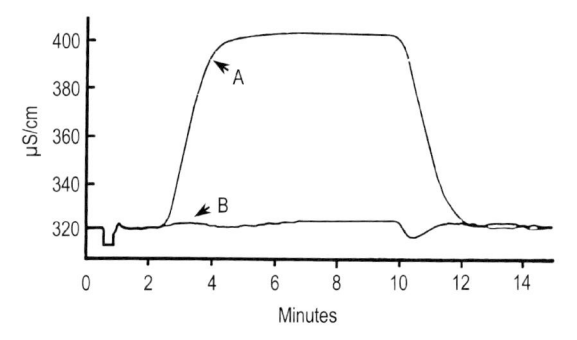

Fig. 3-152. Comparison of the conductivity change resulting from a pure concentration gradient (A) to that obtained under isoconductive conditions (B). Separator column: Waters IC-Pak Anion; eluant: see Table 3-25; (taken from [152]).

3 Anion Exchange Chromatography (HPIC)

Table 3-25. Eluant composition for a concentration gradient (A) and an isoconductive eluant (B) according to Fig. 3-152; (taken from [152]).

Component	A. Eluant 1	Eluant 2	B. Eluant 1	Eluant 2
Boric acid [mol/L]	0.011	0.01375	0.011	0.01375
Gluconic acid [mol/L]	0.00148	0.00185	0.00148	0.00185
Potassium hydroxide [mol/L]	0.00349	0.00436	0.00349	--
Lithium hydroxide [mol/L]	--	--	--	0.00513
Glycerol [mol/L]	0.00069	0.00081	0.00069	0.00081
Acetonitrile [mL/L]	120	120	120	120

The chromatogram of various inorganic anions (Fig. 3-153) obtained with isoconductive eluants clearly shows the limitation of this technique as compared with the concentration gradient concept with subsequent chemical suppression. Although the analysis time for the isoconductive separation is shorter than under isocratic conditions, improved resolution between signals, especially in the first part of the chromatogram, is not achieved. Furthermore, carbonate, which is often present in high concentrations in real-world samples, elutes between fluoride and chloride. In practice, this may lead to significant interferences. Time will tell how this technique develops with new eluant compositions and gradient profiles.

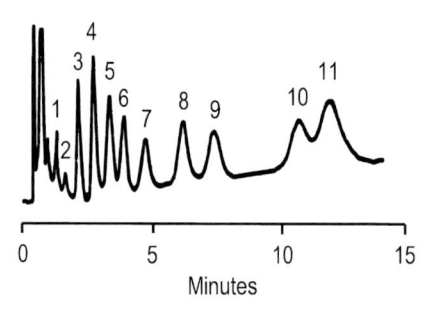

Fig. 3-153. Separation of various inorganic anions with an isoconductive eluant. – Separator column: Waters IC-PAK Anion; eluant: see Table 3-25 (eluant switching at the time of injection); detection: direct conductivity; injection volume: 100 µL; solute concentrations: 1 mg/L fluoride (1), 2 mg/L each of carbonate (2) and chloride (3), 4 mg/L each of nitrite (4), bromide (5), and nitrate (6), 6 mg/L orthophosphate (7), 4 mg/L each of sulfate (8) and oxalate (9), and 10 mg/L each of chromate (10) and molybdate (11); (taken from [152]).

3.10 Carbohydrates

GC and HPLC methods were developed in the past for the chromatographic separation of carbohydrates. Gas chromatographic methods are very time consuming, because carbohydrates, owing to their non-volatile character, have to be derivatized prior to the determination [153]. The LC method using strong-acid cation exchangers in the calcium form with de-ionized water as an eluant [154] is more widespread, but has several drawbacks:

- If the sample to be analyzed contains organic acids, metal ions fixed at the ion-exchange groups can be removed by complexation, which increases the frequency of column regeneration.
- High-capacity cation exchangers exhibit poor selectivity for higher oligosaccharides. Even the separation of short-chain oligosaccharides requires an ion-exchange resin with a low degree of cross-linking.
- Owing to the resin's compressibility (due to the cross-linking), only low flow rates can be used, which results in long analysis times.
- Optimal separation is achieved only at temperatures around 85 °C, a serious problem for refractive index detection of sugars.
- The number of retention-determining parameters is limited, because only water or mixtures of water and organic solvents can be utilized as eluants.

Another possibility is the separation of carbohydrates after complexation with boric acid [155]. Although such separations are characterized by high selectivities, the kinetic of the complex formation is very slow. Thus, low flow rates are required, which again results in long analysis times.

Good separations for mono- and disaccharides are obtained with silica-based, chemically bonded octadecyl [156] and aminopropyl phases [157, 158]. However, larger oligosaccharides are eluted as broad peaks. Such separator columns have a limited lifetime, especially when the samples to be analyzed contain aldehydes or ketones. These compounds, which are often constituents in foods and beverages, react with the amino group to yield a Schiff base (RR'C=NR", azomethine).

Alternatively, carbohydrates can be separated by anion exchange chromatography [159]. Since the pK values for sugars lie between 12 and 14 [160], they can be converted into their anionic form in a superalkaline environment, and may be separated on a strong basic anion exchanger in the hydroxide form. Sodium hydroxide proved to be a suitable eluant in the concentration range between 0.001 mol/L and 0.15 mol/L. The hydroxide ions have two functions: (1) they act as eluant ions and (2) they determine the pH value of the mobile phase. Thus, a change in the hydroxide ion concentration in the mobile phase has two different effects on the retention behavior of carbohydrates. While carbohydrate dissociation and, consequently, retention increases with increasing pH value, the associated higher eluant ion concentration results in a retention decrease. As long as the carbohydrates are not fully dissociated, the two effects compensate each other. At complete dissociation of carbohydrates, a further increase in the hydroxide ion concentration merely results in decreased retention. Carbohydrates with a high affinity towards the stationary phase can be analyzed by adding sodium acetate to the eluant, which is not detected by the pulsed amperometry detection system [161].

In some carbohydrates, for example epimers, co-elution is often observed, because the retention behavior of these compounds does not differ very much. Taking into account the different dissociation behavior, however, the selectivity of the separation system can be enhanced significantly by lowering the mobile phase pH to a value comparable to the analyte pK value.

In preparing sodium hydroxide based eluants, extreme care must be taken that they are carbonate-free. Since carbonate ions exhibit a much higher elution strength than hydroxide ions, the presence of even small amounts of carbonate reduces resolution. Therefore, eluants should be prepared from a 50% NaOH concentrate which contains only small amounts of carbonate. The de-ionized water being used to make up the mobile phase has to be degassed thoroughly with helium prior to adding the concentrate. The eluant itself must be stored under helium. For the separation of monosaccharides and amino sugars, which is usually carried out with hydroxide concentrations $c \ll 0.1$ mol/L, electrolytical hydroxide generation with an Eluant Generator™ is a welcome alternative (see Section 3.9).

Presently, a number of different anion exchangers for carbohydrate analysis are offered by Dionex; their structural-technical properties are summarized in Table 3-26.

Figure 3-154 shows a standard chromatogram with an isocratic separation of various sugar alcohols and saccharides utilizing CarboPac PA1. Although the affinity of carbohydrates towards the stationary phase increases in the following order — sugar alcohols < monosaccharides < oligosaccharides < polysaccharides — it is possible that mixtures of different oligosaccharides will yield overlaps in the elution order. Thus, at temperatures slightly above ambient (as seen in Fig. 3-154), the trisaccharide raffinose and the tetrasaccharide stachyose elute ahead of the disaccharide maltose.

The major advantage of anion exchange chromatography for carbohydrate separations is the ability to experimentally affect both the retention times and the elution order. Figure 3-155 shows the dependence of retention on the hydroxide ion concentration, exemplified by some selected carbohydrates. This figure reveals that optimal resolution between raffinose, stachyose, and maltose is obtained with 0.15 mol/L NaOH.

3.10 Carbohydrates

Table 3-26. Structural-technical properties of anion exchangers suitable for carbohydrate analysis.

Separator	CarboPac PA1	CarboPac PA10	CarboPac PA20	CarboPac PA-100	CarboPac MA1	Metrosep Carb 1
Manufacturer	Dionex	Dionex	Dionex	Dionex	Dionex	Metrohm
Dimensions (Length × i. d.) [mm]	250 x 4 250 x 9	250 × 4 250 × 9	150 × 3	250 × 4	250 × 4	250 × 4.6
Support material	PS/DVB	PS/DVB	Ethylvinyl-benzene/DVB	Ethylvinyl-benzene/DVB	Vinylbenzyl-chloride/DVB	PS/DVB
Particle diameter [µm]	10	10	6.5	10	8.5	5
Degree of cross-linking [%]	5	5	5	6	15	--
Latex particle Diameter [nm]	350	400	130	350	--	--
Capacity [µequiv/column]	100	100	65	90	4500	1530
Recom. flow rate [mL/min]	1	1.5	0.5	1	0.5	1.0
Max. pressure [MPa]	34	27	21	34	13	15
pH stability	0 – 14	0 – 14	0 – 14	0 – 14	0 – 14	0 – 14
Solvent compatibility [%]	0	90	100	100	0	0
Application	Separation of acidic, basic and neutral mono-saccharides	Separation of mono- and disaccharides	Separation of mono- and disaccharides	Separation of complex mixtures of oligosaccharides	Special column for sugar alcohols	Universal column for mono- and disaccharides

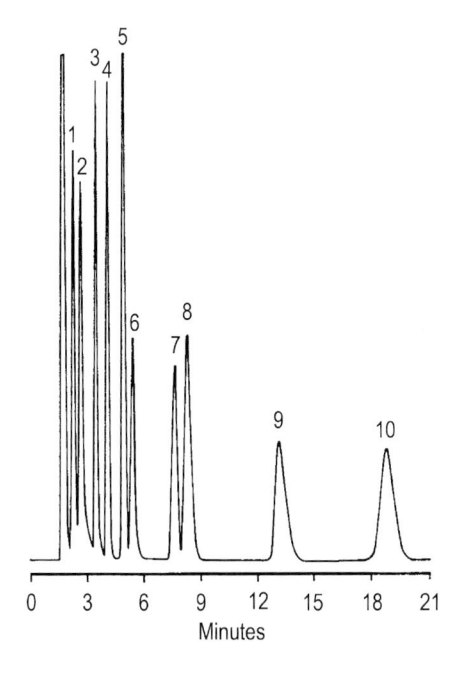

Fig. 3-154. Separation of various sugar alcohols and saccharides. − Separator column: CarboPac PA1; eluant: 0.15 mol/L NaOH; flow rate: 1 mL/min; detection: pulsed amperometry at a Au working electrode; injection volume: 50 µL; solute concentrations: 10 mg/L xylitol (1), 5 mg/L sorbitol (2), 20 mg/L each of rhamnose (3), arabinose (4), glucose (5), fructose (6), and lactose (7), 100 mg/L sucrose (8), 100 mg/L raffinose (9), and 50 mg/L maltose (10).

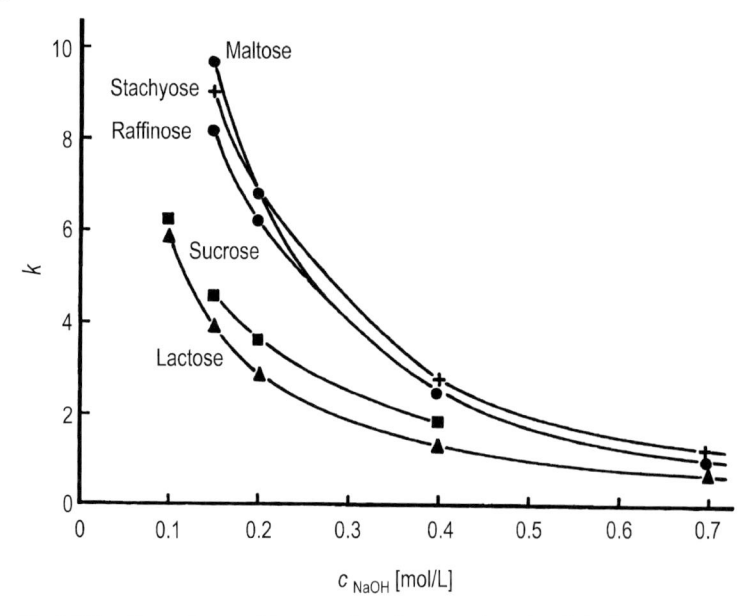

Fig. 3-155. Dependence of the retention of some selected carbohydrates on the hydroxide ion concentration. — Separator column: CarboPac PA1; flow rate: 1 mL/min; detection and injection volume: see Fig. 3-154; solute concentrations: 100 mg/L each of the respective carbohydrates.

A second parameter that affects retention is the column temperature. In general, retention decreases with increasing temperature. The temperature influence on retention time can be correlated with the molecular size of the analyte and decreases in the order: oligosaccharides < sugar alcohols < monosaccharides. The dependence of retention on the column temperature is displayed in Fig. 3-156. It is noteworthy that the elution order of maltose and stachyose changes with increasing temperature. However, the column temperature should not be raised too much. At temperatures above 45 °C, column efficiency decreases, and secondary signals and tailing effects are observed, which may be attributed to the Lobry-de-Bruyn-von-Ekenstein rearrangement. This represents an isomerization occurring predominantly at C-atom 2 of the sugar moiety that leads to a change in configuration via an intermediate endiol:

$$\begin{array}{c}\text{HC}=\text{O}\\|\\\text{HC}-\text{OH}\\|\\\text{HO}-\text{CH}\\|\\\text{HC}-\text{OH}\\|\\\text{HC}-\text{OH}\\|\\\text{H}_2\text{C}-\text{OH}\\\text{D-Glucose}\end{array} \rightleftharpoons \left[\begin{array}{c}\text{HC}-\text{OH}\\||\\\text{C}-\text{OH}\\|\\\text{HO}-\text{CH}\\|\\\text{HC}-\text{OH}\\|\\\text{HC}-\text{OH}\\|\\\text{H}_2\text{C}-\text{OH}\\\text{Endiol}\end{array}\right] \rightleftharpoons \begin{array}{c}\text{H}_2\text{C}-\text{OH}\\|\\\text{C}=\text{O}\\|\\\text{HO}-\text{CH}\\|\\\text{HC}-\text{OH}\\|\\\text{HC}-\text{OH}\\|\\\text{H}_2\text{C}-\text{OH}\\\text{D-Fructose}\end{array} \quad (102)$$

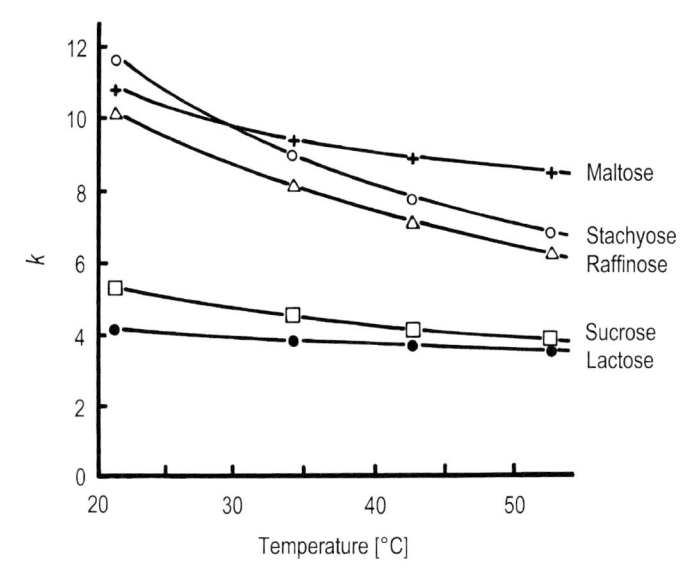

Fig. 3-156. Dependence of the retention of some selected carbohydrates on column temperature. — Separator column: CarboPac PA1; eluant: 0.15 mol/L NaOH; flow rate: 1 mL/min; detection and injection volume: see Fig. 3-154; solute concentrations: 100 mg/L each of the respective carbohydrates.

Therefore, the anion exchange column is operated at temperatures between 20 °C and 40 °C.

The conversion of aldoses into ketoses is also caused by dilute sodium hydroxide solutions. However, these reactions occur very slowly and are not observed in the time period necessary for the chromatography.

3.10.1
Sugar Alcohols

Figure 3-154 reveals that a variety of sugar alcohols, saccharides, and oligosaccharides can be separated under isocratic conditions using 0.15 mol/L sodium hydroxide. It is extremely difficult to separate sugar alcohols; they exhibit a very low affinity towards the stationary phase of a CarboPac PA1 column. Even lowering the sodium hydroxide concentration to 1 mmol/L and applying two separator columns in series allows the baseline-resolved separation only of sugar alcohols with different numbers of C-atoms, e.g. xylitol and sorbitol. A complete separation of the most important hexites such as sorbitol, mannitol, and dulcitol is impossible.

Cyclic polyalcohols (or "cyclitols", according to the nomenclature of sugar alcohols) also have a very low affinity. The most important representative of this class of compounds is inositol, which exhibits a retention behavior comparable to that of pentitols.

Inositol

In its *meso*-configuration, inositol can be phosphorylated. Corresponding procedures for the separation and determination of various inositol phosphates were introduced by Smith and MacQuarrie [162].

Baseline-resolved separations of various sugar alcohols are obtained on Carbo-Pac MA1, which has a much higher ion-exchange capacity as compared to Carbo-Pac PA1. The CarboPac MA1 material consists of a fully aminated, macroporous substrate functionalized with a quaternary alkyl amine. Since this material is not very stable mechanically, the separator column has to be operated at the relatively low flow rate of 0.4 mL/min. At this stationary phase, the most important sugar alcohols present in foodstuffs and physiological samples can be separated with inositol under isocratic conditions (see Fig. 3-157). Due to the extremely high ion-exchange capacity of CarboPac MA1, a highly concentrated NaOH eluant is required. Under these conditions, mono- and di-saccharides are strongly retained, but can be eluted within about one hour using 0.5 mol/L NaOH. Interestingly, resolution between glycerol and inositol can be improved by increasing the NaOH concentration, although the resolution between mannitol and mannose will be sacrificed.

For comparison, Fig. 3-158 shows the separation of sugar alcohols on a conventional polystyrene/divinylbenzene-based cation exchanger in the lead form. The dominating retention mechanism in this case is ligand exchange by interaction of the hydrated metal ions of the stationary phase with the hydroxide groups of the sugar alcohols and saccharides. Usually, de-ionized water is used as a mobile phase. Separator columns of this kind have to be operated above 70 °C to achieve good chromatographic efficiency, which is problematic for the subsequent refractive index detection because of temperature gradients. However, cation exchangers in the calcium or lead form can be connected to the more sensitive and selective pulsed amperometric detector via post-column addition of NaOH concentrate. Figure 3-158 also demonstrates a much higher analysis time for sugar alcohols as compared to CarboPac MA1. This is due to the reversed elution order.

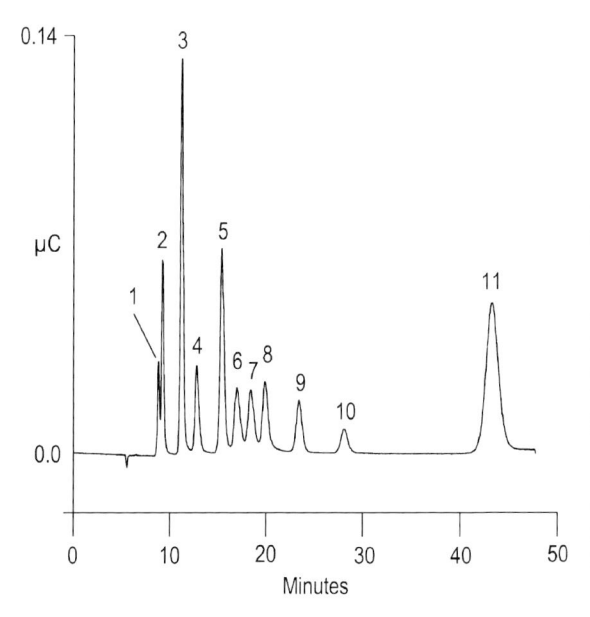

Fig. 3-157. Separation of various sugar alcohols and saccharides on CarboPac MA1. — Eluant: 0.50 mol/L NaOH; flow rate: 0.4 mL/min; detection: pulsed amperometry on a Au working electrode; injection volume: 10 µL; solute concentrations: 2 nmol each of *myo*-inositol (1), glycerol (2), *i*-erythritol (3), xylitol (4), arabitol (5), sorbitol (6), dulcitol (7), mannitol (8), glucose (9), fructose (10), and sucrose (11).

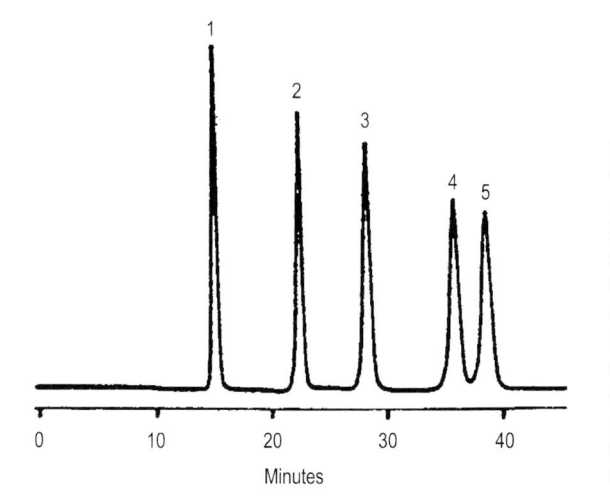

Fig. 3-158. Separation of various sugar alcohols on a conventional cation exchanger in the lead form. — Separator column: Rezex RPM Monosaccharide (Phenomenex); column temperature: 75 °C; eluant: de-ionized water; flow rate: 0.6 mL/min; detection: RI; peaks: (1) glucose, (2), erythritol, (3) mannitol, (4) dulcitol, and (5) sorbitol.

The resolution between glycerol and inositol on CarboPac MA1 in Fig. 3-157 can be improved by applying a gradient elution technique. As illustrated in Fig. 3-159 with cyclites and sugar alcohols present in human urine, even stereo isomers of inositol such as *myo*- and *scyllo*-inositol can be separated. However, this does not apply to all of the eight different stereo isomers of inositol. When working with internal standards on CarboPac MA1, 2-deoxyglucose is recommended instead of rhamnose, because the latter is interfered by galactosamine.

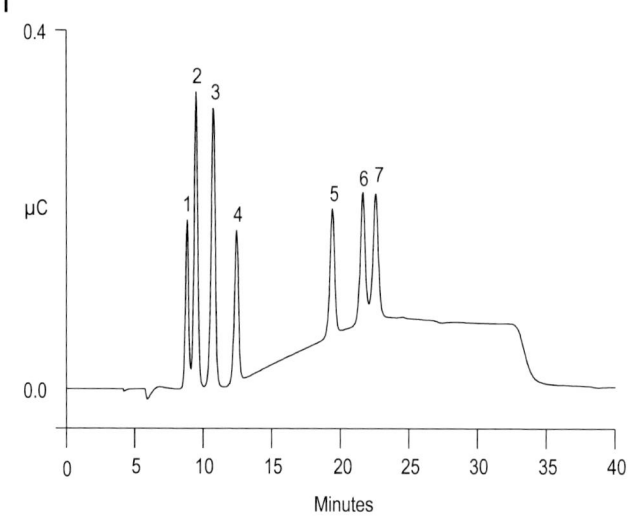

Fig. 3-159. Separation of physiologically relevant cyclites and sugar alcohols on CarboPac MA1. — Eluant: (A) water, (B) 1 mol/L NaOH; gradient: 8% B linearly to 70% B in 18 min; flow rate: 0.4 mL/min; detection: pulsed amperometry on a Au working electrode; injection volume: 25 µL; solute concentrations: 2 nmol each of glycerol (1), *myo*-inositol (2), *scyllo*-inositol (3), erythritol (4), arabitol (5), sorbitol (6), and dulcitol (7).

3.10.2
Monosaccharides

The analysis of the various monosaccharides, which can be separated on CarboPac PA1, PA10, and PA20, is much easier than that for sugar alcohols. This is surprising, because the affinities of these compounds are not very different. However, when the pH value of the mobile phase is lowered to a value corresponding to the pK value of the respective compounds, even small differences in the dissociation behavior contribute to the separation. Thus, the chromatogram in Fig. 3-160 is obtained using a dilute sodium hydroxide solution of $c = 1$ mmol/L. Under these conditions, even epimeric compounds such as glucose, mannose, and galactose are baseline-resolved. Two important aldopentoses — arabinose and xylose — can also be determined in the same run. Remarkable is the elution order, which cannot be correlated with the carbon chain length of the monosaccharides.

As seen in Fig. 3-160, two other compounds belonging to the group of deoxy sugars are eluted prior to aldopentoses and aldohexoses. While fucose (6-deoxy-L-galactose) and other 6-deoxyhexoses (also called methyloses) are abundant in nature in cardio toxic glycosides, 2-deoxy-D-ribose, one of the building blocks of

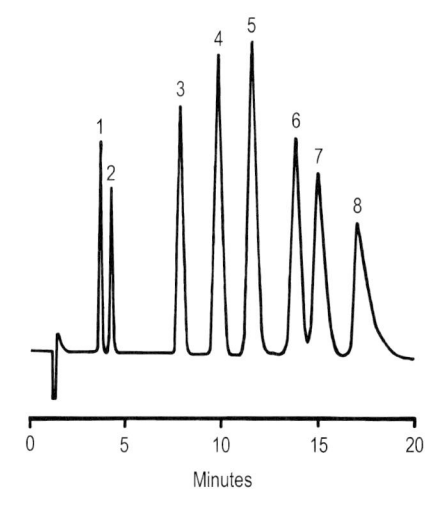

Fig. 3-160. Separation of various monosaccharides on CarboPac PA1. − Eluant: 1 mmol/L NaOH; flow rate: 1 mL/min; detection: pulsed amperometry on a Au working electrode; injection volume: 50 µL; solute concentrations: 25 mg/L each of fucose (1), deoxyribose (2), arabinose (3), galactose (4), glucose (5), and xylose (6), 50 mg/L each of mannose (7) and fructose (8).

deoxyribonucleic acid, is of great biological significance. Rhamnose (6-deoxy-L-mannose), as a fission product of many glycosides, elutes between arabinose and galactose.

The separation of monosaccharides on CarboPac PA1 under isocratic conditions shown above − especially the separation of epimers − can only be carried out with a very dilute NaOH eluant (1 mmol/L). An NaOH concentrate (0.3 mol/L) must be added to the column effluent before it enters the detector cell. Subsequently, pulsed amperometric detection with the conventional pulse sequence of three different potentials is applied. Post-column addition of an NaOH concentrate is necessary to raise the eluant pH to about 13, a value that is optimal for pulsed amperometric detection. The addition of NaOH can be done pneumatically via a metal-free manifold connected to a reaction coil of corresponding dimension in which the column effluent and the NaOH concentrate are mixed. Alternatively, a pump with a corresponding pulse damper can be used. Application of the new quadruple pulse sequence (see Section 7.1.2.2) has rendered the post-column addition of an NaOH concentrate obsolete. However, the linear range is not that large in comparison with the detection at pH 13.

An excellent separation of monosaccharides, including rhamnose and ribose, is obtained with an inverse NaOH gradient [163]. To carry out such a gradient, the separator column is equilibrated with 0.1 mol/L NaOH. After about four minutes, the eluant is switched to pure de-ionized water. Even with a neutral eluant the separation of monosaccharides is possible on a latexed anion exchanger, because the basicity of the ion-exchange groups in the hydroxide form is sufficient to partly convert monosaccharides to the anionic form. Non-ionic interactions also contribute to the retention mechanism. As can be seen from the chromatogram in Fig. 3-161, the analyte pairs arabinose/rhamnose and mannose/

xylose, which can only be separated with a very dilute NaOH eluant under isocratic conditions (see Fig. 3-160), can be separated to baseline with an inverse NaOH gradient. This is of great importance for the analysis of pectin hydrolysates and plant extracts (see Section 9.9).

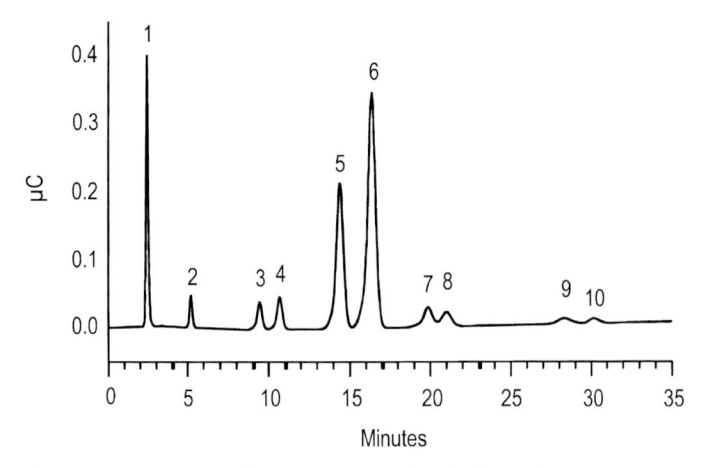

Fig. 3-161. Separation of various monosaccharides by applying an inverse NaOH gradient. — Separator column: CarboPac PA1; eluant: (A) water, (B) 0.1 mol/L NaOH; gradient: 10% B isocratically for 4 min, then switch to 100% A; flow rate: 1 mL/min; detection: see Fig. 3-160; injection volume: 100 µL; solute concentrations: 2 mg/L inositol (1), 0.5 mg/L fucose (2), 1 mg/L rhamnose (3), 1 mg/L arabinose (4), 5 mg/L galactose (5), 10 mg/L glucose (6), 1 mg/L xylose (7), 1 mg/L mannose (8), 1 mg/L fructose (9), and 1 mg/L ribose (10).

The class of amino sugars is obtained by replacing an alcoholic hydroxide group in monosaccharides with an amino group. The two most important representatives are D-glucosamine and D-galactosamine, which can be separated together with their N-acetylated derivatives via anion exchange chromatography. When using CarboPac PA1 as a separator column, 10 mmol/L sodium hydroxide (pH 12) is the recommended eluant. If the concentration falls below this value, a high concentration of sodium hydroxide (0.3 mol/L) must be added to the column effluent before it enters the detector cell. The two chromatograms of various amino sugars in Fig. 3-162 illustrate how important the addition of an NaOH concentrate is for baseline stability. The left-hand chromatogram in Fig. 3-162 clearly demonstrates that peak area evaluation is not possible without post-column addition of a NaOH concentrate because of the baseline shifts.

Post-column addition of concentrated NaOH is not only of significance for the electrochemical detection of amino sugars. For the first time, it allows the separation of carbohydrates with strongly different retention behavior in the same run by applying a gradient elution technique (see Fig. 3-161). As a prerequisite, a constant pH value must be maintained in the detector cell, because in

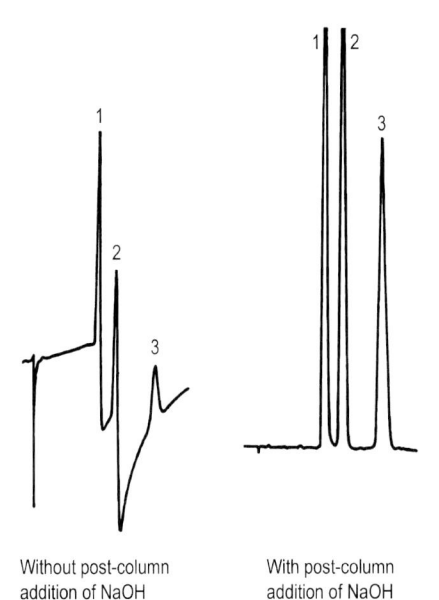

Fig. 3-162. Separation of amino sugars with and without post-column addition of NaOH. — Separator column: CarboPac PA1; eluant: 10 mmol/L NaOH; flow rate: 1 mL/min; detection: see Fig. 3-160; injection volume: 50 µL; solute concentrations: 30 mg/L galactosamine (1), 60 mg/L glucosamine (2), and 30 mg/L N-acetylgalactosamine (3).

electrochemical detection pH variations lead to strong baseline drifts. For gradient elution of mono- and di-saccharides, however, the NaOH concentration in the mobile phase must be increased during the run. The resulting pH change requires post-column addition of a NaOH concentrate in order to ensure baseline stability. The introduction of the gradient technique with pulsed amperometric detection represents one of the greatest innovations in the field of carbohydrate analysis in terms of selectivity and sensitivity.

With the introduction of CarboPac PA10, the analysis of neutral and acidic monosaccharides and amino sugars could be improved significantly. CarboPac PA10 is a solvent-compatible latexed anion exchanger; its selectivity for monosaccharides is optimized by choosing appropriate functional groups. Figure 3-163 shows a selectivity comparison of the two columns taking monosaccharides derived from glycoproteins as an example. As can be seen from the bottom chromatogram in Fig. 3-163, CarboPac PA10 allows a baseline-resolved separation of glucosamine and galactose despite the relatively high eluant concentration (18 mmol/L NaOH). Because CarboPac PA10 is commonly operated with a flow rate of 1.5 mL/min, only about ten minutes are required for an isocratic separation of monosaccharides. Another advantage of the CarboPac PA10 column is the better separation between fucose and by-products formed during the hydrolysis of glycoconjugates. For further improvement of monosaccharide quantitation in samples containing larger amounts of amino acids, an AminoTrap column has been developed, which is used instead of the regular PA10 guard column. Due to its high ion-exchange capacity, the AminoTrap column delays the reten-

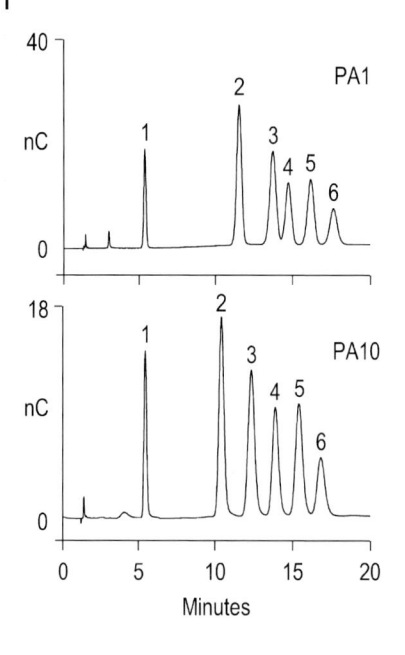

Fig. 3-163. Selectivity comparison between CarboPac PA1 and PA10 exemplified by the separation of monosaccharides derived from glycoproteins. — Eluant: 18 mmol/L NaOH; flow rate: 1 mL/min; detection: see Fig. 3-160; peaks: (1) fucose, (2) galactosamine, (3) glucosamine, (4) galactose, (5) glucose, and (6) mannose.

tion of interfering amino acids, which can also be detected via pulsed amperometry [164]. In contrast, carbohydrates remain practically unaffected in their retention. One of the most strongly interfering amino acids is lysine, which elutes short of galactosamine on CarboPac PA10. With the exception of arginine, which is not retained at all on this stationary phase, all other amino acids elute behind the monosaccharides shown in Fig. 3-163. Figure 3-164 illustrates the influence of the AminoTrap column on the retention of lysine, which was added as an interfering amino acid to the monosaccharide mixture in a 100-fold excess.

Amino acids not only represent a potential interference in monosaccharide separation, they also interfere with pulsed amperometric detection. This is especially true for amine-containing glycoconjugates with a low degree of glycosylation, such as monoclonal antibodies. When an AminoTrap column is combined with CarboPac PA10, the monosaccharide building blocks of the respective MAb hydrolysates can be separated much more effectively (Fig. 3-165). As mentioned above, lysine elutes short of galactosamine under the chromatographic conditions in Fig. 3-165. Since lysine is strongly adsorbed on the surface of the Au working electrode, and its oxidation products desorb very slowly, sensitivity for the subsequently eluting monosaccharides decreases. When using an AminoTrap column, this problem is circumvented, because lysine elutes behind mannose.

The CarboPac PA10 column also solves the problem of oxygen interference in the form of a negative peak, which only appears in the chromatogram at a high sensitivity scale (pmol range). Figure 3-166 again shows a comparison between

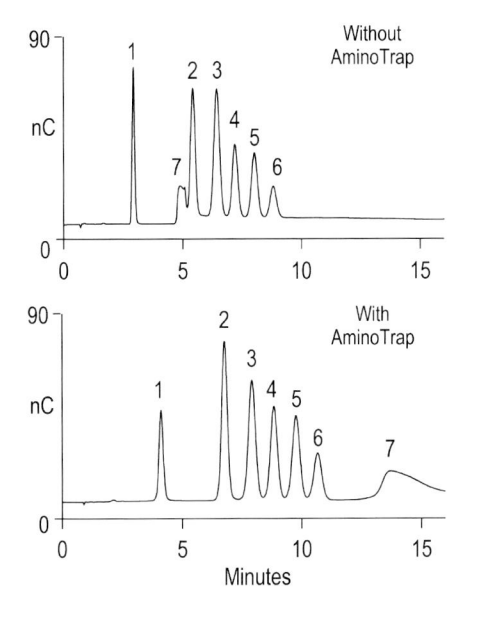

Fig. 3-164. Influence of the AminoTrap column on lysine as an interfering amino acid in monosaccharide analysis. — Separator column: CarboPac PA10; eluant: 18 mmol/L NaOH; flow rate: 1.5 mL/min; detection: see Fig. 3-160; solute concentrations: 1 nmol each of fucose (1), galactosamine (2), glucosamine (3), galactose (4), glucose (5), mannose (6), and 100 nmol lysine (7).

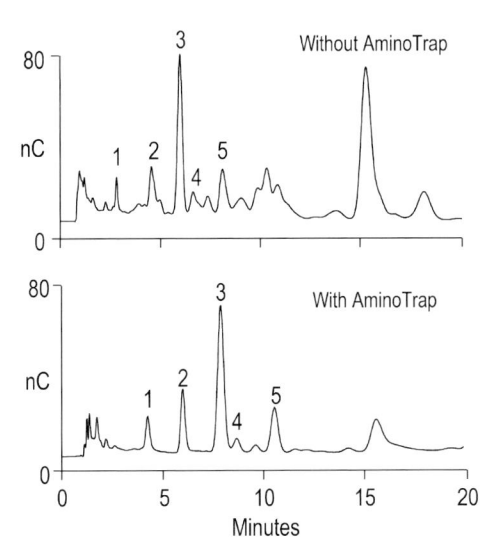

Fig. 3-165. Isocratic separation of a MAb hydrolysate. — Separator column: CarboPac PA10; eluant: 18 mmol/L NaOH; flow rate: 1.5 mL/min; detection: see Fig. 3-160; peaks: (1) fucose, (2) 2-deoxyglucose, (3) glucosamine, (4) galactose, and (5) mannose.

CarboPac PA1 and PA10, with the analysis of monosaccharide from glycoprotein hydrolysates as an example. As can be seen from Fig. 3-166, this negative peak leads to severe interferences in the retention range of galactose, which cannot be quantified under these conditions. The two neighboring peaks are also affected. In contrary, CarboPac PA10 exhibits a higher selectivity for oxygen, which moves the negative peak behind mannose, so that monosaccharide quantitation is not interfered.

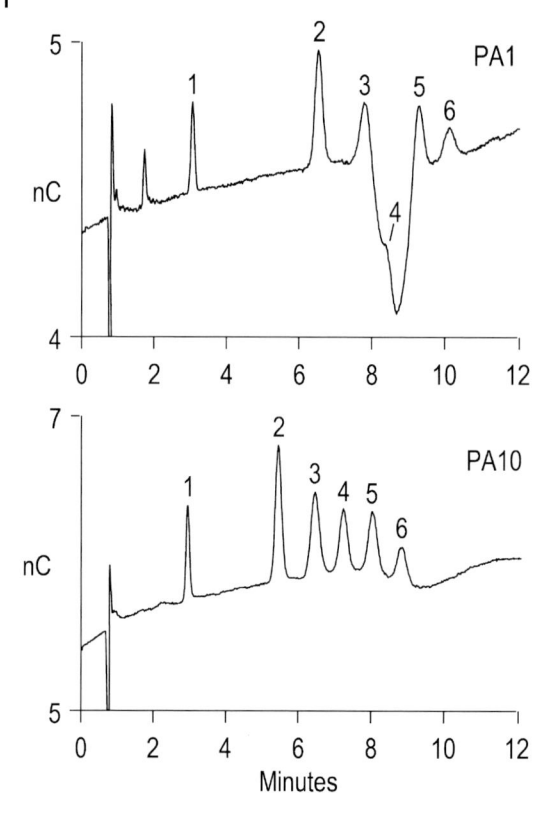

Fig. 3-166. Comparison between CarboPac PA1 and PA10 for monosaccharide analysis in glycoprotein hydrolysates at high sensitivity. — Chromatographic conditions: see Fig. 3-165; solute concentrations: 5 pmol each of fucose (1), galactosamine (2), glucosamine (3), galactose (4), glucose (5), and mannose (6).

Besides amino acids, traces of borate in the mobile phase can also affect monosaccharide analysis due to complexation. Borate ions are among the first to be detected in the slip of water purification systems when their mixed-bed ion exchangers approach the capacity limit. (Borate can be determined with ion-exclusion chromatography and subsequent conductivity detection down to the lowest ng/L level utilizing pre-concentration; see Section 5.5.) If de-ionized water used for preparing the mobile phase contains traces of borate, a significant loss of peak symmetry is observed; this effect is most pronounced with mannose and other reduced saccharides. Even borate concentrations in the lowest mg/L levels are sufficient to cause this effect. With a specially-designed BorateTrap column, which is placed between the analytical pump and the injection valve, borate can be completely removed from the mobile phase.

The ability to separate plant monosaccharides on CarboPac PA10 with pure de-ionized water as an eluant is remarkable. Figure 3-167 shows a baseline-resolved separation of xylose and mannose. However, after every chromatographic run the separator column has to be rinsed with 0.2 mol/L NaOH in order to elute potential matrix components. Pure aqueous eluants tend to lead to retention time variations, which Puls [165] got under control by using a very dilute NaOH eluant (4 mmol/L). In any case, post-column addition of a NaOH concentrate to the column effluent is necessary to ensure sensitive detection.

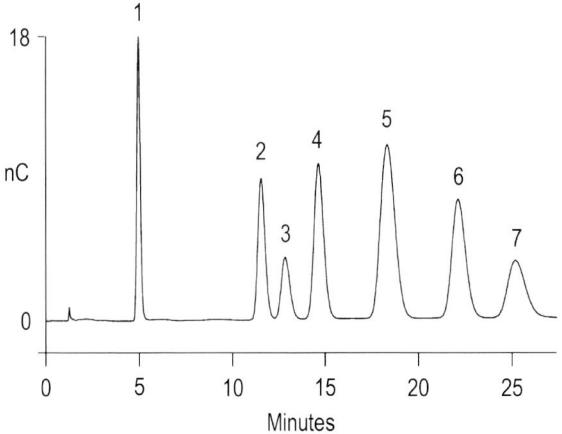

Fig. 3-167. Separation of plant monosaccharides. — Separator column: CarboPac PA10; eluant: de-ionized water (regeneration with 0.2 mol/L NaOH); flow rate: 1.5 mL/min; detection: pulsed amperometry on a gold working electrode with post-column addition of a NaOH concentrate (0.3 mol/L); solute concentrations: 50 mg/L each of fucose (1), arabinose (2), rhamnose (3), galactose (4), glucose (5), xylose (6), and mannose (7).

The latest addition to the CarboPac column family is CarboPac PA20. The CarboPac PA20 column uses 6.5-µm pellicular resin technology to provide improved resolution, peak shapes, and efficiencies for the six common monosaccharides found in glycoprotein hydrolysates. This smaller resin particle is agglomerated with an optimized latex that further improves the column performance by imparting a unique selectivity. This selectivity results in a major improvement in resolution between galactose and glucose, a pair of analytes whose separation has historically been problematic. This optimized anion exchange material is packed in 150 mm × 3 mm PEEK hardware to provide fast separations. The recommended flow rate is 0.5 mL/min, which reduces eluant consumption. Figure 3-168 illustrates the effect of hydroxide concentration on elution times for the six common monosaccharides on CarboPac PA20. In comparison with CarboPac PA10, the retention times of all monosaccharides are approximately 50% shorter despite the use of a less-concentrated hydroxide eluant. A positive side effect of these short retention times is the much higher sensitivity, demonstrated with a 200-fmol standard in Fig. 3-169. (For comparison: Using CarboPac PA10, the limit of determination for standard monosaccharides is on the order of 5 pmol.) As can be seen from this chromatogram, the six monosaccharides can be quantitated at this concentration level.

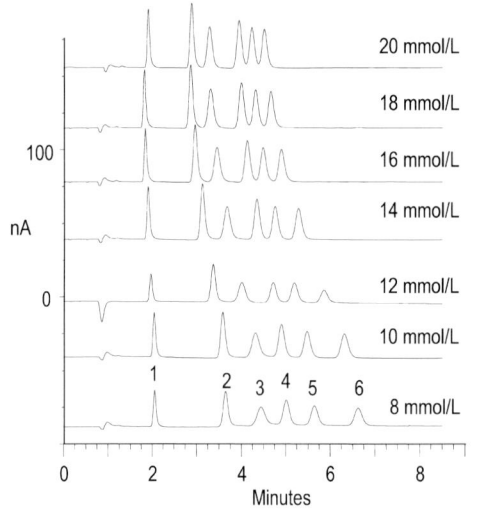

Fig. 3-168. Effect of hydroxide concentration on elution times for the six common monosaccharides found in glycoprotein hydrolysates on CarboPac PA20. — Eluant: 8-20 mmol/L NaOH; flow rate: 0.5 mL/min; detection: pulsed amperometry on a gold working electrode; peaks: (1) fucose, (2) galactosamine, (3) glucosamine, (4) galactose, (5) glucose, and (6) mannose.

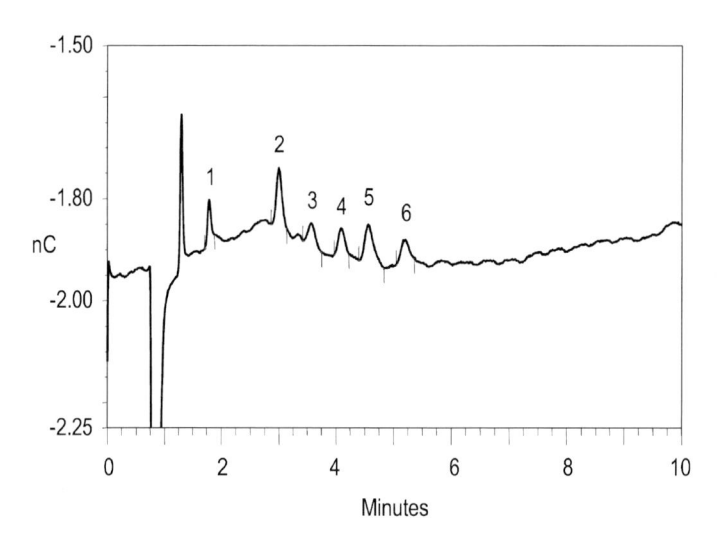

Fig. 3-169. Low level determination of monosaccharides using CarboPac PA20. — Eluant: 12 mmol/L NaOH; flow rate: 0.5 mL/min; detection: pulsed amperometry on a gold working electrode; injection volume: 10 µL; solute concentrations: 200 fmol each of fucose (1), galactosamine (2), glucosamine (3), galactose (4), glucose (5), and mannose (6).

There are two solutions for the problem of changes in selectivity and retention times caused by carbonate impurities when working with hydroxide eluants, especially in the lowest mmol/L concentration range. A very simple and inexpensive solution is provided by adding divalent, electroinactive cations such as Ca(II), Sr(II), or Ba(II) to the mobile phase, which improves peak symmetry for sugar alcohols and, therefore, chromatographic efficiency [166-168]. Moreover, the addition of Ba(II) in form of barium acetate leads to a significant sensitivity increase. Adding an alkaline-earth metal results in a very effective removal of carbonate ions from the alkaline eluant via formation of low solubility salts and via complexation of cyclic and acyclic polyhydroxylated compounds. This phenomenon has been known for a long time and is used for ligand exchange chromatography of mono- and di-saccharides on metal-loaded cation exchangers [154, 169-171]. As an example, Fig. 3-170 compares the separation of rhamnose, arabinose, glucose, ribose, lactose, and cellobiose carried out in a conventional way with 0.1 mol/L NaOH (A) and with 1 mmol/L barium hydroxide added to the NaOH eluant (B).

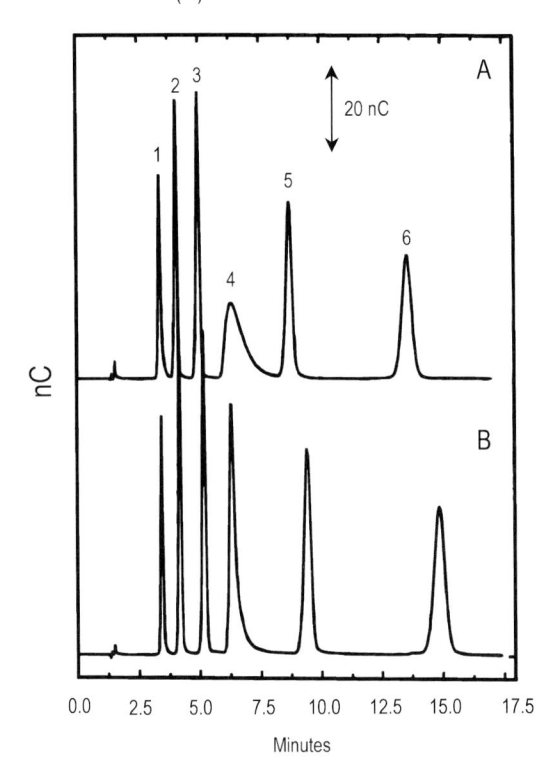

Fig. 3-170. Influence of Ba(II) on carbohydrate separation by anion exchange. — Separator column: CarboPac PA1; eluant: (A) 0.1 mol/L NaOH, (B) 0.1 mol/L NaOH + 1 mmol/L Ba(OH)$_2$, flow rate: 1 mL/min; detection: pulsed amperometry on a Au working electrode; injection volume: 10 µL; solute concentrations: 50 µmol/L each of rhamnose (1), arabinose (2), and glucose (3), 100 µmol/L ribose (4), and 50 µmol/L each of lactose (5) and cellobiose (6); (taken from [168]).

As expected, the addition of barium hydroxide results in a somewhat higher retention of all components; this effect is more pronounced with lactose and cellobiose. The retention increase is attributed to the complete removal of carbonate, which precipitates in the eluant reservoir as low soluble barium carbonate. The improved peak shape for ribose that Angyal [172] attributes to the complexation of sugar molecules by alkaline-earth metals, especially by Sr(II) and Ba(II), is remarkable. Sugar molecules with a six-membered ring and a sequence of an axial (ax), an equatorial (eq) and an axial (ax) hydroxide group bind alkaline-earth metals very strongly. Ribose has such a structure:

α-D-Ribopyranose

(ax1-eq2-ax3) ⇌ (ax2-eq3-ax4)

The sensitivity increase for some selective carbohydrates that is caused by the addition of Ba(II) is shown in Fig. 3-171. While retention of the first three components remains practically the same when adding barium acetate, a slight retention decrease is observed for fructose and sucrose. At first view, this seems surprising; one might suppose that the complete removal of carbonate would lead to a retention increase. However, since in this case Ba(II) has been added as acetate salt, and acetate is a stronger eluant than hydroxide, even traces of acetate in the mobile phase will result in a retention decrease. The sensitivity increase is attributed by Cataldi et al. [166] to the inhibition of the gold oxide formation by alkaline-earth metals in the order Ca(II) > Sr(II) > Ba(II) which, in turn, results in an increase of the electrocatalytic activity of the electrode.

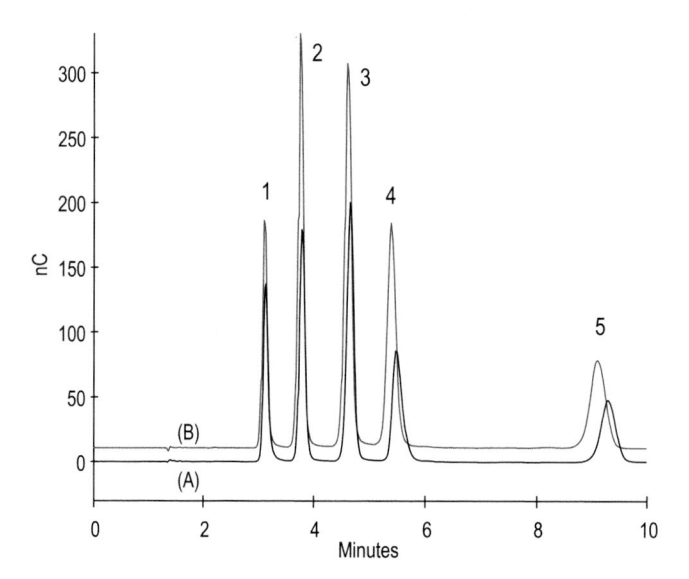

Fig. 3-171. Sensitivity increase for carbohydrates in anion exchange chromatography when adding Ba(II) to the mobile phase. — Separator column: CarboPac PA-100; eluent: (A) 0.1 mol/L NaOH, (B) 0.1 mol/L NaOH + 1 mmol/L Ba(OAc)$_2$; flow rate: 1 mL/min; detection: see Fig. 3-170; injection volume: 25 µL; solute concentrations: 10 mg/L each of rhamnose (1), arabinose (2), glucose (3), fructose (4), and sucrose (5).

3.10 Carbohydrates

The second way to avoid carbonate impurities in hydroxide eluants involves electrolytic hydroxide generation [173]. The operating principle of the Eluant Generator™ was described in Section 3.9. Hydrogen gas that is generated together with KOH at the cathode is removed via a pressure resistant, semipermeable membrane before the eluant leaves the Eluant Generator™ module. However, traces of dissolved hydrogen gas are left in the mobile phase after degassing which, in turn, leads to a significant increase of background current by 20-60 nC in the subsequent pulsed amperometric detection due to the electroactivity of hydrogen. For this reason, the high pressure degas unit is connected to the vacuum degas of the analytical pump according to Fig. 3-172, which increases the degas efficiency.

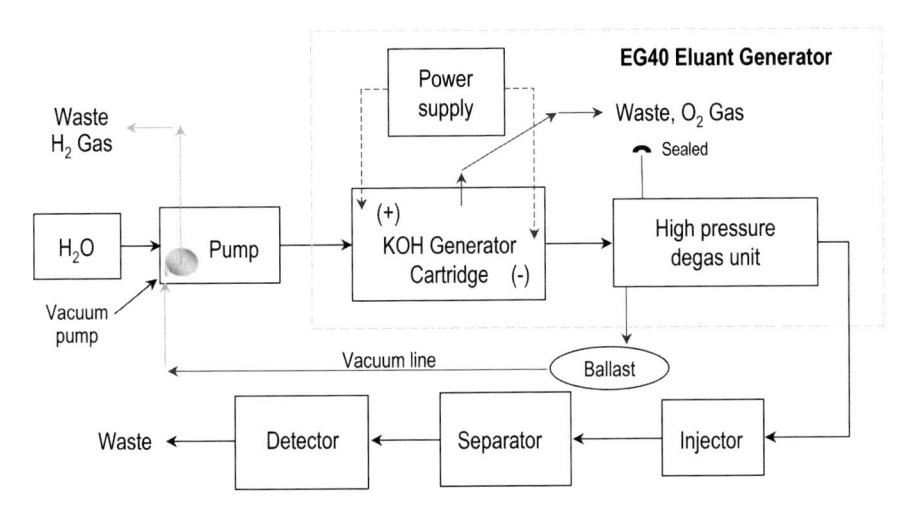

Fig. 3-172. Schematic of a HPAEC-PAD system with vacuum degassed, electrolytically generated hydroxide eluants.

The enormous influence of carbonate impurities in conventionally prepared NaOH eluants on the selectivity of monosaccharide separations is demonstrated in Fig. 3-173. Without rinsing the separator column with a concentrated sodium hydroxide solution after every chromatographic run, the retention time of mannose at the 14th injection becomes identical to that of glucose in the 5th run. So far, acceptable retention time stability could only be achieved by post-chromatographic rinsing of the separator column with a more concentrated NaOH solution. This increases the analysis time per sample to about 50 minutes.

In contrast, Fig. 3-174 shows a series of chromatograms from subsequent injections utilizing an electrolytically generated KOH eluant, starting with a standard (injection #18), followed by an oligosaccharide hydrolysate (injection #19), a protein hydrolysate (injection #20), and another standard (injection #21). After mannose is eluted, hydroxide concentration is raised to $c = 0.08$ mol/L for a short time to remove strongly-retained basic components in protein hydrolysates

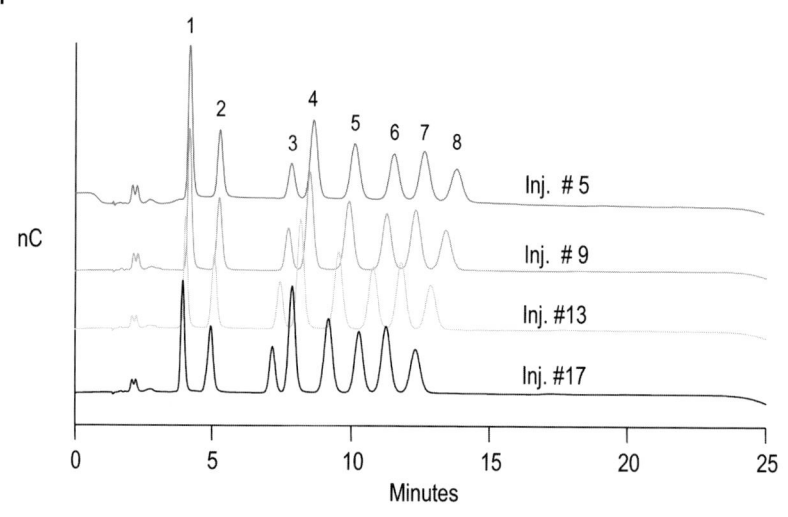

Fig. 3-173. Carbonate-induced retention time decrease for monosaccharides with conventionally prepared NaOH eluants without post-chromatographic column cleanup. — Separator column: CarboPac PA10; eluant: 18 mmol/L NaOH; flow rate: 1 mL/min; detection: see Fig. 3-170; peaks: (1) mannitol, (2) fucose, (3) 2-deoxyglucose, (4) galactosamine, (5) glucosamine, (6) galactose, (7) glucose, and (8) mannose.

from the column. The baseline dip after about 24 minutes results from the reduction of oxygen. Oxygen reaches the column in the form of a minute bubble when using an autosampler to inject small sample volumes. Total analysis time per sample is about 30 minutes. Table 3-27 characterizes the response factors for fucose, galactosamine, and mannose as constant over a longer period of time, without applying post-chromatographic column cleanup.

Mono- and di-phosphoric acid esters of the sugars are particularly important in enzymatic carbohydrate metabolism. The phosphate group is attached to the glycosidic hydroxide group or to a primary alcohol group. As primary esters of phosphoric acid, phosphorylated sugars are very resistant to hydrolysis. Biochemically, they are digested only by enzymes. Figure 3-175 shows a separation of phosphorylated monosaccharides. Again, simultaneous detection of mono- and di-phosphoric acid esters is only possible by applying a gradient elution technique. Because phosphorylated sugars are strongly retained on the stationary phase of an anion exchanger, large amounts of sodium acetate, which has a much higher elution strength than sodium hydroxide, are added to the mobile phase. In this case, sodium hydroxide primarily serves to adjust the pH required for electrochemical detection. It is important to note that it is possible to distinguish between anomers of phosphorylated monosaccharides with anion exchange chromatography. The comparison of the retention behavior of the two anomers **4** and **6** in Fig. 3-175 reveals that an equatorially bound phosphate rest in β-D-glucose-1-phosphate leads to a markedly higher retention than the axially bound rest in α-D-glucose-1-phosphate. The separation of the glucopyranose anomer mixture present in the solution, however, is not possible.

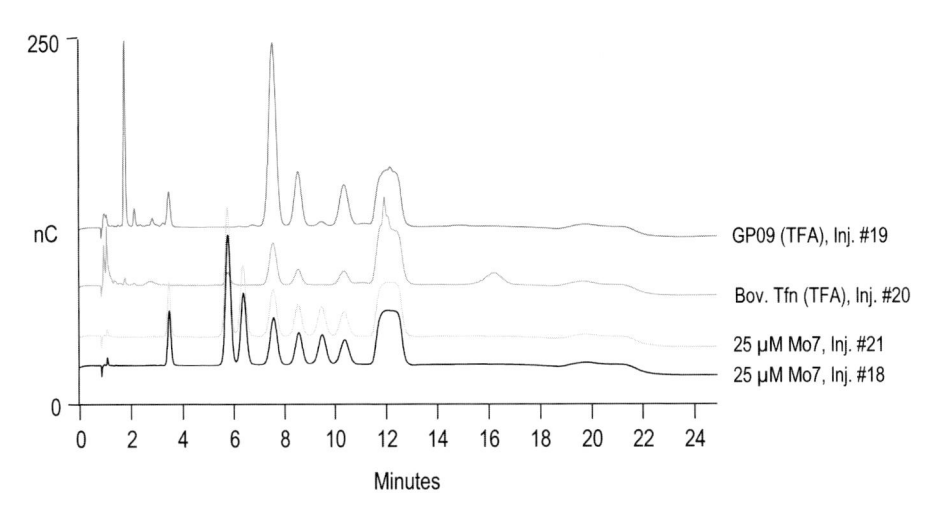

Fig. 3-174. Retention time stability of selected monosaccharides in standards and hydrolysates with electrolytically generated hydroxide eluants (short post-chromatographic column cleanup with 0.08 mol/L KOH after the elution of mannose).

Table 3-27. Stability of the response factors for fucose, galactosamine, and mannose as a function of the number of injections without post-chromatographic column cleanup.

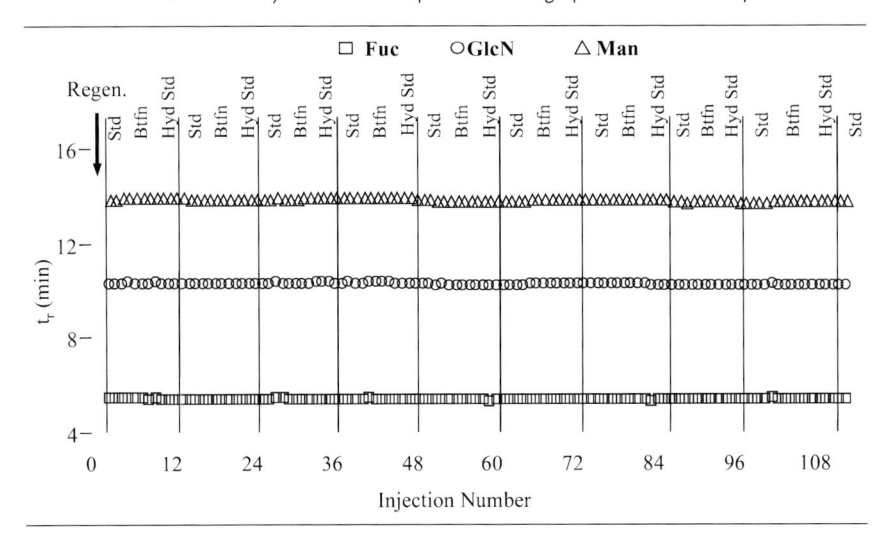

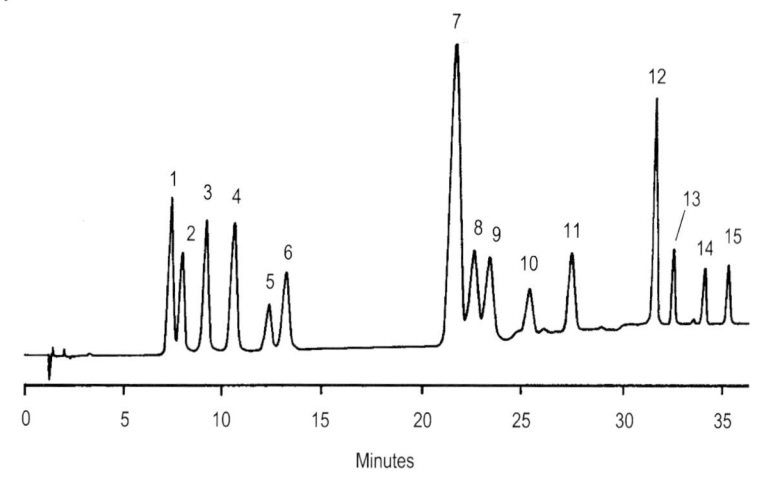

Fig. 3-175. Gradient elution of phosphorylated mono- and di-saccharides. – Separator column: CarboPac PA1; eluant: (A) 0.1 mol/L NaOH, (B) 0.1 mol/L NaOH + 1 mol/L NaOAc; gradient: linear, 10% B to 20% B in 20 min, then to 50% B in 10 min; flow rate: 1 mL/min; detection: pulsed amperometry on a Au working electrode; injection volume: 50 µL; solute concentrations: 22.6 mg/L α-D-galactosamine-1-P (1), 9 mg/L α-D-glucosamine-1-P (2), 35 mg/L each of α-D-galactose-1-P (3) and α-D-glucose-1-P (4), 29.2 mg/L α-D-ribose-1-P (5), 35 mg/L β-D-glucose-1-P (6), 75 mg/L D-glucosamine-6-P (7), 40.8 mg/L D-galactose-6-P (8), 25 mg/L D-glucose-6-P (9), 19.2 mg/L D-fructose-1-P (10), 8.4 mg/L D-fructose-6-P (11), 61.6 mg/L α-D-glucuronic acid-1-P (12), 21.2 mg/L α-D-glucose-1,6-diP (13), 18.4 mg/L each of β-D-fructose-2,6-diP (14), and D-fructose-1,6-diP (15).

3.10.3

Oligosaccharides

Disaccharides are formed by coupling a hemiacetalic hydroxide group of a monosaccharide with any hydroxide group of a second sugar molecule. Trisaccharides result from a glycosidic formation between three molecules and, in an analogous way, tetra-, penta-, and hexa-saccharides are formed.

In disaccharides, one distinguishes between reducing and non-reducing compounds, depending on the type of coupling. Non-reducing disaccharides are characterized by different hexoses coupling across the hemiacetalic hydroxide group. The two most important representatives are trehalose and sucrose. In reducing sugars, the hemiacetalic hydroxide group of one sugar molecule is linked via glycosidic coupling to the alcoholic hydroxide group of the other sugar. These are called maltose-type disaccharides. In the naturally-occurring disaccharides of this group, the reducing sugar molecule is coupled to the glycosidic sugar moiety in 4- or 6-position.

Maltose Isomaltose

Haworth Projections

*without configuration assignment

Reducing and non-reducing disaccharides exhibit different retention behaviors, but can be analyzed in a single run with a gradient elution technique. To reduce the analysis time, small amounts of acetic acid are added to the sodium hydroxide eluant. Figure 3-176 shows a separation of different disaccharides on a CarboPac PA1 anion exchanger.

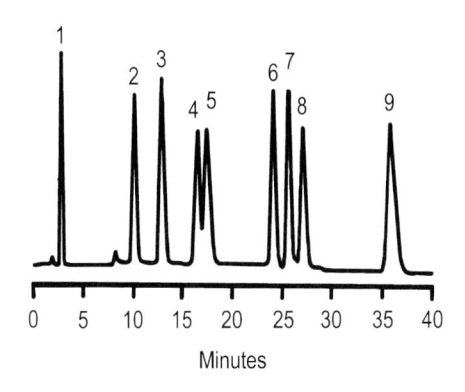

Fig. 3-176. Gradient elution of various disaccharides. – Separator column: CarboPac PA1; eluent: (A) water, (B) 0.05 mol/L NaOH + 1.5 mmol/L acetic acid; gradient: 20% B for isocratically for 5 min, then linearly to 100% B in 15 min; flow rate: 1 mL/min; detection: pulsed amperometry on a Au working electrode; injection volume: 50 µL; solute-concentrations: 10 mg/L trehalose (1), 25 mg/L each of sucrose (2), lactose (3), isomaltose (4), melibiose (5), gentiobiose (6), and cellobiose (7), and 50 mg/L each of turanose (8), and maltose (9).

The separation between the two disaccharides lactose and lactulose, which coelute under the chromatographic conditions in Fig. 3-154, is extremely difficult.

Lactose Lactulose

Separation of these two carbohydrates is of great importance for the dairy industry as well as for clinical analysis. In milk products, for example, a failure in the pasteurization process can force lactose to isomerize to lactulose due to heating.

Because lactulose is not metabolized and ingestion of large amounts of lactulose can produce diarrhoea [174], the lactulose content in milk products is a quality feature for a perfect pasteurization. The fact that lactulose cannot be metabolized makes it valuable for clinical applications. If lactulose is administered to a patient, kidney permeability can be examined via the urinary excretion rate.

A successful isocratic separation of lactose and lactulose was first achieved by Kynaston et al. [175] by adding small amounts of zinc acetate to an NaOH eluant, which resulted in a baseline-resolved separation due to complexation. However, occasional precipitations in the mobile phase are a serious problem; it can only be avoided by adjusting concentration ratio between NaOH and zinc acetate to at least 100:1. Moreover, Bruggink and Bastiaanse [176] observed problems with the pumping system during long-term operation; alkaline eluants containing zinc acetate are not recommended, because the long-term stability of this method cannot be ensured.

A significantly better separation between the two disaccharides is obtained with a more complex NaOH gradient. Instead of zinc acetate, small amounts of sodium acetate are added to the mobile phase in order to elute more strongly-retained saccharides within an acceptable time frame. Figure 3-177 shows a respective chromatogram with carbohydrates, which are relevant for the analysis of urine to assess kidney functionality. In addition to lactose and lactulose, these carbohydrates include *meso*-erythritol, mannitol, sucrose, and turanose. Under these chromatographic conditions, resolution between lactose and lactulose is so large that even an excess of lactose does not prohibit clear quantitation of lactulose.

Modern disaccharide-based artificial sweeteners such as isomaltulose (palatinose) [177], trehalose, and leucrose [178] can also be separated on CarboPac PA1 [179]. Sweeteners of this kind, with their graduated sweetness, broaden the variety of sweeteners, especially for dietary formulations. Of great importance is the artificial sweetener palatinitol, which is synthesized via hydrogenation of isomaltulose. Hydrogenation of isomaltulose leads to an equimolar mixture of α-D-glucopyranosido-1,6-sorbitol (GPS) and -mannitol (GPM). Both compounds can be separated in the same run with sorbitol, mannitol, and isomaltose with a 0.1 mol/L NaOH eluant (Fig. 3-178) [180].

Tri- and tetrasaccharides such as raffinose and stachyose may be separated under the chromatographic conditions given in Fig. 3-154. As with phosphorylated monosaccharides, shorter analysis times are obtained when adding sodium acetate to the eluant. The same holds for the analysis of tri- and tetra-saccharides of the kestose group, which also function as artificial sweeteners.

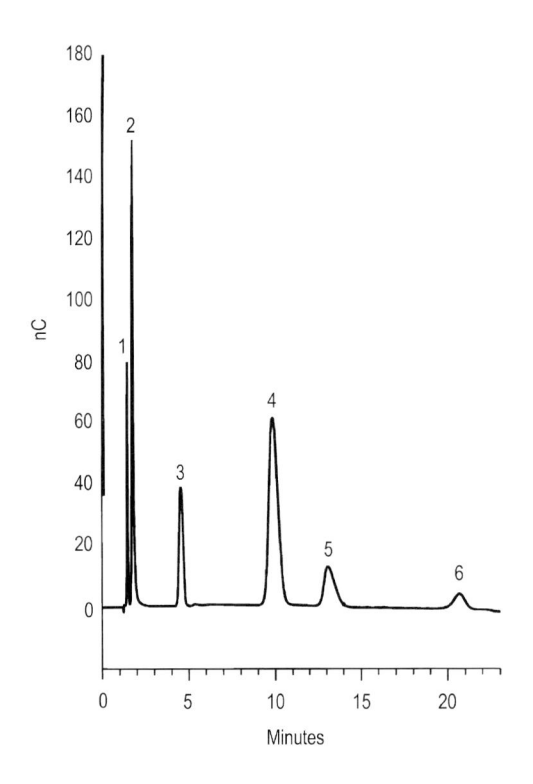

Fig. 3-177. Gradient elution of lactose and lactulose. — Separator column: CarboPac PA1; eluant: (A) 0.125 mol/L NaOH, (B) 0.125 mol/L NaOH + 0.5 mol/L NaOAc, (C) 0.0125 mol/L NaOH, (D) 0.008 mol/L NaOAc; gradient: 70% C + 30% D isocratically for 3.5 min, then linearly to 4% A + 26% C + 70% D in 13 min; hold for 3.5 min, then to 50% A + 50% B in 7.5 min; flow rate: 1 mL/min; detection: pulsed amperometry on a Au working electrode; injection volume: 25 µL; solute concentrations: 5 mg/L *meso*-erythritol (1), 20 mg/L mannitol (2), 20 mg/L sucrose (3), 20 mg/L lactose (4), 5 mg/L lactulose (5), and 5 mg/L turanose (6).

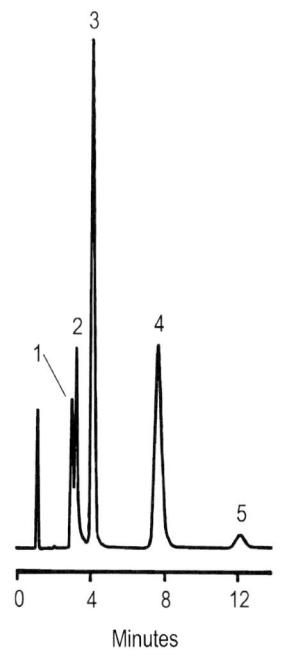

Fig. 3-178. Separation of the artificial sweetener palatinitol. — Separator column: CarboPac PA1; eluant: 0.1 mol/L NaOH; flow rate: 1 mL/min; detection: pulsed amperometry on a gold working electrode; injection volume: 50 µL; peaks: (1) sorbitol, (2) mannitol, (3) 10 mg/L α-D-glucopyranosido-1,6-sorbitol (GPS), 10 mg/L α-D-glucopyranosido-1,6-mannitol (GPM), and (4) isomaltose.

3 Anion Exchange Chromatography (HPIC)

[Structures of 1-Kestose, Nystose, Neokestose, and 6-Kestose]

Their separation is illustrated in Fig. 3-179. The addition of 0.02 mol/L sodium acetate to the NaOH eluant is sufficient for a baseline-resolved separation in less than 15 minutes.

Larger oligosaccharides can also be analyzed by anion exchange chromatography. However, chromatographic analysis within an acceptable time frame requires gradient elution. Figure 3-180 shows the gradient elution of maltose oligomers (DP2 to DP6) as an example.

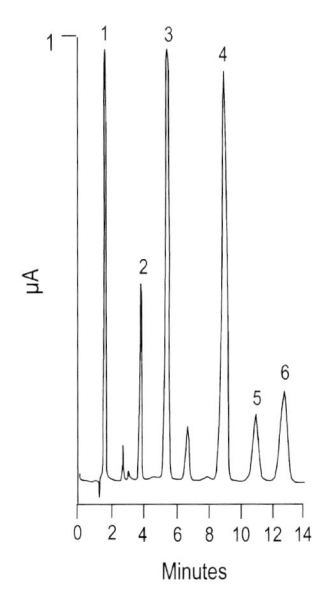

Fig. 3-179. Separation of kestose and related oligosaccharides. – Separator column: CarboPac PA1; eluant: 0.1 mol/L NaOH + 0.02 mol/L NaOAc; flow rate: 1 mL/min; detection: pulsed amperometry on a gold working electrode; peaks: (1) galactinose, (2) sucrose, (3) 1-kestose, (4) 6-kestose, (5) neo-kestose, and (6) nystose.

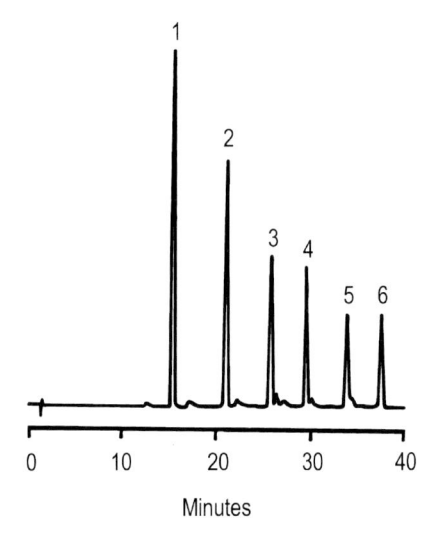

Fig. 3-180. Separation of maltose oligomers. – Separator column: CarboPac PA1; eluant: (A) 0.1 mol/L NaOH, (B) 0.1 mol/L NaOH + 1 mol/L NaOAc; gradient: 100% A isocratically for 2 min, then linearly to 100% B in 200 min; flow rate: 1 mL/min; detection: see Fig. 3-179; injection volume: 50 µL; solute concentrations: 166 mg/L each of maltose (1), maltotriose (2), maltotetraose (3), maltopentaose (4), maltohexaose (5), and maltoheptaose (6).

3.10.4
Polysaccharides

Polysaccharides consist of a number of monosaccharide residues and occur in the vegetable and animal kingdoms as storage and skeletal carbohydrates. In the past, polysaccharides were also separated with chemically bonded octadecyl and aminopropyl phases using water or water/acetonitrile mixtures as an eluant [156, 182-183]. Although these phases exhibit high chromatographic efficiencies, only oligomers with a comparatively low degree of polymerization (<DP15) can be analyzed. This is a result of the detection system being employed. Because until now carbohydrates were detected almost exclusively via refractive index measurements, gradient elution techniques, essential for the separation of higher oligo- and poly-saccharides, could not be applied.

Using anion exchange chromatography with pulsed amperometric detection, polymers up to DP70 may be analyzed. The necessary gradient elution technique is based on the combination of sodium hydroxide and sodium acetate eluants as described above.

The performance of anion exchange chromatography is demonstrated using the separation of cyclodextrins [184] as an example. Cyclodextrins are macrocycles with six to eight glucose residues. Depending on the number of glucose residues, one distinguishes between α-, β-, and γ-cyclodextrin. Isolated from liquified potato starch after fermentation based on the enzyme system of *Bac. macerans*, they are constituents of pharmaceutical and cosmetic products. The separation in Fig. 3-181 reveals that cyclodextrin retention cannot be correlated with its molecular weight.

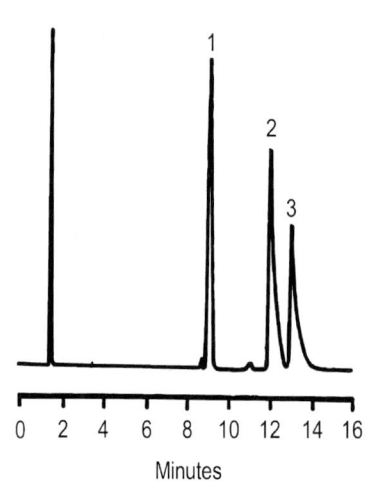

Fig. 3-181. Separation of α-, β-, and γ-cyclodextrin. — Separator column: CarboPac PA1; eluant: (A) 0.1 mol/L NaOH, (B) 0.1 mol/L NaOH + 0.5 mol/L NaOAc; gradient: linear, 100% A to 100% B in 20 min; flow rate: 1 mL/min; detection: pulsed amperometry on a gold working electrode; injection volume: 50 µL; solute concentrations: 500 mg/L each of α- (1), β- (2), and γ-cyclodextrin (3).

In dextrans, in which α-D-glucose residues are linked in 1,6-position, the retention increases with increasing molecular weight. Dextrans are mainly produced extracellularly by lactic acid bacteria. Sucrose, for example, is transformed into glucose and fructose in a fermentation process by *Leuconostoc mesenteroides* with transglucosidase. Since only glucose is polymerized, dextrans are so-called homopolymers (α-1,6-glucans). After partial hydrolysis, their molecular weight ranges from 20,000 to 70,000 g/mol. Figure 3-182 shows the separation of a dextran hydrolysate (Dextran 1000). In addition to the polymers up to DP18, small amounts of glucose, fructose, and sucrose from the microbial process were detected. From Fig. 3-182 it appears that a separation of the α- and β-forms (pair of anomers) of the individual polysaccharides is not possible by anion exchange chromatography. The alkalinity of the mobile phase accelerates the conversion of the anomers into each other and, therefore, the sum of the signals is obtained for each anomeric pair. This is in contrast with the double-peaks, representing the α- and β-forms, respectively, which are observed when separating those compounds on chemically bonded aminopropyl phases.

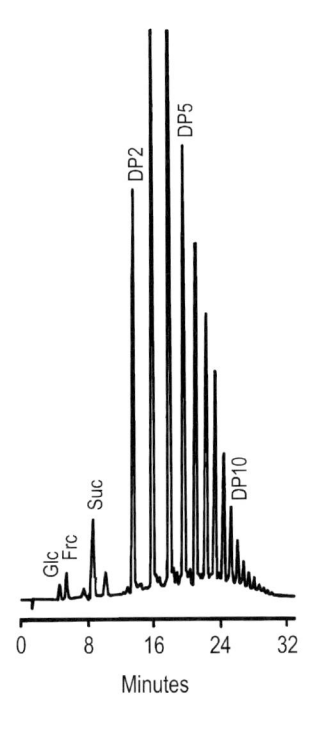

Fig. 3-182. Gradient analysis of a dextran hydrolysate (Dextran 1000). – Chromatographic conditions: see Fig. 3-181; injection: 50 µL of a 0.3% solution.

With the introduction of CarboPac PA-100 (see Table 3-25), resolution between the various oligosaccharides can be improved significantly. For comparison, Fig. 3-183 shows the separation of the same sample under optimized chromatographic conditions utilizing this special anion exchanger. Instead of a linear gradient, a convex gradient profile has been applied, which results in an improved resolution in the higher molecular weight range, especially when analyzing polysaccharides. In this way, linear oligosaccharides present as major components can be separated from branched structures, which elute as small peaks ahead of the major components. Depending on the required resolution, both CarboPac PA1 and PA-100 are suitable for separating oligosaccharide mixtures.

When the slope of the gradient profile is made steeper by adding more sodium acetate per time unit, even higher polymers can be resolved. Figure 3-184 shows this effect with another dextran hydrolysate separation (Dextran 5000), where polymers up to DP50 can be detected. The extremely long analysis time when separating on CarboPac PA1 (Fig. 3-184, bottom chromatogram) is attributed to the application of a linear gradient with a shallow profile in order to obtain

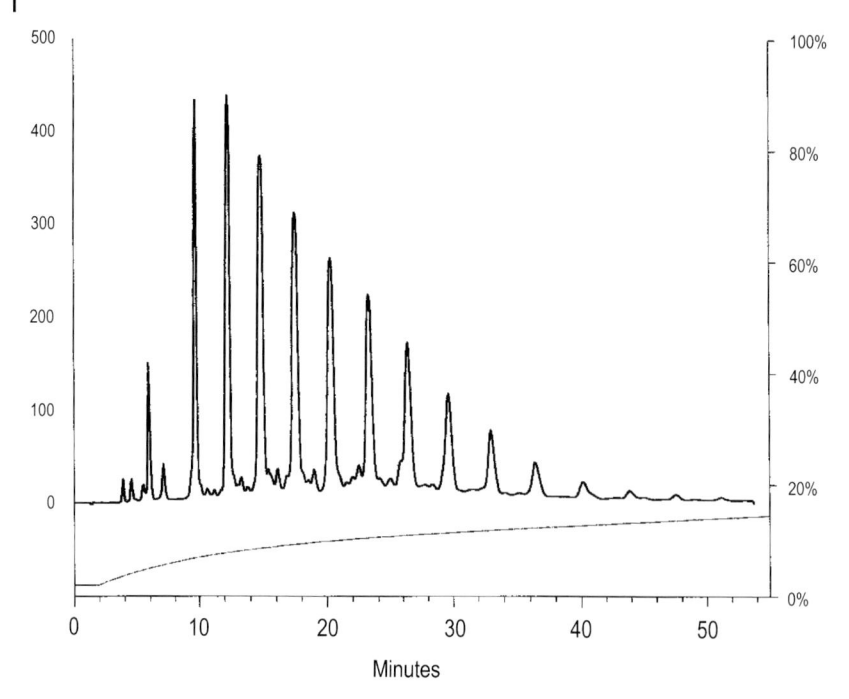

Fig. 3-183. Gradient analysis of a dextran hydrolysate (Dextran 1000) on CarboPac PA-100. – Eluant: (A) 0.1 mol/L NaOH; (B) 0.1 mol/L NaOH + 0.5 mol/L NaOAc; gradient: 2% B isocratically for 2 min, then convexly to 15% B in 58 min; flow rate: 1 mL/min; detection: see Fig. 3-181; injection: 25 µL of a 0.1% solution.

enough resolution in the higher molecular weight range. In contrast, when applying the convex gradient profile in combination with CarboPac PA-100, an almost equivalent resolution is achieved in a much shorter time (Fig. 3-184, top chromatogram).

Storage carbohydrates such as polyfructans may also be analyzed under similar chromatographic conditions. The most important representative of this class of compounds is inulin, which is increasingly being used to diagnose kidney function. The structure of inulin corresponds to a poly-β-(2-1)-fructofuranosan. It contains about 30 D-fructose residues and a terminal sucrose molecule. The separation of a 0.3% inulin solution in 0.1 mol/L NaOH is displayed in Fig. 3-185 exhibiting a polymer distribution between DP10 and DP50. Because the investigated inulin was pretreated with water, polymer residues smaller than DP10 have been washed out and, thus, are not detectable.

Similar chromatographic conditions are used for the analysis of $(1 \rightarrow 4)$-α-D-glucans of a short-chain amylose EX-1 (DP $\approx$ 17) with a chain length of up to DP50. Amylose with a weight distribution of 15% to 20% is one of the two important components of starch. The chromatogram in Fig. 3-186 was obtained on CarboPac PA1 by Koizumi et al. [185] utilizing a modified NaOH/NaOAc gradient.

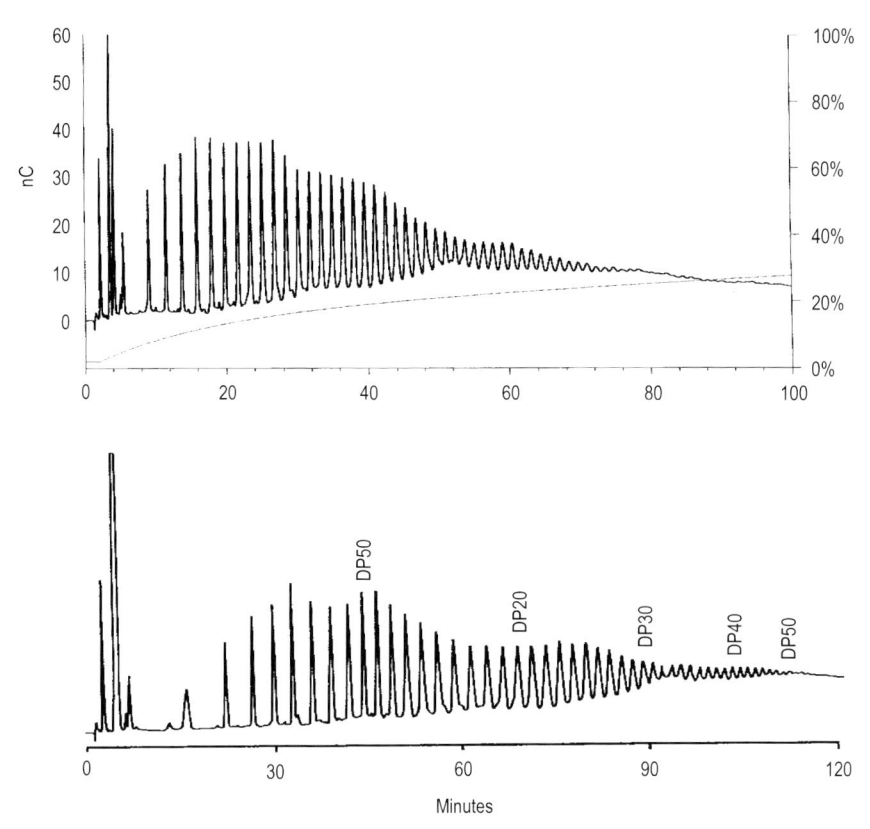

Fig. 3-184. Gradient analysis of a dextran hydrolysate (Dextran 5000). — Separator columns: (top) CarboPac PA-100, (bottom) CarboPac PA1; eluant: see Fig. 3-180; gradient: (top) 2% B isocratically for 2 min, then convexly to 30% B in 118 min, (bottom) linear, 0% B to 17% B in 40 min, then to 100% B in 10 min; flow rates: 1 mL/min; detection: see Fig. 3-182; injection: (top) 25 µL of a 0.1% solution, (bottom) 50 µL of a 0.3% solution.

The major constituent of starch, amylopectin, can also be analyzed after depolymerization [186]. As can be seen from the chromatogram in Fig. 3-187, information on the chain length distribution of amylopectin derived from rice, for example, can be obtained in less than 40 minutes.

The most widely distributed polysaccharide on earth is cellulose, which is built up in plants as a skeletal compound. Cellulose is depolymerized by acetolysis, i.e. by treatment with a mixture of acetic acid anhydride and sulfuric acid. The β-glucopyranose residues in cellulose are linked in 1-4-position. Figure 3-188 shows the chromatogram of an aqueous solution of β-(1-4)-linked D-glucans, which was obtained by Koizumi et al. [185] after partial acetolysis of cellulose and subsequent deacetylation in 0.15 mol/L NaOH. Oligomers up to DP28 were detected.

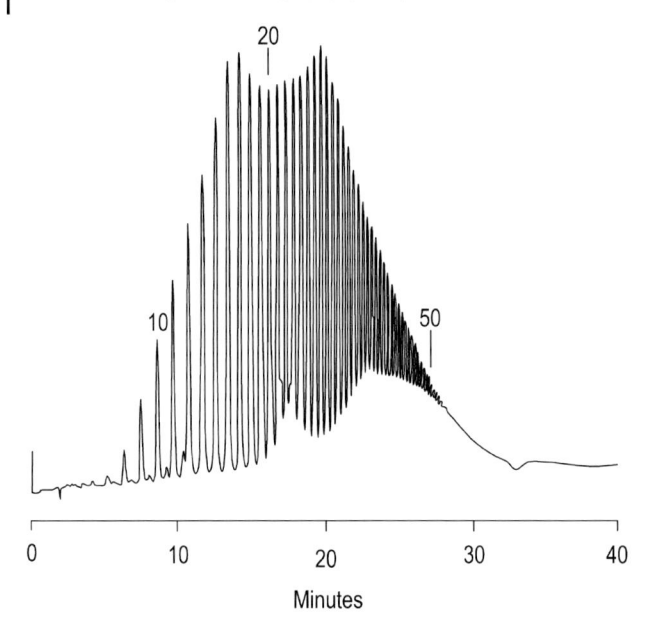

Fig. 3-185. Gradient analysis of an inulin hydrolysate. — Separator column: CarboPac PA1; eluant: (A) 0.1 mol/L NaOH, (B) 0.1 mol/L NaOH + 1 mol/L NaOAc; gradient: linear, 20% B to 60% B in 40 min; flow rate: 1 mL/min; detection: see Fig. 3-182; injection: 50 µL of a 0.3% solution.

Still higher polymers were detected after enzymatic degradation of pullulan, a polysaccharide consisting of α-D-glucan and glycosyl residues joined by 1,4- or 1,6-linkages. Pullulan is produced by the fungus *Aureobasidium pullulans* and can be digested by pullulanase (E.C. 2.2.1.41). This is an enzyme that specifically hydrolyzes α-1,6-linkages. In this reaction, maltotriose residues with different degrees of polymerization are formed; they can be separated chromatographically [184]. The corresponding chromatogram is displayed in Fig. 3-189 and shows the separation of polymers up to DP69. Thus, polysaccharides with a molecular weight of about 10,000 g/mol can be separated and detected via anion exchange chromatography.

3.10.5
Carbohydrates Derived from Glycoproteins

Of particular importance are carbohydrates that are connected to proteins. They are called glycoproteins. Until the 1960s, protein chemists did not give much credit to carbohydrates. Today, we know that most of the eukaryotic proteins exist in a glycosylated form and that 90% of their total molecular weight is attributed to the carbohydrate moiety. Glycoproteins are found in all biological areas

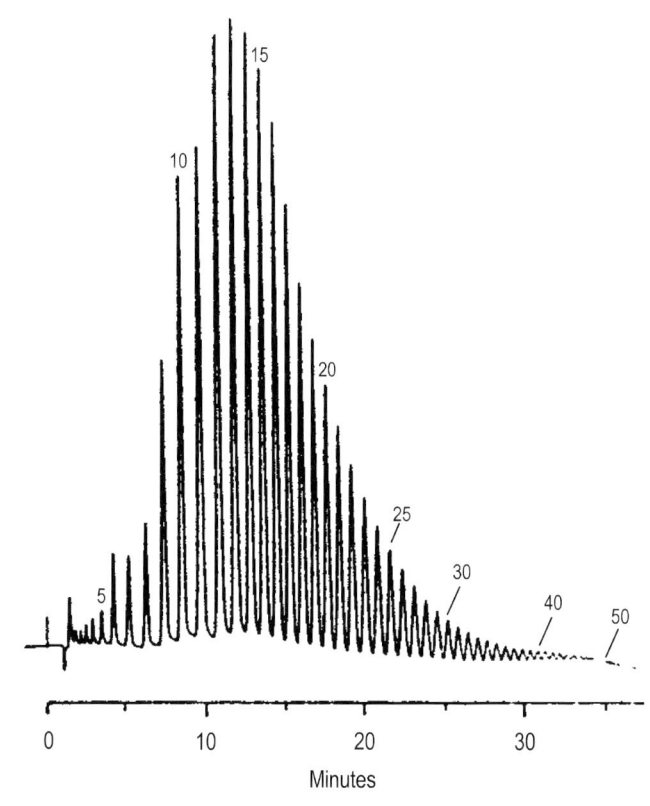

Fig. 3-186. Separation of (1 → 4)-α-D-glucans of a short-chain amylose EX-1. — Separator column: CarboPac PA1; eluant: (A) 0.15 mol/L NaOH, (B) 0.15 mol/L NaOH + 0.5 mol/L NaOAc; gradient: linear, 40% B to 60% B in 10 min, then to 80% B in 20 min; flow rate: 1 mL/min; detection: see Fig. 3-182; injection volume: 50 µL; solute concentration: 3 mg/mL amylose EX-1; (taken from [185]).

— in enzymes, transport and receptor proteins, hormones, structural and other proteins [187]. While the primary structure of a protein is genetically defined, carbohydrate moieties are enzymatically coupled to the protein during or after the translation (co- or post-translational). Thus, glycoproteins possess a microheterogeneous carbohydrate composition that can be responsible for a certain biological or therapeutic activity. The diversity of carbohydrate structures is also observed with other products of metabolism such as glycolipids and proteoglycans and their intermediates [188, 189].

The most important function of carbohydrates is their role as a mediator in cellular communication [190]. They participate in processes such as insemination, differentiation and aggregation of organs as well as in bacterial and viral cell infection [191]. It is obvious that new therapeutic possibilities are opening up in the fight against disease. Free oligosaccharides found in human milk are

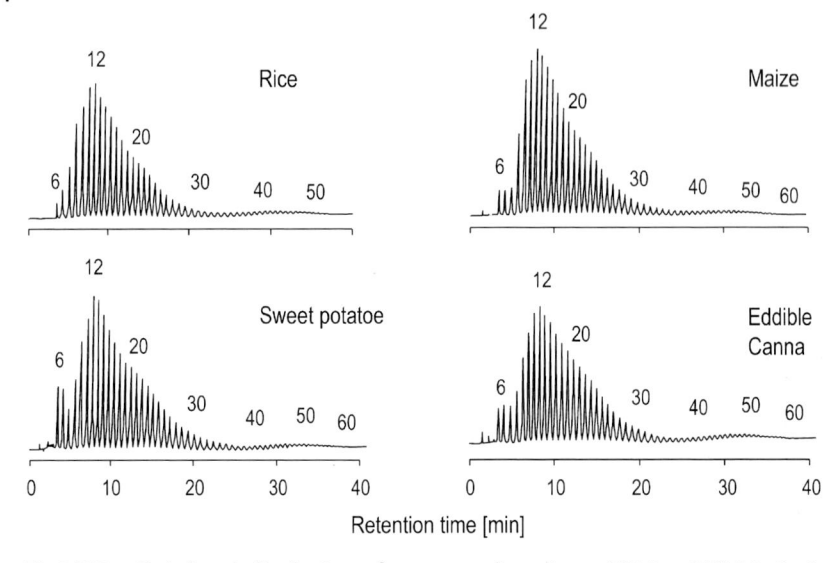

Fig. 3-187. Chain length distributions of amylopectins of different origin after depolymerization. – Separator column: CarboPac PA1; eluant: (A) 0.15 mol/L NaOH, (B) 0.15 mol/L NaOH + 0.5 mol/L NaOAc; gradient: linear, 40% B to 50% B in 2 min, then to 60% B in 8 min, then to 80% B in 30 min; flow rate: 1 mL/min; detection: see Fig. 3-182; injections: 25 µL; (taken from [186]).

also of great importance. Those sugars seem to mask the binding sites of the bacteria on the epithelium cells [192, 193] and, thus, to prophylactically protect the child from bacterial infections. For these reasons, there has been great interest for a number of years in investigating the composition and structure of the carbohydrates.

Oligosaccharides can be linked at different sites of the protein in the N- and/or O-glycosidic form. In the N-glycosidic form, N-acetylglucosamine is bound β-glycosidical to the amide function of asparagine. Hence, more than 50% of all catalogued proteins contain the sequence Asn-X-Ser or Asn-X-Thr. X can be any amino acid with the exception of Pro or Asp [194, 195]. N-glycosidical bound oligosaccharides consist of a core of two N-acetylglucosamine (GlcNAc) and three mannose residues (Man). To those, additional mannose residues (mannose type), lactosamine residues and other carbohydrates (complex type), or different sugars in irregular order (hybrid type), can be linked (Fig. 3-190). Other constituents of N-glycosidical bound oligosaccharides include galactose (Gal), N-acetylneuraminic acid (Neu5Ac), and Fucose (Fuc).

The structural diversity becomes even more complex by:

- Branching isomerism
- Anomeric compounds (α or β) as stereo isomeric forms in the C1 position of the hexose
- Different sugar content in the structure.

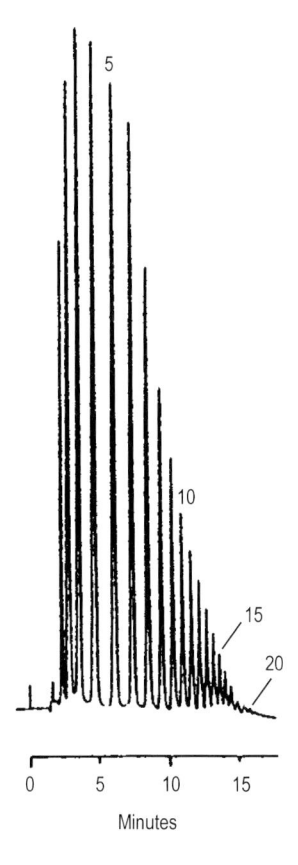

Fig. 3-188. Separation of β-(1-4)-linked D-glucans after partial acetolysis of cellulose. — Separator column: CarboPac PA1; eluant: (A) 0.15 mol/L NaOH, (B) 0.15 mol/L NaOH + 0.5 mol/L NaOAc; gradient: linear, 20% B to 50% B in 10 min, then to 60% B in 20 min; flow rate: 1 mL/min; detection: see Fig. 3-182; injection: 50 µL of a 0.3% aqueous solution of the acetolysate; (taken from [185]).

This diversity determines the three-dimensional structure of the carbohydrates. The most common O-glycosidical fixation is mediated by a core disaccharide, β-galactosyl-(1-3)-α-N-acetylgalactosamine, which is linked α-glycosidical to Ser or Thr. N-acetylgalactosamine is not present in human N-glycosidical bound oligosaccharides and, thus, as a result of compositional analysis it can be interpreted as a first hint to O-glycosidical bonds. In addition, galactose, mannose, and xylose are also sometimes bound to Ser or Thr (Fig. 3-190). However, very little is known about the function of O-glycosidical glycostructures.

The post-translational modification of proteins takes place in the endoplasmatic reticulum and the Golgi apparatus. Usually, it is determined by glycosyltransferases, which are regulated in a cell- and development-specific manner [196]. If all parameters that determine the glycan sequence of a protein are put together, the complexity of the glycan structures becomes apparent. One of simplest and best-investigated glycoproteins is the iron transport protein, transferrin, which circulates in the plasma. In this glycoprotein, the carbohydrate moiety determines the lifetime of transferrin in blood. Other examples of glycoproteins

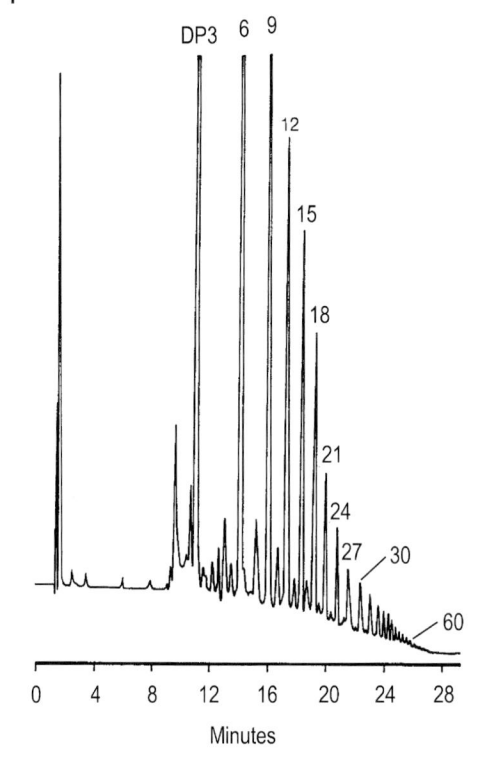

Fig. 3-189. Gradient analysis of pullulan after enzymatic digestion with pullulanase. — Chromatographic conditions: see Fig. 3-182; injection: 50 µL of a 0.2% solution.

include anti-freeze glycoproteins, mucines, immunoglobulins (especially monoclonal antibodies), erythropoietin (rEPO), the tissue plasminogen activator (tPA), ovalbumine, and ribonuclease B.

3.10.5.1 Compositional and Structural Analysis

The aim of oligosaccharide analysis is to provide a first look into the variety of different structural possibilities of the protein glycostructures as quickly and accurately as possible. While the methodology for characterizing proteins and peptides has been continuously improved, identification and characterization of oligosaccharide moieties (and the corresponding sample preparation) remained cumbersome. Conventional methods such as HPLC on aminopropyl-bonded phases or metal-loaded cation exchangers with refractive index and UV detection were insufficient for separating and detecting such complex structures. Other conventional methods such as NMR and MS are especially useful for structural identification, but do not offer a satisfactory alternative for carbohydrate analysis, due to the high demands on the operator. For the different isomers, a single separator column with a subsequent sensitive detection system was required. Besides the above-mentioned methods, capillary electrophoresis and other electrophoretic techniques were developed in recent years which do not require large

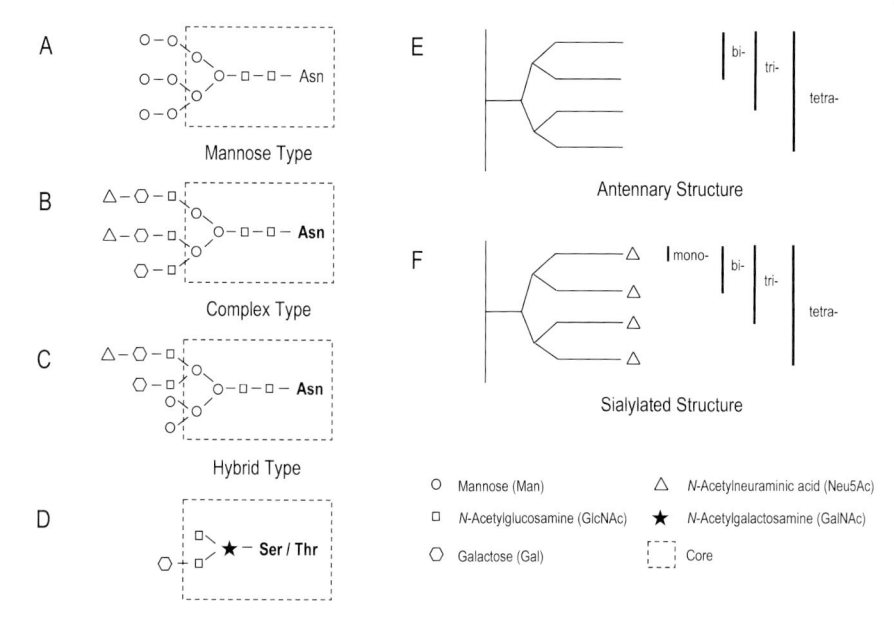

Fig. 3-190. Schematic representation of oligosaccharides derived from human glycoproteins. – In N-glycosidical bound oligomers di-N-acetylchitobiose together with three mannose residues are forming the core of the mannose- (A), the complex- (B), or the hybrid type (C). O-glycosidical bound oligosaccharides (D) usually contain N-acetylgalactosamine. N-glycosidical bound carbohydrates are present as bi-, tri-, or tetra-antennary structures (E) and can be mono-, di-, tri-, or tetra-sialylated (F).

amounts of sample [197], but do require a sophisticated sample derivatization [198, 199]. Oligosaccharides derived from glycoproteins can also be determined immunologically via carbohydrate-binding proteins called the lectines [200]. However, this technique is of rather qualitative character and requires the ability to recognize the epitopes from glycoproteins.

HPAEC-PAD (**H**igh **P**erformance **A**nion **E**xchange **C**hromatography with **P**ulsed **A**mperometric **D**etection) was successfully employed as a simple method for the analysis of free or bound oligosaccharides (see Section 3.10.5). This method can be used for oligosaccharide mapping as well as for the separation and detection of glycopeptides. The separation of oligosaccharides via anion exchange chromatography is by far more selective than other separation techniques. In combination with pulsed amperometric detection, which allows determinations down to the pmol level, direct and sensitive detection of glycans is possible. Similar sensitivities can only be achieved by radiolabelling [201] or derivatizations with chromophoric or fluorophoric compounds [202, 203]. Further technical details are summarized in Technical Notes 40 to 42 from Dionex Corporation [204-206].

Compositional Analysis of Monosaccharides Derived from Glycoproteins

To investigate the monosaccharide composition, the carbohydrate moiety of the glycoprotein has to be hydrolyzed with acid, which results in a quantitative cleavage of the neutral or aminated monosaccharides. A separation of monosaccharides derived from glycoprotein hydrolysates is shown in Fig. 3-191. Both neutral and amino sugars resulting from the hydrolysis appear as sharp and symmetric peaks separated on CarboPac PA1 in less than 15 minutes. For determining recovery values deoxyglucose, which does not exist in glycoproteins, is used as an internal standard. Since deoxyglucose is an acid labile sugar, it is added as a reference compound only after hydrolysis is completed.

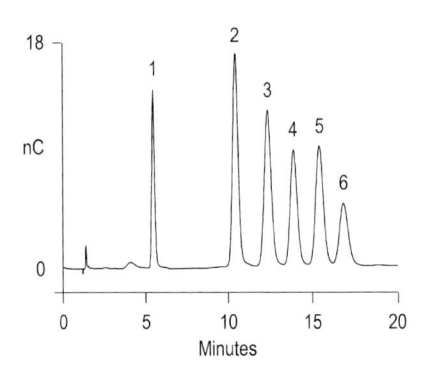

Fig. 3-191. Separation of monosaccharides derived from glycoproteins. – Separator column: CarboPac PA1; eluent: 22 mmol/L NaOH; flow rate: 1 mL/min; detection: pulsed amperometry on a gold working electrode with post-column addition of 0.3 mol/L NaOH; injection volume: 50 µL; solute concentrations: 4 nmol fucose (1), 5 nmol each of galactosamine (2), glucosamine (3), galactose (4), glucose (5), and mannose (6).

A new standard for the quantitative analysis of acidic, neutral, and basic monosaccharides employs CarboPac PA10 [207], the successor to CarboPac PA1. As already outlined in Section 3.10, this separator column facilitates trace level analysis of monosaccharides, because the negative peak resulting from the oxygen interference appears behind mannose, so that all carbohydrates released during hydrolysis can be determined free of any interferences. A selectivity comparison between CarboPac PA1 and PA10 is shown in Fig. 3-163 (Section 3.10.2). CarboPac PA10 is characterized by a number of other advantages including the total analysis time for monosaccharides, which can be decreased to 10 minutes when applying a flow rate of 1.5 mL/min. Also, the resolution between monosaccharides is significantly improved over CarboPac PA1, despite the short analysis time. The solvent compatibility of the CarboPac PA10 column allows it to be regenerated with organic solvents. Minimum detection limits are in the fmol range for monosaccharides obtained with CarboPac PA10.

Sialic acids such as N-acetylneuraminic acid (Neu5Ac) and N-glycolylneuraminic acid (Neu5Gc) can also be separated on CarboPac PA10 with an NaOH eluant [208]. Under identical chromatographic conditions the resulting retention times do not differ very much from CarboPac PA1 used thus far. Hermentin et al. [209] validated the HPAEC-PAD method for quantifying Neu5Ac derived from glycoproteins after sulfuric acid hydrolysis ($c = 0.05$ mol/L).

In general, the HPAEC-PAD method allows free amino sugars and neutral monosaccharides to be analyzed in the same run. Under these conditions, acetylated monosaccharides such as N-acetylgalactosamine and N-acetylglucosamine co-elute with mannose and glucose, respectively. They may need to be determined under different conditions. When hydrolyzing glycostructures at 100 °C for two hours, all amino sugars are completely de-N-acetylated. Therefore, interferences from GalNAc and GlcNAc in the monosaccharide compositional analysis are very unlikely. In case of an enzymatic cleavage, on the other hand, the de-N-acetylated amino sugars stay intact and should be separated from other monosaccharides on a high-capacity separator (e.g., CarboPac MA1).

Although amino acids and peptides can also be electrochemically detected under the conditions being used [164], Hardy et al. [210] could not identify any interfering signals in the monosaccharide retention range when analyzing hydrolysates of bovine fetuin and fibrinogen. UV active components of these protein hydrolysates elute either in the column void volume or far behind the monosaccharides.

The above mentioned proteins are characterized by a marked carbohydrate content (>20%), so sensitive monosaccharide detection at trace level is not really necessary. Relative to their total molecular weight, immunoglobulins of the G-type and recombinant monoclonal antibodies (MAb) have a carbohydrate content of only 2%. In such cases, the high content of amino acids in the hydrolysate interferes with monosaccharide determination. Figure 3-165 (see Section 3.10.2) shows an isocratic separation of monosaccharides from MAb hydrolysate using CarboPac PA10. In this analysis, lysine as an interfering amino acid renders more difficult the quantitative analysis of the subsequently eluting monosaccharides, because the oxidation products of lysine desorb very slowly from the working electrode. This, in turn, results in an insufficient cleaning of the working electrode under the waveform settings optimized for carbohydrates. However, this problem can be solved by using a special AminoTrap column which is inserted instead of the regular CarboPac PA1 guard column [207]. The AminoTrap column causes surface-active amino acids to elute much later, so that interferences in the retention range of the monosaccharides of interest can be excluded. The separation of carbohydrates is not affected by the AminoTrap column.

Derivatized monosaccharide acids belong to another sub-group of monosaccharides existing in glycostructures. For cleaving N-acetylneuraminic acid and N-acetylglycolylneuraminic acid mild hydrolysis conditions are required. The determination of the neuraminic acid content of a glycoprotein — as in the case of transferrin — allows statements about the physical condition of a patient. The separation of these two acidic carbohydrates is achieved under gradient conditions with a somewhat higher concentration of NaOH and NaOAc (Fig. 3-192). Glucuronic acid serves as an internal standard.

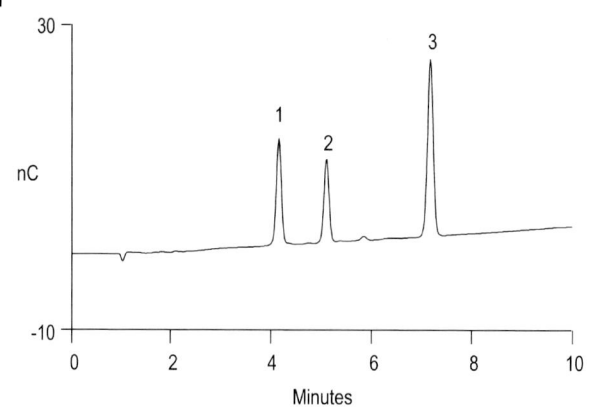

Fig. 3-192. Separation of neuraminic acids derived from glycoproteins. — Separator column: CarboPac PA10; eluant: (A) 0.1 mol/L NaOH, (B) 0.1 mol/L NaOH + 1 mol/L NaOAc; gradient: linear, 7% B to 30% B in 10 min; flow rate: 1.5 mL/min; detection: pulsed amperometry on a Au working electrode; injection volume: 20 µL; solute concentrations: 200 pmol each of N-acetylneuraminic acid (1), KDN (2), and N-glycolyl-neuraminic acid (3).

If oligosaccharides are not cleaved from the protein by acid hydrolysis but by β-elimination for determining monosaccharide composition, as is done for O-glycosidical bound sugars, reduced monosaccharide alditols and reduced oligosaccharides will result. Those sugar alcohols have such a high pK value that they can only be separated on a high-capacity separator under strongly alkaline conditions. Figure 3-193 shows the separation of alditols and aldoses as well as the N-acetylated forms of galactose and glucose on CarboPac MA1. This separator column is also suitable for the isocratic separation of sugar alcohols and their N-acetyl derivatives. The difference in selectivity of CarboPac MA1 compared with PA1 is remarkable (see Fig. 3-191). The separation performance of this column as well as the sensitivity of the detection system makes it possible to investigate the purity and the activity of glucosidases [211].

With the HPAEC-PAD method, the degree of glycosylation and the type of oligosaccharide present can already be judged during the compositional analysis of monosaccharides derived from glycoproteins. If, for example, a glycoprotein is expressed in cell cultures, its quality can be controlled by the carbohydrate composition during the production process with the HPAEC-PAD method.

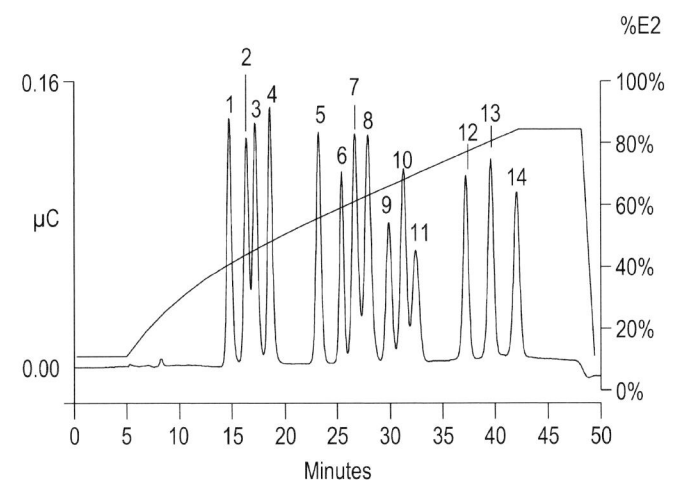

Fig. 3-193. Separation of alditols and aldoses. – Separator column: CarboPac MA1; eluant: 0.75 mol/L NaOH; flow rate: 0.4 mL/min; detection: pulsed amperometry on a gold working electrode; solute concentrations: 2 nmol each of fucitol (1), N-acetylgalactosamine (reduced, 2), N-acetylglucosamine (reduced, 3), xylitol (4), arabitol (5), fucose (6), sorbitol (7), dulcitol (8), N-acetylglucosamine (9), mannitol (10), N-acetylgalactosamine (11), mannose (12), glucose (13), and galactose (14).

Structural Analysis of Oligosaccharides Derived from Glycoproteins

Due to the high number and complexity of potential carbohydrate compounds, a number of methods have been developed that allow identification of carbohydrate structures, depending on the analytical problem and conditions, either directly or in combination with other techniques. In general, oligosaccharide analysis derived from glycoproteins is divided into three phases:

1. Release of oligosaccharides
2. Separation and detection of oligosaccharides
3. Structural analysis of oligosaccharides

Release of Oligosaccharides

After compositional analysis, oligosaccharides are chemically or enzymatically cleaved off the protein. The chemical reaction is carried out by hydrazinolysis, cleaving off O- or N-glycans from the protein, or by β-elimination with subsequent reduction, releasing especially O-glycosidical bound sugars from the protein. When utilizing the chemical method, unwanted side reactions of the oligosaccharides that may destroy the protein cannot be excluded. Maintaining the intact protein, N-glycans are cleaved off the protein enzymatically by means of

N-glycopeptidase F [peptidase-N^4-(N-acetyl-β-glucosaminyl)asparagine-amidase] [212]. A simple method that allows the cleavage of glycans from the protein and the subsequent analysis of those glycans is described in Section 9.9.

Separation and Detection of Oligosaccharides
Neutral oligosaccharides differing in the number, the type and the sequence of sugar residues can be separated both with pellicular separator columns and chemically bonded alkyl- [213] and aminopropyl phases [214]. With this technique, separations between the 1-6-isomers and their 1-2-, 1-3-, and 1-4-analogues can also be carried out as shown for various oligosaccharides [212, 214]. However, successful separations of the 1-2-, 1-3-, and 1-4-isomers on aminopropyl phases are only documented for di-, tri-, and tetra-saccharides [213, 215]. This emphasizes the decisive advantage of using the HPAEC-PAD method with pellicular anion exchangers for separation. As already mentioned in Section 3.10, the high selectivity of these stationary phases allows the separation of structural isomeric oligosaccharides, which only differ in the linkage of two sugar residues. A very impressive example is the separation of the two glycopeptides from asialo fetuin shown in Fig. 3-194.

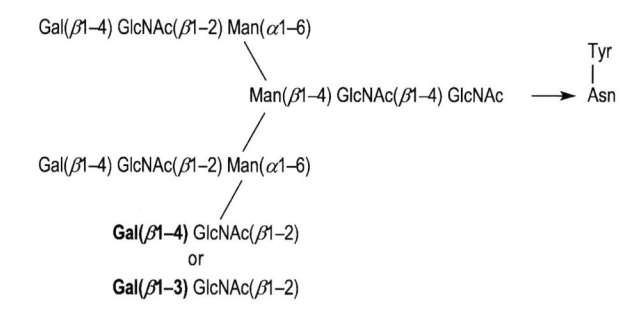

```
Gal(β1–4) GlcNAc(β1–2) Man(α1–6)
                                   \
                                    Man(β1–4) GlcNAc(β1–4) GlcNAc  →  Tyr
                                   /                                    |
Gal(β1–4) GlcNAc(β1–2) Man(α1–6)                                       Asn
                                 /
             Gal(β1–4) GlcNAc(β1–2)
                        or
             Gal(β1–3) GlcNAc(β1–2)
```

The terminal N-acetylneuraminic acid was cleaved off with neuraminidase to yield neutral carbohydrate compounds. The structures shown above only differ in the labeled β-1,3- and β-1,4-linkage between galactose and N-acetylglucosamine in the lactosamine residue of this tri-antennary oligosaccharide [216]. Neither isomers can be separated with conventional RPLC techniques.

Hardy and Townsend [217] attribute the high selectivity of a hydroxide-selective anion exchanger to the different acidity of the individual hydroxide groups of a monosaccharide; they demonstrate this with the separation of carbohydrate isomers. Consequently, the different acidity of the individual hydroxide groups of a monosaccharide implies the different acidity of oligosaccharides with their diverse linkages of the monomeric building blocks. The reducing anomeric hydroxide group of a sugar exhibits the highest acidity, followed by ring hydroxide groups in the order 2-OH >> 6-OH > 4-OH > 3-OH [160]. Correlating the retention behavior of various oligosaccharides with their chemical structure, it becomes evident that the selectivity of the separation is affected by the accessibility

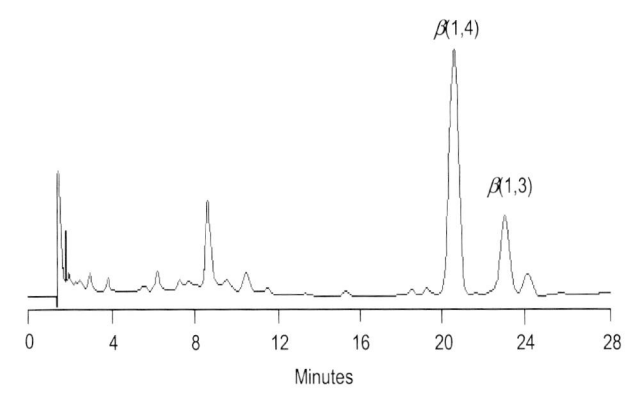

Fig. 3-194. Separation of two structural isomeric undecasaccharides derived from asialo fetuin. — Separator column: CarboPac PA1; eluant: (A) 0.15 mol/L NaOH, (B) 0.15 mol/L NaOH + 1 mol/L NaOAc; gradient: linear, 50% B to 100% B in 20 min; flow rate: 1 mL/min; detection: pulsed amperometry on a gold working electrode; injection volume: 50 µL.

of the fixed stationary phase for the oxyanions formed in the mobile phase. Other specific characteristics for carbohydrates separated on pellicular anion exchangers can be demonstrated with a number of neutral oligosaccharides (see Table 3-28) that resemble N-glycans derived from glycoproteins (Fig. 3-195).

In contrast to the retention behavior of linear homooligomers, retention of an oligosaccharide does not simply increase with the molecular size [218]. In comparison with the nonasaccharide **4**, the fucosylated trisaccharide **1** exhibits a slightly lower retention time. The significantly lower retention time of the fucosylated nonasaccharide **4** as compared with the non-fucosylated heptasaccharide **9** emphasizes again the effect of fucosylation on the retention behavior of oligosaccharides. The shorter retention time can be traced back to the fucosyl substitution at C-3 of the N-acetylglucosamine residue and to the steric arrangement of the α-(1-3)-linkage between fucose and N-acetylglucosamine. The retention behavior of the compounds **3** and **6** reveals that reduction at the anomeric C-1 to the corresponding sugar alcohol leads to a retention time decrease of 15 minutes. This is probably due to a conformational transition of the branching mannose residue from the chair form typical for pyranoses into an open-chain structure. A fucose substitution at C-3 of N-acetylglucosamine also shortens retention. This effect is obvious in the oligosaccharides **9**, **2** and **4**. Compound **9**, for example, has a bi-antennaric structure in which Gal(β1-4) is linked terminally with GlcNAc. When these terminal residues are replaced by Fuc(α1-3) (compound **2**), the retention time is significantly reduced, because the high acidity of the hydroxide groups [219] attached to C-3 of N-acetylglucosamine and at C-6 of galactose is not realized. A similarly strong reduction in the retention time is observed when two Fuc(α1-3) residues are linked with the galactosylated com-

Table 3-28. Chemical structures of oligosaccharides separated in Fig. 3-195.

Peak	Structure	Peak	Structure
1	Fuc(α1–3) GlcNAc(β1–2) Man	7	Gal(β1–3) GlcNAc(β1–3) Gal(β1–4) Glc
2	Fuc(α1–3) GlcNAc(β1–2) Man(α1-6) \\ Man / Fuc(α1–3) GlcNAc(β1–2) Man(α1–3)	8	Gal(β1–4) GlcNAc(β1-6) \\ Man(α1–2) Man / Gal(β1–4) GlcNAc(β1–3)
3	Gal(β1–4) GlcNAc(β1–2) \\ Man-OH / Gal(β1–4) GlcNAc(β1–4)	9	Gal(β1–4) GlcNAc(β1–2) Man(β1-6) \\ Man / Gal(β1–4) GlcNAc(β1–2) Man(α1–3)
4	Fuc(β1–3) \\ Gal(β1–4) GlcNAc(β1–2) Man(α1-6) \\ Man / Gal(β1–4) GlcNAc(β1–2) Man(α1–3) / Fuc(β1–3)	10	Gal(β1–4) GlcNAc(β1–2) Man(α1-6) \\ Man / Gal(β1–4) GlcNAc(β1–2) Man(α1–3) / Gal(β1–4) GlcNAc(β1–4)
5	Gal(β1–4) GlcNAc(β1–3) Gal(β1–4) Glc	11	Gal(β1–4) GlcNAc(β1-6) \\ Gal(β1–4) GlcNAc(β1–2) Man(α1-6) \\ Man / Gal(β1–4) GlcNAc(β1–2) Man(α1–3) / Gal(β1–4) GlcNAc(β1-6)
6	Gal(β1–4) GlcNAc(β1–2) \\ Man / Gal(β1–4) GlcNAc(β1–4)	12	Gal(β1–4) GlcNAc(β1–2) Man(α1-6) \\ Man / Gal(β1–4) GlcNAc(β1–2) Man(α1–3) / Gal(β1–4) GlcNAc(β1–4)

pound **9**. Although this substitution prevents the formation of oxyanions at the C-3 of the two branching GlcNAc residues, the resulting compound **4** has six additional hydroxide groups. The oxyanions located at C-2 of both fucose residues are particularly acidic. However, the interaction of the oxyanion formed at this position is sterically hindered due to the close proximity of the acetamido group, which explains the distinct retention decrease. The retention time comparison of compound **10** and **11** is another example of the high selectivity of this method regarding structural isomers. Since the acidity of the hydroxide group

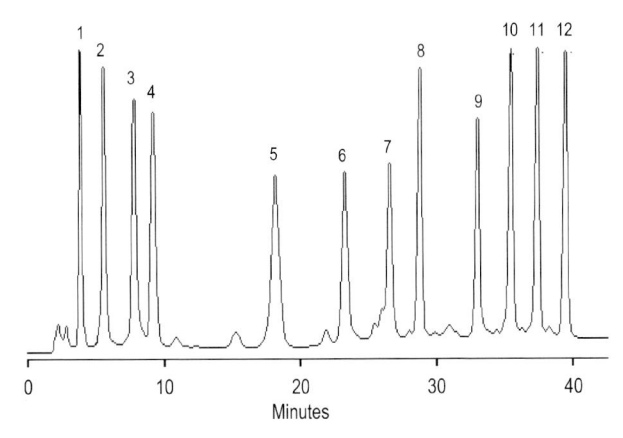

Fig. 3-195. Separation of various neutral oligosaccharides. – Separator column: CarboPac PA1; eluant: (A) 0.1 mol/L NaOH, (B) 0.1 mol/L NaOH + 0.15 mol/L NaOAc; gradient: linear, 100% A isocratically for 10 min, then to 80% B in 60 min; flow rate: 1 mL/min; detection: see Fig. 3-194; injection volume: 50 µL; solute concentrations: 1 nmol each of the structures **1** to **12** (see Table 3-28); (taken from [217]).

at C-6 is higher than that of the hydroxide group attached to C-4, the nonasaccharide **11** with a (β1-6)-linkage is more strongly retained, followed by the tetraantennary oligosaccharide **12**.

The following example emphasizes the selectivity of the HPAEC-PAD method in a different way. Figure 3-196 shows the separation of the oligosaccharides listed in Table 3-29 that can exist in glycoproteins.

Although the structures **3**, **10**, and **12** exhibit different compositions, their hydrodynamic volumes are very similar. The hydrodynamic volume of an oligosaccharide can be determined via the determination of glucose equivalents by means of ion-exclusion chromatography. Structures **10** and **12**, which differ by only 0.1 glucose equivalents, can easily be separated by the HPAEC-PAD method, whereas the separation by ion-exclusion chromatography is extremely difficult.

It can also be seen from Fig. 3-196 that the HPAEC-PAD method can be applied for the separation of linkage isomers of charged oligosaccharides. The trisialylated oligosaccharides **10** and **11**, which differ in the α2-6- and α2-3-linkage, are separated to baseline. Moreover, Fig. 3-197 demonstrates the retention behavior of neutral- and differently-charged oligosaccharides, which are eluted in groups from the column depending on the degree of sialylation. Neutral oligosaccharides elute ahead of monosialylated (not shown here), followed by disialylated **8** and **9**, trisialylated **10** and **11**, and tetrasialylated carbohydrates **12** and **13**. Thus, the elution pattern provides the first evidence as to the sialylation state of a proteoglycan.

Table 3-29. Chemical structures of oligosaccharides separated in Fig. 3-196.

Peak	Structure	Peak	Structure
1	Man\\ 　Man–GlcNAc–GlcNAc（Fuc） Man/	7	Gal–GlcNAc–Man\\ 　Man–GlcNAc–GlcNAc（Fuc） Gal–GlcNAc–Man/
2	Man\\ 　Man–GlcNAc–GlcNAc Man/	8	Gal–GlcNAc–Man\\ 　Man–GlcNAc–GlcNAc Gal–GlcNAc–Man/
3	GlcNAc–Man\\　　　　　Fuc 　Man-GlcNAc-GlcNAc GlcNAc–Man/	9	Gal–GlcNAc–Man\\ 　Man–GlcNAc–GlcNAc Gal–GlcNAc–Man/ Gal–GlcNAc/
4	GlcNAc–Man\\ 　Man–GlcNAc–GlcNAc GlcNAc–Man/	10	GlcNAc–Man\\ 　GlcNAc–Man–GlcNAc–GlcNAc Man–Man/ Man/
5	Man–Man\\ 　Man–GlcNAc–GlcNAc Man–Man/	11	Gal–GlcNAc\\ Gal–GlcNAc–Man\\ 　　Man-GlcNAc-GlcNAc Gal–GlcNAc–Man/ Gal–GlcNAc/
6	GlcNAc–Man\\ 　Man–GlcNAc–GlcNAc GlcNAc–Man/ GlcNAc/	12	Man–Man–Man\\ 　Man–GlcNAc–GlcNAc Man–Man–Man/ Man–Man/

The empirical findings of the retention behavior of oligosaccharides on pellicular anion exchangers such as CarboPac PA1 and PA-100 are summarized in Table 3-30.

Since certain carbohydrates easily undergo base-catalyzed reactions, other peaks could appear in the chromatogram when separating oligosaccharides by anion exchange chromatography in strongly alkaline media. These reactions include the Lobry-de-Bruyn-van-Ekenstein rearrangement, the de-acetylation of

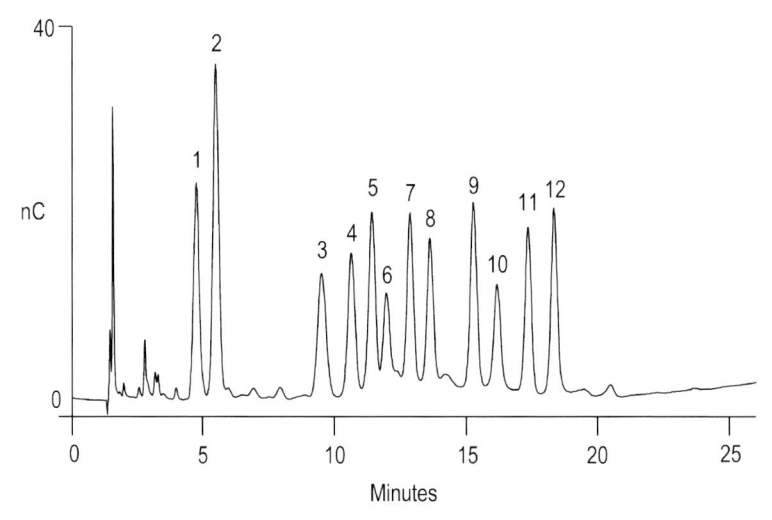

Fig. 3-196. Separation of neutral oligosaccharides derived from glycoproteins. — Separator column: CarboPac PA-100; eluant: (A) 0.25 mol/L NaOH, (B) 0.25 mol/L NaOH + 0.08 mol/L NaOAc; gradient: linear; flow rate: 1 mL/min; detection: see Fig. 3-194; injection volume: 25 µL; sample: structures **1** to **12** (see Table 3-29).

N-acetylated carbohydrates, and the β-elimination of 3-O-substituted compounds. However, in the compounds listed in Tables 3-29 and 3-30, epimerization does not occur within the time frame needed for the chromatographic separation. Epimerization to N-acetylmannosamine (ManNAc) at pH 13 could be expected to a very small extent only in terminal GlcNAc building blocks. Even in separations on CarboPac MA1 under very strong alkaline conditions epimerization does not occur within the time frame needed for chromatography [230].

Structural Analysis of Oligosaccharides
The investigation of a glycoprotein usually starts with the identification of a possible glycosylation; this can be carried out very easily with immunological tests. In contrast, structural analysis of oligosaccharides requires more effort. Complex methods such as methylation analysis with subsequent gas chromatography and mass spectrometry contribute to the identification of glycosidic bonds between the monosaccharides. The molecular weight of an oligosaccharide or its fragments is often determined by FAB (**F**ast **A**tom **B**ombardment)-MS analysis after complete methylation [231, 232]. In this context, MALDI (**Ma**trix **A**ssisted **L**aser **D**esorption **I**onization)-MS is more and more applied for this analysis; it does not require oligosaccharide derivatization, yet it is more sensitive by three to four orders of magnitude compared with FAB-MS [233]. Online mass determination of oligosaccharides can also be automated by hyphenating HPAEC-PAD with a mass spectrometer via a thermospray interface [234-238]. This hyphenation requires complete desalting of the eluant (NaOH and sodium

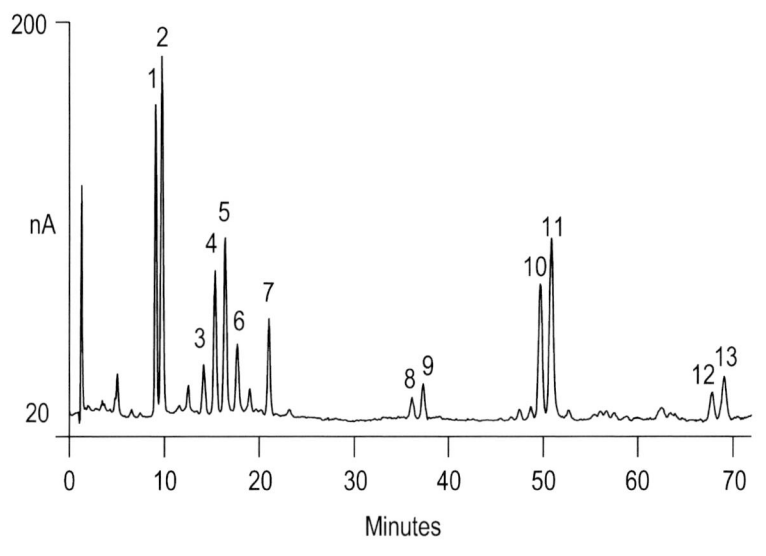

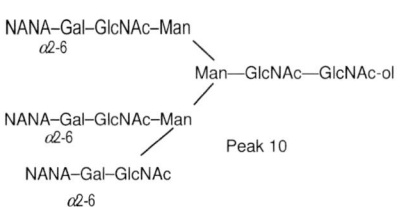

NANA–Gal–GlcNAc–Man
    α2-6
                    ╲
                     Man—GlcNAc—GlcNAc-ol
                    ╱
NANA–Gal–GlcNAc–Man
    α2-6   Peak 10
                    ╱
NANA–Gal–GlcNAc
    α2-6

NANA—Gal—GlcNAc—Man
    α2-3
                    ╲
                     Man—GlcNAc—GlcNAc-ol
                    ╱
NANA—Gal—GlcNAc—Man
    α2-6   Peak 11
                    ╱
NANA—Gal—GlcNAc
    α2-6

Fig. 3-197. Separation of various neutral and sialylated oligosaccharides. – Separator column: CarboPac PA-100; eluant: (A) 0.1 mol/L NaOH, (B) 0.1 mol/L NaOH + 0.25 mol/L NaOAc; gradient: linear, 0% B to 100% B in 110 min; flow rate: 1 mL/min; detection: see Fig. 3-194; injection volume: 25 µL; structures: FucMan$_3$GlcNAc$_2$ (1), Man$_3$GlcNAc$_2$ (2), asialo, agalacto bi, core Fuc (3), asialo, agalacto bi (4), asialo bi, core Fuc (5), Asialo bi (6), Man$_9$GlcNAc$_2$ (7), disialylated, bi-antennary (reduced, 8, 9), trisialylated, tri-antennary (reduced, 10, 11), tetrasialylated, tetra-antennary (reduced, 12, 13). The structures **1** to **7** refer to the structures **1** to **4**, **7**, **8** and **12** in Table 3-29.

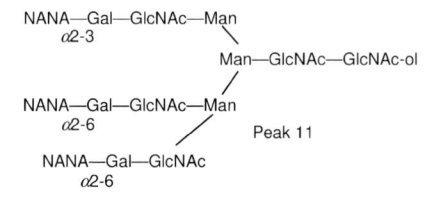

acetate) before it enters the thermospray interface. In 1989, Spellman et al. [239] described the use of a micromembrane suppressor (AMMS) for eluant desalting. For the chemical regeneration of the suppressor, the authors used sulfuric acid (0.03 mol/L). However, with this desalting procedure only neutral oligosaccharides are amenable to the subsequent structural analysis, since the required sodium acetate concentration for eluting sialylated oligosaccharides exceeds the suppression capacity of an AMMS. An increase of both the regenerant concentration and flow rate does not lead to a significant increase in suppression capacity, because sulfate ions – as observed by Rouse and Vath [240] – can permeate through the cation exchange membrane in the AMMS at high regenerant concentration and appear in the oligosaccharide fraction which, in turn, leads to a suppression of the MALDI-TOF detector signal. Thayer et al. [241] solved that

Table 3-30. Empirical findings of the retention behavior of oligosaccharides on pellicular anion exchangers such as CarboPac PA1 and PA-100; (taken from [220]).

1. The larger the oligosaccharide, the higher its retention time (an oligosaccharide with 5 mannose residues, for example, elutes ahead of an oligosaccharide with 9 residues).	Hardy and Townsend, 1988 [217]; Basa and Spellman, 1990 [221]; Hernandez et al., 1990 [222]; Pfeiffer et al., 1990 [223]; Townsend et al., 1991 [224].
2. With increasing negative charge, retention time increases (sialylation, sulfatation, phosphorylation).	Townsend et al., 1988 [225]; 1989 [226]; Hermentin et al., 1992 [227].
3. Fucosylation decreases retention time.	Hardy and Townsend, 1988 [217]; 1989 [228]; Basa und Spellman, 1990 [221]; Pfeiffer et al., 1990 [223].
4. An oligosaccharide with Neu5Ac(α2-3)Gal elutes behind the respective oligosaccharide with Neu5Ac(α2-6)Gal.	Townsend et al., 1988 [225]; 1989 [226].
5. An oligosaccharide with Gal(β1−3)GlcNAc elutes behind the respective oligosaccharide with Gal(β1−4)GlcNAc.	Hardy and Townsend, 1988 [217].
6. Linkage to a GlcNAc branch increases the retention time.	Hermentin et al., 1992 [227].
7. Exchange of Neu5Ac by Neu5Glc increases the retention time.	Hermentin et al., 1992 [227].
8. Linkage of a lactosamine residue to a neutral oligosaccharide increases the retention time.	Hermentin et al., 1992 [227].
9. Linkage of a lactosamine residue to a sialylated oligosaccharide decreases the retention time.	Hermentin et al., 1992 [227].
10. Reduction of the reducing terminal GlcNAc to GlcNAc-ol decreases the retention time.	Hardy and Townsend, 1988 [217].
11. An oligosaccharide with a complete core (Man$_3$GlcNAc$_2$) elutes behind an oligosaccharide with an incomplete core (Man$_3$GlcNAc).	Basa and Spellman, 1990 [221]; Pfeiffer et al., 1990 [223].
12. Sialylated oligosaccharides with Neu5Ac(α2-3)Gal residues together exhibit better resolution at pH 5.	Watson et al., 1992 [229].

problem by using a desalting procedure based on an ASRS self-regenerating suppressor that has been successfully used by other authors [242, 243]. The operating principle of an ASRS has been described in detail in Section 3.6 (Fig. 3-77). Instead of sulfuric acid, Thayer et al. used a trifluoroacetic acid (TFA) regenerant. Since TFA is a volatile acid, possible permeation through the cation exchange membrane does not result in its accumulation in the oligosaccharide fraction after evaporation, so that acid-catalyzed degradation of oligosaccharides is largely eliminated. Due to its high acid strength, TFA is very well-suited as a regenerant and can be employed at higher concentrations. With 0.15 mol/L TFA at a flow rate of 5 mL/min, sodium ions with a concentration of up to 0.35 mol/L can be exchanged for hydronium ions. Under these conditions the desalting efficiency is higher than 99.9%. When applying sodium acetate gradients, the desalting procedure yields acetic acid, which can be evaporated in a second step. The desalting unit has no negative effects on the detector noise. According to Thayer et al. recoveries for a number of neutral and charged oligosaccharides are between 75% and 100%. Although oligosaccharides come into contact with cation exchange membranes in the desalting unit for a short time, desialylation, which is often observed during manual treatment with cation exchange cartridges, does not occur. With high probability this can be attributed to the short time the oligosaccharides stay in the desalting unit. With this inline desalting technique, neutral and sialylated oligosaccharides, following chromatographic purification via HPAEC, can be prepared for subsequent mass spectrometric investigations.

Among chromatographic techniques, the HPAEC-PAD method represents an attractive alternative to structure identification methods. In a short time, it is possible to identify the most important structures by directly comparing the retention times of unknown oligosaccharides with reference glycans. The determined retention coefficients of unknown glycans can be compared with those of the data bank, which leads to an undisputable identification of the oligosaccharides found [227]. This structural analysis can also be supported by sequential enzymatic deglycosilation with subsequent determination of the degraded mono- and oligo-saccharides.

3.10.5.2 Chosen Examples

Until now, the characterization of glycoproteins and the accompanying sample preparation have proven troublesome. Taking human serum transferrin (HST) and recombinant erythropoietin (rEPO) as examples, a possible strategy for determining oligosaccharide structures is shown below.

Human Serum Transferrin (HST)
Transferrin as an iron-binding β-globin protein exists in the serum of many vertebrates. It has two glycosylation sites (Asn 413 and Asn 611), which carry bi-antennary (85%) and tri-antennary (15%) glycans, most of them being sialylated [244]. Differences in the degree of sialylation of transferrin can be correlated with the clinical state of a patient. Patients suffering from cancer or rheumatic arthritis show a remarkably high degree of sialylation of this serum protein. A decrease in the degree of sialylation results in an accelerated exclusion of this protein from the blood circle.

Due to its glycan residues, transferrin can be separated into three differently sialylated sub-populations {mono- (F1), di- (F2), and trisialylated (F3)} on a pellicular anion exchanger (see Section 3.10.1) [245, 246]. After enzymatic cleavage, the respective glycan populations of the F1-, F2-, and F3-fractions can be characterized via HPAEC-PAD. Figure 3-198 shows the glycan profiles of the protein itself and the fractions, which were identified as mono- (**1, 2**), di- (**3, 4, 5**), and trisialylated oligosaccharides (**7, 8, 9**) by HPAEC-PAD.

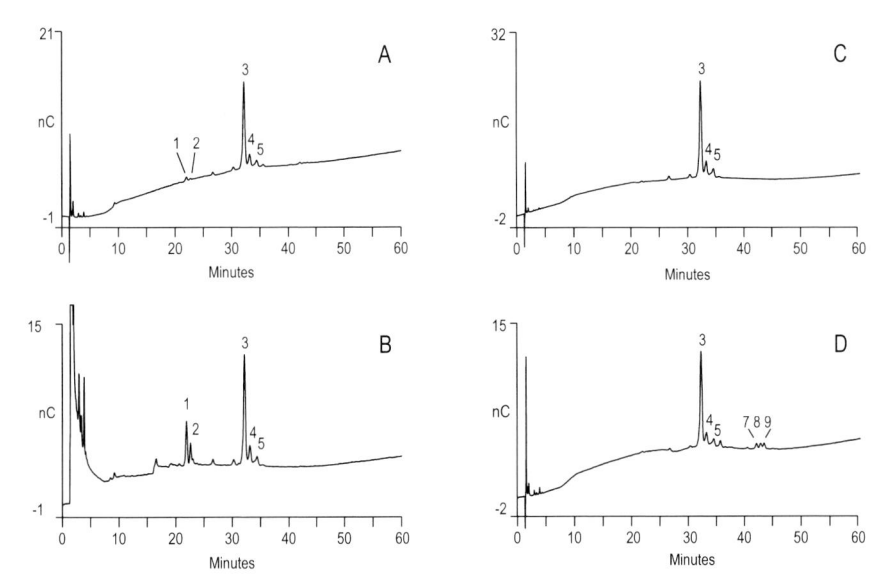

Fig. 3-198. Separation of oligosaccharides from human serum transferrin (HST) cleaved by N-glycopeptidase F. — Separator column: CarboPac PA-100; eluant: (A) 0.1 mol/L NaOH + 0.02 mol/L NaOAc, (B) 0.1 mol/L NaOH + 0.2 mol/L NaOAc; gradient: linear, 0% B to 100% B in 65 min; flow rate: 1 mL/min; detection: pulsed amperometry on a gold working electrode. (A) oligosaccharides of the HST digest (25 µL) prior to fractionation, (B) oligosaccharides of the 1st fraction of the separated HST (25 µL), (C) oligosaccharides of the 2nd fraction of the separated HST (10 µL), (D) oligosaccharides of the 3rd fraction of the separated HST (20 µL). Monosialylated oligosaccharides (**1, 2**), disialylated oligosaccharides (**3, 4, 5**), and trisialylated oligosaccharides (**7, 8, 9**).

Recombinant Erythropoietin (rEPO)
Recently, the HPAEC-PAD method has been extended by a simple sample preparation technique [247]. This technique includes purification of the wanted protein via SDS polyacrylamide gel electrophoresis (SDS-PAGE) [247] or isoelectric focusing (IEF) [248] and electrophoretic transfer to an inert membrane. After identification of the protein, the protein band is cut out of the membrane and hydrolyzed with acid for compositional analysis. Oligosaccharide cleavage is carried out either via enzymatic digest with N-glycopeptidase F [247] or via reductive β-elimination [248]. In both cases, N- and O-glycosidical bound oligosaccharides can be determined via HPAEC-PAD. Figure 3-199 shows the respective HPAEC-PAD monosaccharide profiles of rEPO after acid hydrolysis. The different monosaccharide profiles are attributed to the hydrolytic properties of the two acids

being used as well as to the different hydrolysis conditions. The "oligosaccharide fingerprint" of the membrane-bound rEPO is shown in Fig. 3-200; it does not differ from the profile of the free rEPO. When N-glycopeptidase F is used as a glycerol-stabilized protein, the more electroactive glycerol appears in the chromatogram as a prominent peak ahead of oligosaccharides.

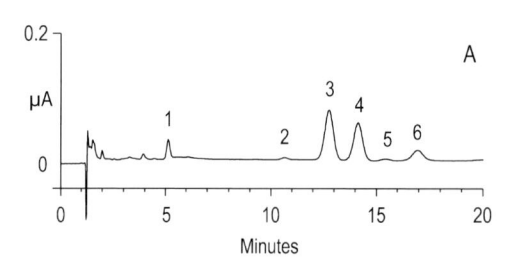

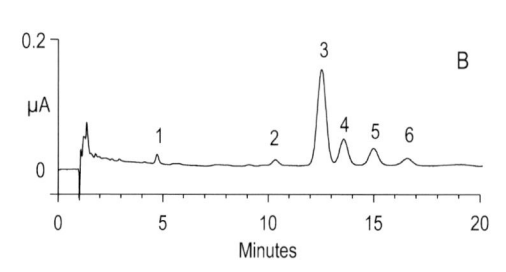

Fig. 3-199. Separation of monosaccharides of the immobilized recombinant erythropoietin (rEPO) after acid hydrolysis. — Separator column: CarboPac PA1; eluant: 0.016 mol/L NaOH; flow rate: 1 mL/min; detection: pulsed amperometry on a gold working electrode. (A) monosaccharides of rEPO after hydrolysis with 2 mol/L TFA at 100 °C, (B) monosaccharides of rEPO after hydrolysis with 6 mol/L HCl at 100 °C. Fucose (1), galactosamine (2), glucosamine (3), galactose (4), glucose (5), and mannose (6).

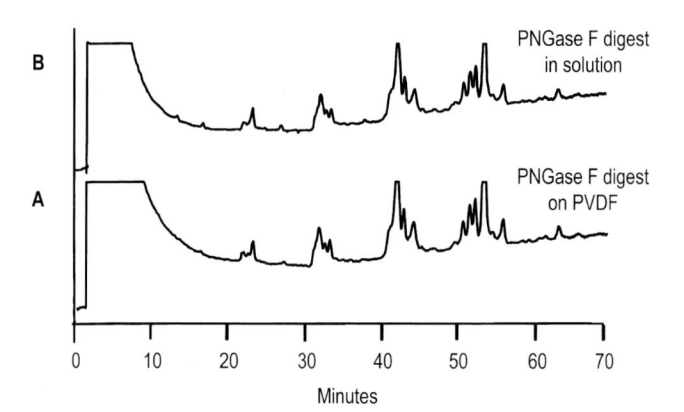

Fig. 3-200. Separation of oligosaccharides of the recombinant erythropoietin (rEPO) after enzymatic digest with N-glycopeptidase F. — Separator column: CarboPac PA-100; eluant: (A) 0.1 mol/L NaOH, (B) 0.1 mol/L NaOH + 0.2 mol/L NaOAc; gradient: linear, 0% B to 100% B in 110 min; flow rate: 1 mL/min; detection: pulsed amperometry on a gold working electrode. (A) oligosaccharides of the free rEPO, (B) oligosaccharides of the immobilized rEPO. (In this case, the enzymatic digest was carried out without triton X-100. The system peak at the beginning of the chromatogram is attributed to glycerol from the protein lot of the PNGase F.)

Thus, the strategy outlined in Fig. 3-201 has a number of advantages: glycoproteins are separated from other proteins via gel electrophoresis. Therefore, they are mostly present in high purity and high concentration. Finally, the protein is transferred onto an inert PVDF (polyvinylene difluoride)-membrane by electrotransfer. On that membrane, glycans – free of any salts or detergents – are amenable to acid or enzymatic hydrolysis. The hydrolysate can then be investigated by HPAEC-PAD. Because this method can be applied for other glycoproteins, it offers the advantage of universal applicability and quality control of recombinant glycoproteins [249].

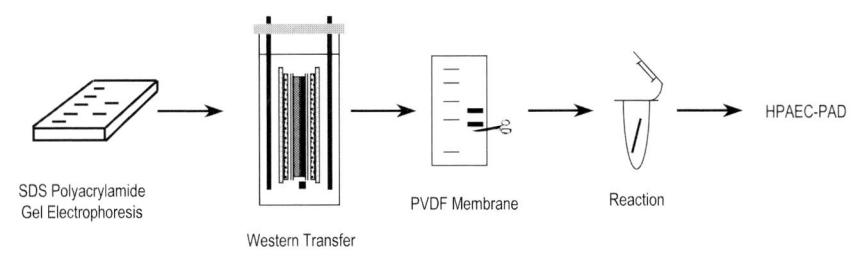

Fig. 3-201. Experimental strategy for investigating carbohydrates of a membrane-bound glycoprotein.

3.11
Amino Acids

Although amino acids are traditionally separated on totally sulfonated cation exchangers (see Section 5.9.1), pellicular anion exchangers are also used today, especially in combination with integrated amperometric detection [250]. (See Section 7.1.2.2)

The AminoPac PA-10 was developed for anion exchange chromatography of amino acids. The resin consists of a polystyrene/divinylbenzene substrate with a particle diameter of 8.5 µm and a degree of cross-linking of 55%. This stationary phase is compatible with typical HPLC solvents, although solvents are not necessary for separating amino acids by anion exchange chromatography. The latex beads which are functionalized with an alkylamine have a diameter of 80 nm and a relatively low degree of cross-linking of 1%. To achieve high mass sensitivity, the AminoPac PA-10 column is only available in the microbore format (250 mm × 2 mm i. d.). Until now, no methods were available for the separation of amino acids by anion exchange, primarily because of the difficulties in manufacturing *surface-aminated* materials with good chromatographic properties.

Latexed anion exchangers are an alternative to cation exchangers. Figure 3-202 illustrates a typical elution profile of hydrolysate amino acids on an AminoPac PA-10. It was obtained in less than 30 minutes with a NaOH/NaOAc gradient utilizing integrated amperometry. The gradient program being used is described in Table 3-31 and graphically depicted in Fig. 3-202. As can be seen from the chromatogram in Fig. 3-202, anion exchange chromatography allows the complete separation of phenylalanine and tyrosine, which is not possible on conventional cation exchange resins. So far, integrated amperometry has been almost exclusively utilized for the detection of carbohydrates, amines, and divalent sulfur compounds. Now it can also be used for direct detection of amino acids. The elution order is reversed in comparison with cation exchange: arginine and lysine elute first and threonine and serine are separated much better (Fig. 3-203).

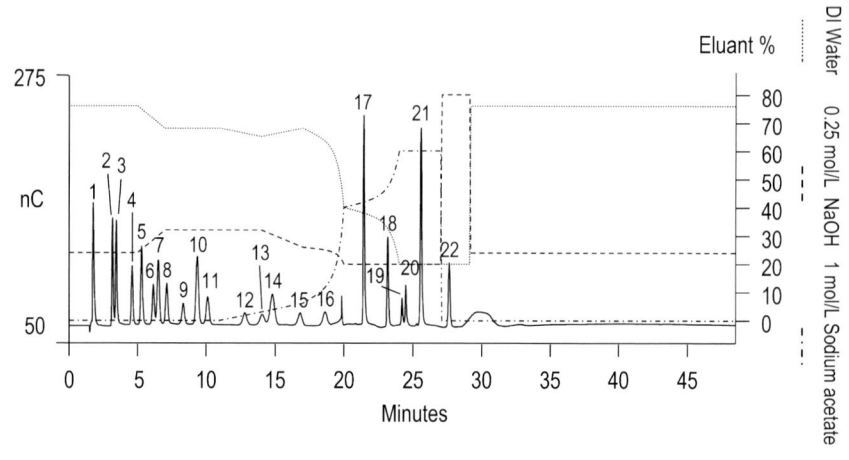

Fig. 3-202. Separation of amino acids on an AminoPac PA-10 latexed anion exchanger (250 × 2 mm i. d.). — Eluant: NaOH — sodium acetate; gradient: see Table 3-31; flow rate: 0.25 mL/min; detection: integrated amperometry on a gold working electrode; analytes: 100 pmol each of arginine (1), ornithine (2), lysine (3), glutamine (4), asparagine (5), alanine (6), threonine (7), glycine (8), valine (9), serine (10), proline (11), isoleucine (12), leucine (13), methionine (14), norleucine (15), taurine (16), histidine (17), phenylalanine (18), glutamate (19), aspartate (20), cystine (21), and tyrosine (22).

Table 3-31. Gradient program for the separation of amino acids in Fig. 3-202.

Time [min]	NaOH [mmol/L]	Sodium acetate [mmol/L]	Gradient type
00.0	40		
02.0	40		
12.0	80		8
16.0	80		
24.0	60	400	8
35.0	60	400	
35.1	200		
37.1	200		
37.2	40		

3.11 Amino Acids

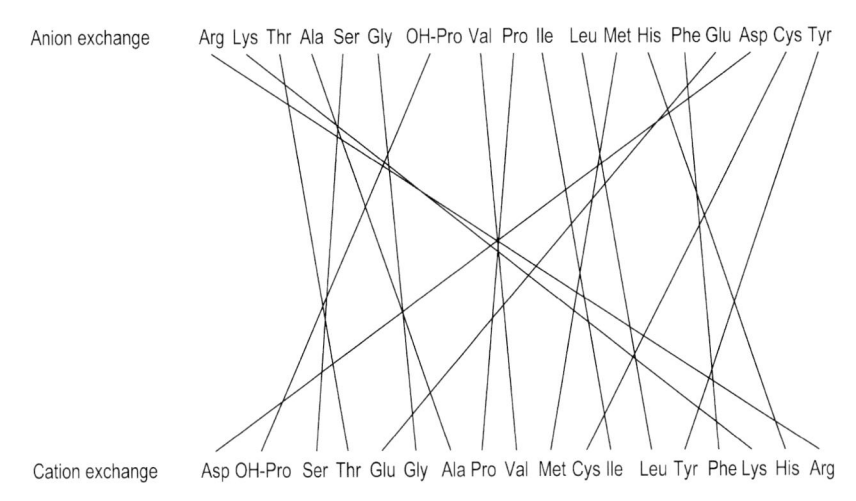

Anion exchange   Arg Lys Thr Ala Ser Gly OH-Pro Val Pro Ile   Leu Met His Phe Glu Asp Cys Tyr

Cation exchange   Asp OH-Pro Ser Thr Glu Gly Ala Pro Val Met Cys Ile   Leu Tyr Phe Lys His Arg

Fig. 3-203. Comparison of the retention of amino acids on anion and cation exchangers, respectively.

Using the pulse sequence employed in Fig. 3-202 to detect amino acids, carbohydrates can also be detected when they are present in approximately equimolar concentrations. This is the case, for example, with cell culture media [251] and protein hydrolysates [252]. Thus, a 5-fold higher glucose concentration interferes with the determination of alanine, while fructose interferes with the determination of threonine and glycine, and sucrose interferes with that of proline. As can be seen from Fig. 3-204, the retention times of amino acids and carbohydrates on an AminoPac PA-10 are dependent on the initial NaOH concentration, so the resolution between amino acids and interfering sugars can be optimized by changing the gradient profile. If the initial NaOH concentration is decreased to 16 mmol/L and the mobile phase pH is increased by a concave gradient over the next 10 minutes, amino acids – starting with alanine – elute behind the most important monosaccharides and amino sugars. Thus, the retention of monosaccharides is less strongly affected by the initial NaOH concentration. Figure 3-205 shows the resulting chromatogram with the graphically depicted gradient profile, which can be utilized for amino acid analysis in samples exhibiting an equimolar carbohydrate content.

If amino acids are to be analyzed in samples containing much higher concentrations of sugars (e. g. vegetable or fruit juices), anion exchange chromatography with integrated amperometric detection must be preceded by a sugar-eliminating step. For this purpose, Jandik et al. [253] developed an automated chromatographic procedure that captures amino acids on a strong acid cation exchange resin with sulfonic acid groups and a highly hydrophobic bead surface under conditions of no retention for carbohydrates. In an initial step, the sample is brought onto a short cation exchange column in hydrogen form. Carbohydrates pass through this column, while amino acids are retained. Subsequently,

3 Anion Exchange Chromatography (HPIC)

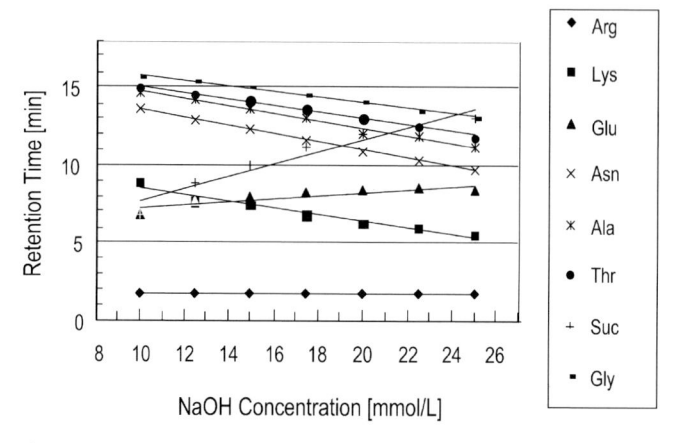

Fig. 3-204. Retention times of various amino acids and carbohydrates on AminoPac PA-10 as a function of the initial NaOH concentration.

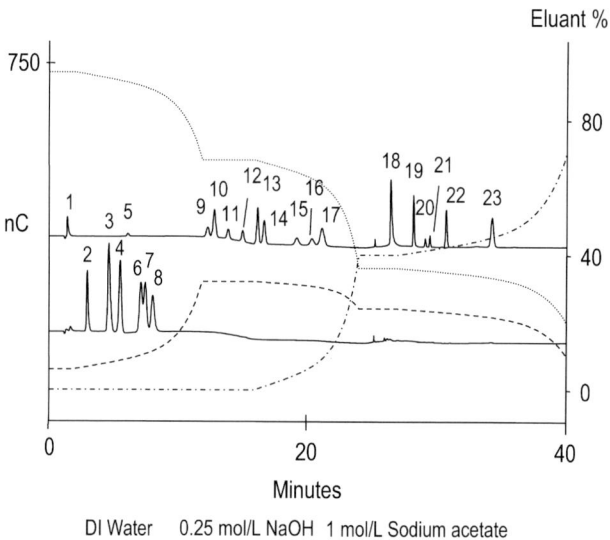

Fig. 3-205. Optimized separation of amino acids and carbohydrates on an AminoPac PA-10 latexed anion exchanger (250 mm × 2 mm i.d.). – Eluant: NaOH – sodium acetate; gradient: graphically depicted; flow rate: 0.25 mL/min; detection: integrated amperometry on a gold working electrode; analytes: (1) arginine, (2) fucose, (3) galactosamine, (4) glucosamine, (5) lysine, (6) galactose, (7) glucose, (8) mannose, (9) alanine, (10) threonine, (11) glycine, (12) valine, (13) serine, (14) proline, (15) isoleucine, (16) leucine, (17) norleucine, (18) histidine, (19) phenylalanine, (20) glutamate, (21) aspartate, (22) cysteine, and (23) tyrosine.

the cation exchange column holding the amino acid fraction is switched inline with the anion exchange separator. The mobile phase used at the beginning of the separation (NaOH, pH 12.7) transfers amino acids from the trap column onto the anion exchange separator. Using a modified NaOH/NaOAc gradient, the separation of amino acids is completed without any interference by carbohydrates. This is shown in Fig. 3-206. Trace 1 is a direct separation of a 125-pmol amino acid standard (SRM 2389, NIST Gaithersburg, MD, USA) to which the authors added 10 µmol/L levels of glucose, fructose, sucrose, and maltotriose. Also present are peaks of maltose, maltotetraose, and maltopentaose, which were introduced as impurities with the maltotriose. Complete co-elution is observed for the fructose/glycine and sucrose/proline peak pairs. The peaks of glucose and alanine, histidine and maltotetraose are incompletely resolved. Trace 2 in Fig. 3-207 is a chromatogram generated under optimized conditions enabled by the sample preparation technique described above. All seven carbohydrates added to the amino acid mixture are completely removed. The authors were able to verify that carbohydrates were completely removed even at a 100-fold molar excess relative to the amino acids. All common amino acids were recovered following the carbohydrate removal step. The average value of their recovery was 88.1%. The mean value of relative standard deviations from five replicants was 3.9%. These data were generated with solutions containing 40 µmol/L glucose.

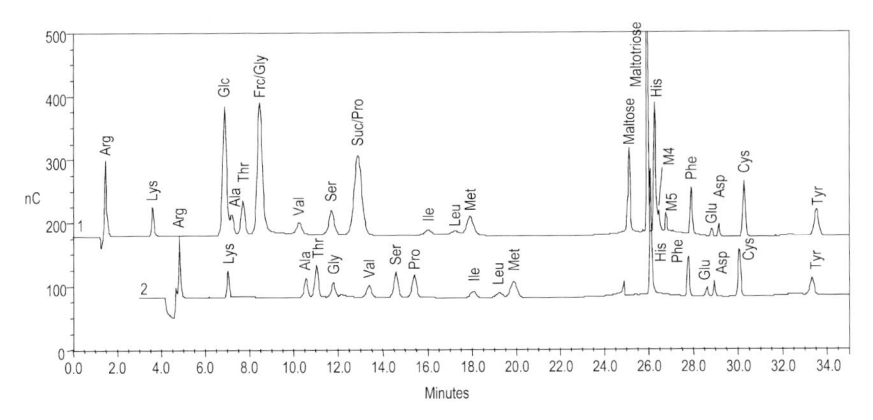

Fig. 3-206. Direct injection of a carbohydrate-amino acid mixture (Trace 1) and a chromatogram obtained with inline sample preparation of the same mixture (Trace 2). – Separator column: AminoPac PA-10 (250 mm × 2 mm i. d.); eluant: NaOH/NaOAc gradient; flow rate: 0.25 mL/min; detection; integrated amperometry on a gold working electrode; solute concentration: 125 pmol of each amino acid + 10 µmol/L of each carbohydrate. Note: In the chromatogram of Trace 1, the sample enters the separator column at time zero of the gradient program. In the chromatogram of Trace 2, the amino acid fraction is transferred to the separator column at 3.0 minutes of the gradient program. The retention times are not directly comparable; however, the selectivity of the separation is almost unchanged. (taken from [253]).

The chromatogram in Fig. 3-202 shows a typical anion exchange separation of an amino acid mixture. It illustrates the weak interaction of arginine with the stationary phase, which places the peak of arginine very close to the system void. Arginine retention is completely unaffected by gradient manipulations. Jandik et al. [254] described a new method of increasing its retention on an anion exchange column; the method relies on the dual ion-exchange functionality of the pellicular packing material. Those particles, by design, carry a residual cation exchange functionality that was demonstrated experimentally by Wimberley [255]. The new method is based on the interactions of the protonated form of arginine with the residual cation exchange groups on the core beads. Although the direct addition of acid is very effective in influencing the retention of arginine, it affects peak shape and retention of other peaks in the chromatographic separation. By co-injecting sulfuric acid using the valve scheme with two guard columns shown in Fig. 3-207, a similar retention enhancement for arginine was achieved with only minimal effect on the rest of the separation. The valve scheme consists of two interconnected six-port valves equipped with 25-µL injection loops. Two AminoPac PA-10 guard columns are installed between the SAMPLE and the ACID valve, and one AminoPac PA-10 analytical column is installed at the outlet of the ACID valve. The SAMPLE INJECT and the timing for the movements of the ACID valve are summarized in Table 3-32.

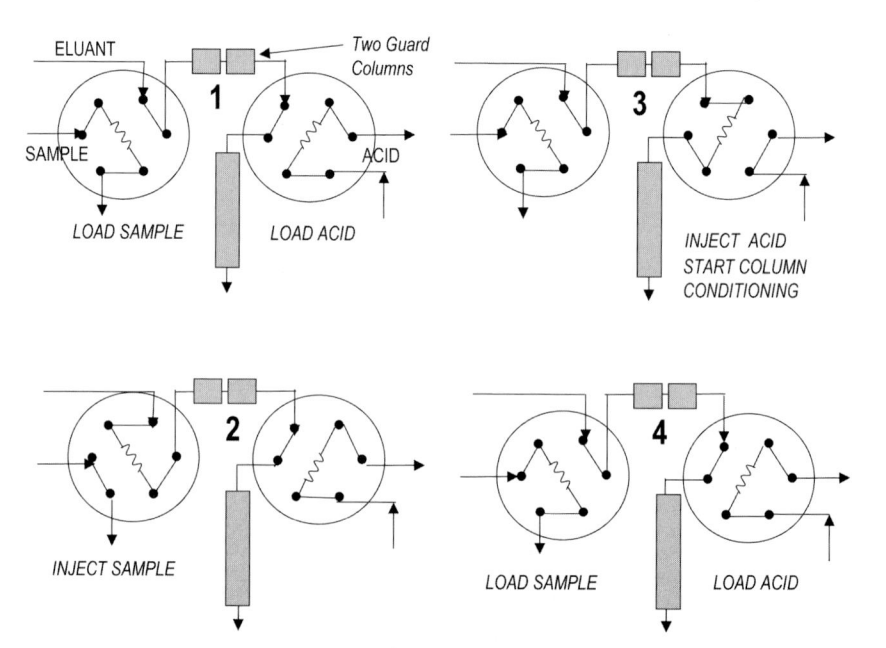

Fig. 3-207. Valve scheme for acid co-injection to selectively enhance retention on a pellicular anion exchanger.

Table 3-32. Gradient conditions and valve movements for the valve scheme in Fig. 3-207.

	Time [min]	NaOH [mmol/L]	NaOAc [mmol/L]	Curve
LOAD ACID	0.0	50		5
INJECT SAMPLE	1.0	50		
INJECT ACID	1.6	50		
LOAD ACID	2.1	50		
	12.0	50		
	16.0	80		5
	24.0	60	400	8
	27.0	60	400	
	40.0	60	400	
	40.1	200		8
	42.0	200		
	42.1	50		8
	75.0	50		8

The effect of sulfuric acid co-injection on arginine retention is illustrated in Fig. 3-208. Tris was added to the amino acid mixture as a model compound for unretained, electroactive compounds interfering with arginine. Without the direct addition of acid or acid co-injection, the arginine peak would be completely obscured by the Tris peak. As can be seen from both chromatograms in Fig. 3-208,

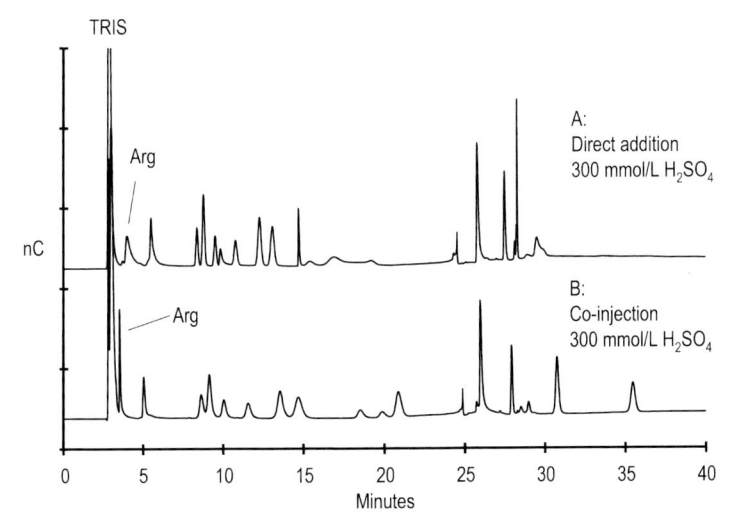

Fig. 3-208. Comparison of direct sulfuric acid addition and sulfuric acid acid co-injection for selectively enhancing the retention of arginine utilizing a pellicular anion exchanger. — Separator column: AminoPac PA-10; eluant: NaOH/NaOAc gradient (see Table 3-32); flow rate: 0.25 mL/min; detection: see Fig. 3-205; injection volume: 25 µL; solute concentrations: 250 pmol of each amino acid, 20 µmol/L Tris; (taken from [254]).

arginine is resolved from Tris. However, the chromatogram in Fig. 3-208B, which results from acid co-injection, is free of the peak distortion that occurs with direct acid addition.

The AminoPac PA-10 anion exchanger can also be used for the separation of O-phosphorylated amino acids such as P-Thr, P-Ser, and P-Tyr, which are barely retained on cation exchangers and are therefore only poorly separated. On AminoPac PA-10, they elute at the end of the hydrolysate program in the retention range of tyrosine and may thus be separated from hydrolysate amino acids (Fig. 3-209). To complete the elution of O-phosphorylated amino acids within 30 minutes, the acetate gradient was started earlier than it was for the standard conditions in Fig. 3-202. The bottom chromatogram in Fig. 3-209 shows the separation of 200 pmol each of the hydrolysate amino acids, the top chromatogram the separation of 50 pmol each of the O-phosphorylated amino acids. The minimum detection limits for O-phosphorylated amino acids are in the upper fmol range.

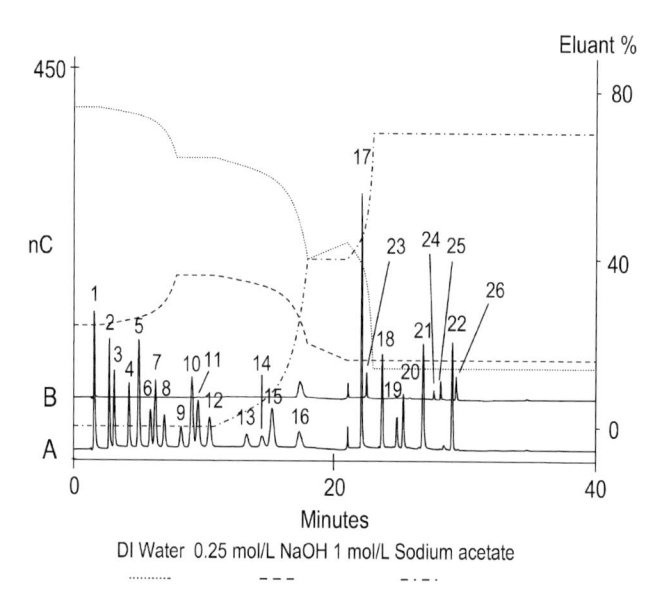

Fig. 3-209. Separation of hydrolysate amino acids and O-phosphorylated amino acids on AminoPac PA-10 (250 × 2 mm i. d.). − Eluant: NaOH − NaOAc; gradient: graphically depicted; flow rate: 0.25 mL/min; detection: see Fig. 3-206; chromatogram A: 200 pmol each of arginine (1), hydroxylysine (2), lysine (3), galactosamine (4), glucosamine (5), alanine (6), threonine (7), glycine (8), valine (9), hydroxyproline (10), serine (11), proline (12), isoleucine (13), leucine (14), methionine (15), norleucine (16), histidine (17), phenylalanine (18), glutamate (19), aspartate (20), cystine (21), and tyrosine (22), chromatogram B: 50 pmol each of P-arginine (23), P-serine (24), P-threonine (25), and P-tyrosine (26).

The separation of amino acids on anion exchangers with integrated amperometric detection is compatible with all common hydrolysis protocols based on hydrochloric acid, performic acid, methanesulfonic acid, and sodium hydroxide (see also Section 5.9.3). Taking soybean hydrolysis as an example, Fig. 3-210 shows the resulting chromatograms of three different hydrolysis protocols. The characteristic common to hydrochloric acid and methanesulfonic acid hydrolyses is the degradation of tryptophane, which can only be hydrolyzed successfully with sodium hydroxide. In this case, sample preparation is confined to sample filtration and dilution, because the separation on an anion exchanger also involves an alkaline eluant. In comparison with pre-column derivatization of amino acids requiring neutralization, this method offers a significant advantage; neutralization happens along with the formation of large amounts of sodium chloride, which impairs the derivatization reaction.

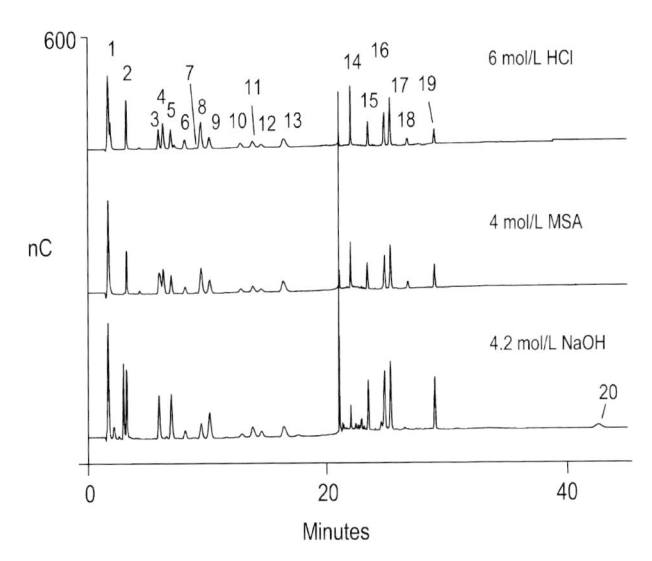

Fig. 3-210. Influence of various hydrolysis techniques on the amino acid distribution taking soybean powder as an example. — Separator column: AminoPac PA-10 (250 × 2 mm i. d.); eluant: NaOH — NaOAc; flow rate: 0.25 mL/min; detection: see Fig. 3-206; hydrolysis protocols: (A) 0.1 g in 5 mL HCl (c = 6 mol/L) at 110 °C over 24 h, dilution: 1:1250, (B) 0.1 g in 5 mL MSA (c = 4 mol/L) at 110 °C over 24 h, dilution: 1:1250, (C) 0.1 g in 5 mL NaOH (c = 4.2 mol/L) at 110 °C over 24 h, dilution: 1:500; analytes: arginine (1), lysine (2), alanine (3), threonine (4), glycine (5), valine (6), hydroxyproline (7), serine (8), proline (9), isoleucine (10), leucine (11), methionine (12), norleucine (13), histidine (14), phenylalanine (15), glutamate (16), aspartate (17), cystine (18), tyrosine (19), and tryptophane (20).

3 Anion Exchange Chromatography (HPIC)

Figure 3-211 shows the chromatogram of a collagen hydrolysate in comparison with a 200-pmol amino acid standard. In conventional amino acid analysis, collagen analysis is considered to be difficult because of the large concentration differences between glycine, proline, hydroxyproline, and alanine and the other amino acids, which usually require two runs with different dilutions. This is the consequence of the limited dynamic range of post-column derivatization with ninhydrin. In contrast, those large concentration ratios can be determined in one run when direct detection via integrated amperometry is applied. The results of both detection methods, shown in Table 3-33 using a collagen hydrolysate as an example, are in good agreement [250].

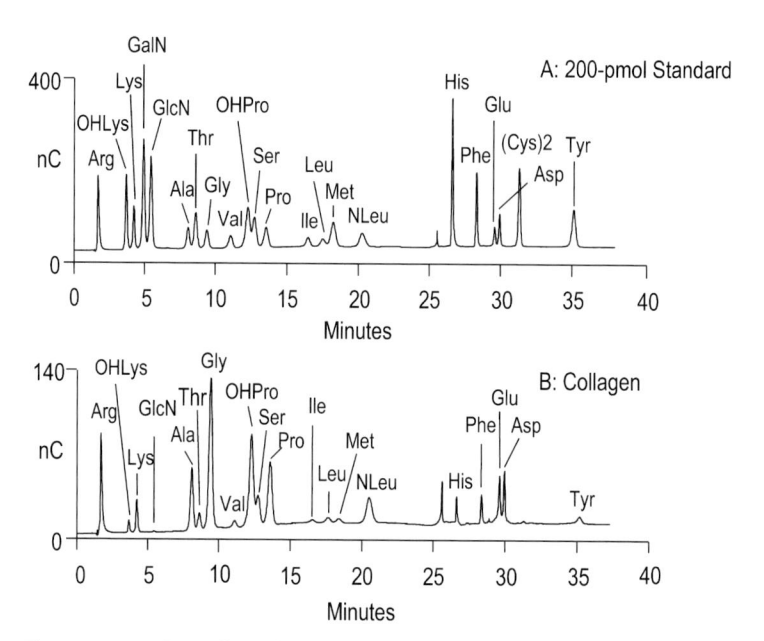

Fig. 3-211. Analysis of a collagen hydrolysate and an amino acid/amino sugar standard with anion exchange chromatography and integrated amperometry. — Chromatographic conditions: see Fig. 3-207; injection volume: 25 µL; (A) 200-pmol standard, (B) 0.5 µg collagen hydrolysate.

Table 3-33. Comparison of integrated amperometry and post-column derivatization with ninhydrin as detection methods for determining amino acids in a collagen hydrolysate.

Amino acid	Integrated Amperometry [pmol]	[mol%]	Ninhydrin method [mol%]
Arg	143	5.3	5.8
OH-Lys	20	0.8	0.8
Lys	74	2.8	2.7
Ala	313	11.6	11.0
Thr	48	1.8	1.7
Gly	86	31.8	32.7
Val	58	2.1	2.1
OH-Pro	264	9.8	9.0
Ser	90	3.4	3.3
Pro	321	11.9	12.4
Ile	31	1.2	1.2
Leu	66	2.4	2.5
His	15	0.6	0.5
Phe	38	1.4	1.3
Glu	178	6.6	7.6
Asp	119	4.4	4.7
Tyr	11	0.4	0.4

Furthermore, anion exchange chromatography is suited for the separation of arogenic acid, an intermediate in the biosynthesis of phenylalanine and tyrosine.

Arogenic acid

Arogenic acid is not stable at acid pH. Thus, it cannot be analyzed by cation exchange chromatography. However, on a latexed anion exchanger with alkaline eluants, the separation of this compound from the amino acids phenylalanine and tyrosine is accomplished without any problem.

3.12 Proteins

Today, proteins are isolated and characterized with different chromatographic techniques. Depending on the protein, ion-exchange-, ion-exclusion-, affinity-, and reversed-phase chromatography may be applied. Traditionally, separation materials with low mechanical stability and limited resolution were used. In recent years, HPLC materials were developed that offered the separation power required by a protein chemist. Now, pellicular ion-exchange resins exhibit high resolution at relatively short retention times, while maintaining the biological activity of the molecule.

Among the pellicular ion exchangers used to separate proteins are latexed ion exchangers and tentacular materials. Due to their pH compatibility, they can be operated with almost all aqueous buffer systems and serve many separation problems, and so they are universally applicable. The general structure of tentacular materials [256] is shown in Fig. 3-212. They consist of a microporous ethylvinylbenzene substrate with a particle diameter of 10 μm that is cross-linked

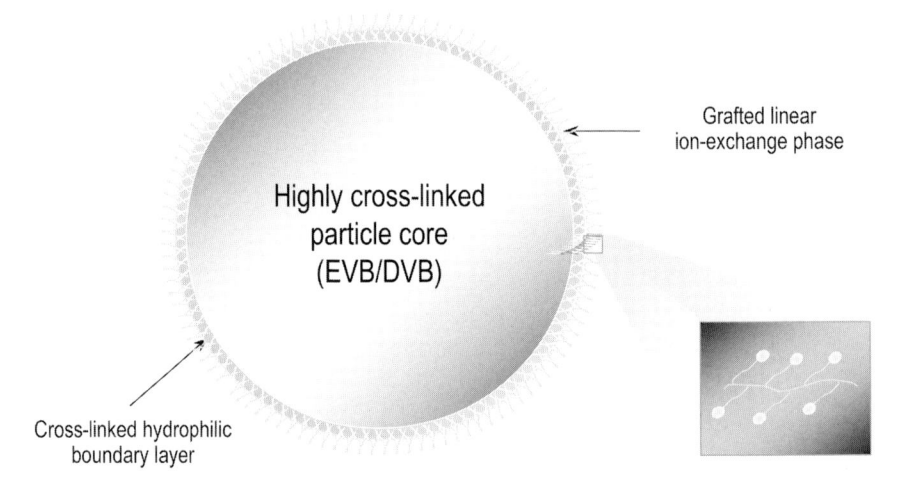

Fig. 3-212. Schematic structure of tentacular materials for protein separation.

with 55% divinylbenzene. The substrate particle is covered with a thin film of covalently bonded polymer chains, whose length and surface distribution are well-defined. Depending on the type of ion exchanger, they carry carboxyl-, sulfonate-, quaternary ammonium-, or amino groups. Stationary phases of this kind are extremely hydrophilic, which inhibits hydrophobic interaction with the protein. This can be derived from the fact that protein retention first decreases with increasing salt content in the mobile phase and then stays constant at higher salt concentration (Fig. 3-213).

Tentacular materials are predominantly used for the following separation problems:
- Basic proteins
- Monoclonal antibodies
- Deamidation products
- N-terminal pyroglutamate variants
- Protein isoforms

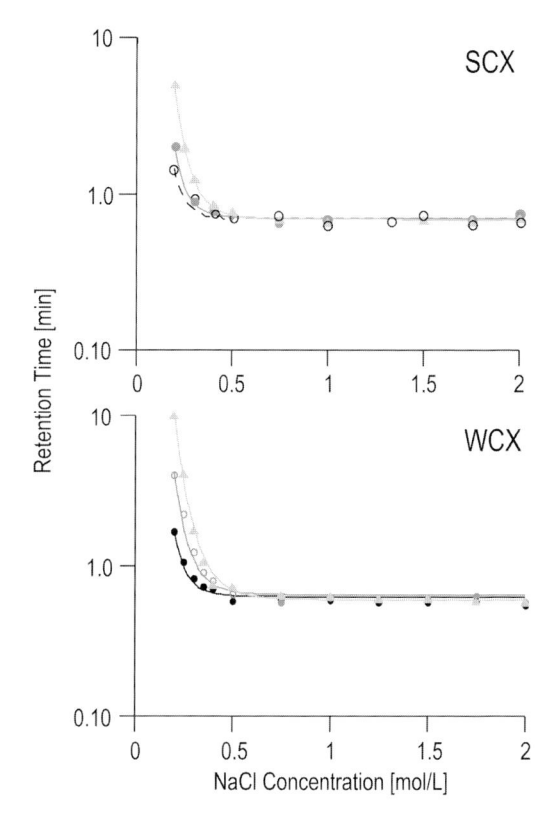

Fig. 3-213. Illustration of the hydrophilic nature of tentacular packing materials for protein separation. — Separator columns: 250 mm × 4 mm i. d. ProPac SCX and WCX; eluant: (A) 10 mmol/L sodium phosphate, pH 7, (B) 10 mmol/L sodium phosphate + 2 mol/L NaCl, pH 7; flow rate: 1 mL/min; detection: UV (254 nm); injection volume: 10 µL; sample: 0.5 mg/mL each of lysozyme (▲), ribonuclease A (○), and cytochrome c (●).

Selectivity differences between strong acid (SCX-10) and weak acid cation exchangers (WCX-10) as a function of the mobile phase pH are impressively demonstrated taking five different proteins as an example (Fig. 3-214).

A special challenge in the development and production of therapeutic proteins is the characterization of the structural variants of monoclonal antibodies. Of the five classes of immunoglobulines in human blood plasma, the IgG globulins are the most common and the best understood. They have a molecular weight of 150,000 and consist of four polypeptide chains: two identical heavy chains with 430 amino acid residues, and two identical light chains with 214 amino acid residues. These chains are tied together by disulfide bridges in a Y-shaped, flexible structure as schematically illustrated in Fig. 3-215. The heavy chains contain covalently-bonded oligosaccharide moieties. Each chain is characterized by a region with a constant amino acid sequence and a region with a variable sequence. The arms of the "Y" can be cleaved by papain, a proteolytic enzyme that delivers the so-called Fab- and Fc-fragments (in the presence of cysteine), or F(ab')-fragments (in the absence of cysteine). A frequent structural variation requiring thorough analysis [257-261] is the C-terminal processing of lysine residues at the heavy chain of monoclonal antibodies. Incomplete protein

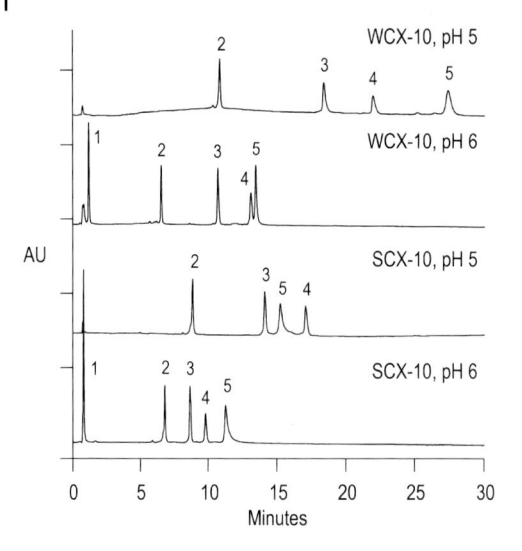

Fig. 3-214. Selectivity differences between strong acid and weak acid cation exchangers as a function of pH, taking certain proteins as an example. — Separator columns: 250 mm × 4 mm i. d. ProPac SCX and WCX; eluant: NaCl gradient in presence of 10 mmol/L sodium phosphate; flow rate: 1 mL/min; detection: UV (254 nm); injection volume: 10 µL; peaks: myoglobine (1), α-chymotrypsinogen (2), cytochrome c (3), ribonuclease A (horse, 4), and lysozyme (5).

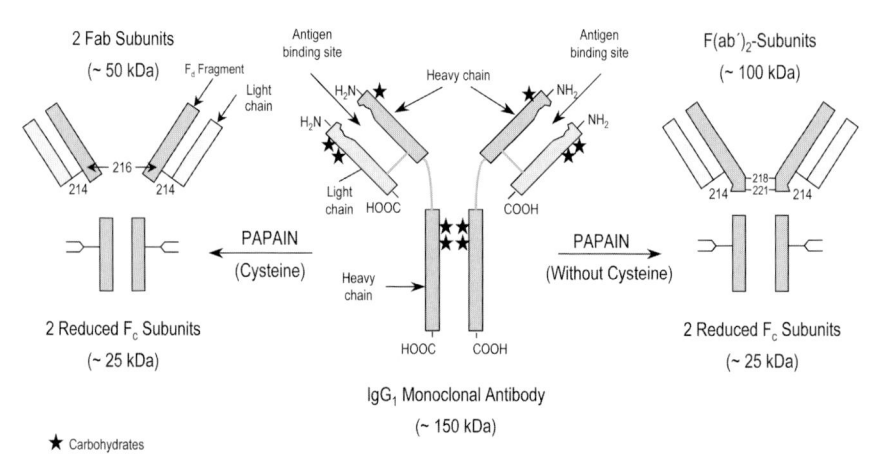

Fig. 3-215. Structure of the IgG immunoglobulin.

processing leads to a charge heterogeneity due to the absence of C-terminal lysine residues, which can be identified by cation exchange chromatography. As can be seen from the top chromatogram in Fig. 3-216, C-terminal lysine variants can be separated from the native IgG$_1$ antibody. Highest resolution is achieved with a weak acid cation exchanger at pH 7. When the IgG$_1$ sample is treated with carboxypeptidase B, an exopeptidase that cleaves residues (especially lysine) at the C-terminal end, peaks **2** and **3** can no longer be detected (bottom chromatogram), while the area of peak **1** increases. This can be taken as a proof that peaks **2** and **3** differ by one or two lysine residues at the C-terminal end of the chain. Thus, cation exchange chromatography can be applied to the quality control of therapeutic proteins.

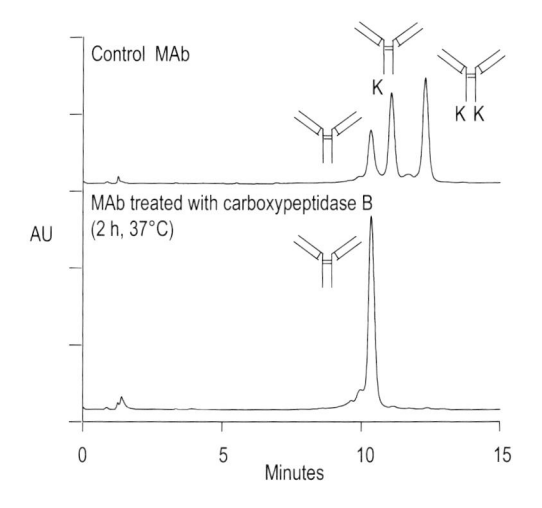

Fig. 3-216. Analysis of IgG$_1$ on a ProPac WCX-10 weak acid cation exchanger before and after treatment with carboxypeptidase B. – Eluant: (A) 0.01 mol/L sodium phosphate, pH 7, (B) 0.01 mol/L sodium phosphate + 1 mol/L NaCl; gradient: linear; 4% B to 15% B in 30 min; flow rate: 1 mL/min; detection: UV (220 nm); sample: IgG$_1$ Mab; peaks: (1) Mab standard, (2) Mab with one lysine residue on the C-terminal end of the heavy chain, and (3) Mab with two lysine residues on the C-terminal end of the heavy chain.

Tentacular cation exchangers have also proved to be good for the determination of the deamidation degree of a protein. Deamidation is a process occurring under physiological conditions that can be taken as a measure for protein half-life in blood. Deamidation occurs along with the loss of a functional amide residue of asparagine, which is metabolized to the respective acidic amino acid:

Modifications of this kind occur in a number of recombinant proteins, such as the human growth hormone, TPA, hirudine, monoclonal antibodies, interleukine-1, and others with various effects on the activity and stability of therapeutic proteins [262-266]. The determination of Asn deamidation residues in recombinant proteins represents a challenge for analytical and protein chemists [267]. It is clear from Fig. 3-217 that the separation of various deamidated isoforms of ribonuclease A from the native protein can be carried out in one run on ProPac PA1, whereas Donato et al. [268] reported a first separation on a cation exchanger

(Mono S) followed by a second separation via hydrophobic interactions for separating the two deamidation variants. Figure 3-218 illustrates the crystal structure of ribonuclease A with the deamidation site at Asn^{67}.

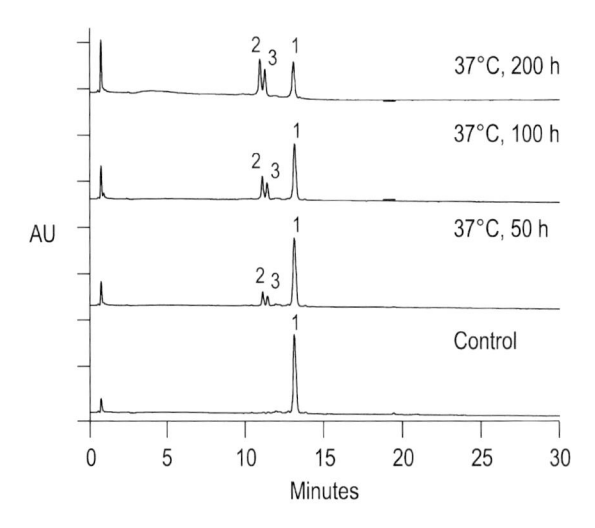

Fig. 3-217. Determination of the deamidation pattern of ribonuclease A after incubation of the protein at 37 °C in ammonium carbonate buffer, pH 8.2. — Separator column: ProPac WCX-10; eluant: (A) 0.01 mol/L sodium phosphate (pH 6), (B) 0.01 mol/L sodium phosphate (pH 6) + 1 mol/L NaCl; gradient: linear, 4% B to 70% B in 30 min; flow rate: 1 mL/min; detection: UV (280 nm); peaks: (1) 3 mg/mL native ribonuclease A, (2) and (3) deamidation products.

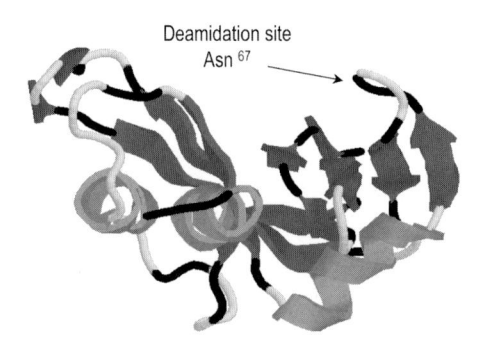

Fig. 3-218. Crystal structure of ribonuclease A with the deamidation site at Asn^{67}. (Source: Protein Database, Brookhaven National Labs, original data from L. Esposito, F. Sica, L. Vitagliano, A. Zagari, and L. Mazzarella.)

Another example of the high resolution of tentacular cation exchangers is the separation of peptides with N-terminal Glu- or Gln residues that tend to undergo cyclisation reactions, by which the N-terminal end will be blocked. This blocked residue can be specifically removed with pyroglutamate aminopeptidase. As can be seen from Fig. 3-219, the ProPac WCX-10 weak acid cation exchanger is suitable for the separation of blocked peptides from the N-1-peptides.

The selectivity differences between weak acid and strong acid cation exchangers are illustrated in Fig. 3-220, taking cytochrome c variants as an example. Cytochrome c is an electron-transporting protein which exists in all animals, plants, and aerobic microorganisms. It consists of a single chain of 104 amino acids in land-living vertebrates. In a detailed study, Margoliash [269] listed vari-

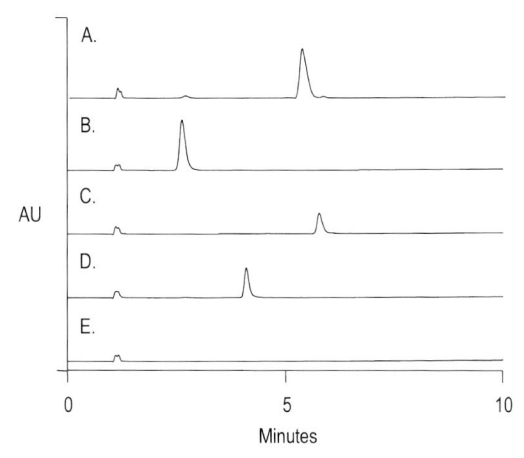

Fig. 3-219. Separation of blocked and N-1-peptides after treatment with pyroglutamate aminopeptidase. – Separator column: ProPac WCX-10; eluant: (A) 0.02 mol/L sodium phosphate, pH 4.8, (B) 0.02 mol/L sodium phosphate, pH 4.8 + 1 mol/L NaCl; gradient: linear, 0% B to 60% B in 15 min; flow rate: 1 mL/min; detection: UV (220 nm); samples: A. pGlu-Ser-Leu-Arg-Trp-amide + pyroglutamate aminopeptidase, B. pGlu-Ser-Leu-Arg-Trp-amide, C. neurotensin + pyroglutamate aminopeptidase, D. neurotensin (pGlu-Leu-Tyr-Gln-Asn-Lys-Pro-Arg-Arg-Pro-Tyr-Ile-Leu), E. buffer + enzyme control (0.1 mmol/L sodium phosphate, 10 mmol/L Na$_2$-EDTA, 5 mmol/L DTT, 5% (v/v) glycerol, pH 8.

ations of the amino acid sequence of cytochrome c in more than 50 different species. In all species, 27 amino acid residues are absolutely invariant and positioned irregularly in the chain. The number of differences in the residues roughly corresponds to the phylogenetic differences between the species. The ones separated in Fig. 3-220 originate from bovine, horse, and rabbit.

Pellicular anion exchangers are predominantly used for separating microheterogeneic proteins, which differ most in the sialylation degree of the glycol moieties. This special application is of great interest, since the sialylation pattern can be responsible for the half-life and the biological activity of a therapeutic protein. The DNAPac PA-100 latexed anion exchanger can be employed for the separation of differently sialylated glycoproteins and glycopeptides [270, 271]. In the case of glycoproteins, not only the portion of neuraminic acids but also the conformation of the carbohydrate part is decisive for the separation. A pre-purification of the protein in its microheterogeneic populations is often carried out prior to the analysis of the glyco moieties. Taking the human serum transferrins as an example, one can obtain three large sub-populations. The respective chromatogram in Fig. 3-221 shows the separation in the mono- (F1), di- (F2), and trisialylated form (F3) of the human serum transferrin.

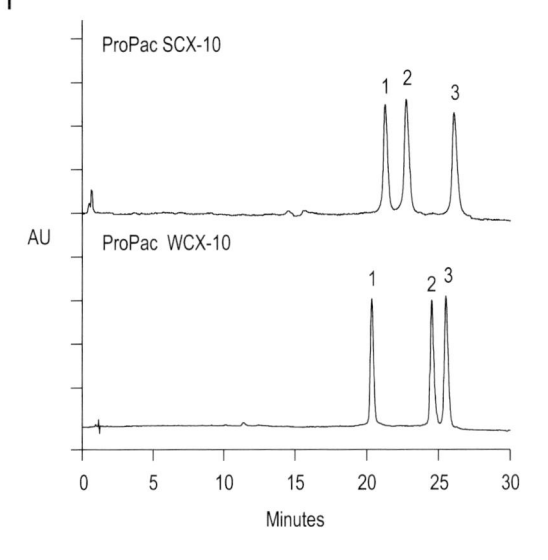

Fig. 3-220. Separation of cytochrome c variants. – Separator columns: ProPac WCX-10 and SCX-10; eluant: (A) 0.02 mol/L sodium phosphate, pH 7, (B) 0.02 mol/L sodium phosphate, pH 7 + 1 mol/L NaCl; gradient: linear, 4% B to 15% B in 30 min; flow rate: 1 mL/min; detection: UV (220 nm); peaks: (1) cytochrome c (bovine), (2) cytochrome c (horse), and (3) cytochrome c (rabbit).

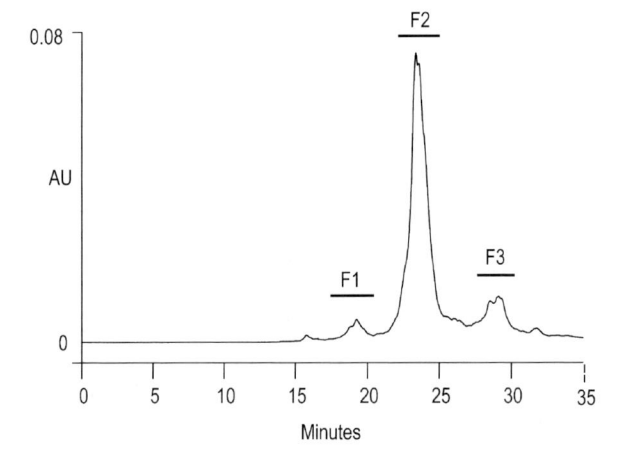

Fig. 3-221. Determination of the sialylation pattern of Human Serum Transferrin (HST). – Separator column: DNAPac PA-100; eluant: (A) 0.025 mol/L NH$_4$Ac, (B) 0.2 mol/L NH$_4$Ac; gradient: linear, 100% A to 100% B in 33 min; flow rate: 1 mL/min; detection: UV (215 nm); F1: monosialylated HST, F2: disialylated HST, F3: trisialylated HST.

Apart from neuraminic acid, other functional groups such as phosphate, sulfate, etc. determine the net charge of a protein, moreso than the amino acid residues do. Ovalbumine is a good example of this. As a phosphoglycoprotein, it is an essential constituent of chicken protein. It carries N-linked oligosaccharides of the hybrid- and mannose type, and phosphate residues on two serine amino acids. Hence, the degree of phosphorylation can vary quite a bit [272]. The ProPac SAX-10 strong basic, tentacular anion exchanger separates ovalbumine in nine prominent peaks (Fig. 3-222), two of which could be identified as the mono- and diphosphorylated form [273]. The other signals probably represent further

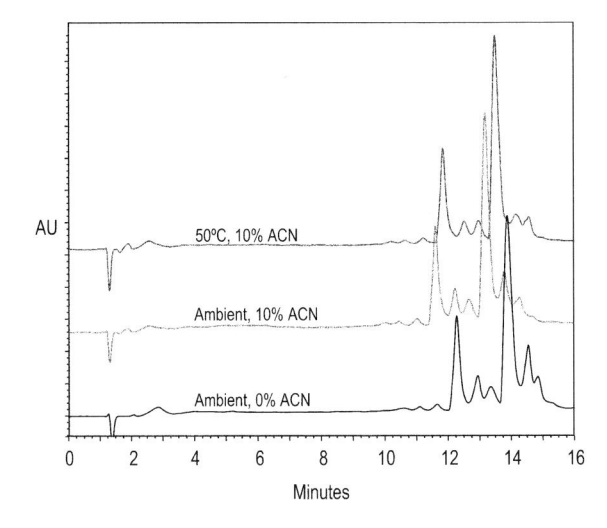

Fig. 3-222.
Determination of the phosphorylation pattern of ovalbumine. — Separator column: ProPac SAX-10; eluant: (A) 0.02 mol/L Tris/HCl — acetonitrile, pH 8.5, (B) 2 mol/L NaCl; gradient: linear, 0% B to 25% B in 15 min; flow rate: 1 mL/min; detection: UV (214 nm); sample: 50 µg ovalbumine.

microheterogeneic forms or ovalbumine variants that differ in amino acid composition [274]. Furthermore, Fig. 3-222 shows the influence of acetonitrile on the elution profile as a function of temperature.

3.13
Nucleic Acids

In recent years, the demand for synthetic oligonucleotides has drastically increased because their medical-therapeutic and molecular-biological use, especially in genetic engineering, has gained more and more importance. Oligonucleotides are used for gene sequencing, as linkers for recombinant DNA technology, as primers for sequencing, as substrates for enzymatic and structural analysis, or as sensors for detecting DNA- and RNA sequences in the diagnosis of genetic defects. Another increasingly important function of oligonucleotides is their use as *antisense* therapeutics for suppressing the expression of cancer- and other genes. *Antisense oligonucleotides* exist in a modified form as phosphorothioates.

The availability of oligonucleotides is provided by the incredible advancement in DNA solid phase synthesis, which is performed in fully automated systems. Optimization of the synthesis is controlled by means of chromatography; the optimization of the chromatographic separation and the preparative purification of the synthetic products are of special interest. Oligonucleotides have a number of properties which can be utilized in their chromatographic separation. As polyvalent anions, they differ in net charge depending on their chain length, so that they can be separated in a simple way on anion exchangers. Their retention behavior towards reversed phase materials is determined by the lipophilic nucleobases. In the case that they significantly differ in size, a separation based on ion-exclusion might also be suitable [275, 276]. All three separation techniques usually employ macroporous resins, which only partly satisfy the demand for resolution and mechanical stability. Yet modern applications of modified oligonucleotides as diagnostic sensors or therapeutic agents require the highest possible resolution (N,N-1 with oligonucleotides >DP30) and short retention times (<30 minutes). Only pellicular ion exchangers can meet these demands, because their ion-exchange groups are directly accessible to the analytes. In this way, their small diffusion rate is compensated; normally, the small diffusion rate has a negative influence on the efficiency of the separation when using macroporous resins.

The DNAPac PA-100 latexed anion exchanger is based on a microporous substrate with a particle diameter of 13 µm, to which 100-nm latex particles are agglomerated electrostatically. This stationary phase exhibits very few hydrophobic properties, so separations of N,N-1 bases up to the higher polymeric range (DP50 to DP70) are possible. Figure 3-223 shows the separation of linear poly(dA) polymers on such a stationary phase.

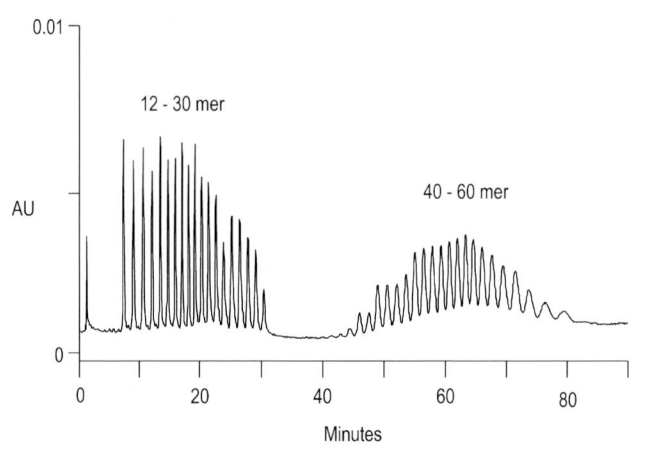

Fig. 3-223. Separation of polydeoxyadenosine with 12 to 30 and 40 to 60 adenosine residues. — Separator column: DNAPac PA-100; eluant: (A) 25 mmol/L Tris-HCl + 7.5 mmol/L NaClO$_4$, pH 8, (B) 25 mmol/L Tris-HCl + 124 mmol/L NaClO$_4$; gradient: non-linear; flow rate: 1.5 mL/min; detection: UV (260 nm).

Baseline-resolved separations, efficiency, and rapid elutions are the predominant characteristics of the pellicular anion exchanger for separations of single-stranded nucleic acids, even when additional agents such as formamide and urea are used. While homopolymers are usually separated under non-denaturing conditions, heteropolymers require denaturing conditions due to their secondary structures (hairpin, palindrome). This can be achieved in a very simple way by separating at high pH or at elevated temperature. Thus, hydrophobic or van-der-Waals interactions of complementary sequences or guanine-rich areas within the single strand are avoided (Fig. 3-224).

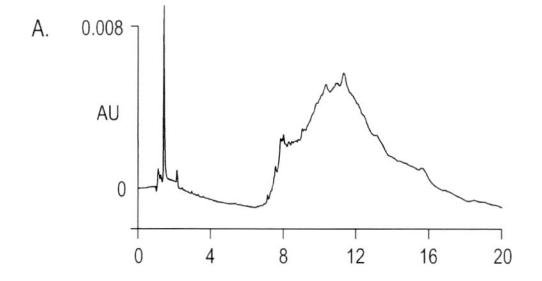

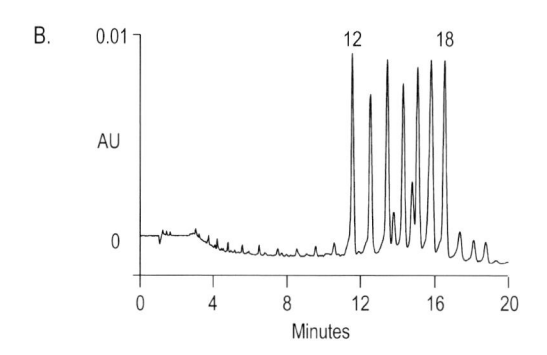

Fig. 3-224. Separation of polydeoxyguanosine with 12 to 18 guanosine residues. — Separator column: DNAPac PA-100; eluant for A.: (A) 0.03 mol/L Tris-HCl + 0.1 mol/L HCl (pH 8.4), (B) 0.03 mol/L Tris-HCl + 0.9 mol/L HCl (pH 8.4); eluant for B.: (A) 0.025 mol/L NaOH + 0.5 mol/L NaCl (pH 12.4), (B) 0.025 mol/L NaOH + 0.9 mol/L NaCl (pH 12.4); flow rate: 1.5 mL/min; detection: UV (260 nm).

Apart from the use of "simple" oligonucleotides for molecular biology purposes, oligonucleotides in a modified form are used as diagnostic sensors or *antisense* therapeutics, which must be separated from their synthetic precursors. Also, their degradation rate in biological fluids is controlled chromatographically [277]. Synthetic RNA oligomers or phosphorothioates are separated analytically or semi-preparatively from their precursors on DNAPac PA-100 (Fig. 3-225). The high capacity of a semi-preparative anion exchanger (250 mm × 22 mm i. d.) allows N,N-1 separations of up to 30 mg sample, wherein the chromatographic conditions developed for the analytical separation are directly transferred to the semi-preparative procedure. Figure 3-226 shows an example for such a transfer of chromatographic conditions from the analytical to the semi-preparative scale.

Ion exchangers based on DEAE belong to the class of non-porous materials. Since these are weak basic ion exchangers, alkaline eluants are not suitable for creating denaturing conditions. The high backpressure of this column material, caused by the small particle diameter of 2 µm to 3 µm, only partly allows it to be used for semi-preparative purposes [278]. Non-porous materials without ion-exchange functional groups separate oligonucleotides based on ion-pair mechanisms [279]. Analytical separations of oligonucleotides are characterized by high resolution. All separation materials mentioned in this Section can also be applied for the analysis of double-stranded nucleic acids.

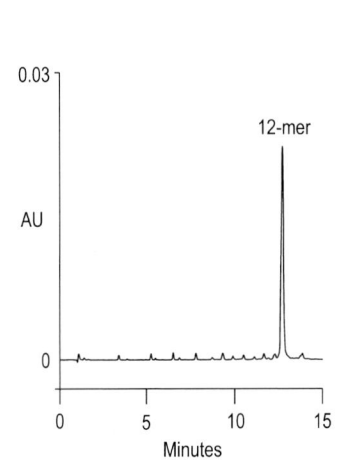

Fig. 3-225. Separation of synthetic RNA (A.) and phosphorothioates (B.) from their substrates. — Separator column: DNAPac PA-100; eluant for A.: (A) 25 mmol/L Tris-HCl + 1.9 mmol/L NaClO$_4$, (pH 8), (B) 25 mmol/L Tris-HCl + 11 mmol/L NaClO$_4$, pH 8; gradient: linear, 0% B to 100% B in 20 min; eluant for B.: (A) 25 mmol/L NaOH + 56 mmol/L NaClO$_4$, pH 12.4, (B) 25 mmol/L NaOH + 330 mmol/L NaClO$_4$, pH 12.4; gradient: 0% B to 100% B in 20 min; flow rate: 1.5 mL/min; detection: UV (260 nm).

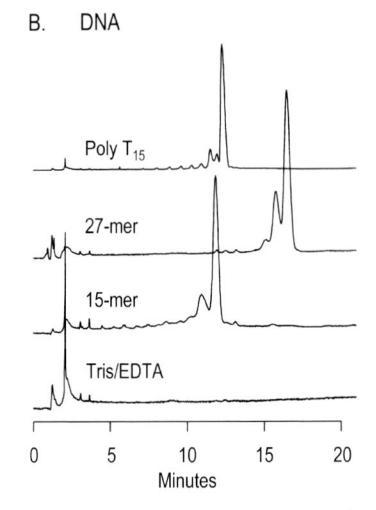

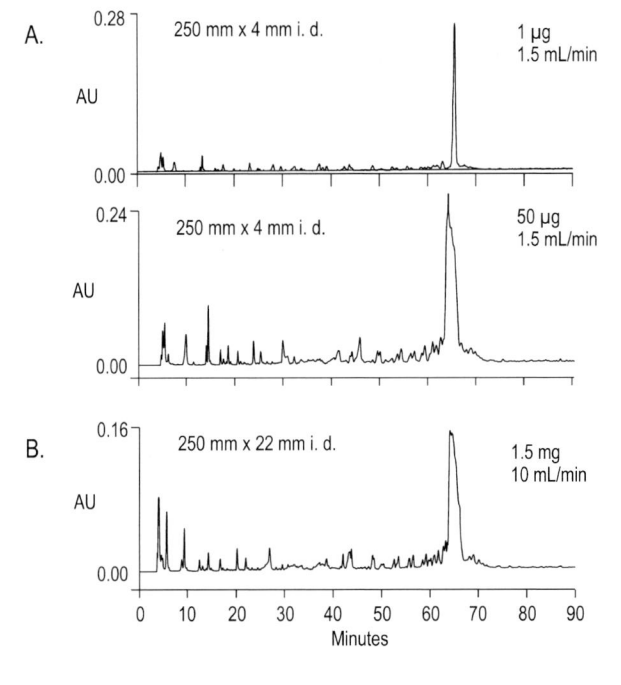

Fig. 3-226. Direct comparison of an oligonucleotide (20mer) separation at analytical (A.) and semi-preparative (B.) scale. — Separator column: DNAPac PA-100; eluant: (A) 25 mmol/L Tris-HCl + 7.5 mmol/L NaClO$_4$, pH 8, (B) 25 mmol/L Tris-HCl + 124 mmol/L NaClO$_4$, pH 8; gradient: non-linear; flow rate: A. 0.33 mL/min, B. 10 mL/min; detection: UV (260 nm).

4
Cation Exchange Chromatography (HPIC)

Small et al. [1] devoted a significant part of their pioneering work in ion chromatography to the separation and determination of cations. The required hardware was identical to that used for anion analysis. It comprised a low-capacity cation exchanger and a suppressor column containing a strong basic anion exchange resin in the hydroxide form. With the exception of the suppressor column, which has since been replaced by modern membrane suppressors, the principle setup remains unchanged. As in anion analysis, the application of a suppressor system is not necessarily a prerequisite for the cation determination by means of conductivity detection.

The exchange reaction with a cation M^+ that occurs at the stationary phase of a cation exchanger can be expressed as follows:

$$\text{Resin} - SO_3^- \, H^+ + M^+ A^- \xrightleftharpoons{K} \text{Resin} - SO_3^- \, M^+ + H^+ A^- \tag{103}$$

The separation of cations is determined by their different affinities towards the stationary phase.

4.1
Stationary Phases

As with anion exchangers, cation exchangers are classified according to the type of substrate material. Although cation exchangers are usually made using organic polymers, a variety of substrate materials are appropriate. Because dilute acids serve as the eluent in cation separations, the stability over the whole pH range (a condition provided by organic polymers) is not required. Therefore, silica-based cation exchangers, which exhibit a significantly higher chromatographic efficiency, are also used.

4.1.1
Polymer-Based Cation Exchangers

4.1.1.1 Styrene/Divinylbenzene Copolymers

Styrene/divinylbenzene copolymers are widely used as substrate materials for the manufacture of cation exchangers. The principle properties of this support material were described in detail in Section 3.4.1. The material is surface-sulfonated via the reaction with concentrated sulfuric acid. The ion-exchange capacity is determined by the degree of sulfonation. The latter depends essentially on the reaction time and the temperature program during the reaction [2]. Typical ion-exchange capacities are between 0.005 and 0.1 mequiv/g [3]. A surface-sulfonated cation exchange material is depicted schematically in Fig. 4-1. The diffusion of totally dissociated ions such as sodium, potassium, and magnesium into the interior of the stationary phase can be ignored because of the interior's comparatively strong hydrophobic character. This results in short diffusion pathways and, thus, in a higher chromatographic efficiency as compared with conventional totally sulfonated cation exchangers.

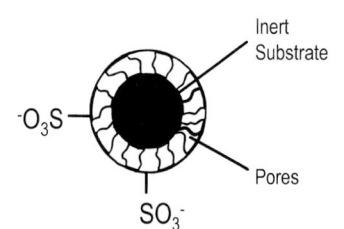

Fig. 4-1. Schematic representation of a surface-sulfonated cation exchange material.

Surface-sulfonated styrene/divinylbenzene copolymers are offered by a number of manufacturers. The structural and technical properties of these columns are summarized in Table 4-1. Because the ion-exchange functional groups in all strong acid cation exchangers are sulfonate groups, columns with packing materials differ considerably only in their chromatographic efficiency and, consequently, in column dimensions. The IonPac CS1 separator column, manufactured by Dionex (Sunnyvale, CA, USA), is still commercially available, but is only of historical significance due to the large particle diameter of 20 μm. All other separator columns listed in Table 4-1 are still in use, mainly for the sequential analysis of alkali- and alkaline-earth metals, because a simultaneous analysis of mono- and di-valent cations is not possible with surface-sulfonated cation exchangers.

As a representative for all separator columns of this type, Fig. 4-2 shows a chromatogram with a separation of alkali metals; suppressed conductivity detection was used.

Mineral acids such as hydrochloric acid or nitric acid are employed as eluants, regardless of whether the background conductivity is chemically suppressed or electronically compensated. While alkali metals are eluted within 10 minutes

Table 4-1. Structural and technical properties of surface-sulfonated styrene/divinylbenzene copolymers.

Separator	Manufacturer	Dimensions [Length × i.d.] [mm]	Max. flow rate [mL/min]	Max. pressure [MPa]	Solvent stability [%]	Capacity [mequiv/g]	Particle diameter [µm]
IonPac CS1	Dionex	200 × 4.6	3	5	5	0.01....0.05	20
IonPac CS2	Dionex	250 × 4.6	3	10	5	0.01....0.05	15
LCA-K01	Sykam	125 × 4	3	20	5	0.05	10
MCI Gel SCK01	Mitsubishi Kasei	150 × 4.6	2	8	5	—[a]	10
PRP-X200	Hamilton	250 × 4	8	30	100	0.035	10
Shimpack IC-C1	Shimadzu	150 × 5	2	5	10	—[a]	10
TSK Gel IC Cation	Toyo Soda	50 × 4.6	2	5	10	0.012	10

[a] no details

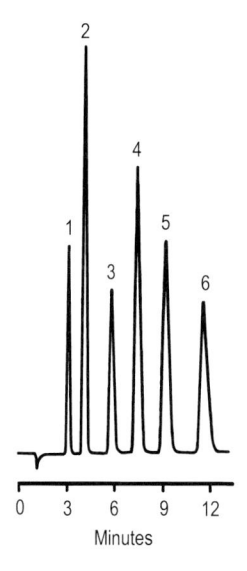

Fig. 4-2. Separation of alkali metals at a surface-sulfonated styrene/divinylbenzene copolymer. — Separator column: IonPac CS2; eluant: 30 mmol/L HCl; flow rate: 1 mL/min; detection: suppressed conductivity; injection volume: 50 µL; solute concentrations: 5 mg/L lithium (1), 5 mg/L sodium (2), 10 mg/L ammonium (3), 10 mg/L potassium (4), 20 mg/L rubidium (5), and 30 mg/L cesium.

under the chromatographic conditions in Fig. 4-2, alkaline-earth metals are strongly retained due to their high affinity towards the stationary phase. When the samples to be analyzed contain alkaline-earth metals, too, alkali metal analysis will be interfered with by late eluting alkaline-earth metals.

Generally, divalent eluant ions are used for eluting alkaline-earth metals from surface-sulfonated cation exchangers. It is true that alkaline-earth metals can also be eluted by increasing the acid strength in the mobile phase, but the resulting background conductivity would be too high, especially when using non-suppressed conductivity detection. For this reason, ethylenediamine in combination with a complexing agent such as tartaric acid is used as an eluant for non-suppressed conductivity detection of alkaline-earth metals. In suppressed conductivity detection, however, 2,3-diaminopropionic acid (DAP) in combination with hydrochloric acid has been successfully used as an eluant.

$$\begin{bmatrix} \overset{+}{H_3N} \ \overset{+}{NH_3} \\ H_2C-CH-COO^- \end{bmatrix}^+ \underset{-H^+}{\overset{+H^+}{\rightleftharpoons}} \begin{bmatrix} \overset{+}{H_3N} \ \overset{+}{NH_3} \\ H_2C-CH-COOH \end{bmatrix}^{2+} \quad (104)$$

$pK_a \quad -COO^- \approx 1.3$
$ \quad -NR_3^+ \approx 6.7$

The big advantage of this eluant is that the net charge of 2,3-diaminopropionic acid and, thus, the elution strength, is solely determined by the HCl concentration. In the early days of cation exchange chromatography, *m*-phenylenediamine was recommended as an eluant for alkaline-earth metals, but it isn't used anymore because it is not available in the required purity. Moreover, *m*-phenylenediamine tends to undergo polymerization at the stationary phase, which leads to a significant and irreparable loss of separation power.

4.1.1.2 Ethylvinylbenzene/Divinylbenzene Copolymers

As mentioned above, alkali- and alkaline-earth metals can only be analyzed sequentially on strong acid cation exchangers due to their widely different affinities towards the stationary phase. This is also true for the significantly more efficient latexed cation exchangers (see Section 4.1.2), when the latex particles are functionalized with sulfonate groups. The dream of the inorganic analytical chemist – the simultaneous analysis of alkali- and alkaline-earth metals – can only be realized with weak acid cation exchangers using carboxyl groups as ion-exchange functionality. Schomburg et al. [4] were first to report of an isocratic separation of mono- and di-valent cations within an acceptable time frame of 20 minutes using a silica phase coated with poly(butadiene-maleic acid) (PBDMA); (see Section 4.1.3). Because a weak acid complexing agent is used as an eluant, cation exchange interactions as well as complexation processes contribute to the retention mechanism. The complexation processes strongly depend on sample pH. As with many polymer-coated silica columns, reproducible manufacturing is extremely difficult, so that the eluant composition has to be adapted when using columns from different production lots. The same applies to the separator columns derived from the original concept [5, 6] of coating silica with polymers. This stability problem was solved with the development of a cation exchanger based on a polymeric substrate material. This column was introduced in 1993 under the trade name IonPac CS12 [7]. It consists of a highly cross-linked mesoporous ethylvinylbenzene/divinylbenzene copolymer with a particle diameter of 8 μm and a specific surface of 450 m²/g, which is grafted at the surface with a thin film (500 nm to 1000 nm) of an ion-exchange polymer. This ion-exchange polymer contains the ion-exchange functional groups, namely carboxyl groups (pK < 3). The resulting high ion-exchange capacity of 2.8 mequiv/column (250 mm × 4 mm i.d.) is necessary, because in carboxylate-based weak acid cation exchangers only a fraction of the ion-exchange groups are available for the cation exchange process due to the low acid strength. Because such a high ion-exchange capacity cannot be realized with latexed cation exchangers and their microporous substrates, mesoporous substrates with high specific surfaces are now used.

The big advantage of such a development concept is that mono- and di-valent cations, in contrary to latexed cation exchangers such as the IonPac CS10 (see Section 4.1.2), can be eluted with mineral acids such as H_2SO_4 or organic acids such as methanesulfonic acid. As can be seen from Fig. 4-3, the five most important cations – sodium, ammonium, potassium, magnesium, and calcium – can be separated to baseline under isocratic conditions within 10 minutes. Barium that is strongly retained on conventional cation exchangers and, thus, exhibits significant tailing, already elutes just after 15 minutes on the CS12 column under the conditions given in Fig. 4-3. Therefore, a minimum detection limit of about 0.1 mg/L can be achieved without any problem when applying suppressed conductivity detection and a larger injection loop of 100 µL.

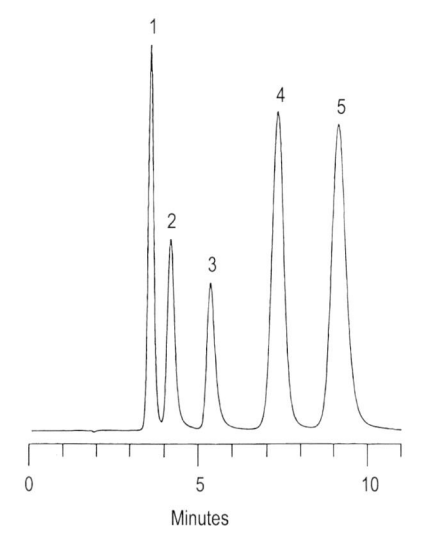

Fig. 4-3. Simultaneous analysis of alkali metals, alkaline-earth metals, and ammonium on IonPac CS12. – Eluant: 20 mmol/L methanesulfonic acid; flow rate: 1 mL/min; detection: suppressed conductivity; injection volume: 25 µL; solute concentrations: 5 mg/L sodium (1), 5 mg/L ammonium (2), 5 mg/L potassium (3), 5 mg/L magnesium (4), and 10 mg/L calcium (5).

A quality feature of all cation exchangers suitable for simultaneous analysis of mono- and di-valent cations is the maximal concentration ratio between sodium and ammonium that allows a separation of both components. Under isocratic conditions with 10 mmol/L methanesulfonic acid at 2 mL/min, the IonPac CS12 separates ammonium in the presence of a 250-fold to 500-fold amount of sodium. Higher concentration differences are only applicable when using a step gradient or a continuous concentration gradient (see Section 4.7).

A flaw of weak acid cation exchangers is the dependence of the dissociation of carboxyl groups on the sample pH. Polymer-coated silica columns such as the one introduced by Schomburg et al. [4] require pH adjustment of the samples with nitric acid or sodium hydroxide. In addition to the time necessary for this kind of sample preparation, the likely possibility that cationic impurities in the chemicals will contaminate the samples is problematic. Conversely, even strongly acidic samples up to pH 1.3 (corresponding to 50 mmol/L acid) can be injected into the IonPac CS12 without observing any negative effect on peak shape or separation power. This is attributed to the relatively small pK_a value of the carboxyl ion-exchange groups ($pK_a < 3$). Regarding the robustness towards extreme sample pH, the IonPac CS12 is very similar to a sulfonated cation exchanger.

As mentioned above, the support material of the IonPac CS12 is highly cross-linked to ensure solvent compatibility. While the addition of organic solvents to the mobile phase is of great importance for the analysis of organic cations such as organic amines [8-10], the retention behavior of inorganic cations is not significantly affected by organic solvents. This is especially true of strong acid cation exchangers. However, the selectivity of weak acid cation exchangers should be affected by aprotic solvents such as acetonitrile, because they prevent the dissociation of the carboxyl groups which, in turn, decreases the ion-exchange capacity. As can be seen from Fig. 4-4, the addition of acetonitrile to the mobile phase indeed leads to a retention decrease, especially for divalent cations. The addition of 10% (v/v) acetonitrile, for example, decreases retention of magnesium and calcium by 45%. The effect of linear convex and concave gradients has also been investigated (see Section 4.7). Such gradients are suitable for samples contaminated by organics, because in this way the separator is more or less continuously rinsed with an organic solvent.

Of increasing interest, above all for the power generating industry (see Section 9.2), is the chromatographic analysis of manganese together with alkali- and alkaline-earth metals. Manganese(II) is regarded as a corrosion indicator in boiler waters. Thus, it is of great importance to be able to analyze manganese together with standard inorganic cations at the highest possible sensitivity. This separation problem can be solved in a simple way with the IonPac CS10 latexed cation exchanger (see Section 4.1.2), with manganese(II) eluting just ahead of calcium. In real-world samples with high calcium and low manganese content, a separation of both cations is not possible. In principle, one can solve that problem by adding a small amount of pyrophosphoric acid to the mobile phase. Pyrophosphoric acid complexes manganese by reducing its positive net charge, so that it elutes earlier. By optimizing the pyrophosphoric acid concentration, manganese retention can be adjusted in the way so that it elutes precisely between potassium and magnesium. The resulting low peak efficiency for manganese is a disadvantage; so is the fact that pyrophosphoric acid is not available in reagent-grade purity. Moreover, dilute solutions of pyrophosphoric acid at low

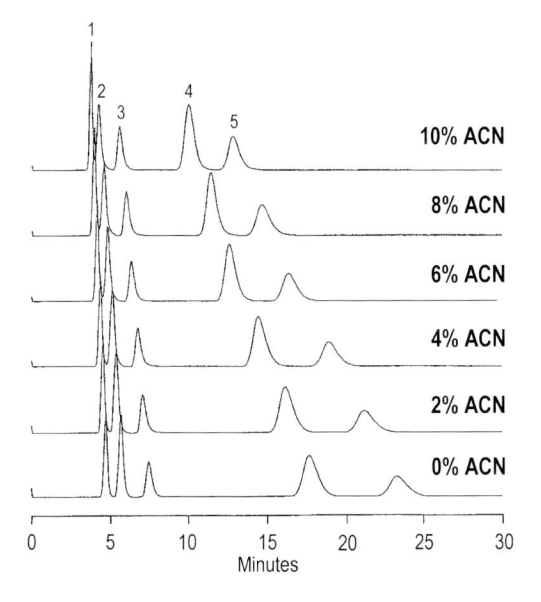

Fig. 4-4. Influence of acetonitrile on the retention of alkali- and alkaline-earth metals employing the IonPac CS12. — Eluant: 10 mmol/L methanesulfonic acid — acetonitrile; detection: suppressed conductivity; injection volume: 25 µL; solute concentrations: 10 mg/L each of sodium (1), ammonium (2), potassium (3), magnesium (4), and calcium (5).

pH are very unstable. The only alternative would be the daily adjustment of the pyrophosphoric acid concentration via titration, which is rather impractical. This problem was solved with the development of the IonPac CS12A [11], which contains phosphonate and carboxyl groups in the ion-exchange polymer. The selectivity comparison shown in Fig. 4-5, which employs a sulfuric acid eluant, reveals that the influence of the phosphonate groups on the retention of the six standard cations is minimal. Alkaline-earth metals are slightly more strongly retained on IonPac CS12A than on IonPac CS12.

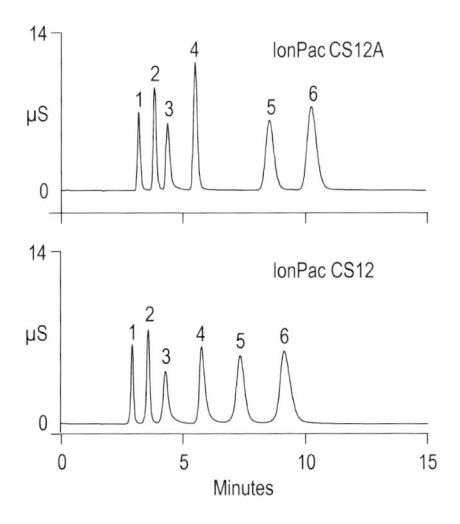

Fig. 4-5. Selectivity comparison between IonPac CS12 and IonPac CS12A. — Eluant: 11 mmol/L H_2SO_4; flow rate: 1 mL/min; injection volume: 25 µL; detection: suppressed conductivity; solute concentrations: 0.5 mg/L lithium (1), 2 mg/L sodium (2), 2.5 mg/L ammonium (3), 5 mg/L potassium (4), 2.5 mg/L magnesium (5), and 5 mg/L calcium (6).

In the case that all analyte cations in the samples to be analyzed are present in a comparable concentration range, the sulfuric acid concentration can be increased to 15.5 mmol/L because of the high resolution between the peaks. This method would result in a total analysis time of less than eight minutes (Fig. 4-6).

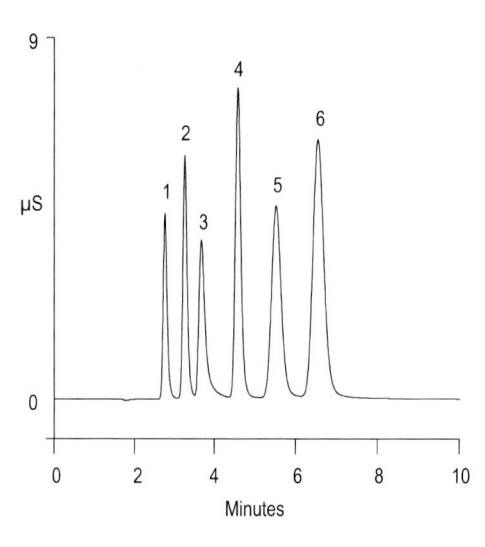

Fig. 4-6. Fast separation of alkali metals, alkaline-earth metals, and ammonium on IonPac CS12A. — Eluant: 15.5 mmol/L H_2SO_4; flow rate: 1 mL/min; injection volume: 25 µL; detection: suppressed conductivity; solute concentrations: see Fig. 4-5.

The separation of alkali metals, alkaline-earth metals, and manganese is shown in Fig. 4-7. Interesting in this respect is the fact that an increase of the phosphonate concentration in the ion-exchange polymer of the IonPac CS12A results in an opposite effect as compared to the pyrophosphoric acid addition to the mobile phase: manganese is more strongly retained than alkaline-earth metals. As can be seen from Fig. 4-7, the carboxylate/phosphonate ratio in the ion-exchange polymer has been optimized to have manganese eluting right between magnesium and calcium. The high resolution between magnesium and calcium that is attributed to the influence of the phosphonate groups is a great advantage. However, optimal separations are only achieved with a methanesulfonic acid eluant. If a sulfuric acid eluant of equal hydronium ion concentration is used, the resulting separation between magnesium and manganese is insufficient due to the latent complexing properties of sulfate for alkaline-earth metals and manganese. Since the manganese complex is more stable, the retention decreasing effect is more pronounced for manganese. The same counts for the analysis of strontium and barium, for which a methanesulfonic acid eluant is recommended, too.

Figure 4-8 shows the respective separation of all alkali- and alkaline-earth metals under the chromatographic conditions of Fig. 4-7. In comparison with IonPac CS12, divalent cations are eluted with higher peak efficiencies, which can even be improved by increasing column temperature. In general, a decrease in retention of monovalent cations and an increase in retention of divalent cations are

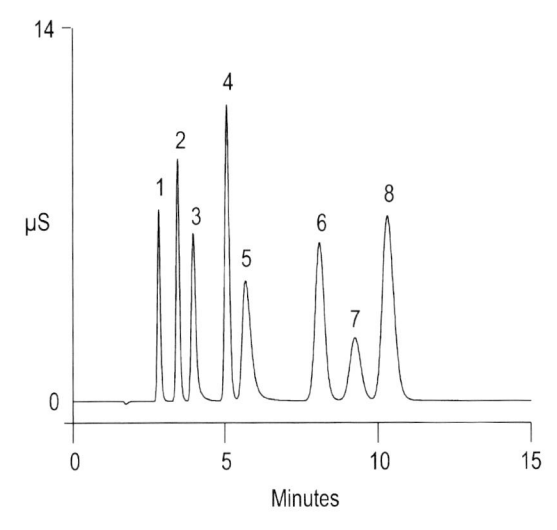

Fig. 4-7. Separation of alkali metals, alkaline-earth metals, ammonium, and manganese on IonPac CS12A. – Eluant: 20 mmol/L methanesulfonic acid; flow rate: 1 mL/min; injection volume: 25 µL; detection: suppressed conductivity; solute concentrations: 0.5 mg/L lithium (1), 2 mg/L sodium (2), 2.5 mg/L ammonium (3), 5 mg/L potassium (4), 10 mg/L diethylamine (5), 2.5 mg/L magnesium (6), 2.5 mg/L manganese (7), and 10 mg/L calcium (8).

observed with increasing column temperature. Thus, by optimizing eluant concentration and column temperature, the analysis time for a baseline-resolved separation of the six standard cations can be reduced down to five minutes.

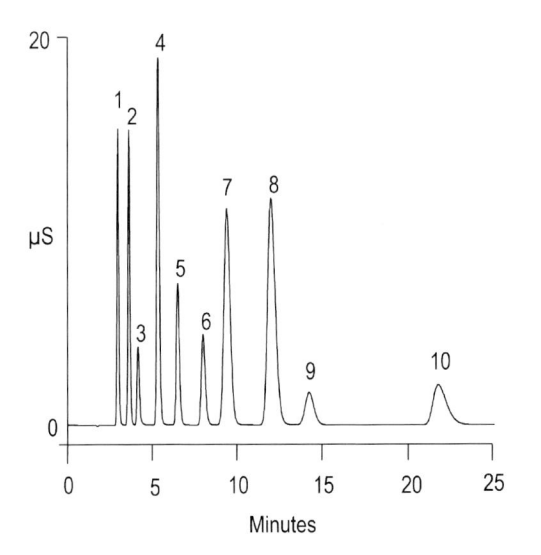

Fig. 4-8. Separation of all alkali- and alkaline-earth metals and ammonium on IonPac CS12A. – Chromatographic conditions: see Fig. 4-7; solute concentrations: 1 mg/L lithium (1), 4 mg/L sodium (2), 5 mg/L ammonium (3), 10 mg/L potassium (4), 10 mg/L rubidium (5), 10 mg/L cesium (6), 5 mg/L magnesium (7), 10 mg/L calcium (8), 10 mg/L strontium (9), and 10 mg/L barium (10).

Even faster separations are obtained with the 5-µm IonPac CS12A, which is commercialized in the format of 150 mm × 3 mm i.d. with an ion-exchange capacity of 0.94 mequiv. The chromatographic efficiency of this column is so high that it can be operated with an acid concentration of 33 mmol/L, what reduces the total analysis time for the six standard cations to less than three minutes (Fig. 4-9). Even with a smaller acid concentration of 20 mmol/L, the separation of all alkali- and alkaline-earth metals is completed in eight minutes.

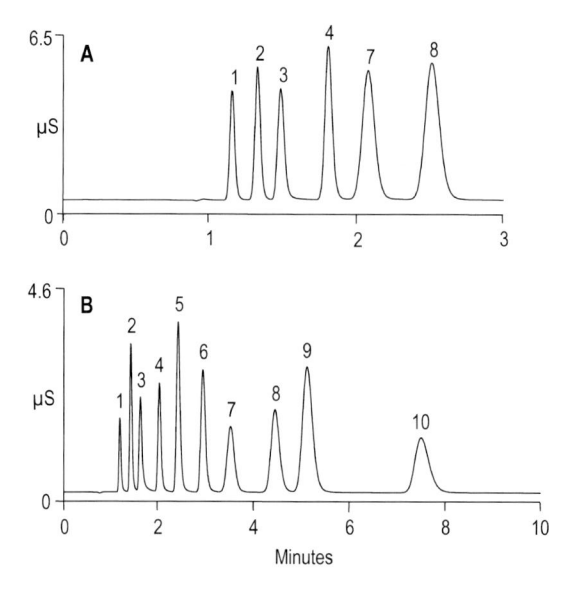

Fig. 4-9. Fast separation of alkali- and alkaline-earth metals and ammonium on the 5-μm IonPac CS12A. − Eluant: (A) 33 mmol/L methanesulfonic acid, (B) 20 mmol/L methanesulfonic acid; flow rates: (A) 0.8 mL/min, (B) 1 mL/min; detection: suppressed conductivity; injection volume: 25 μL; analytes: (1) lithium, (2) sodium, (3) ammonium, (4) potassium, (5) rubidium, (6) cesium, (7) magnesium, (8) calcium, (9) strontium, and (10) barium.

In addition to alkali- and alkaline-earth metals, a number of aliphatic amines can also be separated on IonPac CS12A. In most cases, gradient elution techniques (see Section 4.7) have to be employed due to the widely different retention behavior of these compounds. This is especially true when amines are to be analyzed together with alkali- and alkaline-earth metals. An exception is morpholine, which is used for conditioning boiler waters in the power generating industry. As demonstrated by the respective chromatogram in Fig. 4-10, morpholine can be separated from alkali- and alkaline-earth metals under isocratic conditions with a sulfuric acid eluant. On IonPac CS12A the separation is carried out with a purely aqueous eluant; for the same separation on IonPac CS12 an organic solvent has to be added to the mobile phase in order to increase peak efficiency.

The IonPac CS14 is specifically designed for the analysis of aliphatic amines. The structure of this carboxylate-based weak acid cation exchanger is very similar to the IonPac CS12. A significant difference between the separators is the ion-exchange capacities (see also Table 4-2). With 1.3 mequiv/column, the ion-exchange capacity of the IonPac CS14 is only half of the capacity of the IonPac CS12. As expected, the required eluant concentration of 10 mmol/L methanesulfonic acid for a baseline-resolved separation of mono- and di-valent cations is respectively lower, too. The separator columns also differ in the surface properties of their ion-exchange groups. Because even short-chain aliphatic amines exhibit a strong tailing when separated on IonPac CS12, one can conclude that

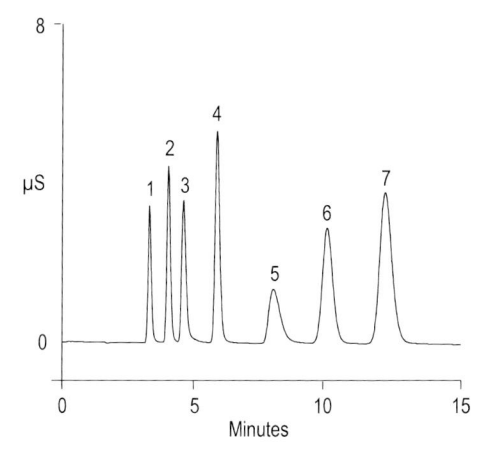

Fig. 4-10. Separation of alkali metals, alkaline-earth metals, and morpholine on IonPac CS12A. — Eluant: 10 mmol/L H_2SO_4; flow rate: 1 mL/min; injection volume: 25 µL; detection: suppressed conductivity; solute concentrations: 0.5 mg/L lithium (1), 2 mg/L sodium (2), 2.5 mg/L ammonium (3), 5 mg/L potassium (4), 25 mg/L morpholine (5), 2.5 mg/L magnesium (6), and 5 mg/L calcium (7).

the ion-exchange functionality in the ion-exchange polymer of the IonPac CS14 is much more hydrophilic. The respective comparison shown in Fig. 4-11 takes the separation of methylamines as an example. Although both columns are operated with optimized eluant concentrations, aliphatic amines elute significantly faster and with better peak symmetries on IonPac CS14. The hydrophobic interactions responsible for the tailing on IonPac CS12 are reduced to a minimum on IonPac CS14.

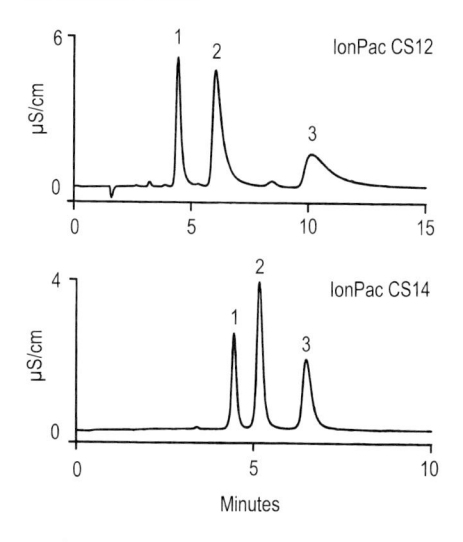

Fig. 4-11. Peak symmetry comparison between IonPac CS12 and CS14 for methylamines. — Eluants: IonPac CS12: 20 mmol/L MSA, IonPac CS14: 10 mmol/L MSA; flow rates: 1 mL/min; detection: suppressed conductivity; injection volumes: IonPac CS12: 25 µL, IonPac CS14: 18 µL; solute concentrations: 10 mg/L monomethylamine (1), 20 mg/L dimethylamine (2), and 30 mg/L trimethylamine (3).

The isocratic separation between monoethanolamine and ammonium is difficult. However, potassium, ammonium, and monoethanolamine are more strongly retained when a crown ether is added to the mobile phase; in this way, they can be separated from other mono- and di-valent cations. 18-crown-6 is especially suitable for this purpose. The chromatogram shown in Fig. 4-12 illustrates that under these chromatographic conditions, resolution between sodium

4 Cation Exchange Chromatography (HPIC)

and ammonium is so high that even large concentration differences between the cations do not prevent their quantitation. The long analysis time for potassium (more than 40 minutes) is a disadvantage.

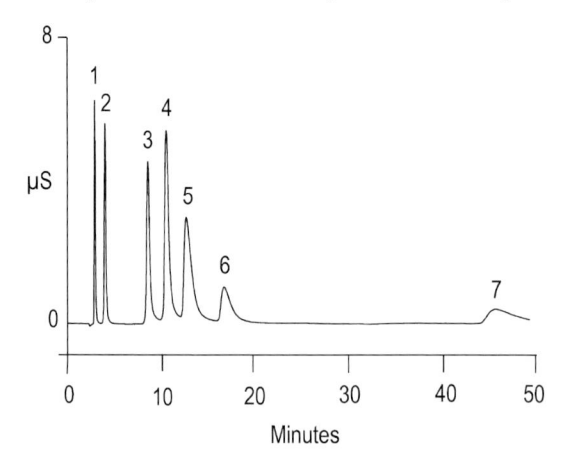

Fig. 4-12. Isocratic separation of alkali metals, alkaline-earth metals, and monoethanolamine on IonPac CS14. – Eluant: 10 mmol/L trifluoroacetic acid + 8 mmol/L 18-crown-6 + 12 mL/L acetonitrile; flow rate: 1 mL/min; detection: suppressed conductivity; injection volume: 18 µL; solute concentrations: 1 mg/L lithium (1), 4 mg/L sodium (2), 5 mg/L magnesium (3), 10 mg/L calcium (4), 5 mg/L monoethanolamine (5), 5 mg/L ammonium (6), and 10 mg/L potassium (7).

Amino alcohols such as 2-ethylaminoethanol and 2-diethylaminoethanol can also be separated on IonPac CS14, utilizing a trifluoroacetic acid eluant. The respective chromatogram in Fig. 4-13 shows that the retention behavior of those compounds is very similar to alkali- and alkaline-earth metals. Again, small amounts of acetonitrile are added to the mobile phase to improve peak efficiencies. Instead of trifluoroacetic acid, methanesulfonic acid can also be used as an eluant. Because the latter one is the stronger acid, its concentration has to be decreased accordingly. Under the chromatographic conditions given in Fig. 4-13, triethanolamine can also be analyzed; it has the same retention time as 2-ethylaminoethanol.

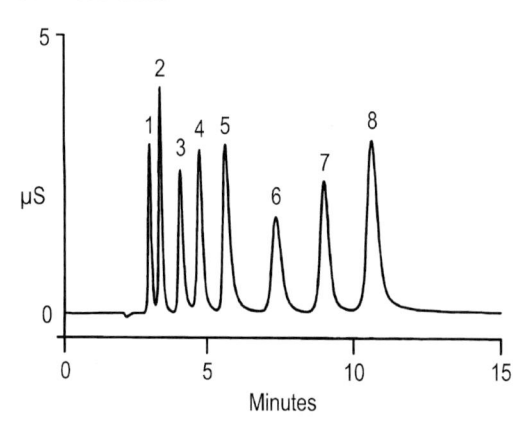

Fig. 4-13. Isocratic separation of alkali metals, alkaline-earth metals, and amino alcohols on IonPac CS14. – Eluant: 10 mmol/L trifluoroacetic acid + 15 mL/L acetonitrile; flow rate: 1 mL/min; detection: suppressed conductivity; injection volume: 18 µL; solute concentrations: 0.5 mg/L lithium (1), 2 mg/L sodium (2), 2.5 mg/L ammonium (3), 10 mg/L 2-ethylaminoethanol (4), 5 mg/L potassium (5), 20 mg/L 2-diethylaminoethanol (6), 2.5 mg/L magnesium (7), and 5 mg/L calcium (8).

Furthermore, the IonPac CS14 can be used for the analysis of a number of other amines when gradient elution techniques are employed. Examples of such analyses are presented in Section 4.7.

As shown in Fig. 4-12, a significantly better resolution between sodium, ammonium, and monoethanolamine is achieved when a crown ether is added to the mobile phase. A more elegant way to separate those cations is by functionalization of the ion-exchange polymer with crown ethers, implemented in the development of the IonPac CS15 [12]. The primary structural and technical properties of this separator column are identical to those of the IonPac CS12A. Only the ion-exchange polymer differs; the CS15 column contains additional crown ether groups besides the carboxyl- and phosphonate groups of the CS12A column. The resulting selectivity is a good fit for trace analysis of ammonium in environmental samples and sodium in amine-containing waters. The respective example chromatogram in Fig. 4-14 shows the separation of ammonium and monoethanolamine in presence of alkali- and alkaline-earth metals. Elevating the column temperature to 40 °C and adding of small amounts of acetonitrile to the mobile phase are necessary to improve the peak efficiency of the late-eluting potassium.

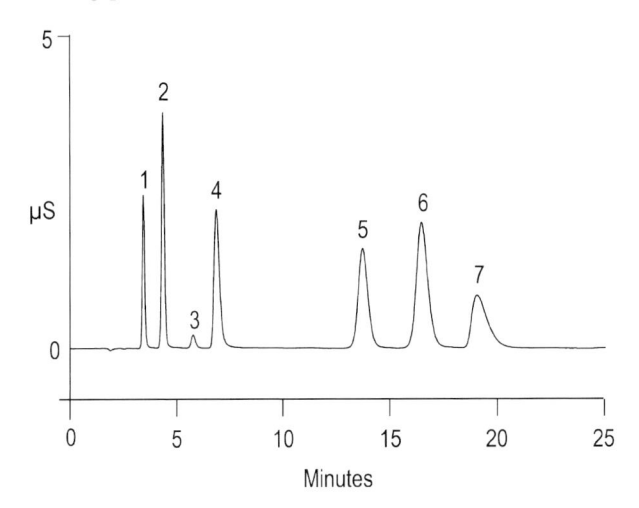

Fig. 4-14. Isocratic separation of alkali metals, alkaline-earth metals, ammonium, and monoethanolamine on IonPac CS15. – Eluant: 5 mmol/L H_2SO_4 – acetonitrile (91:9 v/v); flow rate: 1.2 mL/min; column temperature: 40 °C; detection: suppressed conductivity; injection volume: 25 µL; solute concentrations: 1 mg/L lithium (1), 4 mg/L sodium (2), 3 mg/L monoethanolamine (3), 5 mg/L ammonium (4), 5 mg/L magnesium (5), 10 mg/L calcium (6), and 10 mg/L potassium (7).

The influence of the column temperature on the selectivity of the CS15 column when using a purely aqueous eluant is illustrated in Fig. 4-15. When operating it at 25 °C (top chromatogram), the potassium peak exhibits strong tailing. The increase of column temperature to 40 °C in combination with a higher eluant concentration (bottom chromatogram) results in a certain improvement of the peak shape and a significantly decreased retention time, but these measures are not sufficient to optimize the separation. In comparison, Fig. 4-16 illustrates the temperature influence when small amounts of acetonitrile are added to the mobile phase. As can be seen from the top chromatogram, the addition of an organic solvent to the eluant almost completely eliminates the tailing of the potassium peak. However, the resulting total analysis time of 25 minutes is far too high for routine analyses. Only the combination of solvent addition and elevated column temperature leads to the desired result (Fig. 4-16, bottom chromatogram). When the concentrations of the analyte cations are comparable, the total analysis time can be decreased to about 13 minutes by increasing eluant concentration and acetonitrile content (Fig. 4-17).

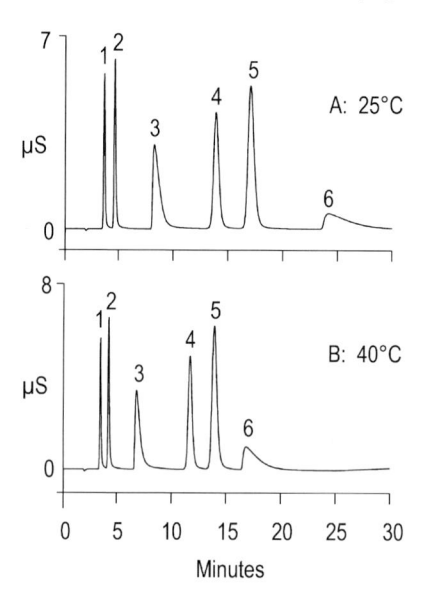

Fig. 4-15. Temperature influence on the selectivity of the IonPac CS15 when utilizing a purely aqueous eluant. – Eluant: (A) 6.5 mmol/L H_2SO_4, (B) 7.5 mmol/L H_2SO_4; flow rate: 1.2 mL/min; detection: suppressed conductivity; injection volume: 25 µL; solute concentrations: 1 mg/L lithium (1), 4 mg/L sodium (2), 10 mg/L ammonium (3), 5 mg/L magnesium (4), 10 mg/L calcium (5), and 10 mg/L potassium (6).

Like the IonPac CS14, the IonPac CS15 is also suitable for gradient elution of amines. To illustrate this, the analysis of ethanolamines together with alkali- and alkaline-earth metals is also presented in Section 4.7.

Another development in carboxylate-based weak acid cation exchangers is the IonPac CS16, which is a unique hydrophilic, high-capacity carboxylate functionalized cation exchange column that provides excellent peak shape for alkali- and alkaline-earth metals and amines. The IonPac CS16 column is available in two column formats: 250 mm × 3 mm i.d. and 250 mm × 5 mm i.d. The column packing consists of 5-µm diameter macroporous particles that are 100% solvent compatible due to the 55% degree of cross-linking. The substrate is functionalized with a hydrophilic carboxylic acid layer that permits the simultaneous elution of mono- and di-valent cations using a dilute acid eluant such as methanesulfonic acid. The use of a smaller resin particle size improves peak efficiencies, while a new grafting technology allows higher capacity (8.4 mequiv for the 250 mm × 3 mm i.d. column) via the incorporation of a much higher number of carboxylic acid cation exchange sites. Common inorganic cations and

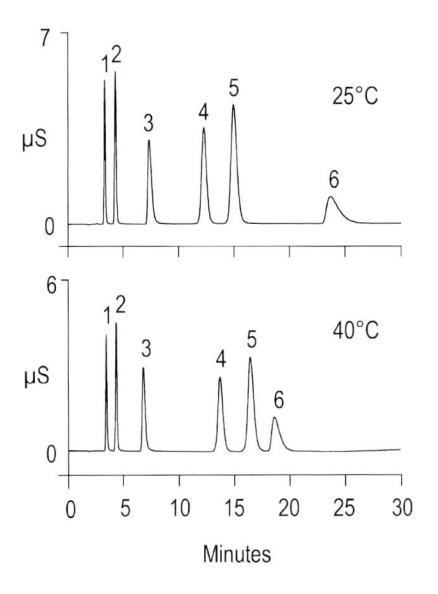

Fig. 4-16. Temperature influence on the selectivity of the IonPac CS15 when utilizing a solvent-containing eluant. − Eluant: 5 mmol/L H_2SO_4 − acetonitrile (91:9 v/v)); other chromatographic conditions: see Fig. 4-15.

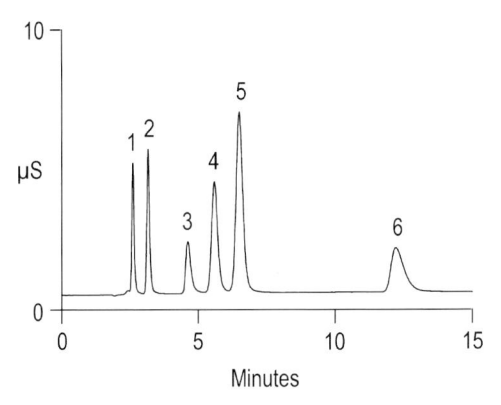

Fig. 4-17. Fast isocratic separation of alkali metals, alkaline-earth metals, and ammonium on IonPac CS15. − Eluant: 9 mmol/L H_2SO_4 − acetonitrile (87:13 v/v); flow rate: 1.2 mL/min; other chromatographic conditions: see Fig. 4-15.

ammonium can be resolved in about 20 minutes using a 30 mmol/L methanesulfonic acid eluant, as shown in Fig. 4-18. The acid concentration can be optimized for the fast determination of mono- and di-valent cations by using 48 mmol/L MSA; this is also illustrated in Fig. 4-18.

The IonPac CS16 is suitable for the determination of low concentrations of ammonium in environmental waters [13]. Figure 4-19 shows a respective chromatogram of a tap water sample with a typical content of about 65 µg/L ammonium. Under standard chromatographic conditions with 30 mmol/L MSA, the separation between sodium and ammonium is baseline-resolved.

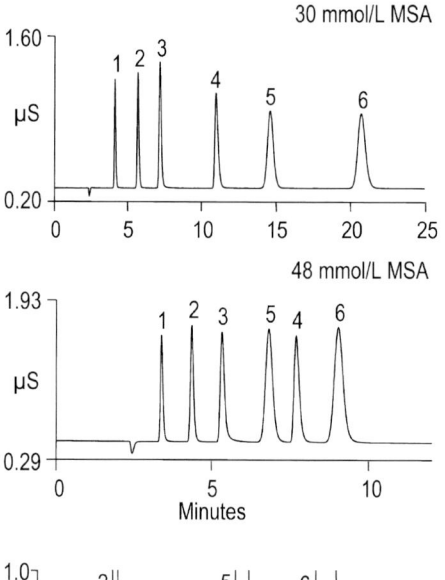

Fig. 4-18. Isocratic separation of alkali- and alkaline-earth metals and ammonium on 250 mm × 5 mm i.d. IonPac CS16. — Eluant: see chromatogram; flow rate: 1 mL/min; detection: suppressed conductivity; injection volume: 25 µL; solute concentrations: 0.1 mg/L lithium (1), 0.4 mg/L sodium (2), 0.5 mg/L ammonium (3), 1 mg/L potassium (4), 0.5 mg/L magnesium (5), and 1 mg/L calcium (6).

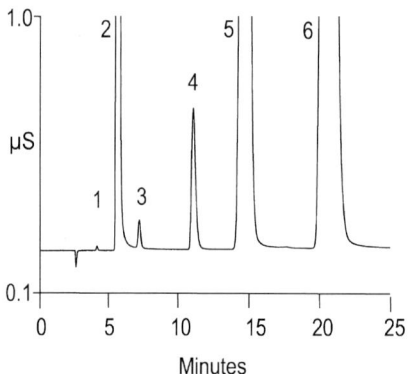

Fig. 4-19. Isocratic separation of alkali- and alkaline-earth metals and ammonium in a tap water sample on IonPac CS16. — Eluant: 30 mmol/L MSA; flow rate: 1 mL/min; detection: suppressed conductivity; injection volume: 25 µL; solute concentrations: 0.002 mg/L lithium (1), 19.7 mg/L sodium (2), 0.065 mg/L ammonium (3), 0.98 mg/L potassium (4), 7.2 mg/L magnesium (5), and 18.5 mg/L calcium (6).

The IonPac CS16, which was specifically designed for high-to-low-ratios of sodium to ammonium in diverse sample matrices, is an alternative to the IonPac CS15. It provides improved resolution of sodium from ammonium and alkanolamines, even for high ionic strength samples. Figure 4-20A illustrates the determination of trace level ammonium in the presence of high sodium concentrations, at ratios up to 10,000:1. This is accomplished with an isocratic MSA eluant on the 5-mm CS16 column at an elevated temperature of 40 °C, utilizing suppressed conductivity detection. Ratios up to 20,000:1 can be resolved using a MSA gradient with the CS16 column in the 3-mm format, as illustrated in Fig. 4-20B.

The IonPac CS16 can also be used to monitor the amine content in the quality control of chemical additives, process solutions, plating baths, and scrubber solutions, as well as for the determination of trace level sodium in ammonium and amine-treated cooling waters (see Section 9.2).

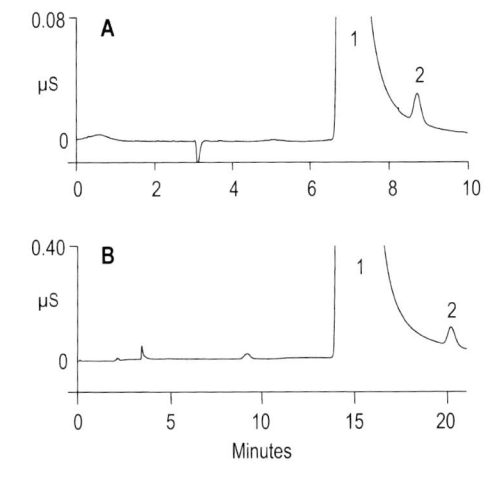

Fig. 4-20. Determination of trace level ammonium in presence of high sodium concentrations. − Separator columns: (A) 250 mm × 5 mm i.d. IonPac CS16, (B) 250 mm × 3 mm i.d. IonPac CS16; eluants: (A) 30 mmol/L MSA, (B) 7 mmol/L MSA isocratically for 23 min, then linearly to 52 mmol/L in 5 min; flow rates: (A) 1 mL/min, (B) 0.43 mL/min; detection: suppressed conductivity; injection volume: 25 µL; solute concentrations: (A) 100 mg/L sodium (1), 0.01 mg/L ammonium (2), (B) 500 mg/L sodium (1), 0.025 mg/L ammonium (2).

The latest development in carboxylate-based weak acid cation exchangers is the IonPac CS17, which was designed for the determination of hydrophobic and polyvalent amines, including biogenic amines, alkylamines, and diamines using simple aqueous eluants and elevated column temperature. The CS17 column is also ideal for separating certain hydrophilic amines, for example, alkanolamines. It offers improved performance for most IonPac CS14, CS15, and CS16 applications. The IonPac CS17 is a unique, hydrophilic, moderate capacity, carboxylate functionalized cation exchanger that provides excellent selectivity and peak shape for amines. The IonPac CS17 resin bead consists of a highly cross-linked core. This core makes it solvent compatible, so that the column can easily be cleaned-up after the analysis of complex sample matrices. The substrate for the CS17 column is a macroporous resin bead with a 7-µm particle diameter consisting of ethylvinylbenzene cross-linked with 55% divinylbenzene. The CS17 resin bead is produced using a novel grafting technology, where a non-functional coating is first grafted to the resin surface and pores, and then the cation exchange polymer is grafted onto this non-functional coating. This novel grafting technology reduces hydrophobic interactions between the analytes and the resin surface resulting in excellent mass-transfer characteristics and, consequently, high peak efficiencies.

Figure 4-21 illustrates the improved resolution between adjacent peaks when step-gradient elution is used instead of isocratic elution. Standard inorganic cations and alkylamines are used for this example. Step-gradient elution improves the resolution between the first five eluting peaks (the four monovalent cations and dimethylamine). When using an Eluant Generator and a continuously regenerated trap column (CR-CTC), the baseline shift is minimized by stepping from a low to a high eluant concentration; this simplifies peak integration and analyte quantitation. Moderately hydrophobic amines such as triethylamine are eluted efficiently without adding an organic solvent to the mobile phase.

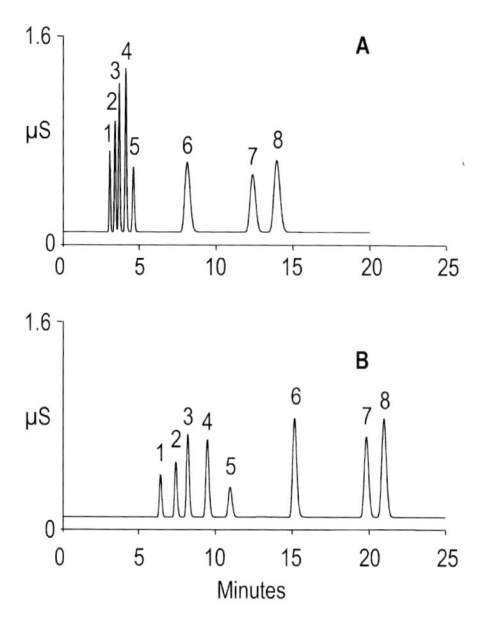

Fig. 4-21. Separation of alkali metals, alkaline-earth metals, ammonium and alkylamines using the IonPac CS17. — Column temperature: 30 °C; eluant: (A) 6 mmol/L methanesulfonic acid, (B) 2 mmol/L methanesulfonic acid for 11 min isocratically, then step change to 7 mmol/L and isocratically for 11 min; flow rate: 1 mL/min; detection: suppressed conductivity; injection volume: 25 µL; solute concentrations: 0.1 mg/L lithium (1), 0.4 mg/L sodium (2), 0.5 mg/L ammonium (3), 1 mg/L potassium (4), 1 mg/L dimethylamine (5), 7 mg/L trimethylamine (6), 0.5 mg/L magnesium (7), and 1 mg/L calcium (8).

A typical gradient application of the IonPac CS17 is the analysis of alkanolamines, including mono-, di-, and tri-ethanolamine. These three alkanolamines are most commonly used, individually or in combination, to optimize the efficiency of the scrubber treatment for a specific chemical process. The CS17 column resolves all combinations of these scrubber amines using a methanesulfonic acid gradient and elevated column temperature, as illustrated in Fig. 4-22.

The structural-technical properties of all carboxylate-based weak acid cation exchangers are summarized in Table 4-2.

4.1.1.3 Polymethacrylate and Polyvinyl Resins

Polymethacrylate and polyvinyl resins play only a secondary role in the manufacturing of cation exchangers. Currently, the only polymethacrylate-based cation exchanger on the market is offered by Sykam (Gauting, Germany) under the trade name LCA K02. This column differs from its PS/DVB-analogue (see Table 4-1) only in particle size (5-µm) and ion-exchange capacity (0.4 mequiv/g). Used with a tartaric acid eluant, this phase is preferred for the analysis of transition metals.

The only polyvinyl-based cation exchanger available today, the ION-200 (Interaction Chemicals, Mountain View, CA, USA), exhibits a poor chromatographic efficiency in comparison to PS/DVB copolymers. The 50 mm × 3 mm i.d. column can be employed for the separation of inorganic and organic cations. With a maximum flow rate of 2.5 mL/min, this column may be operated at a maximum

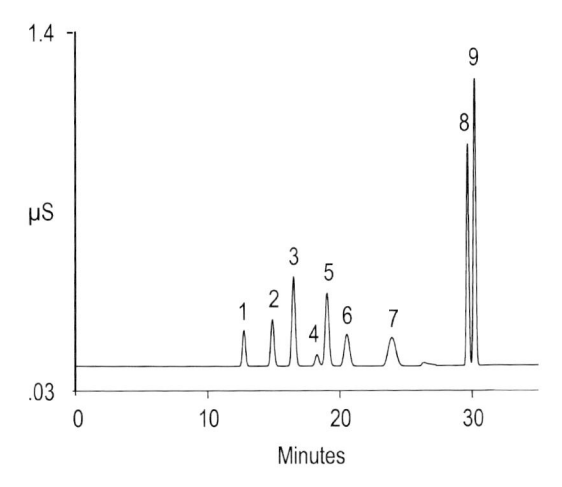

Fig. 4-22. Separation of alkali metals, alkaline-earth metals, ammonium and alkanolamines using the IonPac CS17. — Column temperature: 30 °C; eluant: 0.5 mmol/L methanesulfonic acid linearly to 0.8 mmol/L in 25 min, then step change to 9 mmol/L and isocratically for 10 min; flow rate: 1.4 mL/min; detection: suppressed conductivity; injection volume: 25 µL; solute concentrations: 0.1 mg/L lithium (1), 0.4 mg/L sodium (2), 0.5 mg/L ammonium (3), 0.5 mg/L monoethanolamine (4), 1 mg/L potassium (5), 1 mg/L diethanolamine (6), 18 mg/L triethanolamine (7), 0.5 mg/L magnesium (8), and 1 mg/L calcium (9).

Table 4-2. Structural and technical properties of carboxylate-based weak acid cation exchangers.

Column	IonPac CS12	IonPac CS12A	IonPac CS14	IonPac CS15	IonPac CS16	IonPac CS17
Manufacturer	Dionex	Dionex	Dionex	Dionex	Dionex	Dionex
Dimensions [length × i.d.] [mm]	250 × 2 250 × 4	250 × 2 250 × 4 150 × 3	250 × 2 250 × 4	250 × 2 250 × 4	250 × 3 250 × 5	250 × 2 250 × 4
Degree of cross-linking	55	55	55	55	55	55
Maximum pressure [MPa]	27.6	27.6	27.6	27.6	27.6	27.6
Solvent compatibility	100*)	100*)	100*)	100*)	100*)	100*)
Capacity [mequiv/column]	0.7 2.8	0.7 2.8	0.325 1.3	0.7 2.8	3.0 8.4	0.363 1.45
Particle diameter [µm]	8.5	8.5 8.5 5	8.5	8.5	5	7
Functionality	Carboxyl groups	Carboxyl and phosphonate groups	Carboxyl groups	Carboxyl, phosphonate and crown ether groups	Carboxyl groups	Carboxyl groups

*) aprotic solvents such as acetonitrile

pressure of 15 MPa. An ion-exchange capacity of 0.1 mequiv/g is stated for this column packing, the particle size of which is 5 µm. The corresponding chromatogram of an alkali metal separation is shown in Fig. 4-23.

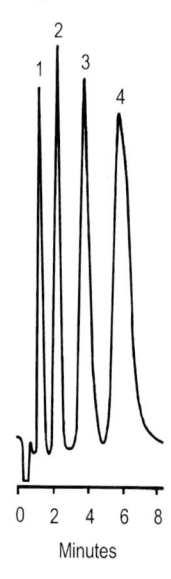

Fig. 4-23. Separation of alkali metals on a surface-sulfonated polyvinyl resin. — Separator column: ION 200; eluant: 2 mmol/L picolinic acid, pH 2.0; flow rate: 2.6 mL/min; detection: direct conductivity; peaks: (1) lithium, (2) sodium, (3) ammonium, and (4) potassium.

4.1.2
Latexed Cation Exchangers

While the latex concept for anion exchangers (see Section 3.4.2) was realized with the introduction of ion chromatography, it was not until 1986 that its usefulness for cation exchangers was recognized. The reason for the long development time was the lack of suitable surface-aminated substrates. The manufacturing of those substrates via *direct* amination was impractical because of the poor reproducibility of this method. Based on experiences gained in the manufacturing of latexed anion exchangers, it eventually became possible to cover the anion exchange beads with a second layer of totally sulfonated latex beads. These materials, developed by Dionex Corp., are called latexed cation exchangers. Their structure is schematically depicted in Fig. 4-24.

Latexed cation exchangers consist of a weakly sulfonated polystyrene/divinylbenzene substrate with a particle size of, for example, 10 µm. Totally aminated latex beads with a much smaller diameter (about 50 nm) are agglomerated on its surface by both electrostatic and van-der-Waals interactions. The anion exchange substrate that results is covered by a second layer of latex beads which carry the actual ion-exchange functional groups, namely sulfonate groups. In the IonPac CS3 latex cation exchanger, which is offered in the standard dimensions of 250 mm × 4.6 mm i.d., the sulfonated latex beads have a diameter of about

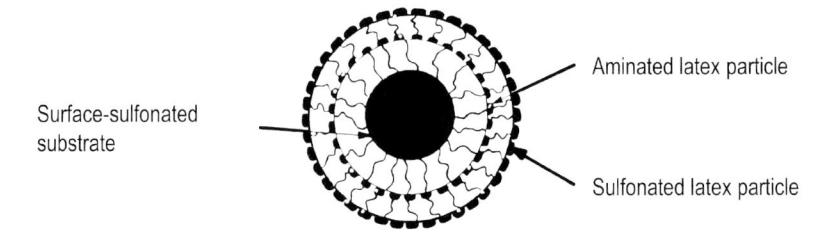

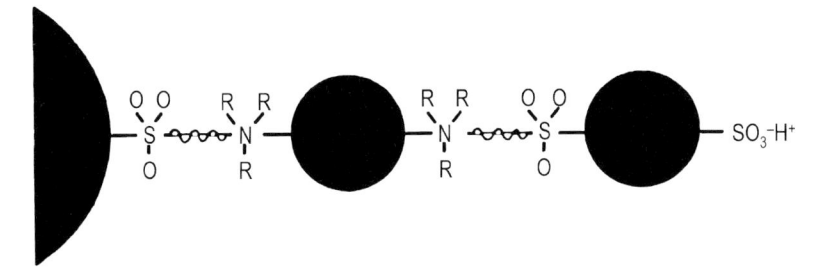

Fig. 4-24. Schematic representation of a latexed cation exchange particle.

250 nm and a degree of cross-linking of 5 %. In contemporary practice, such a separator column is only used for the sequential analysis of alkali- or alkaline-earth metals because it provides a relatively high resolution between the individual components.

The separator column introduced as the Fast-Sep Cation has similar physical properties. This column was developed primarily for the fast separation of alkali- or alkaline-earth metals. In combination with a column-switching technique, it is also suitable for the simultaneous analysis of mono- and di-valent cations (see Section 4.4). However, in terms of separation power and peak efficiencies, the separations obtained in this way are inferior to those obtained with carboxylate-based weak acid cation exchangers (described in Section 4.1.1). The dimensions of the Fast-Sep Cation column are 250 mm × 4 mm i.d. The 13-µm particle diameter of the substrate is slightly higher than that of the CS3 column. The sulfonated latex beads are cross-linked with 4 % DVB and have a diameter of about 225 nm.

In contrast to surface-sulfonated materials, latex cation exchangers exhibit a significantly higher chromatographic efficiency. This is illustrated by the chromatogram in Fig. 4-25 which was obtained using the chromatographic conditions outlined in Fig. 4-2. Latexed cation exchangers may be operated at a flow rate of 2 mL/min without significant loss in separation efficiency. These conditions enable a baseline-resolved separation of sodium, ammonium, and potassium within three minutes.

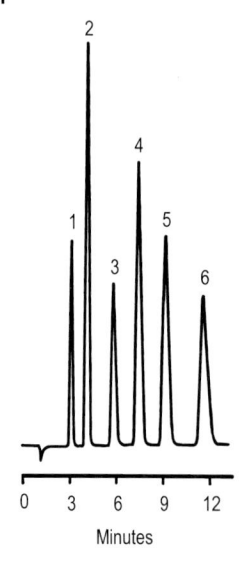

Fig. 4-25. Separation of alkali metals on a latexed cation exchanger. – Separator column: Fast-Sep Cation I; chromatographic conditions: see Fig. 4-2.

A special latexed cation exchanger was introduced in the early 1990's for the simultaneous analysis of alkali- and alkaline-earth metals under isocratic conditions. Its support material is composed of a highly cross-linked ethylvinylbenzene/divinylbenzene copolymer with a particle size of 8.5 µm. It is offered in the standard dimensions 250 mm × 4.6 mm i.d. under the trade name IonPac CS10. The support material is manufactured as follows: At the time of polymerization, the individual particles have a reactive surface to which – in contrast to the conventional latex structure shown in Fig. 4-24 – a monolayer of a totally aminated colloidal polymer particles are covalently bound. However, those particles only act as anchor groups for the second layer of latex beads, which carry the actual ion-exchange functional groups, namely sulfonate groups. In this way, the densitiy of sulfonate groups at the surface of the stationary phase is lessened. As schematically depicted in Fig. 4-26, this causes divalent cations to behave somewhat like monovalent ones during the cation exchange process; consequently, they elute much faster. The covalent bond of the latex beads to the support material enables the high mechanical and chemical stability of this stationary phase. A new feature of latexed cation exchangers is the 100% solvent stability, which results from the high degree of cross-linking of the microporous support. This polymeric stationary phase can, therefore, be cleaned using commercial HPLC solvents such as methanol, acetonitrile, etc. However, for the simultaneous analysis of alkali- and alkaline-earth metals, it is not necessary to put organic additives in the mobile phase (see Fig. 4-27). A mixture of 2,3-diaminopropionic acid (DAP) and hydrochloric acid was used as an eluant.

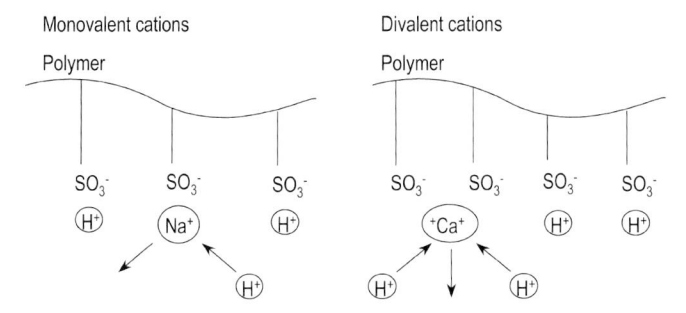

Fig. 4-26. Schematic representation of the cation exchange processes for mono- and di-valent cations on IonPac CS10.

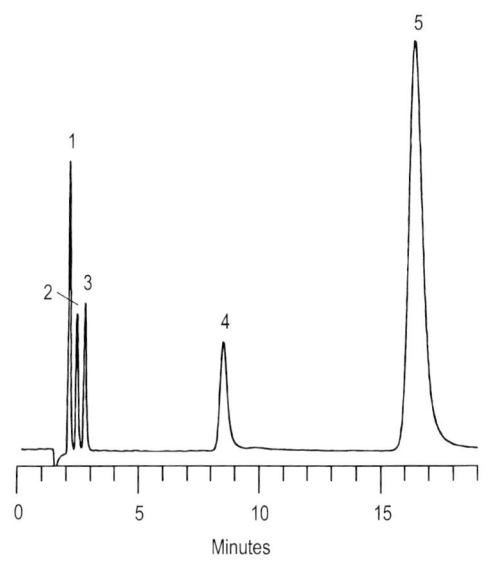

Fig. 4-27. Simultaneous separation of alkali- and alkaline-earth metals and ammonium on IonPac CS10. — Eluant: 40 mmol/L HCl + 4 mmol/L 2,3-diaminopropionic acid (DAP); flow rate: 1 mL/min; detection: suppressed conductivity; injection volume: 25 µL; solute concentrations: 1 mg/L sodium (1), 0.5 mg/L ammonium (2), 1 mg/L potassium (3), 1 mg/L magnesium (4), and 10 mg/L calcium (5).

Separations on the IonPac CS10 are characterized by a relatively limited resolution between monovalent cations, which elute within a short time due to the low ion-exchange capacity of the column (0.08 mequiv). Nevertheless, concentration ratios of sodium to ammonium as high as 100:1 can be quantified. The extreme resolution between divalent cations is another characteristic of this column; the resolution is much higher than it is with weak acid cation exchangers (see Section 4.1.1). It can be somewhat decreased by increasing the DAP concentration in the mobile phase. However, for the analysis of manganese, this high resolution between divalent cations is advantageous, because manganese elutes just between potassium and magnesium under the chromatographic conditions outlined in Fig. 4-27. Even in presence of a large excess of alkali- and alkaline-earth metals, manganese can be detected very sensitively. Because of the introduction of weak acid cation exchangers, the IonPac CS10 is no longer used for the simultaneous analysis of mono- and divalent cations. The IonPac CS10 is still used for the analysis of soil samples, where the soil is extracted with, for example, strontium chloride.

The analysis of amines is an application area of great significance [8]. As an important object of recent application developments, for example, Fig. 4-28 shows the isocratic elution of monovalent cations and ethanolamines on IonPac CS11, the microbore version of the IonPac CS10 with a slightly higher ion-exchange capacity. Under the chromatographic conditions used, a baseline-resolved separation of mono-, di-, and tri-ethanolamine as well as N-methyldiethanolamine is achieved within 30 minutes.

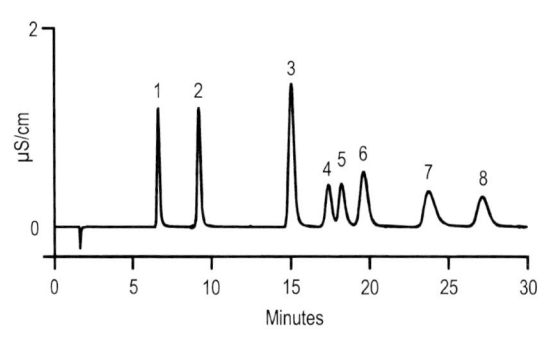

Fig. 4-28. Isocratic separation of ethanolamines and alkali metals on IonPac CS11. – Eluant: 35 mmol/L methanesulfonic acid; flow rate: 0.25 mL/min; detection: suppressed conductivity; solute concentrations: 0.25 mg/L lithium (1), 1 mg/L sodium (2), 2 mg/L ammonium (3), 2 mg/L monoethanolamine (4), 1 mg/L potassium (5), 10 mg/L diethanolamine (6), 100 mg/L triethanolamine (7), and 10 mg/L N-methyldiethanolamine (8).

This is a typical application used in the petrochemical industry [14]. Wastewaters from oil refineries, for example, contain these amines, which can be present in widely different concentration levels. Also, the building material industry is interested in this separation, because ethanolamines are used as grinding additives for cement to increase the efficiency of the mills [15]. While alkylamines can be detected very sensitively with suppressed conductivity detection, the equivalent conductances of alkanolamines drastically decrease from monoalkanolamines to trialkanolamines. On the other hand, trialkanolamines can be detected by UV detection at 210 nm, which is more sensitive than suppressed conductivity detection by a factor of 40. Integrated pulsed amperometry on a gold working electrode is an interesting alternative (see Section 7.1.2.2) to using two different detectors for sensitive detection of alkanolamines. Using this method, the response factors of primary, secondary, and tertiary amines do not differ that much as compared with suppressed conductivity and UV detection.

Another important application for the IonPac CS10 is the separation of biologically relevant amines such as choline and acetylcholine, which previously have been used to be separated by ion-pair chromatography (see Section 6.4).

Cation exchange chromatography with a methanesulfonic acid eluant is a much simpler alternative. For optimizing peak efficiencies, 1% (v/v) methanol is added to the mobile phase. As an alternative to the IonPac CS10, an OmniPac PCX-100 (see Section 6.7) can also be used. The latter one has very similar separation properties, but it is characterized by a lower hydrophobicity of the cation exchange functionality, which improves peak efficiency for choline and its derivatives even more.

4.1.3
Silica-Based Cation Exchangers

Silica-based cation exchangers with sulfonic acid groups as ion-exchange sites are typically produced via silanization of silica with reagents such as $R-Si(CH_3)_2-C_3H_6-C_6H_5-SO_3H$ (R = H, Cl). Such cation exchangers are only of minor importance despite their comparatively high chromatographic efficiency. Vydac 400 IC 405, manufactured by The Separations Group (Hesperia, CA, USA), is a stationary phase that is best suited for the analysis of alkali metals and small aliphatic amines.

In contrast, the two 5-µm cation exchangers, Nucleosil 5 SA by Macherey & Nagel (Düren, Germany) with the dimensions of 125 mm × 4 mm i.d., and TSK Gel IC Cation SW from Toyo Soda (Tokyo, Japan) with the dimensions 50 mm × 4.6 mm i.d., are designed for the analysis of divalent cations. All of these stationary phases exhibit a relatively high ion-exchange capacity of 0.5 mequiv/g and are resistant to all water-miscible organic solvents. Figure 4-29 shows an example of the separation of alkaline-earth metals on Nucleosil 5 SA, which is characterized by high resolution and excellent peak symmetries.

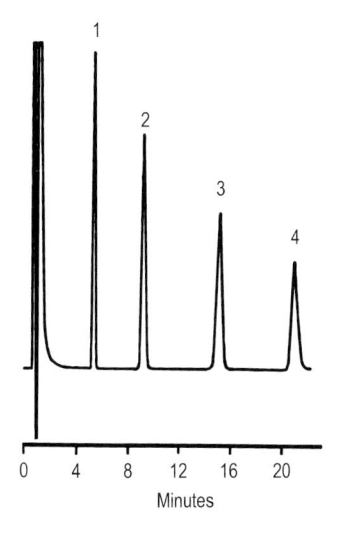

Fig. 4-29. Separation of alkaline-earth metals on a silica-based cation exchanger. – Separator column: Nucleosil 5 SA; eluant: 3.5 mmol/L oxalic acid + 2.5 mmol/L ethylenediamine + 50 mL/L acetone, pH 4; flow rate: 1.5 mL/min; detection: direct conductivity; injection volume: 100 µL; solute concentrations: 2.5 mg/L magnesium (1), 5 mg/L calcium (2), 20 mg/L strontium (3), and 40 mg/L barium (4).

A great deal of interest has been shown in a polymer-coated silica-based cation exchange phase introduced by Schomburg et al. [4] in 1987. It belongs to the class of weak acid cation exchangers. The coating of the silica is performed with "prepolymers" which are synthesized in a separate step and then applied to the support material and immobilized. For a number of years, this method has been successfully employed in the synthesis of various stationary phases based on silica and alumina, creating resins with remarkable selectivities. Further information on this subject may be found in the review by Schomburg [16].

The prepolymer used for manufacturing this novel cation exchanger consists of a copolymer that is derived from a mixture of butadiene and maleic acid in equal parts:

Poly(butadiene-maleic acid), PBDMA

The structural formula reveals that this polymer contains two different types of carboxyl groups that have different dissociation constants. The first dissociation step is characterized by a pK value of 3.4; the pK value of the second step is about 7.4. These pK values were determined via titration of the prepolymer with sodium hydroxide solution. The ion-exchange capacity of the finished stationary phase is directly proportional to its polymer content. The ion-exchange capacity can be calculated in advance, because the concentration of the ion-exchange groups in the prepolymer is known based on the chemical composition.

PBDMA-coated silica is suitable for the simultaneous analysis of alkali- and alkaline-earth metals. The first coating experiments with 40% PBDMA resulted in a more group-wise separation of both classes of compounds when formic acid was used as an eluant. However, the separation between the most important components was significantly improved by slightly increasing the thickness of the layer and by changing the eluant. PBDMA-coated silica is offered, for example, by Metrohm (Herisau, Switzerland) under the trade name Metrosep Cation 1-2 (125 mm × 4 mm i.d.). Figure 4-30 shows an example chromatogram obtained under standard chromatographic conditions with an eluant mixture of tartaric acid and pyridine-2,6-dicarboxylic acid (dipicolinic acid). Under these chromatographic conditions, calcium elutes ahead of magnesium, followed by strontium and barium. This unusual retention behavior can be attributed to the complexing properties of pyridine-2,6-dicarboxylic acid. With a pure tartaric acid eluant, alkaline-earth metals elute in the usual order, with a much greater resolution between mono- and divalent cations (Fig. 4-31).

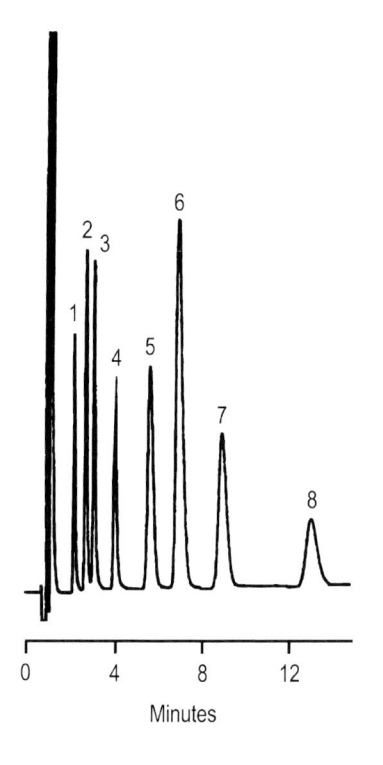

Fig. 4-30. Simultaneous separation of alkali metals, alkaline-earth metals, and ammonium on PBDMA-coated silica with a tartaric acid/pyridine-2,6-dicarboxylic acid eluant. − Separator column: Metrosep Cation 1-2; eluant: 5 mmol/L tartaric acid + 0.75 mmol/L pyridine-2,6-dicarboxylic acid; flow rate: 1 mL/min; detection: direct conductivity; injection volume: 10 µL; solute concentrations: 1 mg/L lithium (1), 5 mg/L sodium (2), 5 mg/L ammonium (3), 10 mg/L potassium (4), 10 mg/L calcium (5), 10 mg/L magnesium (6), 20 mg/L strontium (7), and 20 mg/L barium (8).

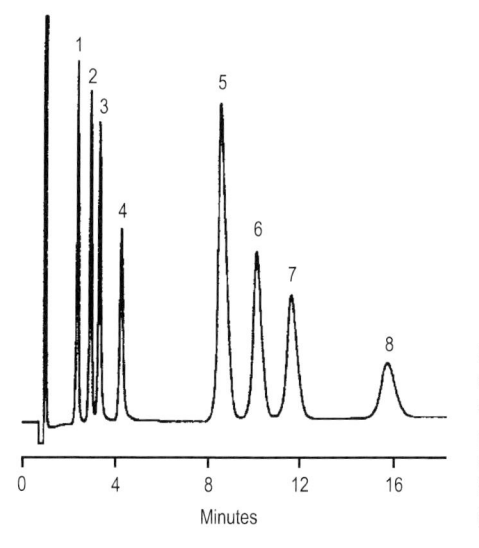

Fig. 4-31. Simultaneous separation of alkali metals, alkaline-earth metals, and ammonium on PBDMA-coated silica with a pure tartaric acid eluant. − Separator column: Metrosep Cation 1-2; eluant: 5 mmol/L tartaric acid; other chromatographic conditions: see Fig. 4-30.

A characteristic feature of separations on PBDMA-coated silica is the unusually short analysis time for barium − of less than 15 minutes under standard chromatographic conditions − without compromising the resolution between other mono- and di-valent cations. The only exception is the resolution between sodium and ammonium, which is significantly lower than that obtained with polymer-based weak acid cation exchangers such as IonPac CS12A within a total analysis time of approximately 20 minutes.

E. Merck (Darmstadt, Germany) also offers a 5-µm PBDMA-coated silica column under the trade name LiChrosil® IC CA. This analytical column in the 100 mm × 4.6 mm i.d. format is also used with an eluant mixture of tartaric acid and dipicolinic acid. In terms of peak efficiency and resolution, this column is comparable to the Metrosep Cation 1-2. Detection is carried out with non-suppressed conductivity. Under the chromatographic conditions given in Fig. 4-30, transition metals could potentially interfere. While the majority of transition metals are not retained under these conditions, nickel elutes between potassium and calcium, thus interfering with the determination of rubidium and cesium. Nickel, zinc, and cobalt can be separated together with alkali- and alkaline-earth metals using a mixture of tartaric acid and oxalic acid as an eluant, but only in the non-suppressed conductivity detection mode (Fig. 4-32). Total analysis time, however, is significantly longer in this case.

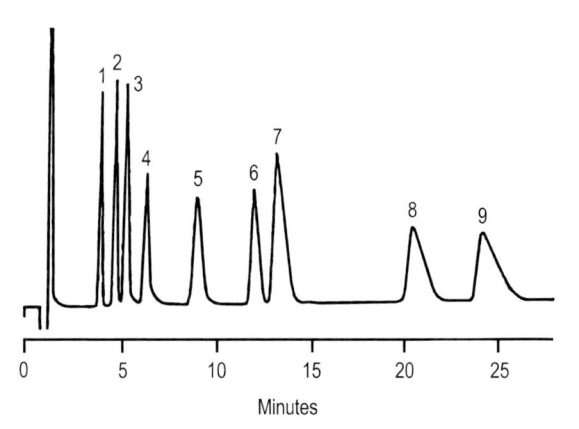

Fig. 4-32. Separation of transition metals together with alkali- and alkaline-earth metals on PBDMA-coated silica. – Separator column: LiChrosil® IC CA; column temperature: 40 °C; eluant: 2 mmol/L tartaric acid + 0.5 mmol/L oxalic acid; flow rate: 1 mL/min; detection: direct conductivity; injection volume: 15 µL; solute concentrations: 1 mg/L lithium (1), 5 mg/L sodium (2), 5 mg/L ammonium (3), 10 mg/L potassium (4), 10 mg/L nickel (5), 10 mg/L zinc (6), 10 mg/L cobalt (7), 10 mg/L magnesium (8), and 20 mg/L calcium (9).

The LiChrosil® IC CA column has been successfully employed in the analysis of ethanolamines, which can be eluted with high efficiency in less than seven minutes with a tartaric acid/dipicolinic acid eluant. The chromatogram in Fig. 4-33 shows the three ethanolamines eluting with high chromatographic efficiencies within seven minutes.

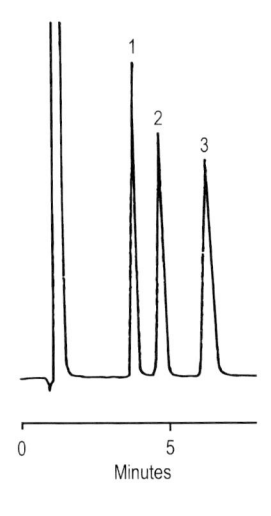

Fig. 4-33. Separation of ethanolamines on PBDMA-coated silica. – Separator column: LiChrosil® IC CA; column temperature: 40 °C; eluant: 5 mmol/L tartaric acid + 1 mmol/L pyridine-2,6-dicarboxylic acid; flow rate: 1 mL/min; detection: direct conductivity; injection volume: 15 µL; solute concentrations: 30 mg/L monoethanolamine (1), 50 mg/L diethanolamine (2), and 100 mg/L triethanolamine (3).

The third product of this kind is offered by Alltech (Deerfield, IL, USA) under the trade name Universal Cation [17]. While the column dimensions of Universal Cation and LiChrosil® IC CA are identical, the 7-µm particle diameter of the Universal Cation substrate is slightly larger. Remarkably, with a pure methanesulfonic acid eluant and consequent use of a suppressor system, it is possible to elute alkali- and alkaline-earth metals. As shown in Fig. 4-32, some transition metals can be separated with this stationary phase, if an eluant mixture of two chelating agents is used [18].

In addition to strong acid and weak acid cation exchangers, cross-linked polymers carrying cyclic polyethers as anchor groups have been used for the separation of alkali- and alkaline-earth metals. The structure of these stationary phases, which are also based on silica, was described in detail in Section 3.4.4. A reasonable separation of alkali metals was obtained by Kimura et al. [19] on silica modified with poly(benzo-15-crown-5) using de-ionized water as an eluant (Fig. 4-34). This separation requires a long time, although it can be significantly shortened by coating ODS (octadecyl silica) with lipophilic crown ether derivatives [19]. However, the chromatographic efficiency of these separations does not meet today's requirements. The only advantage of this method is the ability to perform a very sensitive detection via electrical conductivity, because the eluant being used does not have an intrinsic conductance.

In contrast to conventional cation exchangers, a reversed elution order is observed with crown ether phases. This elution order is mainly determined by the size ratio between the crown ether ring and the alkali metal ion. Due to the high affinity of poly(benzo-15-crown-5) towards potassium and rubidium ions, these are more strongly retained than lithium, sodium, and cesium ions. However, the complexing properties of crown ethers also depend on the counter ion being used. In potassium salts, for example, an increase in retention in the order KCl < KBr < KI is observed with an increasing size of the counter ion.

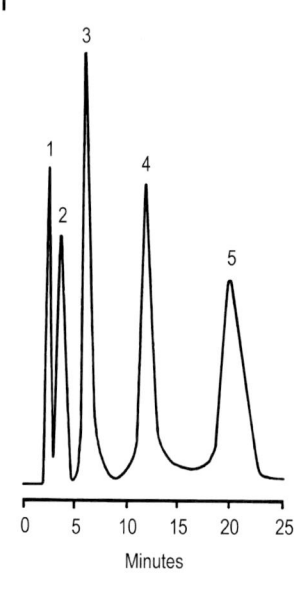

Fig. 4-34. Separation of alkali metals on silica modified with poly(benzo-15-crown-5). — Eluant: water; flow rate: 1 mL/min; detection: direct conductivity; injection volume: 1 µL; solute concentrations: 13.6 g/L LiBr (1), 14.4 g/L NaBr (2), 26.2 g/L KBr (3), 36.4 g/L RbBr (4), and 34 g/L CsBr (5); (taken from [19]).

Alkaline-earth metal ions, on the other hand, elute from a crown ether phase in the normal elution order ($Mg^{2+} < Ca^{2+} < Sr^{2+} < Ba^{2+}$). Such a separation is of pure academic interest, since the resolution between magnesium and calcium is extraordinarily poor due to the low interactions of both ions with crown ethers.

4.2
Eluants in Cation Exchange Chromatography

As in anion exchange chromatography, the type of eluants for cation exchange chromatography depends on the detection method being used. Classification was not necessary in former times due to the limited number of stationary phases. However, the confusing number of strong acid and weak acid cation exchangers today requires a more systematic handling of this subject.

For the separation of alkali metals, ammonium, and small aliphatic amines on strong acid cation exchangers, mineral acids such as hydrochloric acid, sulfuric acid, or nitric acid are typically used as eluants, independent of whether the subsequent conductivity detection is carried out with or without chemical suppression. Depending on the type of separator column, the concentration range lies between 2 mmol/L and 40 mmol/L. Organic eluants such as methanesulfonic acid are employed only, if an electrolytically regenerated suppressor system such as CSRS (Dionex, Sunnyvale, CA, USA) is used. For the operation of separator columns such as ION-200 and Vydac 400 IC 405, the respective manufacturers recommend pyridinecarboxylic acids such as dipicolinic and isonicotinic acid. Alkali metals can also be detected via indirect fluorescence detection. Bächmann et al. [20] employed cerium(III) nitrate in very low concentrations as an eluant for this method.

Divalent cations such as alkaline-earth metals exhibit a much higher affinity towards the stationary phase of strong acid cation exchangers and, thus, cannot be eluted with dilute mineral acids. However, increasing the acid concentration in the mobile phase is impossible because of the following reasons:

- In a system *without* chemical suppression, the extremely high background conductivity would render impossible the sensitive detection of alkaline-earth metals via electrical conductivity.
- In a system *with* chemical suppression, the required reduction of the background conductivity would only be possible by using high-capacity membrane-based suppressor systems.

In his pioneering paper, Small [1] suggested silver nitrate, which also exhibits a high affinity towards the stationary phase, as an eluant for alkaline-earth metals. However, this system required a suppressor column in the chloride form to precipitate the silver as insoluble silver chloride. The application of this eluant quickly proved to be disadvantageous, because peak broadening and high backpressure from the suppressor column are associated with the precipitation of silver chloride.

The mixture of *m*-phenylenediamine-dihydrochloride and hydrochloric acid used later did not gain acceptance. The solution turns purple upon light absorption because of a slow dimerization of the amine and, thus, has only a limited stability.

In combination with a membrane-based suppressor (see Section 4.3) a mixture of 2,3-diaminopropionic acid (DAP) and hydrochloric acid is frequently used for the separation of alkaline-earth metals. Using this eluant provides the advantage of being able to adjust the elution power via the dissociation equilibrium of 2,3-diaminopropionic acid; see Eq. (104) in Section 4.1.1. With $pK_a = 1.33$ for the dissociation of the carboxyl group, DAP may exist in the mobile phase as a monovalent cation, a divalent cation, or a mixture of both. In any case, the product of the suppressor reaction is the zwitterionic form, which has no intrinsic conductance:

$$\underset{H_2C-CH-COO^-}{H_2N\ \overset{+}{N}H_3}$$

In non-suppressed ion chromatography systems, mixtures of ethylenediamine and aliphatic dicarboxylic acids such as tartaric acid or oxalic acid are typically employed as an eluant for the separation of alkaline-earth metals. Those mixtures have a relatively high intrinsic conductance, so that chromatographic signals are registered as negative peaks.

For the simultaneous analysis of alkali metals, alkaline-earth metals, and ammonium on carboxylate-based weak acid cation exchangers, methanesulfonic acid or sulfuric acid are usually recommended as eluants. Depending on the type of separator column, eluant concentration varies, but it rarely exceeds 20 mmol/L. If a polymer-coated silica is used as the stationary phase, the respective manufacturers recommend a mixture of pyridine-2,6-dicarboxylic acid (dipicolinic acid) and tartaric acid as standard eluant for this same application, although the elution order of magnesium and calcium is reversed. In many application examples, pure tartaric acid eluants are used as well. Under these conditions, alkaline-earth metals are more strongly retained than alkali metals, but they elute in the usual order, according to the periodic table of elements. Mixtures of pyridine-2,6-dicarboxylic acid and oxalic acid have proved to be suitable for analyzing some transition metals together with alkali- and alkaline-earth metals. However, non-suppressed conductivity detection must be applied when using complexing agents in the mobile phase in those cases.

Transition metals are usually analyzed as a separate class of compounds, so that a higher selectivity can be achieved by choosing a more suitable mobile phase and detection system. This is necessary, because in real-world samples transition metals are present at much lower concentrations than alkali- or alkaline-earth metals. Independent of the type of stationary phase being used, weak organic acids are used as complexing agents for the elution of transition metals. Because they are separated as anionic or neutral complexes, the type and concentration of the organic acid(s) depend on the separation method. Pyridine-2,6-dicarboxylic acid and oxalic acid are, again, the eluants most widely used in this application. Further details are described in Section 4.5.3.

4.3
Suppressor Systems in Cation Exchange Chromatography

4.3.1
Suppressor Columns

In the past, packed bed suppressor columns containing an ion-exchange resin were also employed in cation exchange chromatography. The resin was a strong basic microporous polystyrene/divinylbenzene-based anion exchange resin such as Dowex 1 × 10. It was functionalized with quaternary ammonium bases and had a relatively high degree of cross-linking of about 10%, which resulted in a high mechanical stability and prevented the resin material from swelling in water-miscible organic solvents. In contrast, macroporous strong basic anion exchangers did not gain acceptance as packing material for suppressor columns, because the high specific surface of the resin (200 to 800 m^2/g) leads to undesired adsorption effects with weak electrolytes.

The function of a suppressor column was already described in detail in Section 3.6. The disadvantages of a conventional suppressor column, also discussed in that section, are also extant to cation analysis. Moreover, the ion-exchange functional groups of the suppressor column typically exist in the hydroxide form, converting weakly dissociated cations into the corresponding free bases. The ammonium ion, for example, is converted into the free base NH_3. Because the latter is non-ionic, it is not subject to Donnan exclusion. Therefore, weakly dissociated cations may interact with the anion exchange resin of the suppressor column, resulting in considerably longer retention times.

In HPLC, peak dispersion is significantly affected by the particle size of the resin. Thus, the band broadening in a suppressor column decreases as the particle diameter of the resin being used decreases. Similarly, the void volume of the suppressor column should be as small as possible in order to reduce band broadening effects. Thus, an optimal suppressor column would have a very low volume and would contain an ion-exchange resin with a very small particle diameter. Because the resulting suppression capacity of such a suppressor column is very small, it can only be employed, if it is regenerated after every chromatographic run. The other option, using a suppressor column with a much higher volume, has all the negative consequences for peak dispersion.

According to the procedure described in Section 3.6.1, regeneration of conventional suppressor columns such as CSC-1 or CSC-2 is performed using sodium hydroxide solution with a concentration of 0.5 mol/L.

4.3.2
Hollow Fiber Suppressors

Today, conventional suppressor columns have only historical significance. The above-mentioned disadvantages of suppressor columns were overcome with the development of the CFS hollow fiber suppressor suited for cation exchange chromatography. The schematics of this suppressor corresponds precisely to the respective system for anion exchange chromatography described in Section 3.6.2. Instead of sulfonate groups, the membrane carries quaternary ammonium bases in the hydroxide form in order to exchange anions which are present as counter ions, for hydroxide ions. A dynamic equilibrium is established at constant operating conditions, so that the detection sensitivity for weakly dissociated cations remains unaffected. In comparison to suppressor columns, hollow fiber suppressors have a void volume which is more than one order of magnitude smaller, significantly reducing peak broadening and, thus, increasing sensitivity.

Either 0.04 mol/L potassium hydroxide or 0.02 mol/L tetramethylammonium hydroxide may be used for continuous regeneration of this suppressor. The choice of regenerant depends on the kind of analyte. A potassium hydroxide regenerant is recommended for the analysis of alkali metals. If ammonium is also to be analyzed, it should be noted that the linear range for the determination of this ion is very small when using a potassium hydroxide regenerant. This is because of the equilibrium between the suppressor product NH_4OH and the free base NH_3.

The regenerant flow rate should be 2 to 3 mL/min. With eluant concentrations up to 5 mmol/L, a background conductivity of 10 µS/cm at maximum is ensured.

Potassium hydroxide solution may also be used as a regenerant for the analysis of alkaline-earth metals, as long as a detection sensitivity in the mg/L range is sufficient.

4.3.3
Micromembrane Suppressors

Although hollow fiber suppressors are rather effective in increasing sensitivity, their application in cation analysis limits the concentration of the acid eluant. As mentioned in Section 3.6.2, a higher ionic strength eluant results in a higher background conductivity, because it is not possible to increase the regenerant concentration accordingly. In such a case, the comparatively small potassium ions would surmount the Donnan exclusion forces and would diffuse into the interior of the membrane. If the application of concentrated eluants is indispensible, as in modern latexed cation exchangers, a micromembrane suppressor with significantly higher suppression capacity must be employed.

The schematics and operation of the CMMS micromembrane suppressor suitable for cation exchange chromatography corresponds precisely to the AMMS suppressor described in Section 3.6.3, which was developed for anion exchange chromatography. The high capacity of this micromembrane suppressor allows the use of high-ionic strength eluants, which leads to shorter analysis times. The low void volume of the suppressor, which has little effect on the efficiency of the separation, also contributes to the sensitivity increase.

The CMMS micromembrane suppressor also allows the application of concentration gradients without sacrificing conductivity detection, which is irreplaceable for cation detection. Depending on the type of stationary phase being employed, a mixture of 2,3-diaminopropionic acid and hydrochloric acid suitable for chemical suppression or pure acid (hydrochloric acid, sulfuric acid, methanesulfonic acid) with or without an organic solvent is used as an eluant. Due to the lack of suitable stationary phases, in the past gradient techniques played only a secondary role in cation analysis, all the more so, because alkali- and alkaline-earth metals can be analyzed in one run under isocratic conditions. Also, gradient techniques cannot be applied at all for transition metal analysis, for which different chromatographic conditions are required; conductivity detection has to be ruled out as detection system due to its low selectivity and sensitivity. On the other hand, with an increasing number of applications for aliphatic amines, gradient techniques have become increasingly important in cation exchange chromatography (see Section 4.7).

The regeneration of a CMMS micromembrane suppressor is not fundamentally different from that of a hollow fiber suppressor, excepting only that tetrabutylammonium hydroxide (TBAOH) is used as a regenerant. Depending on the concentration of the acid(s) employed as an eluant, regenerant concentrations from 0.04 to 0.1 mol/L are sufficient for isocratic conditions. Upon application of the gradient technique, the flow rate must be adjusted in the way so that a sufficiently low background conductivity is ensured at the maximum eluant concentration being used. Thus, the regenerant should be recirculated according to Fig. 3-75 (see Section 3.6.3). In order to maintain a constant hydroxide ion concentration over a longer period of time, the regenerant is passed through an ion-exchange cartridge containing a strong basic anion exchanger in the hydroxide form, before it is delivered to the suppressor system via a small pump. In this way, tetrabutylammonium chloride formed in the suppressor is converted to the corresponding base again. The lifetime of such a cartridge with a stated capacity of about 350,000 µequiv depends, of course, on the concentration and flow rate of the eluant being used. To calculate the cartridge lifetime, use Eq. (69) in Section 3.6.3.

4.3.4
Self-Regenerating Suppressors

The schematics and operation of the CSRS self-regenerating suppressor suitable for cation exchange chromatography corresponds to the ASRS, which was developed for anion exchange chromatography (see Section 3.6.4). Both suppressor systems were introduced in summer 1992 [21, 22]. Since then, dynamic suppression capacity has been improved; they have developed into the SRS-Ultra type suppressors now being available on the market. The principle schematic of a CSRS only differs from a CMMS in the platinum electrodes for water electrolysis, which are placed at the top and at the bottom, between the respective regenerant screen and the adjacent enclosure. After suppression, the de-ionized effluent of the conductivity cell is recycled back into the regenerant screens and, thus, serves as the water source for electrolysis.

The neutralization reactions taking place inside a CSRS are schematically illustrated in Fig. 4-35. Hydroxide ions formed at the cathode permeate the anion exchange membrane and neutralize the eluant, which is methanesulfonic acid in this case. Methanesulfonate counter ions are attracted by the anode, permeate through the anodic regenerant chamber and combine with hydronium ions, which are generated there for maintaining electroneutrality. At both electrodes, a little bit of gas is generated, which goes to waste together with the liquid reaction products.

In contrast to a CMMS, the suppressor reaction inside a CSRS is directed by the electrodes. In a CMMS, for instance, the chemical regenerant is directed countercurrent through both regenerant chambers. Therefore, regenerant ions

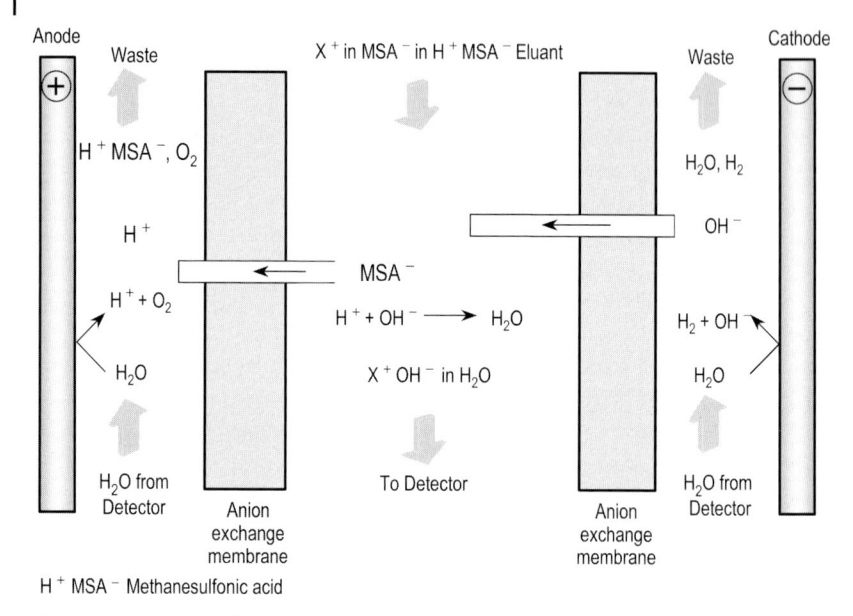

Fig. 4-35. Schematic illustration of the neutralization reactions inside a CSRS.

required for neutralization are provided to the eluant chamber through both membranes. However, in a CSRS the hydroxide ions required for the suppression reaction are exclusively generated at the cathode, so that only the anion exchange membrane in the cathodic regenerant chamber is permeable for hydroxide ions. Conversely, the exchange of counter ions is only carried out at the membrane in the anodic regenerant chamber, where they associate with hydronium ions being generated therein.

Depending on the application, a self-regenerated suppressor can also be operated in three different modes, which were described in Section 3.6.4. The most common mode of operation for the CSRS is *AutoSuppression* in the Recycle Mode, in which the conductivity cell effluent is used as the source for the required de-ionized water (see Fig. 3-78 in Section 3.6.4). Because in this case an external chemical regenerant such as tetrabutylammonium hydroxide is unnecessary, cost of ownership is significantly reduced and suppressor operation is simplified. However, this mode of operation can only be used for applications that use pure aqueous eluants.

In case organic solvents need to be added to the mobile phase, *AutoSuppression* with *External Water Supply* is recommended as the mode of operation, to avoid the recycling of degradation products from organic solvents under electrolysis conditions; this would increase noise and background conductivity. But even under those conditions, the organic solvent content in the mobile phase is limited to 40% (v/v). The CSRS self-regenerating suppressor can be operated at temperatures up to 40°C.

It should also be mentioned that a CSRS, like a CMMS, can be regenerated in a conventional way with tetrabutylammonium hydroxide. There are no special applications for it, but the noise level would be lower because the electrolysis process is omitted.

As outlined above, alkali- and alkaline-earth metals are converted to the metal hydroxides in the suppressor system. Since those metal hydroxides are strong bases and totally dissociated, they give a linear response over a wide concentration range. In contrast, weak bases such as ammonium and amine hydroxides formed in the suppressor, are only partially dissociated and, thus, give a non-linear response as the analyte concentration increases. For most applications, this non-linear response of weak bases can be managed easily. Many applications do not require a wide calibration range. In any case, a quadratic fit of the calibration curve should be used to produce accurate, reliable data. However, the non-linear response can be overcome by converting the weak base analyte to a totally ionized salt form, thus extending the linear response over three orders of magnitude. A modified version of the CSRS, the CSRS-SC, has been designed to make this conversion.

The CSRS-SC is a post-column electrolytic eluant suppressor package consisting of an Eluant Suppressor component (ES) and an Analyte Converter component (AC) that enables the suppressed conductivity detection of low-level ammonium and other amines. Figure 4-36 illustrates the ion movement through a CSRS-SC in detail. The eluant and the ammonium analyte (an example of the cation analytes) exit the analytical column as methanesulfonic acid (H^+MSA^-) and ammonium methanesulfonate ($NH_4^+MSA^-$). In the Eluant Suppressor component, the majority of the eluant is suppressed to water and the analyte is converted to the base form, $NH_4^+OH^-$. However, a small amount of MSA^- re-enters the eluant chamber near the exit, which converts the analyte back to the salt form, so that the analyte exits the Eluant Suppressor component as $NH_4^+MSA^-$ in a background of dilute methanesulfonic acid. In the Analyte Converter (AC) component, the NH_4^+ analyte cation is exchanged for a hydronium ion, resulting in conversion of the analyte to methanesulfonic acid, H^+MSA^-. The eluant remains unchanged as methanesulfonic acid of low concentration. In the conductivity detector, the analyte is detected as totally dissociated methanesulfonic acid in a low background of methanesulfonic acid. This means that at the time the analyte passes through the detector cell, an increase in methanesulfonic acid concentration is observed that is proportional to the analyte concentration.

When using the CSRS-SC, the response at varying concentrations is linear, because the analyte is detected as totally dissociated methanesulfonic acid. Figure 4-37 shows the linear calibration plot for ammonium, comparing the CSRS-SC with the conventional CSRS over three orders of magnitude (0.1 mg/L to 100 mg/L). Table 4-3 lists the comparison of correlation coefficients for ammonium and a broad range of aliphatic amines using the CSRS-SC and the conventional CSRS suppressors. Remarkably, the correlation coefficient for the weak base, triethanolamine, is excellent in the concentration range of 0.1 mg/L to 100 mg/L, when using the CSRS-SC suppressor.

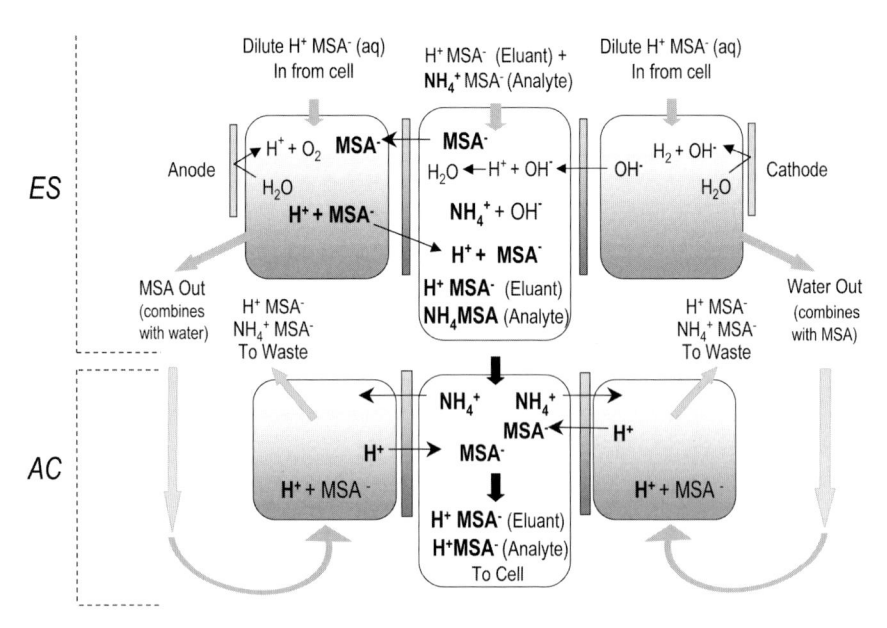

Fig. 4-36. Illustration of the ion movement through a CSRS-SC.

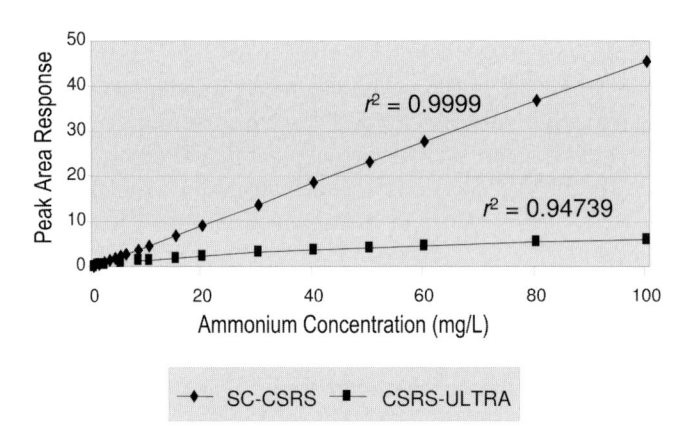

Fig. 4-37. Comparison of response of ammonium using CSRS-SC and CSRS-Ultra.

The CSRS-SC significantly improves the response factors of ammonium and amines when using conductivity detection, because the analytes are actually detected as totally dissociated methanesulfonic acid, which has a higher equivalent

Table 4-3. Linearity comparison for ammonium and aliphatic amines using the CSRS-SC and the conventional CSRS in the concentration range of 0.1 mg/L to 100 mg/L.

Amine	pK_a	Correlation coefficients		CSRS-SC
		CSRS		
		Quadratic fit	Linear fit	Linear fit
Triethanolamine	7.92	0.9977	0.9789	0.9994
Morpholine	8.33	0.9978	0.9778	1.0000
Ammonium	9.25	0.9946	0.9644	1.0000
Trimethylamine	9.76	1.0000	0.9986	0.9994
Monomethylamine	10.62	0.9998	0.9967	1.0000
Monoethylamine	10.7	0.9999	0.9960	1.0000
Diethylamine	11.09	1.0000	0.9990	1.0000

conductance (399 S cm^2 mol^{-1}). In contrast, sodium detected as Na$^+$OH$^-$ has an equivalent conductance of only 249 S cm^2 mol^{-1} and ammonium detected as NH$_4^+$OH$^-$ has an equivalent conductance of only 274 S cm^2 mol^{-1}. In Fig. 4-38, separations of standard inorganic cations, ammonium, and amines using a conventional CSRS are compared to the same separations using a CSRS-SC.

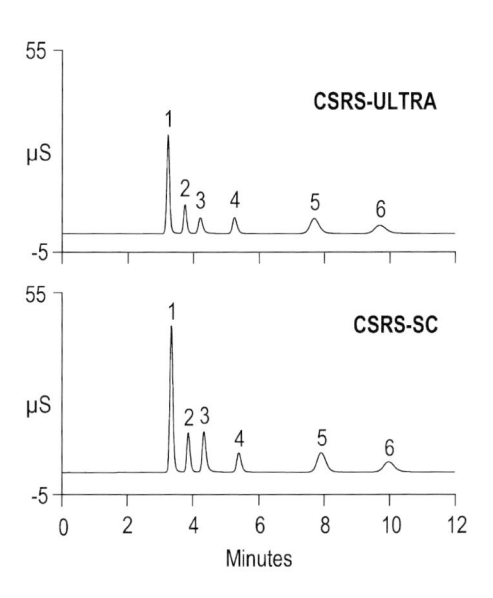

Fig. 4-38. Comparison of standard suppression and the converter mode for the separation of inorganic cations. – Separator column: IonPac CS12A; column temperature: 30°C; eluant: 20 mmol/L methanesulfonic acid; flow rate: 1 mL/min; injection volume: 25 µL; detection: suppressed conductivity; solute concentrations 4 mg/L each of lithium (1), sodium (2), ammonium (3), potassium (4), magnesium (5), and calcium (6).

4.3.5
Suppressors with Monolithic Suppression Beds

In 2001, a third type of suppressor – the Atlas Electrolytic Suppressor (AES) – was introduced. The Atlas suppressor is a continuously regenerated suppressor operated in the recycle mode and designed for optimal performance with conventional methanesulfonic acid (MSA) eluants. Able to suppress up to 25 mmol/L MSA at 1 mL/min, it can be used in combination with all weak acid cation exchangers except the high-capacity column IonPac CS16 (see Section 4.1.1).

In contrast to the membrane-based micromembrane and self-regenerated suppressor devices, the suppression bed of an Atlas suppressor is made of an ion-exchange monolith. As previously illustrated in the respective cross-section in Fig. 3-79 (see Section 3.6.5), the electrodes are placed on both sides of the monolithic suppression bed, which is approximately 1 cm long. The length of this monolithic bed is a compromise between suppression capacity and dead volume. To ensure maximum penetration of the monolith with the eluant, it is cut in slices (MonoDiscs), which are separated from each other by flow distribution discs. The small boreholes in those discs alternate between top and bottom, so that the eluant is forced to flow through every ion-exchange MonoDisc. This configuration facilitates an efficient exchange of eluant MSA ions for regenerant hydroxide ions generated at the anode.

The new design of the Atlas suppressor enables a much faster daily startup times after an overnight shutdown, which improves the throughput of routine sample analysis. A stable baseline is observed after less than 30 minutes, which is by far shorter than that with an SRS device. The Atlas suppressor also improves the analysis of standard cations by significantly lowering the noise, which is only slightly higher than that observed for a micromembrane suppressor in the chemical suppression mode. Like self-regenerated suppressors, the Atlas suppressor is not compatible with organic solvents, which is not a real drawback, because organic solvents are not necessary for routine MSA-based separations.

4.4
Cation Exchange Chromatography of Alkali Metals, Alkaline-Earth Metals, and Amines

Typically, alkali metals and ammonium are separated using a dilute mineral acid or a strong organic acid such as methanesulfonic acid as an eluant. Sulfonate-based strong acid cation exchangers and carboxylate-based weak acid cation exchangers are used as stationary phases. A representative example for the sequential separation of alkali metals and ammonium on a latexed cation exchanger is shown in Fig. 4-25 (see Section 4.1.2). This figure reveals that retention of alkali

4.4 Cation Exchange Chromatography of Alkali Metals, Alkaline-Earth Metals, and Amines

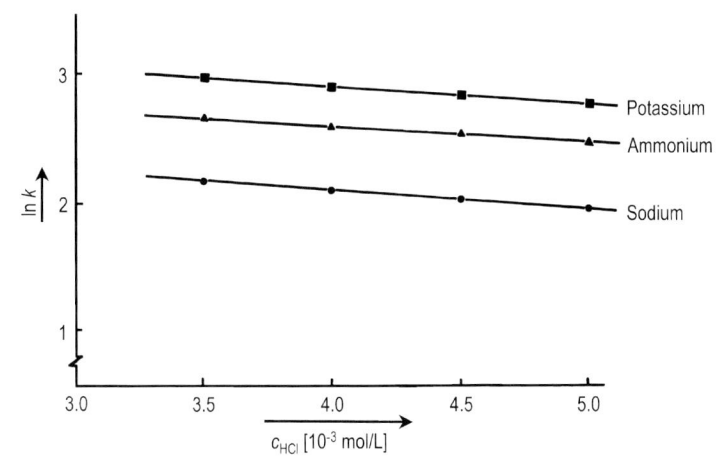

Fig. 4-39. Monovalent cation retention as a function of the eluant ionic strength. — Separator column: IonPac CS1; eluant[1]: HCl; flow rate: 2.3 mL/min; detection: suppressed conductivity.

metals increases with increasing ionic radius. Compared to conventional instrumental analysis methods, the advantage of ion chromatography is the simultaneousness of the method. Without any doubt, the key ion in this chromatogram is ammonium, which elutes between sodium and potassium on strong acid cation exchangers. Sensitive detection of ammonium by other methods is very difficult. The greatest difference among the numerous stationary phases that can be used for this separation problem is the resolution between sodium and ammonium. When analyzing alkali metals sequentially, this resolution is always larger than it is in the simultaneous analysis of alkali- and alkaline-earth metals.

The sole parameter that determines retention in the separation of alkali metals is the concentration of the acid eluant. Linear relationships are obtained for the different solute ions when the logarithm of the capacity factors is plotted as a function of the ionic strength (see Fig. 4-39).

Many aliphatic amines also elute under the chromatographic conditions that are suitable for the separation of alkali metals. This includes hydroxylamine, which can be separated as a cation after protonation in the mobile phase and detected via direct conductivity detection. However, a much more selective method is amperometric detection, using a platinum working electrode at which hydroxylamine is oxidized at a working potential above +0.8 V. The corresponding chromatogram is displayed in Fig. 4-40. Hydroxylammonium ions exhibit about the same retention behavior as ammonium ions. If one component is to

---

1 HCl should only be used as an eluant in combination with chemical suppression.

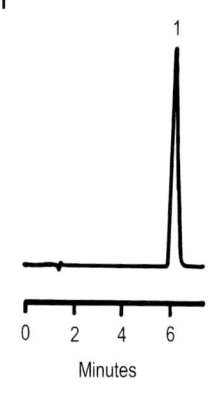

Fig. 4-40. Amperometric detection of hydroxylamine after cation exchange separation. — Separator column: IonPac CS3; eluant: 30 mmol/L HCl; flow rate: 1 mL/min; detection: DC amperometry on a platinum working electrode; oxidation potential: +0.85 V; injection volume: 50 µL; solute concentration: 10 mg/L hydroxylamine (1).

be detected in presence of the other, the necessary selectivity may only be obtained by applying different detection methods; the separator column effluent is passed first through the amperometric cell and then through the suppressor system into the conductivity cell. Since hydroxylamine cannot be detected via suppressed conductivity and — vice versa — ammonium cannot be oxidized on a platinum electrode, both detection methods are highly selective. The respective signals may be acquired with a dual-channel data system. Another way to differentiate between the two analytes is offered by fluorescence detection. While ammonium ions may be detected with a fluorescence detector after derivatization with o-phthaldialdehyde/2-mercaptoethanol, hydroxylammonium ions do not undergo this reaction.

Conductivity detection in *both* of its modes of application is also suitable for the detection of primary, secondary, and tertiary alkylamines. This is illustrated in Fig. 4-41; three methylamines are separated on Vydac 400 IC 405 with an isonicotinic acid eluant. Interferences are possible if the sample contains potassium, rubidium, and/or cesium in addition to amines. However, the separation can be optimized by adding a small amount of an organic solvent to the mobile phase, which mainly affects the retention of alkylsubstituted cations. In comparison to Fig. 4-41, Fig. 4-42 shows the separation of the corresponding ethylamines on a latexed cation exchanger using hydrochloric acid as an eluant. If necessary, the higher retention times may be shortened significantly by adding 2,3-diaminopropionic acid.

Compounds having several amino functions can also be analyzed under the chromatographic conditions given in Fig. 4-42. This includes both the formamidinium and the guanidinium cations:

Formamidinium Cation   Guanidinium Cation

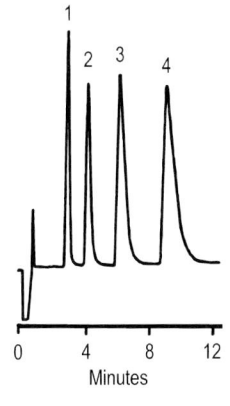

Fig. 4-41. Separation of monomethylamine, dimethylamine, and trimethylamine on Vydac 400 IC 405. — Eluant: 5 mmol/L isonicotinic acid + 2.5 mmol/L HNO_3 + 100 mL/L methanol; flow rate: 2 mL/min; detection: direct conductivity; injection volume: 100 µL; solute concentrations: 1 mg/L ammonium (1), 2 mg/L monomethylamine (2), 5 mg/L dimethylamine (3), and 10 mg/L trimethylamine (4).

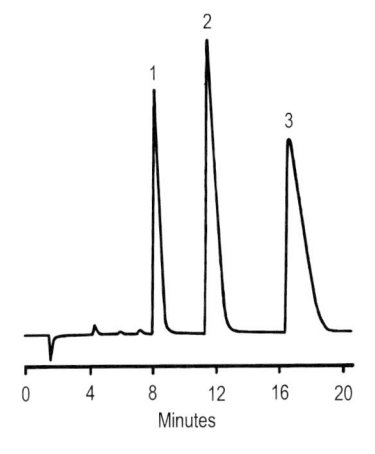

Fig. 4-42. Separation of monoethylamine, diethylamine, and triethylamine on a latexed cation exchanger. — Separator column: IonPac CS3; eluant: 40 mmol/L HCl; flow rate: 1 mL/min; detection: suppressed conductivity; injection volume: 50 µL; solute concentrations: 20 mg/L monoethylamine (1), 50 mg/L diethylamine (2), and 100 mg/L triethylamine (3).

The latter is a urea derivative in which the three NH_2 groups are positioned symmetrically and equidistantly to the central C-atom. Both cations may be detected by either suppressed or non-suppressed conductivity; this is shown in Fig. 4-43.

Guanylurea with four amino groups has an even higher retention time and may be determined either by direct conductivity detection or by measuring the light absorption of the carbonyl group at a wavelength of 215 nm.

$$H_2N-\underset{\underset{NH}{\|}}{C}-NH-\underset{\underset{O}{\|}}{C}-NH_2$$

Guanylurea

As an example, Fig. 4-44 shows the respective chromatogram obtained upon application of UV detection.

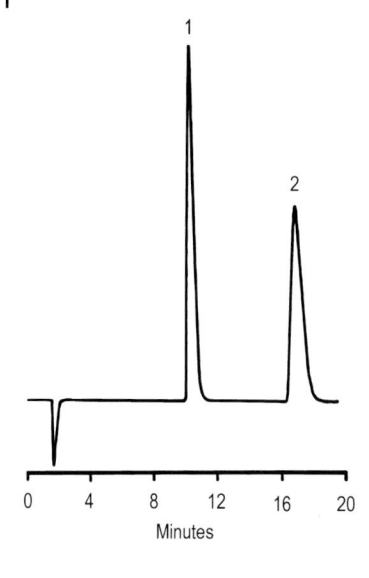

Fig. 4-43. Separation of formamidinium and guanidinium cations. − Chromatographic conditions: see Fig. 4-42; solute concentrations: 50 mg/L each of formamidinium acetate (1) and guanidinium sulfate (2).

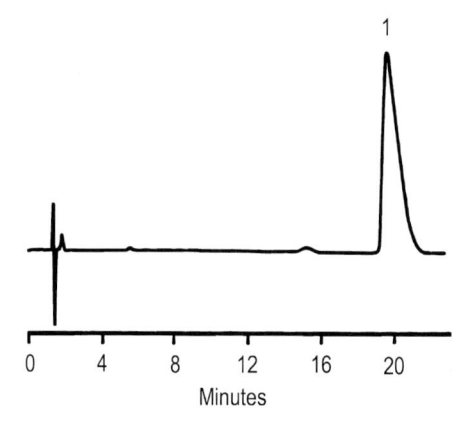

Fig. 4-44. Analysis of guanylurea. − Separator column, eluant, and flow rate: see Fig. 4-42; detection: UV (215 nm); injection volume: 50 µL; solute concentration: 100 mg/L guanylurea (sulfate salt) (1).

The sequential analysis of alkaline-earth metals, which elute in the order $Mg^{2+} < Ca^{2+} < Sr^{2+} < Ba^{2+}$ on strong acid cation exchangers, can also be performed using both conductivity detection modes. An eluant mixture of hydrochloric acid and 2,3-diaminopropionic acid is used for suppressed conductivity detection; ethylenediammonium ions are suitable for direct conductivity detection. Figure 4-45 shows a separation of alkaline-earth metals on Shimpack IC-C1 obtained with this eluant. Because of the high elution power of the mobile phase, all monovalent cations present in the sample are eluted as one peak within the void volume of the column. As an alternative to strong eluants, shorter separator columns may be employed to reduce the retention of cations that have high affinities towards the stationary phase.

4.4 Cation Exchange Chromatography of Alkali Metals, Alkaline-Earth Metals, and Amines

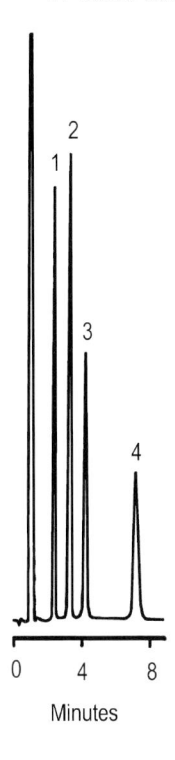

Fig. 4-45. Separation of alkaline-earth metals on Shimpack IC-C1. – Eluant: 4 mmol/L tartaric acid + 2 mmol/L ethylenediamine; flow rate: 1.5 mL/min; detection: direct conductivity; injection volume: 20 µL; solute concentrations: 5 mg/L magnesium (1), 10 mg/L calcium (2), 20 mg/L strontium (3), and 36 mg/L barium (4).

The simultaneous analysis of the most important alkali- and alkaline-earth metals used to be impossible due to their markedly different retention behavior. Once only been a dream of the inorganic analytical chemist, such an analysis is considered a routine nowadays. One method involves using a silica-based cation exchanger modified with poly(butadiene-maleic acid), which was introduced in Section 4.1.3. As it was shown in Fig. 4-30, the most important alkali- and alkaline-earth metals and ammonium can be analyzed in a single run with direct conductivity detection using a tartaric/dipicolinic acid eluant mixture. The extremely short time required for such a separation is quite impressive.

Strong acid cation exchangers are not suited for this type of analysis. A column switching technique [23] introduced years ago uses two latexed cation exchangers of different length (Fast Cation I and II). The sequence of the columns is changed via valve switching upon sample injection while maintaining the flow direction. However, it was never really embraced by the marketplace. Although the peak efficiencies are very high for monovalent cations passing through both columns, the strong tailing of divalent cations passing through the shorter column with its low chromatographic efficiency is a disadvantage. The calcium signal was particularly affected; it could only be reproducibly analyzed with a precise programming of the integration parameters. The relatively short analysis time of about 10 minutes did not compensate for all these disadvantages.

Some certain improvement was achieved with the IonPac CS10 separator column (introduced in Section 4.1.2) because divalent cations pass through the guard *and* the separator column. As can be seen from the corresponding chromatogram in Fig. 4-27, all components exhibit an almost symmetrical peak shape. However, the limited resolution between the monovalent cations and the unreasonably high resolution between the divalent cations are problematic. Although total analysis time may be shortened by increasing the 2,3-diaminopropionic acid concentration in the mobile phase, resolution in the front part of the chromatogram would decrease even more. Interestingly, the transition metal manganese can also be analyzed together with alkali- and alkaline-earth metals, although it is generally assumed that the transition metal hydroxides formed in the suppressor system will precipitate and, thus, may not be detected by suppressed conductivity detection. However, the kinetics of this reaction is not the same for all metals, especially because a weakly acidic milieu (pH 6) prevails in the suppressor after the suppressor reaction, but many transition metals only precipitate at pH values above 7.

The real breakthrough in the simultaneous analysis of alkali- and alkaline-earth metals was the development of a carboxylate-based weak acid cation exchanger with a polymeric backbone, which is significantly more stable than PBDMA-coated silica. This separator, based on Schomburg's idea, was developed in the early 1990's. Columns of that type have been improved over the past several years by modifying the ion-exchange polymer with other functional groups such as phosphonate or crown ether moieties. As shown in the Figures 4-3 and 4-7 through 4-17, the separator columns listed in Table 4-2 are suitable for the simultaneous analysis of alkali- and alkaline-earth metals as well as for a large number of aliphatic amines. To a certain extent, those amines can be analyzed together with inorganic cations, often by using gradient techniques (see Section 4.7). Excellent separations can also be obtained with simple step gradients utilizing organic solvents. An example of this technique, illustrating the separation of ethylamines together with standard inorganic cations, is shown in Fig. 4-46. The stepwise increase of the methanesulfonic acid concentration during the run results in a comparatively high resolution between early eluting species, without compromising on the total analysis time. Furthermore, the slight increase of the acetonitrile content in the mobile phase improves the peak form of the late eluting components.

With the development of carboxylate-based weak acid cation exchangers, sequential analyses of alkali- and alkaline-earth metals became much less common.

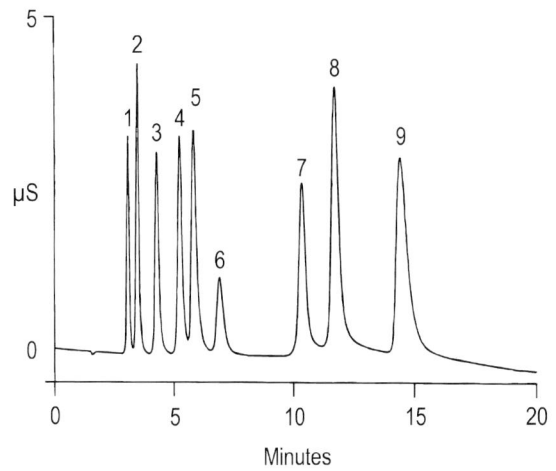

Fig. 4-46. Separation of ethylamines together with alkali- and alkaline-earth metals utilizing a step gradient. – Separator column: IonPac CS14; eluant: 9 mmol/L methanesulfonic acid + 5 mL/L acetonitrile for 6 min, then to 11 mmol/L methanesulfonic acid + 15 mL/L acetonitrile; flow rate: 1 mL/min; injection volume: 18 µL; detection: suppressed conductivity; solute concentrations: 0.5 mg/L lithium (1), 2 mg/L sodium (2), 2.5 mg/L ammonium (3), 10 mg/L monoethylamine (4), 5 mg/L potassium (5), 10 mg/L diethylamine (6), 2.5 mg/L magnesium (7), 5 mg/L calcium (8), and 50 mg/L triethylamin (9).

4.5 Transition Metal Analysis

As early as 1939, Samuelson [24] reported the successful separation of transition metals by means of ion-exchange chromatography. This method employed an ion-exchange resin to separate the metal ions. Quantitation was achieved via manual spectrophotometry of the individual fractions. The ion chromatographic determination of transition metals was significantly improved by the introduction of highly efficient low-capacity ion exchangers by Fritz et al. [25] in 1974 and the use of a continuous solute-specific photometric detection after derivatization of the column effluent. In 1979, Cassidy et al. [26] successfully analyzed lanthanides by ion chromatography. The commercialization of this method since 1982 is attributed to the work of Riviello et al. [27]. By introducing membrane reactors, he succeeded in significantly improving reagent delivery for post-column derivatization. Following the introduction of highly selective pellicular ion exchangers with defined anion and cation exchange capacity [28], ion chromatography gained general acceptance as a fast and sensitive multi-element detection method for the determination of transition metals, allowing detection limits in the lower µg/L range and analysis times of about 15 minutes.

4.5.1
Basic Theory

The separation of transition metals with ion exchangers requires a complexation of the metal ions in the mobile phase to reduce their effective charge density. Monovalent cations such as Na^+ or H^+ are unsuitable as eluants. Because the selectivity coefficients for transition metals of the same valency are so similar, a selectivity change is obtained only by the introduction of a secondary equilibrium, such as a complexation equilibrium, which is established by adding appropriate complexing agents to the mobile phase. Common complexing agents include weak organic acids that preferably form anionic or neutral complexes with metal ions such as citric acid, oxalic acid, tartaric acid, or pyridine-2,6-dicarboxylic acid (dipicolinic acid). Such complexes may be separated on anion or cation exchangers. Nowadays, stationary phases with a defined anion *and* cation exchange capacity are used. The separations obtained with aliphatic or aromatic carboxylic acids are clearly superior to those obtained with ammonia or ethylenediamine; the latter forms predominantly neutral complexes. Often, two different organic acids are used as complexing agents to optimize a particular separation. For simplicity, however, only a single ligand is considered in the following kinetic treatment.

Imagine a solution containing a multiply-charged transition metal ion M^{n+} in low concentration and a monovalent cation A^+ (Na^+ or H^+) in a significantly higher concentration. When this solution is in contact with a cation exchanger in the A^+-form, the following equilibrium is established:

$$M^{n+} + n \cdot RA \rightleftharpoons R_nM + n \cdot A^+ \qquad (105)$$

R Resin

According to the law of mass action the equilibrium constant or selectivity coefficient is expressed as:

$$K_{MA} = \frac{[R_nM] \cdot [A^+]^n}{[RA]^n \cdot [M^{n+}]} \qquad (106)$$

If we are dealing with a pure ion-exchange mechanism, the distribution coefficient D_M is given by the ratio of the analyte concentration M^{n+} in the stationary and in the mobile phase at equilibrium:

$$\begin{aligned}D_M &= \frac{[R_nM]}{[M^{n+}]} \\ &= K_{MA} \cdot \frac{[RA]^n}{[A^+]^n}\end{aligned} \qquad (107)$$

The metal ion concentration is very low relative to the eluant concentration, hence:

$$[M^{n+}] \ll [A^+] \quad \text{and} \quad [R_nM] \ll [RA] \tag{108}$$

It follows that K_{MA} is constant for any given eluant concentration. Because, in good approximation, [RA] is also constant and represents the resin ion-exchange capacity C, the distribution coefficient may be expressed as:

$$D_M \sim \frac{K_{MA} \cdot C^n}{[A^+]^n} \tag{109}$$

In logarithmic form one obtains:

$$\log D_M \sim \log K_{MA} + n \cdot \log C - n \cdot \log[A^+] \tag{110}$$

However, the capacity factor, k, according to Eq. (7) in Section 2.1, is usually measured instead of the distribution coefficient D_M.

From Eq. (110) it follows that the logarithm of the distribution coefficient, D_M, depends linearly on the logarithm of the cation concentration $[A^+]$ in the mobile phase. Plotting $\log D_M$ versus $\log [A^+]$ yields a straight line with the slope n, which corresponds to the charge number of the ion.

The selectivity of the separation of metal ions with the same charge number is exclusively determined by their K_M-values. The concentration of eluant cations $[A^+]$ does not contribute to the selectivity increase, because the differences in the logarithms of the distribution coefficients remain unchanged with different eluant concentrations. Adequate separations are only obtained if the selectivity coefficients of two transition metal ions M_1^{n+} and M_2^{n+} differ significantly. The only exceptions are metal ions of different charge numbers, whose separation depends on the eluant concentration.

If the eluant contains complexing agents such as citric acid or oxalic acid, then the equation for the distribution coefficient of the metal ion, M^{n+}, must also take into account the competing equilibrium of the complex formation. The complex formation coefficient may be expressed as follows:

$$a_{M(L)} = 1 + [L]\, B_{ML} + [L]^2\, B_{ML2} + \cdots \tag{111}$$

$a_{M(L)}$ Complex formation coefficient
[L]   Concentration of anionic ligand
B_{ML} Formation constant for the equilibrium
     $M^{n+} + L^{x-} \rightleftharpoons ML^{n-x}$
B_{ML2} Formation constant for the equilibrium
     $M^{n+} + 2\,L^{x-} \rightleftharpoons ML_2^{n-2x}$

If the complexing form of the ligand is a sufficiently strong conjugated base, protonation of the non-complexing ligands must be taken into account for the calculation of $a_{M(L)}$. Because under normal chromatographic conditions the concentration of the complexing form of the ligand, L^{x-}, is much higher than that of the metal ion, the complex formation equilibrium is barely affected by a partial protonation of the ligand. Thus, the distribution coefficient of the metal ion in the presence of complexing agents can be expressed as follows:

$$\log D_M = \log K_{MA} - \log a_{M(L)} + n \log C - n \log [A^+] \quad (112)$$

The separation of the various transition metals may be optimized by varying the pH value. If weak organic acids are used as complexing agents, lowering pH leads to a decrease in the effective ligand concentration. Therefore, an increase in the proton concentration results in a longer retention time. This is easily recognized when putting the dissociation equilibrium for [L] into Eq. (111) and substituting it into Eq. (112). If the effective ligand concentration is raised by a higher concentration of the organic acid, this leads to a higher concentration of the respective counter ion (Na^+ or H^+) as well. This, in turn, accelerates the displacement of metal ions from the cation exchange functional groups and results in a severe loss of separation efficiency.

When low-capacity cation exchangers are used, the ion-exchange capacity depends on the kind of counter ion. Accordingly, lithium hydroxide is recommended for adjusting the pH value, because lithium ions exhibit a low affinity towards the stationary phase.

The ability of metals to form anionic complexes is used for the separation of transition metals on anion exchangers. Such separations are based on (1) the significant differences in the stability of these complexes with anionic complexing agents and (2) the different affinity of anionic complexes towards the stationary phase of anion exchangers. The most important parameter determining retention is the valency of the complex, which, in turn, depends on the charge numbers of the central metal ion and the ligand, as well as on the coordination number of the complex.

The distribution coefficient of a metal ion with a monovalent anionic ligand between the mobile and the stationary phase is given by:

$$D_M = \frac{[R_{n-x} \cdot ML_x]}{[M^{n+}]}$$

$$= K_M \cdot \Phi_x \cdot \left(\frac{[RL]}{[L^-]}\right)^{n-x} \quad (113)$$

K_M is called the selectivity coefficient for the anions ML^{n-x} and L^-. It can be described as:

$$K_M = \frac{[R_{n-x} \cdot ML_x] \cdot [L^-]^{n-x}}{[RL]^{n-x} \cdot [ML_x^{n-x}]} \quad (114)$$

Φ_x is the molar portion of the complex ML_x^{n-x} in solution, which may be calculated via Eq. (115):

$$\Phi_x = \frac{[L^-] \cdot \beta_{ML_x}}{1 + [L^-] \cdot \beta_{ML} + [L^-]^2 \cdot \beta_{ML_2} + \cdots} \tag{115}$$

Under normal chromatographic conditions, however, it holds

$$[M^{n+}] \ll [L^-]$$

so that the concentration of the free ligand ions $[L^-]$ may be substituted by the total ligand concentration $[L_t]$. When [RL] is approximated via the ion-exchange capacity C, one obtains for D_M

$$D_M = K_M \, \Phi_x \, C^{n-x} \, [L^-]^{n-x} \tag{116}$$

or in a logarithmic form

$$\log D_M = \log K_M + \log \Phi_x + (n-x) \log C + (n-x) \log [L^-] \tag{117}$$

In cation exchange chromatography of transition metals, an increase in the concentration of the complexing ligand reduces the resolution. In contrast, in anion exchange chromatography, an optimal separation is obtained at a ligand concentration in which predominantly neutral complexes prevail. If the concentration is further raised, the retention also increases because of the increased formation of anionic complexes. This effect is partly compensated, however, because the free anionic ligands are responsible for the elution of anionic metal complexes. Excellent separations were accomplished on pellicular anion exchangers using an oxalic acid eluant [29].

When selecting a ligand for the separation and elution of transition metal ions from *cation exchangers* the following guidelines should be taken into account:
- Metal ions and ligands must form neutral or anionic complexes.
- Different complex formation constants for the various metals increase the selectivity.
- The transition metal complexes being formed should be thermodynamically stable and kinetically labile; that is, the complex should have a high energetic and entropic formation tendency and the thermodynamic equilibrium should be established quickly and without restriction. This requirement is discussed in the following examples:
  a) If a transition metal complex has a high formation constant but is kinetically stable, then the cation exchange process is impossible; that is, the anionic complex is not retained due to Donnan exclusion.
  b) If the formation constant is small but the complex is kinetically unstable, the retention mechanism is dominated by cation exchange. Correspondingly long retention times could result.

c) If the formation constant is small and the complex is kinetically stable, the chromatographic signals are often tainted by a strong tailing.
- If the transition metal complexes are detected by post-column derivatization with a suitable metallochromic indicator, the constant of the formation reaction should be high. Also, the indicator complex, MeIn, should be kinetically stable.

$$\text{MeL}_x + \text{In} \rightleftharpoons \text{MeIn} + x \cdot \text{L} \tag{118}$$

Me   Metal
L   Ligand
In   Indicator

This also applies for the separation of anionic transition metal complexes on anion exchangers. Here, the ligand molecules should be relatively small, so that the resulting complex is similar in size to the hydrated metal ion. In metal ions of the same valency that share the same coordination number and geometry with any given ligand, large ligands cause a loss in efficiency if the formation constants of these anionic transition metal complexes do not differ significantly.

4.5.2
Transition Metal Analysis with Non-Suppressed Conductivity Detection

While both conductivity detection modes are suitable for the detection of alkali- and alkaline-earth metals, only non-suppressed conductivity detection can be used for the analysis of transition metals. Upon application of the suppressor technique, transition metals would be mostly transformed by the suppressor reaction into the insoluble hydroxides. Direct conductivity detection is also impeded by the presence of complexing agents in the mobile phase; these agents are required for separating transition metals. In 1983, Sevenich and Fritz [30] found an eluant mixture suited for direct conductivity detection. It is comprised of ethylenediamine and tartaric acid. Ethylenediamine serves as an eluant ion, because it is already fully protonated (EnH_2^{2+}) at pH values ≤ 5. The selection of tartrate as a complexing anion is based on the final degree of complexation of the metal ions to be analyzed. The metal ions should be only partly complexed by the selected ligand, because a complete complexation would result in a loss of selectivity. In the example in Fig. 4-47, both eluant components are employed in almost equimolar concentrations for the separation of various divalent metal ions on a surface-sulfonated 20-µm cation exchanger. Again, the pH value of the mobile phase is the experimental parameter determining retention. It primarily affects the complexing properties of tartrate. Its complexing ability increases with increasing pH, which shifts the equilibrium to the right and, consequently decreases retention.

4.5 Transition Metal Analysis

$$\text{Resin} - M^{2+} + EnH_2^{2+} \rightleftarrows \text{Resin} - EnH_2^{2+} + M^{2+}$$

$$\updownarrow L^{2-} \qquad (119)$$

$$ML$$

Besides tartrate, α-hydroxyisobutyrate can be used as a complexing anion. It was employed by Cassidy et al. [26] for the separation of lanthanides. In comparison to tartrate, no meaningful advantages for the separation of divalent metal ions are revealed.

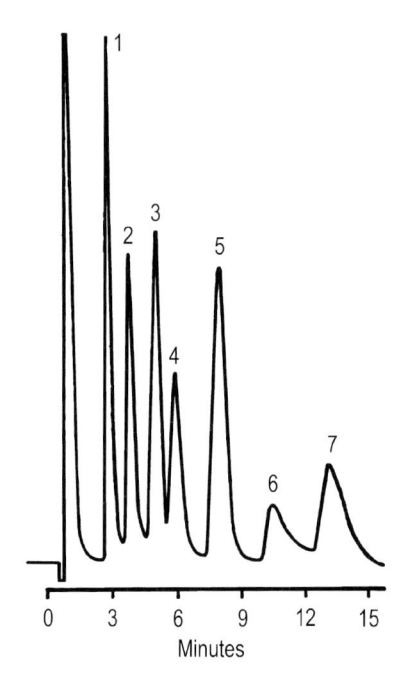

Fig. 4-47. Separation of various divalent cations utilizing direct conductivity detection. — Separator column: surface-sulfonated cation exchanger (Benson Co., Reno, USA); eluant: 1.5 mmol/L ethylenediamine + 2 mmol/L tartaric acid, pH 4; flow rate: 0.85 mL/min; injection volume: 100 µL; solute concentrations: 10.3 mg/L Zn^{2+} (1), 9.1 mg/L Co^{2+} (2), 16 mg/L Mn^{2+} (3), 16.1 mg/L Cd^{2+} (4), 17.1 mg/L Ca^{2+} (5), 16 mg/L Pb^{2+} (6), and 20.3 mg/L Sr^{2+} (7); (taken from [30]).

In most applications, the method of cation exchange with direct conductivity detection does not have the required specificity for the analysis of transition metals, because cationic matrix components that are often present in high amounts are also detected.

4.5.3 Transition Metal Analysis with Spectrophotometric Detection

According to Fig. 4-48, both cation and anion exchange processes contribute to the separation of transition metals with complexing agents in the mobile phase. Hence, before the introduction of ion exchangers with defined anion and cation exchange capacities, surface-sulfonated cation exchangers as well as latexed anion exchangers were used; these two types exhibit totally different selectivities [31, 32].

$$M^{n+} + C^{x-} \rightleftharpoons MC^{n-x} + MC_2^{n-2x} + \ldots$$

Anion exchange / Cation exchange

Resin–$SO_3^-M^{n+}$

or

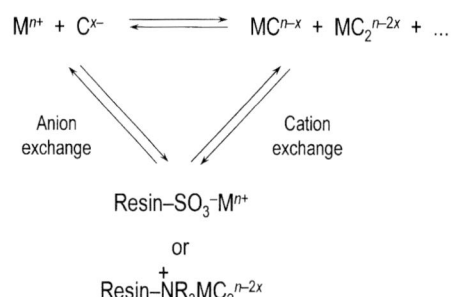

Resin–$\overset{+}{N}R_3MC_2^{n-2x}$

Fig. 4-48. Overview of the equilibria involved in transition metal separations.

The aggravating problem of separating transition metals on surface-sulfonated cation exchangers, typically carried out with an eluant mixture of oxalic acid and citric acid, was always the lack of resolution between iron(III), copper, and nickel, which are not very well retained under these chromatographic conditions. The complete exclusion of iron(III) can be attributed to the formation of $Fe(Ox)_3^{3-}$ which is kinetically very stable. Due to the unusually high stability of this trivalent complex (log K = 18.5 [33]) it does not exhibit any interaction with the stationary phase and, consequently, elutes in the void volume. Moreover, adjusting the sample pH to pH 1.5 proved to be of crucial importance in accomplishing acceptable separations between copper and nickel. On the other hand, cadmium and manganese are strongly retained with an oxalic/citric acid eluant mixture, but may be eluted more rapidly with a mixture of citric acid and tartaric acid. In any case, detection is carried out via post-column derivatization of the column effluent with 4-(2-pyridylazo)resorcinol (PAR) and subsequent photometric detection of the resulting chelate complexes at 520 nm (see Section 7.2.1).

Significantly better separations between iron(III), copper, and nickel are accomplished on a polymethacrylate-based cation exchanger using a pure tartaric acid eluant. As can be seen from the respective chromatogram in Fig. 4-49, iron(III) still elutes very close to the void volume, which makes it difficult to quantify this signal. The addition of ZnEDTA to the PAR reagent provides a big advantage, because it enables the simultaneous detection of alkaline-earth metals [34]. This is accomplished via the release of zinc ions from the ZnEDTA complex that results from the substitution reaction with alkaline-earth metals, whose EDTA complexes are characterized by a higher stability. The zinc ions, in turn, form chelate complexes with PAR, which can then be detected spectrophotometrically in the usual way.

The introduction of the IonPac CS5 separator column [28] in the mid-1980s finally remedied the problems associated with the determination of iron(III). The CS5 column is a latexed anion exchanger with mixed anion *and* cation

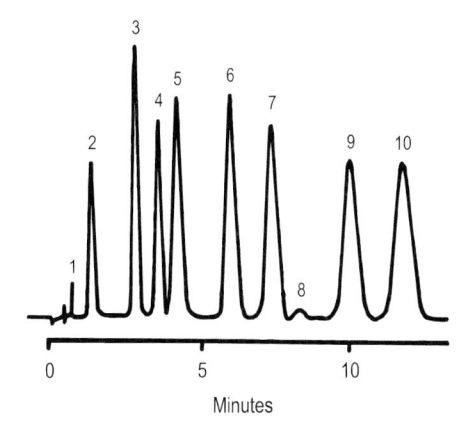

Fig. 4-49. Separation of transition metals on a polymethacrylate-based cation exchanger. — Separator column: Sykam LCA K02; eluant: 0.1 mol/L tartaric acid, pH 2.95 with NaOH; flow rate: 2 mL/min; detection: photometry at 500 nm after reaction with PAR and ZnEDTA; injection volume: 100 µL; solute concentrations: 2 mg/L Fe^{3+} (1), 2 mg/L Cu^{2+} (2), 4 mg/L Pb^{2+} (3), 1 mg/L Zn^{2+} (4), 2 mg/L Ni^{2+} (5), 2 mg/L Co^{2+} (6), 4 mg/L Cd^{2+} (7), 1.8 mg/L Fe^{2+} (8), 1 mg/L Ca^{2+} (9), and 1 mg/L Mg^{2+} (10).

exchange capacity in the dimensions 250 mm × 4 mm i.d. and a surface-sulfonated microporous PS/DVB substrate with a particle size of 13 µm. The electrostatically agglomerated, totally aminated latex beads have a diameter of 150 nm and a degree of cross-linking of about 2%. Because not all sulfonate groups on the substrate surface carry positively charged latex beads, a certain cation exchange capacity results, which can be defined by the degree of latex coverage on the substrate surface. For the first time, pyridine-2,6-dicarboxylic acid (PDCA) has been used as a complexing agent in the mobile phase. As revealed by the respective chromatogram in Fig. 4-50, iron(III) is more strongly retained when using PDCA than when using an oxalic/citric acid eluant mixture. The retention behavior of the other metal ions is characterized by differences in the geometry, charge density, and hydrophobicity of the metal-PDCA complexes, the structure of which are illustrated in Fig. 4-51. The limited resolution between copper, nickel, cobalt, and zinc is a drawback and cannot be improved very much, even under optimized chromatographic conditions. Anion exchange is the dominating separation mechanism, because the stability constants of the metal-PDCA complexes are very high (see Table 4-4). Lead elutes under these chromatographic conditions, but cannot be detected because the lead-PDCA complex is more stable than the corresponding chelating complex with the PAR reagent. In comparison with surface-sulfonated cation exchangers, this stationary phase exhibits a higher chromatographic efficiency; this can be solely attributed to the smaller particle diameter of the substrate material.

A reversal of the elution order for metals such as lead, cobalt, zinc, and nickel is obtained with the use of oxalic acid as a single complexing agent. Under these chromatographic conditions, the metal separation is controlled by anion *and* cation exchange processes. The degree to which each mechanism contributes to the separation process depends on the stability of the oxalate complexes (see Table 4-5) and is, therefore, different for each metal ion. The anion exchange

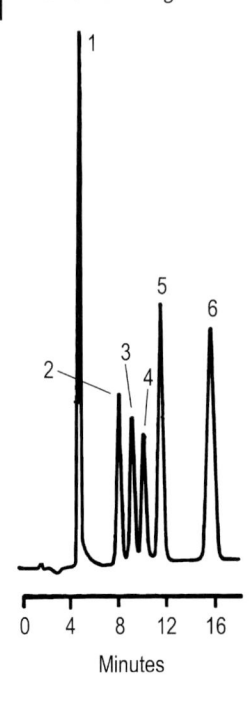

Fig. 4-50. Separation of transition metals on an ion exchanger with both anion *and* cation exchange capacity. — Separator column: IonPac CS5; eluant: 6 mmol/L pyridine-2,6-dicarboxylic acid, pH 4.8 with LiOH; flow rate: 1 mL/min; detection: photometry at 520 nm after derivatization with PAR; injection volume: 50 µL; solute concentrations: 1 mg/L Fe^{3+} (1), 1 mg/L Cu^{2+} (2), 3 mg/L Ni^{2+} (3), 4 mg/L Zn^{2+} (4), 2 mg/L Co^{2+} (5), and 3 mg/L Fe^{2+} (6).

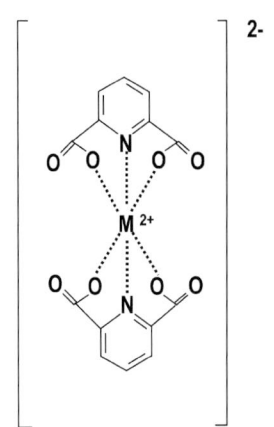

Fig. 4-51. Schematic representation of the structure of metal-PDCA complexes.

mechanism predominates where stable anionic oxalate complexes are formed. In metal ions which do not form stable oxalate complexes, the cation exchange mechanism is favored. A corresponding separation is displayed in Fig. 4-52. Under these chromatographic conditions, iron(II) and iron(III) cannot be eluted. The high affinity of iron(III) towards the stationary phase is attributed to the formation of the stable $Fe(Ox)_3^{3-}$ complex, which cannot be eluted by the divalent oxalate.

Table 4-4. Stability constants of various metal-PDCA complexes.

Metal ion	Complex	log K
Fe^{3+}	$Fe(PDCA)_2^-$	17.1
Cu^{2+}	$Cu(PDCA)_2^{2-}$	16.5
Ni^{2+}	$Ni(PDCA)_2^{2-}$	13.5
Zn^{2+}	$Zn(PDCA)_2^{2-}$	11.8
Co^{2+}	$Co(PDCA)_2^{2-}$	12.7
Cd^{2+}	$Cd(PDCA)_2^{2-}$	11.2
Mn^{2+}	$Mn(PDCA)_2^{2-}$	8.5
Fe^{2+}	$Fe(PDCA)_2^{2-}$	10.4

Table 4-5. Stability constants of various metal-oxalate complexes.

Metal ion	Complex	log K
Pb^{2+}	$Pb(Ox)_2^{2-}$	5.8
Cu^{2+}	$Cu(Ox)_2^{2-}$	9.2
Cd^{2+}	$Cd(Ox)_2^{2-}$	5.8
Mn^{2+}	$Mn(Ox)_2^{2-}$	4.4
Co^{2+}	$Co(Ox)_2^{2-}$	5.6
Zn^{2+}	$Zn(Ox)_2^{2-}$	6.4
Ni^{2+}	$Ni(Ox)_2^{2-}$	7.6

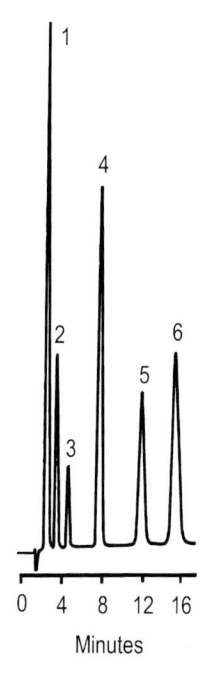

Fig. 4-52. Separation of some selected transition metals with oxalic acid as a complexing agent. — Separator column: IonPac CS5; eluant: 50 mmol/L oxalic acid, pH 4.8 with LiOH; flow rate, detection and injection volume: see Fig. 4-50; solute concentrations: 4 mg/L Pb^{2+} (1), 0.5 mg/L Cu^{2+} (2), 4 mg/L Cd^{2+} (3), 2 mg/L Co^{2+} (4), 2 mg/L Zn^{2+} (5), and 4 mg/L Ni^{2+} (6).

For a long time, post-column derivatization with PAR was limited to the analysis of iron, cobalt, nickel, copper, cadmium, manganese, zinc, lead, uranium, and lanthanides, primarily due to the kinetics of complex formation with PAR, which is hindered by the complexing agents present in the mobile phase. Recent investigations have revealed that the ligand exchange between the transport complex and PAR is aided by the adding of ligands such as phosphate, carbonate, or alkylamines to the eluant. In this way, transition metals that are masked by the presence of complexing agents in the eluant can also be detected sensitively with PAR.

Another limiting factor of complexation with PAR is the pH value of the reagent. As the pH value of the reagent increases, so does the PAR dissociation, which is necessary for the complexation of metals. However, at higher reagent pH values, metal ions hydrolyze, which inhibits their complexation with PAR. By lowering the pH value of the PAR reagent, transition metals sensitive to hydrolysis can be detected. On the basis of these new insights, the composition of the mobile phase was modified. It is now possible to determine nine different metals within approximately 15 minutes (Fig. 4-53). If the pH value of the PAR reagent is lowered to pH 8.8 by adding a phosphate buffer, gallium(III) and vanadium(IV)/(V) can be detected; they can be separated from other transition metals with the PDCA eluant [35]. Figure 4-54 shows the separation of vanadium(V) in the form of ammonium(meta)vanadate, NH_4VO_3. Under these conditions, vanadium(IV) as vanadium sulfate ($VOSO_4$) elutes after about 17 minutes, so that it is easy to distinguish between the two most important oxidation states of vanadium. Gallium(III), of particular importance for the semiconductor industry, elutes near the void volume.

The simultaneous analysis of chromium(III)/(VI) is still a difficult task. Numerous ion chromatographic methodologies for this analytical problem were described in the past, but all of them turned out to be rather impractical. Although chromium(III) forms a chelate complex with PDCA, its ligand-exchange kinetics is too slow for post-column derivatization. The method once described by Heberling [36], namely the separation of chromium(III) as $Cr(PDCA)_2^-$ complex and chromium(VI) as chromate, CrO_4^{2-}, on a suitable anion exchanger, suffered from exactly that problem. To accelerate the ligand-exchange process, a pre-column derivatization with PDCA at elevated temperature was suggested. Detection was carried out photometrically after derivatization of the column effluent with 1,5-diphenylcarbazide (DPC) at 520 nm (see Section 7.2.1). With this detection method, chromate can be analyzed very sensitively, while the chromium(III)-PDCA complex is detected directly at 520 nm. The method described by Heberling works very well with standards, but difficulties were reported with pre-column derivatization of chromium(III) in real-world samples from the electroplating industry. While this method does not appear to be suited for the simultaneous analysis of chromium(III) and chromium(VI), it is unmatched for trace analysis of chromium(VI) in pure chromium(III) utilizing post-column derivatization with 1,5-DPC (no pre-column derivatization with PDCA). Based on an injection volume of 50 μL, it is possible to determine 20 μg/L chromium (VI) without any problem. Eluant pH should be kept neutral to ensure that all chromium (VI) exists in the chromate form.

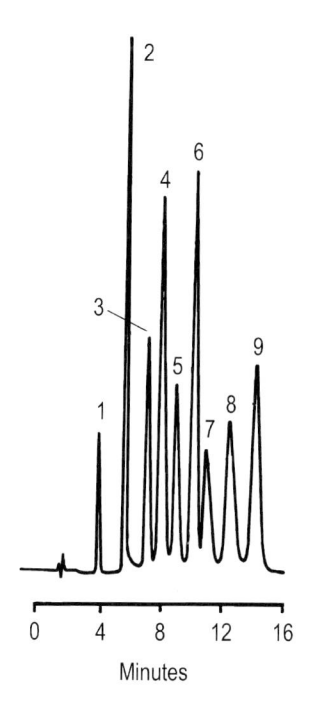

Fig. 4-53. Simultaneous analysis of nine different transition metals. − Separator column: IonPac CS5; eluant: 4 mmol/L pyridine-2,6-dicarboxylic acid + 15 mmol/L NaCl, pH 4.8 with LiOH; flow rate, detection and injection volume: see Fig. 4-50; solute concentrations: 10 mg/L Pb^{2+} (1), 1 mg/L each of Fe^{3+} (2), Cu^{2+} (3), Ni^{2+} (4), Zn^{2+} (5), and Co^{2+} (6), 3 mg/L Cd^{2+} (7), 1 mg/L Mn^{2+} (8), and 1 mg/L Fe^{2+} (9).

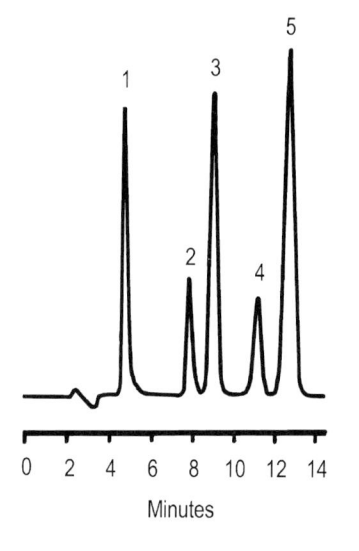

Fig. 4-54. Analysis of vanadium(V). − Separator column: IonPac CS5; eluant: 6 mmol/L pyridine-2,6-dicarboxylic acid, pH 4.8 with LiOH; flow rate: 1 mL/min; detection: photometry at 520 nm after reaction with PAR in a phosphate buffer; injection volume: 50 μL; solute concentrations: 0.5 mg/L Fe^{3+} (1), 10 mg/L V^{5+} (as VO_3^-) (2), 2 mg/L Cu^{2+} (3), 2 mg/L Zn^{2+} (4), and 2 mg/L Co^{2+} (5).

The best ion chromatographic method for the simultaneous analysis of chromium(III) and chromium(VI) at present is based — according to Somerset [37] — on the ion-exchange separation of both species with post-chromatographic oxidation of chromium(III) to chromium(VI), which can then be detected photometrically at 365 nm. The oxidation is carried out with potassium peroxodisulfate, using silver nitrate as a catalyst. Complete conversion is achieved within a short time by raising the reaction temperature to 80 °C.

The limited resolution between copper, nickel, cobalt, and zinc that was discussed in connection with transition metal separations on IonPac CS5 (see Fig. 4-50) can be solved by optimizing the stationary phase. In contrast to the IonPac CS5, the IonPac CS5A commercialized in 1996 is a bifunctional latexed cation exchanger. The separate steps in the manufacturing of such a stationary phase are illustrated in Fig. 4-55. A 55% cross-linked microporous DVB substrate with a particle diameter of 9 µm serves as a starting material. As in carboxylate-based weak acid cation exchangers (see Section 4.1.1), the substrate is coated at the surface with an anion exchange polymer. This positively charged surface is then electrostatically agglomerated with 10% cross-linked sulfonated latex beads with a particle diameter of 140 nm. The third layer consists of 2% cross-linked aminated latex beads with a particle diameter of 76 nm.

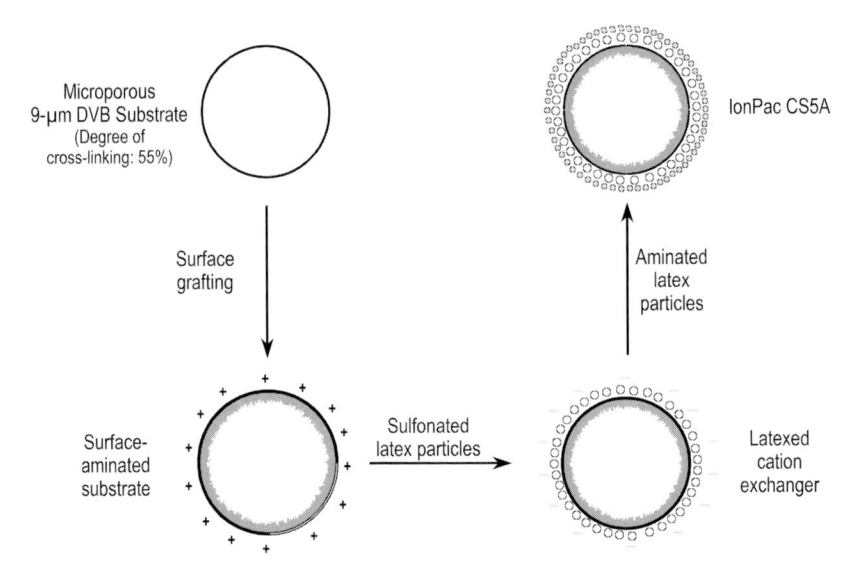

Fig. 4-55. Schematic representation of the separate steps in the manufacturing of the IonPac CS5A bifunctional latexed cation exchanger.

The finished ion exchanger (250 mm × 4 mm i.d.) has a cation exchange capacity of 20 µequiv and an anion exchange capacity of 40 µequiv. Due to the high degree of cross-linking of the substrate material, typical HPLC solvents up to 50% (v/v) can be added to the mobile phase. The pressure stability of the separator column was determined to be 17.3 MPa (2500 psi). In addition to optimizing the stationary phase, eluants have also been optimized to improve the selectivity, buffer capacity, and total analysis time. In addition, the PAR reagent formulation used so far (0.12 g/L PAR in 3 mol/L NH_4OH + 1 mol/L CH_3COOH) has also been changed to increase the solubility, reactivity, and stability of the PAR reagent. The new formulation consists of

0.12 g/L PAR (for metal ion concentrations > 0.5 mg/L) or
0.06 g/L PAR (for metal ion concentrations < 0.5 mg/L) in
1 mol/L 2-Dimethylaminoethanol +
0.5 mol/L NH_4OH +
0.3 mol/L $NaHCO_3$

and has a pH of 10.4. Figure 4-56 illustrates the separation of eight different transition metals with pyridine-2,6-dicarboxylic acid as a complexing agent. In comparison with the corresponding separation on the IonPac CS5, the resolution between metal ions was increased significantly while the total analysis time was shortened by about three minutes. A similar result was achieved for the separation of transition metals with oxalic acid as a single complexing agent; the separation is performed in a shorter period of time than on IonPac CS5 without sacrificing separation power (Fig. 4-57).

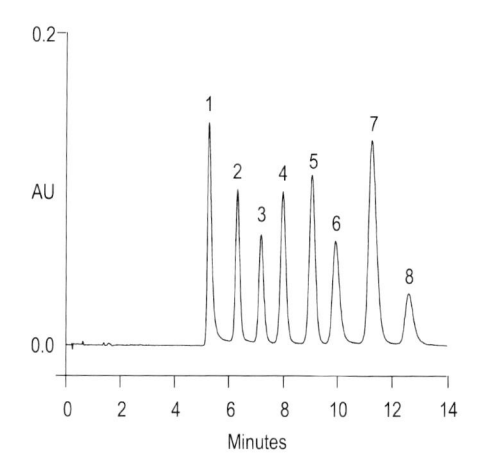

Fig. 4-56. Separation of transition metals on the IonPac CS5A bifunctional latexed cation exchanger. — Eluant: 7 mmol/L pyridine-2,6-dicarboxylic acid + 66 mmol/L KOH + 74 mmol/L formic acid + 56 mmol/L K_2SO_4; flow rate: 1.2 mL/min; detection: photometry at 520 nm after reaction with PAR; injection volume: 50 µL; solute concentrations: 1.3 mg/L Fe^{3+} (1), 1.3 mg/L Cu^{2+} (2), 2.6 mg/L Ni^{2+} (3), 1.3 mg/L Zn^{2+} (4), 1.3 mg/L Co^{2+} (5), 6 mg/L Cd^{2+} (6), 2.6 mg/L Mn^{2+} (7), and 1.3 mg/L Fe^{2+} (8).

Cation exchange chromatography in combination with a post-column derivatization is also suitable for analyzing uranium and thorium [38]. Because they are actinides and in close relation with lanthanides, they are always associated with them. Due to the lack of simple and sensitive colorimetric methods, the trace analysis of these elements represents a challenge. Although alternative analytical methods for these elements such as neutron activation analysis [39] and ICP-MS [40] are described in literature, they are not suitable for routine analysis

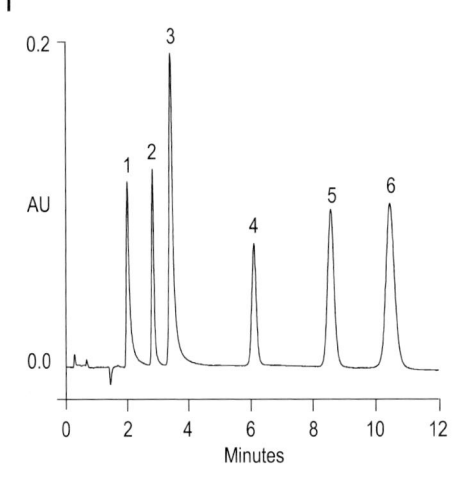

Fig. 4-57. Separation of transition metals on the IonPac CS5A bifunctional latexed cation exchanger with oxalic acid as a complexing agent. – Eluant: 80 mmol/L oxalic acid + 50 mmol/L KOH + 0.1 mol/L TMAOH; flow rate: 1.2 mL/min; detection: photometry at 520 nm after reaction with PAR; injection volume: 50 µL; solute concentrations: 5 mg/L Pb^{2+} (1), 0.7 mg/L Cu^{2+} (2), 3.3 mg/L Cd^{2+} (3), 0.7 mg/L Co^{2+} (4), 1.3 mg/L Zn^{2+} (5), and 2 mg/L Ni^{2+} (6).

and are prone to interferences. Major interferences include alkali metals, alkaline-earth metals, and iron, which in real-world samples (e.g., digested ores and minerals) differ in concentration by several orders of magnitude. Cation exchange chromatography of uranium and thorium is hampered by their widely different affinities towards the stationary phase. While uranium usually exists as a divalent cation, UO_2^{2+}, thorium is tetravalent. Thus, uranium cations can be eluted from a cation exchanger with relatively low concentrations of a mineral acid such as hydrochloric acid or nitric acid, whereas an acid concentration of $c > 3$ mol/L is required for the elution of thorium. In principle, gradient elution is an option, but fails because of the pH dependence of the background absorbance of the reagent suitable for post-column derivatization. However, the affinity of thorium towards the stationary phase can be decreased with sulfate, due to the formation of a thorium-sulfate complex, $ThSO_4^{2+}$. Thus, both elements can be eluted within ten minutes with a sulfate gradient, keeping the acid concentration in the mobile phase constant. Also, both elements are separated from potential interferences such as Fe^{3+}, Ca^{2+}, Hf^{4+}, ZrO^{3+}, and lanthanides. Because the separation is carried out at constant pH, the baseline remains stable during the gradient run. Uranium and thorium can be determined down to the µg/L range, using post-column derivatization with subsequent spectrophotometry of the metal complexes that formed. Arsenazo III has proved to be a suitable metallochromic reagent [41-43]. It is specifically employed as reagent for the detection of thorium and uranium, because it forms colored complexes with a number of metals. Iron, calcium, hafnium, zirconium, and lanthanides, to name the most important elements, can also be detected with Arsenazo III. When an ion chromatographic separation is performed prior to detection, such interferences are mostly eliminated. It is advantageous that the selectivity of Arsenazo III increases with decreasing pH, and that in an acidic milieu, the number of metals that form stable complexes with Arsenazo III also decreases. Therefore, iron and calcium, which in real samples are present at much higher concentrations relative to uranium and thorium, do not represent an insuperable interference because of their lower response factors in an acidic pH range. Metal complexes formed with Arsenazo III absorb at $\lambda = 660$ nm, whereas free Arsenazo III only weakly absorbs at this wavelength, especially at low pH [44]. In addition to Arsenazo III, the derivatization reagent contains acetic acid and Triton X-100 as a nonionic surfactant. Both components stabilize the reagent and help to avoid adsorption effects inside the reaction loop. The chromatogram in Fig. 4-58 shows the separation of uranium and thorium on IonPac CS2.

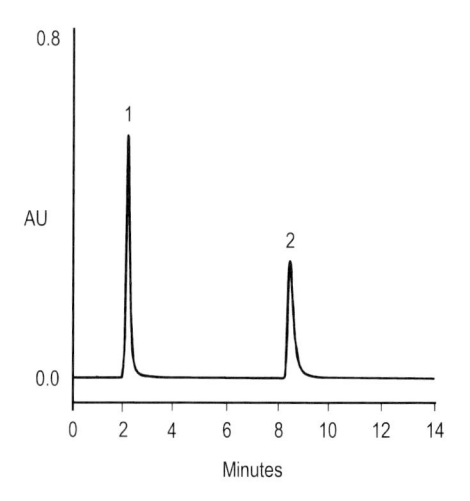

Fig. 4-58. Separation of uranium and thorium on IonPac CS2. — Eluent: (A) 0.6 mol/L HCl, (B) 0.6 mol/L HCl + 0.5 mol/L Na_2SO_4; gradient: 100% A to 100% B in 15 min; flow rate: 1 mL/min; detection: photometry at 660 nm after derivatization with Arsenazo III; injection volume: 50 µL; solute concentrations: 40 mg/L uranium (as UO_2^{2+}) (1) and 20 mg/L thorium (2).

Alternatively, uranyl cations can also be separated on an IonPac CS3 latexed cation exchanger utilizing a buffer mixture of ammonium sulfate and sulfuric acid as an eluent. Detection is performed as it is for transition metals, via derivatization with PAR.

The buffer mixture of ammonium sulfate and sulfuric acid can also be employed for the elution of aluminum. The specific detection of aluminum is carried out via derivatization with Tiron (4,5-dihydroxy-1,3-benzenedisulfonic acid disodium salt), with subsequent photometry of the complex formed at a wavelength of 313 nm (Fig. 4-59). In an acidic sulfate medium, aluminum ions may exist as $AlSO_4^+$ ions (depending on the hydrogen ion and sulfate ion concentration). In this case, the retention-determining parameter is the ammonium sulfate concentration.

Transition metals may also be separated via ion-pair chromatography on macroporous PS/DVB resins or chemically bonded silica [45]. The mobile phase contains complexing agents and a corresponding ion-pair reagent. If these columns are equilibrated with a surface-active acid such as octanesulfonic acid, metal ions such as copper, nickel, zinc, and cobalt elute in the same order as they do on surface-sulfonated cation exchangers. This suggests that in this case,

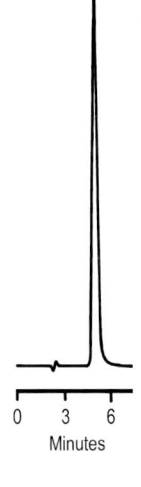

Fig. 4-59. Separation of aluminum on a latexed cation exchanger. — Separator column: IonPac CS3; eluent: 0.2 mol/L $(NH_4)_2SO_4$ + 0.01 mol/L H_2SO_4; flow rate: 1 mL/min; detection: photometry at 313 nm after derivatization with Tiron; injection volume: 50 µL; solute concentration: 5 mg/L Al^{3+} (1).

the ion-exchange process is the dominating separation mechanism followed by ion-pairing. The separation results obtained with MPIC phases are clearly inferior to those obtained with modern ion exchangers.

The methods used in anion exchange chromatography for transition metal analysis can also be applied to ion-pair chromatography. With oxalic acid as a complexing agent and tetrabutylammonium hydroxide as an ion-pair reagent, the elution order is opposite to that obtained in cation exchange analysis. The separation of anionic metal complexes on MPIC phases occurs exclusively via anion exchange processes. Because of the process by which it is manufactured, the IonPac CS5A ion exchanger features anion exchange and cation exchange capacities. Due to the separation mechanisms being involved, this characteristic is of decisive importance for the selective separation of transition metals.

Analysis of Lanthanides
The separation of lanthanides is of great importance for many industries. In the nuclear power industry, for instance, lanthanide fission products indicate how many nuclear fuel rods are already burned off. In the mining industry, the trace analysis of lanthanides supports the prospecting of lanthanide deposits. Moreover, it contributes to the understanding of a number of geological processes. Because the lanthanide concentration in rock samples remains constant over time, geological correlations can be investigated via "fingerprint" analyses.

In the past, lanthanides were almost exclusively analyzed with element-specific methods such as AAS and ICP. However, spectral interferences are observed in trace analysis of lanthanides in geological samples, so that time-consuming extractions or group separations on ion exchangers have to be carried out first. In many cases, pre-concentrations are necessary, because the detection limits for lanthanides obtained with AAS and ICP are only in the lower mg/L to sub-mg/L range. For this reason, traces of lanthanides in geological samples are now analyzed in the United States by means of neutron activation analysis (NAA).

As an alternative to the above-mentioned analytical methods, lanthanides can also be analyzed by anion or cation exchange chromatography. In their pioneering 1979 paper, Cassidy et al. [26] presented successful separations on silica-based cation exchangers. In analogy to transition metal analysis, they used an organic complexing agent such as α-hydroxyisobutyric acid (HIBA) as the eluant to enhance the selectivity of the separation. As trivalent cations, lanthanides are strongly hydrated in aqueous solution and differ very little in the physical-chemical properties that are of importance in the separation. Therefore, conventional methods of cation exchange chromatography do not furnish satisfactory results. However, the various lanthanides differ in their complexing behavior, so that gradient elution is indispensable for fast and efficient separations. Analysis times of less than 15 minutes may be realized, for example, on a modern IonPac CS5A latexed cation exchanger by using a linear concentration gradient based on oxalic acid and diglycolic acid. As can be seen from the corresponding chromatogram in Fig. 4-60, the elution starts with lanthanum and ends with lutecium; however, lutecium is not completely separated from ytterbium. Detection limits in the lower µg/L range are realized by means of photometric detection after reaction with PAR. Similarly, good separations were accomplished by Knight and Cassidy [46] on dynamically coated ODS phases. They employed this procedure for the analysis of rare-earth elements in burned-up nuclear fuel rods.

Lanthanides may be eluted in reverse order with α-hydroxyisobutyric acid on a cation exchanger. In this case, linear composition gradients are usually applied.

When determining lanthanides in geological samples, co-elution problems occur with some transition metals, because transition metals are present in much higher concentrations. For an improved separation between these classes of compounds the different complexing behavior of transition metals and rare-earth elements with pyridine-2,6-dicarboxylic acid [47] can be exploited. While transition metals form stable monovalent and divalent complexes with PDCA, the lanthanides complexes with PDCA are trivalent:

$$M^{3+} + 2\ PDCA^{2-} \rightarrow [M(PDCA)_2]^- \tag{120}$$
$$M^{3+}\quad Fe^{3+},\ Ga^{3+},\ Cr^{3+}$$

$$M^{2+} + 2\ PDCA^{2-} \rightarrow [M(PDCA)_2]^{2-} \tag{121}$$
$$M^{2+}\quad Cu^{2+},\ Ni^{2+},\ Zn^{2+}$$

$$L^{3+} + 3\ PDCA^{2-} \rightarrow [L(PDCA)_3]^{3-} \tag{122}$$
$$L^{3+}\quad La^{3+},\ Ce^{3+},\ Pr^{3+}$$

Based on this charge difference, transition metals can be eluted with PDCA, while lanthanides are retained at the beginning of the column. After the transition metals are completely eluted, the second step switches (as described above) to a mixture of oxalic and diglycolic acid, which elutes the rare-earth elements. After careful optimization of this technique, it is possible to analyze transition metals and lanthanides in the same run (see Fig. 4-61). Watkins et al. [48] successfully applied this method for the analysis of silicate rocks.

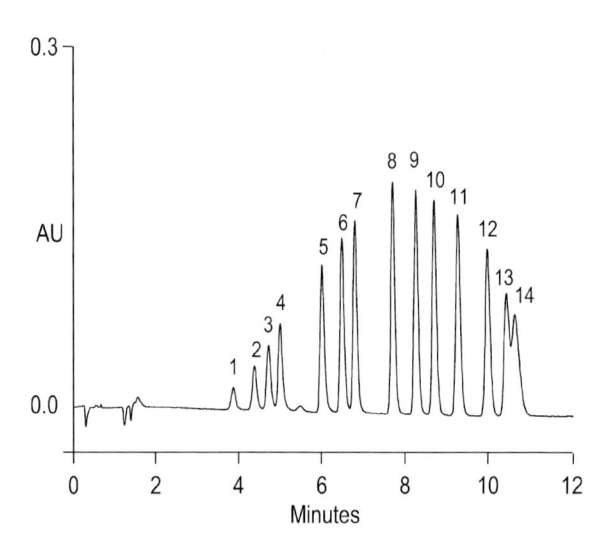

Fig. 4-60. Separation of lanthanides on IonPac CS5A. — Eluant: oxalic acid / diglycolic acid gradient; flow rate: 1.2 mL/min; detection: photometry at 520 nm after reaction with PAR; injection volume: 50 µL; solute concentrations: 5 mg/L each of lanthanum (1) cerium (2), praseodymium (3), neodymium (4), samarium (5), europium (6), gadolinium (7), terbium (8), dysprosium (9), holmium (10), erbium (11), thulium (12), ytterbium (13), and lutecium (14).

4.6 Analysis of Polyamines

In the past, the separation of polyamines was a big problem. These compounds exhibit a very high affinity towards the stationary phase of a conventional cation exchanger and, thus, can only be eluted with high ionic strength eluants. Totally sulfonated cation exchange materials are not suitable for analyzing polyamines because they would shrink in the high-ionic strength environment which, in turn, would lead to a drastic loss in the separation efficiency. For this reason, only pellicular cation exchangers are used for the analysis of polyamines.

The first successful polyamine separations were accomplished in the mid 1970s after the introduction of surface-sulfonated cation exchangers. Figure 4-62 displays a standard chromatogram with the separation of putrescine (1,4-diaminobutane) and cadaverine (1,5-diaminopentane) as well as spermidine (*N*-(3-aminopropyl-1,4-diaminobutane) and spermine (*N*, *N'*-Bis-(3-aminopropyl)-1,4-diaminobutane). The required eluant was comprised of several buffer solutions. Their composition is listed in Table 4-6. Owing to the high ionic

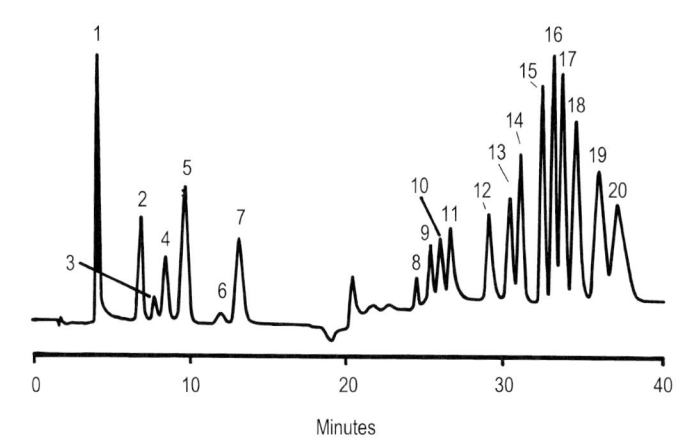

Fig. 4-61. Simultaneous analysis of transition metals and lanthanides. — Separator column: IonPac CS5; eluant: (A) water, (B) 6 mmol/L PDCA + 50 mmol/L NaOAc + 50 mmol/L HOAc, (C) 0.1 mol/L oxalic aicd + 0.19 mol/L LiOH, (D) 0.1 mol/L diglycolic acid + 0.19 mol/L LiOH; flow rate: 1 mL/min; detection: see Fig. 4-60; injection volume: 50 µL; solute concentrations: 2 mg/L Fe^{3+} (1), 1 mg/L Cu^{2+} (2), 3 mg/L Ni^{2+} (3), 4 mg/L Zn^{2+} (4), 2 mg/L Co^{2+} (5), 1 mg/L Mn^{2+} (6), 3 mg/L Fe^{2+} (7), and 7 mg/L each of La^{3+} (8), Ce^{3+} (9), Pr^{3+} (10), Nd^{3+} (11), Sm^{3+} (12), Eu^{3+} (13), Gd^{3+} (14), Tb^{3+} (15), Dy^{3+} (16), Ho^{3+} (17), Er^{3+} (18), Tm^{3+} (19), and Yb^{3+} (20); gradient program:

Time [min]	A [%]	B [%]	C [%]	D [%]
0	0	100	0	0
12	0	100	0	0
12.1	100	0	0	0
17	100	0	0	0
17.1	40	0	60	0
21	40	0	60	0
21.1	20	0	80	0
30	51	0	26	23

strength of these buffer solutions, detection cannot be carried out by means of conductivity measurements. Polyamines carry terminal NH_2 groups, thus, fluorescence detection after reaction with o-phthaldialdehyde is a suitable and very sensitive detection method.

Following the introduction of highly efficient latex-based cation exchangers such as the IonPac CS3, the eluant composition could be simplified considerably. Ethylenediamine, for example, can be eluted from this stationary phase within an acceptable time frame utilizing an eluant mixture of hydrochloric acid and 2,3-diaminopropionic acid (DAP) and subsequently detected via suppressed conductivity detection. Today, diamines can be analyzed together with alkali- and alkaline-earth metals using weak acid cation exchangers in combination with gradient elution techniques (see Section 4.7). Instead of divalent eluant ions

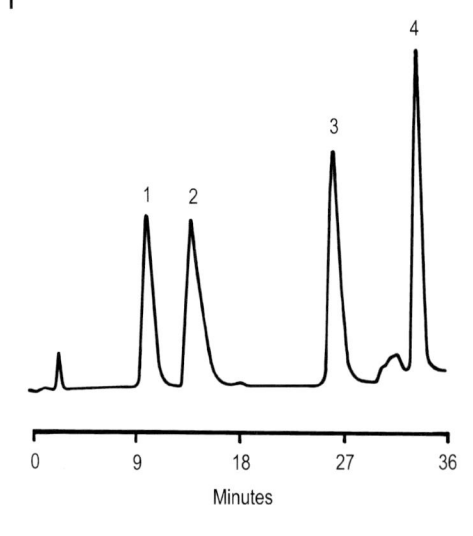

Fig. 4-62. Separation of various polyamines. — Separator column: IonPac CS1; eluant: see Table 4-6; flow rate: 0.6 mL/min; detection: fluorescence after reaction with o-phthaldialdehyde; injection volume: 20 µL; solute concentrations: 4.4 mg/L putrescine (1), 5.1 mg/L cadaverine (2), 7.3 mg/L spermidine (3), and 10.1 mg/L spermine (4).

Table 4-6. Eluant composition for the separation of polyamines on a surface-sulfonated cation exchanger.

Composition	Buffer type			
	A	B	C	Regenerant
pH value	5.80	5.55	5.55	–
Trisodium citrate [g/L]	0.98	1.97	7.84	–
Sodium chloride [g/L]	16.94	42.63	83.00	5.84
Phenol [g/L]	1.0	1.0	1.0	–
EDTA [g/L]	–	–	–	0.25
Sodium hydroxide [g/L]	–	–	–	4.01

such as DAP, monovalent methanesulfonic acid, with or without aprotic solvents such as acetonitrile, is used as an eluant. Figure 4-63 illustrates such a separation on IonPac CS14; the separation was obtained with a combined concentration and solvent gradient. As can be seen from this chromatogram, the retention of diamines increases with decreasing carbon chain length between the terminal amino groups. The observed tailing of the diamine peaks can be minimized by increasing column temperature to 30-40 °C.

The trend, however, in cation exchange stationary phase design is to develop weak acid cation exchange materials with such a hydrophilic surface that organic solvents in the mobile phase would become obsolete. An impressive example is the separation of biogenic amines such as putrescine, cadaverine, spermine, and spermidine together with alkali- and alkaline-earth metals on IonPac CS17 using an aqueous eluant without any organic solvent (Fig. 4-64). The biogenic amines elute with good peak efficiencies and symmetries when the column is operated with a simple acid gradient at elevated temperature. Using suppressed conductivity detection, these biogenic amines can easily be determined in complex food matrices. Alternatively, amperometric detection provides the advantage of high specificity for oxidizable amines, including the biogenic amines. Higher concentrations of inorganic cations do not interfere with their quantitation.

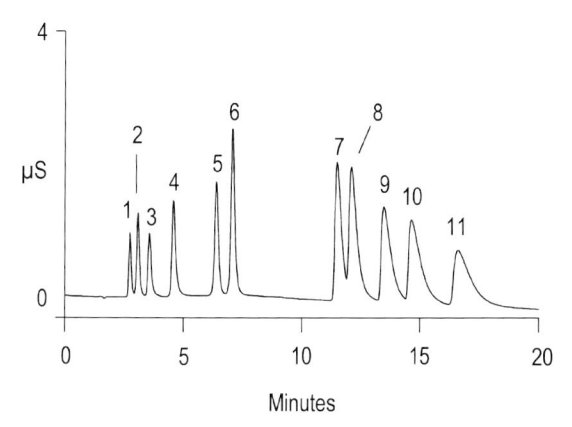

Fig. 4-63. Separation of diamines on IonPac CS14. — Column temperature: 23 °C; eluant: (A) 10 mmol/L methanesulfonic acid — acetonitrile (95:5 v/v), (B) 20 mmol/L methanesulfonic acid — acetonitrile (85:15 v/v); gradient: 100% A to 100% B in 10 min; flow rate: 1 mL/min; detection: suppressed conductivity; injection volume: 25 µL; solute concentrations: 0.5 mg/L lithium (1), 2 mg/L sodium (2), 2.5 mg/L ammonium (3), 5 mg/L potassium (4), 2.5 mg/L magnesium (5), 5 mg/L calcium (6), 50 mg/L each of cadaverine (7), putrescine (8), 1,2-diaminopropane (9), and ethylenediamine (10), and 100 mg/L N,N-diethylethylenediamine (11).

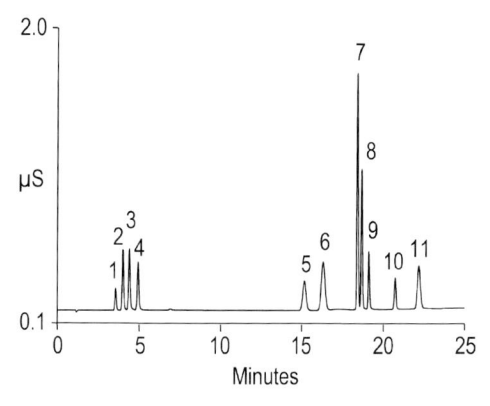

Fig. 4-64. Determination of biogenic amines together with common inorganic cations on IonPac CS17. — Separator column: IonPac CS17 (2-mm); column temperature: 40 °C; eluant: 3 mmol/L methanesulfonic acid isocratically for 3.5 min, then linear gradient to 6 mmol/L methanesulfonic acid in 8.5 min, isocratically for 3 min, linear gradient to 40 mmol/L methanesulfonic acid in 5 min, isocratically for 4 min; flow rate: 0.4 mL/min; injection volume: 25 µL; detection: suppressed conductivity; solute concentrations: 10 µg/L lithium (1), 40 µg/L sodium (2), 50 µg/L ammonium (3), 100 µg/L potassium (4), 50 µg/L magnesium (5), 100 µg/L calcium (6), 1000 µg/L putrescine (7), 600 µg/L cadaverine (8), 600 µg/L histamine (9), 200 µg/L spermidine (10), and 400 µg/L spermine (11).

Higher polyamines such as triamines and tetramines, however, cannot be analyzed under these chromatographic conditions. Excellent separations for higher polyamines with remarkably good peak symmetry are obtained with an eluent mixture of potassium chloride and EDTA and the application of a linear concentration gradient; this eluent was developed for polyphosphate analysis (see Section 3.7.4). Such a separation of diethylenetriamine and triethylenetetramine, which were available as relatively pure reference compounds, is illustrated in Fig. 4-65. The signals in the chromatogram of the next higher tetraethylenepentamine shown in Fig. 4-66 cannot be unequivocally assigned, because this product could only be obtained in technical grade quality; thus, it contains more than one major component. It is even more difficult to interpret the chromatograms of higher polyamines. For their elution, the potassium chloride concentration in the mobile phase has to be increased to 2 mol/L.

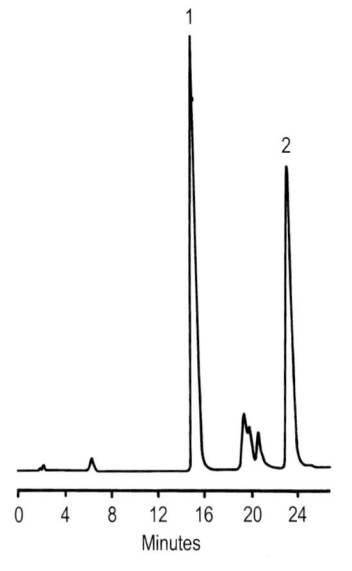

Fig. 4-65. Separation of diethylenetriamine and triethylenetetramine. – Separator column: IonPac CS3; eluent: (A) 0.2 mol/L KCl + 3.2 mmol/L EDTA, pH 5.1, (B) 1 mol/L KCl + 3.2 mmol/L EDTA, pH 5.1; gradient: linear, 100% A for 2 min isocratically, then to 100% B in 20 min; flow rate: 1 mL/min; detection: see Fig. 4-62; injection volume: 50 µL; solute concentrations: 10 mg/L each of diethylenetriamine (1) and triethylenetetramine (2).

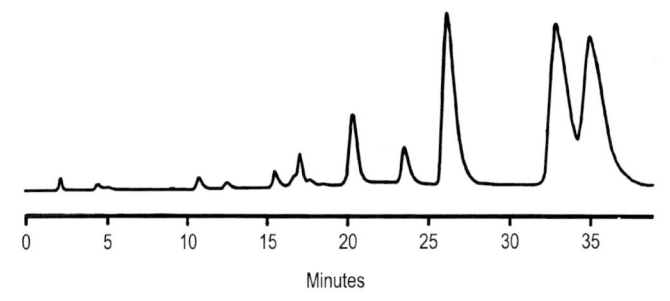

Fig. 4-66. Separation of tetraethylenepentamine (technical grade). – Chromatographic conditions: see Fig. 4-65; gradient: linear, 30% B for 2 min isocratically, then to 100% B in 15 min; injection volume: 50 µL; solute concentration: 20 mg/L.

4.7
Gradient Techniques in Cation Exchange Chromatography of Inorganic and Organic Cations

Gradient elution techniques in cation exchange chromatography are less common than in anion exchange chromatography. The primary reason for this is the availability, since the late 1980's, of separation materials that allow the simultaneous analysis of alkali- and alkaline-earth metals. Those separation materials are silica- or polymer-based weak acid cation exchangers which allow baseline-resolved separations of alkali- and alkaline-earth metals in less than 15 minutes. Thus, the application of a gradient elution technique would not lead to a significant decrease in analysis time. This is especially true because the time required for column re-equilibration has to be taken into account.

The situation is completely different if amines have to be analyzed together with alkali- and alkaline-earth metals. In the case of morpholine as an amine component, analysis is possible under isocratic conditions (see Fig. 4-10). Selected amino alcohols can also be eluted together with alkaline-earth metals by adding small amounts of acetonitrile to the mobile phase to increase peak efficiencies (see Fig. 4-13). In general, these are individual cases because usually, gradient elution techniques have to be applied for the analysis of amines together with mono- and di-valent inorganic cations. Often, an easy-to-perform step gradient is sufficient to solve the analytical problem. This is exemplified in Fig. 4-67, which shows the separation of the six standard cations and 3-dimethylaminopropylamine, which serves as a corrosion inhibitor in cooling water circuits. Without the application of a step gradient, which in this case is based on dilute sulfuric acid, the amine would elute much later and exhibit significant tailing. If several amines are used simultaneously as corrosion indicators, as it is often the case in steam generators and boiler feed waters, a combined step

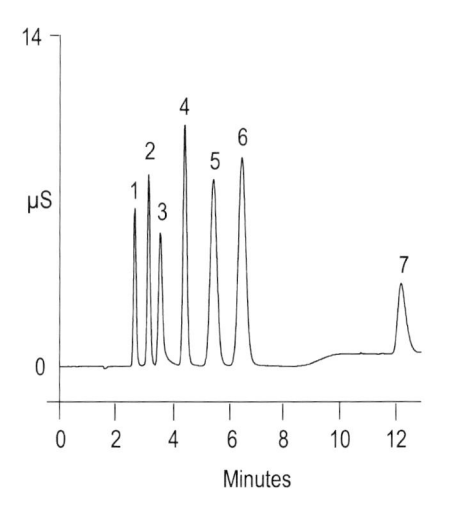

Fig. 4-67. Separation of the six standard inorganic cations and 3-dimethylaminopropylamine. – Separator column: IonPac CS12A; step gradient: after 5.1 min from 15 mmol/L to 35 mmol/L H_2SO_4; flow rate: 1 mL/min; detection: suppressed conductivity; injection volume: 25 µL; solute concentrations: 0.5 mg/L lithium (1), 2 mg/L sodium (2), 2.5 mg/L ammonium (3), 5 mg/L potassium (4), 2.5 mg/L magnesium (5), 5 mg/L calcium (6), and 10 mg/L 3-dimethylaminopropylamine (7).

gradient can help to solve the problem. Such an example is illustrated in Fig. 4-68 with the separation of morpholine, 2-diethylaminoethanol, cyclohexylamine, and the six standard inorganic cations [11]. Total analysis time can be reduced to about 20 minutes by gradually increasing the sulfuric acid and acetonitrile concentration in the mobile phase. Under these chromatographic conditions and the elevated column temperature of 40 °C, even the surface-active cyclohexylamine elutes as a completely symmetric peak. By further increasing column temperature to 60 °C, the addition of organic solvents can be omitted.

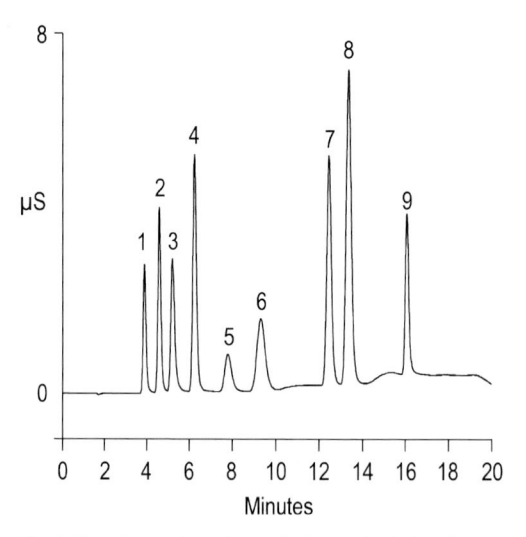

Fig. 4-68. Separation of morpholine, 2-diethylaminoethanol, cyclohexylamine, and the six standard inorganic cations. — Separator column: IonPac CS12A; column temperature: 40 °C; step gradient: after 7 min from 8 mmol/L H_2SO_4-acetonitrile (98:2 v/v) to 14 mmol/L H_2SO_4-acetonitrile (96:4 v/v), then after 11 min to 25 mmol/L H_2SO_4-acetonitrile (95:5 v/v); flow rate: 1 mL/min; detection: suppressed conductivity; injection volume: 25 µL; solute concentrations: 0.5 mg/L lithium (1), 2 mg/L sodium (2), 2.5 mg/L ammonium (3); 5 mg/L potassium (4), 10 mg/L morpholine (5), 10 mg/L 2-diethylaminoethanol (6), 2.5 mg/L magnesium (7), 5 mg/L calcium (8), and 15 mg/L cyclohexylamine (9).

Applications in the field of amine analysis are predominantly based on methanesulfonic acid or sulfuric acid concentration gradients. When the components to be analyzed are surface-active amines, combined acid and organic solvent gradients are used. The separation of methylamines together with the six standard inorganic cations is an interesting application for a pure sulfuric acid concentration gradient. As can be seen from the respective chromatogram in Fig. 4-69, ammonium and methylamine cannot be separated at ambient temperature despite the low acid concentration in the mobile phase. The elevation of the column temperature to 60 °C is required for a successful separation; it not only leads to an increased peak efficiency but also affects the selectivity of the separation in a significant way.

4.7 Gradient Techniques in Cation Exchange Chromatography of Inorganic and Organic Cations

The special selectivity of the IonPac CS15 with its crown ether functional groups is suitable for the isocratic separation of standard cations and monoethanolamine, which has already been shown in Fig. 4-14. Di- and tri-ethanolamine, however, would be difficult to separate from sodium under these chromatographic conditions. If a purely aqueous methanesulfonic acid gradient is applied, the separation of all three ethanolamines can be carried out without any problem (Fig. 4-70).

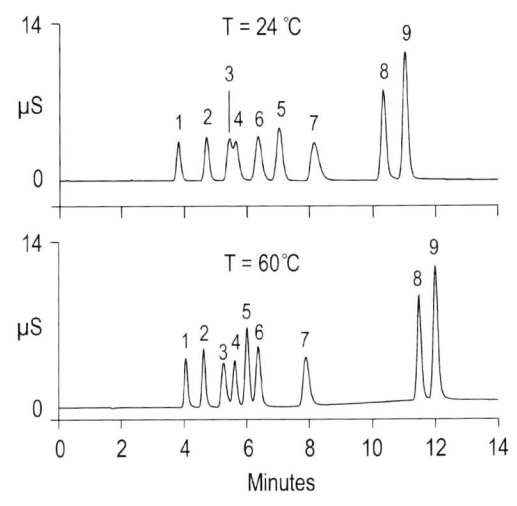

Fig. 4-69. Gradient elution of methylamines together with the six standard inorganic cations. — Separator column: IonPac CS12A; eluant: H_2SO_4; gradient: 8 mmol/L isocratically for 4 min, then to 20 mmol/L in 4 min; flow rate: 1 mL/min; detection: suppressed conductivity; injection volume: 25 µL; solute concentrations: 0.5 mg/L lithium (1), 2 mg/L sodium (2), 2.5 mg/L ammonium (3), 5 mg/L monomethylamine (4), 5 mg/L potassium (5), 10 mg/L dimethylamine (6), 15 mg/L trimethylamine (7), 2.5 mg/L magnesium (8), and 5 mg/L calcium (9).

The separation of ethylamines together with the standard cations can also be performed in a simple way with a linear acid gradient by exploiting the selectivity change caused by the crown ether functional groups of the IonPac CS15. To improve the peak form, small amounts of acetonitrile are added to the mobile phase. Shown in the chromatogram in Fig. 4-71, monoethylamine elutes way behind di- and tri-ethylamine due to its higher affinity towards the crown ether. Even long-chain aliphatic amines can be eluted from an IonPac CS12A column with a purely aqueous acid gradient. However, this requires a relatively high column temperature of 60 °C, which limits the lifetime of an electrolytically operated suppressor system. Chemically regenerated micromembrane suppressors, on the other hand, withstand elevated temperatures. As can be seen from the respective chromatogram illustrated in Fig. 4-72, the two last eluting components, 1,2-dimethylpropylamine and di-*n*-propylamine, exhibit a signifi-

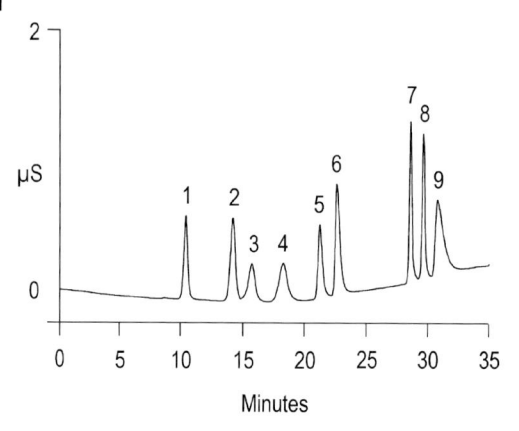

Fig. 4-70. Gradient elution of ethanolamines together with the six standard inorganic cations. — Separator column: IonPac CS15 (2-mm); column temperature: 40 °C; eluant: methanesulfonic acid; gradient: 2 mmol/L isocratically for 14 min, then to 27 mmol/L in 16 min; flow rate: 0.3 mL/min; detection: suppressed conductivity; injection volume: 2.5 µL; solute concentrations: 0.5 mg/L lithium (1), 2 mg/L sodium (2), 10 mg/L diethanolamine (3), 100 mg/L triethanolamine (4), 5 mg/L monoethanolamine (5), 2.5 mg/L ammonium (6), 1 mg/L magnesium (7), 1.5 mg/L calcium (8), and 5 mg/L potassium (9).

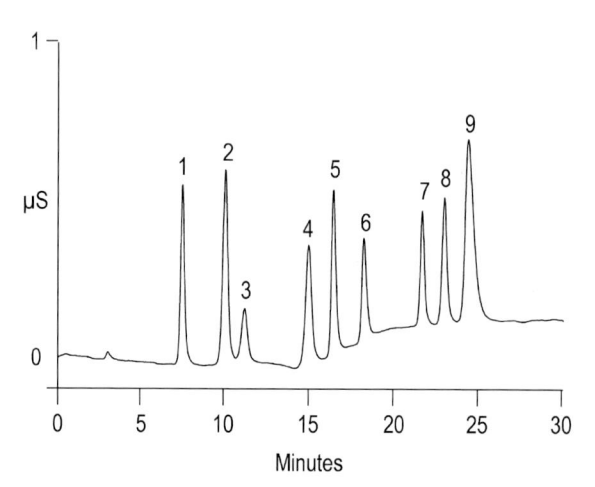

Fig. 4-71. Gradient elution of ethylamines together with the six standard inorganic cations. — Separator column: IonPac CS15 (2-mm); column temperature: 40 °C; eluant: methanesulfonic acid-acetonitrile (94:6 v/v); gradient: 2.5 mmol/L isocratically for 10 min, then to 14 mmol/L in 5 min; flow rate: 0.3 mL/min; detection: suppressed conductivity; injection volume: 2.5 µL; solute concentrations: 0.5 mg/L lithium (1), 2 mg/L sodium (2), 2.5 mg/L diethylamine (3), 10 mg/L triethylamine (4), 5 mg/L ammonium (5), 2.5 mg/L monoethylamine (6), 0.5 mg/L magnesium (7), 1 mg/L calcium (8), and 5 mg/L potassium (9).

4.7 Gradient Techniques in Cation Exchange Chromatography of Inorganic and Organic Cations

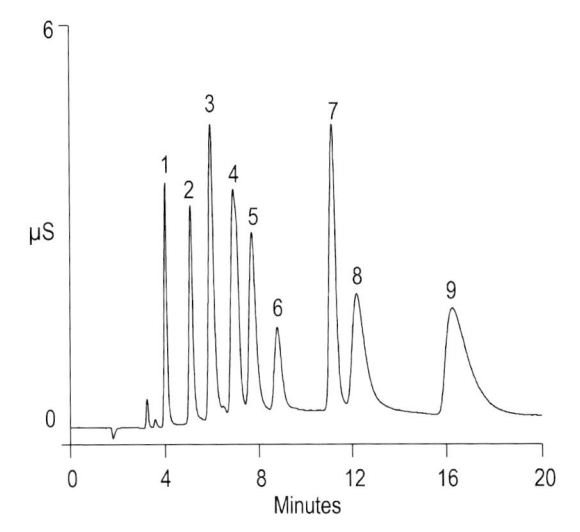

Fig. 4-72. Gradient elution of aliphatic amines. – Separator column: IonPac CS12A; column temperature: 60 °C; eluant: H_2SO_4; gradient: 15 mmol/L to 25 mmol/L in 10 min; flow rate: 1 mL/min; detection: suppressed conductivity; injection volume: 25 µL; solute concentrations: 5 mg/L monoethylamine (1), 7.5 mg/L n-propylamine (2), 12.5 mg/L tert-butylamin (3), 12.5 mg/L sec-butylamine (4), 12.5 mg/L iso-butylamin (5), 20 mg/L n-butylamin (6), 20 mg/L 1,2-propanediamine (7), 20 mg/L 1,2-dimethylpropylamine (8), and 40 mg/L di-n-propylamine (9).

cant tailing under these chromatographic conditions. This is ameliorated by decreasing column temperature to 40 °C. The higher retention that results can be compensated by adding small amounts of acetonitrile to the mobile phase.

Long-chain diamines exhibit an even higher affinity towards the stationary phase of a conventional cation exchanger. Thus, their elution usually requires the addition of an organic solvent to the mobile phase. Because early generations of cation exchangers were not solvent compatible, such compounds used to be separated by ion-pair chromatographic methods. Modern cation exchangers, however, are compatible with common HPLC solvents, so that even diamines can easily be separated by cation exchange chromatography. A combined acid and solvents gradient is required for their separation. As an example, Fig. 4-73 illustrates the separation of standard inorganic cations together with a number of aliphatic diamines. Due to hydrophobic interactions, the retention of the aliphatic diamines increases with increasing chain length between the terminal amino groups. Interferences between inorganic and organic cations are not observed, because even the smallest member of the diamine series, the 1,2-propanediamine, is more strongly retained than, for example, calcium. To improve the peak form, especially of the late eluting components, the column temperature is raised to 40 °C.

4 Cation Exchange Chromatography (HPIC)

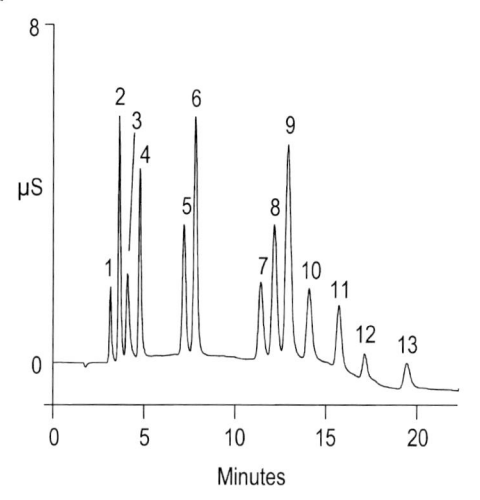

Fig. 4-73. Gradient elution of aliphatic diamines. — Separator column: IonPac CS12A; column temperature: 40 °C; eluant: H_2SO_4-acetonitrile; gradient: linear, from 11 mmol/L H_2SO_4-acetonitrile (98:2 v/v) to 22 mmol/L H_2SO_4-acetonitrile (85:15 v/v) in 10 min, then to 25 mmol/L H_2SO_4-acetonitrile (70:30 v/v) in 14 min; flow rate: 1 mL/min; detection: suppressed conductivity; injection volume: 25 µL; solute concentrations: 0.2 mg/L lithium (1), 0.8 mg/L sodium (2), 1 mg/L ammonium (3), 2 mg/L potassium (4), 1 mg/L magnesium (5), 2 mg/L calcium (6), and 8 mg/L each of 1,2-propanediamine (7), 1,6-hexanediamine (8), 1,7-heptanediamine (9), 1,8-octanediamine (10), 1,9-nonanediamine (11), 1,10-decanediamine (12), and 1,12-dodecanediamine (13).

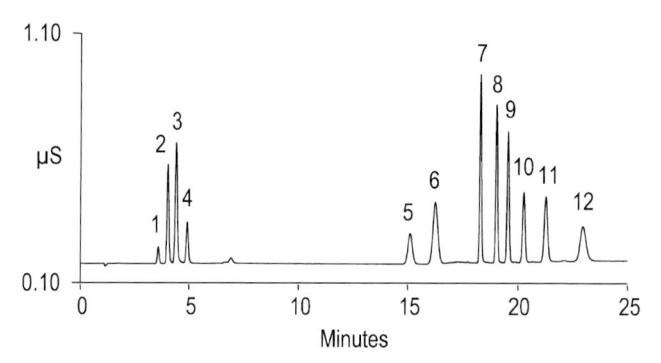

Fig. 4-74. Gradient elution of aliphatic diamines. — Separator column: IonPac CS17 (2-mm); column temperature: 40 °C; eluant: 3 mmol/L methanesulfonic acid isocratically for 3.5 min, linear gradient to 6 mmol/L in 8.5 min, isocratically for 3 min, linear gradient to 40 mmol/L in 5 min, isocratically for 4 min; flow rate: 0.4 mL/min; detection: suppressed conductivity; injection volume: 25 µL; solute concentrations: 5 µg/L lithium (1), 65 µg/L sodium (2), 89 µg/L ammonium (3), 47 µg/L potassium (4), 24 µg/L magnesium (5), 56 µg/L calcium (6), 500 µg/L 1,2-propanediamine (7), 250 µg/L 1,6-hexanediamine (8), 125 µg/L 1,7-heptanediamine (9), 250 µg/L 1,8-octanediamine (10), 375 µg/L 1,9-nonanediamine (11), and 375 µg/L 1,10-decanediamine (12).

4.7 Gradient Techniques in Cation Exchange Chromatography of Inorganic and Organic Cations

Today, the unique hydrophilic cation exchange surface of the IonPac CS17 allows the elution of standard inorganic cations and aliphatic diamines with a solvent-free eluant. The diamines elute after the inorganic cations with good efficiencies and peak symmetries when the column is operated with an acid gradient at elevated temperature. The selectivity of the IonPac CS17 for diamines is illustrated in Fig. 4-74. Despite the concentration change from 3 mmol/L to 40 mmol/L methanesulfonic acid, no baseline shift is observed. Aliphatic diamines larger than 1,10-decanediamine are not soluble in aqueous solutions. When they are present in the sample, the addition of an organic solvent to the eluant is required.

The IonPac CS12A cation exchanger can also be used to separate quaternary ammonium compounds (Fig. 4-75). The chromatographic conditions are very similar to those described in Fig. 4-73. Because quaternary ammonium compounds have a much larger carbon backbone, hydrophobic interactions with the stationary phase are so strong that an increase in eluant acid concentration does not have an influence on their retention behavior. Therefore, a solvent gradient

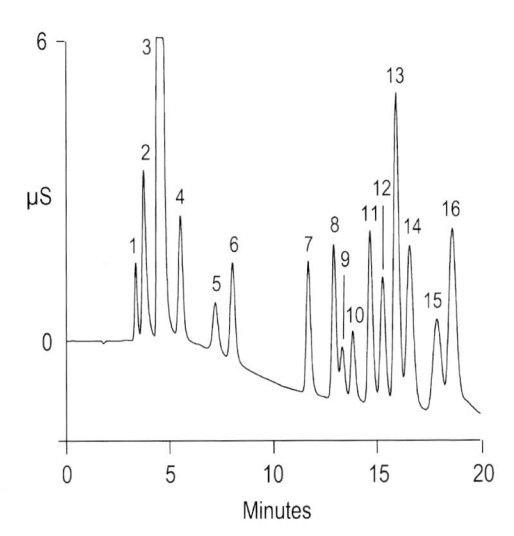

Fig. 4-75. Gradient elution of quaternary ammonium compounds. – Separator column: IonPac CS12A; eluant: H_2SO_4-acetonitrile; gradient: 11 mmol/L H_2SO_4-acetonitrile (90:10 v/v) to 11 mmol/L H_2SO_4-acetonitrile (20:80 v/v) in 15 min; flow rate: 1 mL/min; detection: suppressed conductivity; injection volume: 25 µL; solute concentrations: 0.3 mg/L sodium (1), 2 mg/L ammonium (2), 5 mg/L potassium (3), 5 mg/L tetramethylammonium (4), 8 mg/L calcium (5), 20 mg/L tetraethylammonium (6), 25 mg/L tetrapropylammonium (7), 50 mg/L each of tributylmethylammonium (8), heptyltriethylammonium (9), and tetrabutylammonium (10), 50 mg/L decyltrimethylammonium (11), 50 mg/L tetrapentylammonium (12), 100 mg/L dodecyltrimethylammonium (13), 100 mg/L tetrahexylammonium (14), 100 mg/L tetraheptylammonium (15), and 100 mg/L hexadecyltrimethylammonium (16).

that maintains a constant acid concentration is applied. The negative baseline drift during the gradient run is attributed to the high solvent content in the mobile phase towards the end of the gradient run.

Conventional cation exchangers can also be used for the analysis of aromatic amines, in which hydrophobic interactions dominate the separation mechanism. As with the separation of quaternary ammonium compounds, the elution of anilines shown in Fig. 4-76 is also carried out with a solvent gradient at constant acid concentration. Anilines are detected by UV detection at 210 nm. A slight elevation of the column temperature — as already discussed for the analysis of aliphatic diamines — has a positive effect on the peak form of the late-eluting components.

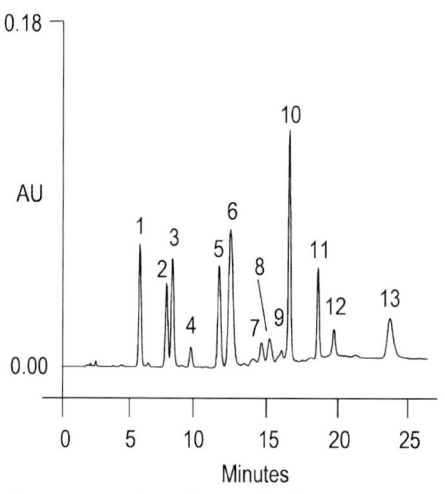

Fig. 4-76. Gradient elution of anilines. — Separator column: IonPac CS12A; column temperature: 40 °C; eluent: H_2SO_4-acetonitrile; gradient: 20 mmol/L H_2SO_4-acetonitrile (95:5 v/v) to 20 mmol/L H_2SO_4-acetonitrile (75:25 v/v) in 10 min, then to 20 mmol/L H_2SO_4-acetonitrile (40:60 v/v) in 17 min; flow rate: 1 mL/min; detection: UV (210 nm); injection volume: 25 µL; analytes: (1) aniline, (2) N-methylaniline, (3) 3-toluidine, (4) N,N'-dimethylaniline, (5) N,N'-diethylaniline, (6) 4,4'-methylenedianiline, (7) not identified, (8) not identified, (9) not identified, (10) 4-nitroaniline, (12) 2-nitroaniline, and (13) 2,6-dichloro-4-nitroaniline.

The separation of pyridines using a combined acid and solvent gradient is more strongly affected by column temperature. As can be seen from the respective chromatogram in Fig. 4-77, 2-aminopyridine and 4-picoline cannot be separated at room temperature. However, when increasing column temperature to 60 °C, a baseline-resolved separation of both compounds is obtained.

4.7 Gradient Techniques in Cation Exchange Chromatography of Inorganic and Organic Cations

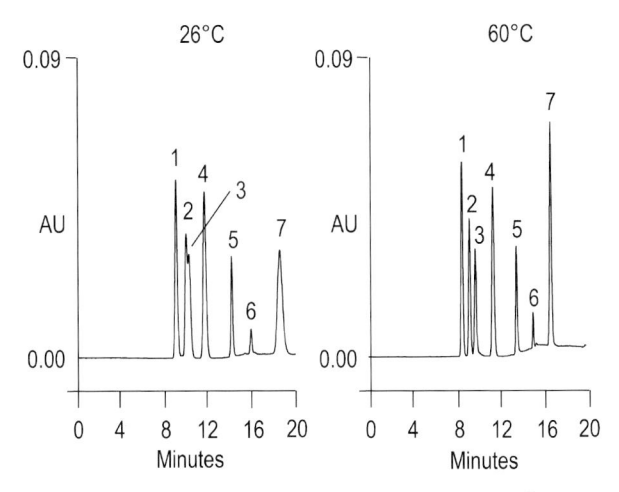

Fig. 4-77. Gradient elution of pyridines. – Separator column: IonPac CS12A; eluant: H_2SO_4-acetonitrile; gradient: 5 mmol/L H_2SO_4-acetonitrile (90:10 v/v) to 9 mmol/L H_2SO_4-acetonitrile (75:25 v/v) in 10 min, then to 20 mmol/L H_2SO_4-acetonitrile (50:50 v/v) in 13 min; flow rate: 1 mL/min; detection: UV (254 nm); injection volume: 25 µL; solute concentrations: 20 mg/L each of pyridine (1), 2-aminopyridine (2), 4-picoline (3), 2-dimethylaminopyridine (4), 2,2'-bipyridine (5), 4-benzylpyridine (6), and 2-(2-aminoethyl)pyridine (7).

The last two examples show that, in a number of cases, cation exchange chromatography can be regarded as an alternative to reversed-phase chromatography on chemically bonded silica phases, especially because the characteristic interactions between the basic compounds and the silanol groups of silica are not observed with polymer-based cation exchangers.

5
Ion-Exclusion chromatography (HPICE)

The introduction of ion-exclusion chromatography is attributed to Wheaton and Bauman [1]. It is primarily employed for the separation of weak inorganic and organic acids. In addition, ion-exclusion chromatography can be used for the separation of alcohols, aldehydes, amino acids, and carbohydrates. Due to Donnan exclusion, fully dissociated acids are not retained at the stationary phase, eluting therefore within the void volume as a single peak. Undissociated compounds, however, can diffuse into the pores of the resin, since they are not subject to Donnan exclusion. In this case, separations are based on non-ionic interactions between the solute and the stationary phase.

Ion-exclusion chromatography in combination with ion-exchange chromatography (HPICE/HPIC coupling) makes it possible to separate a wealth of inorganic and organic anions within 30 minutes in a single run.

Detection is usually carried out by measuring the electrical conductivity. When combined with a suppressor system, this detection method is superior to all other detection methods (for example, refractive index or UV detection at low wavelengths) with regard to specificity and sensitivity.

5.1
The Ion-Exclusion Process

Typically, HPICE separator columns contain a totally sulfonated high-capacity cation exchange resin. The separation mechanism occuring at this stationary phase is based on three phenomena:
- Donnan exclusion
- Steric exclusion
- Adsorption

Figure 5-1 represents a schematic picture of the separation process on an HPICE column. It shows the resin surface with its bonded sulfonic acid groups. If pure water is passed through the separator column, a hydration shell is formed around the sulfonic acid groups. Therefore, some of the water molecules are therefore in a higher state of order compared to the water molecules in the bulk mobile phase.

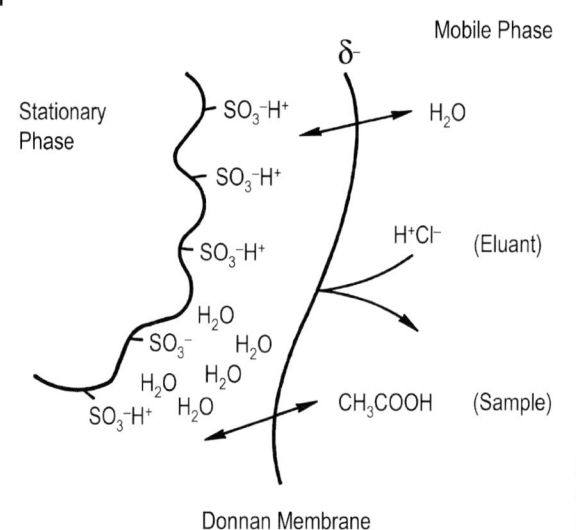

Fig. 5-1. Schematic representation of the separation process on a HPICE column.

In this retention model, a negatively charged layer analogous to the Donnan membrane characterizes the interface between the hydration shell, which is only permeable by undissociated compounds, and the bulk mobile phase. Fully dissociated acids, such as hydrochloric acid acting as the eluant, cannot penetrate this layer because of the negative charge of the chloride ion. Thus, such ions are excluded from the stationary phase. Their retention volume is called the *exclusion volume* V_e. On the other hand, neutral water molecules may diffuse into the pores of the resin and back into the mobile phase. The volume corresponding to the "retention time" of water is called the *totally permeated volume* V_p. Depending on the eluant pH, a weak organic acid (e.g., acetic acid) may be present after injection in partly undissociated form, which is not subject to Donnan exclusion. Although both acetic acid and water may interact with the stationary phase, a retention volume is observed for acetic acid that is higher than V_p. This phenomenon can only be explained by adsorption occuring on the surface of the stationary phase. The separation mechanism in the case of aliphatic monocarboxylic acids, therefore, is determined by Donnan exclusion and adsorption. Retention time increases with increasing length of the alkyl chain of the acid. By adding organic solvents such as acetonitrile or 2-propanol, the retention of aliphatic monocarboxylic acids may be lowered. This is due to: (1) adsorption sites being blocked by solvent molecules and (2) the solubility in the eluant being enhanced.

Di- and tri-carboxylic acids such as oxalic acid, citric acid, etc. elute between the exclusion volume and the totally permeated volume. Apart from Donnan exclusion, the predominant separation mechanism is, in this case, mainly steric exclusion. The retention is determined by the size of the sample molecule. Because the pore volume of the resin is determined by its degree of cross-linking, resolution can only be improved by utilizing another column or by coupling with another column.

In general, organic acid separations can be optimized by changing the pH, because the eluant pH influences the degree of dissociation and, consequently, the solute retention.

A thermodynamic treatment of the processes occurring at the stationary phase of ion-exclusion columns has recently been published (in Italian) by Sarzanini et al. [2] in their book on ion chromatography.

5.2
Stationary Phases

The selection of stationary phases for ion-exclusion chromatography is relatively limited. Typically, totally sulfonated polystyrene/divinylbenzene-based cation exchangers in the hydrogen form are used. The percentage of divinylbenzene — expressed as the degree of cross-linking — is of particular importance for the retention behavior of organic acids. Depending on the degree of cross-linking, these acids may diffuse into the interior of the stationary phase to a greater or lesser degree, resulting in different retention times. In their investigations into the retention behavior of various inorganic and organic acids, Harlow and Morman [3] found out that resins with a relatively high degree of cross-linking of 12% are suitable for the separation of weakly dissociated organic acids. On the other hand, more strongly dissociated acids are best separated on resins with a low degree of cross-linking of 2%. Most of the materials offered today have a degree of cross-linking of 8%. This compromise takes into account the different retention behavior of strong and weak acids. Instead of organic polymers, silica-based cation exchangers are sometimes used for ion-exclusion chromatography [4]. These materials, however, are only of secondary importance due to their lack of pH stability.

Comparably small particle diameters between 5 µm and 15 µm, which enable fast diffusion processes, are characteristic of the stationary phases that are used today in ion-exclusion chromatography. The structural-technical properties of these phases are listed in Table 5-1.

Dionex currently offers two ion-exclusion columns: IonPac ICE-AS1 and AS6. The former is a moderately hydrophilic, microporous polystyrene/divinylbenzene-based cation exchanger with a particle diameter of 7.5 µm and an ion-exchange capacity of 27 mequiv/column, functionalized with sulfonate groups. The IonPac ICE-AS1 is primarily used for the separation of weak inorganic acids, aliphatic monocarboxylic acids, and alcohols. Difficulties are encountered, however, in the separation of aliphatic di- and tri-carboxylic acids. These acids elute from such stationary phases within the totally permeated volume. The selectivity of the separation in this retention range is usually very poor. Since the totally permeated volume is determined by the degree of cross-linking of the resin, selectivity can only be improved marginally by altering the chromatographic conditions. For example, Fig. 5-2 shows a separation of short-chain fatty acids on IonPac ICE-AS1 utilizing suppressed conductivity detection.

Table 5-1. Structural-technical properties of various polystyrene/divinylbenzene-based ion-exclusion columns.

Column	Manufacturer	Dimensions [Length × i.d.] [mm]	Particle diameter [µm]	Max. pressure [MPa]	Ion-exchange functionality	Application
IonPac ICE-AS1	Dionex	250 × 9	7.5	9.6	Sulfonate groups	High-performance separator for weak inorganic acids and short-chain fatty acids
IonPac ICE-AS6	Dionex	250 × 9	8	5.6	Sulfonate and carboxyl groups	High-performance separator for aliphatic carboxylic acids and alcohols
PRP-X300	Hamilton	250 × 4.1	10		Sulfonate groups	High-performance separator for aliphatic carboxylic acids
ORH-801	Interaction Chemicals	300 × 6.5	8		Sulfonate groups	High-performance separator for aliphatic carboxylic acids
ION-300	Interaction Chemicals	300 × 7.8	8		Sulfonate groups	Separator for Krebs cycle acids

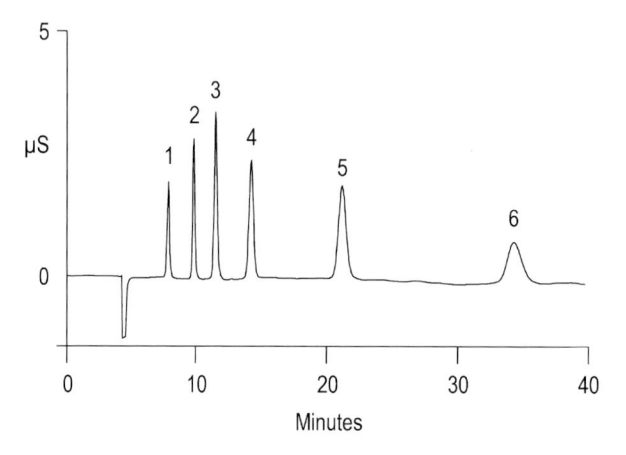

Fig. 5-2. Separation of short-chain fatty acids on IonPac ICE-AS1. – Eluant: 0.4 mmol/L heptafluorobutyric acid; flow rate: 1 mL/min; detection: suppressed conductivity; injection volume: 50 µL; analytes: (1) formic acid, (2) acetic acid, (3) propionic acid, (4) butyric acid, (5) valeric acid, and (6) caproic acid.

The packing material of the IonPac ICE-AS1 exhibits a 50% solvent compatibility towards common HPLC solvents such as methanol, 2-propanol, and acetonitrile, which can be used to drastically decrease retention of strongly hydrophobic compounds.

The IonPac ICE-AS6 also contains a microporous resin with a particle diameter of 8 µm. In comparison to the ICE-AS1 column it is more hydrophobic; this is due to the incorporation of additional polymethacrylates in the substrate, which is functionalized with sulfonate- and carboxyl groups. The carboxyl groups are responsible for a higher selectivity which can be attributed to hydrogen bonding between the resin and the analyte acid. Thus, this stationary phase is very well suited for the analysis of aliphatic hydroxycarboxylic acids, which are not well separated on a conventional ion-exclusion column. Because non-ionic adsorption interactions also contribute to retention, mono- and di-carboxylic acids with hydrophobic segments such as propionic acid, butyric acid, fumaric acid and succinic acid are very strongly retained. The separation of aliphatic carboxylic acids and hydroxycarboxylic acids illustrated in Fig. 5-3 impressively demonstrates the unique selectivity of this separator column. Analyte pairs such as tartaric acid/citric acid, glycolic acid/lactic acid, lactic acid/formic acid, and succinic acid/formic acid, which are poorly separated or not at all separated on conventional ion-exclusion columns, are separated to baseline on IonPac ICE-AS6 under standard chromatographic conditions.

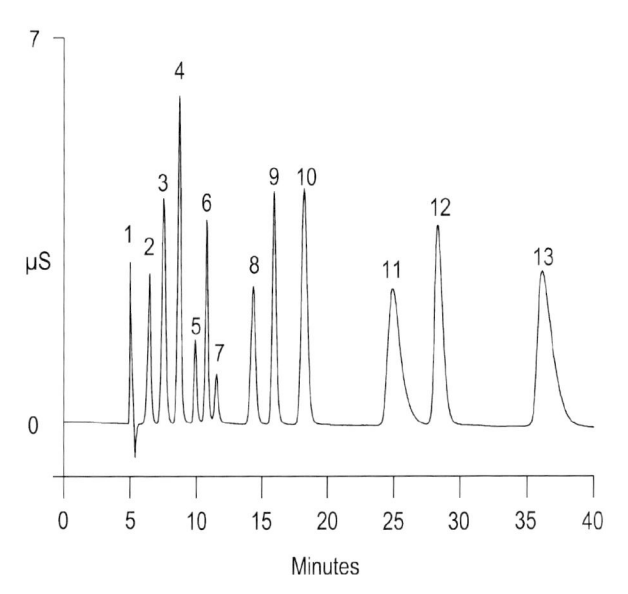

Fig. 5-3. Separation of carboxylic acids and hydroxycarboxylic acids on IonPac ICE-AS6. – Eluant: 0.4 mmol/L heptafluorobutyric acid; flow rate: 0.4 mL/min; detection: suppressed conductivity; injection volume: 50 µL; solute concentrations: 5 mg/L oxalic acid (1), 10 mg/L tartaric acid (2), 15 mg/L citric acid (3), 20 mg/L malic acid (4), 10 mg/L glycolic acid (5), 10 mg/L lactic acid (7), 30 mg/L hydroxyisobutyric acid (8), 25 mg/L acetic acid (9), 25 mg/L succinic acid (10), 35 mg/L fumaric acid (11), 50 mg/L propionic acid, (12), and 40 mg/L glutaric acid (13).

The IonPac ICE-AS6 also has a limited solvent compatibility of 20% towards methanol, 2-propanol, and acetonitrile. As shown below (Fig. 5-14 in Section 5.6), the addition of 10% solvent is sufficient to reduce, for example, the retention of acrylic acid by more than 40% at this stationary phase.

Polystyrene/divinylbenzene-based ion-exclusion columns are also offered by Hamilton Co. (Reno, NV, USA) under the trade name PRP-X300. This is a 10-µm material with an ion-exchange capacity of 0.2 mequiv/g [5]. This column material is obtained by sulfonation of PRP-1, a macroporous PS/DVB polymer with reversed-phase properties. Figure 5-4 shows a separation of various organic acids on this stationary phase using a dilute sulfuric acid eluant. The much higher retention of succinic acid as compared to acetic acid reveals that the retention of organic acids is characterized, apart from reversed-phase effects, by the formation of hydrogen bonds.

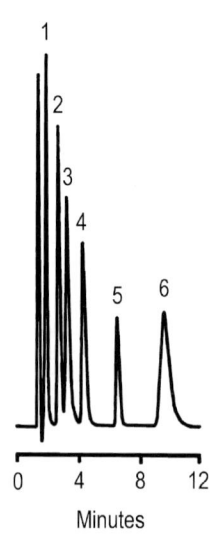

Fig. 5-4. Separation of organic acids on PRP-X300. – Eluant: 0.5 mmol/L H_2SO_4; flow rate: 1 mL/min; detection: direct conductivity; injection volume: 100 µL; solute concentrations: 4 mg/L tartaric acid (1), 7.5 mg/L malic acid (2), 7.5 mg/L citric acid (3), 10 mg/L lactic acid (4), 25 mg/L acetic acid (5), and 40 mg/L succinic acid (6).

Interaction Chemicals also offers two ion-exclusion columns for the separation of organic acids under the trade names ORH-801 and ION-300. Both stationary phases contain a 8-µm substrate which was optimized for its respective purpose by slight modifications of the chemical structure. While the separation properties of the ORH-801 column are very similar to those of the Dionex IonPac ICE-AS1, the ION-300 column was especially developed for the analysis of the Krebs cycle acids. Citric and pyruvic acid, succinic and lactic acid, as well as oxalacetic and α-ketoglutaric acid may also be successfully separated on this stationary phase. Such separations are of importance for the analysis of fruit juices, foods, and physiological samples. Figure 5-5 shows a respective separation of a standard mixture of organic acids, carbohydrates and alcohols, which resembles the matrix of a variety of beverages.

A common characteristic of all polystyrene/divinylbenzene-based ion-exclusion phases is the high retention of aromatic carboxylic acids. This is due to π-π interactions between the aromatic ring systems of the polymer and the solute. The separation of aromatic carboxylic acids, therefore, is more elegantly accomplished by reversed-phase chromatography.

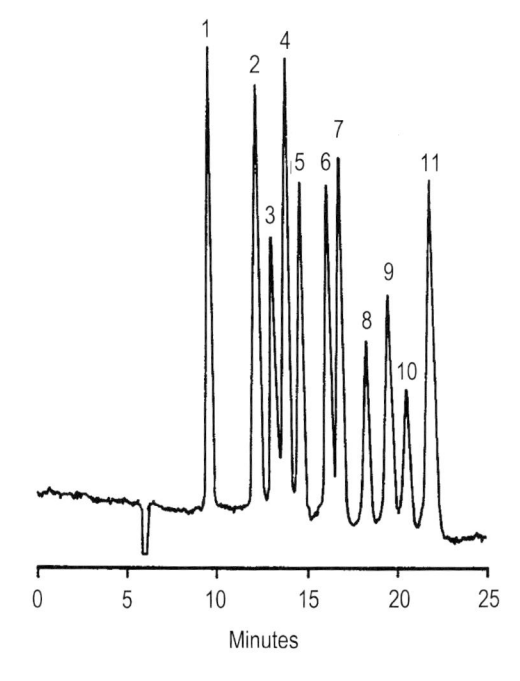

Fig. 5-5. Separation of a standard mixture of organic acids, carbohydrates, and alcohols on ION-300. – Column temperature: 30 °C; eluant: 5 mmol/L H_2SO_4; flow rate: 0.5 mL/min; detection: RI; analytes: (1) sucrose, (2) glucose, (3) citric acid, (4) fructose, (5) tartaric acid, (6) malic acid, (7) glycerol, (8) acetic acid, (9) lactic acid, (10) methanol, and (11) ethanol.

5.3 Eluants in Ion-Exclusion Chromatography

In ion-exclusion chromatography, the selection of eluants is very limited. In the most simple case, pure de-ionized water can be utilized. However, the work of Turkelson and Richard [6] on the separation of organic acids occuring in the citric acid cycle using an Aminex 50W-X4 cation exchanger (30 µm to 35 µm), has shown that when pure water is used, the peak form of the acids is characterized by a large half width and a strong tailing. This phenomenon probably occurs because organic acids are present in their molecular as well as in their anionic form. Acidifying the mobile phase suppresses dissociation, thus significantly improving peak shape. Nowadays, pure de-ionized water is recommended only for the analysis of carbonate, which can easily be determined in this way.

For the separation of organic acids, mineral acids and long-chain aliphatic carboxylic acids are usually employed as eluants. Among mineral acids, hydrochloric acid and sulfuric acid are the most commonly used. When applying direct UV detection, sulfuric acid is usually used as the eluant [4, 7], while hydrochloric acid eluants are considered outdated. They were used in the late 1970's, when cation exchangers in the silver form were used as suppressor columns in ion-exclusion chromatography. In those suppressor columns, chloride precipitated prior to entering the conductivity cell (see Section 5.4). Although hydrochloric acid eluants are compatible with modern membrane suppressors, the high equivalent conductance of the chloride ion results in an unnecessarily high background conductivity. Good separations are obtained with tridecafluoroheptanoic acid (perfluoroheptanoic acid), which is applied (like all other eluants) in a concentration range between $c = 0.5$ mmol/L and 10 mmol/L. Octanesulfonic acid exhibits similar elution properties, and is also suited for the analysis of borate and carbonate. For the sensitive detection of boric acid by electrical conductivity measurement, a mixture of octanesulfonic acid and mannitol should be used, because boric acid is only weakly dissociated. The sugar alcohol complexes boric acid, which leads to a significantly higher conductivity. On the other hand, for an IonPac ICE-AS6 separator column, short-chain perfluorated fatty acids such as heptafluoropropanoic acid (perfluorobutyric acid) are recommended as the eluants.

The high retention of aliphatic monocarboxylic acids ($n_c > 4$) and aromatic carboxylic acids can be lowered by adding small amounts of an organic solvent (10 to 30 mL/L) to the eluant. This blocks adsorption sites on the surface of the stationary phase [8]. Acetonitrile, 2-propanol, or ethanol are particularly suitable. If possible, methanol should not be used, because the stationary phase is subject to strong volume changes when using this solvent.

According to Tanaka and Fritz [9], comparably low background conductivities were obtained with a benzoic acid eluant, as benzoic acid is only partly dissociated. The acid strength of a dilute benzoic acid solution is sufficient to elute aliphatic carboxylic acids with good peak shapes.

5.4
Suppressor Systems in Ion-Exclusion Chromatography

Because the detection of aliphatic carboxylic acids is usually performed by measuring the electrical conductivity, suppressor systems are also used in ion-exclusion chromatography to chemically reduce the background conductivity of the acid eluant. In the past, modern membrane suppressors were not available, so packed bed suppressor columns were used; such columns contained a cation exchange resin in the silver form. With a dilute hydrochloric acid eluant, the suppressor reaction is described as follows:

$$\text{Resin-SO}_3^- \text{Ag}^+ + \text{H}^+\text{Cl}^- \rightarrow \text{Resin-SO}_3^- \text{H}^+ + \text{AgCl} \downarrow \tag{123}$$

Because the organic acids to be analyzed do not precipitate with silver, they reach the conductivity cell unchanged. However, this type of suppression has a number of disadvantages:
- With increasing formation of silver chloride in the suppressor column, a pressure increase and a significant peak broadening were observed. The signals could only be evaluated by comparing the peak areas with those of reference compounds of known concentrations.

- In theory, a periodical regeneration of the suppressor column is possible with a concentrated ammonium hydroxide solutions. However, it does not make much sense since it is not practical because it is so time-consuming.
- A continuous regeneration of the suppressor column – a standard feature in anion exchange chromatography for years – was impossible with the suppressor columns used for ion-exclusion chromatography.

The only way to limit the steadily increasing back pressure caused by the continuing precipitation of silver chloride in the suppressor column was to cut off the exhausted part of the cartridge after some days of operation.

A cation exchange membrane was developed to overcome these disadvantages. This membrane allows continuous regeneration and consists of a sulfonated polyethylene derivative. It is compatible with water-miscible organic solvents and exhibits a high permeability for quaternary ammonium bases such as tetrabutylammonium hydroxide. Hence, the suppression mechanism differs substantially from the process for anion exchange chromatography described in Section 3.6.

In the regenerated state, the membrane exists in the tetrabutylammonium form. The oxonium ions of the organic acids that flow through the interior of the membrane are replaced by tetrabutylammonium ions during the suppressor reaction. Because of the different equivalent conductances of oxonium and tetrabutylammonium ions, the salt that is formed as the suppressor product has a markedly lower conductivity than the corresponding acid form. In the zone of dynamic equilibrium, the neutralization reaction between H^+ ions and OH^- ions forms water as the suppressor product; the water is discharged from the suppressor together with an excess of tetrabutylammonium hydroxide. The higher background conductivity that results when hydrochloric acid is used as the eluant is due to the fact that the chloride ion has a higher equivalent conductance than the octanesulfonate ion.

Eluants and regenerants suitable for the analysis of organic acids when applying an AFS-2 hollow fiber membrane suppressor are listed in Table 5-2. For the analysis of borate and carbonate with an octanesulfonic acid eluant, an ammonium hydroxide solution with a concentration $c = 10$ mmol/L can also be used as the regenerant.

A micromembrane suppressor for ion-exclusion chromatography has been introduced under the trade name AMMS-ICE. Its structure corresponds to the systems developed for anion and cation exchange chromatography (see Sections 3.6.3 and 4.3.3). However, in its mode of operation, it corresponds to the AFS-2 hollow fiber suppressor. An AMMS-ICE micromembrane suppressor also contains membranes that are compatible with water-miscible organic solvents. Therefore, it is used for the analysis of long-chain fatty acids, which are separated on a non-polar stationary phase in a weakly acidic medium with methanol or acetonitrile as mobile phase components. In this case, a dilute potassium hydroxide solution is used as the regenerant. With respect to the ion-exchange capacity, no significant differences exist between the two types of membrane suppressors. The distinctly lower dead volume of the micromembrane suppressor, however, contributes to the sensitivity increase. The regeneration of an AMMS-ICE micromembrane suppressor is no different than that of the corresponding hollow fiber suppressor. It is usually carried out with tetrabutylammonium hydroxide with a concentration $c = 10$ mmol/L (see Table 5-2).

Table 5-2. Eluants and regenerents in ion-exclusion chromatography.

Eluant	Concentration [10^3 mol/L]	Regenerant[a] [mol/L]	Background conductivity[b] [µS/cm]
HCl	0.5 ... 1		50 ... 1000
Perfluorobutyric acid	0.5 ... 1	TBAOH 0.005 ... 0.01	20 ... 45
Perfluoroheptanoic acid	0.5 ... 1		20 ... 40
Octanesulfonic acid	0.5 ... 1		25 ... 45

a) The regenerant concentration should be about 10 times as high as the eluant concentration, based on a flow rate of 2 mL/min.

b) The values indicated for the resulting background conductivity refer to an eluant flow rate of 0.8 mL/min.

5.5
Analysis of Inorganic Acids

The separator columns listed in Table 5-1 (see Section 5.1) allow the separation of a variety of inorganic acids. While mineral acids elute within the void volume because of Donnan exclusion, weak inorganic acids are more strongly retained. This includes the barely dissociated boric acid ($pK_a = 9.23$), which is detected by measuring the electrical conductivity. The analysis of boric acid in the mg/L range is carried out with a purely aqueous octanesulfonic acid eluant ($c = 1$ mmol/L) in combination with a membrane-based suppressor system. A chromatogram from such an analysis is shown in Fig. 5-6. The chromatogram illustrates that the more strongly retained carbonate can be analyzed under the same chromatographic conditions. The minimum detection limits for both compounds are about 1 mg/L.

For sensitive borate detection, a large excess of mannitol is added to a methanesulfonic acid eluant to increase the borate conductance via complexation. Recently, a special concentrator column (Trace Borate Concentrator, TBC-1) was developed for the ultra-trace analysis of borate. The stationary phase of this concentrator is functionalized with a *cis*-diol, on which borate is selectively retained. Minimum detection limits in the lower ng/L range are obtained when pre-concentrating large sample volumes, so that this method is suitable for the trace analysis of borate in de-ionized water. As an example, Fig. 5-7 shows the chromatogram of a 500-ng/L borate standard, which was obtained upon pneumatic preconcentration of 160 mL of the respective standard solution. An IonPac ICE-Borate column was used as the separator, which is an IonPac ICE-AS1 column being especially conditioned for borate analysis.

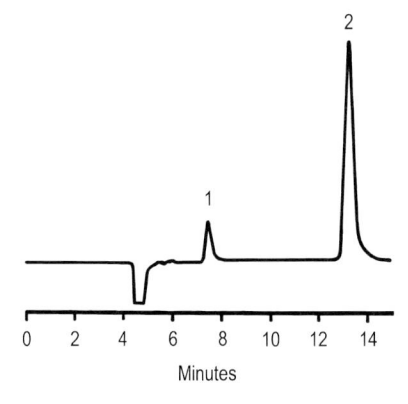

Fig. 5-6. Separation of borate and carbonate by ion-exclusion chromatography. – Separator column: IonPac ICE-AS1; eluant: 1 mmol/L octanesulfonic acid; flow rate: 1 mL/min; detection: suppressed conductivity; injection volume: 50 µL; solute concentrations: 10 mg/L borate (1) and 50 mg/L carbonate (2).

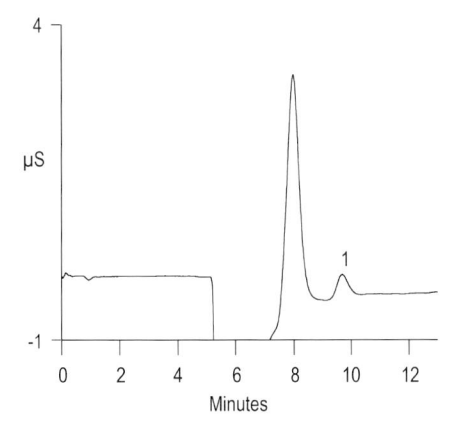

Fig. 5-7. Trace analysis of borate. – Separator column: IonPac ICE-Borate; eluant: 2.5 mmol/L methanesulfonic acid + 60 mmol/L mannitol; flow rate: 1 mL/min; detection: suppressed conductivity; regenerant: 25 mmol/L TMAOH + 15 mmol/L mannitol; concentrator column: TBC-1; pre-concentrated volume: 160 mL; solute concentration: 500 ng/L borate (as boron) (1).

In addition to borate and carbonate, fluoride can also be analyzed by ion-exclusion chromatography. Again, both conductivity detection modes may be employed. Because fluoride is clearly separated from the mineral acids that elute within the void volume, ion-exclusion chromatography with its different selectivity for fluoride is a welcomed alternative to anion exchange chromatography. Ion-exclusion chromatography is also suited for the trace analysis of anionic impurities in hydrofluoric acid [10, 11]. Using an ICE column, anionic impurities are pre-separated from the HF matrix and then concentrated on an anion exchange concentrator column (IonPac AC10); (see Section 5.7). In a second step, the pre-concentrated anions are eluted from the concentrator column with sodium hydroxide and separated on an analytical anion exchanger. Such a coupling of HPIC and HPICE is also known as *two-dimensional ion chromatography*

[10]. Only this method allows the determination of anionic impurities in concentrated hydrofluoric acid in the lowest µg/L range. This level of sensitivity is now required by the semiconductor industry for process chemicals used in the manufacturing of integrated circuits (see also Section 9.4).

In combination with an amperometric detection, ion-exclusion chromatography is suitable for the determination of sulfite and arsenite. Both ions can be oxidized at a platinum working electrode and, thus, be detected selectively and very sensitively. Introduced in 1986 by Kim et al. [12], DC amperometry of sulfite is widely used in the food industry and accepted as a standard method by AOAC (Association of Official Analytical Chemists) [13]. The method is highly selective, so that a time-consuming sample preparation is not necessary. The only disadvantage is that the electrode surface changes its characteristics relatively quickly, resulting in a 40% decrease of the response factor within a time frame of about eight hours [14]. For this reason, the working electrode has to be polished frequently, and a check standard has to be chromatographed after each sample to ensure exact quantitation. When applying pulsed amperometry of sulfite [15] instead of D.C. amperometry, the surface characteristics and, consequently, the response factor remains constant due to the fast and repeating sequence of oxidizing and reducing potentials. Figure 5-8 displays the chromatogram of a sulfite standard which was obtained with a dilute sulfuric acid eluant. To avoid rapid autoxidation of sulfite to sulfate, mannitol that elutes ahead of sulfite was added to the standard solution. The minimum detection limit for sulfite under these chromatographic conditions was calculated to be 40 µg/L.

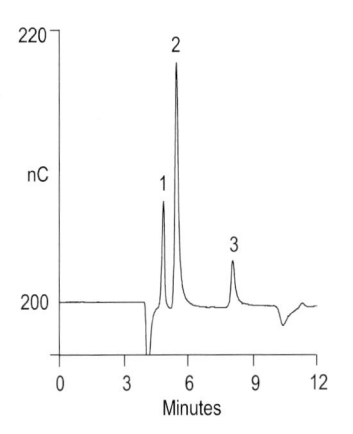

Fig. 5-8. Analysis of sulfite utilizing pulsed amperometric detection. — Separator column: IonPac ICE-AS1; eluant: 0.01 mol/L sulfuric acid; flow rate: 1 mL/min; detection: pulsed amperometry on a platinum working electrode; injection volume: 50 µL; analytes: (1) unknown, (2) mannitol, and (3) sulfite.

For the amperometric detection of arsenite on the platinum working electrode the applied oxidation potential can be kept constant. Figure 5-9 shows a respective chromatogram of a 10-mg/L arsenite standard that was obtained with an oxidation potential of +0.95 V.

Ion-exclusion chromatography is also suited for the determination of cyanide, which can be oxidized on a silver working electrode at low potential. Nitric acid ($c = 0.1$ mol/L) is used as the eluant. Because an alkaline medium is required for the subsequent oxidation reaction, an AMMS-ICE micromembrane suppressor, regenerated with 0.5 mol/L NaOH, is used for the post-chromatographic pH adjustment. Sodium ions migrate through the cation exchange membranes into the eluant chamber and, thus, neutralize the strongly acidic eluant. Figure 5-10 shows the respective chromatogram of a 10-mg/L cyanide standard.

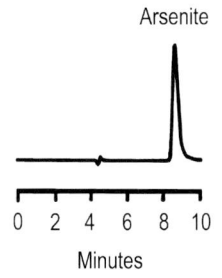

Fig. 5-9. Analysis of arsenite with amperometric detection. – Separator column: IonPac ICE-AS1; eluant: 0.5 mmol/L HCl; flow rate: 1 mL/min; detection: D.C. amperometry on a platinum working electrode; oxidation potential: +0.95 V; injection volume: 50 µL; solute concentration: 10 mg/L arsenite.

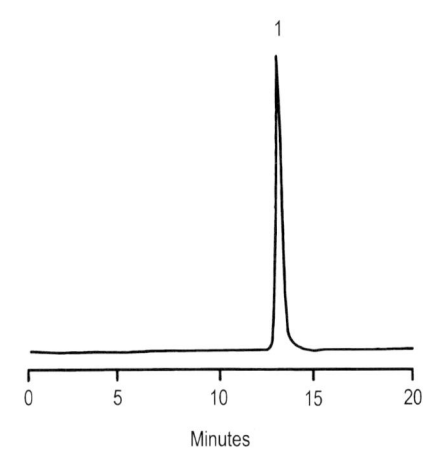

Fig. 5-10. Ion-exclusion chromatography of cyanide with amperometric detection. – Separator column: IonPac ICE-AS1; eluant: 0.1 mol/L nitric acid; flow rate: 0.8 mL/min; detection: amperometry on a silver working electrode; oxidation potential: 0 V; suppressor: AMMS-ICE; regenerant: 0.5 mol/L NaOH; injection volume: 50 µL; solute concentration: 10 mg/L cyanide (1).

High selectivities are also obtained by combining ion-exclusion chromatography with photometric detection after derivatization of the column effluent with suitable reagents. For a typical example, Fig. 5-11 shows the chromatogram of a 20-mg/L silicate standard obtained by derivatization with sodium molybdate in acidic solution and subsequent photometric detection at 410 nm. Orthophosphate may also be detected under these conditions. Because of its different acid strength it elutes *prior* to silicate, so that this method is very selective for the determination of both anions.

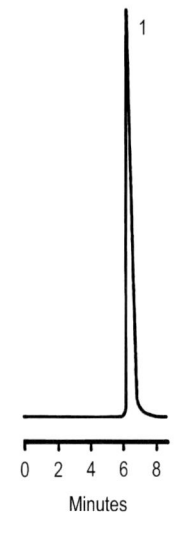

Fig. 5-11. Analysis of silicate after derivatization with sodium molybdate. — Separator column: IonPac ICE-AS1; eluant: 0.5 mmol/L HCl; flow rate: 1 mL/min; detection: photometry at 410 nm after reaction with sodium molybdate; injection volume: 50 µL; solute concentration: 20 mg/L Si (as $SiCl_4$, 1).

5.6 Analysis of Organic Acids

In addition to a couple of inorganic acids, a wealth of organic acids may be separated by using the separator columns listed in Table 5-1. A selection of the analyzable compounds with their pK_a values is listed by elution order in Table 5-3.

Table 5-3. Elution order of selected organic acids that can be analyzed by ion-exclusion chromatography on a totally sulfonated polystyrene/divinylbenzene-based cation exchanger.

Acid		pK_a	Acid		pK_a
Maleic acid	K_1	1.83	Glycolic acid		3.83
	K_2	6.07	Lactic acid		3.08
Oxalic acid	K_1	1.23	Formic acid		3.75
	K_2	4.19	Adipic acid	K_1	4.43
Citric acid	K_1	3.14		K_2	5.41
	K_2	4.77	Fumaric acid	K_1	3.03
	K_3	6.39		K_2	4.44
Pyruvic acid		2.49	Acetic acid		4.75
Tartaric acid	K_1	2.98	Propionic acid		4.87
	K_2	4.34	Acrylic acid		4.25
Malonic acid	K_1	2.83	iso-Butyric acid		4.84
	K_2	5.69	n-Butyric acid		4.81
Succinic acid	K_1	4.16	Mandelic acid		3.85
	K_2	5.61			

5.6 Analysis of Organic Acids

As mentioned in Section 5.5, fully dissociated acids elute within the void volume due to Donnan exclusion. The retention behavior of weak organic acids, on the other hand, may be predicted based on the following criteria:

- Members of a homologous series elute in the order of decreasing acid strength and decreasing water solubility, that is, increasing solvophobicity. Taking aliphatic monocarboxylic acids as an example, the elution order, according to Fig. 5-12, is formic acid > acetic acid > propionic acid, and so forth. The characteristics of such separations are similar to those of reversed-phase chromatography. Van-der-Waals forces between the solute and the polymeric resin material (mainly benzene rings) as well as the decrease in solubility of the solutes in the eluant influence the distribution of the solutes between the stationary and the mobile phases.

- Di-basic acids are eluted prior to the corresponding mono-basic acids because of their higher solubility in polar eluants. Thus, it is observed that oxalic acid elutes prior to acetic acid, and malonic acid prior to propionic acid.

- Organic acids with a branched carbon skeleton such as iso-butyric acid are generally eluted prior to the non-branched analogues such as *n*-butyric acid. Again, this corresponds to the retention behavior in RPLC.

- A double-bond in the carbon skeleton leads to a significantly higher retention because of π-π interactions with the aromatic rings of the polymer: acrylic acid is eluted *after* propionic acid.

- Aromatic acids are strongly retained because of the interactions described above. Thus, they should not be analyzed by ion-exclusion chromatography. The high retention of unsaturated and aromatic moieties on ICE stationary phases is unlike the behavior on ODS phases, where π-π interactions are not possible and where only the enhanced solubility in the eluant comes to fruition.

In the analysis of di- and poly-basic acids (see Fig. 5-3), the acid concentration in the mobile phase determines the retention. It affects the degree of dissociation and, thus, the retention time of the carboxylic acid to be analyzed. In general, a higher resolution is observed with increasing acid concentration. This effect is illustrated in Fig. 5-13, which shows the separation of organic acids with low pK_a values such as glucuronic acid, tartaric acid, citric acid, and malonic acid. They are only separated to baseline at higher acid concentrations in the mobile phase. The retention of short-chain fatty acids, on the other hand, is not significantly affected by changes made to the mobile phase pH. Because these acids are mainly retained by non-ionic interactions resembling reversed-phase effects, retention times can only be lowered by adding small amounts of 2-propanol or acetonitrile (10 mL/L up to 30 mL/L). The two chromatograms in Fig. 5-14 demonstrate how strongly the separation of various organic acids is affected by the addition of organic solvents. Later eluting acids such as shikimic acid, itaconic acid, and acrylic acid are most strongly affected by solvent addition: the addition of 10% (v/v) acetonitrile decreases their retention by 40%.

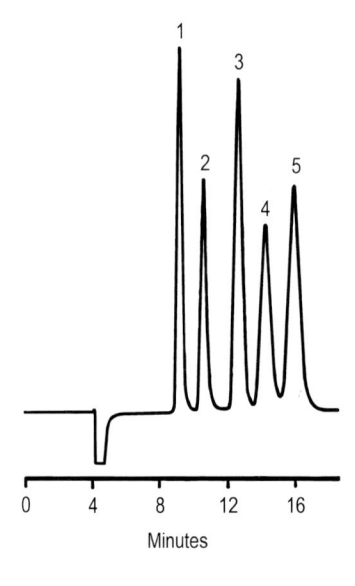

Fig. 5-12. Separation of aliphatic monocarboxylic acids. − Separator column: IonPac ICE-AS1; eluant: 1 mmol/L octanesulfonic acid; flow rate: 1 mL/min; detection: suppressed conductivity; injection volume: 50 µL; solute concentrations: 20 mg/L formic acid (1), 20 mg/L acetic acid (2), 40 mg/L propionic acid (3), 40 mg/L iso-butyric acid (4), and 60 mg/L n-butyric acid (5).

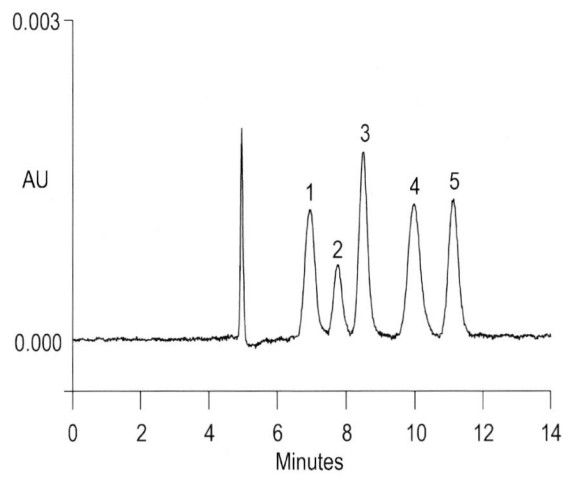

Fig. 5-13. Separation of organic acids with low pK_a values. − Separator column: IonPac ICE-AS6; eluant: 0.01 mol/L sulfuric acid; flow rate: 1 mL/min; detection: UV (210 nm); analytes: (1) glucoronic acid, (2) galacturonic acid, (3) tartaric acid, (4) citric acid, and (5) malonic acid.

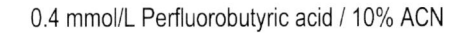

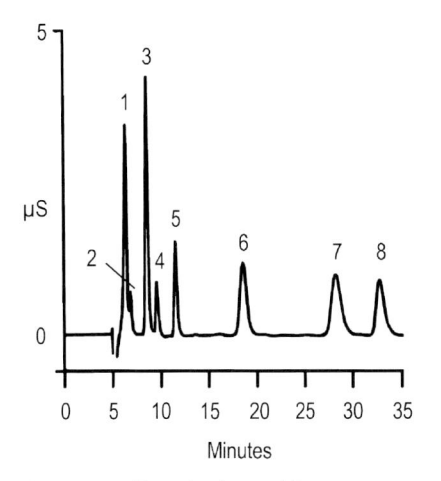

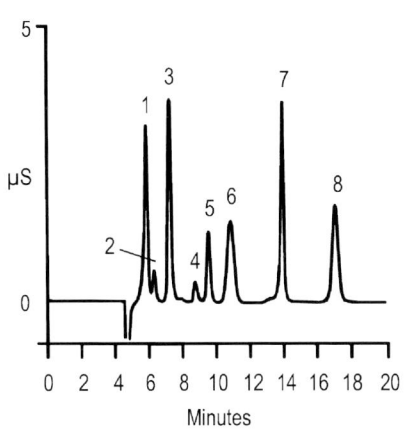

Fig. 5-14. Effect of solvent addition on organic acid separation. — Separator column: IonPac ICE-AS6; eluant: 0.4 mmol/L heptafluorobutyric acid — acetonitrile; flow rate: 1 mL/min; detection: suppressed conductivity; analytes: (1) tartaric acid, (2) galacturonic acid, (3) malic acid, (4) boric acid, (5) lactic acid, (6) shikimic acid, (7) itaconic acid, and (8) acrylic acid.

The IonPac ICE-AS6 separator column has proved to be best suited for the separation of aliphatic hydroxycarboxylic acids. See Fig. 5-3 above for the example chromatogram of a standard solution containing various organic acids. Perfluorobutyric acid is employed as the eluant. In this stationary phase, the acid concentration in the eluant is one of the most important parameters determining retention. Furthermore, the selectivity of the IonPac ICE-AS6 is affected by the column temperature. Hydrogen bonding contributes to the solute retention at this stationary phase, therefore an increase in column temperature will result in a general retention decrease, according to Eq. (40) (Section 3.2). An additional loss in selectivity will result from hydrogen bonds breaking at the elevated column temperature.

5.7 HPICE/HPIC-Coupling

The concept of combining ion-exclusion and ion-exchange chromatography was introduced by Rich et al. [16]. It improves the selectivity of the chromatographic separation of inorganic and organic acids in complex matrices. Its mode of action is schematically depicted in Fig. 5-15.

5 Ion-Exclusion chromatography (HPICE)

Fig. 5-15. Schematic representation of the HPICE/HPIC-coupling.

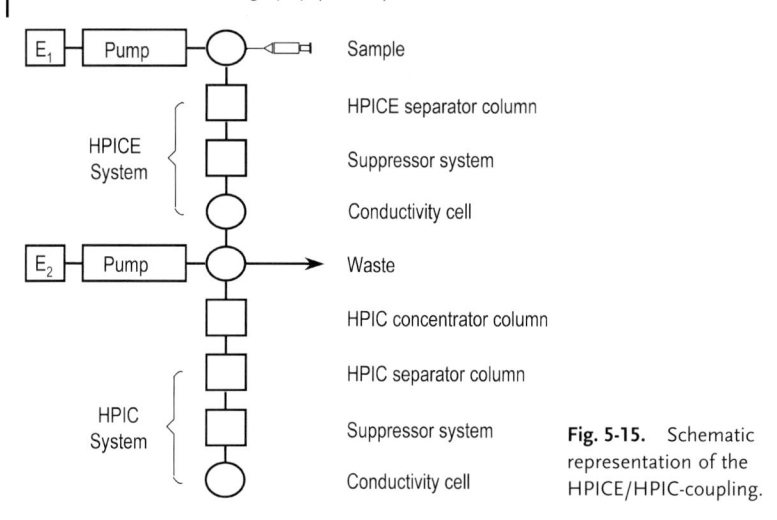

After its injection, the sample to be analyzed reaches an ion-exclusion column which causes all mineral acids to be separated, as a one peak, from weak inorganic or organic acids. The column effluent is then directed into the suppressor system and passed on to the conductivity cell. A column-switching technique connects the cell outlet with a concentrator column that contains an anion exchange resin. With this concentrator column, one (or several) of the peaks separated by HPICE is collected and then separated into its components on a subsequent anion exchanger. Inorganic and organic anions can be co-determined with minimum sample preparation when the appropriate eluants are used. In addition, single anions can be separated from others that are present in much larger concentrations.

The combination of HPICE/HPIC was successfully applied by Rich et al. [17] to the separation of pyruvic acid and lactic acid in blood sera. At that time, a separation of organic acids in biological liquids could not be accomplished solely by ion-exchange chromatography. Under isocratic conditions, the retention behavior of the organic acids of interest was too similar to that of certain inorganic anions under isocratic conditions. The application of HPICE/HPIC solved that problem. In addition, organic acids with identical retention on HPICE phases are often well separated on anion exchangers. Characteristic examples are α-ketoisovaleric acid and pyruvic acid. These organic acids cannot be separated on a conventional ion-exclusion phase; however, their capacity factors differ significantly upon application of an anion exchanger.

Since the introduction of gradient techniques in anion exchange chromatography (see Section 3.9), HPICE/HPIC applications have come to play only a secondary role. Today, most, if not all, of these problems can be solved by gradient elution techniques using a sodium hydroxide eluant. Thus, the HPICE chromatogram of a human serum depicted in Fig. 5-16 is of purely historical interest.

It was obtained using a suppressor column in the silver form that was commonly employed in the beginning of the 1980s. The individual signals, marked by letters, represent sum peaks of organic acids with similar pK values and comparable hydrophobicity. The chromatogram in Fig. 5-17 shows the separation of the peak marked "A" in Fig. 5-16 into its components nitrate and sulfate. This separation was obtained with the IonPac AS1 anion exchanger used at that time, after concentrating the HPICE effluent on a corresponding pre-column.

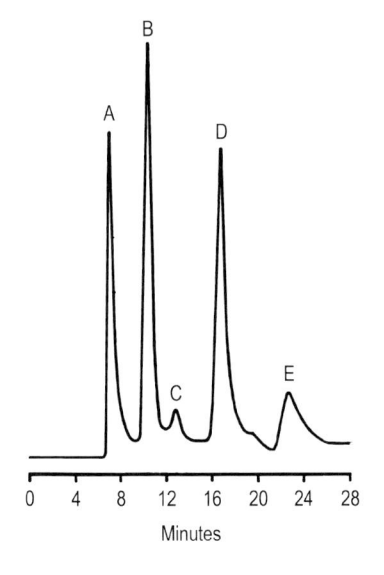

Fig. 5-16. HPICE chromatogram of a human serum. — Separator column: IonPac ICE-AS2 (precursor of IonPac ICE-AS1); eluant: 0.01 mol/L HCl; flow rate: 0.8 mL/min; detection: suppressed conductivity (cation exchanger in Ag form); (A) strong inorganic acids, (B) phosphoric acid group, (C) pyruvic acid group, (D) lactic acid group, (E) hydroxybutyric acid group.

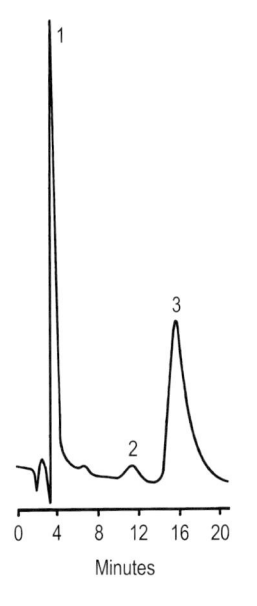

Fig. 5-17. HPIC chromatogram of the peak marked "A" in Fig. 5-16. — Separator column: IonPac AS1; eluant: 3 mmol/L $NaHCO_3$ + 2.4 mmol/L Na_2CO_3; flow rate: 2.3 mL/min; detection: suppressed conductivity (packed-bed suppressor ASC-2); analytes: (1) chloride, (2) nitrate, and (3) sulfate.

On the other hand, HPICE/HPIC is almost indispensable for the analysis of anionic impurities in ultra-pure chemicals [10, 11, 18] such as concentrated hydrofluoric acid and orthophosphoric acid, very important to the semiconductor industry. The purity of those chemicals is of vital importance for proper functioning of semiconductor devices. The increasing integration density of those devices, therefore requires analytical methods that can determine ionic impurities in ultra-pure acids down to the lower µg/L range. Such a high sensitivity cannot be achieved by diluting the acid or by treating it with sample preparation cartridges, because sample dilution decreases sensitivity and sample preparation cartridges might contaminate the sample. On the other hand, the sensitivity being required today cannot be achieved with other forms of matrix elimination. Siriraks et al [19], for example, report on the enrichment of an HF sample on a high-capacity concentrator column, which is then rinsed with methanol to remove the excess of HF from the concentrator prior to eluting the retained anionic impurities. The elution of fluoride from the concentrator with pure methanol is possible, because other anions are more strongly retained. With this method, the authors obtained minimum detection limits of 250 µg/L for simple inorganic anions in HF. Chen et al. [11] report minimum detection limits of 2 µg/L for chloride and sulfate, 5 µg/L for nitrate, and 10 µg/L for orthophosphate in 24.5% HF (w/w) after a pre-separation on an ICE column. An IonPac ICE-AS1 or AS6 is suited for this pre-separation, using pure de-ionized water as the eluant, which is directed through an IonPac AC10 trap column prior to entering the chromatographic system. As exemplified in Fig. 5-18 with the separation of 10 mg/L chloride and 100 mg/L fluoride, the ICE column effluent containing the analyte anions and a little bit of fluoride is fractionated by means of a switching valve, pre-concentrated, and then separated into its constituents on an anion exchanger. Further details regarding the switching technique are outlined in Section 9.10. Either carbonate/bicarbonate [18] or hydroxide [10] is a suitable eluant for the anion exchange separation.

As an example, Fig. 5-19 shows the separation of inorganic anions in 24.5% HF (w/w) on an IonPac AS9-HC following a pre-separation on an IonPac ICE-AS6. The concentrated hydrofluoric acid has to be diluted with ultra-pure de-ionized water, because repeated injections of a 49% HF (w/w) lead to column bleeding. As can be seen from Fig. 5-19, the fluoride matrix is so much impoverished after the pre-separation on an ICE column that the analysis of the anionic impurities can be done without interferences. A satisfactory separation between chloride and carbonate, which elute immediately after fluoride, is also ensured under these chromatographic conditions. The method of standard addition is recommended for calibration. The relative standard deviation is less than 10% in the concentration range between 10 and 50 µg/L.

In a very similar way, anionic impurities can also be determined in concentrated orthophosphoric acid [20], because orthophosphate is also separated from mineral acids on an ICE column. A hydroxide-selective stationary phase, such

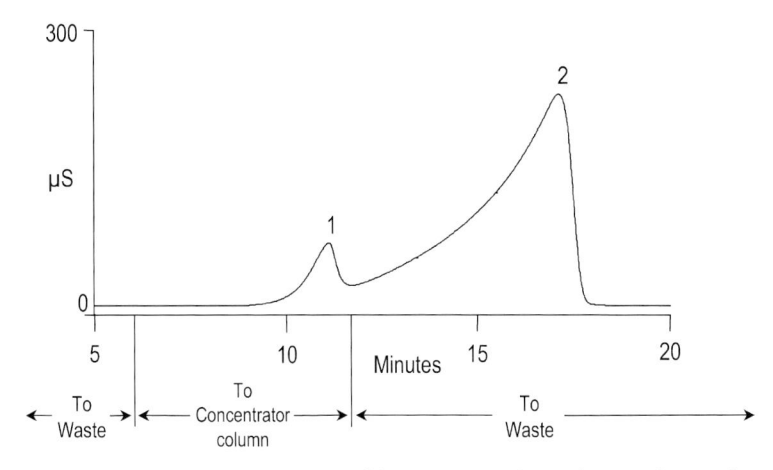

Fig. 5-18. Schematic representation of the fractionation for the separation of chloride and fluoride on an ion-exclusion column with subsequent pre-concentration on an anion exchanger. – Separator column: IonPac ICE-AS6; eluant: de-ionized water; flow rate: 0.55 mL/min; detection: direct conductivity; sample volume: 750 µL; solute concentrations: 10 mg/L chloride (1) and 100 mg/L fluoride (2); (taken from [18]).

as IonPac AS11-HC, with a 20 mmol/L hydroxide eluant is recommended for separating anionic impurities such as chloride, nitrate, and sulfate. Under these conditions, orthophosphate is strongly retained and, therefore, very well separated from the other anions. Because a sufficient separation between mineral acids and orthophosphoric acid is only possible, if the latter one is present in a concentrated form (85%) and, consequently, completely protonated, the 200-µL injection valve has to be filled pneumatically. This is done to avoid the formation of air bubbles which form due to the high sample viscosity. A representative chromatogram (see Fig. 5-20) depicts the separation of chloride, nitrate, and sulfate in 85% orthophosphoric acid; it shows that orthophosphate remaining in the sample after the pre-separation is well separated from the analyte anions. The relative standard deviation of this method with minimum detection limits for chloride, nitrate, and sulfate is less than 6%.

5.8 Analysis of Alcohols and Aldehydes

Totally sulfonated cation exchangers are suitable not only for the separation of weak inorganic and organic acids, but can also be used for the analysis of short-chain monohydric and polyhydric alcohols and aldehydes. For the separation of monohydric alcohols, Jupille et al. [21] used Aminex HPX-85H (Bio-Rad, Richmond, VA, USA) as the stationary phase and dilute sulfuric acid as the eluant. As

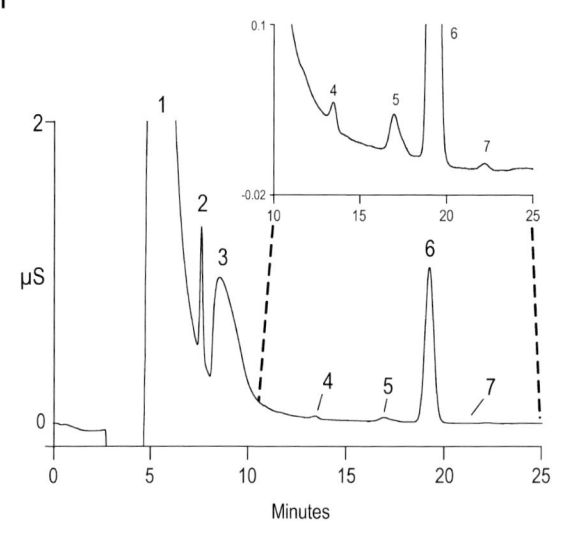

Fig. 5-19. Separation of chloride, nitrate, and sulfate in 24.5% hydrofluoric acid (w/w) after pre-separation on an IonPac ICE-AS6. – Eluant for the pre-separation: de-ionized water; flow rate: 0.55 mL/min; concentrator column: IonPac AG9-HC (4-mm); separator column: IonPac AS9-HC (2-mm); eluant: 8 mmol/L Na_2CO_3 + 1.5 mmol/L NaOH; flow rate: 0.25 mL/min; injection volume: 750 µL; solute concentrations: fluoride (1), 7.9 µg/L chloride (2), carbonate (3), 0.9 µg/L nitrate (4), unknown (5), 10.1 µg/L sulfate (6), and 2.4 µg/L orthophosphate (7).

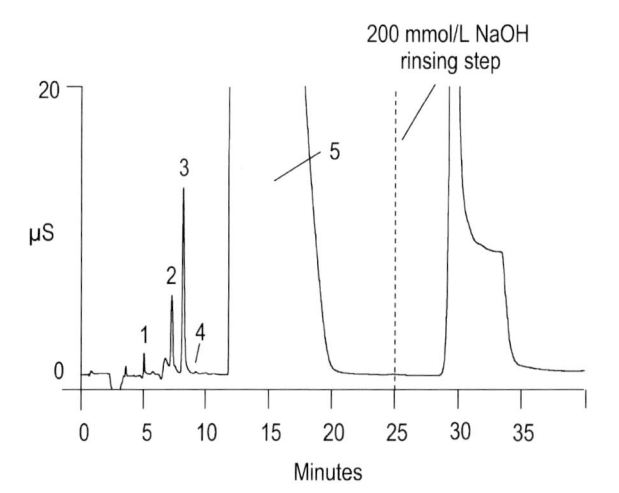

Fig. 5-20. Separation of chloride, nitrate, and sulfate in 85% orthophosphoric acid after a pre-separation on an IonPac ICE-AS6. – Eluant for the pre-separation: de-ionized water; sample volume: 200 µL; flow rate: 0.5 mL/min; concentrator column: IonPac AG11-HC (4-mm); separator column: IonPac AS11-HC (2-mm); eluant: 0.02 mol/L NaOH; flow rate: 0.38 mL/min; injection volume: 200 µL; solute concentrations: 36 µg/L chloride (1), not identified (2), 750 µg/L sulfate (3), 15 µg/L nitrate (4), and 15 µg/L orthophosphate (5).

can be seen from the corresponding chromatogram in Fig. 5-21, the investigated compounds elute in the order of increasing sorption area, just as they do on a chemically bonded reversed phase, which points to similarities in the retention mechanisms. Detection was carried out in this case by measuring the change in refractive index.

5.8 Analysis of Alcohols and Aldehydes

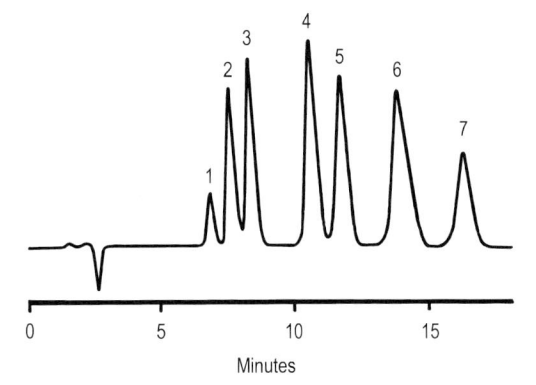

Fig. 5-21. Separation of monohydric aliphatic alcohols on Aminex HPX-85H. — Eluant: 5 mmol/L sulfuric acid; flow rate: 0.4 mL/min; detection: RI; injection volume: 20 µL; solute concentrations: 0.5 % (w/w) each of methanol (1), ethanol (2), 2-propanol (3), 2-methylpropanol-1 (4), n-butanol (5), 3-methyl-butanol-1 (6), and n-pentanol (7); (taken from [21]).

In comparison, Fig. 5-22 shows the simultaneous separation of alcohols and aldehydes on IonPac ICE-AS1, using pure de-ionized water as the mobile phase.

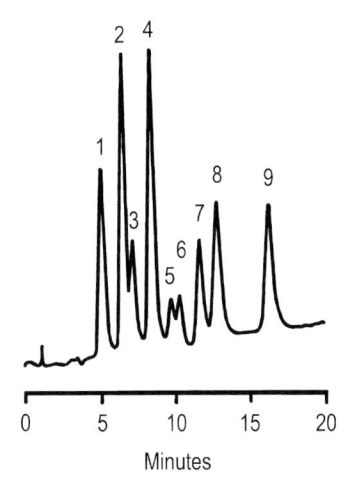

Fig. 5-22. Simultaneous separation of alcohols and aldehydes on IonPac ICE-AS1. — Eluant: de-ionized water ; flow rate: 1 mL/min; detection: RI; injection volume: 50 µL; solute concentrations: 100 mg/L each of glyoxal (1), glycerol (2), formaldehyde (3), ethylene glycol (4), glutaric dialdehyde (5), methanol (6), ethanol (7), 2-propanol (8), and n-propanol (9).

Pulsed amperometry on a platinum working electrode is an alternative to RI detection for the detection of alcohols [22]; it results in a higher sensitivity and selectivity. Figure 5-23 shows a standard chromatogram with a simultaneous separation of monohydric and dihydric alcohols. Dilute perchloric acid is used as the eluant because of its electrochemical inertness. Eluant degassing is extremely important, because oxygen is reduced at the potential that is suitable for the oxidation of alcohols and aldehydes. Online vacuum degassing has proved to be extremely effective. Without that, a high negative background signal is observed which, in turn, results in a strong baseline drift. Theoretically, eluant degassing can also be carried out offline. However, this form of degassing is not very effective, because oxygen is able to permeate through the Teflon tubing that connects the pump with the eluant reservoirs. Oxygen that is dissolved in the samples is retained at the stationary phase and elutes in the totally permeated volume. A negative peak for oxygen is observed in the chromatogram because the mobile phase is usually degassed but the samples are not. An IonPac NG1 pre-column is put inline with the ion-exclusion column [23] to avoid the interference of this negative peak with the sample components of interest, such as propylene glycol and acetaldehyde, that elute in the same retention range. This pre-column contains a highly cross-linked, non-modified divinylbenzene polymer that increases the totally permeated volume and, consequently, the retention time of oxygen. Hence, the retention times of other sample components are not influenced by the NG1 pre-column.

Clearly, glycerol is the key substance in Fig. 5-23, because it cannot be detected by gas chromatography due to its high boiling point. Although corresponding RPLC methods are available for the analysis of monohydric alcohols, ion-exclusion chromatography offers the advantage of the simultaneous detection of polyhydric alcohols. Because the IonPac ICE-AS6 is more hydrophobic than the ICE-AS1 column, it is the preferred column for the separation of short-chain alcohols and aldehydes because it is more selective. Mono-, di-, and tri-ethylene glycol can also be separated to baseline on IonPac ICE-AS6 (Fig. 5-24). Only long-chain alcohols, such as n-butanol and others that are strongly retained on the ICE-AS6 column, elute in a much shorter time on the ICE-AS1 column.

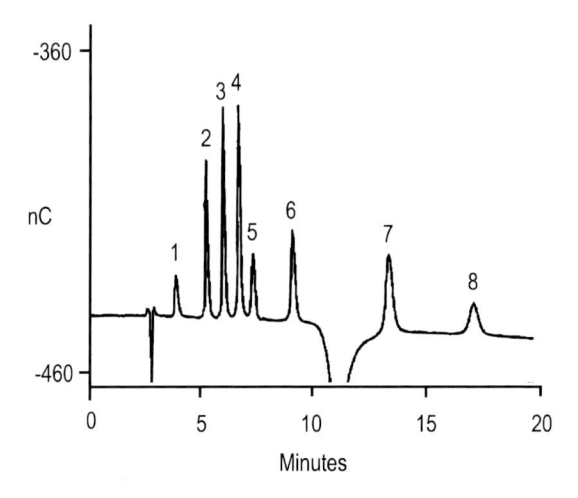

Fig. 5-23. Simultaneous separation of mono- and poly-hydric alcohols on IonPac ICE-AS6. — Eluant: 0.1 mol/L $HClO_4$; flow rate: 2 mL/min; detection: pulsed amperometry on a platinum working electrode; injection volume: 25 µL; solute concentrations: 10 mg/L each of glucose (1), glycerol (2), ethylene glycol (3), methanol (4), propylene glycol (5), and ethanol (6), 100 mg/L 2-propanol (7) and 10 mg/L n-propanol (8).

5.9
Amino Acid Analysis

The separation of amino acids on totally sulfonated cation exchangers is still one of the most commonly used methods for amino acid analysis. A milestone in the development of this method is the paper by Spackman, Stein, and Moore [24], published in 1958, wherein an automated quantitative detection of physiological amino acids is described for the first time. This method, for which the

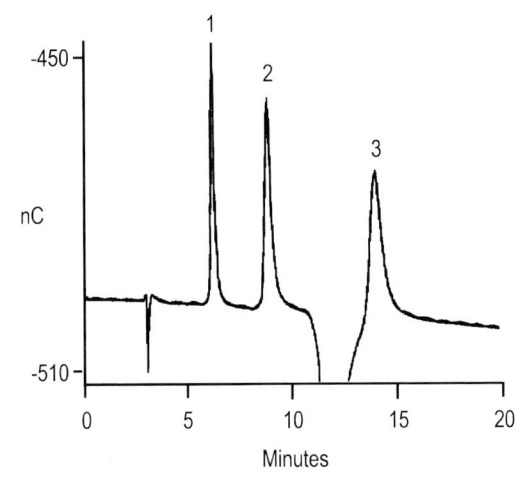

Fig. 5-24. Simultaneous separation of mono-, di-, and triethylene glycol on IonPac ICE-AS6. – Chromatographic conditions: see Fig. 5-23; solute concentrations: 10 mg/L monoethylene glycol (1), 20 mg/L diethylene glycol (2), and 30 mg/L triethylene glycol (3); (taken from [23]).

authors received the Nobel prize in 1972, utilized two ion-exchange columns; acidic and neutral amino acids were separated in one column, while the other column was used for basic amino acids. The separated compounds were then reacted with ninhydrin post-chromatographically. The resulting derivatives were detected by measuring their light absorption. At that time, an analysis took about two days and required a sample amount of 4 mL of plasma. Today, a 50-µL sample is sufficient for a single run using a commercial amino acid analyzer based on an ion-exchange separation with subsequent post-column derivatization with ninhydrin. Analysis times are about 30 minutes for protein hydrolysates and about 2 hours for physiological samples.

Although conventional amino acid analysis is extremely reliable, it does not satisfy today's demands for both higher sensitivity (fmol range) and shorter analysis times (10 to 15 min) along with the same high resolution. These requirements can only be met by HPLC methods which have since been developed; such applications involve a pre-column derivatization with either o-phthaldialdehyde (OPA) [25, 26], fluorenylmethyloxycarbonylchloride (FMOC) [27, 28], phenylthiohydantoin (PTH) [29-31], phenylisothiocyanate (PITC) [32, 33], or dansyl chloride (Dns) [34, 35]. However, these methods will not be discussed here. The methodological variety that presently characterizes amino acid analysis, might appear confusing but should not be regarded as a disadvantage. Currently, there is no method that is capable of analyzing all of the more than 300 known amino acids and their metabolites. A review of the various methods has been published by Ogden [36].

5.9.1
Separation of Amino Acids

Amino acids are amphoteric compounds; that is, they may exist in anionic, cationic, and zwitterionic form, depending on the pH:

$$\underset{\text{zwitterionic}}{\overset{R}{\underset{R}{H_3\overset{+}{N}-C-COO^-}}} \longleftrightarrow \underset{\text{anionic}}{\overset{R}{\underset{R}{H_2N-C-COO^-}}} \longleftrightarrow \underset{\text{cationic}}{\overset{R}{\underset{R}{H_3\overset{+}{N}-C-COOH}}}$$

According to the structure of the residue R, one distinguishes between "acidic", "neutral", and "basic" amino acids. In buffer solutions of low pH the equilibrium is shifted to the right-hand side.

$$\overset{R}{\underset{R}{H_2N-C-COOH}} + H^+ \rightleftharpoons \overset{R}{\underset{R}{H_3\overset{+}{N}-C-COOH}} \tag{124}$$

Amino acids may interact in their cationic form with the negatively charged sulfonate groups of a cation exchanger. When proton concentrations in the eluant are high, the equilibrium represented by Eq. (124) is shifted to the right-hand side. This also increases the concentration of exchangeable cations. Therefore, amino acids are strongly retained on the stationary phase at very low pH. If the eluant pH is raised, the equilibrium is shifted more and more to the left-hand side. Amino acid molecules without any positive charge are formed. They are not retained on the stationary phase and, thus, elute very quickly. Conversely, amino acids can interact with the quaternary ammonium groups of an anion exchanger when the pH in the mobile phase is high.

Stationary Phases
Totally sulfonated polystyrene/divinylbenzene-based cation exchangers are generally used as stationary phases. The technical specifications of a few selected materials are listed in Table 5-4. Totally sulfonated cation exchangers are not very rugged and have low permeabilities. This results in high back pressure of the separator columns, so that only low flow rates can be selected. If a flow rate is accidentally adjusted too high, the quality of the packing material deteriorates, resulting in a loss in separation efficiency. Shorter analysis times can only be realized by selecting suitable eluants. Totally sulfonated cation exchangers have a disadvantage in that the kinetic processes are slow due to the diffusion of sample molecules into the interior of the stationary phase.

Table 5-4. Characteristic structural-technical properties of some selected totally sulfonated cation exchangers for amino acid separations.

Designation	Manufacturer	Particle diameter [µm]	Degree of cross-linking [%]
DC 6A	Dionex	11	8
DC 4A	Dionex[a]	8	8
BTC 2710	Mitsubishi[b]	7	10

a) Can also be purchased from Pharmazia/LKB.
b) Can also be purchased from Eppendorf-Biotronik.

Thus, pellicular ion exchangers fulfilled an important need. Two important objectives were realized with their development:

a) With a pellicular ion exchanger, analysis times for hydrolysates could be reduced to 30-40 minutes.
b) At a flow rate of 1 mL/min, the pressure drop along the separator column does not exceed 19.5 MPa (3000 psi).

To realize these objectives, two different methods were originally applied to the separation of amino acids: by anion exchange *and* cation exchange, respectively. The latexed cation exchanger, introduced a few years ago under the trade name AminoPac PC-1, did not gain recognition and, therefore, is no longer commercially available. It consisted of a polystyrene/divinylbenzene substrate with a particle diameter of 10 µm. In contrast to the columns employed in conventional cation analysis, these latex beads were not made of the same material but of methylmethacrylate and styrene. In comparison with totally sulfonated cation exchangers, this latexed cation exchanger had a significantly lower capacity. This was determined by the size of the latex beads, which was reported to be approximately 200 nm. The AminoPac PC-1 exhibited a very high selectivity for early eluting amino acids, which was mainly because of the polymer that was used to synthesize the latex beads. The separation of amino acids using latex exchangers was performed in an isocratic mode of operation at ambient temperature. The chromatographic efficiency was high, as it usually is on latex exchangers; the separation was carried out with dilute nitric acid (pH 3.2) in less than 15 minutes. (In comparison: this run required almost 30 minutes on conventional exchangers!) Mixtures of nitric acid and oxalic acid were used to elute the other amino acids.

The corresponding AminoPac PA-1 latexed anion exchanger, introduced at the same time as the AminoPac PC-1, has been replaced meanwhile by a more modern product, the AminoPac PA-10 (see Section 3.11). The baseline-resolved separation of 17 amino acids contained in a hydrolysate standard requires about 40 minutes, which is significantly longer than with a latex cation exchanger.

Eluants

A series of buffer solutions with increasing cation concentration and increasing pH value are generally used for the separation of amino acids on conventional cation exchangers. Dus et al. [37], for example, employed four different citrate buffer solutions with pH values between 3.2 and 10.9 and sodium concentrations between 0.2 mol/L and 1.4 mol/L. Hare [38] developed an elution procedure based on a buffer system of constant cation concentration ($c(Na^+)$ = 0.2 mol/L). The pH value is varied between 3.25 and 10.1. Finally, in his Pico-Buffer® system, Benson [39] used a series of buffer solutions that have similar pH values but different cation concentrations. Compared to systems involving pH gradients, this method, originally introduced by Piez and Morris [40], offers the advantage of a more stable baseline.

The Hi-Phi buffer solutions recommended in the past by Dionex for separations on totally sulfonated cation exchangers were similar in their composition to those employed by Dus et al. The sodium-based eluants are intended for the separation of hydrolysates. The separation of asparagine, glutamine, and glutamic acid, important for physiological samples, can only be accomplished with buffer solutions in the lithium form.

The chemicals from which the various buffer solutions are prepared should be of high purity. Of particular concern are ammonia and other amines present in laboratory air which dissolve very well in the buffer solutions. Ammonia is eluted only very slowly as ammonium ion. If the sample to be analyzed contains ammonia in appreciable quantities, it elutes as a sharp signal in the retention range of strong basic amino acids. As a contaminant in the buffer solution however, ammonium ions are continuously bound to the stationary phase. Because ammonium ions exhibit a very high affinity towards the stationary phase, they pass very slowly through the column, thereby leading to a broad and diffuse signal that disturbs the baseline. For these reasons, the eluants required for amino acid analysis should be stored under inert gas. To prevent the growth of diverse microorganisms, a very common problem, a concentration c = 1 g/L phenol is added to the buffer solution. Out of historic reason, Fig. 5-25 displays the chromatogram of a hydrolysate standard obtained with citrate buffers on a totally sulfonated AminoPac Na-1 cation exchanger. Hydrolysates from connective tissue samples, which also contain hydroxyproline and hydroxylysine, can also be analyzed under the same chromatographic conditions. The only disadvantage is the long analysis time of about 70 minutes, which is mainly due to the low number of buffer solutions used for elution. When the number of citrate/borate buffers is increased from three to five, slightly shorter analysis times with significantly improved peak shapes are obtained. This is impressively illustrated in Fig. 5-26, which shows the separation of a hydrolysate standard on a totally sulfonated BTC 2710 cation exchanger (Eppendorf-Biotronik, Hamburg, Germany).

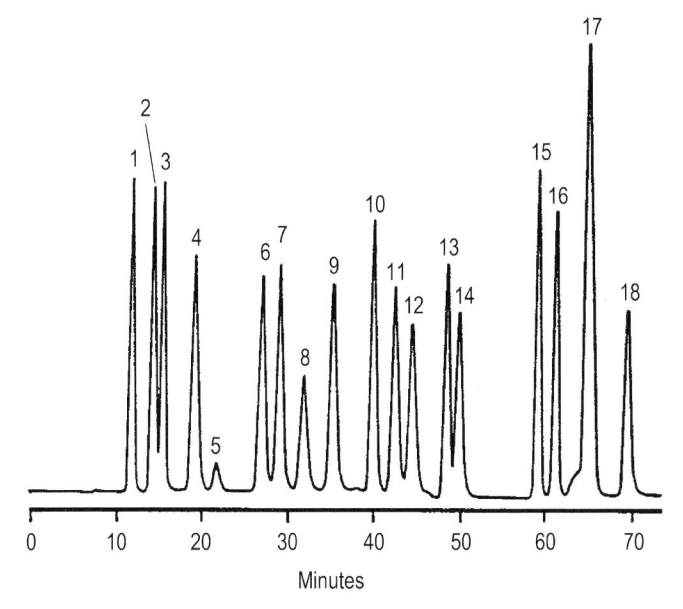

Fig. 5-25. Separation of a hydrolysate standard on a totally sulfonated cation exchanger AminoPac Na-1. — Column temperature: 50 °C; eluant: Hi-Phi buffer A, B and C; flow rate: 0.3 mL/min; detection: photometry with NIN-filter after reaction with ninhydrin at 130 °C; injection volume: 50 µL; solute concentrations: 10 nmol each of ASP (1), THR (2), SER (3), GLU (4), PRO (5), GLY (6), ALA (7), CYS (8), VAL (9), MET (10), ILE (11), LEU (12), TYR (13), PHE (14), HIS (15), LYS (16), NH$_3$ (17), and ARG (18).

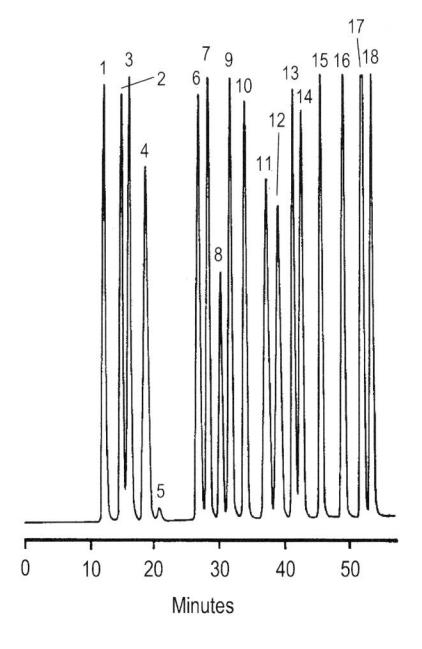

Fig. 5-26. Separation of a hydrolysate standard on a totally sulfonated cation exchanger BTC 2710. — Column temperature: four-step temperature program from 48 °C to 70 °C; eluant: 5 sodium citrate/borate buffers; flow rate: 0.3 mL/min; all other chromatographic conditions and elution order: see Fig. 5-25.

The determination of amino acids in physiological samples is one of the most demanding tasks in the field of amino acid analysis. Often, more than 40 components have to be quantified in complex matrices such as serum or urine. For such separations, lithium citrate buffers are still used as the eluant. The separation of asparagine, glutamic acid, and glutamine, important for physiological samples, is only possible on totally sulfonated cation exchangers in the lithium form. As a representative example of this technique, Fig. 5-27 shows the optimized separation of 40 physiologically relevant amino acids on BTC 2710. The separation has been obtained with five different lithium citrate buffers and a four-step temperature program. The analysis time is more than two hours, but may be reduced significantly if one can dispense with the separation of sarcosine, homocystine, ethanolamine, hydroxylysine, and anserine. The gradient program must be adapted accordingly for the analysis of free amino acids in fruit juices, beer, wine, and milk.

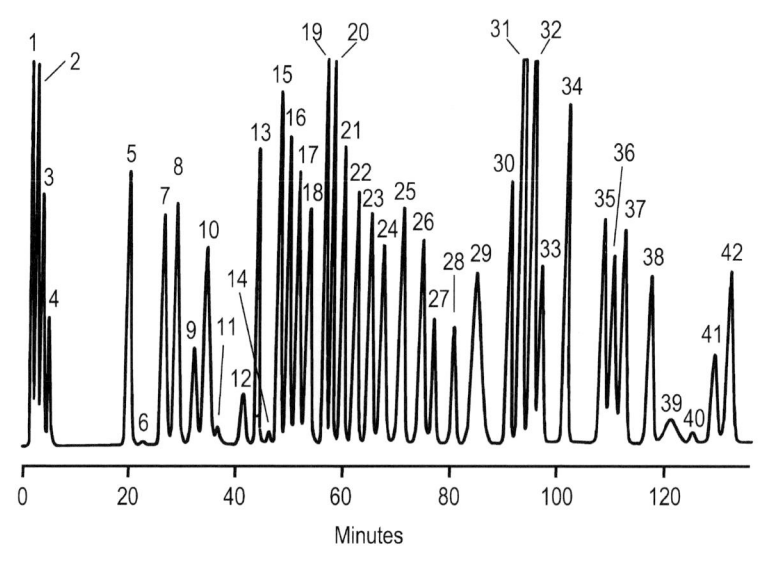

Fig. 5-27. Separation of a standard of physiologically relevant amino acids on a totally sulfonated cation exchanger BTC 2710. — Column temperature: four-step temperature program from 48 °C to 70 °C; eluant: 5 lithium citrate buffer; flow rate: 0.3 mL/min; detection: see Fig. 5-25; solute concentrations: 10 nmol each of Pser (1), Tau (2), PETA (3), Urea (4), Asp (5), OH-Pro (6), Thr (7), Ser (8), Asn (9), Glu (10), Gln (11), Sarc (12), AAAA (13), Pro (14), Gly (15), Ala (16), Citr (17), AABA (18), Val (19), Cys (20), Met (21), Cyst (22), Ile (23), Leu (24), Tyr (25), Phe (26), B-Ala (27), B-AIBA (28), Homocys (29), GABA (30), ETA (31), NH_3 (32), OH-Lys (33), Orn (34), Lys (35), 1-Meth (36), His (37), 3-Meth (38), Trp (39), Anser (40), Carn (41), and Arg (42).

5.9.2
Post-Column Derivatizations of Amino Acids

Ninhydrin

Even today, the post-column derivatization with ninhydrin introduced by Spackman, Stein, and Moore [24] represents the most common detection method for quantitative amino acid analysis. As a strong oxidant, ninhydrin reacts with the α-amino groups of eluting amino acids at temperatures around 130 °C, according to Eq. (125), releasing ammonia and carbon dioxide.

$$\text{Ninhydrin} + \text{Amino acid} \longrightarrow R-\overset{O}{C}-H + NH_3 + CO_2 + \text{Hydrindantin}$$

$$\text{Hydrindantin} \xrightarrow[+ \text{Ninhydrin}]{-H_2O,\ +NH_3} \text{Ruhemann's Purple} \tag{125}$$

An aldehyde shorter by one C-atom and hydrindantin, the reduced form of ninhydrin, are formed in this reaction. Hydrintantin reacts with ammonia and a second molecule of ninhydrin to form a red dye called Ruhemann's purple. This dye has an absorption maximum at 570 nm (see Fig. 5-28).

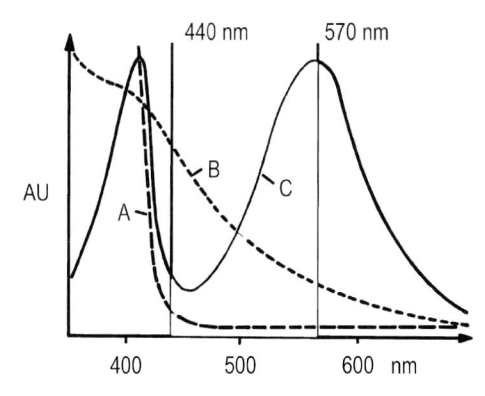

Fig. 5-28. Absorption spectrum of ninhydrin and its reaction products: (A) ninhydrin, (B) derivatized imino acid, (C) derivatized amino acid.

Secondary amino acids, so-called "imino acids" such as proline and hydroxyproline, do not possess an α-amino group, and react with ninhydrin to form a yellow product which is usually detected at 440 nm. Therefore, amino acid analyzers are equipped with a photometric detector capable of measuring at two different wavelengths (570 nm and 440 nm). The sensitivity of this detection method is about 200 pmol.

o-Phthaldialdehyde

o-phthaldialdehyde (OPA) is the reagent most widely used to convert primary amino acids into fluorescing derivatives. The reaction with OPA, first described by Roth [41] in 1971, quickly proved to be a more sensitive alternative to the ninhydrin method [42-44]. Initially, OPA was used exclusively for post-column derivatization. Today, it is increasingly employed for pre-column derivatizations. In an alkaline medium, o-phthaldialdehyde reacts with primary amino acids, according to Eq. (126), to form a strongly blue-fluorescing adduct, which can be detected in the lower pmol range. The reaction requires only a few seconds, even at room temperature.

$$\text{o-phthaldialdehyde} + R-\underset{H}{\underset{|}{C}}(NH_2)-COOH + HS-CH_2-CH_2-OH \longrightarrow \text{adduct} + H_2O \quad (126)$$

Postulated adduct

The subsequent fluorescence detection is carried out at an excitation wavelength of 340 nm; the emission is measured at 455 nm. The addition of thiols such as 2-mercaptoethanol increases the fluorescence yield. However, OPA only reacts with primary amino acids. Secondary amino acids can only be detected after their oxidation (for example with hypochlorite or chloramine T) [45]. In practical applications, this creates significant difficulties.

Other Derivatization Methods

An interesting derivatization method was recently described by Jenke and Brown [46]. They mixed the column effluent with a buffered solution of 5,5'-dithiobis(2-nitrobenzoic acid) (DTNB), yielding a strongly yellow-colored chromophore, which can be detected photometrically at 412 nm:

$$\text{(127)}$$

DTNB   Mixed disulfide   Chromophore

As an alternative to derivatization techniques, amino acids can also be detected directly via pulsed amperometry. Further details are described in Section 6.3.2.2.

5.9.3
Sample Preparation

Sample preparation is the key step in an accurate and reproducible amino acid analysis. In principle, physiological samples should be processed as quickly as possible to avoid contaminations by the sample handling or by the reagents being used. In addition, metabolic activities may lead to false results.

Investigations into urine and plasma samples must consider that certain compounds such as antibiotics are not metabolized after its administration. Antibiotics pass through the body without changing their structure and, finally, get into the urine or plasma and interfere with the amino acid analysis via reaction with ninhydrin. Perry et al. [47] reported this problem for the first time. In urine samples oxidized with hydrogen peroxide they detected a signal in the retention range of cysteic acid and homocystine, which they attributed to D-penicillamine-sulfonic acid. This compound was formed by oxidation of the antibiotic D-penicillamine, which was administered to the patient whose urine was investigated.

If free amino acids are to be analyzed in foods or animal feeds, it is advisable to remove matrix components such as lipids, carbohydrates, and nucleic acids from the sample. Of course, the necessary extraction steps are time-consuming, but higher concentrations of such compounds may interfere severely with the amino acid analysis. After hydrolysis of protein-free mixtures of nucleosides, nucleotides, and nucleic acids, Paddock et al. [48] even observed the formation of amino acids, predominantly glycine.

To analyze free amino acids in plasma or tissue homogenates, it is necessary to remove proteins and peptides present in solution. The most widely used deproteinization method is precipitation with 5-sulfosalicylic acid followed by centrifugation for separating the precipitate. In comparison to other precipitation agents such as trichloroacetic acid, perchloric acid, picrinic acid, or acetonitrile, the best results with respect to completeness of precipitation are obtained with 5-sulfosalicylic acid [49]. Other deproteinization methods comprise ultrafiltration and ultracentrifugation [50], which have only recently been considered as sample preparation methods for amino acid analysis.

Cleavage of the proteins into their amino acid moieties is typicallly accomplished via acid hydrolysis at 110 °C for 24 hours with hydrochloric acid in a concentration $c = 6$ mol/L. Only 17 of the standard amino acids survive this procedure without degradation. Glutamine and asparagine are transferred into the corresponding acids glutamic acid and aspartic acid. Sulfur-containing amino acids such as cysteine, cystine, and methionine have to be oxidized prior to hydrolysis; while isoleucine, valine, threonine, and serine require different hydrolysis times. To protect tryptophan, 2% thioglycolic acid is added to the sample [51]. To remove hydrochloric acid, the hydrolysates are lyophilized or vacuum-dried in a centrifuge, then resuspended in small amounts of water, and dried again. The samples are taken up in special buffer solutions for subsequent amino acid analysis. Quantitative results for sulfur-containing amino acids are obtained via performic acid hydrolysis. However, methionine is oxidized with this method to the respective sulfoxide or sulfone. The best hydrolysis technique that preserves tryptophane is the alkaline hydrolysis with NaOH. This technique is not compatible with the classical methods of amino acid analysis via pre- or post-column derivatization without preceding neutralization, because large amounts of sodium chloride are formed during the hydrolysis. When separating amino acids by anion exchange chromatography, a neutralization is not necessary, because alkaline eluants based on NaOH are used.

Because of the high sensitivity of modern amino acid analyzers, the purity of eluants becomes increasingly important. A possible source for contamination is the de-ionized water that is used to prepare the eluants. It may be contaminated with amino acid traces from the mixed-bed ion exchangers. This problem can be circumvented by using reversed osmosis for water de-ionization. Microbial contaminations in the finished buffer solutions are best avoided by cooling the eluants and by adding phenol. Reagents used to make eluants can also contribute to contaminations. According to one chemical manufacturer (Pierce, Rockford, IL, USA), it is citric acid which is primarily contaminated with traces of amino acids.

Further information about problems that may occur in amino acid analysis by ion-exchange chromatography can be found in the review paper published by Williams [52] in 1986.

6
Ion-Pair Chromatography (MPIC)

Ion-pair chromatography provides a useful alternative to ion-exchange chromatography. The selectivity of the separation in ion-pair chromatography is mainly determined by the type of the mobile phase, so both anionic and cationic compounds can be separated. This universal applicability has helped ion-pair chromatography reach its present significance.

Haney et al. [1, 2], Waters Associates [3], and Knox et al. [4, 5] – to name just a few – found that by adding lipophilic ions such as alkanesulfonic acid or quaternary ammonium compounds to the mobile phase, solute ions of opposite charge can be separated on a chemically bonded reversed phase. The term "Reversed-Phase Ion Pair Chromatography" (RPIPC) has generally been adopted for this technique. The term "Mobile Phase Ion Chromatography" (MPIC) describes a method which combines the major elements of RPIPC with suppressed conductivity detection, previously described. Besides the above-mentioned chemically bonded reversed phases, neutral divinylbenzene resins featuring a high surface area and a weakly polar character are also used as stationary phases.

The physical-chemical phenomena of RPIPC, the basis for the retention mechanism, are still not fully understood. This mechanistic uncertainty is reflected by the many terms proposed for this kind of separation method in the past. Here, we discuss two hypotheses, although both of them lack an unequivocal experimental basis. Horvath et al. [6, 7] take the view that solute ions form neutral ion pairs with the lipophilic ions in the aqueous mobile phase. These neutral ion pairs are retained at the non-polar stationary phase. In contrast, Huber, Hoffmann, and Kissinger [8-10] support the ion-exchange model, where the lipophilic reagent first adsorbs at the surface of the stationary phase, giving it an ion-exchange character. Both hypotheses represent limiting cases. It is not to be expected that the retention process is fully described by just one of the two limiting cases. One must also consider that it is impossible to develop a retention model, that accounts for all experimental data in the same way, is impossible on the basis of chromatographic data only. A deeper insight into the very complex mechanistic events is severely obscured, first by the lack of physical-chemical data, and second by the difficulty in obtaining accurate values for the equilibrium constants of interest.

6.1
Survey of Existing Retention Models

According to the model of *ion-pair formation* [6, 11], the solute ion E interacts with the lipophilic ion H to form a complex, EH. This complex can be reversibly bound at the non-polar surface of the stationary phase L (chemically bonded reversed phase or divinylbenzene resin), giving LEH. The equilibria thus established may be described as follows:

$$E + H \xrightleftharpoons{K_1} EH \tag{128}$$

$$EH + L \xrightleftharpoons{K_2} LEH \tag{129}$$

The respective equilibrium constants can be expressed by Eqs. (130) and (131). The concentrations of individual species in the mobile and the stationary phases are indicated by the indices m and s, respectively.

$$K_1 = [EH]_m / [E]_m \cdot [H]_m \tag{130}$$

$$K_2 = [LEH]_s / [EH]_m \cdot [L]_s \tag{131}$$

$[L]_s$ is defined as the available free surface area of the stationary phase.

According to the *ion-exchange model* [8-10], the lipophilic ion is adsorbed at the surface of the stationary phase, forming LH. This renders the non-polar resin into a dynamic ion exchanger. The solute ion E may subsequently interact with LH:

$$H + L \xrightleftharpoons{K_3} LH \tag{132}$$

$$LH + E \xrightleftharpoons{K_4} LEH \tag{133}$$

The equlibria constants are defined in an analogous way:

$$K_3 = [LH]_s / [H]_m \cdot [L]_s \tag{134}$$

$$K_4 = [LEH]_s / [LH]_s \cdot [E]_m \tag{135}$$

Under the condition that solute ions and lipophilic ions can interact completely independently with the stationary phase, one obtains, according to Horvath [6], a very complicated equilibrium system (see Fig. 6-1).

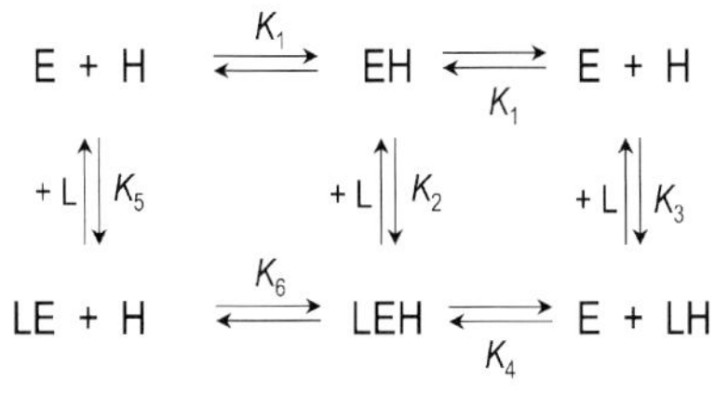

$$
\begin{array}{ccccc}
E + H & \underset{K_1}{\overset{K_1}{\rightleftarrows}} & EH & \rightleftarrows & E + H \\
{\scriptstyle +L} \updownarrow {\scriptstyle K_5} & & {\scriptstyle +L} \updownarrow {\scriptstyle K_2} & & {\scriptstyle +L} \updownarrow {\scriptstyle K_3} \\
LE + H & \underset{K_4}{\overset{K_6}{\rightleftarrows}} & LEH & \rightleftarrows & E + LH
\end{array}
$$

Fig. 6-1. Schematic representation of the equilibrium system in ion-pair chromatography on non-polar stationary phases with lipophilic ions in the mobile phase.

Hence the equilibrium system must be extended as follows:

$$[E] + [L] \overset{K_5}{\rightleftarrows} [LE] \tag{136}$$

$$[LE] + [H] \overset{K_6}{\rightleftarrows} [LEH] \tag{137}$$

and

$$K_5 = [LE]_s/[E]_m \cdot [L]_s \tag{138}$$

$$K_6 = [LEH]_s/[LE]_s \cdot [H]_m \tag{139}$$

With the definition of the capacity factor, k,

$$k = \Phi \cdot K \tag{140}$$

Φ Phase volume ratio
K Distribution coefficient

the capacity factor for the solute ion E is expressed as

$$k = \Phi \cdot \frac{[LEH]_s \cdot [LE]_s}{[E]_m \cdot [EH]_m} \tag{141}$$

When the stationary and mobile phases are in equilibrium, the concentration of lipophilic ions is constant. With

$$[E] \ll [H]$$

only a small fraction of lipophilic ions is complexed, so the concentration of these ions remains virtually unchanged:

$$[H]_m \equiv [H] \tag{142}$$

Assuming further that the fraction of solute ions adsorbed on the stationary phase is very small and that the total surface area L_T of the stationary phase is fixed, it holds that:

$$[L]_T = [L]_s + [LH]_s \equiv [L] \tag{143}$$

If lipophilic ions and solute ions in the mobile phase form ion pairs, and if those ion pairs interact with the stationary phase, Eq. (141) can be expressed with the expressions in Eqs. (130), (131), (134), (135), (142), and (143) as a combination of the various equilibrium constants in dependence on [H] and [L]. For this, Eqs. (130), (131), (134), and (135) are transformed as follows:

$$[LE]_s = K_1 \cdot [E]_m \cdot [L]_s \tag{144}$$

$$[E]_m = [EH]_m / K_2 \cdot [H]_m \tag{145}$$

$$[L]_s = [LH]_s / K_3 \cdot [H]_m \tag{146}$$

$$[LEH]_s = K_4 \cdot [EH]_m \cdot [L]_s \tag{147}$$

Inserting Equations (144) to (147) into Eq. (141) and putting $[L]_s/[E]_m$ outside the brackets, one obtains:

$$k = \Phi \cdot \frac{[L]_s}{[E]_m} \cdot \frac{K_4 \cdot [EH]_m + K_1 \cdot [E]_m}{1 + K_2 \cdot [H]_m} \tag{148}$$

This expression is expanded with $(1 + K_3 \cdot [H]_m)$:

$$k = \Phi \cdot \frac{[L]_s}{[E]_m} \cdot \frac{(1 + K_3 \cdot [H]_m)(K_4 \cdot [EH]_m + K_1 \cdot [E]_m)}{(1 + K_2 \cdot [H]_m)(1 + K_3 \cdot [H]_m)} \tag{149}$$

With Eqs. (145) and (146) it follows:

$$k = \Phi \cdot \frac{K_2 \cdot [LH]_s}{K_3 \cdot [EH]_m} \cdot \frac{(1 + K_3 \cdot [H]_m)(K_4 \cdot [H]_m + K_1 \cdot [E]_m)}{(1 + K_2 \cdot [H]_m)(1 + K_3 \cdot [H]_m)} \tag{150}$$

$$k = \Phi \cdot [LH]_s \cdot \frac{\left(\dfrac{K_2}{K_3} + K_2 \cdot [H]_m\right) \cdot \left(K_4 + \dfrac{K_1 \cdot [E]_m}{[EH]_m}\right)}{(1 + K_2 \cdot [H]_m)(1 + K_3 \cdot [H]_m)} \tag{151}$$

Inserting Eq. (145) and subsequent multiplication of the bracketed expression gives:

$$k = \Phi \cdot [LH]_s \cdot \frac{\dfrac{K_2 \cdot K_4}{K_3} + \dfrac{K_1}{K_3 \cdot [H]_m} + K_2 \cdot K_4 \cdot [H]_m + K_1}{(1 + K_2 \cdot [H]_m)(1 + K_3 \cdot [H]_m)} \tag{152}$$

$$k = \Phi \cdot \frac{[LH]_s \cdot \left(\dfrac{K_2 \cdot K_4}{K_3} + \dfrac{K_1}{K_3 \cdot [H]_m}\right) + [LH]_s \cdot (K_2 K_4 \cdot [H]_m + K_1)}{(1 + K_2 \cdot [H]_m)(1 + K_3 \cdot [H]_m)} \tag{153}$$

Inserting Eq. (146) it follows:

$$k = \Phi \cdot \frac{K_2 K_4 [H]_m \cdot [L]_s + K_1 \cdot [L]_s + [LH]_s \cdot (K_2 K_4 \cdot [H]_m + K_1)}{(1 + K_2 \cdot [H]_m)(1 + K_3 \cdot [H]_m)} \quad (154)$$

After putting $[L]_s$ outside the brackets and considering Eq. (143) the intended dependence is derived:

$$k = \Phi \cdot [L] \frac{K_1 + K_2 K_4 \cdot [H]}{(1 + K_2 \cdot [H])(1 + K_3 \cdot [H])} \quad (155)$$

According to the model of dynamic ion-exchange, the following expression for the capacity factor, k, is obtained in much the same way using the expressions of Eqs. (130), (131), (134), (138), and (141) to (143):

$$k = \Phi \cdot [L] \frac{K_1 + K_3 K_5 \cdot [H]}{(1 + K_2 \cdot [H])(1 + K_3 \cdot [H])} \quad (156)$$

Equations (155) and (156) show the dependence of the capacity factor, k, on the concentration of lipophilic ions. In a general form, this dependence can be described as

$$k = (K_0 + B \cdot [H])/(1 + K_2 \cdot [H]) \cdot (1 + K_3 \cdot [H]) \quad (157)$$

K_0 is the capacity factor of the solute ion if no lipophilic ions take part in the retention process. According to this model, K_2 is the formation constant for the respective ion pair, and K_3 is a constant characterizing the bonding between the lipophilic ions and the stationary phase. According to Eqs. (155) and (156), constant B represents the product of two equilibrium constants.

Plotting k versus $[H]$, a parabolic curve is obtained when, according to Eq. (157), $1/K_3 \cdot [H] \ll 1/K_2$. Such dependences have been observed by Knox et al. [5, 12].

Finally, in 1979, Bidlingmeyer et al. [13, 14] introduced a third model which they termed the *ion interaction model*. It is based on conductivity measurements, the results of which rule out the formation of ion pairs in the mobile phase. This retention model, also used by Pohl [15] to interpret the retention mechanism on a MPIC phase, neither presupposes the formation of ion pairs nor is it based on classical ion-exchange chromatography.

According to the ion interaction model, a high surface tension is generated between the non-polar stationary phase and the polar mobile phase. From this, the stationary phase obtains a high affinity towards those components of the mobile phase which are able to reduce the high surface tension. Such components include, for example, polar organic solvents, surfactants with their respective counter ions, and quaternary ammonium bases. Moreover, the model concept of ion interaction provides for an electrically charged double layer at the surface of the stationary phase. This phenomenon is schematically represented in Fig. 6-2 taking the analysis of surface-inactive anions as an example.

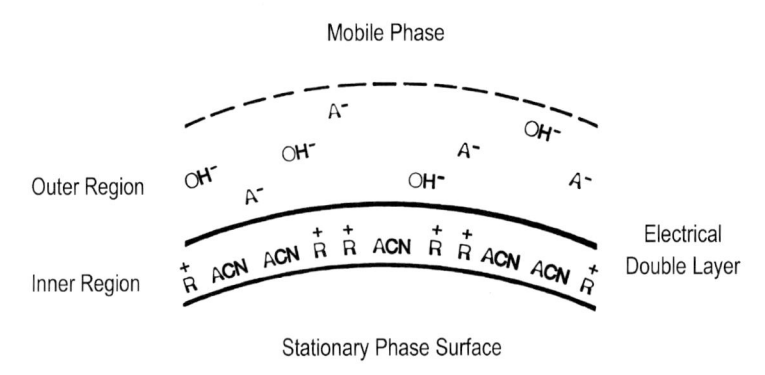

Fig. 6-2. Schematic representation of the electrically charged double layer when separating surface-inactive anions.

As can be seen from Fig. 6-2, the lipophilic ions (e.g., tetrabutylammonium cations) and acetonitrile (as the organic modifier) are adsorbed in the inner region at the surface of the non-polar stationary phase. As all lipophilic cations are equally charged, the surface can be only partly covered with such ions because of the repulsive forces between these charges. The corresponding counter ions (typically OH^- ions when applying conductivity detection), as well as the analyte ions A^-, are found in the diffuse outer region. When the lipophilic ion concentration in the mobile phase is increased, the concentration of ions adsorbed to the surface will also increase because of the dynamic equilibrium between the mobile and the stationary phases. The transfer of a solute ion through the electrical double layer is, therefore, a function of electrostatic and van-der-Waals forces. If a solute ion with an opposite charge is attracted by the charged surface of the stationary phase, retention is a result of Coulomb attractive forces and additional adsorptive interactions between the lipophilic part of the solute ion and the non-polar surface of the stationary phase. Adding a negative charge to the positively charged inner region of the double layer is tantamount to removing one charge from this region. To re-establish electrostatic equilibrium, another lipophilic ion can be adsorbed at the surface. Finally, two oppositely charged ions (not necessarily an ion pair) are adsorbed at the stationary phase.

The separation of surface-inactive cations can be interpreted analogously. In this case, lipophilic anions are adsorbed at the resin surface, while the analyte cations are retained in the outer region of the double layer.

Unlike normal solute ions, surface-active ions may penetrate the inner region of the double layer, where they are adsorbed at the surface of the stationary phase. With this class of compounds, retention depends on the carbon chain length and, thus, on the degree of hydrophobicity. Retention increases with growing chain length. Acetonitrile as an organic modifier is also adsorbed at the resin surface, and therefore is involved in a competing equilibrium with the lipophilic ions. When both surface-active and surface-inactive ions are analyzed, the organic modifier serves to shorten retention by blocking adsorption sites on the resin surface. In the case of surface-active ions, this is achieved by direct competition; for surface-inactive ions it is achieved via competition with $R\text{-}SO_3^-$ and R_4N^+, the solvophobic counter ions.

6.2 Suppressor Systems in Ion-Pair Chromatography

As with ion-exchange and ion-exclusion chromatography, the background conductivity in ion-pair chromatography was also chemically decreased to enable a sensitive conductivity detection. Conventional suppressor columns such as the ASC-1 (for anion determinations) and CSC-1 (for cation determinations) were initially used. After an average operation lasting about six hours, these columns had to be regenerated by the procedure described in the Sections 3.6.1 and 4.3.1.

The hollow fiber suppressors, developed for ion-exchange chromatography at the beginning of the 1980s, could not be employed in ion-pair chromatography, because the occasionally high concentrations of organic solvents in the mobile phase were detrimental to the membrane material. Also, the membranes were not permeable by the ion-pair reagents being used. To alleviate the problems with the conventional suppressor columns, that were discussed in Section 3.6.1, a sulfonated membrane was developed for the ion-pair chromatographic determination of anions. The membrane featured good permeabilities for quaternary ammonium bases and was characterized by a high resistance to organic solvents. Such hollow fiber suppressors were introduced under the trade name AFS-2. Dilute sulfuric acid with $c = 0.01$ mol/L is used as a regenerant with the AFS-2 for ion-pair chromatography, just as it is with the AFS-1 for anion exchange chromatography. This means that in the suppressor reaction, tetrabutylammonium cations are exchanged for regenerant hydronium ions. As in the process described in Section 3.6.2, water is the reaction product; its formation provides the driving force for the hydronium ion diffusion through the membrane wall. The analyte anions are converted into their respective acids, thereby enabling a more sensitive and selective detection. Hollow fiber suppressor are employed in ion-exclusion chromatography and in the ion-pair chromatography of anions; in the case of alternate operation, it is advisable to condition the suppressor overnight with the appropriate regenerant.

The CFS hollow fiber suppressor (see Section 4.3.2) that was developed for cation exchange chromatography can also be applied to cation analysis via ion-pair chromatography. It features good solvent stability and sufficient permeability for the anionic ion-pair reagent. This suppressor is regenerated with tetramethylammonium hydroxide using at a concentration of $c = 0.04$ mol/L.

As described in Sections 3.6.2 and 4.3.2, hollow fiber suppressors no longer represent the state-of-the-art. A micromembrane suppressor was introduced under the trade name AMMS-MPIC for the ion-pair chromatography of anions. Its structure corresponds to the systems developed for ion-exchange and ion-exclusion chromatography. Like the AFS-2, the AMMS-MPIC micromembrane suppressor contains a solvent-resistant membrane that is permeable to quaternary ammonium bases. Regarding the ion-exchange capacity, there is no difference with the AFS-2; a significantly smaller void volume contributes to the sensitivity increase. This micromembrane suppressor is regenerated in the same manner as the related hollow fiber suppressor, it is also performed with 0.01 mol/L sulfuric acid.

The solvent stability of the CMMS micromembrane suppressor, introduced in Section 4.3.3, is sufficient to employ such suppressors in the ion-pair chromatography of cations. In this case, the suppressor is regenerated with tetrabutylammonium hydroxide. The regenerant concentration adheres to the concentration of the ion-pair reagent. A reagent concentration of $c = 2$ mmol/L, for example, requires a TBAOH concentration of $c = 0.04$ mol/L.

6.3
Experimental Parameters that Affect Retention

The primary advantage of ion-pair chromatography over ion-exchange chromatography is its great flexibility, which allows the chromatographic conditions to be adjusted for a given separation problem. This flexibility results from the great variety of experimental parameters that affect retention. Thus, ion-pair chromatographic separations of ions on a MPIC phase are affected by the following parameters:
- Type of lipophilic counter ion in the mobile phase
- Concentration of lipophilic counter ion in the mobile phase
- Type of organic modifier
- Concentration of organic modifier in the mobile phase
- Type and concentration of inorganic additives
- Eluant pH
- Column temperature

The effects on the retention process achieved by varying those individual parameters are discussed below.

6.3.1
Type and Concentration of Lipophilic Counter Ions in the Mobile Phase

For the ion-pair chromatographic separation of *anions*, quaternary ammonium bases are preferably added as lipophilic ions to the mobile phase. According to the literature [16-18], there are basically two types of stationary phases that can be used – permanently coated and dynamically coated stationary phases. Dynamic coating is realized with reagents of low hydrophobicity such as tetraalkylammonium compounds. Permanent coating is accomplished with strongly hydrophobic reagents such as cetyltrimethylammonium salts. After coating the stationary phase with these strongly hydrophobic salts, tetramethylammonium compounds are generally used to elute the solutes. As in liquid-liquid partition chromatography with pre-coated support material, these "permanently" coated phases are not stable for long periods of time, because even the strongly adsorbed cetyl compounds are slowly washed off.

It is imperative to consider the kind of counter ion in an ion-pair reagent for selecting the appropriate detection method. If suppressed conductivity detection is applied, the ion-pair reagent is used in its hydroxide form. With direct conductivity detection, salicylate is preferred as the counter ion for tetraalkylammonium cations [19, 20], since these salts exhibit a lower background conductivity in aqueous solution. According to Wheals [21], eluants such as cetyltrimethylammonium bromide in combination with citric acid at pH 5.5 have proved suitable for UV-, RI-, and amperometric detection, as well as for direct conductivity detection. This example is an impressive illustration of the versatility of ion-pair chromatography.

The same applies to the ion-pair chromatographic separation of *cations*, which is performed either with long-chain alkanesulfonic acids or, in the simplest manner, with mineral acids. A survey of the most commonly used reagents is listed in Table 6-1 in the order of increasing hydrophobicity. A comparison by Cassidy and Elchuk [18] of chemically bonded reversed phases with organic divinylbenzene-based (Dionex IonPac NS1) versus phases with polystyrene/divinylbenzene-based polymers (Hamilton PRP-1) revealed that the latter have a lower chromatographic efficiency, but are characterized by a higher affinity towards certain ion-pair reagents.

Table 6-1. Commonly used reagents for ion-pair chromatography on MPIC phases, in the order of increasing hydrophobicity.

Anion analysis	Cation analysis
Ammonium hydroxide	HCl, HClO$_4$
Tetramethylammonium hydroxide	Hexanesulfonic acid
Tetrapropylammonium hydroxide	Heptanesulfonic acid
Tetrabutylammonium hydroxide	Octanesulfonic acid

(Hydrophobicity increasing downward)

The choice of lipophilic ion depends solely on the degree of hydrophobicity of the analyte ion. For the separation of surface-inactive ions, a hydrophobic reagent is necessary; on the other hand, the separation of ions with long alkyl chains requires a strongly hydrophilic reagent. The hydrophobicity of an ion-pair reagent is generally determined by the number of its carbon atoms and, above all, by the length of its alkyl groups. This is illustrated in Fig. 6-3, which depicts the dependence of the k value for diphenhydramine on the chain length of the ion-pair reagent, when using a polymer phase. The sodium salts of various alkanesulfonic acids were employed as ion-pair reagents. It appears from Fig. 6-3 that the diphenhydramine retention significantly increases with the increasing chain length of the ion-pair reagent. According to the ion-pair model, such a dependence may be accounted for by the fact that the hydrophobic character of the formed ion pairs increases with increasing chain length of the ion-pair reagent. This results in both a stronger interaction between the ion pair and the stationary phase and a reduced solubility in the eluant. A variation in the chain length of the ion-pair reagent affects the various solute ions in different ways. This parameter enables, therefore, the very efficient adjustment of the selectivity of the chromatographic system. With alkaloids such as atropine and papaverine, Jost et al. [22] even observed a change in the retention order with the increasing chain length of the alkanesulfonates that were used as the ion-pair reagent.

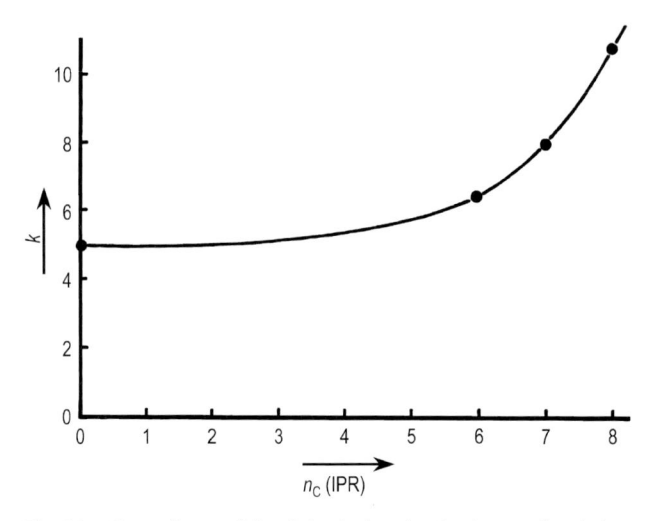

Fig. 6-3. Dependence of the diphenhydramine k value on the chain length n_C of the ion-pair reagent. — Separator column: IonPac NS1 (10-µm); eluant: 0.05 mol/L KH_2PO_4 (pH 4.0) / acetonitrile (70:30 v/v) + 0.005 mol/L IPR; flow rate: 1 mL/min; detection: UV (220 nm); injection volume: 50 µL; solute concentration: 40 mg/L diphenhydramine hydrochloride.

Another method for controlling retention is to vary the concentration of the ion-pair reagent. Figure 6-4 shows the dependence of the k value on the concentration of the ion-pair reagent, again taking diphenhydramine as an example. A DVB resin was employed as the stationary phase, the eluant was comprised of a mixture of acetonitrile and a phosphate buffer, and sodium octanesulfonate

was used as the ion-pair reagent. As seen in Fig. 6-4, in the investigated concentration range between 0 and 0.02 mol/L, the k value depends almost linearly on the concentration. For some compounds, such curves may pass a maximum [23]. The concentration of lipophilic ions to be used is limited by two factors:
- With increasing reagent concentration, the surface of the stationary phase is increasingly blocked by undissociated molecules of the ion-pair reagent or by ion pairs from the ion-pair reagent and buffer ions.
- When applying conductivity detection, an increase in reagent concentration results in an increase in background conductivity.

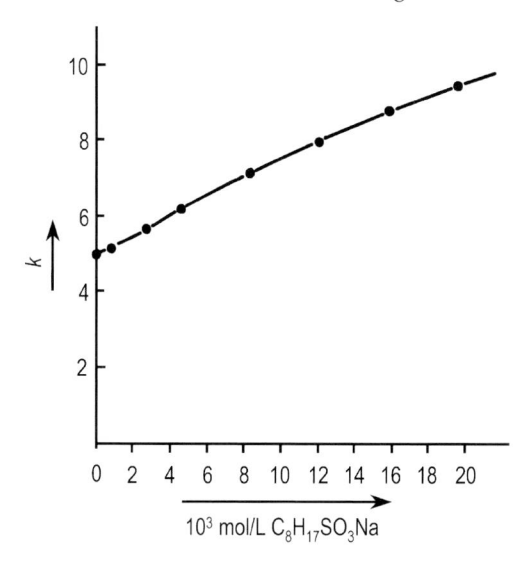

Fig. 6-4. Dependence of the diphenhydramine k value on the ion-pair reagent concentration. — Separator column: IonPac NS1 (10-μm); eluant: 0.05 mol/L KH_2PO_4 (pH 4.0)/acetonitrile (70:30 v/v) + $C_8H_{17}SO_3Na$; flow rate: 1 mL/min; all other chromatographic conditions: see Fig. 6.3.

Therefore, the compounds listed in Table 6-1 are used in a concentration range between $5 \cdot 10^{-4}$ and 10^{-2} mol/L. The most significant changes in the retention values in dependence on the ion-pair reagent concentration occur in this range. In many cases, $c = 2$ mmol/L has proved to be an appropriate reagent concentration.

6.3.2
Type and Concentration of the Organic Modifier

In analogy to reversed-phase chromatography, in ion-pair chromatography, organic solvents such as acetonitrile or methanol are added to the aqueous mobile phase as organic modifiers. Because the solvent molecules are adsorbed at the surface of the stationary phase, they are in a competing equilibrium with lipophilic ions, vying for active centers on the stationary phase available for adsorption. This effect is illustrated by Fig. 6-5. The degree to which the retention of two aromatic sulfonic acids (p-toluenesulfonic acid and naphthalene-2-sulfonic acid) depends on the water content of the mobile phases was quantified in this investigation. An acetonitrile/water mixture was used, to which tetrabutylammonium hydroxide (TBAOH) was added as the ion-pair reagent. Plotting ln k versus the acetonitrile content in the mobile phase yields the parabolic dependence shown in Fig. 6-5.

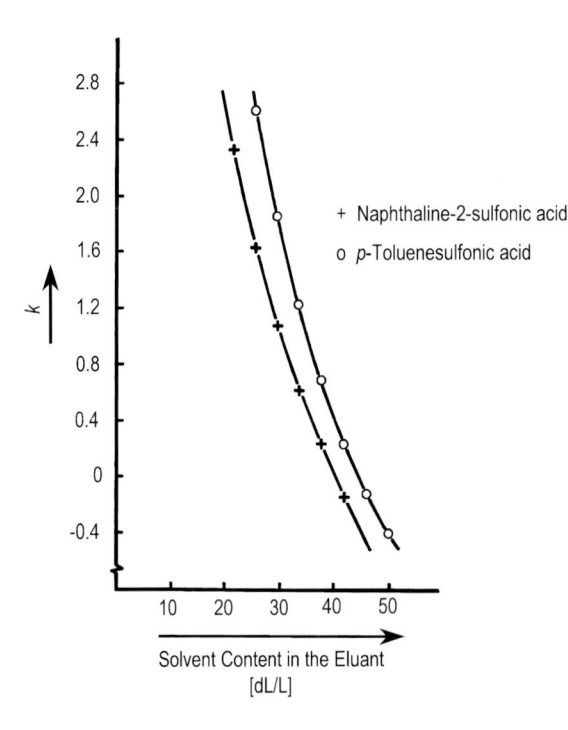

Fig. 6-5. Dependence of the retention of aromatic sulfonic acids on the organic solvent content in the mobile phase. — Separator column: IonPac NS1 (10-µm); eluant: 2 mmol/L tetrabutyl-ammonium hydroxide/acetonitrile; flow rate: 1 mL/min; detection: UV (254 nm); injection volume: 50 µL; solute concentrations: 40 mg/L each of p-toluenesulfonic acid monohydrate and naphthaline-2-sulfonic acid.

When replacing acetonitrile with methanol, the solvent content in the mobile phase must be enhanced to obtain comparable retention times. This effect can be illustrated by using the separation of naphthalene-2-sulfonic acid as an example. An aqueous TBAOH solution with a volume fraction of 500 mL/L acetonitrile as the organic modifier was selected to elute the acid. When this fraction is gradually replaced by methanol, a significant retention increase results. This is graphically depicted in Fig. 6-6 as a function of the percentage distribution between methanol and acetonitrile in the volume fraction of the organic solvent. The selectivity difference obtained with methanol as the organic modifier is due to the capability of methanol to form hydrogen bonds. Compared with acetonitrile, the higher viscosity of methanol is a disadvantage, because it leads to a larger pressure drop along the column.

From the dependences shown in the Figs. 6-3, 6-4, and 6-5, it is obvious that the majority of analyte ions may be separated under different chromatographic conditions; i.e., with different ion-pair reagents and solvent contents. This is impressively demonstrated by the analysis of alkanesulfonates with medium

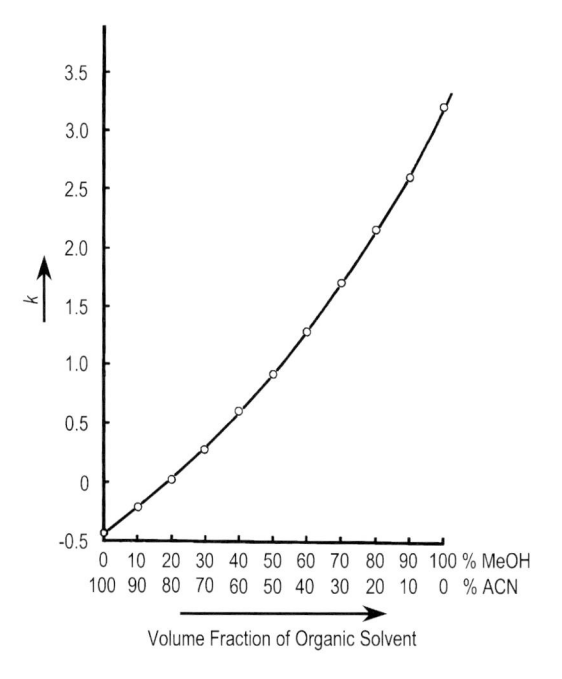

Fig. 6-6. Dependence of the naphthalene-2-sulfonic acid retention on the percentage distribution between methanol and acetonitrile in the volume fraction of the organic modifier. – Separator column: IonPac NS1 (10-µm); eluant: 2 mmol/L TBAOH – acetonitrile/methanol (50:50 v/v); flow rate: 1 mL/min; detection: UV (254 nm).

chain lengths (C_5 to C_8). Because these compounds have a hydrophobic character, the hydrophilic ammonium hydroxide suggests itself as the ion-pair reagent. The corresponding chromatogram is shown in Fig. 6-7a. It was obtained under isocratic conditions with an acetonitrile content of 150 mL/L in the mobile phase. All compounds are separated to baseline, but the heptane- and octane-sulfonic acid peaks are characterized by a strong tailing. If an ion-pair reagent of a higher hydrophobicity, such as tetramethylammonium hydroxide, is used, the resolution between the different signals is enhanced. The retention time increase may be compensated, as shown in Fig. 6-7b, by raising the acetonitrile content in the eluant. The resulting chromatogram is achieved within comparable analysis times but the tailing of the two late-eluting components is distinctly reduced. This effect can be heightened by selecting tetrabutylammonium hydroxide as the ion-pair reagent. Figure 6-7c shows the separation that is optimized with regard to analysis time, resolution, and peak shape.

This example corroborates the above statement that analyte compounds can be separated by several different methods. In many cases, the sample matrix will determine the selection of the chromatographic conditions.

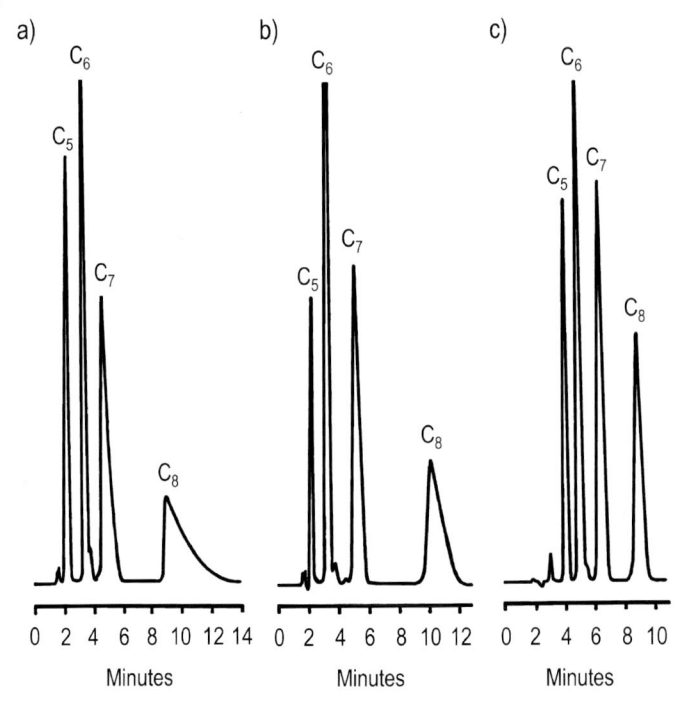

Fig. 6-7. Dependence of the separation of alkylsulfonates with medium chain lengths on the type of lipophilic ion and the concentration of organic modifier. — Separator column: IonPac NS1 (10-µm); eluant: a) 2 mmol/L NH$_4$OH/acetonitrile (85:15 v/v); b) 2 mmol/L TMAOH/acetonitrile (82:18 v/v), c) 2 mmol/L TBAOH/acetonitrile (63:37 v/v).

6.3.3
Inorganic Additives

Inorganic additives such as sodium carbonate are added to the mobile phase to control the retention of di- and multi-valent anions and to improve their peak shape. The sodium carbonate effect on retention is most simply illustrated by using the analysis of inorganic anions as an example. Note the chromatogram in Fig. 6-8a, which depicts the separation of fluoride, chloride, bromide, nitrate, and sulfate. It was obtained using tetrabutylammonium hydroxide as the ion-pair reagent and a small amount of acetonitrile as the organic modifier. Under these chromatographic conditions, the monovalent anions are separated to baseline and elute within ten minutes. For the divalent sulfate, however, an extremely long retention time and a pronounced peak broadening is observed. Orthophosphate, which also exists as a divalent anion under these conditions, is even more strongly retained. After adding 1 mmol/L sodium carbonate to this eluant, the chromatogram depicted in Fig. 6-8b is obtained. It reveals that the retention decrease is much higher for divalent anions than for monovalent ones, which avoids the unnecessarily high resolution between anions of different valency.

The addition of even minute amounts of sodium carbonate has a particularly strong effect on the retention behavior of multivalent anions. The two iron cyanide complexes, $Fe(CN)_6^{3-}$ and $Fe(CN)_6^{4-}$, are a good example; their separation is obtained with an eluant containing only $3 \cdot 10^{-4}$ mol/L sodium carbonate (see Fig. 6-9), in addition to tetrabutylammonium hydroxide and acetonitrile. Lowering the acetonitrile content in favor of sodium carbonate, the resolution between both signals will decrease drastically, although the peak shape of the iron(II) complex will be significantly improved.

The effect of sodium carbonate as an inorganic additive is mechanistically not completely clear. According to the dynamic ion-exchange model, it is to be assumed that carbonate ions are found in a competing equilibrium with solute ions for the ion-exchange groups that are adsorbed at the surface of the stationary phase. This is a plausible explanation for the strong effect of carbonate on the retention of divalent species.

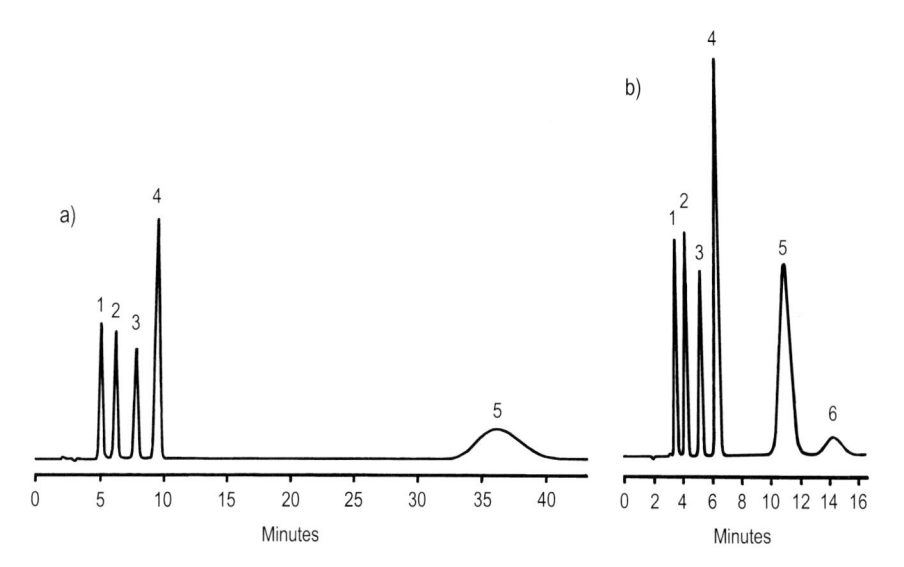

Fig. 6-8. Illustration of the sodium carbonate influence on retention exemplified with an inorganic anion separation. — Separator column: IonPac NS1 (10-μm); eluent:
a) 2 mmol/L TBAOH/acetonitrile (82:18 v/v),
b) 2 mmol/L TBAOH + 1 mmol/L Na_2CO_3/acetonitrile (82:18 v/v); flow rate: 1 mL/min; detection: suppressed conductivity; injection volume: 50 μL; solute concentrations: 3 mg/L fluoride (1), 4 mg/L chloride (2), 10 mg/L bromide (3), 20 mg/L nitrate (4), 25 mg/L sulfate (5), and 10 mg/L orthophosphate (6).

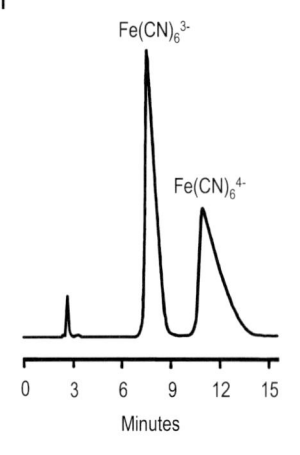

Fig. 6-9. Separation of the two iron cyanide complexes. — Separator column: IonPac NS1 (10-µm); eluant: 2 mmol/l TBAOH + 0.3 mmol/L Na_2CO_3/acetonitrile (60:40 v/v); flow rate: 1 mL/min; detection: suppressed conductivity; injection volume: 50 µL; solute concentrations: 40 mg/L each of $Fe(CN)_6^{3-}$ and $Fe(CN)_6^{4-}$.

6.3.4
pH Effects and Temperature Influence

When analyzing multivalent ions it is often necessary to change the pH value of the mobile phase by adding appropriate acids or bases. As the retention of multivalent ions increases with the degree of dissociation, the pH value affects retention by determining the degree of dissociation. This effect is illustrated by the analysis of thioglycolic acid. This compound can be separated by using tetrabutylammonium hydroxide as the ion-pair reagent and adding an appropriate amount of acetonitrile to the mobile phase. When detecting thioglycolic acid by suppressed conductivity detection, no chromatogram is obtained under the above-mentioned chromatographic conditions (pH 10.8). However, if the eluant pH is lowered to 7.25 by adding crystalline boric acid, the chromatogram in Fig. 6-10 results.

Boric acid is especially suitable for lowering the pH value. Because of its low degree of dissociation, it contributes only marginally to the increase in background conductivity. In general, an increase in the eluant pH is accomplished with sodium hydroxide, because it suppresses to water and hardly affects the background conductivity.

In addition to controlling the degree of dissociation of the analyte species, eluant pH changes are necessary to avoid unwanted side reactions in an acidic or alkaline medium. This applies, for example, to mercaptans, which may react to form disulfides in alkaline medium.

Unlike the eluant pH value, the column temperature is seldom relevant for optimizing the separation. Retention can be somewhat reduced by raising the column temperature. Generally speaking, the viscosity of the mobile phase will be reduced and the chromatographic efficiency will be increased when the column temperature is raised. For mechanistic investigations, however, a variation in column temperature offers the ability to determine the dependence of retention on the temperature and to derive important thermodynamic quantities such as the sorption enthalpies; (see also Section 3.2).

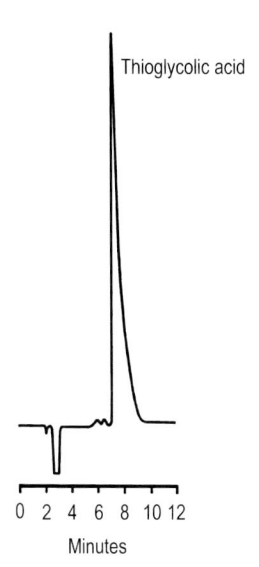

Fig. 6-10. Separation of thioglycolic acid. — Separator column: IonPac NS1 (10-µm); eluant: 2 mmol/L TBAOH/acetonitrile (80:20 v/v), pH 7.25 with H_3BO_3; flow rate: 1 mL/min; detection: suppressed conductivity; injection volume: 50 µL; solute concentration: 50 mg/L.

6.4 Analysis of Surface-Inactive Ions

In the field of *anion analysis*, ion-pair chromatography is a significant alternative to ion-exchange chromatography. If a co-elution of two anions is suspected with one of the two methods, it is often possible to solve the separation problem with the other method; two different compounds rarely show the same retention behavior under completely different chromatographic conditions.

A typical example is the analysis of nitrate and chlorate. In the past, it was not possible to separate the two on conventional anion exchangers such as IonPac AS4A(-SC) (Dionex Corp.), which made it harder to accurately interpret the chromatogram. The only way to distinguish between nitrate and chlorate was to make use of their different absorption characteristics: chlorate is UV-transparent while nitrate can be detected at a wavelength of 215 nm. If both species were present in solution, determinations were only be possible via differential detection.

On the other hand, with ion-pair chromatography, nitrate and chlorate are resolved using tetrabutylammonium hydroxide as the ion-pair reagent, with nitrate eluting prior to chlorate (Fig. 6-11). Pursuing the hypothesis that the stationary phase is transformed into a dynamic ion exchanger by adding lipophilic ions, this different selectivity (as compared to conventional ion exchangers) seems to be related to the different type of ion-exchange selectivity provided by the lipophilic ions.

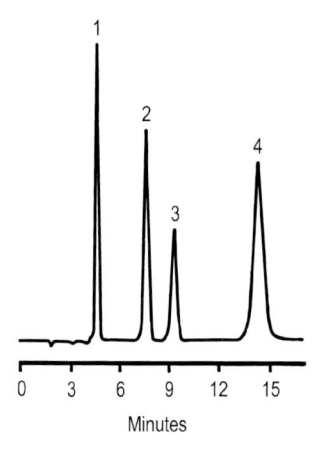

Fig. 6-11. Ion-pair chromatographic separation of nitrate and chlorate. — Separator column: IonPac NS1 (10-µm); eluant: 2 mmol/L TBAOH + 1 mmol/L Na_2CO_3/acetonitrile (85:15 v/v); flow rate: 1 mL/min; detection: suppressed conductivity; injection volume: 50 µL; solute concentrations: 5 mg/L chloride (1), 10 mg/L nitrate (2), 10 mg/L chlorate (3), and 15 mg/L sulfate (4).

Another example is the analysis of benzoic acid, which elutes on a conventional anion exchanger directly after nitrite. Interferences in the determination of this compound caused by a high electrolyte content in the sample cannot be ruled out until the more selective UV detection is applied. With ion-pair chromatography, benzoic acid exhibits a completely different retention behavior. This is due to π-π interactions between the aromatic benzoic acid ring and the aromatic skeleton of the stationary phase, so that benzoic acid is more strongly retained. Under the chromatographic conditions given in Fig. 6-12, it elutes long after the mineral acids.

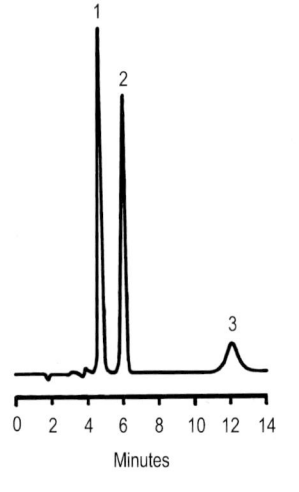

Fig. 6-12. Ion-pair chromatographic separation of nitrite, nitrate, and benzoate. — Separator column: IonPac NS1 (10-µm); eluant: 2 mmol/L TBAOH + 1 mmol/L Na_2CO_3/acetonitrile (82:18 v/v); flow rate: 1 mL/min; detection: suppressed conductivity; injection volume: 50 µL; solute concentrations: 10 mg/L nitrite (1), 10 mg/L nitrate (3), and 20 mg/L benzoate (3).

Tetraalkylammonium salts that are suitable ion-pair reagents for ion-pair chromatography of simple inorganic anions are predominantly employed in combination with polymer phases [24, 25]. However, they may also be used when applying chemically bonded reversed phases [16, 26]. The separation of inorganic anions on LiChrosorb RP18 shown in Fig. 6-13 is an example. In contrast to polymer phases, the fluoride determination is a problem with chemically bonded reversed phases because fluoride is poorly retained. If direct conductivity detection is applied, a quantitative evaluation of the fluoride signal is difficult.

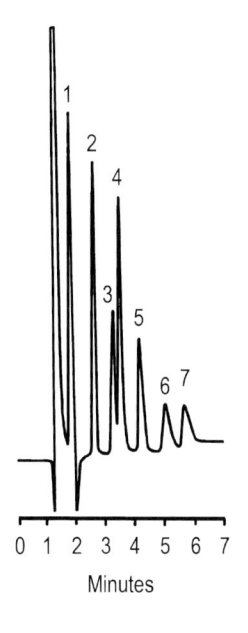

Fig. 6-13. Ion-pair chromatographic separation of inorganic anions on a chemically bonded reversed phase. – Separator column: LiChrosorb RP 18 (10-µm); eluant: 2 mmol/L TBAOH + 0.05 mol/L phosphate buffer (pH 6.7); flow rate: 2 mL/min; detection: direct conductivity; injection volume: 20 µL; solute concentrations: 1000 mg/L each of fluoride (1), chloride (2), sulfate (3), nitrite (4), bromide (5), dichromate (6), and nitrate (7); (taken from [26]).

If the counter ion of the ion-pair reagent exhibits suitable absorption characteristics, indirect photometric detection is feasible. Based on a 1982 paper by Dreux et al. [27], Bidlingmeyer et al. [28] developed a procedure for separating inorganic anions with tetrabutylammounium salicylate as the eluant. This method is suitable for the simultaneous application of indirect photometric and direct conductivity detection. At a measuring wavelength of 288 nm, the authors obtained the chromatogram depicted in Fig. 6-14 using a radially compressed ODS cartridge as the stationary phase.

Furthermore, ion-pair chromatographic separations of inorganic anions are performed on chemically bonded cyanopropyl phases [17, 29]. Figure 6-15 shows a typical result for a separation on such a stationary phase.

To analyze bromide and nitrate in foodstuff, Leuenberger et al. [30] used a chemically bonded aminopropyl phase with a phosphate buffer eluant. Cortes [31] applied this method successfully to the separation of other inorganic anions. The chromatogram depicted in Fig. 6-16 confirms the extraordinary selectivity of ion-pair chromatography for monovalent anions. A final assessment of the applicability of cyano- and amino-propyl phases in the separation of inorganic anions is not presently feasible, as only UV absorbing species have been investigated so far. Although such stationary phases exhibit the high chromatographic efficiency as expected, they have a limited pH stability because of their silica structure.

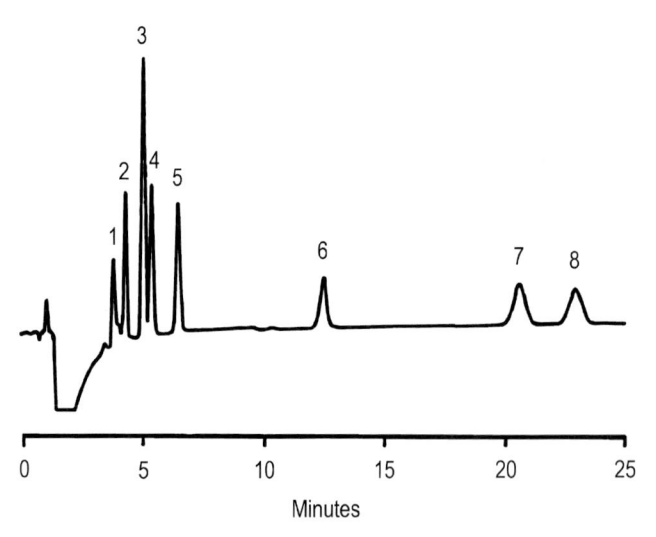

Fig. 6-14. Ion-pair chromatographic separation of inorganic anions with indirect photometric detection. — Separator column: Waters C18 Radial-PAK (5-μm); eluant: 0.4 mmol/L tetrabutylammonium salicylate (pH 4.62); flow rate: 2 mL/min; detection: indirect UV (288 nm); injection volume: 50 μL; solute concentrations: 4 mg/L orthophosphate (1), 2 mg/L chloride (2), 4 mg/L each of nitrite (3), bromide (4), nitrate (5), and iodide (6), 6 mg/L sulfate (7), and 6 mg/L thiosulfate (8); (taken from [28]).

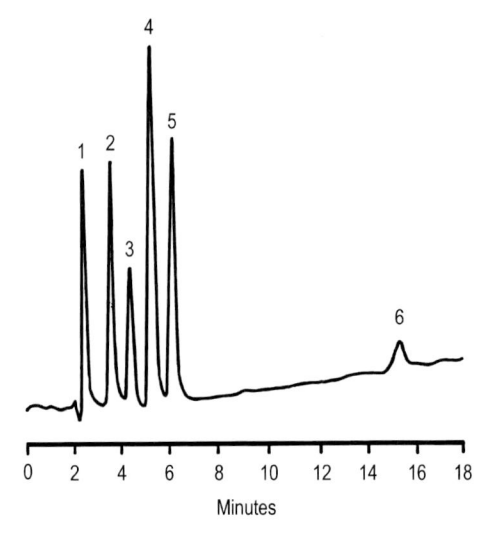

Fig. 6-15. Ion-pair chromatographic separation of inorganic anions on a chemically bonded cyanopropyl phase. — Separator column: Polygosil-60-D-10 CN; eluant: 0.1 mol/L Na_2HPO_4 + 0.1 mol/L KH_2PO_4 + 1 g/kg cetyltrimethylammonium chloride/ acetonitrile (75:25 v/v); flow rate: 1.5 mL/min; detection: UV (205 nm); injection volume: not given; peaks: (1) iodate, (2) bromate, (3) nitrite, (4) bromide, (5) nitrate, and (6) iodide; (taken from [29]).

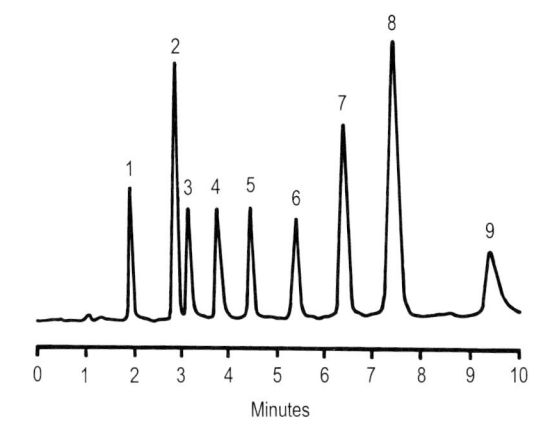

Fig. 6-16. Separation of inorganic anions on a chemically bonded aminopropyl phase. — Separator column: Zorbax NH_2; eluant: 0.03 mol/L H_3PO_4 (pH 3.2 with NaOH); flow rate: 2 mL/min; detection: UV (205 nm); injection volume: 20 µL; solute concentrations: 25 to 100 mg/L each of acetate (1), acrylate (2), glycolate (3), formate (4), nitrite (5), bromide (6), nitrate (7), iodide (8), and dichloroacetate (9); (taken from [31]).

All the aforementioned examples indicate the significance that ion-pair chromatography has gained in solving certain separation problems, even with the separation of simple inorganic anions.

Ion-pair chromatography assumes a special position in the analysis of strong polarizable anions. On conventional anion exchangers, they can only be eluted with strong eluants. The class of polarizable anions comprises perchlorate and citrate, oxidic sulfur anions, and metal complexes. For the ion-pair chromatographic analysis of these compounds, it generally suffices to increase the acetonitrile content in the mobile phase. This is indicated in Fig. 6-17, which shows a perchlorate separation that uses TBAOH as the ion-pair reagent and a comparably high solvent content of 340 mL/L. Remarkably, the trivalent citrate elutes prior to the monovalent perchlorate. The broader peak width of the citrate signal is caused by the higher valency of this compound.

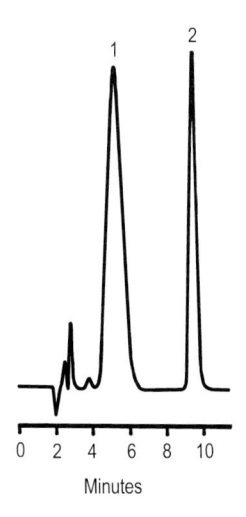

Fig. 6-17. Analysis of citrate and perchlorate. — Separator column: IonPac NS1 (10-µm); eluant: 2 mmol/L TBAOH + 1 mmol/L Na_2CO_3/acetonitrile (66:34 v/v); flow rate: 1 mL/min; detection: suppressed conductivity; injection volume: 50 µL; solute concentrations: 50 mg/L citrate (1) and 10 mg/L perchlorate (2).

6 Ion-Pair Chromatography (MPIC)

In the field of inorganic sulfur compounds, ion-pair chromatography is applied to the analysis of dithionate [32], $S_2O_6^{2-}$, peroxodisulfate [33], $S_2O_8^{2-}$, and polythionates [34, 35], $S_nO_6^{2-}$. In all cases, TBAOH is suited as the ion-pair reagent. Because the above-listed compounds are divalent anions, retention may be reduced with sodium carbonate. Another parameter affecting retention is the acetonitrile content in the mobile phase. The chromatogram in Fig. 6-18 shows the separation of sulfate, dithionate, and tetrathionate. Under these chromatographic conditions, peroxodisulfate elutes shortly before tetrathionate.

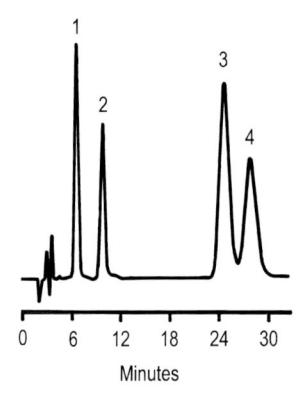

Fig. 6-18. Ion-pair chromatographic separation of sulfate, dithionate, peroxodisuolfate, and tetrathionate. — Separator column: IonPac NS1 (10-µm); eluant: 2 mmol/L TBAOH + 1 mmol/L Na_2CO_3/acetonitrile (77:23 v/v); flow rate: 1 mL/min; detection: suppressed conductivity; injection volume: 50 µL; solute concentrations: 5 mg/L sulfate (1), 10 mg/L dithionate (2), 20 mg/L peroxodisulfate (3), and 20 mg/L tetrathionate (4).

In contrast, disulfite, $S_2O_5^{2-}$, and disulfate, $S_2O_7^{2-}$, decompose in aqueous solution according to

$$S_2O_5^{2-} + H_2O \rightleftharpoons 2 HSO_3^- \tag{158}$$

$$2 S_2O_7^{2-} \rightleftharpoons 2 SO_4^{2-} + SO_3 \tag{159}$$

and thus cannot be analyzed by liquid chromatography. To analyze polythionates, $S_nO_6^{2-}$ ($n > 2$), the acetonitrile content in the mobile phase must also be enhanced. A separation of higher polythionates (n_S = 5 to 11) which was obtained under isocratic conditions by Steudel et al. [35] is shown in Fig. 6-19. Polythionates are sulfur chains bearing terminal sulfonate groups, so measuring the light absorption at 254 nm can be used as the detection method. Plotting the ln k values of the investigated polythionates versus the sulfur atom number yields the parabolic dependence represented in Fig. 6-20. It is obvious that higher polythionates can be detected with this method when a gradient technique is applied.

Ion-pair chromatography is also suited for the analysis of metal complexes. To be separated chromatographically, the complexes must be thermodynamically *and* kinetically stable. This means that complex formation must be thermodynamically possible and furthermore an irreversible process. Metal-ETDA and metal-DTPA complexes exhibit similary high stabilities. To separate the Gd-DTPA complex (Fig. 6-21), which is of great relevance in the pharmaceutical industry, TBAOH was used as the ion-pair reagent [36]. Detection was carried out by measuring the electrical conductivity in combination with a suppressor system.

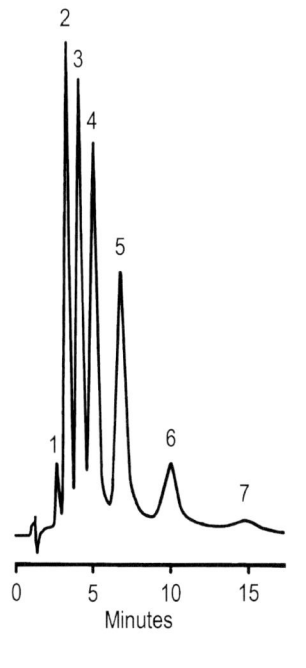

Fig. 6-19. Ion-pair chromatographic separation of higher polythionates, $S_nO_6^{2-}$ (n_S = 5 to 11). – Separator column: IonPac NS1 (10-µm); eluant: 2 mmol/L TBAOH + 1 mmol/L Na_2CO_3/acetonitrile (60:40 v/v); flow rate: 1 mL/min; detection: UV (254 nm); peaks: (1) $S_5O_6^{2-}$, (2) $S_6O_6^{2-}$, (3) $S_7O_6^{2-}$, (4) $S_8O_6^{2-}$, (5) $S_9O_6^{2-}$, (6) $S_{10}O_6^{2-}$, (7) $S_{11}O_6^{2-}$; (taken from [35]).

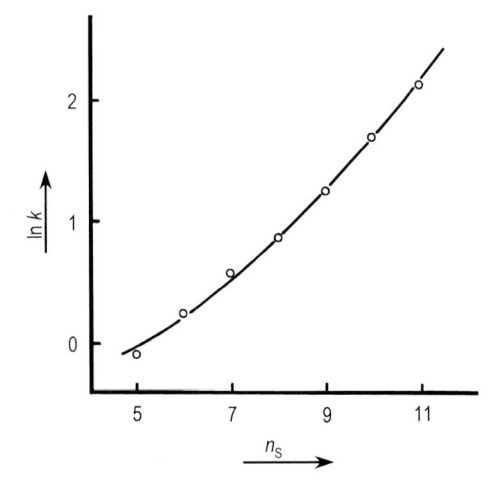

Fig. 6-20. Dependence of ln k values for polythionates, $S_nO_6^{2-}$ (n_S = 5 to 11) on the sulfur atom number. – For chromatographic conditions: see Fig. 6-19; (taken from [35]).

In the electroplating industry, applications were specifically developed to analyze metal-cyanide complexes by ion-pair chromatography, which is interesting because the oxidation state of the metal can be determined via its complexation with cyanide. The two iron cyanide complexes, for example, are eluted in the

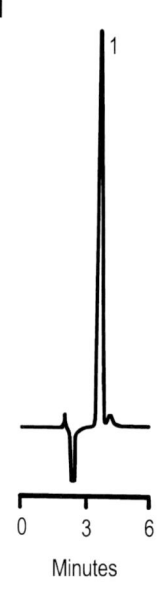

Fig. 6-21. Separation of Gd-DTPA. — Separator column: IonPac NS1 (10-µm); eluant: 2 mmol/L TBAOH + 1 mmol/L Na_2CO_3/acetonitrile (75:25 v/v); flow rate: 1 mL/min; detection: suppressed conductivity; injection: 50 µL sample (1:1000 diluted) with Gd-DTPA (1).

order of increasing charge number (see Fig. 6-9 in Section 6.3.3 for the corresponding chromatogram). In combination with TBAOH as the ion-pair reagent, the addition of small amounts of sodium carbonate is of crucial importance. In contrast, for the monovalent gold complexes, $Au(CN)_2^-$ and $Au(CN)_4^-$, the different coordination numbers and, thus, the different spatial arrangements, are responsible for the separation of both complexes (Fig. 6-22). The cyano-complexes of iron, cobalt, and gold exhibit a very high stability and therefore exist as anions. The cyano-complexes of nickel, copper, and silver cannot be quantitated by means of chromatography because of the low formation constants of these complexes, which lead to a slight dissociation into metal and ligand. By adding small amounts of potassium cyanide to the mobile phase, the equilibrium is shifted to the complex side, rendering chromatographic analysis possible. Figure 6-23 illustrates a chromatogram of kinetically stable and unstable metal-cyanide complexes. Be advised, however, that the suppression product of an eluant containing potassium cyanide is HCN, and that for safety, the effluent from the chromatographic system should be collected in a waste container containing a strongly basic solution.

The analysis of thio- and seleno-metalates [37] is an interesting application of ion-pair chromatography. These compounds play a significant role in some bioinorganic, nutritional physiological, and veterinary medical problems [38]. Thio- and seleno-metalate ions are formed from electron-deficient transition metals such as vanadium, niobium, tantalum, molybdenum, tungsten, and rhenium in their highest oxidation states. Multimetal complexes with remarkable electronic properties can be obtained in this manner. It is worth noting that poly(thiometalates) with mixed valencies are formed from thiometalates by novel redox processes with simultaneous condensation.

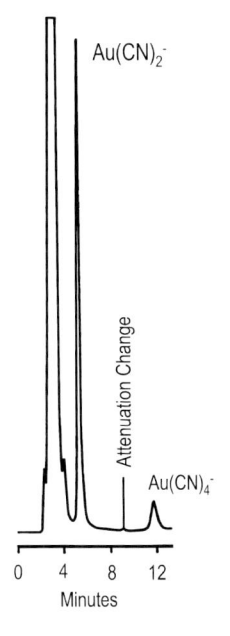

Fig. 6-22. Separation of gold(I) and gold(III) as cyanide complexes. — Separator column: IonPac NS1 (10-µm); eluant: 2 mmol/L TBAOH + 2 mmol/L Na_2CO_3/acetonitrile (60:40 v/v); flow rate: 1 mL/min; detection: suppressed conductivity.

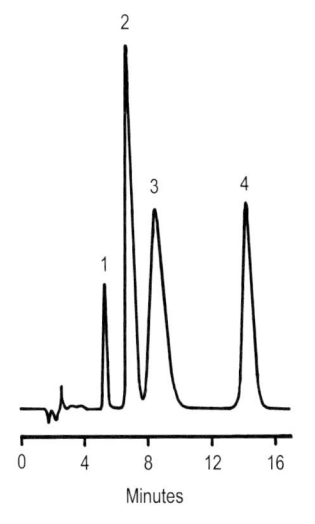

Fig. 6-23. Separation of kinetically stable and unstable metal-cyanide complexes. — Separator column: IonPac NS1 (10-µm); eluant: 2 mmol/L TBAOH + 1 mmol/L Na_2CO_3 + $2 \cdot 10^{-4}$ mol/L KCN/acetonitrile (70:30 v/v); flow rate: 1 mL/min; detection: suppressed conductivity; injection volume: 50 µL; solute concentrations: 80 mg/L $KAg(CN)_2$ (1), 40 mg/L $K_2Ni(CN)_4$ (2), 40 mg/L $K_3Co(CN)_6$ (3), and 80 mg/L $KAu(CN)_2$ (4); (taken from [36]).

The starting compounds, tungstate and molybdate, can be separated by anion exchange chromatography (see Section 3.7.4). Substituting oxygen atoms with sulfur or selenium, however, causes such an increase in affinity towards the stationary phase that even monothio- or monoseleno-metalates are no longer eluted within an acceptable time frame with the usual carbonate/bicarbonate mixtures. As illustrated in Fig. 6-24, the thiomolybdates series, $MoO_nS_{4-n}^{2-}$ (n = 0 to 4), may be separated in a single run via ion-pair chromatography under isocratic conditions. The parabolic retention increase observed with the increased substitution of oxygen atoms with sulfur or selenium was explained by Weiß et al. [37] on the basis of the retention model proposed by Bidlingmeyer et al. [13].

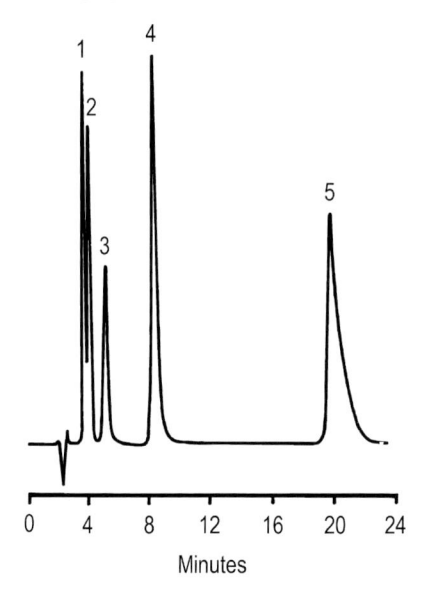

Fig. 6-24. Separation of molybdate and its thia-substituted derivatives. — Separator column: IonPac NS1 (10-µm); eluant: 2 mmol/L TBAOH + 1 mmol/L Na_2CO_3/acetonitrile (75:25 v/v); flow rate: 1 mL/min; detection: suppressed conductivity; injection volume: 50 µL; solute concentrations: 50 mg/L $(NH_4)_2MoO_4$ (1), $(NH_4)_2MoO_3S$ (2) as an impurity, 200 mg/L each of $(NH_4)_2MoO_2S_2$ (3), $(NH_4)_2MoOS_3$ (4), and $(NH_4)_2 MoS_4$ (5); (taken from [37]).

When the acetonitrile content in the mobile phase is further increased, the poly(thiometalate), $[Mo_2^VO_2S_2(S_2)_2]^{2-}$, is also accessible for analysis. This binuclear complex with η-S_2^{2-} ligands is formed from dithiomolybdate by intramolecular redox processes. Figure 6-25 shows the chromatogram of the tetramethylammonium salt of this compound. Obviously, molybdenum sulfur clusters can also be detected under the same chromatographic conditions. These include the binuclear cluster, $[Mo_2(S_2)_6]^{2-}$, in which molybdenum is only surrounded by disulfido ligands, as well as the later eluting trinuclear cluster, $[Mo_3S(S_2)_6]^{2-}$, which functions as a model compound for crystalline molybdenum(IV) sulfide. (MoS_2 catalysts are utilized in desulfurization of mineral oil). This is an impressive example of the potential of ion-pair chromatography to separate and detect inorganic anions with complex structures.

In the field of *cation analysis*, ion-pair chromatography is the preferred method for the separation of all types of amines. While short-chain aliphatic amines (C_1 to C_3) and some smaller aromatic amines [39] can also be separated on surface-sulfonated cation exchangers, ion-pair chromatographic applications have been developed for the separation of structurally isomeric amines, alkanolamines, quaternary ammonium compounds as well as arylalkylamines, barbiturates, and alkaloids.

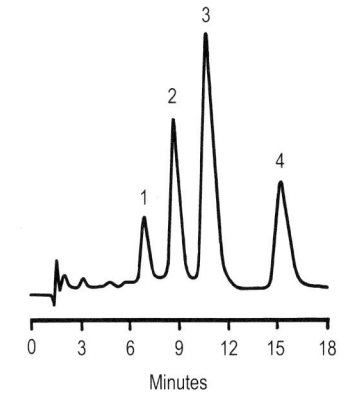

Fig. 6-25. Separation of various molydbenum disulfido complexes. — Separator column: IonPac NS1 (10-µm); eluent: 2 mmol/L TBAOH + 1 mmol/l Na_2CO_3/acetonitrile (50:50 v/v); flow rate: 1 mL/min; detection: UV (254 nm); injection volume: 50 µL; solute concentrations: impurity (1), 50 mg/L $(NH_4)_2[Mo_2S_{12}]$ (2), 50 mg/L $[(CH_3)_4N]_2[Mo_2O_2S_2(S_2)_2]$ (3), and 100 mg/L $(NH_4)_2[Mo_3S_{13}]$ (4); (taken from [37]).

As an example, Fig. 6-26 shows the separation of mono-, di-, and tri-ethylamine, accomplished by using octanesulfonic acid as the ion-pair reagent. The less hydrophobic hexanesulfonic acid is used in combination with boric acid as the eluent for the separation of ethanolamines, as shown in Fig. 6-27. These compounds are detected by measuring the electrical conductivity, so the background conductivity is generally lowered with a membrane suppressor. The addition of boric acid to both the eluant and the regenerant serves to enhance the sensitivity for di- and tri-ethanolamine.

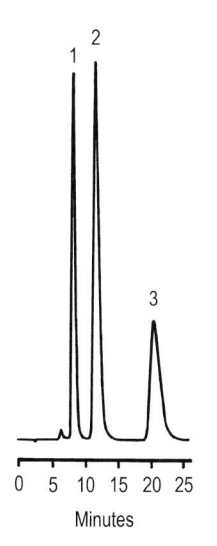

Fig. 6-26. Separation of mono-, di-, and tri-ethylamine. — Separator column: IonPac NS1 (10-µm); eluent: 2 mmol/L octanesulfonic acid/acetonitrile (92: 8 v/v); flow rate: 1 mL/min; detection: suppressed conductivity; injection volume: 50 µL; solute concentrations: 50 mg/L MEA (1), 100 mg/L DEA (2), and 100 mg/L TEA (3).

Quaternary ammonium compounds can be analyzed under similar chromatographic conditions, but without the addition of boric acid. This includes choline (2-hydroxyethyl-trimethylammonium hydroxide), which is commonly found in fauna and flora as a basic constituent of lecithin-type phospholipids. The ion-pair chromatographic separation of choline requires an eluant, in addition to

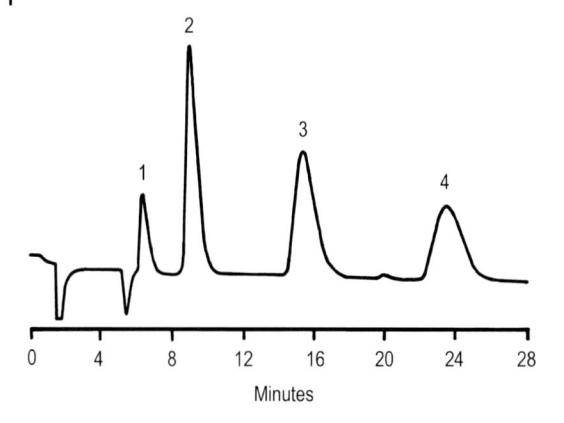

Fig. 6-27. Separation of ethanolamines. — Separator column: IonPac NS1 (10-µm); eluant: 4 mmol/L hexanesulfonic acid + 0.04 mol/L H_3BO_3; flow rate: 1 mL/min; detection: suppressed conductivity; injection volume: 100 µL; solute concentrations: ammonium (1), 2.5 mg/L monoethanolamine (2), 5 mg/L diethanolamine (3), and 10 mg/L triethanolamine (4).

hexanesulfonic acid as the ion-pair reagent, that contains a small amount of acetonitrile as an organic modifier. Important choline derivatives such as acetylcholine and chlorocholine are more strongly retained under these chromatographic conditions and, thus, can be separated from choline. As an example, Fig. 6-28 shows the separation of choline and chlorocholine, which differ only in the substitution at the alkyl side-chain end:

$$[HO-CH_2-CH_2-\overset{+}{N}(CH_3)_3]\ OH^- \qquad [Cl-CH_2-CH_2-\overset{+}{N}(CH_3)_3]\ OH^-$$

Choline   Chlorocholine

Much higher amounts of organic solvents are required for the elution of tetraalkylammonium compounds, which are used as reagents in ion-pair chromatography of anions. If the analysis times are kept constant, an amount of 60 mL/L acetonitrile suffices to elute tetramethylammonium hydroxide at a hexanesulfonic acid concentration of $c = 1$ mmol/L (Fig. 6-29), 280 mL/L is needed for the elution of tetrapropylammonium ions, and 480 mL/L for the even more

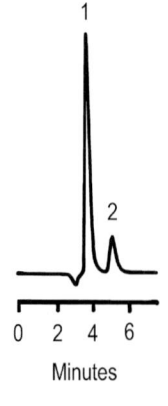

Fig. 6-28. Analysis of choline and chlorocholine. — Separator column: IonPac NS1 (10-µm); eluant: 2 mmol/L hexanesulfonic acid/acetonitrile (97:3 v/v); flow rate: 1 mL/min; detection: suppressed conductivity; injection volume: 50 µL; solute concentrations: 10 mg/L each of choline chloride (1) and chlorocholine chloride (2).

hydrophobic tetrabutylammonium ions. Because of the drastic increase in retention time with a larger carbon skeleton, a simultaneous analysis of these compounds is only accomplished by applying a gradient technique.

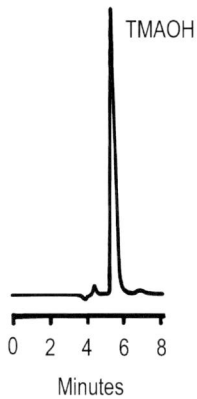

Fig. 6-29. Separation of tetramethylammonium hydroxide. – Separator column: IonPac NS1 (10-µm); eluant: 1 mmol/L hexanesulfonic acid/acetonitrile (94:6 v/v); flow rate: 1 mL/min; detection: suppressed conductivity; injection volume: 50 µL; solute concentration: 20 mg/L.

The various arylalkylamines and alkaloids can be separated with ion-pair chromatography by coating the stationary phase with sodium octanesulfonate. The effect of the type and concentration of the ion-pair reagent on the retention of these compounds has been described in detail in Section 6.3.1, taking diphenhydramine as an example. Figure 6-30 reveals that under similar chromatographic conditions, other arylalkylamines such as epinephrine and ephedrine can be separated in the presence of opium alkaloids such as morphine and codeine (3-monomethylester of morphine).

Morphine

For the determination of ephedrines and other catecholamines in biological samples, the UV detection that was used in Fig. 6-30, does not exhibit the required sensitivity and selectivity. Much better results are obtained with amperometric detection on a glassy carbon electrode. Figure 6-31 shows a representative separation of various catecholamines on an ODS phase with butanesulfonic acid as the ion-pair reagent.

Under the chromatographic conditions used in Fig. 6-30, atropine and cocaine can also be analyzed. However, the acetonitrile content in the mobile phase has to be slightly increased (Fig. 6-32).

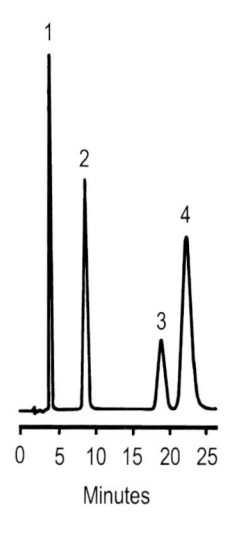

Fig. 6-30. Separation of epinephrine, ephedrine, and opium alkaloids. — Separator column: IonPac NS1 (10-μm); eluant: 5 mmol/L sodium octanesulfonate + 0.05 mol/L KH_2PO_4 (pH 4.0) / acetonitrile (89: 11 v/v); flow rate: 1 mL/min; detection: UV (220 nm); injection volume: 50 μL; solute concentrations: 10 mg/L epinephrine (1), 10 mg/L morphine sulfate (2), 20 mg/L ephedrine hydrochloride (3), and 20 mg/L codeine phosphate (4).

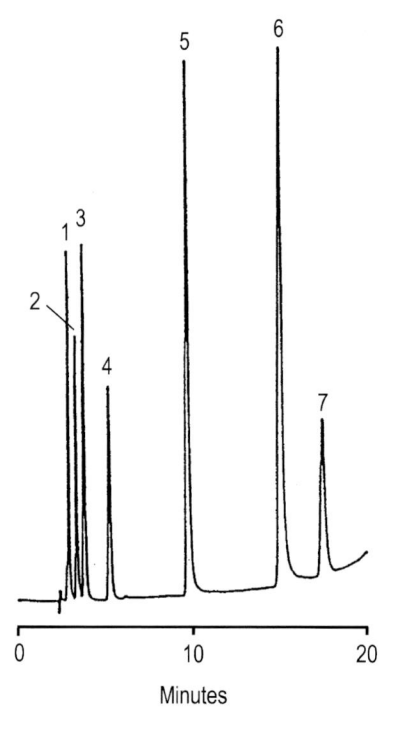

Fig. 6-31. Separation of catecholamines. — Separator column: Zorbax ODS; eluant: (A) 70 mmol/L KH_2PO_4 + 5 mmol/L triethylamine + 1 mmol/L butanesulfonic acid (pH 3), (B) methanol + 5 mmol/L triethylamine; gradient: linear, 5% B to 56% B in 20 min; flow rate: 1 mL/min; detection: DC amperometry on a glassy carbon electrode; solute concentrations: 200 ng each of norepinephrine (1), epinephrine (2), 3,4-dihydroxybenzylamine (3), dopamine (4), serotonin (5), 5-hydroxyindolacetic acid (6), and homovanillic acid (7).

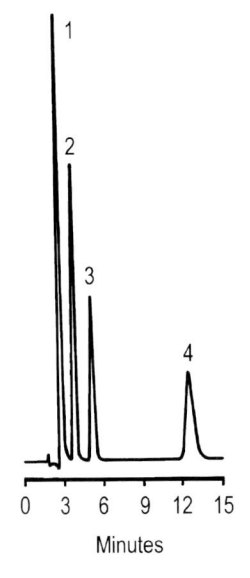

Fig. 6-32. Separation of atropine and cocaine. — Separator column: IonPac NS1 (10-µm); eluant: 5 mmol/L sodium octanesulfonate + 0.05 mol/L KH_2PO_4 (pH 4.0) / acetonitrile (77:23 v/v); flow rate: 1 mL/min; detection: UV (220 nm); injection volume: 50 µL; solute concentrations: 10 mg/L morphine sulfate (1), 20 mg/L ephedrine hydrochloride (2), 20 mg/L atropine sulfate (3), and 20 mg/L cocaine hydrochloride (4).

Atropine

(2R,3S)-(−)-Cocaine

A much higher acetonitrile content is required to elute the most important barbiturates: i.e., barbital, phenobarbital, and hexobarbital. Their separation is depicted in Fig. 6-33. Barbiturates are derived from barbituric acid (pK_a = 3.9 [40]), which is quite acidic because of its activated methylene group in 5-position.

Barbituric acid

Substituting the H-atoms of this group with various alkyl groups lowers the acidity of the resulting derivative. The acid strength of barbital (5,5-diethylbarbituric acid), for example, is given as pK_a = 7.89 [40]. The retention of barbiturates is completely unaffected by the type and concentration of lipophilc anions, such as octanesulfonate, present in the mobile phase. Nevertheless, it is logical to add sodium octanesulfonate to the eluant used for barbiturates, because barbiturates and alkaloids can then be analyzed in a single run. This applies, for example, to papaverine, which has a retention time of about six minutes under the chromatographic conditions given in Fig. 6-33.

Papaverine

Because arylalkylamines as well as alkaloides and barbiturates have an aromatic skeleton, they can all be detected very sensitively by measuring the light absorption at a wavelength of 220 nm.

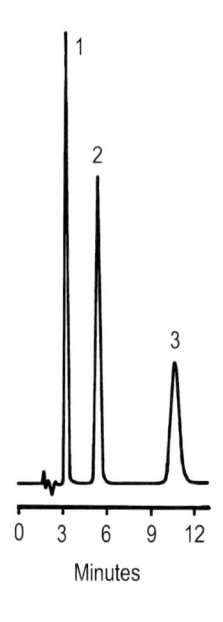

Fig. 6-33. Separation of various barbiturates. — Separator column: IonPac NS1 (10-µm); eluant: 5 mmol/L sodium octanesulfonate + 0.05 mol/L KH_2PO_4 (pH 4.0) / acetonitrile (64:36 v/v); flow rate: 1 mL/min; detection: UV (220 nm); injection volume: 50 µL; solute concentrations: 10 mg/L each of barbital (1), phenobarbital (2), and hexobarbital (3).

6.5
Analysis of Surface-Active Ions

The analysis of surface-active *anions* comprises the determination of simple aromatic sulfonic acids, hydrotropes (toluene-, cumene-, and xylene sulfonates), alkane- and alkene sulfonates, fatty alcohol ether sulfates, alkylbenzene sulfonates, and α-sulfofatty acid methyl esters. Such compounds are relevant predominantly in the detergent and cleansing industry.

Surface-active anions with aromatic backbone have long been separated by RPIPC and detected sensitively via their UV-absorption [41]. However, a chromatographic determination of the other compound classes mentioned above is only feasible by conductivity detection.

The retention behavior of surface-active anions depends on the alkyl side chain length and, thus, on their hydrophobicity, which increases with increasing chain length. In general, the hydrophilic ammonium hydroxide is employed as the ion-pair reagent, because the analyte compounds already have a strongly hydrophobic character. Hydrophobicity depends, apart from the alkyl chain length, on the kind and number of ionic and/or non-ionic substituents in the carbon skeleton. Figure 6-34 shows the different retention behavior of alkyl sulfates and alkyl sulfonates of equal carbon chain length. It is obvious that alkyl sulfates are more strongly retained than the corresponding alkyl sulfonates. This effect is directly connected with a phenomenon observed during the investigation of the retention behavior of thioalkanes in reversed-phase chromatography. According to Fig. 6-35, each CH_2-S or CH-S entity in dialkyl sulfides and in alkane sulfonates represents a local polar center that allows a stronger interaction with the polar eluant [42] via solvation. Thus, for each of these entities, a retention decrease is observed which, depending on the water content of the mobile phase, corresponds to the loss of 1 to 3 methylene groups. Moreover, the effect that a methylene group has on retention is reduced by neighboring sulfur atoms. However, this effect does not arise in alkyl sulfates, which explains the higher retention of these compounds.

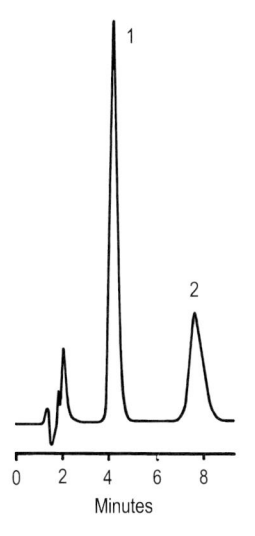

Fig. 6-34. Separation of dodecyl sulfonate and dodecyl sulfate. – Separator column: IonPac NS1 (10-μm); eluant: 10 mmol/L NH_4OH / acetonitrile (65:35 v/v); flow rate: 1 mL/min; detection: suppressed conductivity; injection volume: 50 μL; solute concentration: 100 mg/L each of dodecylsulfonate (1) and dodecylsulfate (2).

Replacing a hydrogen atom with a hydroxide group in any of the methylene groups significantly reduces the retention time. The position of the hydroxide group also exerts a noticeable influence on the resulting retention time. Figure 6-36 shows the separation of two C_{16}-hydroxyalkane sulfonates, which are hydroxy-substituted in the 2- and 3-position of the alkyl chain. In comparison to non-substituted alkane sulfonates, the retention decrease corresponds to the loss of 2 to 3 methylene groups from the solvophobic alkyl groups.

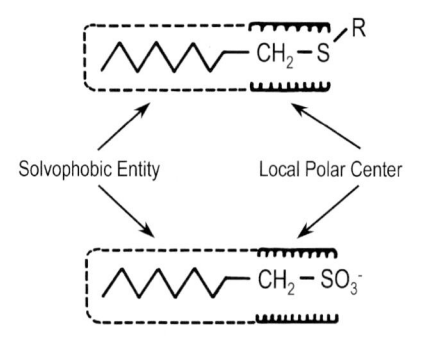

Fig. 6-35. Schematic representation of the loss of molecular surface area due to solvation of local polar centers comprising of a sulfur atom and neighboring CH-fragments.

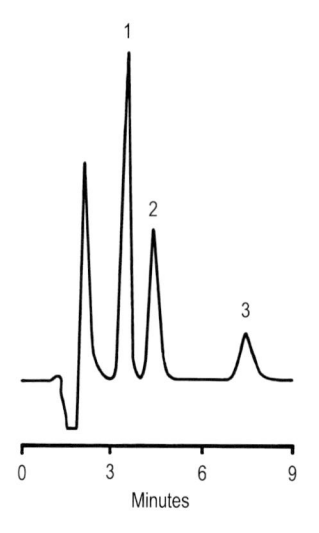

Fig. 6-36. Separation of two hydroxyalkane sulfonates with different positions of the hydroxide group. – Separator column: IonPac NS1 (10-µm); eluant: 10 mmol/L NH$_4$OH/ acetonitrile (62:38 v/v); flow rate: 1 mL/min; detection: suppressed conductivity; injection volume: 50 µL; solute concentrations: 100 mg/L each of C$_{14}$-alkane sulfonate (2-hydroxy) (1), C$_{16}$-alkane sulfonate (3-hydroxy) (2), and C$_{16}$-alkane sulfonate (2-hydroxy) (3).

The analysis of an olefin sulfonate, which is rarely a pure product, is much more difficult. The different positions of the double bond in the alkyl chain and the cis/trans isomerism generally lead to a product mixture of numerous compounds. Thus, satisfactory separations are only obtained if the spatial arrangement of the alkyl groups at the double bond is defined; the position of the double bond must also be known. The chromatogram of a C$_{16}$-olefin sulfonate (trans-3-en) is shown in Fig. 6-37. The C$_{16}$-olefin sulfonate (trans-3-en) has a slightly shorter retention time than the alkane sulfonate of equal alkyl chain length. This difference is explained by the enhanced solubility of the unsaturated compound in the polar eluent.

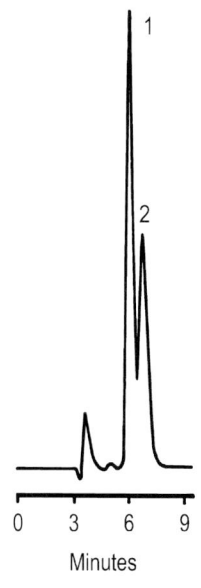

Fig. 6-37. Separation of alkane- and olefin sulfonates of equal carbon chain length. – Separator column: IonPac NS1 (10-µm); eluant: 10 mmol/L NH_4OH / acetonitrile (65:35 v/v); flow rate: 1 mL/min; detection: suppressed conductivity; injection volume: 50 µL; solute concentration: 100 mg/L each of C_{16}-olefine sulfonate (trans-3-en) (1) and C_{16}-alkane sulfonate (2).

The surface-active anions investigated so far include the class of *aryl sulfonates*. A respective chromatogram the separation of benzene-, toluene-, xylene-, and cumene sulfonates is shown in Fig. 6-38. They are eluted in this order according to the number of carbon atoms of their substituents. To obtain high selectivity, tetrabutylammonium hydroxide was used as the ion-pair reagent. Due to the comparably small retention differences between the various aryl sulfonates, they can be eluted under isocratic conditions. Detection of aryl sulfonates is carried out by applying either the different modes of conductivity detection or UV detection at a wavelength of 254 nm, as shown in Fig. 6-38. The choice of the detection system appropriate for the analysis usually depends on the nature of the sample matrix.

The chromatographic separation of *aliphatic sulfonic acids* requires the application of a gradient technique. Retention of these compounds increases exponentially with increasing alkyl chain length. The separation of a standard mixture with chain lengths between C_1 and C_{10} is shown in Fig. 6-39. Because short-chain sulfonic acids are only moderately hydrophobic, the relatively hydrophobic tetrabutylammonium hydroxide was used as the ion-pair reagent. In contrary, long-chain fatty alcohol sulfates are separated with ammonium hydroxide as the ion-pair reagent. Fatty alcohol sulfates are constituents of many cosmetics, so their determination is commercially important. In Fig. 6-40 the separation of a standard mixture containing octyl-, decyl-, dodecyl-, tetradecyl-, and hexadecyl sulfate is juxtaposed to the analysis of a raw material sample (Rewopol NLS 28, Rewo, Steinau, Germany). In addition to the major component, lauryl sulfate, this product contains small amounts of tetradecyl- and hexadecyl sulfate. Because aliphatic sulfonic acids and fatty alcohol sulfates are non-chromophoric, sensitive detection of these compounds is only feasible via suppressed conductivity detection. The negative baseline drift during the gradient run is a result of the rapidly increasing acetonitrile concentration in the mobile phase. It can be offset either by baseline subtraction or by increasing the concentration of the ion-pair reagent during the run.

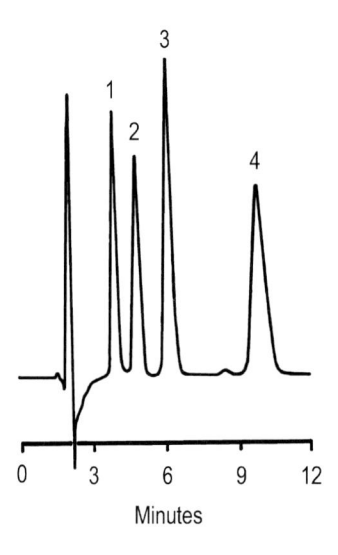

Fig. 6-38. Separation of various aryl sulfonates. — Separator column: IonPac NS1 (10-µm); eluant: 2 mmol/L TBAOH / acetonitrile (70:30 v/v); flow rate: 1 mL/min; detection: UV (254 nm); injection volume: 50 µL; solute concentration: 50 mg/L benzene sulfonate (1), 50 mg/L toluene sulfonate (2), 100 mg/L xylene sulfonate (3), and 100 mg/L cumene sulfonate (4).

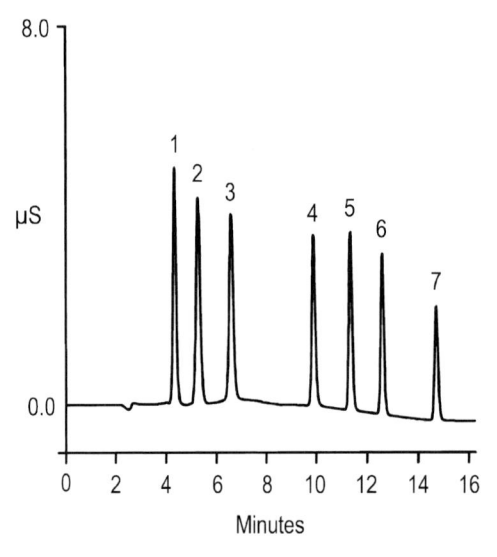

Fig. 6-39. Separation of aliphatic sulfonic acids. — Separator column: IonPac NS1 (10-µm); eluant: (A) 2 mmol/L TBAOH − acetonitrile (76:24 v/v), (B) 2 mmol/L TBAOH − acetonitrile (52:48 v/v); gradient: linear, 100% A to 100% B in 10 min; flow rate: 1 mL/min; detection: suppressed conductivity; injection volume: 50 µL; solute concentrations: 5 mg/L methanesulfonic acid (1), 8.6 mg/L propanesulfonic acid (2), 8.7 mg/L butanesulfonic acid (3), 8.8 mg/L hexanesulfonic acid (4), 8.9 mg/L heptanesulfonic acid (5), 8.9 mg/L octanesulfonic acid (6), and 9.1 mg/L decanesulfonic acid (7).

Ethoxylation of fatty alcohols prior to their sulfonation leads to a class of compounds called *fatty alcohol ether sulfates*, which are contained in a variety of detergents and cleansing agents as active compounds.

$R\text{-}CH_2\text{-}O\text{-}(C_2H_4O)_n\text{-}SO_3Na$ $R = C_{11}$ to C_{13}
$n = 1$ to 5

Fatty alcohol ether sulfate

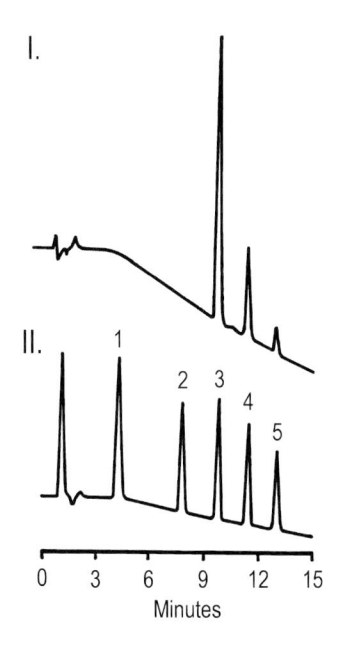

Fig. 6-40. Separation of fatty alcohol sulfates with various chain lengths. − Separator column: IonPac NS1 (10-µm); eluent: (A) 20 mmol/L NH_4OH / acetonitrile (80:20 v/v), (B) 20 mmol/L NH_4OH / acetonitrile (20:80 v/v); gradient: linear 100% A to 100% B in 25 min; flow rate: 1 mL/min; detection: suppressed conductivity; injection volume: 50 µL; solute concentrations: I. 80 mg/L Rewopol NLS 28; II: 80 mg/L each of octyl sulfate (1), decyl sulfate (2), dodecyl sulfate (3), tetradecyl sulfate (4), and hexadecyl sulfate (5).

Depending on their degree of ethoxylation, fatty alcohol ether sulfates are extremely complex mixtures, for which the separation efficiency of a polymer phase is not sufficient. Good separations are obtained with silica-based, chemically bonded, reversed phases. The chromatographic conditions have to be adjusted accordingly. The free base ammonium hydroxide cannot be used as the ion-pair reagent because of the pH limitation of modified silica, so suppressor systems cannot be used for the subsequent conductivity detection. Sodium acetate has proved to be a suitable ion-pair reagent for direct conductivity detection. It exhibits a sufficiently low background conductivity at the required concentration of $c = 1$ mmol/L. In combination with a solvent gradient, fatty alcohol ether sulfates can be separated according to their alkyl chain length and degree of ethoxylation. Figure 6-41 illustrates this with the analysis of Texapon N25 (Henkel KGaA, Düsseldorf, Germany), a raw material that contains dodecyl ether sulfate and the corresponding tetradecyl homologues. Thus, the resulting peak pattern consists of two overlapping peak series of ethoxylated compounds whose concentration decreases as the EO content increases. However, even the high separation efficiency of the stationary phase (Hypersil 5 MOS) is insufficient to resolve all the compounds in this mixture. Co-elution is observed for tetradecyl sulfate, indicated as "C_{14}", and the higher ethoxylated dodecyl compound. Also, only a small retention difference exists between an alkyl ether sulfate with just one EO group and a pure alkyl sulfate with equal carbon chain length. This means that the retention increase, which usually goes along with chain elongation by two methylene groups, is compensated by solvation because another polar center is introduced around the oxygen atom in the EO group. The retention-increasing effect of an EO group is noticed only at higher degrees of ethoxylation.

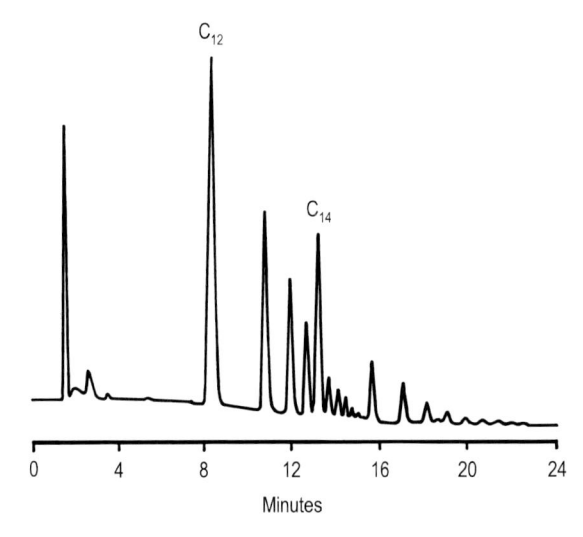

Fig. 6-41. Separation of a lauryl ether sulfate (Texapon N 25). – Separator column: Hypersil 5 MOS; eluant: (A) 1 mmol/L NaOAc / acetonitrile (70:30 v/v), (B) 1 mmol/L NaOAc / acetonitrile (60:40 v/v); gradient: linear, 100% A to 100% B in 10 min; flow rate: 1 mL/min; detection: direct conductivity; injection volume: 50 µL; solute concentration: 500 mg/L of the raw material.

One of the most important anionic surfactants is the linear *alkylbenzene sulfonate* (LAS), which can be analyzed under similar chromatographic conditions.

$$H_3C-(CH_2)_n-\underset{|}{\overset{H}{C}}-(CH_2)_m-CH_3$$

$m+n$ 7 to 10

Alkylbenzene sulfonate

Alkylbenzene sulfonate is yet another complex mixture of compounds; the aromatic ring may be attached to any non-terminal C-atom of the alkyl chain. Thus, the product distribution is statistical. Furthermore, the chain length may differ (C_{10} to C_{14}). Therefore, a complete separation of all components is not to be expected. However, the optimized separation of an LAS raw material sample, Marlon AFR (Chem. Werke Hüls, Marl, Germany) shows that a separation into homologues (C_{10} to C_{13}) is possible, although the respective major components of the different peak groups represent sum signals of the various isomers. The separation is shown in Fig. 6-42. Valuable information for the quality control of raw materials can be inferred from such chromatograms. As an alternative to the direct conductivity detection used in Fig. 6-42, UV detection in the wavelength region around 225 nm [41] can also be applied.

Another constituent of many detergents and cleansing agents are sulfosuccinic acid esters.

$$\begin{array}{l} \quad\;\; SO_3Na \\ \quad\;\; | \\ \quad HC-COOR \\ \quad\;\; | \\ \quad H_2C-COOR \end{array} \qquad R = C_{10} \text{ to } C_{14}$$

Sulfosuccinic acid ester

Because of their good wetting properties, they are often a constituent of many quick-assay strips used in clinical chemistry. The ion-pair chromatographic separation of these compounds on Nucleosil 10 C_8 was recently described by Steinbrech et al. [43], who used tetrabutylammonium hydrogensulfate as the ion-pair reagent and methanol as an organic modifier. Because sulfosuccinic acid esters do not have appreciable chromophores, this class of compound was detected refractometrically. Figure 6-43 shows the chromatogram of a raw material sample Rewopol SBF12 (Rewo, Steinau, Germany), a mixture of dialkyl sulfosuccinates of different carbon chain lengths. As shown, the C_{12}-fraction is the major component. The authors indicated a detection limit of 0.5 µg for this raw material.

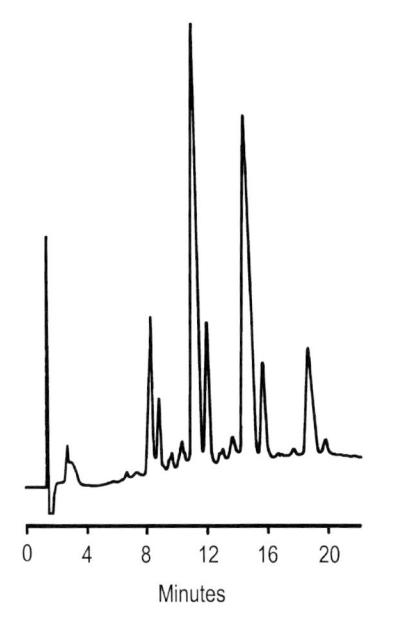

Fig. 6-42. Separation of a linear alkylbenzene sulfonate (Marlon AFR). – Separator column: Hypersil 5 MOS; eluant: (A) 1 mmol/L NaOAc / acetonitrile (80:20 v/v), (B) 1 mmol/L NaOAc / acetonitrile (60:40 v/v); gradient: linear, 60% B to 100% B in 15 min; flow rate: 1 mL/min; detection: direct conductivity; injection volume: 50 µL; solute concentrations: 1000 mg/L of the raw material.

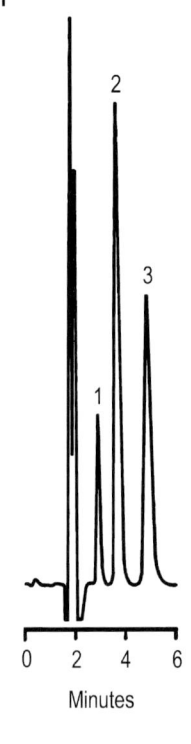

Fig. 6-43. Separation of fatty alcohol sulfosuccinates of different carbon chain length. — Separator column: Nucleosil 10 C_8; eluant: 10 mmol/L tetrabutylammonium hydrogensulfate (pH 3.0) / methanol (23:77 v/v); flow rate: 2 mL/min; detection: RI; injection: 50 µL of a Rewopol-SBF-12 solution with C_{10}-sulfosuccinate (1), C_{12}-sulfosuccinate (2), and C_{14}-sulfosuccinate (3); (taken from [43]).

While sulfosuccinic acid esters with chain lengths between C_{10} and C_{14} can be analyzed under isocratic conditions with the procedure described by Steinbrech et al., the separation of *fatty alcohol polyglycolether sulfosuccinates* requires the application of a gradient technique.

$$R-(OC_2H_4)_n-O-CO-CH_2-CH_2\text{-COONa} \qquad R = C_8$$
$$|$$
$$SO_3Na$$

Fatty alcohol polyglycolether sulfosuccinate

Fatty alcohol polyglycolether sulfosuccinates are used in many cosmetics because of their good compatibility with skin and mucous membrane. Due to the introduction of ethyleneoxide groups, the water solubility of these compounds is better than that of non-ethoxylated sulfosuccinates. Like fatty alcohol ether sulfates and alkylbenzene sulfonates, fatty alcohol polyglycolether sulfosuccinates are highly complex component mixtures. Technical products of this kind are traded under the name Rewopol SBFA 30 (Rewo, Steinau, Germany). A remarkable chromatogram of this raw material was obtained by Janssen [44] by applying an acetonitrile gradient on Hypersil 5 MOS, with sodium acetate as the ion-pair reagent. Although, as in the previous examples, not all compounds are baseline-resolved, the resolution obtained suffices to characterize this raw material unequivocally.

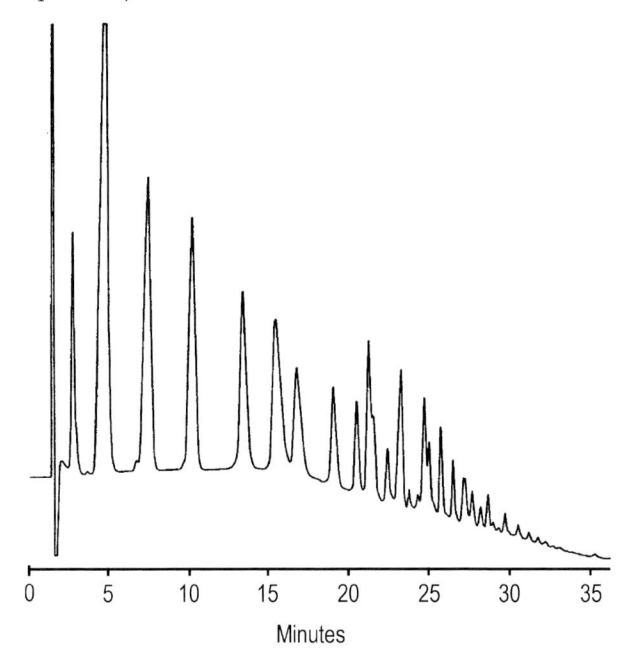

Fig. 6-44. Separation of a fatty alcohol polyglycolether sulfosuccinate (Rewopol SBFA 30). – Separator column: Hypersil 5 MOS; eluant: (A) 1 mmol/L NaOAc / acetonitrile (78:22 v/v), (B) 1 mmol/L NaOAc / acetonitrile (60:40 v/v); gradient: linear, 100% A isocratic for 10 min, then in 20 min to 100% B; flow rate: 1 mL/min; detection: direct conductivity; injection volume: 50 µL; solute concentration: 1000 mg/L of the raw material; (taken from [44]).

In the field of *surface-active cations*, ion-pair chromatography is predominantly applied to the analysis of quaternary ammonium compounds, pyridine-, pyrrolidine-, and piperidine quaternisates, and for sulfonium-, phosphonium-, ammonium- and hydrazinium salts.

The compound class of quaternary alkylammonium salts includes *alkyltrimethylammonium* and *dialkyldimethylammonium compounds*, which are found in low concentrations in many cosmetic products.

$$\left[\begin{array}{c} CH_3 \\ | \\ R-\overset{+}{N}-CH_3 \\ | \\ CH_3 \end{array}\right] Cl^- \qquad \left[\begin{array}{c} R \diagdown \overset{+}{N} \diagup CH_3 \\ R \diagup \diagdown CH_3 \end{array}\right] Cl^- \qquad R = C_{12} \text{ to } C_{18}$$

Alkyltrimethyl-ammonium chloride

Dialkyldimethyl-ammonium chloride

In the past, this type of cationic surfactants has played only a secondary industrial role, but it is being increasingly employed. Not many liquid chromatographic methods for separating such compounds are found in the literature. Because these quaternary ammonium compounds are comparably hydrophobic, hydrochloric acid is used as the ion-pair reagent in their separation. Perfluorinated carboxylic acids are also used from time to time. Because of the lack of suitable chromophores, quaternary ammonium compounds can only be sensitively detected by measuring the electrical conductivity. Figure 6-45 shows a separation of a number of aliphatic quaternary ammonium compounds; it was obtained under gradient conditions with acetonitrile as the organic modifier. For the elution of long-chain dialkyldimethylammonium compounds, the acid concentration and the organic modifier fraction in the mobile phase have to be increased.

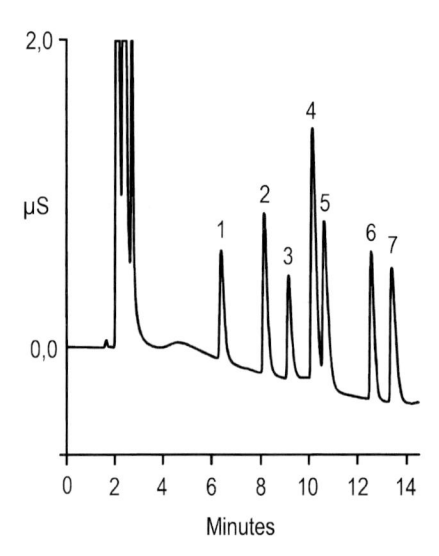

Fig. 6-45. Gradient elution of various aliphatic quaternary ammonium compounds. – Separator column: IonPac NS1 (10-µm); eluant: (A) 2 mmol/L perfluorobutyric acid – acetonitrile (80:20 v/v), (B) 2 mmol/L perfluorobutyric acid – acetonitrile (20:80 v/v); gradient: linear, 100% A to 100% B in 10 min; flow rate: 1 mL/min; detection: suppressed conductivity; injection volume: 25 µL; solute concentrations: 25 mg/L tetrapropylammonium chloride (1), 50 mg/L tributylmethylammonium chloride (2), 25 mg/L decyltrimethylammonium chloride (3), 50 mg/L tetrabutylammonium chloride (4), 50 mg/L dodecyltrimethylammonium chloride (5), 100 mg/L tetrapentylammonium chloride (6), and 100 mg/L hexadecyltrimethylammonium chloride (7).

Alkyl-dimethyl-benzylammonium chlorides are constituents of many disinfectants because of their antibacterial activity.

$$\begin{bmatrix} & CH_3 & \\ & | & \\ R - \overset{+}{N} - CH_2 - C_6H_5 \\ & | & \\ & CH_3 & \end{bmatrix} Cl^- \quad R = C_6 \text{ to } C_{18}$$

Alkyl-dimethyl-benzyl-
ammonium chloride

Figure 6-46 shows the chromatogram of a corresponding Benzalkon A raw material. The gradient technique was applied because of the carbon chain distribution between C_6 and C_{12} in this product. The two main components were identified as the C_6 and C_8 fractions by comparing the raw material with a standard consisting of decyl- and dodecyl-dimethyl-benzylammonium chloride. Accordingly, the two minor components are the C_{10} and C_{12} fractions. Remarkably, the peak symmetry of the individual signals can be significantly improved by raising the hydrochloric acid concentration to c = 20 mmol/L. Because alkyl-dimethyl-benzylammonium chlorides contain an aromatic ring system, they can be detected with sufficient sensitivity by measuring the light absorption at 215 nm.

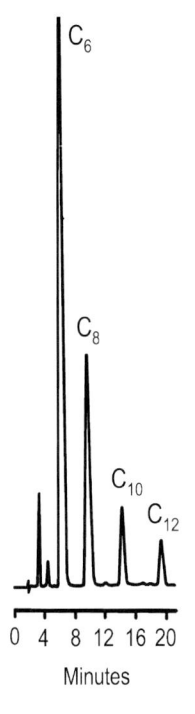

Fig. 6-46. Separation of an alkyl-dimethyl-benzylammonium chloride (Benzalkon A). — Separator column: IonPac NS1 (10-µm); eluant: (A) 20 mmol/L HCl / acetonitrile (50:50 v/v), (B) 20 mmol/L HCl / acetonitrile (20:80 v/v); gradient: linear, from 30% B to 100% B in 20 min; flow rate: 1 mL/min; detection: UV (215 nm); injection volume: 50 µL; solute concentration: 100 mg/L of the raw material.

Derivatives of these compounds, such as *dichlorobenzyl-alkyl-dimethylammonium chloride*, may also be separated and detected under similar chromatographic conditions. As an example, Fig. 6-47 shows the analysis of the raw material Preventol RB50 (Henkel KGaA, Düsseldorf, Germany) which is an exceptionally pure product with a chain length C_{12}. A slightly shorter retention under these chromatographic conditions exhibits *diisobutyl-[2-(2-phenoxyethoxy)ethyl-]dimethylbenzylammonium chloride*, which is utilized under the trade name Hyamine 1622 in analytical chemistry for the two-phase titration of anionic surfactants.

Hyamine 1622

It is also occasionally employed in the pharmaceutical and cosmetics industry. The chromatogram of this compound is illustrated in Fig. 6-48.

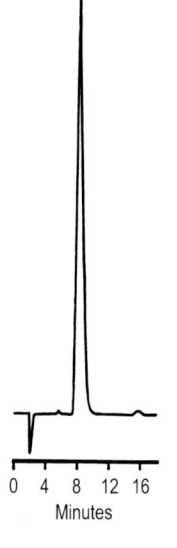

Fig. 6-47. Separation of a dichlorobenzyl-alkyl-dimethylammonium chloride (Preventol RB50). — Separator column: IonPac NS1 (10-µm); eluant: 5 mmol/L HCl / acetonitrile (38:62 v/v); flow rate: 1 mL/min; detection: UV (215 nm); injection volume: 50 µL; solute concentration: 100 mg/L raw material.

Cyclic alkylammonium compounds resemble the linear products in their properties. They are widely found in antiseptic solutions, cremes, shampoos, and mouthwashes. As an example of this class of compounds, Fig. 6-49 displays the chromatogram of a dodecylpyridinium chloride. Its individual components, with the exception of the main component, exhibit a similar retention behavior to that of Benzalkon A. The minor components of this sample are presumably tetradecyl- and hexadecyl-pyridinium chloride.

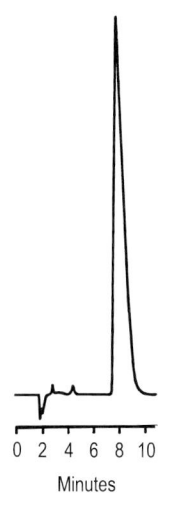

Fig. 6-48. Separation of Hyamine 1622. – Separator column: IonPac NS1 (10-µm); eluant: 5 mmol/L HCl / acetonitrile (42:58 v/v); flow rate: 1 mL/min; detection: UV (215 nm); injection volume: 50 µL; solute concentration: 100 mg/L raw material.

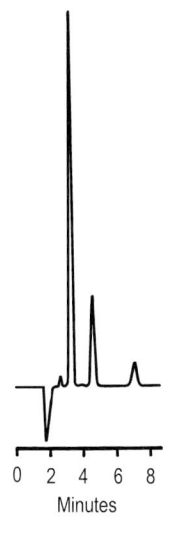

Fig. 6-49. Separation of dodecylpyridinium chloride. – Separator column: IonPac NS1 (10-µm); eluant: 20 mmol/L HCl / acetonitrile (24:76 v/v); flow rate: 1 mL/min; detection: UV (215 nm); injection volume: 50 µL; solute concentration: 100 mg/L raw material.

In principle, modified silica may also be applied in all these separations. However, the high surface area of such materials require a much higher organic solvent fraction in the mobile phase to elute the compounds in a comparable length of time.

As is the case for cationic surfactants, very few liquid chromatographic methods are described in the literature for characterizing *sulfonium salts*. In a recent paper, Anklam et al. [45] showed that five- and six-membered cyclic sulfonium salts of the following structure are easily analyzed via ion-pair chromatography:

Structure of five- and six-membered sulfonium salts

The authors utilized a highly deactivated and stable ODS material, Inertsil ODS II (5-µm), as the stationary phase. The chromatogram shown in Fig. 6-50 was obtained for six-membered sulfonium compounds bearing different alkyl rests R with hexanesulfonic acid as the ion-pair reagent and acetonitrile as the organic modifier. Five-membered analogues show a correspondingly shorter retention. As expected, retention increases as the carbon chain length of the alkyl group increases. This is predicted by the theory of RPLC [47], because the molecular surface area and, thus, the solvophobic character of the solute increases with a larger R. Electrical conductivity measurements achieved with a membrane-based suppressor system are a suitable detection method for non-chromophoric sulfonium salts.

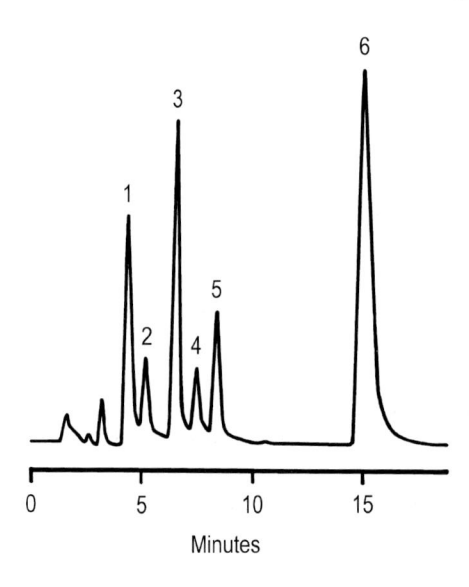

Fig. 6-50. Separation of various six-membered sulfonium salts on Inertsil ODS II (5-µm). — Eluant: 2 mmol/L hexanesulfonic acid / acetonitrile (90:10 v/v); flow rate: 0.8 mL/min; detection: suppressed conductivity; compounds: R = Me (1), R = Et (2), R = i-Pr (3), R = n-Pr (4), R = t-Bu (5), and R = n-Bu (6); (taken from [46]).

As an example for the ion-pair chromatographic analysis of *phosphonium compounds*, the separation of triphenyl-mono(β-jonylidene-ethylene)-phosphonium chloride is illustrated in Fig. 6-51.

Triphenyl-mono(β-jonylidene-ethylene)-phosphonium chloride

This compound is a vitamin A precursor with a strong hydrophobic character. Therefore, sodium perchlorate was employed as the ion-pair reagent. This phosphonium compound contains three conjugated double bonds, so UV detection is much more sensitive than suppressed conductivity detection. Figure 6-51 reveals that the resulting signal is characterized by a strong tailing because of the π-π-interactions with the aromatic backbone of the DVB resin that is used as the stationary phase. Aced [46] obtained much better peak shapes using the already-mentioned Inertsil ODS II column, with which she investigated the retention behavior of various triphenylphosphonium salts $[(Ph)_3P\text{-}R]^+X^-$. Figure 6-52 illustrates this by showing the retention of some n-alkyl-triphenylphosphonium salts of different carbon chain lengths. Again, as with sulfonium salts, a retention increase is observed with increasing chain length. Comparing triphenylphosphonium compounds with the corresponding *arsonium compounds* $[(Ph)_3As\text{-}R]^+X^-$, the latter exhibit a higher retention because of their larger surface area. It is noteworthy that the retention of triphenylphosphonium compounds that was observed by Aced [46] increases when hydrochloric acid is replaced by HBr or HI as the ion-pair reagent.

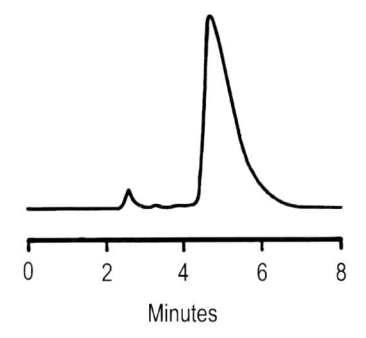

Fig. 6-51. Ion-pair chromatographic separation of triphenyl-mono(β-jonylidene-ethylene)-phosphonium chloride. — Separator column: IonPac NS1 (10-μm); eluant: 5 mmol/L $HClO_4$ / methanol (5:95 v/v); flow rate: 1 mL/min; detection: UV (280 nm).

Finally, note the class of N, N'-substituted *hydrazinium salts*, which may also be investigated by ion-pair chromatography. These compounds do not carry chromophores, thus, a sensitive detection is only possible via suppressed or non-suppressed conductivity detection. As an illustration for the great number of compounds that have already been investigated, Fig. 6-53 displays the chromatograms of two compounds that have the following structures:

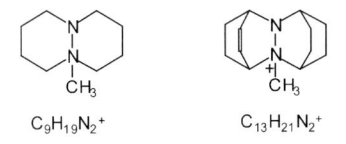

$C_9H_{19}N_2^+$ $C_{13}H_{21}N_2^+$

Although the carbon content of these compounds is fairly high, the bicyclic structure results in a comparatively low sorption area. In this example, hydrochloric acid was used as the ion-pair reagent at a low organic solvent content.

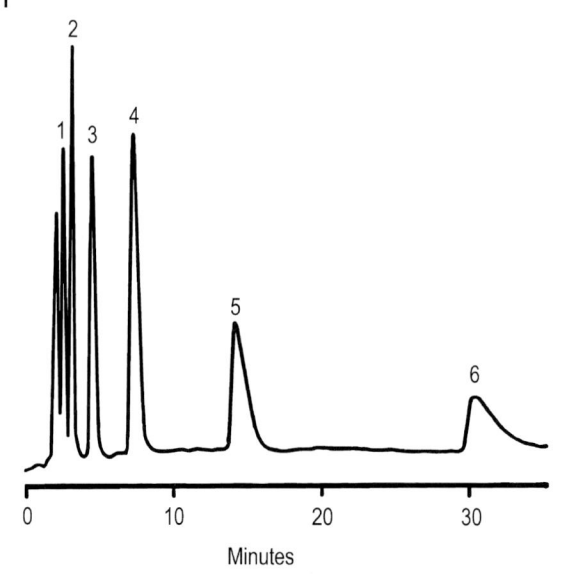

Fig. 6-52. Separation of various n-alkyl triphenylphosphonium compounds. – Separator column: Inertsil ODS II (5-μm); eluant: 2 mmol/L HCl / acetonitrile (72:28 v/v); flow rate: 0.8 mL/min; detection: suppressed conductivity; compounds: R = Me (1), R = Et (2), R = i-Pr (3), R = n-Pr (4), R = t-Bu (5), and R = n-Bu (6); (taken from [46]).

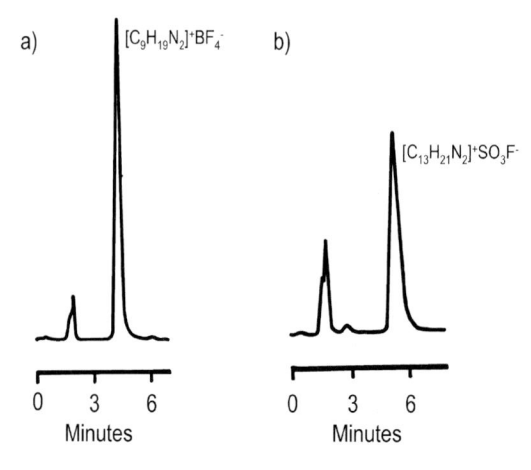

Fig. 6-53. Separation of two different N,N'-substituted hydrazinium compounds. – Separator column: IonPac NS1 (10-μm); eluant: (A) 2 mmol/L HCl / acetonitrile (98:2 v/v), (B) 2 mmol/L HCl / acetonitrile (93:7 v/v); flow rate: 1 mL/min; detection: suppressed conductivity; compounds: a) $[C_9H_{19}N_2]^+BF_4^-$, b) $[C_{13}H_{21}N_2]^+SO_3F^-$.

6.6
Applications of the Ion-Suppression Technique

Another application for neutral, non-polar stationary phases based on silica or organic polymers is the separation of weak acids or bases in their molecular form. Their dissociation is suppressed by choosing the appropriate pH value. The solute interactions with the stationary phase are determined solely by their adsorption and distribution behavior. The term *ion-suppression mode* has been coined for this technique. In some areas of analysis, it provides an alternative to ion-exchange chromatography.

One of the best known and most widely used applications of this technique is the separation of phenols, which are eluted on a non-polar phase using acetonitrile/water- or methanol/water mixtures. A phosphate buffer is added to the solution to suppress the small dissociation of phenols into phenolate anions. Potassium dihydrogenphosphate at concentrations of about $c = 0.01$ mol/L are typically employed. Figure 6-54 shows a separation of various mono-, di-, and trivalent phenols on an organic polymer.

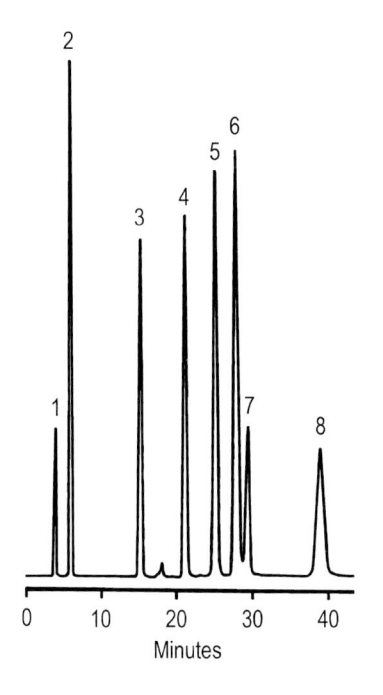

Fig. 6-54. Separation of various mono- and polyvalent phenols. — Separator column: IonPac NS1 (10-µm); eluant: (A) 10 mmol/L KH_2PO_4 (pH 4.0) / acetonitrile (90:10 v/v), (B) 10 mmol/L KH_2PO_4 (pH 4.0) / acetonitrile (20:80 v/v); gradient: linear, 15% B in 20 min to 55% B; flow rate: 1 mL/min; detection: UV (280 nm); injection volume: 50 µL; solute concentrations: 100 mg/L each of pyrogallic acid (1), resorcinol (2), phenol (3), o-cresol (4), 2,4-dimethylphenol (5), α-naphthol (6), 2,4-dichloro-3-nitrophenol (7), and thymol (8).

An ODS phase, with its high chromatographic efficiency can also be used to separate barbiturates, which were also discussed in Section 6.4 (see Fig. 6-33). Under the conditions used in Figure 6-55, the dissociation of barbiturates is suppressed with acetic acid.

Another application of the ion-suppression technique is the separation of long-chain fatty acids described by Slingsby [48]. While short-chain monocarboxylic acids up to valeric acid may be separated by ion-exclusion chromatography (see Section 5.6), long-chain fatty acids exhibit unacceptably long retention times under these conditions. In a protonated state, they can easily be separated on a non-polar stationary phase with a solvent/water-mixture. Figure 6-56 illustrates a respective separation of various fatty acids (butyric acid to palmitic acid); the separation was achieved on an organic polymer by applying a gradient technique.

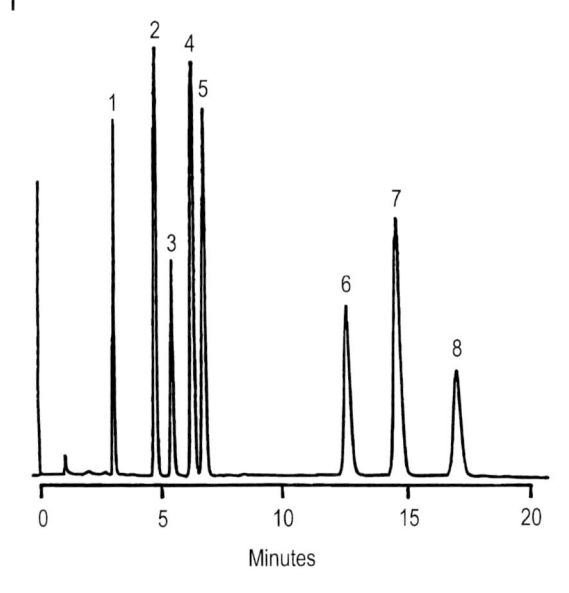

Fig. 6-55. Separation of various barbiturates on an ODS phase. — Separator column: RP-C18 (5-µm); eluant: methanol-water-acetic acid (38:62:1 v/v/v); flow rate: 1.5 mL/min; detection: UV (215 nm); solute concentrations: 50 ng each of barbital (1), allobarbital (2), aprobarbital (3), phenobarbital (4), butabarbital (5), amobarbital (6), mephobarbital (7), and secobarbital (8).

To ensure sufficient solubility of long-chain fatty acids in the mobile phase, a solvent mixture composed of acetonitrile *and* methanol was used. Small amounts of hydrochloric acid in the eluant served to suppress the dissociation of the compounds. If chemically modified silica (e.g. ODS) is used instead of an organic polymer, the enhanced solute interactions with such a material have to be compensated for. This is achieved both by raising the column temperature to 40 °C and by adding 2-propanol or THF to the mobile phase. The detection of long-chain fatty acids depends on the kind of carbon skeleton. If conjugated double bonds are present, as in the case of sorbic acid, a sensitive UV-detection is feasible (Fig. 6-57). On the other hand, if the carbon chain is purely paraffinic, as in the case of the compounds represented in Fig. 6-56, conductivity detection must be utilized, because aliphatic fatty acids do not exhibit any appreciable absorption even at low wavelengths around 200 nm. A prerequisite for conductivity detection is the transformation of fatty acids that are separated as molecular compounds into their dissociated form. For this, the separator column effluent is passed through an AMMS-ICE micromembrane suppressor (see also Section 5.4) before entering the conductivity cell. The fatty acid oxonium ions are exchanged for potassium ions. The solutes enter the detector cell as fully dissociated potassium salts and may be detected conductometrically. In addition, the background conductivity of the hydrochloric acid eluant is also reduced by converting it into the potassium salt. This is advantageous, because it significantly improves the signal-to-noise ratio for the analyte compound. According to Slingsby [48], detection limits for this method are between 50 µg/L for butyric acid and 50 mg/L for stearic acid.

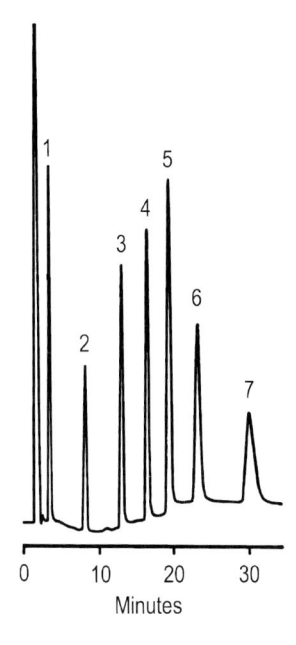

Fig. 6-56. Separation of long-chain fatty acids utilizing an ion-suppression technique. — Separator column: IonPac NS1 (10-µm); eluant: (A) $3 \cdot 10^{-5}$ mol/L HCl / acetonitrile / methanol (70:24:6 v/v/v), (B) $3 \cdot 10^{-5}$ mol/L HCl / acetonitrile / methanol (16:60:24 v/v/v); gradient: linear, 100% A in 15 min to 100% B; flow rate: 1 mL/min; detection: suppressed conductivity; injection volume: 50 µL; solute concentrations: 100 mg/L butyric acid (1), 100 mg/L caproic acid (2), 200 mg/L caprylic acid (3), 200 mg/L capric acid (4), 300 mg/L lauric acid (5), 300 mg/L myristic acid (6), and 400 mg/L palmitic acid (7).

A remarkable application of the ion-suppression technique is the analysis of aminopolycarboxylic acids. In general, NTA and EDTA are separated on an anion exchanger and detected photometrically after derivatization with iron(III) [49]. Compounds such as iminodiacetic acid (IDA), ethylenediaminetriacetic acid (EDTriA), and ethylenediaminediacetic acid (EDDA), however, cannot be detected under these chromatographic conditions, as they do not form the desired complexes with iron(III). This is not surprising, since EDTriA does not exist as an open-chain compound with complex-forming properties in acid pH range. Instead, it exists as a cyclic piperazine, which cannot be detected via post-column derivatization. Thus, the detection of all three complexing agents — IDA, EDTriA, and EDDA — is only possible via direct UV detection of the carboxylate group at low wavelengths around 215 nm. The first successful separations of these complexing agents were accomplished on a chemically bonded ODS phase by means of ion-suppression. Due to the high polarity of these compounds, their retention is very small when the usual methanol/water-mixtures are used. The retention is markedly increased by coating the phase with an anionic surfactant. A good resolution of all three compounds within an acceptable total analysis time is obtained with sodium octylsulfate and methanol in sulfuric acid solution. Mikropak MCH-10 (Varian Co.), for example, with a particle diameter of 10 µm, is a suitable stationary phase. Figure 6-58 shows the corresponding chromatogram of a standard with a baseline-resolved separation of all three components.

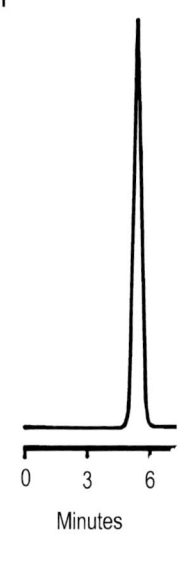

Fig. 6-57. Separation of sorbic acid. — Separator column: IonPac NS1 (10-μm); eluant: $5 \cdot 10^{-5}$ mol/L HCl / acetonitrile / methanol (57:34:9 v/v/v); flow rate: 1 mL/min; detection: UV (250 nm); injection volume: 50 μL; solute concentration: 5 mg/L.

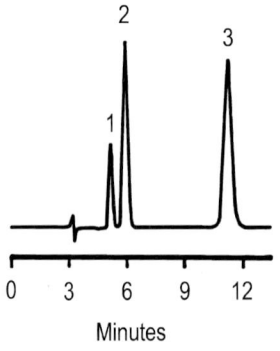

Fig. 6-58. Separation of iminodiacetic acid, ethylenediaminetriacetic acid, and ethylenediaminediacetic acid. — Separator column: Mikropak MCH-10 (Varian); eluant: 0.2 g/L sodium octylsulfate + 0.5 mL/L H_2SO_4 / methanol (92:8 v/v); flow rate: 1 mL/min; detection: UV (215 nm); injection volume: 50 μL; solute concentrations: 500 mg/L IDA (1), 50 mg/L EDTriA (2), and 50 mg/L EDDA (3).

6.7
Applications of Multi-Dimensional Ion Chromatography on Multimode Phases

The term *Multi-Dimensional Ion Chromatography* summarizes separation techniques in which the sample is separated to its components on two or several stationary phases of different properties, using valve switching- or column switching techniques, a change in the type of eluant, or even a change in eluant flow direction. Umile et al., for example, applied these techniques for analyte pre-concentration in samples with complex matrices [50] and for eluting ions with a widely different retention behavior under isocratic conditions [51, 52]. In the following, the term *Multi-Dimensional Ion Chromatography* is used to describe separations, in which ion-exchange- and reversed-phase interactions occur at the same stationary phase and, thus, contribute to the separation of ionic and strongly polar species [53]. The respective stationary phases are commercialized under the trade name OmniPac (Dionex Corporation, Sunnyvale, CA, USA).

6.7 Applications of Multi-Dimensional Ion Chromatography on Multimode Phases

They are also termed *multimode phases*, because they allow separations based on ion-exchange-, ion-pair-, reversed-phase-, and ion-suppression interactions, depending on the type of eluant. Any combinations of these interactions, at the same time or in succession, can also be employed to separate analyte species. Thus, OmniPac columns are probably the most fascinating separation materials in ion chromatography today.

The principle structure of multimode phases illustrated in Fig. 6-59 is very similar to that of a latexed ion exchanger. In contrary to classical latexed ion exchangers (see Sections 3.4.2 and 4.1.2), in which the latex particles are electrostatically agglomerated on the surface-functionalized substrate, the highly cross-linked and therefore solvent-compatible ethylvinylbenzene/divinylbenzene substrate of mutimode phases has a particle diameter of 8.5 µm and is coated with a second polymer containing carboxyl groups. The carboxyl groups, in turn, serve as anchor groups for the oppositely charged latex particles. Even though electrostatic attractive forces initiate the agglomeration procedure, the true bonding of the latex particles is based on pure adsorption. In contrast to traditional latexed anion exchangers, OmniPac anion exchangers do not exhibit any cation exchange capacity. This excludes the enrichment of metal ions in the substrate and consequent precipitations of, for example, orthophosphate when injecting a respective sample. The absence of charges on the substrate surface can also prove to be a disadvantage. This is especially true for some macromolecules, whose peak deformations can be attributed to adsorption processes at the substrate surface. Another disadvantage of the classical latex setup is that functionalization is limited to aqueous systems with hydrophilic reagents in order to maintain the colloidal stability of the latex emulsion during functionalization. Therefore, phosphonate- or iminodiacetic acid groups cannot be used for functionalization due to these restrictions. Because with OmniPac columns the functionalization of the colloidal polymer is carried out after the agglomeration, the functionalization process is also possible in non-aqueous systems and is therefore much more flexible.

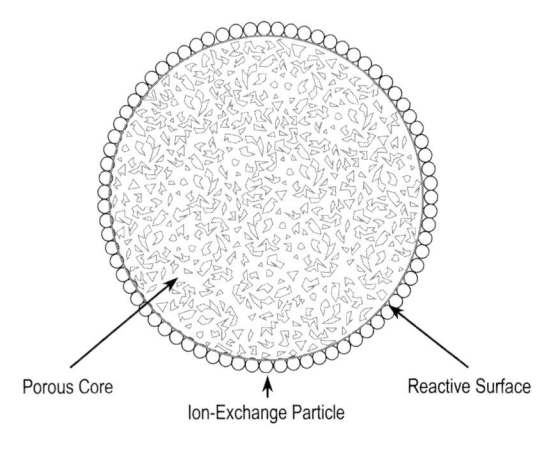

Fig. 6-59. Schematic structure of a multimode phase.

The various columns of the OmniPac series differ in the porosity of the substrate material. When latex particles are bonded to the outer surface of a microporous, highly cross-linked substrate resin particle (specific surface area $< 1 \text{ m}^2/\text{g}$), a pellicular packing material is produced that exclusively allows ion-exchange interactions in a solvent compatible matrix. Separators of that kind are available as anion and cation exchanger under the trade names OmniPac PAX-100 and PCX-100, respectively. When a macroporous substrate resin particle is used, the resulting packing material allows ion-exchange interactions at the substrate surface as well as adsorption processes in the interior of the core material due to the relatively high specific surface area of about $300 \text{ m}^2/\text{g}$ (OmniPac PAX-500 and PCX-500). Because the latex particles with a diameter of 60 nm are significantly larger than the pores of the substrate material (pore size: $\sim$ 6 nm), they cannot penetrate the interior of the stationary phase. Therefore, only the substrate surface is coated with the latex colloid, as shown in Fig. 6-59. The structural-technical properties of the OmniPac columns are summarized in Table 6-2.

Due to their ion-exchange- and reversed-phase properties, multimode phases offer high flexibility for solving complex separation problems. In the past, differentiation between ion-exchange- and reversed-phase chromatography was forcibly given, because the different packing materials did not exhibit similar properties. However, when using multimode phases, ion-exchange-, ion-pair-, and pure reversed-phase interactions can contribute to the separation process at the same time, complimenting each other in an excellent way. The strength of ion-exchange chromatography clearly lies in the separation of multivalent ions, especially if they exhibit the same valency. Usually, ion exchangers with different selectivities have to be used to cover a broad spectrum of analytes. On the other hand, ion-exchange chromatography is not suitable for analyzing surface-active ions. Inversely, the strength of ion-pair chromatography lies in the separation of monovalent and surface-active ions, whereas polyvalent ions are poorly separated. Figure 6-60 illustrates a direct comparison between ion-exchange chromatography on a pure anion exchanger, ion-pair chromatography on a neutral DVB resin, and the combination of both separation mechanisms on a multimode phase. All three separator columns were operated with the same eluant, consisting of sodium hydroxide for anion exchange, tetrabutylammonium hydroxide as ion-pair reagent and acetonitrile as organic modifier. A mixture of various inorganic and organic anions served as the analyte test mixture. As expected, bromide and nitrate cannot be separated on OmniPac PAX-100 (Fig. 6-60, left chromatogram) under the given chromatographic conditions, because adsorptive interactions do not contribute to the separation process on this stationary phase. Resolution between chloride and nitrite is also very poor. The ion-pair chromatographic separation on IonPac NS1 (Fig. 6-60, center chromatogram) shows a markedly better resolution between the monovalent anions, but this compromises the resolution between divalent anions. Generally, a modification of the chromatographic conditions might improve the resolution between mono- and di-valent anions, however an ion-pair chromatographic separation of the divalent anions is impossible. The right-hand chromatogram in Fig. 6-60 impressively demonstrates that all components of the test mixture can be separated on an OmniPac PAX-500 by combining ion-exchange- and ion-pair interactions.

Table 6-2. Structural-technical properties of the OmniPac separator columns.

Separator	PAX-100	PAX-500	PCX-100	PCX-500
Substrate				
Porosity	microporous	macroporous	microporous	macroporous
Pore size [Å]	–	60	–	60
Particle diameter [µm]	8.5	8.5	8.5	8.5
Degree of cross-linking [% DVB]	55	55	55	55
Spec. surface area [m²/g]	<1	300	<1	300
Latex particles				
Particle diameter [nm]	60	60	200	200
Degree of cross-linking [% DVB]	4	4	5	5
Capacity [µequiv/column]	40	40	120	120
Functionality	Quaternary alkanolamine	Quaternary alkanolamine	Sulfonate groups	Sulfonate groups

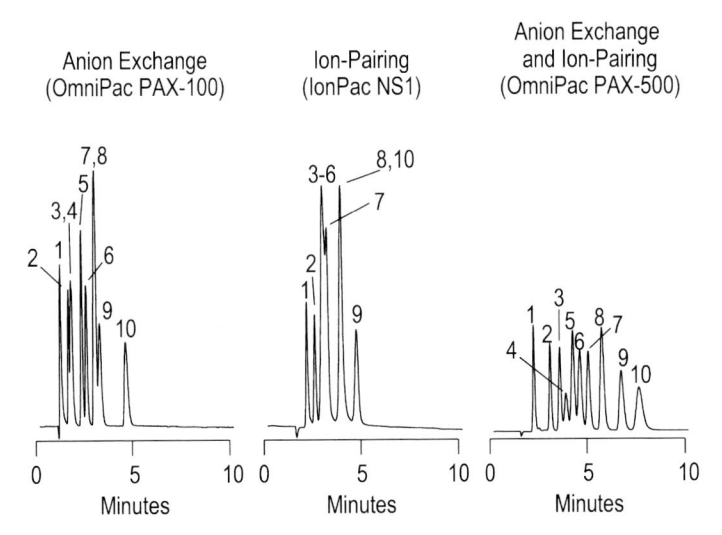

Fig. 6-60. Selectivity comparison between ion-exchange chromatography, ion-pair chromatography and the combination of both on a multimode phase. – Separator columns: OmniPac PAX-100, IonPac NS1 and OmniPac PAX-500; eluent: 42 mmol/L NaOH + 1 mmol/L TBAOH – acetonitrile (83:17 v/v); flow rate: 1 mL/min; detection: suppressed conductivity; analytes: (1) fluoride, (2) chloride, (3) nitrite, (4) succinate, (5) sulfate, (6) oxalate, (7) bromide, (8) nitrate, (9) chlorate, and (10) orthophosphate.

The solvent compatibility of OmniPac columns allows organic solvents to be added to the mobile phase for modifying the selectivity. Organic solvents affect the degree of hydration of solute ions as well as ion-exchange groups. In general, retention times decrease with an increase in the amount of organic solvents in the mobile phase. Because a similar effect is achieved by decreasing the degree of cross-linking of the latex polymer, the addition of an organic solvents to the mobile phase is synonymous with a change of the *effective* degree of cross-linking of the latex polymer (see also Section 3.7.2). This effect can be utilized for separating aliphatic and aromatic carboxylic acids on OmniPac PAX-100. In combination with suppressed conductivity detection, sodium hydroxide is used for gradient elution (Fig. 6-61). If the analyte components carry chromophores that allow UV detection a small amount of sodium hydroxide in the mobile phase is sufficient for complete dissociation of the solutes. Solute elution can then be carried out with a salt gradient (Fig. 6-62). In both cases, a constant amount of an organic solvent is added to the mobile phase.

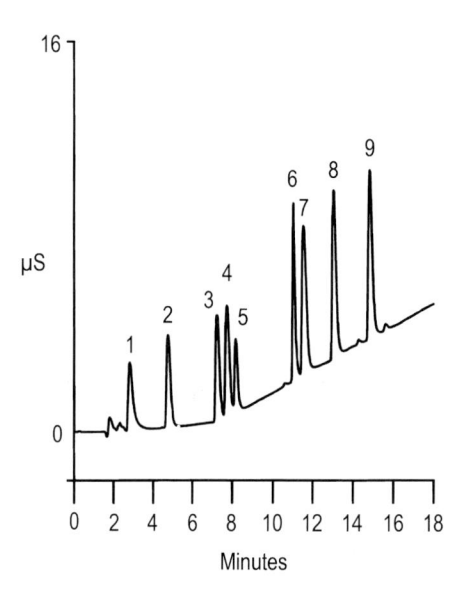

Fig. 6-61. Separation of aliphatic and aromatic carboxylic acids on OmniPac PAX-100 (250 x 2 mm i.d.) with a NaOH concentration gradient. — Eluant: NaOH — methanol (95:5 v/v); gradient: linear, 40 mmol/L NaOH to 280 mmol/L in 15 min; flow rate: 0.25 mL/min; detection: suppressed conductivity; analytes: (1) benzoate, (2) phthalate, (3) citrate, (4) 1,2,3-propanetricarboxylate, (5) isocitrate, (6) phytate, (7) 1,3,5-benzenetricarboxylate, (8) pyromellitate, and (9) mellitate.

As can be seen from Fig. 6-62, very small and symmetric peaks result when a salt gradient is applied. Separations of this kind are not possible with either the ion-suppression technique or with ion-pair chromatography. This is especially true for aromatic polycarboxylic acids, which elute from a multimode phase such as OmniPac PAX-500 according to their valency (Fig. 6-63, bottom chromatogram). When separating on a chemically bonded silica under ion-suppression conditions (Fig. 6-63, top chromatogram), a shorter analysis time is observed but not all components of the test mixture are separated. Moreover, penta- and hexacarboxylates are hardly retained due to their high polarity and, consequently, elute close to the system void.

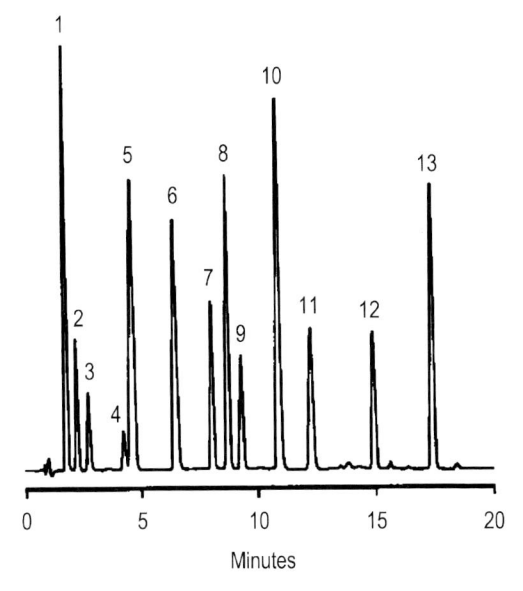

Fig. 6-62. Separation of aliphatic and aromatic carboxylic acids on OmniPac PAX-100 with a salt gradient. – Eluant: NaCl/NaOH – acetonitrile (80:20 v/v); gradient: linear, 0.2 mmol/L NaOH + 50 mmol/L NaCl to 1.6 mmol/L NaOH + 400 mmol/L NaCl in 20 min; flow rate: 1 mL/min; detection: UV (254 nm); analytes: (1) benzoate, (2) benzenesulfonate, (3) toluenesulfonate, (4) p-chlorobenzenesulfonate, (5) p-bromobenzoate, (6) 3,4-dinitrobenzoate, (7) phthalate, (8) terephthalate, (9) p-hydroxybenzoate, (10) p-hydroxybenzenesulfonate, (11) gentisate, (12) trimesate, and (13) pyromellitate.

In a similar way, Fig. 6-64 shows the comparison of ion-pair- and multi mode selectivities for the separation of phenylphosphonate and aromatic sulfonic acids. While all components of the test mixture can be separated to baseline on a multimode phase, the separation of di- and tri-sulfonic acids with ion-pair chromatography is not possible at all. Both examples clearly demonstrate that for an optimal separation of ionic and strongly polar compounds, the combination of ion-exchange- and reversed-phase interactions is of great importance.

The versatility of multimode columns can also be applied for the simultaneous separation of molecular and ionic organic compounds. An impressive example is shown in Fig. 6-65. During the first ten minutes, neutral species, free of any interferences by ionic components, are eluted with an acetonitrile/water mixture. After all neutral compounds are eluted, the organic solvent content in the mobile phase is dropped and ionic components are eluted with a salt gradient. This mode of operation can, of course, be reversed. Figuratively, one can think of two-dimensional chromatography at one and the same stationary phase. An interesting practical application of this technique is on-column sample preparation, which possibly might replace the treatment with extraction cartridges. Slingsby and Rey [54] report the separation of acebutulol in rat urine on an OmniPac PAX-500. As a prerequisite, matrix components and analytes must differ in their hydrophobicity and their degree of dissociation. If this is the case, the gradient is programmed in such a way that the interfering matrix component elutes prior to the analyte. In the above-mentioned example, the matrix components of urine that are mostly of ionic nature are eluted with a high carbonate concentration

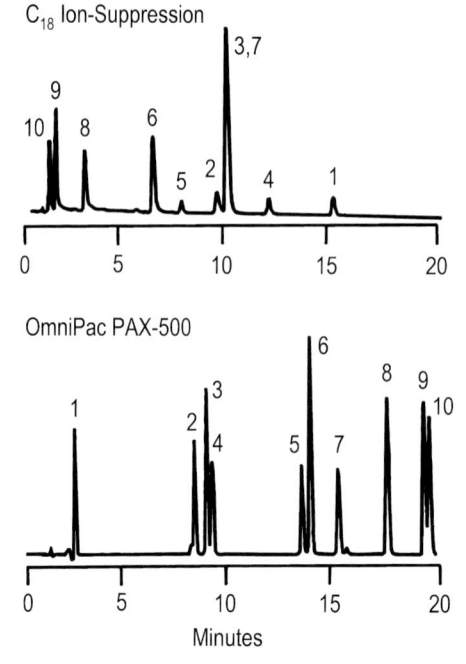

Fig. 6-63. Comparison of ion-suppression- and multimode selectivities exemplified with the separation of aromatic carboxylates. — (C_{18} ion suppression) Separator column: chemically bonded silica (ODS); eluant: 20 mmol/L H_3PO_4 + 20 mmol/L HOAc – acetonitrile; gradient: linear, 5% acetonitrile to 30% in 20 min, (multimode) separator column: OmniPac PAX-500; eluant: KOH/KCl – acetonitrile (80:20 v/v); gradient: linear, 0.5 mmol/L KOH + 50 mmol/L KCl to 4 mmol/L KOH + 400 mmol/L KCl in 20 min; flow rates: 1 mL/min; detection: UV(254 nm); analytes: (1) benzoate, (2) 1,2-benzenedicarboxylate, (3) 1,4-benzenedicarboxylate, (4) 1,3-benzenedicarboxylate, (5) 1,2,3-benzenetricarboxylate, (6) 1,2,4-benzenetricarboxylate, (7) 1,3,5-benzenetricarboxylate, (8) 1,2,4,5-benzenetetracarboxylate, (9) benzenepentacarboxylate, and (10) benzenehexacarboxylate.

after injection of the filtrated sample. Thereafter, the carbonate concentration drops drastically, followed by the elution of the drug component with an organic solvent gradient.

Multimode phases with anion exchange properties are suitable for the analysis of a number of polar and ionic organic species, including the compound class of barbiturates (see also Fig. 6-33 in Section 6.4), which can be separated on an OmniPac PAX-500 with a combined sodium carbonate/acetonitrile gradient and UV detection at 254 nm. Figure 6-66 shows a respective separation of various barbiturates. Their structures are summarized in Table 6-3. Compared to the separation under ion-suppression conditions (see Fig. 6-55) a retention reversal between butabarbital and phenobarbital is observed. When using an OmniPac

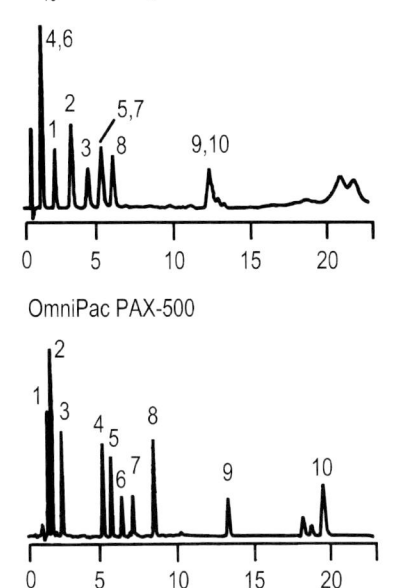

Fig. 6-64. Comparison of ion-pair- and multimode selectivities exemplified with the separation of phenylphosphonate and aromatic sulfonic acids. – (C_{18} ion-pair) separator column: chemically bonded silica (ODS); Eluant: 4 mmol/L H_3PO_4 + 10 mmol/L tetrabutylammonium phosphate (pH 7.5) – acetonitrile; gradient: linear, 25% acetonitrile to 60% in 20 min, (multimode) separator column: OmniPac PAX-500; eluant: KOH/KCl – acetonitrile (75:25 v/v); gradient: linear, 0.5 mmol/L KOH + 50 mmol/L KCl to 7.5 mmol/L KOH + 750 mmol/L KCl in 20 min; flow rates: 1 mL/min; detection: UV (254 nm); analytes: (1) benzenesulfonate, (2) toluenesulfonate, (3) p-chlorobenzenesulfonate, (4) phenylphosphonate, (5) o-benzenedisulfonate, (6) p-hydroxybenzenesulfonate, (7) m-benzenedisulfonate, (8) 5-sulfosalicylate, (9) 1,3,5-benzenetrisulfonate, and (10) 1,3,6-naphtholenetrisulfonate.

PAX-100 instead of the PAX-500 column, the elution order also differs because of the different retention mechanism, especially for the more strongly adsorbing barbiturates.

The separation of anti-inflammatory drugs (see Table 6-4) on OmniPac PAX-500 shown in Fig. 6-67 is also based on the combination of anion exchange- and reversed-phase interactions. It represents an interesting alternative to the ion-suppression technique, because a number of these compounds bear carboxyl groups and are strongly hydrophobic due to their aromatic character. The structural versatility of compounds, which can be separated within 20 minutes with a combined sodium carbonate/acetonitrile gradient, is characteristic for such a separation. In principle, this separation could also be carried out with a chloride salt in the mobile phase, but the concentration required for elution would be higher than that required for carbonate.

In addition, excellent separations of acidic azo dyes such as Methyl Orange, Methyl Red, β-Naphthol Orange, Orange I, Orange II, etc. can be obtained on OmniPac PAX-500. They can be eluted with acetonitrile in an alkaline environment and detected by measuring the light absorption at 254 nm (see also Fig. 6-73).

Among organic cations, OmniPac PCX-100 is predominantly used for analyzing biogenic amines such as catecholamines, choline, and acetylcholine. In comparison with separations on ODS phases, the OmniPac separations are characterized by a different selectivity. Choline and acetylcholine, for example, are more

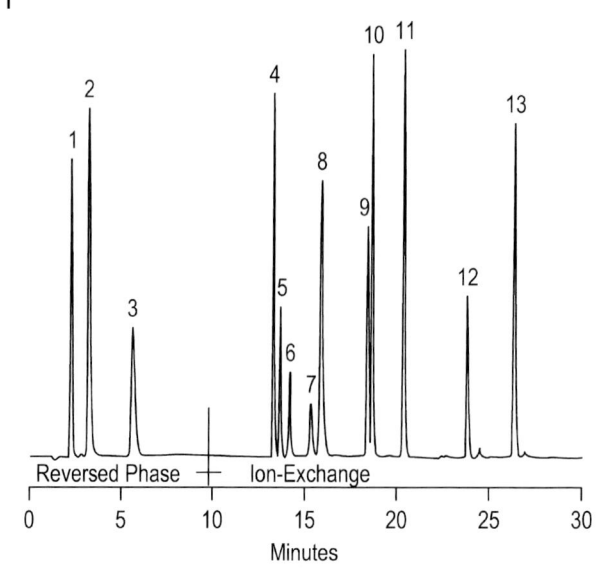

Fig. 6-65. Simultaneous analysis of molecular and ionic organic compounds on OmniPac PAX-500. – Eluant: NaOH/NaCl – acetonitrile; gradient: acetonitrile – water (80:20 v/v) for 10 min isocratically, then linearly from 0.2 mmol/L NaOH + 50 mmol/L NaCl – acetonitrile (80:20 v/v) to 1.6 mmol/L NaOH + 400 mmol/L NaCl – acetonitrile (80:20 v/v) in 10 min; flow rate: 1 mL/min; detection: UV (254 nm); analytes: (1) benzylalcohol, (2) diethyltolueneamide, (3) benzene, (4) benzoate, (5) benzenesulfonate, (6) toluenesulfonate, (7) p-chlorobenzenesulfonate, (8) p-bromobenzenesulfonate, (9) phthalate, (10) terephthalate, (11) p-hydroxybenzenesulfonate, (12) 1,3,5-benzenetricarboxylate, and (13) 1,2,4,5-benzenetetracarboxylate (pyromellitate).

strongly retained by cation exchange interactions and, thus, better separated from the system void. Figure 6-68 shows a respective comparison between the classical HPLC method and the separation on an OmniPac PCX-100 column. Using conductivity detection in combination with the OmniPac column is preferable to using UV detection, because it is more sensitive and more selective. Food products often contain non-ionic, UV-absorbing compounds that do not interfere with conductivity detection. Moreover, inorganic cations such as sodium and potassium can be analyzed in the same run. Comparable separations of choline and acetylcholine are also carried out on conventional cation exchangers such as IonPac CS10 (see Section 4.1.2).

The solvent compatibility of the OmniPac PCX-100 is the prerequisite for the separation of ephedrines. Because ephedrine has two asymmetric C-atoms, two mirrored image isomeric ephedrines exist in addition to the pseudoephedrines, another antipodal pair. The two forms only differ in the configuration at the secondary alcohol group. Figure 6-69 illustrates the influence of the mobile phase ionic strength on the separation of the two pseudoephedrines and their methyl-substituted analogues.

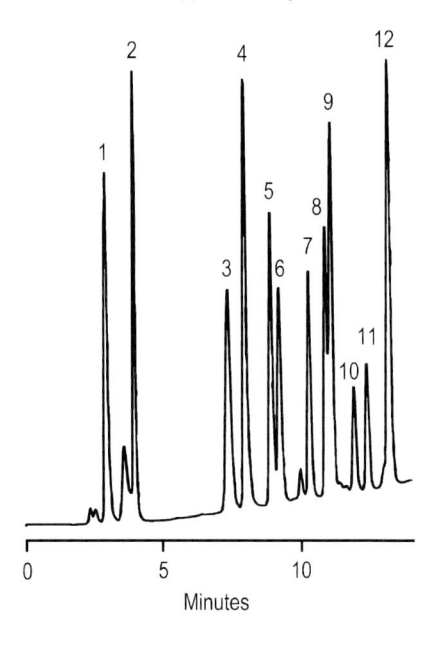

Fig. 6-66. Separation of barbiturates on OmniPac PAX-500. – Eluant: Na_2CO_3 – acetonitrile; gradient: linear, 5 mmol/L Na_2CO_3 – acetonitrile (91:9 v/v) to 20 mmol/L Na_2CO_3 – acetonitrile (80:20 v/v) in 10 min; flow rate: 1 mL/min; detection: UV (254 nm); injection volume: 20 µL; solute concentrations: 3.15 mg/L barbituric acid (1), 12.5 mg/L barbital (2), 8.75 mg/L allobarbital (3), 9.4 mg/L metharbital (4), 6.25 mg/L butabarbital (5), 6.25 mg/L phenobarbital (6), 6.25 mg/L amobarbital (7), 6.25 mg/L mephobarbital (8), 6.25 mg/L secobarbital (9), 3.15 mg/L methohexital (10), 12.5 mg/L diphenylhydantoin (11), and 37.5 mg/L thiamylal (12).

Diphenylhydantoin

Thiamylal

(1R,2R)-(-)-Pseudoephedrin [(-)-Ψ-Ephedrin]

(1S,2S)-(+)-Pseudoephedrin [(+)-Ψ-Ephedrin]

(1S,2R)-(+)-N-Methylephedrin

As can be seen from this chromatogram, a baseline-resolved separation of all three components is only achieved by adding significant amounts of chloride to the mobile phase. In contrast, a high acid concentration does not give the same result, because peaks **2** and **3** co-elute.

The respective multimode phase with cation exchange properties is suitable for the analysis of a number of compound classes. The most important ones include alkaloids, water-soluble vitamins, sulfonamides, cephalosporins, cat-

Table 6-3. Structures of some selected barbiturates.

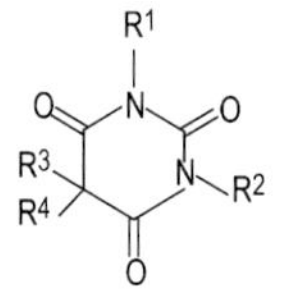

Barbiturate	R¹	R²	R³	R⁴
Metharbital	CH_3	H	C_2H_5	C_2H_5
Allobarbital	H	H	$CH_2=CHCH_2$	$CH_2=CHCH_2$
Amobarbital	H	H	C_2H_5	$(CH_3)_2CHCH_2CH_2$
Mephobarbital	CH_3	H	C_6H_5	C_2H_5
Phenobarbital	H	H	C_2H_5	C_6H_5
Barbituric acid	H	H	H	H
Barbital	H	H	C_2H_5	C_2H_5

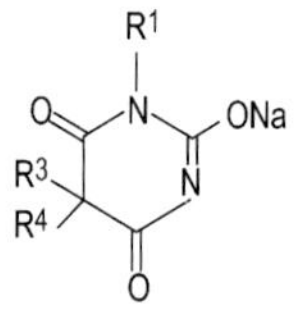

Barbiturate	R¹	R²	R³	R⁴
Secobarbital Na	H	–	$CH_2=CHCH_2$	$CH_3CH_2CH_2CHCH_3$
Methohexital Na	CH_3	–	$CH_2=CHCH_2$	$CH_3CH_2C{\equiv}CCHCH_3$
Butabarbital Na	H	–	CH_3CH_2	$CH_3CH_2CHCH_3$

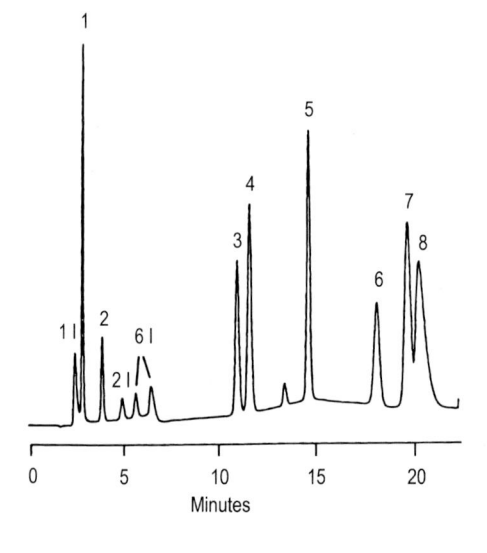

Fig. 6-67. Separation of anti-inflammatory drugs on OmniPac PAX-500. – Eluant: Na_2CO_3 – acetonitrile; gradient: linear, 10 mmol/L Na_2CO_3 – acetonitrile (84:16 v/v) to 50 mmol/L Na_2CO_3 – acetonitrile (70:30 v/v) in 10 min; flow rate: 1 mL/min; detection: UV (254 nm); analytes: (1) aspirin, (2) ibuprofen, (3) tolmetin, (4) naproxen, (5) fenbufen, (6) indomethacin, (7) carprofen, and (8) diflunisal; (peaks marked with "I" are impurities of the indicated drug standard).

Table 6-4. Structures of the compounds separated in Fig. 6-67.

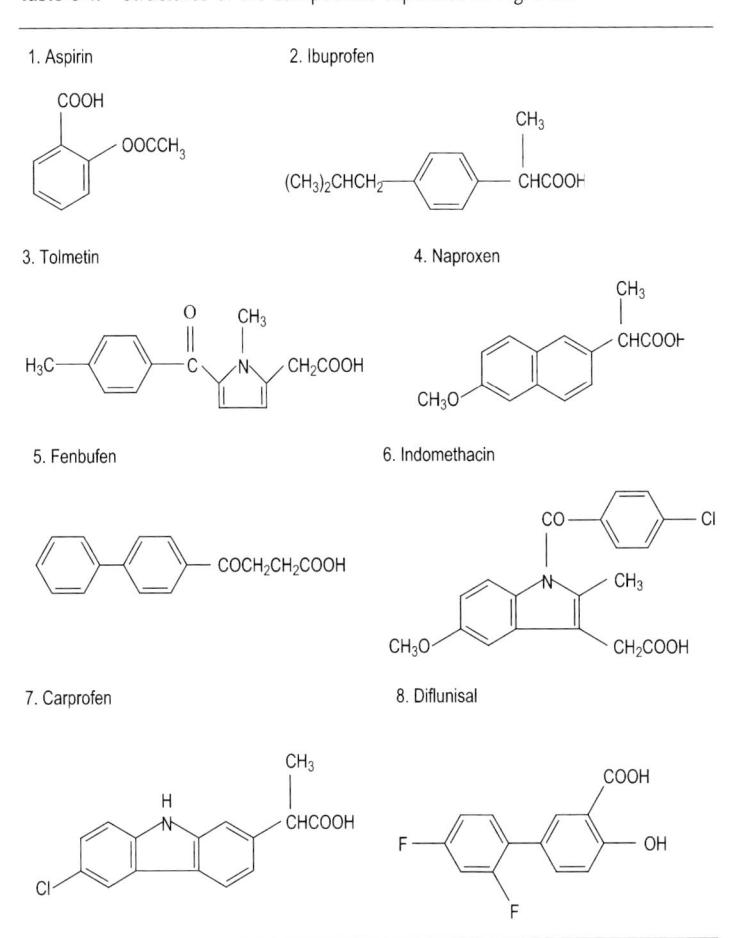

1. Aspirin
2. Ibuprofen
3. Tolmetin
4. Naproxen
5. Fenbufen
6. Indomethacin
7. Carprofen
8. Diflunisal

echolamines, diuretics, and xanthines, as well as a number of basic dyes. Figure 6-70 shows the separation of ten different alkaloids, of which theophylline, theobromine, and caffeine are genuine constituents in beverages such as coffee and tea. In comparison with the classical separation on ODS phases, potentially interfering matrix components are better separated on a multimode phase.

The ability to apply a combination of UV- and conductivity detection to analyze inorganic cations and vitamins in the same run is an advantage for the determination of water-soluble vitamins with a multimode column.

Sulfonamides are effective chemotherapeutics against coccus infections. The most effective sulfonamides include N^1-substituted derivatives of sulfanilamide. Derivatives of carbonic acid as well as heterocyclic rings usually serve as substituents. The other part of sulfanilic acid, H_2N-C_6H_4-SO_2-, is abbreviated as "sulfa". Consequently, the names of sulfonamides are composed of the prefix

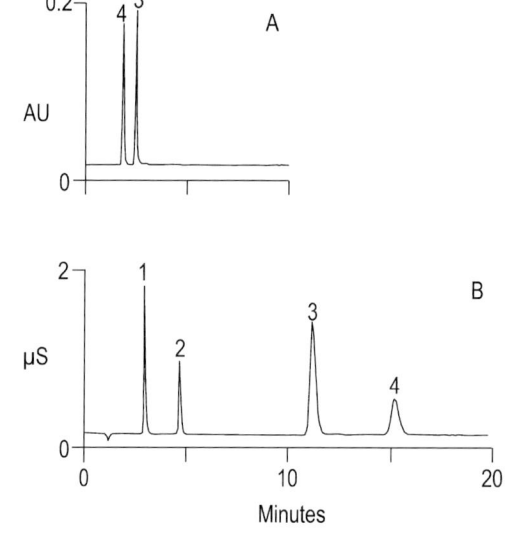

Fig. 6-68. Selectivity comparison between an ODS phase and an OmniPac PCX-100 exemplified with the separation of choline and acetylcholine. — (A) Separator column: Lichrosorb® RP-18; eluant: 5 mmol/L heptanesulfonic acid (pH 4) — acetonitrile (99:1 v/v); flow rate: 1 mL/min; injection volume: 10 µL; (B) Separator column: OmniPac PCX-100; eluant: 75 mmol/L HCl — methanol (99:1 v/v); flow rate: 1 mL/min; injection volume: 10 µL; solute concentrations: 1 mg/L sodium (1), 1 mg/L potassium (2), 10 mg/L choline (3), and 10 mg/L acetylcholine (4).

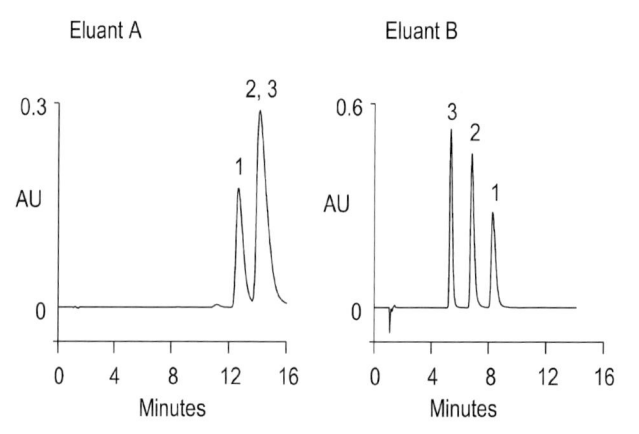

Fig. 6-69. Influence of the mobile phase ionic strength on the separation of ephedrines on OmniPac PCX-100. — Eluant: (A) 80 mmol/L methanesulfonic acid — acetonitrile (80:20 v/v), (B) 20 mmol/L methanesulfonic acid + 50 mmol/L CsCl — acetonitrile (80:20 v/v); flow rate: 1 mL/min; detection: UV (210 nm); analytes: (1) (1R,2R)-(−)-pseudoephedrine, (2) (1S,2S)-(+)-pseudoephedrine, (3) (1S,2R)-(+)-N-methylephedrine.

"sulfa" and the basic group at the S-atom. In practice, sulfonamides can be separated on all OmniPac columns, because their pK_a values allow anion exchange interactions. On the other hand, the amine component of sulfonamides makes cation exchange possible. Because all sulfonamides possess an aromatic ring system, reversed-phase interactions also contribute to the retention mechanism. Due to the huge selectivity differences between the various separator columns, the choice of stationary phase very much depends on the sample matrix.

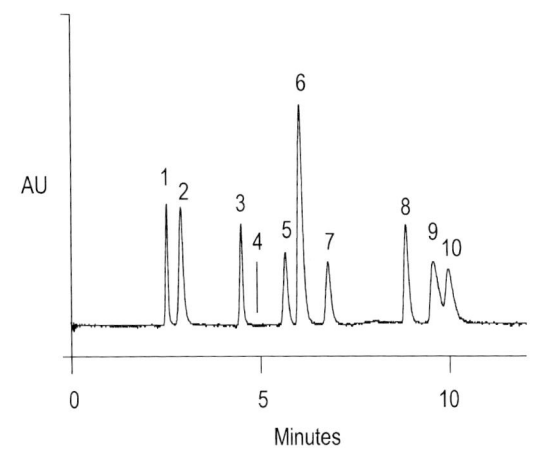

6.7 Applications of Multi-Dimensional Ion Chromatography on Multimode Phases

Fig. 6-70. Separation of alkaloids on OmniPac PCX-500. — Eluant: HCl — KCl — acetonitrile; gradient: stepwise from 25 mmol/L HCl + 0.1 mol/L KCl — acetonitrile (87:13 v/v) after 1 min to 50 mmol/L HCl + 0.1 mol/L KCl — acetonitrile (64:36 v/v), after 3 min to 50 mmol/L HCl + 0.1 mol/L KCl — acetonitrile (52:48 v/v), after 0.1 min to 50 mmol/L HCl + 0.3 mol/L KCl — acetonitril (46:54 v/v); flow rate: 1 mL/min; detection: UV (254 nm); analytes: (1) theobromine, (2) theophylline, (3) caffeine, (4) morphine, (5) colchicine, (6) strychnine, (7) papaverine, (8) nicotine, (9) cinchonine, and (10) quinine.

As an example, Fig. 6-71 illustrates the separation of sulfonamides on OmniPac PCX-500 with a combined sodium acetate/acetonitrile eluant under gradient conditions. Perchlorate anions added to the mobile phase form stable ion pairs, so that those interactions also contribute to the retention mechanism in this case.

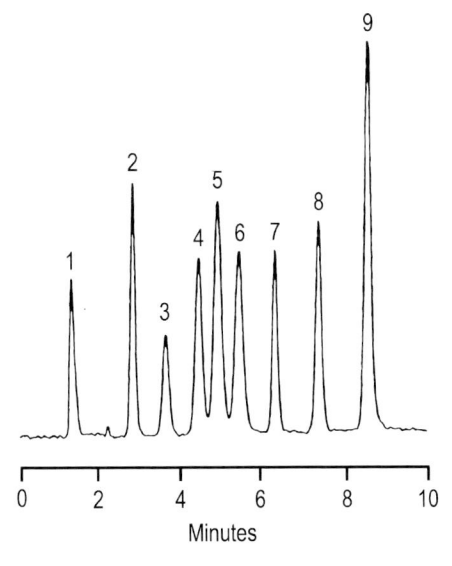

Fig. 6-71. Separation of sulfonamides on OmniPac PCX-500. — Eluant: $HClO_4$ — NaOAc — acetonitrile; gradient: linear, 10.8 mmol/L $HClO_4$ + 20 mmol/L NaOAc — acetonitrile (73:27 v/v) for 2 min isocratically, then to 10.8 mmol/L $HClO_4$ + 0.104 mol/L NaOAc — acetonitrile (44:56 v/v) in 3 min; flow rate: 1.5 mL/min; detection: UV (254 nm); solute concentrations: 0.5 mg/L each of sulfanilic acid (1), sulfanilamide (2), sulfathiazole (3), sulfadiazine (4), sulfamerazine (5), sulfamethazine (6), sulfisoxazole (7), trimethoprim (8), and sulfadimethoxin (9).

Cephalosporins belong to the group of β-lactam antibiotics. The ring system is based on a thiazolidine ring, which is condensated with a β-lactam ring. Cephalosporins synthesized so far only differ in the two substituents R^1 and R^2.

Cephalosporin

Because these compounds are less stable in a basic environment than in an acidic one, a multimode phase with cation exchange properties is again a suitable separation material. Thus, the separation of various cephalosporins with an acetonitrile gradient in the presence of perchlorate anions, as shown in Fig. 6-72, is based on ion-pairing- and cation exchange interactions. The relatively high concentration of perchloric acid in the mobile phase ensures complete protonation of the carboxyl groups that are present in the compounds under investigation. If hydrochloric acid is used instead of perchloric acid, a different selectivity results, as expected. A very remarkable feature of this chromatogram is the fact that some of the precursors of cephalosporin such as D-hydroxyphenylglycine, 7-aminocephalosporanic acid, and cephalosporin C can be analyzed in the same run.

At last, Fig. 6-73 shows a chromatogram with the separation of various dyes. The separation is based on reversed-phase-, cation exchange-, and ion-pair interactions. While some of the late-eluting species bear quaternary ammonium

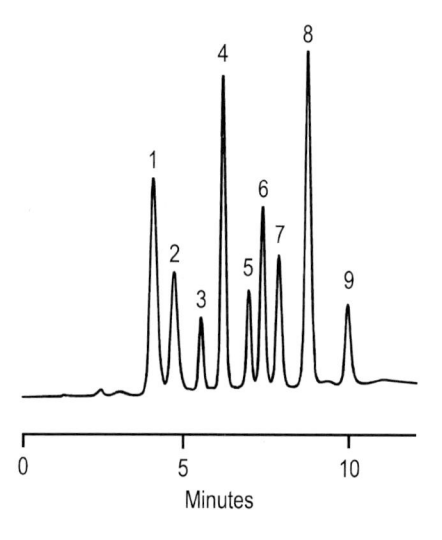

Fig. 6-72. Separation of cephalosporins on OmniPac PCX-500. — Eluant: $HClO_4$ — acetonitrile; gradient: linear, 0.12 mol/L $HClO_4$ — acetonitrile (86:14 v/v) to 0.12 mol/L $HClO_4$ — acetonitrile (46:54 v/v) in 8 min; flow rate: 1 mL/min; detection: UV (215 nm); analytes: (1) D-hydroxyphenylglycine, (2) 7-aminocephalosporanic acid, (3) cephalosporin C, (4) cefadroxil, (5) cefazolin, (6) cephalexin, (7) cefotaxime, (8) cephaloridine, and (9) cephalotin.

groups, the early-eluting components are of an anionic nature. The latter ones are exclusively retained at the stationary phase by hydrophobic interactions. A mixture of perchloric acid, sodium perchlorate, and acetonitrile is used as the mobile phase. With the exception of perchloric acid, its concentration was increased step-wise during the run.

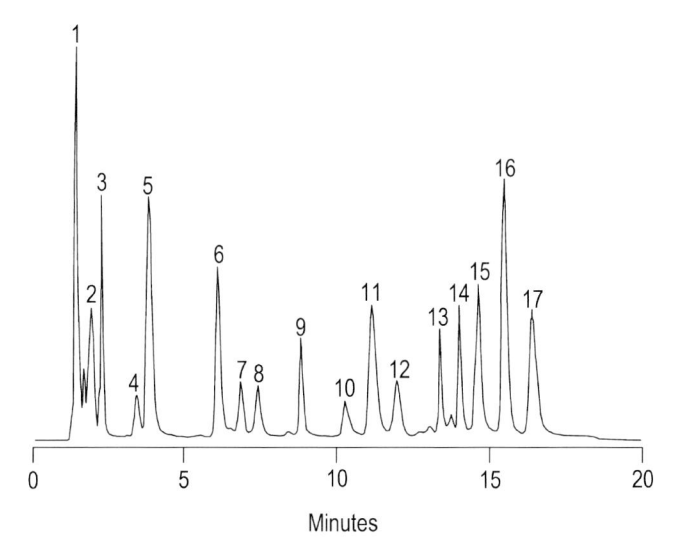

Fig. 6-73. Separation of dyes on OmniPac PCX-500. — Eluant: $HClO_4$ — $NaClO_4$ — acetonitrile step-gradient; flow rate: 1 mL/min; detection: UV (254 nm); analytes: (1) indigocarmine, (2) orange G, (3) tropaeolin O, (4) orange I, (5) alizarin red S, (6) orange II, (7) chromeazurol S, (8) bromocresol purple and acid blue 40, (9) thymol blue, (10) acid blue 113, (11) fluorescein, (12) methyl green, (13) methylene blue and acid red 114, (14) acridine orange, (15) nile blue, (16) rhodamine B, and (17) malachite green.

7
Detection Methods in Ion Chromatography

Detection methods applied in ion chromatography are divided into electrochemical and spectrometric methods. Conductometric and amperometric detection are electrochemical methods, while the spectrometric methods include UV/Vis, fluorescence, and refractive index detection. In addition, there are various application forms of these detection methods; these are described in detail below.

In most cases, the choice of a suitable detection mode depends on the separation method and the corresponding eluants. If detection is to be carried out by direct measurement of a physical property of the solute ion (e.g., UV absorption), the solute ion must differ substantially in this property from eluant ions which are present in much higher concentration. However, eluant and solute ions often exhibit similar properties, so direct detection is only feasible where selective detection of a limited number of solute ions is desirable.

A much broader range of applications employ detection methods that measure changes in a certain physical property of the eluant (e.g., conductance) that are caused by the elution of the solute ion. As a prerequisite, the values of this property for eluant and solute ions must differ. Most of the detection methods applied in ion chromatography are based on this technique. In the following discussion, a further subdivision into direct and indirect methods is made. Direct detection methods are those, in which eluant ions exhibit a much smaller value than solute ions for the property to be measured. On the other hand, detection methods are called indirect, if eluant ions exhibit a much higher value for the property to be measured than do solute ions.

7.1
Electrochemical Detection Methods

7.1.1
Conductivity Detection

As a universal method for the detection of ionic species, conductometric detection has the highest significance in ion chromatography. The fundamental theoretical principles of this detection method are summarized in the following.

7.1.1.1 Theoretical Principles

Electric conductivity of electrolyte solutions [1]
The electric resistance of an electrolyte solution is described by Ohm's law

$$R = \frac{U}{I} \tag{160}$$

- R Resistance
- U Voltage
- I Current strength

Because the resistance depends on the type of conductor, the resistivity ϱ is defined as a material-specific quantity as follows:

$$\varrho = \frac{A \cdot R}{l} \tag{161}$$

- A Cross section of conductor
- l Length of conductor

The electric conductivity $\varkappa$ with unit $S\ cm^{-1}$ represents the reciprocal of the resistivity.

$$\varkappa = \frac{1}{\varrho} \tag{162}$$

The electric conductivity of electrolytes is strongly dependent on concentration. To compare the conducting power of different electrolyte solutions, the electric conductivity is divided by the equivalent concentration, c_{ev}, yielding the equivalent conductance, Λ ($S\ cm^2\ val^{-1}$):

$$\Lambda = \frac{\varkappa}{c_{ev}} \tag{163}$$

When two electrodes immersed in an electrolyte solution are connected to a power supply, an electric field of strength E is created between them. In this field, a directed mass transport occurs. Anions drift to the positive pole while cations drift to the negative pole. Because the ion velocity, v, is strictly proportional to the electric field strength, the mobility of ions, u ($cm^2\ V^{-1}\ s^{-1}$), is an independent characteristic:

$$u = \frac{v}{E} \tag{164}$$

7.1 Electrochemical Detection Methods

In a strong electrolyte, cations and anions are present in the solution in the concentration c_+ and c_-, respectively. The following applies:

$$c = c_+ = c_- \tag{165}$$

When the electrolyte is in an electric field, the number of cations and anions passing the given cross-section A of the electrolyte within the time t is given by $N_A \cdot c_+ \cdot v_+ \cdot A \cdot t$ and $N_A \cdot c_- \cdot v_- \cdot A \cdot t$, respectively. Because each ion carries an electric charge, the directed motion is associated with a charge transport. The ratio of the sum of charges and time represents the electric current strength, I:

$$I = I_+ + I_- \tag{166}$$

$$I = e \cdot N \cdot c_+ \cdot v_+ \cdot A + e \cdot N \cdot c_- \cdot v_- \cdot A \tag{167}$$

N Loschmidt number
e Elemental charge

Using the Faraday constant, $F = N \cdot e$, and taking into account the ion mobility according to Eq. (164), one obtains:

$$I = F \cdot c \cdot A \cdot E \, (u_+ + u_-) \tag{168}$$

Because:

$$\varkappa = \frac{I}{A \cdot E} \tag{169}$$

one obtains with Eq. (168) the electric conductivity for the solution of a strong electrolyte:

$$\chi = c \cdot F \, (u_+ + u_-) \tag{170}$$

Correspondingly, for a polyvalent electrolyte

$$\chi = c \cdot z \cdot F \, (u_+ + u_-) \tag{171}$$

where z is the number of charges. In case of incomplete dissociation, the degree of dissociation also has to be taken into account:

$$\chi = \alpha \cdot c \cdot z \cdot F \, (u_+ + u_-) \tag{172}$$

Thus, the electric conductivity increases
- with increasing ion concentration,
- with increasing charge numbers of the ions, and
- with increasing ion mobility.

With Equations (163) and (171) one obtains:

$$\Lambda = \frac{\varkappa}{z \cdot c} = F \cdot (u_+ + u_-) \tag{173}$$

According to Kohlrausch, the quantities $F \cdot u_+$ and $F \cdot u_-$ are the equivalent ionic conductances Λ_+ and Λ_- for anions and cations, respectively. Thus, the conductivity is a summation:

$$\Lambda = \Lambda_+ + \Lambda_- \tag{174}$$

The equivalent conductance of strong electrolytes always decreases continuously with increasing concentration; with decreasing electrolyte concentration it approaches a material-specific limiting value, the conductivity at infinite dilution, Λ_∞. This behavior may be attributed to the decrease in ion mobility with increasing concentration. At increasing electrolyte concentration, the ions draw so near to one another that they affect each other electrostatically. On time average, Coulomb interaction increases with the shortening distance between ions. This interionic interaction is further enhanced at high electrolyte concentrations because the dielectric constant of the solvent is decreased by the electrolyte. This electrostatic interaction may even cause solvated ions to form ion pairs which do not contribute to the conductivity [2].

Kohlrausch observed the following empirical relation describing the concentrational dependence of the conductivity of strong electrolytes:

$$\Lambda = \Lambda_\infty - k \cdot \sqrt{c} \tag{175}$$

(Kohlrausch square root law)

Here k depends on the charge number of the ion. Plotting Λ versus $\sqrt{c}$ yields a straight line, the slope of which depends on the electrochemical valency of the electrolyte. Equation (175) only applies to low concentrations ($c < 10^{-2}$ mol/L).

An increasing electric conductivity is usually observed with increasing temperature, as the viscosity of the solution decreases exponentially with rising temperature.

Conductivity measurements only define the sum of the equivalent conductances of the ions; no information about their individual values can be derived. However, it is known that the equivalent conductances of ions may differ significantly. According to Kohlrausch's law of independent ion drift, all ions move independently of each other in an infinitely diluted solution. Because the equivalent conductances of ions differ, they contribute differently to the current transport. The contribution of an ionic species, i, to the total current is called the transport number, t_i:

$$t_i = \frac{I_i}{I} \tag{176}$$

With Equations (166) and (167) it applies:

$$I_+ = F \cdot c \cdot A \cdot E \cdot u_+ \tag{177}$$

The result of Eq. (168) is the contribution to the current that is apportioned to the cations. It is denoted as cation transport number, t_+:

$$\begin{aligned} t_+ &= \frac{I_+}{I} \\ &= \frac{\Lambda_+}{\Lambda_+ + \Lambda_-} \\ &= \frac{\Lambda_+}{\Lambda} \end{aligned} \tag{178}$$

Similarly, it applies for the anion transport number, t_-:

$$\begin{aligned} t_- &= \frac{I_-}{I} \\ &= \frac{\Lambda_-}{\Lambda_+ + \Lambda_-} \\ &= \frac{\Lambda_-}{\Lambda} \end{aligned} \tag{179}$$

The equivalent conductances of anions and cations are typically between 35 S cm² val⁻¹ and 80 S cm² val⁻¹ (see Table 7-1). The hydronium ion with 350 S cm² val⁻¹ and the hydroxide ion with 198 S cm² val⁻¹ (25 °C) are the only exceptions. Because these values are that high only in aqueous solution, it has to be assumed that there is a special mechanism for the transport of H^+- and OH^- ions that is correlated with the water structure. Hydrogen bonds between associated water molecules allow the exchange of hydronium- and hydroxide ions over a long chain of water molecules with no actual migration of hydrated ions. The high equivalent conductances of H^+ and OH^- are especially useful in the suppressor reaction to convert the salts under investigation into higher conductive species while the eluant is converted into a less conductive form (see Sections 3.6 and 4.3).

Table 7-1. Equivalent conductances of some selected anions and cations.

Anions	Λ_- [S cm² val⁻¹]	Cations	Λ_+ [S cm² val⁻¹]
OH⁻	198	H⁺	350
F⁻	54	Li⁺	39
Cl⁻	76	Na⁺	50
Br⁻	78	NH₄⁺	73
I⁻	77	K⁺	74
NO₂⁻	72	Mg²⁺	53
NO₃⁻	71	Ca²⁺	60
HPO₄²⁻	57	Sr²⁺	59
SO₄²⁻	80	Ba²⁺	64
Benzoate	32	CH₃NH₃⁺	58
Phthalate	38	N(CH₃CH₂)₄⁺	33

Interionic interactions

For dilute electrolyte solutions, Lewis and Randall observed that the mean activity coefficient of a strong electrolyte does not depend on the kind of ion, but only on the concentration and charge numbers of all ions present in solution. Thus, the individual properties of the ions are not decisive for interionic interactions in dilute electrolyte solutions. These observations paved the way for the introduction of the concept of ionic strength, I:

$$I = \frac{1}{2} \sum_i c_i \cdot z_i^2 \tag{180}$$

This is a very functional quantity, as the mean activity coefficient of an electrolyte may easily be described with it. Lewis discovered the following empirical relation:

$$\log f_\pm = -A \cdot z_+ \cdot |z_-| \cdot \sqrt{I} \tag{181}$$

$f_\pm$ Mean activity coefficient

When $\log f_\pm$ is plotted versus $\sqrt{I}$, a straight line is obtained, with a slope that only depends on the charge numbers of ions present in the electrolyte.

According to the theory of Debye and Hückel, it is assumed that the mutual influence of ions present in the crystal lattice of the solid electrolyte is not fully abolished in electrolyte solution. Accordingly, owing to electrostatic attractive forces, each ion in solution is surrounded by ions of opposite charge. Although these ions are subjected to thermal motion, in time average a positive ion, for example, will form the central ion of an oppositely charged ion cloud. Each ion in this ion cloud interacts with the central ion. The total charge of the ion cloud equals that of the central ion because of the electroneutrality condition.

Following Debye and Hückel, the distribution of ions can be calculated via the Boltzmann energy distribution. The application of this law is based on the concept that the ion cloud represents a space charge, which is most dense in the vicinity of the central ion and decreases with increasing distance from the central ion. A number of simplifying assumptions concerning the state of ions are made:
- The electrolyte is fully dissociated.
- Only electrostatic forces acting between ions are responsible for interionic interactions.
- The electrostatic interaction energy is small compared to the thermal energy.
- The ions are regarded as point charges with an electric field of spherical symmetry and, thus, are non-polarizable.
- The dielectric constant of the electrolyte solution is equal to that of the pure solvent.

With increasing electrolyte concentration, these conditions become more and more inadequate:
- Strong electrolytes form ion pairs at high concentration.
- Apart from electrostatic interactions, ion-molecule interactions also occur at high electrolyte concentrations; such interactions affect the solvation state of the ions and the solvent structure.
- The electrostatic interaction energy becomes so high at high electrolyte concentration, that ions no longer execute an unhindered thermal motion.
- With increasing electrolyte concentration, ions may polarize each other due to close encounters.
- Interactions of ions with the solvent change the dielectric constant of the solvent.

According to Lewis, the constant A in Eq. (181) had to be determined empirically. On the other hand, with the aid of the Debye-Hückel theory, it can be based on physical quantities and, thus, may be calculated. The theoretically derived dependence

$$\log f = -A \cdot z^2 \cdot \sqrt{I} \tag{182}$$

allows the following conclusions:
- At finite ionic strength $\log f$ always has a negative value; that is, the activity coefficient is <1 in the validity range of the equation.
- The logarithm of the activity coefficient decreases from its limiting value $\log f = 0$ (for $I = 0$) with the square root of the ionic strength.
- Ions with high charge numbers deviate more strongly from the ideal behavior than do the ions with small charge numbers.
- The activity coefficient is strongly affected by the temperature and the dielectric constant.

Equation (182) only represents a limiting law owing to the reduced validity of the prerequisites on which the Debye-Hückel theory is based. Thus, Eq. (182) is applicable to monovalent 1-1 electrolytes in aqueous solution up to a maximum ionic strength of 10^{-2} mol/L.

Also, with the Debye-Hückel theory the effect of the ion cloud on ion mobility can be quantified.

When a central ion moves within an electric field, the ion cloud surrounding the ion is permanently renewed. This requires a certain amount of time called the relaxation time. Therefore, as illustrated in Fig. 7-1, the charge density around the central ion is no longer symmetrical, but is lower in front of the central ion than behind it. This dissymmetry in charge distribution leads to an electrostatic deceleration of the central ion, which reduces the mobility of the ion.

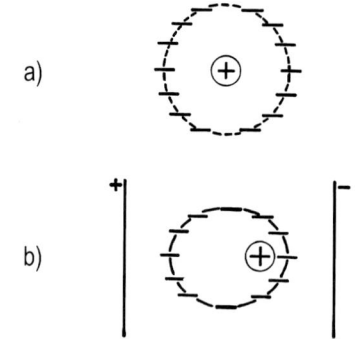

Fig. 7-1. Schematic representation of a central ion with its ion cloud: a) without an external electric field, b) with an external electric field.

The ion cloud, with the solvation shells of its ions, moves in a direction opposite to the central ion. Therefore, the central ion does not move relative to a resting medium but rather against a solvent flow. The resulting reduction in the mobility is called the electrophoretic effect. Both effects become more important with increasing electrolyte concentration and result in a decrease of the equivalent conductance.

According to Onsager the quantitative analysis yields the following equation for the conductivity of a 1-1 electrolyte:

$$\Lambda = \Lambda_\infty - \underbrace{\frac{8.2 \cdot 10^5}{\left(\varepsilon \cdot \frac{T}{K}\right)^{3/2}} \cdot \Lambda_\infty \cdot \sqrt{\frac{c}{\text{mol L}^{-1}}}}_{\text{Relaxation Effect}}$$

$$- \underbrace{\frac{82.48 \text{ cm}^2 \text{ S mol}^{-1}}{\frac{\eta}{\text{g cm}^{-1}\text{s}^{-1}} \cdot \left(\varepsilon \cdot \frac{T}{K}\right)^{1/2}} \cdot \sqrt{\frac{c}{\text{mol L}^{-1}}}}_{\text{Electrophoretic Effect}} \qquad (183)$$

$\Lambda = \Lambda_\infty -$ (Relaxation Effect) $-$ (Electrophoretic Effect)

The lower the dielectric constant, ε, of the solvent, the stronger the interionic interactions and thus, the stronger the relaxation and electrophoretic effects. For the latter, solvent viscosity also plays a decisive role. When both the relaxation and the electrophoretic constants are known, the conductivity coefficient is obtained:

$$f_\Lambda = \frac{\Lambda}{\Lambda_\infty} = 1 - \left(A' + \frac{B}{\Lambda_\infty}\right) \cdot \sqrt{c} \tag{184}$$

A' Relaxation constant
B Electrophoretic constant

7.1.1.2 Application Modes of Conductivity Detection

When passing the separator column effluent into the conductivity cell *without* applying a suppressor system, the electric conductance χ of a solution, according to Fritz et al. [3], is given by:

$$\chi = \frac{(\Lambda_+ + \Lambda_-) \cdot c \cdot \alpha}{10^{-3} K} \tag{185}$$

Λ_+, Λ_- Equivalent conductances of cations and anions
c Concentration
α Dissociation constant of the eluant
K Cell constant

When the cell constant and the equivalent conductances of eluant anions and cations are known, Eq. (185) makes it possible to calculate the conductivities of eluants that are typical for this kind of detection method. To determine the cell constant, the conductivity of a potassium chloride solution with a defined concentration is usually measured, as the equivalent conductances of potassium and chloride ions are known to be 74 S cm^2 val^{-1} and 76 S cm^2 val^{-1}, respectively (see Table 7-1).

When the eluant of the concentration c_E contains cations E$^+$ and anions E$^-$, the background conductivity, χ_E, may be calculated via Eq. (185) as follows:

$$\chi_E = \frac{(\Lambda_{E^+} + \Lambda_{E^-}) \cdot c_E \cdot \alpha_E}{10^{-3} K} \tag{186}$$

When the concentration of solute ions passing the detector is denoted as c_S and their degree of dissociation as α_S, the eluant concentration in the measuring cell during the elution of solute ions is given by $(c_E - c_S \cdot \alpha_S)$. Hence, in this instance, the measured conductivity is caused by eluant and solute ions as well as by the eluant counter ions that are required to maintain electroneutrality. In anion exchange, solute counter ions do not have to be taken into account, because they are not retained at the anion exchanger. The conductivity resulting from the elution of a solute ion is given by:

$$\varkappa_S = \frac{(\Lambda_{E^+} + \Lambda_{E^-})(c_E - c_S \cdot \alpha_S)\alpha_E}{10^{-3} K} + \frac{(\Lambda_{E^+} + \Lambda_{S^-}) c_S \cdot \alpha_S}{10^{-3} K} \qquad (187)$$

The change in conductivity associated with the elution of a solute ion is:

$$\Delta\varkappa = \varkappa_S - \varkappa_E$$
$$= \left[\frac{(\Lambda_{E^+} + \Lambda_{E^-})\alpha_S - (\Lambda_{E^+} + \Lambda_{E^-})\alpha_E \cdot \alpha_S}{10^{-3} K}\right] c_S \qquad (188)$$

In principle, Eq. (188) applies to all ion chromatographic methods. It reveals that the detector signal not only depends on the solute ion concentration, but also on the equivalent conductances of eluant cations and eluant- and solute anions, as well as on their degree of dissociation. The degree of eluant- and solute ion dissociation is determined by the pH value of the mobile phase.

Interestingly, the degree of eluant dissociation significantly affects the detector signal. The amount that sensitivity increases with decreasing degree of eluant dissociation is derived from Eq. (188). This can be confirmed by the work of Fritz and Gjerde [4], who obtained a much higher sensitivity with pure boric acid eluant than with sodium benzoate of comparable elution strength.

According to the above definition, direct detection is feasible when using carefully selected eluants such as phthalate [5] or benzoate [6], which exhibit a low equivalent conductance (see Table 7-1). Using such eluants results in a conductivity increase when solute ions pass the conductivity cell.

Alternatively, a strongly conducting eluant may be used. In this case, elution of the solute ions is associated with a negative conductivity change. This indirect detection method is applied to the separation of anions with a potassium hydroxide eluant [7]. A corresponding chromatogram is displayed in Fig. 7-2. This indirect detection method is also utilized in the analysis of mono- and di-valent cations, which are eluted by dilute nitric acid or nitric acid/ethylenediamine mixtures.

Upon application of a suppressor system, the observed sensitivity enhancement is caused by two processes. On one hand, the eluant is converted into a lower conductive form in the suppressor system which, according to Eq. (188), results in a sensitivity increase. A further enhancement is obtained by converting the solute ions into their corresponding acids or bases. In anion analysis, the associated conductivity change, according to Eq. (189), is to be attributed mainly to the presence of strongly conductive hydronium ions:

$$\Delta \varkappa = \left[\frac{(\varLambda_{H^+} + \varLambda_{S^-}) - (\varLambda_{H^+} + \varLambda_{E^-})}{10^{-3} K} \right] c_S \qquad (189)$$

Finally, it should be pointed out that chemical suppression in form of protonation reactions is also applicable to zwitterionic eluants [8, 9].

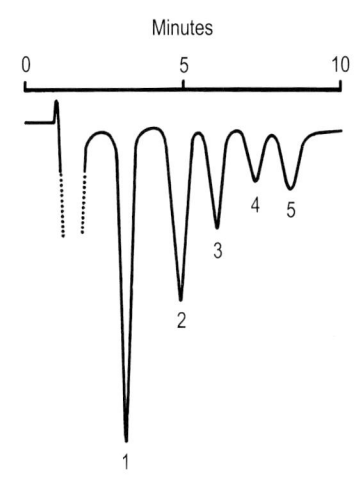

Fig. 7-2. Separation of various inorganic anions with indirect conductometric detection. – Separator column: TSK Gel 620 SA; eluant: 2 mmol/L KOH; flow rate: 1 mL/min; injection volume: 100 µL; solute concentrations: 5 mg/L each of fluoride (1), chloride (2), nitrite (3), bromide (4), and nitrate (5); (taken from [7]).

Comparison of Suppressed and Non-Suppressed Conductivity Detection

Towards the end of the 1970s, the growing use of ion chromatography inspired an argument about necessity of using a suppressor system for sensitive and selective detection of ions via conductivity detection. At this point, an attempt is made to transform the partially very emotional discussion back to a more empirical one.

Without doubt, both application forms of conductivity detection have a number of things in common, while at the same time they bear significant differences. The things in common include the trivial fact that with both application forms the electric conductivity of the analytes is measured. On the other hand, the most obvious difference is that one method uses a suppressor system of whatever kind and the other one does not. There are other distinctions which are less obvious but nonetheless very significant. Two of them are described here in detail: sensitivity and dynamic range. The term sensitivity elicits by far the most misunderstanding in the discussion about the two application forms of conductivity detection. The lack of systematic experimental studies about this subject was corrected by Small [10] with a series of computer simulations that dealt with the relative sensitivity of each application form as used in the chromatographic analysis of sodium and potassium. By intention, Small chose the analysis of cations for his studies because of the great degree of overlap in the two methods as commonly practiced. Both methods use a low-capacity cation exchanger as a stationary phase and a dilute mineral acid such as hydrochloric or nitric acid as a mobile phase. Although stationary phases and eluants have changed over the years, the principle difference between the methods is the same up to the present day. For his hypothetical experiments, Small kept constant the volume of the stationary phase, the ion-exchange capacity of the separator column, the selectivity coefficients for sodium and potassium relative to the hydronium ion, and the injection volume. With these values and the known acid concentration in the mobile phase, it is possible to calculate the elution volumes of sodium and potassium. To further simplify the calculation of the elution profiles, the chromatographic peaks are assumed to be symmetrical, so that they can be described by a Gaussian curve. One can further assume that the membrane-based suppressor system exhibits a very small dead volume and, therefore, subtracts negligibly from the efficiency of the separator column, which is estimated to be 3,000 theoretical plates.

In calculating conductivities, Kohlrausch's law of independent ion mobilities is assumed to be applicable, so conductivities may be calculated by summing up the equivalent conductances of the separate anions and cations in the system. In the non-suppressed mode, the solutes are revealed by the decrements in conductivity that they cause when the hydronium ions are replaced by the less conductive sodium and potassium ions:

$$\Delta L_{Na} = 1000 \, c_{Na} \, (\Lambda_{Na} - \Lambda_{H}) \tag{190}$$

L Conductivity
c_{Na} Sodium ion concentration at any point in the sodium band

For practical reasons, the polarity of the signal is reversed, so that the conductivity values decrease in the positive direction of the ordinate.

In the suppressed mode, the solutes appear as bands of sodium and potassium hydroxides in essentially de-ionized water. The positive deflection in conductivity caused by the sodium band, for example, is calculated according to Eq. (191):

$$\Delta L_{Na} = 1000 \, c_{Na} \, (\Lambda_{Na} + \Lambda_{OH}) \tag{191}$$

In a hypothetical suppressed mode, one could in principle assume the extremely low conductivity of pure water, but a real system is more fairly represented by choosing a value of about 2 µS/cm, because inevitable impurities in the eluant and any slight leakage from the suppressor system have to be taken into account.

Last but not least, the contribution of noise is important. The most significant source of noise is the imperfect control of the temperature of the column effluent reaching the conductivity cell. Even with excellent temperature control a tolerance of $\pm 0.005\,°C$ has to be assumed. Taking the temperature coefficient of conductivity of about 2%/°C into account, the uncertainty in conductivity is $\pm 0.01\%$. The intrinsic noise of the detector is estimated to be 0.01 µS/cm.

The main purpose of this hypothetical experiment is to examine the detectability of sodium and potassium with the two application forms of conductivity detection. If the acid concentration in the eluant is 5 mmol/L, the background conductivity in non-suppressed conductivity detection is roughly 2,100 µS/cm with a noise level of approximately 0.2 µS/cm due to temperature variations. For the suppressed mode, because of the much lower background conductivity, the noise that is relevant is the intrinsic noise of 0.01 µS/cm. Thus, the detectability of the two solute ions essentially depends on the solute ion concentration. At a solute ion concentration of 10 mg/L, the application forms do not differ much with respect to the detectability. However, as the solute ion concentration is lowered to 1 mg/L, the problem of noise clearly begins to affect the accuracy of measurement in the non-suppressed mode. When the solute ion concentration is further lowered to 0.1 mg/L, the signals can hardly be evaluated in the non-suppressed mode (Fig. 7-3A), while baseline noise in the suppressed mode does not significantly influence the evaluation of the sodium and potassium peaks (Fig. 7-3B).

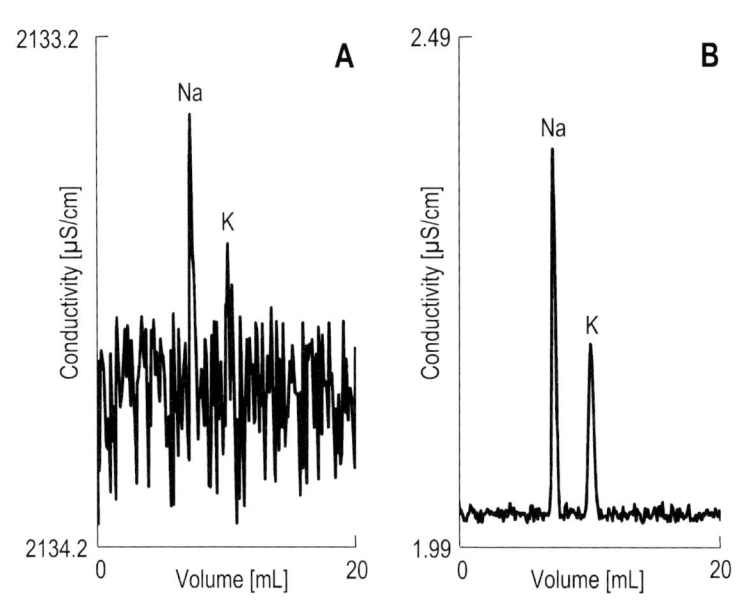

Fig. 7-3. Simulated comparison of the detectability of sodium and potassium with the two application forms of conductometric detection at a solute ion concentration of 0.1 mg/L. – (A) non-suppressed conductivity detection, (B) suppressed conductivity detection. Chromatographic conditions: see text; (taken from [10]).

The large difference in noise between the two application forms is of course caused by the extreme difference in background conductivity. It is clear, therefore, that lowering the acid concentration in the eluant will lower the background conductivity and hence the noise in the non-suppressed mode, but the ion-exchange capacity of the stationary phase also has to be lowered proportionally in order to obtain comparable separations. This measure, in turn, involves a compromise, because lowering the eluant strength and the resin capacity impairs the system's ability to resist column overloading. In contrast, the suppressed system, with a tenfold greater resin capacity, has a tenfold advantage over the non-suppressed system in overload capacity. This will be reflected in a greater dynamic range for the suppressed method.

The ability to handle a wide range of analyte ions is one of the ways in which analytical techniques are evaluated. In this regard, the simulated comparison described above impressively demonstrates the superiority of suppressed conductivity detection. It also shows that the dynamic range of a method does not only depend on the construction of the conductivity detector and cell but also on the type of stationary phase and eluant.

Moreover, the calculation above reveals the pitfall of using an overly restrictive definition of *sensitivity*. If only the peak heights (or peak areas) that result from the two application forms of conductivity detection are compared, it is apparent that the responses in the non-suppressed mode are in fact somewhat greater than in the suppressed mode. When noise is considered alongside response, as it properly should be, then one must conclude that eluant suppression leads to superior sensitivity.

7.1.2
Amperometric Detection

Amperometric detection is generally used for the analysis of solutes with pK values above 7, which, owing to their low dissociation, can hardly be detected or are not at all detected by suppressed conductivity.

Conventional amperometric detectors employ a three-electrode detector cell consisting of a working electrode, a reference electrode, and a counter electrode. The electrochemical reaction at the working electrode is either an oxidation or a reduction. The required potential is applied to the working electrode. The Ag/AgCl reference electrode represents an electrode of the second kind. In this type of electrode, a second solid phase in the form of a sparingly soluble salt contributes to the electrode reaction, in addition to the element and the electrolyte solution. Thus, the activity of the potential-determining cation depends on the activity of the anion involved in the formation of the sparingly soluble salt, via the solubility product. The Ag/AgCl electrode is utilized as a reference electrode because it is characterized by a good potential constancy at current flow. The purpose of the counter electrode, which is usually made of glassy carbon, is to maintain the potential. Furthermore, it inhibits a current flow at the reference electrode that could destroy it. When an electroactive species passes the detector cell it is partly oxidized or reduced. This reaction results in an anodic or cathodic current that is proportional to the concentration of the species over a certain range, and that may be represented as a chromatographic signal.

Such detectors are employed for analyzing a wealth of inorganic and organic ions in the µg/L range. This includes environmetally relevant anions such as sulfide and cyanide [11, 12], as well as arsenic(III) [13], halide ions, oxyhalides [14], nitrite [14], thiosulfate [14], hydrazine, and phenols. A survey of electrochemically active compounds and the working electrodes and potentials is given in Table 7-2.

Table 7-2. Electroactive compounds and the required working electrodes and potentials.

Compound	Working electrode	Working potential [V]
HS^-, CN^-	Ag	0
Br^-, I^-, SCN^-, $S_2O_3^{2-}$	Ag	0.2
SO_3^{2-}	Pt	0.7
OCl^-	Pt	0.2
AsO_2^-	Pt	0.85
N_2H_4	Pt	0.5
NO_2^-	CP[a]	1.1
ClO_2^-	CP	1.1
$S_2O_3^{2-}$	CP	1.1
Phenols	GC[b]	1.2

a) Carbon Paste
b) Glassy Carbon

7.1.2.1 Fundamental Principles of Voltammetry

Information about suitable working potentials for the amperometric detection of electroactive species is obtained in voltammetric experiments. The term "voltammetry" refers to the investigation of current-voltage curves in dependence on the electrode reactions, and the concentrations, and the exploitation of these dependences for analytical chemistry. Among the different types of voltammetry, information from the hydrodynamic and pulsed voltammetry can be best applied to amperometry. In both cases, the analyte ions are dissolved in a supporting electrolyte that has several functions:

- Lowering the resistance of the solution ensures that the voltage drop $i \cdot R_L$ is kept low. This is the case for electrolyte concentrations of 0.1 mol/L.
- Preventing *depolarisators* (anions or cations) which pick up or emit electrons during their reduction or oxidation from reaching the electrode by migration due to the potential gradient. Therefore, the concentration of the supporting electrolyte should exceed that of the depolarisators by a factor of 50 to 100.

The chlorides, chlorates, and perchlorates of alkali- and alkaline-earth metals, alkali hydroxides and carbonates as well as quaternary ammonium compounds are utilized as supporting electrolytes.

In general, the following reaction occurs in an amperometric detection:

$$A \rightleftharpoons B + n \cdot e \tag{192}$$

A transfers n electrons to the working electrode and is oxidized to B. The relation between the concentration of oxidized and reduced species and the applied potential is described by the Nernst equation:

$$E = E_0 + \frac{0.059}{n} \cdot \log \frac{[B]}{[A]} \tag{193}$$

E Working potential
$[A], [B]$ Equilibrium concentrations of both species at the electrode surface
E_0 Standard potential, at which the concentrations of A and B are equal. (Because chromatography is carried out at low concentrations, concentrations may be used instead of activities in Eq. (193) in good approximation.)

The working potential must be selected so that A is fully oxidized to B, in order to yield as high a current as possible. Therefore, according to Eq. (193), the working potential, E, must always be higher than the standard potential, E_0.

A voltammogram of the oxidation of A is recorded by means of hydrodynamic voltammetry. This is done by pumping the substance dissolved in the supporting electrolyte through a flow cell, which contains the working electrode. The potential applied to the electrodes is continuously raised while the current flow is registered. Figure 7-4 shows the sigmoidal curve, which is characteristic for substance A. The curve may be interpreted in terms of the concentration of species A at the electrode surface (Fig. 7-5). Only a small current flows, denoted as residual current, until potential E_A is reached. This is caused by the charging of the double layer at the electrode surface and the reaction of impurities in the solution at the electrode. The concentration of A at the electrode surface is equal to that in the bulk solution. At potential E_A, the concentration ratio B to A increases until at the standard potential, E_0; the concentrations of A and B at the electrode surface are the same. However, a few micrometers away from the electrode surface, the concentration of A remains constant. The distance between the electrode and the point at which the concentration is equal to that in the bulk solution is known as the thickness δ of the diffusion layer. Because of the concentration gradient, the transport of A to the electrode surface occurs only by diffusion within this distance. When the potential is increased in the direction of E_B, the current strength reaches a maximum at E_B and remains nearly constant because A is immediately oxidized to B after it has diffused to the electrode surface. Thus, the current strength is limited by the speed of diffusion because the transport process is the rate-determining step. The resulting current is called the diffusion current. According to Eq. (194), this diffusion current is proportional to the concentration of A in the bulk solution.

$$I = \frac{n \cdot F \cdot A \cdot D \cdot c}{\delta} \tag{194}$$

- F Faraday constant
- A Electrode surface
- D Diffusion coefficient of A

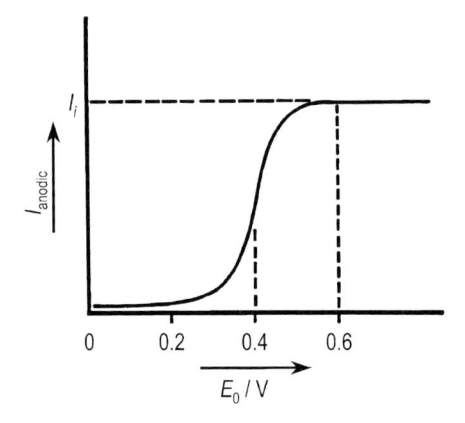

Fig. 7-4. Voltammogram of the oxidation of species A.

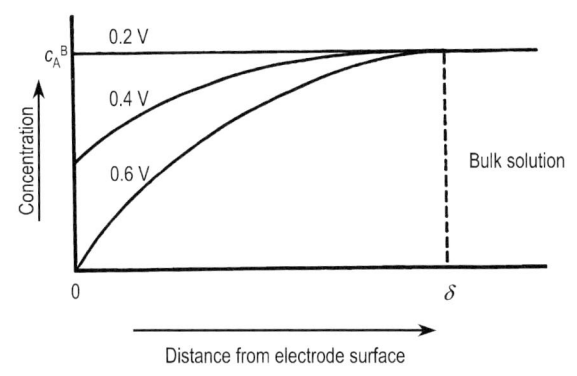

Fig. 7-5. Dependence of solute concentration, c, on the distance to the electrode surface at various working potentials.

Pulsed amperometry in a three-electrode cell is carried out in a non-flowing solution, in which solute ions are dissolved in the supporting electrolyte. A pulsed potential is applied and increased stepwise after each pulse. The resulting current strength is measured at each step. The voltammogram does not differ from that shown in Fig. 7-4. Immediately after a pulse in the diffusion-controlled plateau region the measured current strength is very high, because the molecules A close to the electrode are immediately oxidized. Molecules that are further away from the electrode reach the electrode surface only by diffusion and are then oxidized. As a result, a drop in the current strength is observed (Fig. 7-6).

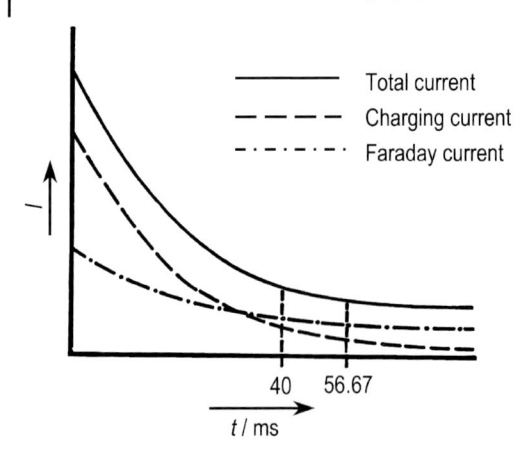

Fig. 7-6. Drop of the total current, the Faraday current, and the charging current after applying a pulse in the diffusion-controlled plateau region.

Two kinds of current are generated during the oxidation of A to B: the charging current and the Faraday current. The charging current results from the charging of the interface between the electrode and the bulk solution, which acts as a condensator. The Faraday current represents the electron transfer in the oxidation of A to B. As illustrated in Fig. 7-6, the drop in the charging current follows an exponential time law. The drop in the Faraday current over time is described by the equation of Cotrell:

$$I(t) = \frac{n \cdot F \cdot A \cdot D^{1/2} \cdot c}{\pi^{1/2} \cdot t^{1/2}} \tag{195}$$

t Time

Although the current strength is a function of $t^{-1/2}$, a direct proportionality to the solute ion concentration exists. The sum of both currents is measured as the total current. Because the charging current is much larger than the Faraday current, measurement of the total current is delayed by more than 40 ms after application of the pulse. During this time, the ratio of Faraday current to charging current is enhanced as the latter drops much faster with time.

7.1.2.2 Amperometry

While the working potential required for the desired electrochemical reaction may be determined with voltammetric experiments, amperometry is used as a detection method in ion chromatography. A distinction is made between pulsed amperometry and amperometry with constant working potential.

Amperometry with Constant Working Potential

This kind of amperometry is the most widely used electrochemical detection method in liquid chromatography. A constant DC potential is continuously applied to the electrodes of the detector cell. The theory of amperometry with constant working potential does not differ from the theory of hydrodynamic voltammetry, even though the applied potential remains constant.

The working potential is chosen to be in the diffusion-controlled plateau region for the analyte ion. When several ions with different standard potentials are to be detected in the same run, the working potential must be high enough to cover the plateau regions of all ions to be analyzed. The amperometric detection with constant working potential is routinely applied to the species listed in Table 7-2. Figure 7-7 shows the application of this detection method to the analysis of iodide in a highly concentrated sodium chloride solution. To eliminate the interfering effects of the chloride matrix, the iodide oxidation is carried out at a platinum working electrode and not, as usually done, at a silver working electrode. The required sensitivity for this application is obtained by conditioning the electrode for one hour with a saturated potassium iodide solution.

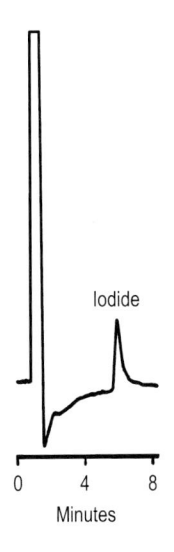

Fig. 7-7. Analysis of iodide in a concentrated NaCl solution. — Separator column: IonPac AS7; eluant: 0.2 mol/L HNO_3; flow rate: 1.5 mL/min; detection: DC amperometry at a platinum working electrode; oxidation potential: +0.8 V; injection: 50 µL of a 10-fold diluted saturated NaCl solution with 24 µg/L iodide.

Environmentally relevant anions such as sulfide and cyanide are detected on a silver working electrode under strongly alkaline conditions (see Fig. 3-86 in Section 3.7.2). The carbon paste electrode is used for detecting nitrite and chlorite under acidic conditions and at neutral pH. Nitrite and chlorite can also be oxidized at a glassy carbon electrode, but the resulting sensitivity is significantly lower because the surface is less electrochemically active.

Conductivity detection and DC amperometry can be used simultaneously. In a non-suppressed system the conductivity cell effluent is directed to the amperometric cell; in a suppressed system the positioning of the amperometric cell depends on the electrode material. The carbon paste electrode, for example, is positioned between the suppressor and the conductivity cell, so that electroactive anions can be detected together with standard anions [14]. An example of applications involving a carbon paste electrode is shown in Fig. 7-8. The separation of nitrite, thiosulfate, and iodide is performed in an acidic medium with a

7 Detection Methods in Ion Chromatography

phthalic acid eluant. The optimal working potential is +1.1 V and the detection limits for the anions shown in this chromatogram are in the lowest µg/L range (based on an injection volume of 100 µL).

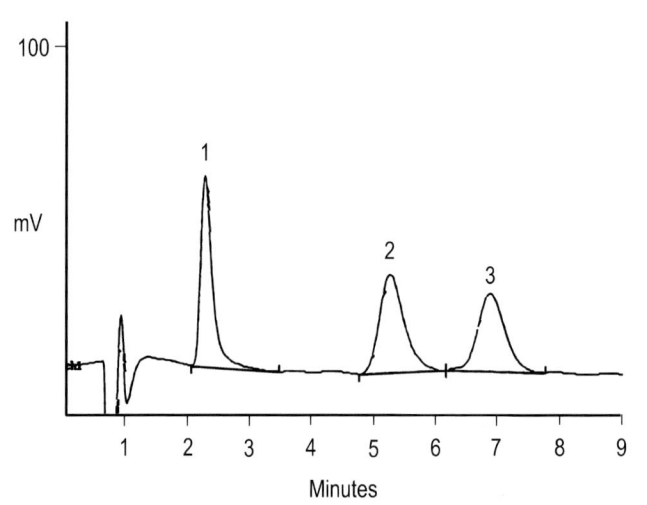

Fig. 7-8. Amperometric detection of nitrite, thiosulfate, and iodide on a carbon paste electrode. – Separator column: Supersep Anion; eluant: 3.3 mmol/L phthalic acid – acetonitrile (90:10 v/v), pH 4.5 with Tris; detection: DC amperometry on a carbon paste working electrode; oxidation potential: +1.1 V; injection volume: 100 µL; solute concentrations: 10 µg/L each of nitrite (1), thiosulfate (2), and iodide (3); (taken from [14]).

Pulsed Amperometry (PAD)

Amperometric detection of electroactive species requires that reaction products from the oxidation or reduction of solutes do not precipitate at the electrode surface. The surface characteristics of electrodes will change if contaminated, thus leading to an enhanced baseline drift, increased background noise, and a constantly changing response. This behavior is particularly pronounced in the amperometric detection of carbohydrates.

Pulsed amperometric detection, on the other hand, utilizes a rapidly repeating sequence of three different working potentials, $E1$, $E2$, and $E3$, which are applied for the times t_1, t_2, and t_3 to the gold working electrode. In contrast to conventional amperometry, the resulting current is only registered in short time intervals. By applying additional potentials (more positive and more negative), oxidizable and reducible species may be removed from the electrode surface. The advantage of this technique is illustrated in Fig. 7-9, which is a comparison of the two amperometric detection methods, taking the chromatographic separation of sugars in chocolate milk as an example.

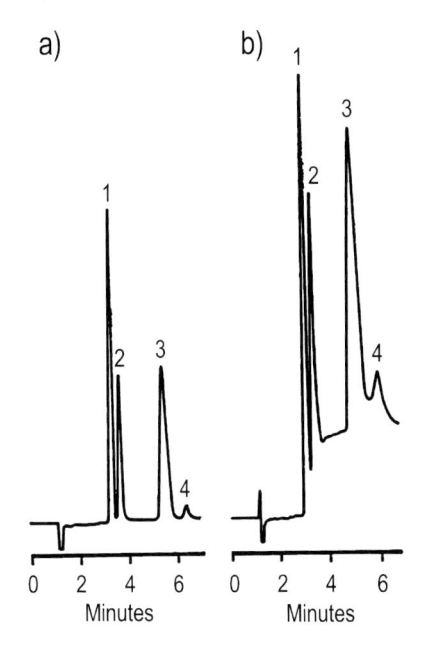

Fig. 7-9. Separation of monosaccharides in a chocolate milk on a latexed anion exchanger: a) with pulsed amperometric detection at a gold working electrode, b) with conventional amperometric detection applying a constant working potential. Analytes: (1) glucose, (2) fructose, (3) lactose, and (4) sucrose.

In their original publication Johnson et al. [15] described the use of a platinum working electrode. Although satisfactory results were achieved with this electrode material, reduction of oxygen is likely to happen at the oxidation potential of -0.4 V, which was used at the time. For this reason, gold was chosen as an alternative electrode material. The required working potentials and the pulse sequence are determined by means of cyclic voltammetry. As an example, Fig. 7-10 shows the cyclic voltammogram of glucose on a gold working electrode [16]. The scattered line illustrates the scan for 0.1 mol/L sodium hydroxide, which serves as a supporting electrolyte. From this scan, the irreversible oxidation of gold starting at 0.25 V can be read. The cathodic peak at 0.1 V represents the reduction of the oxidic gold surface to the pure metal. When glucose is added to that solution, oxidation starts at -0.5 V on the positive branch of the curve (straight line). Until -0.15 V the current remains constant and increases thereafter to a peak at 0.26 V. The sequel of the curve shows a change from a cathodic to an anodic current, which goes along with the reduction of gold oxide. This means that the oxidation of glucose is hindered by the formation of gold oxide. With the beginning of the gold oxide reduction, the oxidation of glucose molecules that diffused to the electrode during the oxidic covering recommences. Comparing cyclic voltammograms of glucose with those of other saccharides shows that the only difference is the height of the anodic peak at 0 V.

According to Johnson et al. [18] the reaction mechanism of the oxidation of carbohydrates on a gold working electrode under alkaline conditions is a sequential process. For example, Fig. 7-11 illustrates the reactions involved in the oxidation of glucose. In a first and extremely fast step, the C_1-aldehyde group is

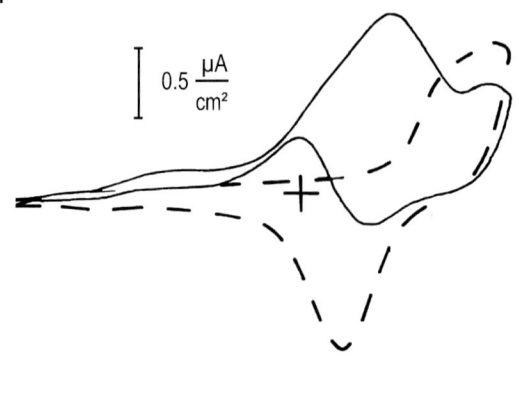

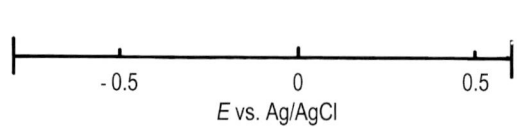

Fig. 7-10. Cyclic voltammogram of glucose (c = 166 mg/L) on a gold working electrode. (Scattered line: supporting electrolyte, 0.1 mol/L NaOH; sweep rate: 200 mV/s; taken from [17]).

oxidized to form a gluconate anion. The second reaction step, the oxidation of the C_6-alcohol group, is also fast, although it is slower than the preceding one. The subsequent decarboxylation occurs with the formation of electroinactive formate and the corresponding dicarboxylate anion. The degree of decarboxylation depends on the amount of time the carbohydrate is in contact with the electrode surface which is, in turn, a function of the type and concentration of the supporting electrolyte.

Fig. 7-11. Reaction mechanism of the oxidation of glucose on a gold working electrode under alkaline conditions; (taken from [18]).

The pulse sequence with three different potentials usually applied for the detection of carbohydrates is shown in Fig. 7-12A. However, the cleaning potential E2 being applied for 200 ms leads to an electrode recession that enlarges the cell volume as shown in Fig. 7-13. This, in turn, lowers the linear speed of the liquid through the detector cell which consequently results in a decrease of the response factor for a given analyte concentration.

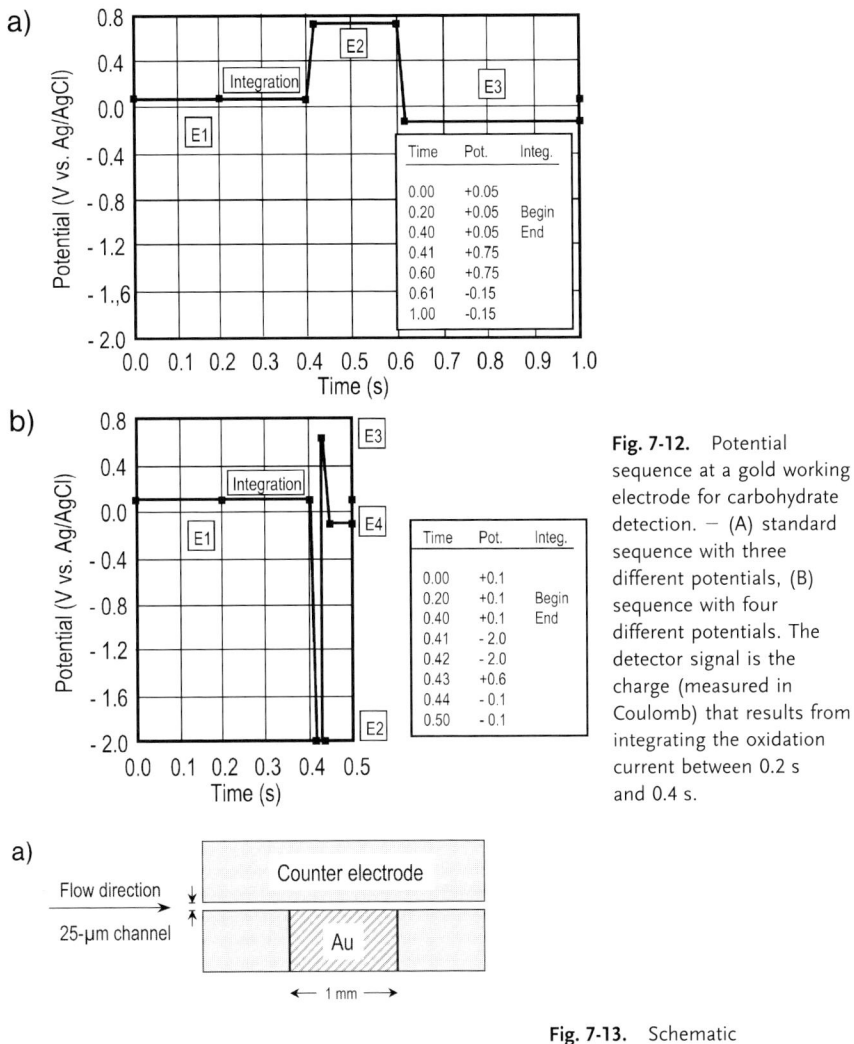

a)

Time	Pot.	Integ.
0.00	+0.05	
0.20	+0.05	Begin
0.40	+0.05	End
0.41	+0.75	
0.60	+0.75	
0.61	−0.15	
1.00	−0.15	

b)

Time	Pot.	Integ.
0.00	+0.1	
0.20	+0.1	Begin
0.40	+0.1	End
0.41	−2.0	
0.42	−2.0	
0.43	+0.6	
0.44	−0.1	
0.50	−0.1	

Fig. 7-12. Potential sequence at a gold working electrode for carbohydrate detection. – (A) standard sequence with three different potentials, (B) sequence with four different potentials. The detector signal is the charge (measured in Coulomb) that results from integrating the oxidation current between 0.2 s and 0.4 s.

a) Flow direction, 25-μm channel, Counter electrode, Au, 1 mm

b) Counter electrode, Au

Fig. 7-13. Schematic representation of the electrode block of an amperometric cell. – (A) new cell with a leveled gold working electrode, (B) used cell with a larger cell volume due to recession of the working electrode.

Electrode recession can be minimized by avoiding a strongly positive potential, although the removal of the top layer is a prerequisite for a clean, catalytically active electrode surface. Just recently, Johnson et al. [19] were able to show that the oxidation products of glucose can also be removed from the electrode surface by applying a strongly negative potential. Based on this work, Rocklin et al. [20] developed a sequence with four different potentials for the detection of carbohydrates (see Fig. 7-12B) that resulted in a constant response factor and better long-term stability. All the potentials and the pulse sequence were optimized for a maximal signal-to-noise ratio.

Every potential marked in Fig. 7-12B has a different meaning. At the potential $E1$ the current resulting from the oxidation of the carbohydrate is integrated. It can be revealed from cyclovoltammetric experiments that the maximal current yield for sugars is achieved at oxidation potentials between 0.1 V and 0.2 V. Because real-world samples often contain amines, which are also oxidized at potentials >0.15 V, the oxidation potential $E1$ was set to 0.1 V, so that amines will not interfere with carbohydrate detection. There is no optimal value for the integration time. In general, higher response factors are observed with increasing integration times; however, noise is not affected at all. Thus, an improved signal-to-noise ratio is obtained with increasing integration time. Integration time cannot be increased indefinitely, because the whole pulse sequence has to be limited in time in order to reproducibly evaluate highly efficient peaks with a chromatography data system. In the end, an integration time of about 200 ms has proved to be suitable. Similar considerations have to be taken into account for the delay time to adjust the potential, at which the integration of the oxidation current can begin. A long delay time lowers noise and does not significantly affect the signal. Normally, the delay time is also recommended to be 200 ms. When integrating highly efficient peaks with a respectively high data acquisition rate, the delay time can be slightly shortened without a significant increase in noise.

In the above mentioned pulse sequence the cleaning potential, $E2$, is strongly negative. The optimal potential for removing adsorbed oxidation products is -1.5 V. However, noise is a problem, because it increases with increasing negative potential. If the sample to be analyzed contains amino acids or small peptides, they can adsorb on the electrode surface after leaving the separator column. This leads to lower response factors for subsequently eluting carbohydrates [21]. With repeatedly injected fetuin hydrolysates it could be shown that matrix components can only be removed effectively at a potential of -2 V. This potential is applied for 10 ms. If the samples do not contain components that foul the surface of the working electrode, the potential $E2$ can be increased to -1.5 V without a noticeable increase in noise.

The potential $E3$ (+0.6 V) activates the electrode surface. It is assumed that the catalytically active sites are created via the formation and subsequent reduction of gold oxide [22]. When investigating the dependence of the response factor as a function of $E3$, it is observed that the response factor remains constant between $+0.4$ V and $+0.9$ V. The suitable value for $E3$ was defined as a result of a long-term study. When glucose and sucrose are injected repeatedly at a potential of $+0.4$ V, constant response factors are observed for both sugars over a period of ten days. Afterwards, the response factors rapidly decrease without electrode recession. Apparently, the formation rate for catalytically active sites is not sufficient at this potential. Only at a potential of $+0.6$ V do the response factors remain constant over a period of more than two weeks. If this value is further increased beyond $+0.6$ V, significant electrode recession is observed. Electrode recession at a potential of $+0.6$ V is avoided as much as possible by lowering $E3$ to -0.1 V immediately after the adjustment at $+0.6$ V. In this way, a catalytically active electrode surface is maintained without electrode recession.

With the potential $E4$ at -0.1 V the oxide that is formed at positive potential is reduced again. The value of -0.1 V is not further lowered in order to mostly avoid the reduction of oxygen present in the samples. Oxygen reduction is responsible for the negative peak, which appears at approximately 14 minutes on CarboPac PA1 when analyzing monosaccharides (see Section 3.10.1.1). Because the reduction of gold oxide is very fast, timing for this pulse is not critical. It has been set so that the time for the whole pulse sequence is 0.5 s.

The detection limits for carbohydrates are in the fmol range with either one of the two pulse sequences; the linear range, depending on the sugar, is two to three orders of magnitude. Concentrations above 1 nmol should be calibrated with several points applying a quadratic curve fitting. Diluting the sample into the workable range is an alternative.

Pulsed amperometry not only allows the detection of carbohydrates (see Section 3.10); it also enables the detection of alcohols and aldehydes. Short-chain primary, secondary, and tertiary alcohols can be separated on an ion-exclusion phase with a perchloric acid eluant (see Section 5.8). Although conventional HPLC methods exist for the analysis of primary alcohols, the advantage of an ion chromatographic separation lies in the ability to simultaneously analyze secondary and tertiary alcohols. Detection of alcohols is carried out on a platinum working electrode. It is assumed that the resulting anodic current is generated via oxidation of hydrogen atoms, which are formed by heterogeneous dehydrogenation of the adsorbed organic molecules [23]. The organic residue remains adsorbed at the electrode surface and hinders the adsorption of other molecules from the bulk solution. Working potentials and pulse sequences can be derived from the cyclic voltammograms obtained by Johnson et al. [24]. At the potential $E1$ ($+0.3$ V), alcohols are oxidized anodically by means of surface-catalyzed dehydrogenation; the electrode surface must be free of platinum oxide. The oxidation products that are strongly adsorbed at the electrode surface are oxidized presumably to carbon dioxide at the potential $E2$ ($+1.3$ V) with simultaneous formation of platinum oxide. The reduction of platinum oxide occurs at the potential $E3$ (<+0.6 V). At concentrations <1 mmol/L the anodic current resulting from the oxidation reaction is proportional to the solute concentration.

In the past, the ion chromatographic analysis of short-chain aliphatic aldehydes was carried out after oxidation to the corresponding carboxylic acid. In contrast, pulsed amperometry allows the simultaneous determination of short-chain aldehydes and formic acid, which can be interfered by methanol and other alcohols. Again, good separations are obtained with ion-exclusion phases in the potassium form. The underlying retention mechanism is a combination of adsorption and weak interactions between the oxygen-containing functional groups of the solutes and the potassium counter ions of the sulfonate groups at the stationary phase.

The amperometric oxidation of aldehydes is also carried out on a platinum working electrode under acidic conditions. Figure 7-14 shows the cyclic voltammogram of formaldehyde, which does not differ much from the one of formic acid. Adsorption of formaldehyde occurs at negative potential; oxidation to carbon dioxide and water occurs at increased potential. Formation of platinum oxide starts at a potential of +0.7 V, which interrupts the further oxidation of solutes. The oxidation of solutes sets in again at potentials >1 V. After reversing the direction of the potential, a change from a cathodic to an anodic current is observed at 0.2 V. The resumption of solute oxidation is attributed to the beginning reduction of platinum oxide and, consequently, the activation of the electrode surface. The cyclic voltammogram of propionaldehyde differs significantly from those of other smaller aldehydes. Although propionaldehyde is also oxidized on a platinum working electrode, the retention mechanism is most probably different than it is for smaller aldehydes. To find out the optimal oxidation potentials for detecting formaldehyde, formic acid, acetaldehyde, and propionaldehyde

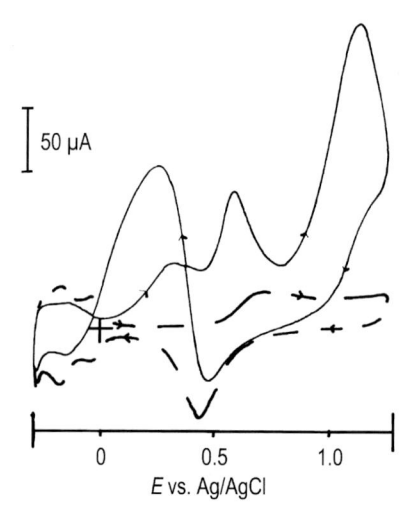

Fig. 7-14. Cyclic voltammogram of formaldehyde (c = 0.02 mol/L) on a platinum working electrode. — Scattered line: 0.05 mol/L H_2SO_4 + 0.05 mol/L K_2SO_4 as supporting electrolyte; sweep rate: 200 mV/s.

the resulting peak heights are plotted versus the oxidation potential E1 (Fig. 7-15). Considering the lowest possible background current, the optimal oxidation potential is +0.2 V.

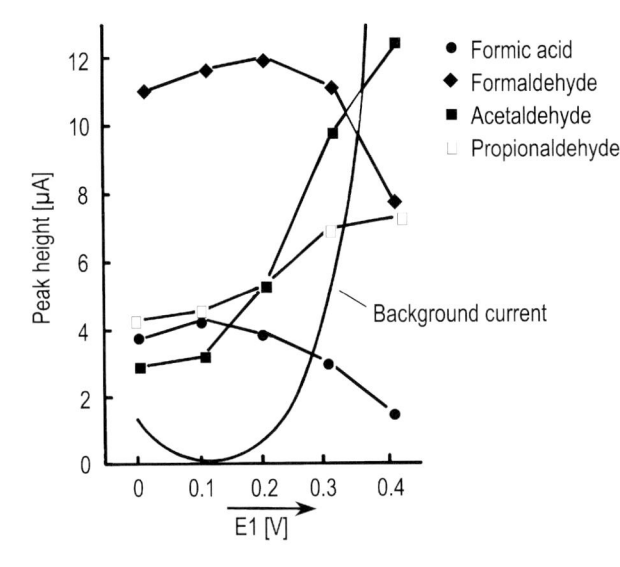

Fig. 7-15. Peak height as a function of the oxidation potential $E1$ for formic acid (100 mg/L), formaldehyde (100 mg/L), acetaldehyde (300 mg/L), and propionaldehyde (300 mg/L); continuous line: background current.

Integrated Amperometry

Integrated amperometry is a variant of pulsed amperometry. It is predominantly used for the detection of amino acids, amines, and organic sulfur compounds. Their oxidation on metal electrodes is catalyzed by metal oxides. When integrated amperometry is employed, baseline disturbances caused by pH gradients, solvent gradients, ionic strength variations, and metal oxide formation are minimized. In pulsed amperometry, the resulting current from the oxidation reaction is measured at fixed oxidation potential after the application of the pulse and the decline of the charging current. In integrated amperometry, the potential $E1$ is not kept constant, but alternated between a high and a low value. Thus, analyte and metal oxidation occur simultaneously at the high potential. However, the metal oxide formed at this potential is immediately afterwards reduced again at the lower potential. Because the oxidation of the electrode surface is a reversible process, while the oxidation of analytes is not, the resulting signal is mainly characterized by the contribution of the analyte oxidation. When integrating the current yield during this cycle, the net signal for the respective analyte is obtained. Positive and negative cleaning potentials are part of the pulse sequence following the integration step. A schematic example of such a sequence is illustrated in Fig. 7-16.

The advantage of integrated amperometry lies in the coulometric compensation of the charges resulting from the formation and subsequent reduction of the metal oxide. Thus, baseline drifts and baseline disturbances caused by small

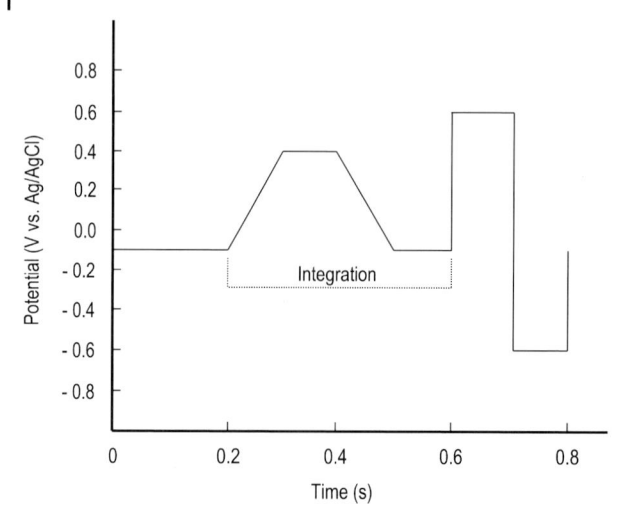

Fig. 7-16. Example of a pulse sequence in integrated amperometry of amines on a gold working electrode.

variations in the mobile phase composition are eliminated. Moreover, the whole system is less sensitive to variations in pH, which influences the potentials for the formation and reduction of the metal oxide. The potential at which the metal electrode surface is oxidized decreases with increasing pH.

LaCourse and Owens [25] showed the superiority of integrated amperometry over conventional pulsed amperometric detection in their work on organic sulfur compounds under reversed-phase conditions. Pulsed amperometry allows direct detection of thio-redox systems ($-SH/-S-S-$), provided they carry a pair of free electrons at the sulfur atom. As an illustration, the chromatogram in Fig. 7-17 displays the separation of lipoic acid, which carries a disulfide bridge as a structural element.

By means of pulsed amperometry, this compound can be directly oxidized without prior conversion into the dihydrolipoic acid. The sensitivity of this method is in the lower pmol range for such compounds.

For the detection of sulfur-containing antibiotics such as cephapirin and ampicillin Dasenbrock et al. [26] developed multicyclic pulse sequences, which result in higher sensitivity and selectivity. In this type of pulse sequence, the potential is cycled between high and low potentials during the integration period (see

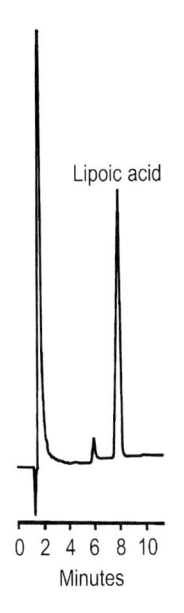

Fig. 7-17. Analysis of lipoic acid utilizing pulsed amperometric detection. — Separator column: CarboPac PA1; eluant: 0.1 mol/L NaOH + 0.5 mol/L NaOAc; flow rate: 1 mL/min; detection: pulsed amperometry on a gold working electrode; injection volume: 50 µL; solute concentration: 40 mg/L.

Fig. 7-18, left illustration) leading to an improved signal-to-noise ratio. Theoretically, the signal-to-noise ratio increases with $n^{1/2}$ for n cycles. However, in practice this value is slightly lower because the frequency of the cycle is so high that only a limited number of analyte molecules can adsorb on the electrode surface between the particular cycles. The pulse sequence developed by Dasenbrock et al. was modified by Hanko and Rohrer [27] who increased the number of cycles from four to ten to achieve a more stable response (see Fig. 7-18, right illustration). For example, Fig. 7-19 shows the separation of ampicillin on a Vydac separator utilizing integrated amperometry with this modified pulse sequence, in comparison with UV detection at 215 nm.

Ampicillin

Although a slightly higher sensitivity is achieved with UV detection at 254 nm, integrated amperometry is five times more sensitive than UV detection at optimized wavelength.

A much larger sensitivity difference between these two detection methods is obtained for lincomycin (see Fig. 7-20).

7 Detection Methods in Ion Chromatography

Lincomycin

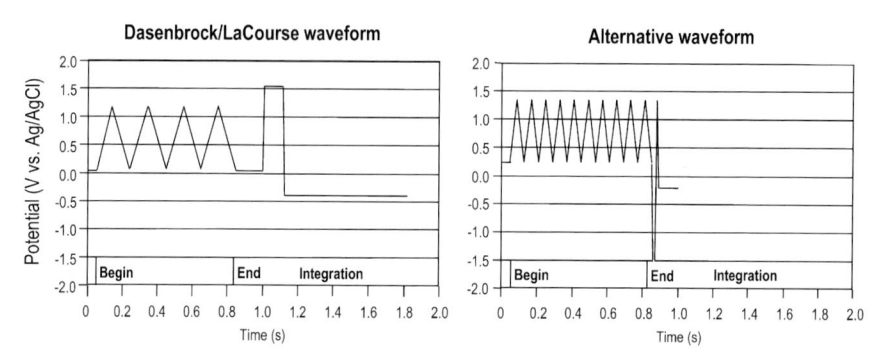

Fig. 7-18. Multicyclic pulse sequences for integrated amperometry of sulfur-containing antibiotics.

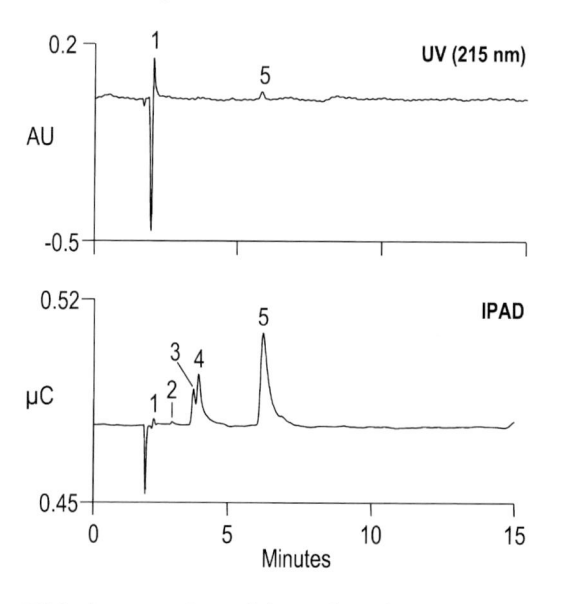

Fig. 7-19. Separation of ampicillin utilizing integrated amperometry with a multicyclic pulse sequence in comparison with UV detection. — Separator: 150 mm × 4 mm i.d. Vydac C8 (208TP5451); column temperature: 30 °C; eluant: 0.1 mol/L NaOAc – acetonitrile (94:6 v/v), pH 3.75; flow rate: 1 mL/min; analytes: (1) to (4) unknown impurities, (5) 100 mg/L ampicillin.

With the exception of the carbonyl group, lincomycin is non-chromophoric. In this case, UV detection is very insensitive.

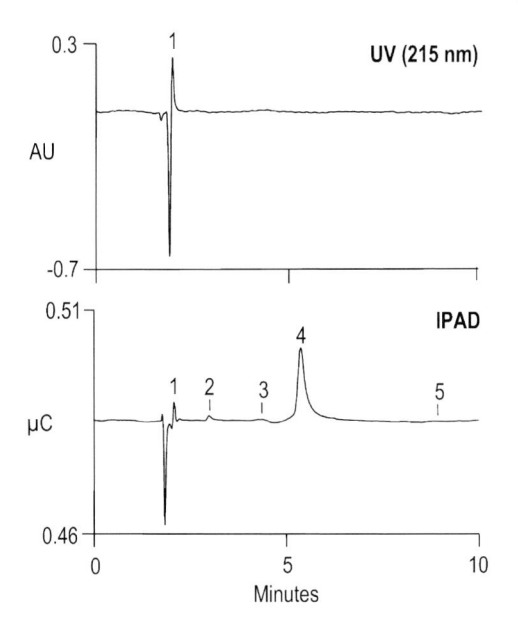

Fig. 7-20. Separation of lincomycin utilizing integrated amperometry with a multicyclic pulse sequence in comparison to UV detection. − Separtor column: 150 mm × 4 mm i.d. Vydac C8 (208TP5451); column temperature: 30 °C; eluant: 0.1 mol/L NaOAc − acetonitrile (91:9 v/v), pH 3.75; flow rate: 1 mL/min; peaks: (1) − (3), (5) unknown impurities, (4) 10 mg/L lincomycin.

In cation analysis, integrated amperometry is predominantly used for detecting primary, secondary, and tertiary amines, which can be oxidized on a gold working electrode at high pH. While low-molecular weight alkylamines can also be detected sensitively and selectively by suppressed conductivity, the sensitivity of conductivity detection for alkanolamines decreases rapidly from monoalkanolamines to trialkanolamines. On the other hand, alkanolamines can also be detected by UV detection at 215 nm, where a sensitivity increase is observed in the same order. In the past, conductivity and UV detectors were used in series to compensate for these sensitivity differences. In integrated amperometry, large sensitivity differences are not observed, so this detection method is an excellent alternative to conductivity and UV detection. The only problem is the fact that the separation of amines via cation exchange requires an eluant with low pH, which is not compatible with the criterion for amperometric detection (pH > 11) mentioned above. To solve this problem, the effluent of the separator column is directed through a CSRS self-regenerating suppressor operated in the recycle mode (see Section 4.3.4) prior to entering the amperometric cell. The reactions occurring in such a suppressor are illustrated in Fig. 7-21. Using the hydroxide ions generated at the cathode, the acid eluant is converted to water, the salt to the respective hydroxide, and the alkanolammonium ions to the free amine. In this way, pH is raised to a value of 13, which is sufficient for amperometric detection. Figure 7-22 shows the optimal pulse sequence for ethanolamines and the chromatogram of a corresponding standard. The chromatogram was obtained with a mixture of lithium sulfate and sulfuric acid on IonPac CS10. Detection limits for ethanolamines with integrated amperometry are between 10 μg/L and 50 μg/L.

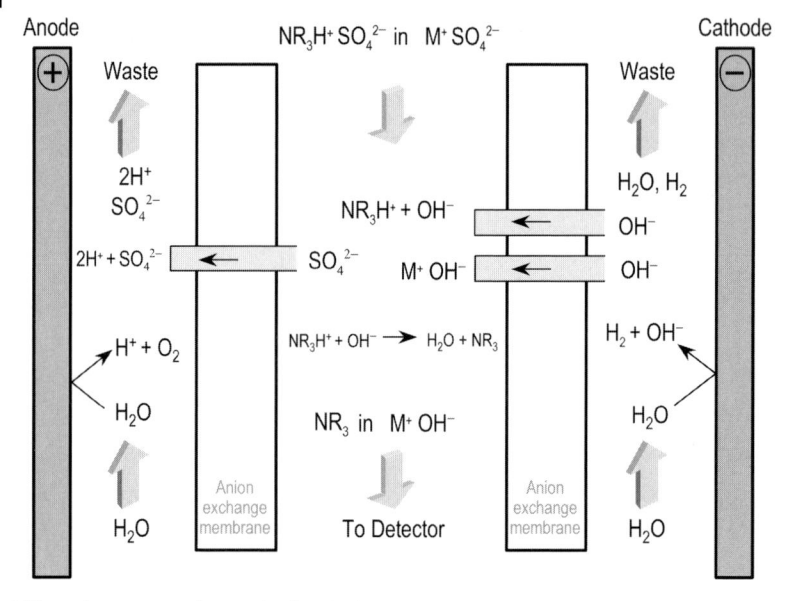

Fig. 7-21. Ion-exchange reactions inside a CSRS as preparation for integrated amperometry of amines.

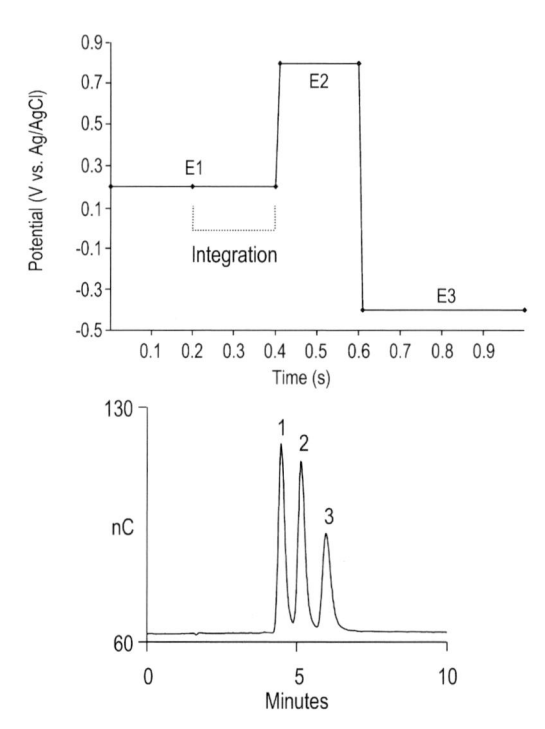

NR_3: primary, secondary, or tertiary amines

Fig. 7-22. Pulse sequence for integrated amperometry of ethanolamines and their separation on a latexed cation exchanger. — Separator: IonPac CS10; eluant: 2.5 mmol/L H_2SO_4 + 100 mmol/L Li_2SO_4; flow rate: 1 mL/min; detection: integrated amperometry on a gold working electrode; suppressor system: CSRS, recycle mode; injection volume: 25 µL; solute concentrations: 1.7 mg/L monoethanolamine (1), 2.8 mg/L diethanolamine (2), and 5.6 mg/L triethanolamine (3).

Under similar chromatographic conditions, but with a modified pulse sequence, primary alkylamines can be analyzed. An IonPac CS14 weak acid cation exchanger served as a stationary phase for the separation shown in Fig. 7-23, which was obtained with a mixture of sodium sulfate and sulfuric acid. A small amount of acetonitrile was added to this mixture in order to improve the peak form. To shorten the analysis time of the strongly retained butylamine, the solvent content is increased stepwise from 5% (v/v) to 40% (v/v) after eight minutes. However, the detection limits (100 µg/L to 1000 µg/L) are somewhat higher relative to the alkanolamines.

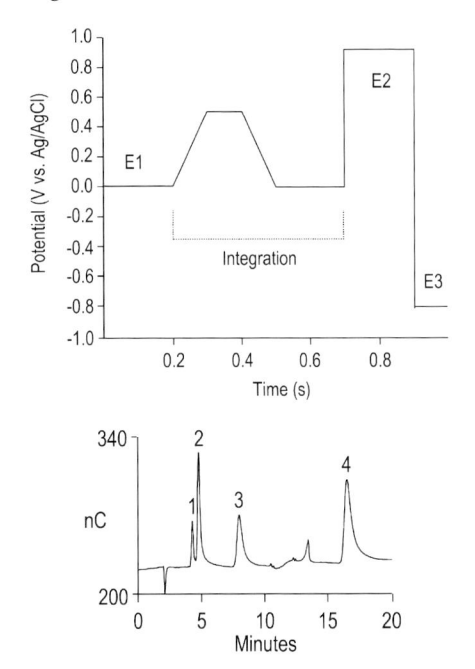

Fig. 7-23. Pulse sequence for integrated amperometry of primary alkylamines and their separation on a weak acid cation exchanger. — Separator column: IonPac CS14; eluant: 2.5 mmol/L H_2SO_4 + 100 mol/L Na_2SO_4 — acetonitrile (5% (v/v) to 40% (v/v) after 8 min); flow rate: 1 mL/min; detection: integrated amperometry on a gold working electrode; suppressor system: CSRS, recycle mode; injection volume: 25 µL; solute concentrations: 2.6 mg/L methylamine (1), 9.1 mg/L ethylamine (2), 0.3 mg/L propylamine (3), and 51 mg/L butylamine (4).

Integrated amperometry is also a welcomed alternative to derivatization techniques for the detection of amino acids. Using this method, first described in 1983 by Johnson et al. [28], amino acids are anodically oxidized in an alkaline medium. Johnson et al. used a platinum working electrode; the use of a gold electrode in connection with the separation of hydrolyzed ribonuclease A on a pellicular anion exchanger was first reported by Frankenberger et al. [29] in 1992. The measured amino acid composition showed good agreement with previously published data. The maximal current yield in the oxidation of the amino group on a gold working electrode, however, occurs at a potential which is high enough for the oxidation of the gold surface itself. The current resulting from this process contributes to noise and to a certain baseline instability. Johnson et al. solved this problem with the introduction of integrated amperometry [30], with amplifies the signal from the amine oxidation and suppresses the signal from the gold oxidation.

Figure 7-24 shows the pulse sequence developed by Avdalovic et al. [31] for integrated amperometry of amino acids. It is very similar to the one introduced by Johnson et al. [30], but optimized for good linearity, minimal baseline drift during the gradient, and long-term stability. The pulse sequence introduced by Avdalovic et al. employs a cleaning step at negative potential which ensures a clean and active electrode surface at constant sensitivity.

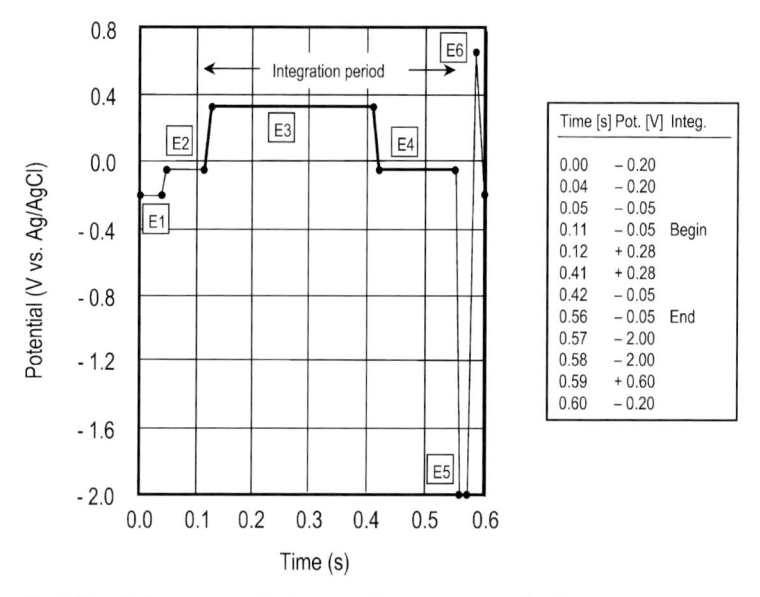

Time [s]	Pot. [V]	Integ.
0.00	−0.20	
0.04	−0.20	
0.05	−0.05	
0.11	−0.05	Begin
0.12	+0.28	
0.41	+0.28	
0.42	−0.05	
0.56	−0.05	End
0.57	−2.00	
0.58	−2.00	
0.59	+0.60	
0.60	−0.20	

Fig. 7-24. Pulse sequence for integrated amperometry of amino acids. The detector signal is the charge (measured in Coulomb) that results from the integration of the current yield from the oxidation of the amino group between 0.11 s and 0.56 s.

The pulse sequence illustrated in Fig. 7-24 can be divided into three distinct regions:
- Adsorption/Initiation $E1, E2$
- Current integration $E3, E4$
- Cleaning/Activation $E5, E6$

From cyclovoltammetric data it is known that the sensitivity for the adsorbing amino acids can be enhanced by an appropriate potential $E1$ (usually negative) as part of the detection waveform. Because the sensitivity for non-adsorbing amino acids is not enhanced, the result can be a divergence of response factors. Moreover, an enhanced adsorption comes at the cost of limiting the linear range for basic amino acids, so that the duration for the adsorption potential must be limited to 40 ms. Cyclovoltammetric data also demonstrated a dependence of amino acid adsorption on the electrode potential. As $E1$ decreases from +0.05 V to −0.2 V, the response increases for all amino acids except arginine, lysine, and methionine. When $E1$ is smaller than −0.2 V, the response for all amino acids decreases. The response for sulfur-containing amino acids such as methionine and cysteine is largely independent of $E1$ until -0.8 V, reflecting a strong chemisorption of sulfur species at the gold electrode. Basic amino acids such as arginine and lysine maintain significant adsorption even at potentials lower than -0.6 V, because they are less anionic (at the pH of the mobile phase) and the rate of their adsorption is less sensitive to changes in the net negative charge on the electrode. The purpose of the delay step $E2$ is to provide a potential at which integration of the current can begin. When increasing the potential from -0.2 V to -0.05 V, a charging current results that almost completely decays during the pulse duration of 60 ms. This short delay time greatly reduces the magnitude of the chromatographic baseline displacement caused by the acetate gradient.

At the beginning of the integration period, the potential $E3$ is swept to a value that is positive enough to oxidize amino acids and form a surface oxide. An examination of the cyclic voltammogram for glycine reveals that a value greater than $+0.20$ V is necessary. In general, the response for amino acids increases with increasing positive potential to a maximum signal at about $+0.30$ V. At potentials higher than $+0.30$ V, the analyte signal for most amino acids decreases. The optimum value for $E3$ is $+0.28$ V. After holding at $E3$ to increase the total charge from the oxidized amino acids, the potential ($E4$) is returned to its initial value of -0.05 V, so that charge from the gold oxide reduction will cancel charge from the gold oxidation. Because the reduction of gold oxide occurs more rapidly than oxidation, the duration of $E4$ is shorter than that of $E3$, therefore allowing a greater proportion of the integration period to be used to measure the analyte signal while still achieving adequate background correction. The optimized value for the total integration time is ~ 450 ms; 290 ms is spent at $E3$, and only 140 ms is spent integrating at $E4$. The integration asymmetry of the waveform maximizes the analyte response without any effect on noise.

The cleaning and activation potentials $E5$ and $E6$ originate from the new quadruple-potential waveform, which was recently introduced for the detection carbohydrates with improved long-term reproducibility [20]. This approach utilizes a cathodic cleaning of the electrode at -2 V ($E5$) and activation of the catalytic surface via transient oxide formation at $+0.6$ V. The potential is then returned to the initial value of -0.2 V for the purpose of reducing the gold oxide.

As can be seen from the chromatogram of an amino acid standard in Fig. 3-200 (see Section 3.11), sensitivity for the various amino acids differs significantly when using integrated amperometry. High sensitivity is obtained for amino acids such as arginine and lysine that elute very early as sharp signals; likewise for aromatic amino acids, which also elute with relatively small peak width due to the focusing power of the acetate gradient. Although the peak areas of acidic amino acids such as glutamate and aspartate are smaller than those of other amino acids, detection limits below 1 pmol are obtained. In general, detection limits for the various amino acids vary between several hundred fmol and 4 pmol (for leucine). With very few exceptions, almost all standard amino acids exhibit a linear response in the concentration range of 1-100 pmol.

With the development of the pulsed amperometric detector (PAD) [32] a new detector cell was also designed. It is schematically shown in Fig. 7-25. To facilitate the replacement of the working electrode, the detector cell consists of two blocks separated by a spacer. One part houses the reference electrode and the capillary connections, the other houses the working electrode. This second cell block may be replaced as a whole. Two stainless steel connectors at the inlet and outlet boreholes serve as a counter electrode.

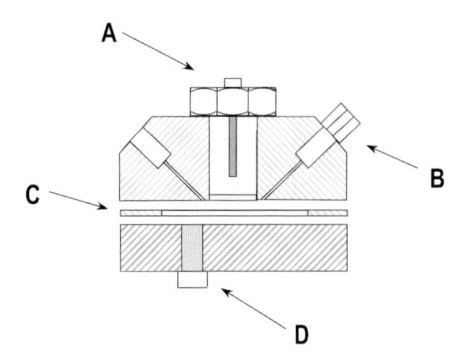

Fig. 7-25. Schematic representation of the detector cell for pulsed amperometry.

The choice of the working electrode suitable for a given application is determined by the following factors:
- Potential limit for the working electrode in the eluant
- Possibility of oxidation of the working electrode by formation of complexes with the solute
- Kinetics of the electrochemical reaction
- Background noise.

Potential Limit

The applied potential is limited in the negative range by the value at which the supporting electrolyte or the solvent is reduced. The limiting factor in the positive potential range is the value at which the solvent, the working electrode itself, or the supporting electrolyte is oxidized. The potential limits for gold-, silver-, platinum-, and glassy carbon working electrodes in acidic and alkaline solution are listed in Table 7-3.

The potential limits are particularly dependent on the eluant pH. In principle, more negative potentials may be applied in alkaline solution than in acidic solution and, conversely, in acidic solution more positive potentials may be selected than in alkaline solution. For oxidation reactions at high positive potentials, glassy carbon- and platinum electrodes are suited; reduction reactions at very negative potentials can be performed at glassy carbon-, silver-, and gold electrodes.

Table 7-3. Potential limits in acidic and alkaline eluants (reference electrode: Ag/AgCl).

Working electrode	Eluant [0.1 mol/L]	Potential range [V]
Gold (Au)	KOH	−1.25 .... 0.75
	$HClO_4$	−0.35 .... 1.10
Silver (Ag)	KOH	−1.20 .... 0.10
	$HClO_4$	−0.55 .... 0.40
Platinum (Pt)	KOH	−0.90 .... 0.65
	$HClO_4$	−0.20 .... 1.30
Glassy Carbon (GC)	KOH	−1.50 .... 0.60
	$HClO_4$	−0.80 .... 1.30

Possibility of Complex Formation

At positive potentials, metals such as silver and gold may form complexes with suitable anions in the analyte solution. This may be a disadvantage because it lowers the positive potential limit. For the analysis of complex-forming anions such as sulfide and cyanide, on the other hand, this is advantageous.

Kinetics

If the reaction from A to B is reversible according to Eq. (192), the concentration ratio of A to B at the electrode surface is in the equilibrium state, as described by the Nernst equation (193). As a prerequisite for this, the electron transfer between the electrode and the solute ions must be kinetically favored. Therefore, a certain species may be oxidized or reduced with different speed at various electrode materials. If an electrode material is chosen at which the electrochemical reaction is very slow, an acceleration of the reaction is only possible by raising the working potential.

The deviations from the equilibrium Galvani potentials[1] and the respective electrode potentials are denoted as overvoltage. If the transport of reaction partners to and from the electrode is slower than the passage reaction, it is called concentration overvoltage. The cause for the passage overvoltage is a slow passage reaction (i.e., the transfer of species from one side to the other side requires a certain amount of activation energy). If the overvoltage exceeds the permissible potential limit for a chosen electrode material, a different material must be selected.

---

[1] The term galvani potential denotes the potential difference between two electrically conducting phases. The equilibrium galvani potential is the galvani potential at electrochemical equilibrium.

Background Noise

The background noise is typically caused by small fluctuations in the eluant flow rate, vortexing in the detector cell, or small temperature variations. Electrode materials with slow kinetics naturally produce less background than do fast-kinetic materials because their response times with regard to interferences are also slower.

7.2
Spectrometric Detection Methods

7.2.1
UV/Vis Detection

7.2.1.1 Direct UV/Vis Detection

In contrast to RPLC, UV detection is of little importance in ion chromatography, but is considered a welcome supplement to conductivity detection. It is a disadvantage of direct UV detection that most inorganic anions do not possess an appropriate chromophore. Thus, they generally absorb at wavelengths below 220 nm [33]. This was corroborated by the works of Reeve [34] and Leuenberger et al. [35], who separated inorganic anions on a chemically bonded cyano- or amino-propyl phase and detected them at a wavelength of 210 nm. Recent work by Williams [36] demonstrated the advantages of simultaneous UV- and conductivity detection in combination with a suppressor system. The absorption wavelength of 195 nm chosen by Williams, however, is only applicable to samples with comparatively simple matrices. This method is only of academic interest; in the area of surface water and wastewater analysis, interference is possible if several organic species absorb at 195 nm.

Direct UV detection gained great significance in the determination of nitrite and nitrate [37, 38], as well as bromide and iodide in the presence of high chloride concentrations. The optimal measuring wavelengths for the determination of those anions are listed in Table 7-4. Figure 7-26 illustrates the superiority of direct UV detection over conductivity detection with a nitrite determination in the presence of a 100-fold chloride excess, even though resolution between these two anions is much higher on modern anion exchangers than on IonPac AS4A(-SC), which was used in this particular example. Determinations of this kind may be performed in all saline samples such as body fluids, sea water, meat products, sausages, etc. It is also worth mentioning that metal-cyano- and metal-chloro complexes [39] can be detected at a wavelength of 215 nm.

Table 7-4. Optimal UV measuring wavelengths for some selected inorganic anions.

Anion	Measuring wavelength [nm]
Bromate	200
Bromide	200
Chromate	365
Iodate	200
Iodide	227
Metal-chloro complexes	215
Metal-cyano complexes	215
Nitrate	202
Nitrite	211
Sulfide	215
Thiocyanate	215
Thiosulfate	215

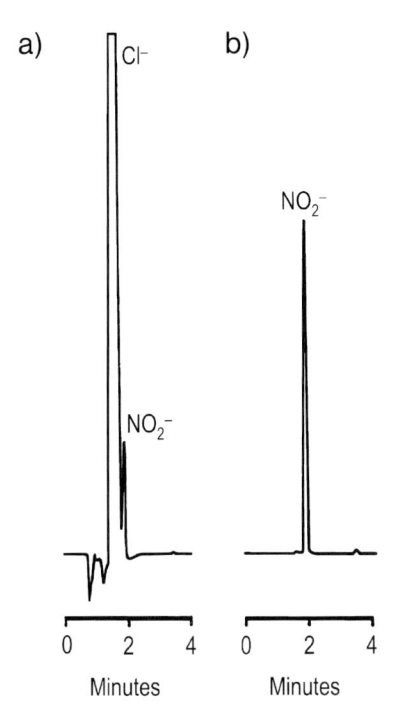

Fig. 7-26. Comparison of direct UV detection and suppressed conductivity detection in the nitrite determination at 100-fold excess of chloride. – Separator column: IonPac AS4A(-SC); eluant: 1.7 mmol/L NaHCO$_3$ + 1.8 mmol/L Na$_2$CO$_3$; flow rate: 2 mL/min; detection: (A) suppressed conductivity (sensitivity: 10 µS/cm), (B) UV detection at 215 nm (sensitivity: 0.05 aufs); injection volume: 50 µL; solute concentrations: 100 mg/L chloride and 1 mg/L nitrite.

7.2.1.2 UV/Vis Detection in Combination with Derivatization Techniques

One of the most important applications of UV/Vis detection is photometric determination after derivatization of the column effluent. First of all, this includes the determination of transition metals after reaction with 4-(2-pyridylazo)-resorcinol (PAR).

PAR

The metal ions are separated on an ion exchanger as oxalate- or PDCA complexes (PDCA: pyridine-2,6-dicarboxylic acid). They are then mixed with the PAR reagent and form chelate complexes which absorb in the wavelength range between 490 nm and 530 nm. PAR is only stable in presence of an ammonia solution at pH > 9. It reacts with a number of metals [40], although usually only Mn, Fe(II/III), Co, Ni, Cu, Zn, Cd, Pb, and lanthanides are analyzed with PAR as a derivatization reagent. When Zn-EDTA is added to the PAR reagent, the application range of this derivatization reaction can be expanded to include alkaline-earth metals [41, 42]. The combination of Zn-EDTA and PAR is frequently cited in the literature, but the kinetics of this reaction are not yet completely understood. Lucy et al. [43] studied in detail the mechanism of reactions in the Zn-EDTA-PAR system and contributed a lot to understanding it. Although zinc forms a stable complex with PAR (log β_2 = 17.1 [44]), the dominating process in the presence of EDTA is the formation of a Zn-EDTA chelate complex (log K_f = 16.44 [45]):

$$Zn(PAR)_2 + EDTA^{4-} \rightleftharpoons 2PAR^- + Zn(EDTA)^{2-} \tag{196}$$

When alkaline-earth metal ions, M^{2+}, are added to this system, zinc ions are released from the Zn-EDTA and subsequently form the chelate complex $Zn(PAR)_2$ with PAR:

$$2PAR^- + M^{2+} + Zn(EDTA)^{2-} \rightleftharpoons Zn(PAR)_2 + M(EDTA)^{2-} \tag{197}$$

Although zinc ions form a more stable complex with EDTA than, for example, magnesium ions, the equilibrium of the reaction described in Eq. (197) is clearly on the right side due to the stability of the $Zn(PAR)_2$ chelate complex. By adding Zn-EDTA to the PAR reagent equal response factors for analyte metal ions are obtained. The respective chelate complex can be detected at its absorption maximum of 490 nm, which leads to a slightly higher sensitivity relative to the pure PAR reagent. It should be noted that alkaline-earth metals such as magnesium and calcium that are present in real samples in much higher concentrations can interfere the analysis of transition metals.

The reagent may be delivered in two ways. Truly pulsation-free delivery is achieved pneumatically by supplying the pressurized reagent to the column effluent through a capillary. Alternatively, a pump with a pulse damper may be employed. When the reagent is delivered via a manifold, the reaction with the reagent occurs in an appropriately dimensioned reaction coil. This is typically filled with chemically inert plastic beads to reduce band spreading and to optimize the mixing process.

Reagent delivery is improved significantly upon application of a membrane reactor. This contains a semipermeable membrane which is permeable to certain reagents. The operation of a membrane reactor requires a pressurized container, from which the reagent solution reaches the membrane reactor pulsation-free. In the membrane reactor, the reagent diffuses into the interior of the membrane, where it mixes with the column effluent. This kind of reagent delivery significantly reduces the short-term noise in the subsequent photometric detection. It is particularly noticeable when the reagent being used exhibits a high background absorption at the measuring wavelength. A much smaller reagent volume is delivered when using a membrane reactor, so a significant sensitivity increase is observed because of the smaller dilution factor. Chromatograms exemplifying the analysis of transition metals using derivatization with PAR are depicted in Section 4.5.3.

A post-column derivatization with subsequent photometric detection has also been developed for the determination of aluminum. Using a mixture of ammonium sulfate and sulfuric acid, aluminum is separated as $AlSO_4^+$ ion on a cation exchanger. It forms a stable complex with the disodium salt of 4,5-dihydroxy-1,3-benzenedisulfonic acid (Tiron) at pH 6.2 which can be detected at 313 nm.

Tiron

At this pH value Tiron only reacts with aluminum and iron(III), hence this method for aluminum determination is very selective. The only disadvantage is the comparatively low extinction coefficient, $\varepsilon = 6000$ L/(mol cm), of the resulting complex at 310 nm. A corresponding standard chromatogram of the separation of aluminum is illustrated in Fig. 4-59 (see Section 4.5.3).

In the area of cation analysis, the extremely selective method of post-column derivatization of alkaline-earth metals with Arsenazo I [o-(1,8-dihydroxy-3,6-disulfo-2-naphthylazo)-benzenearsonic acid] is particularly suited for the determination of these elements in presence of high concentrations of alkali metals [46].

Arsenazo I

Characteristic applications include the determination of strontium in sea water and the determination of magnesium and calcium in ultra-pure brine solutions used for the electrolysis of alkali metal chlorides. With alkaline-earth metals, Arsenazo I forms red-purple complexes that absorb at a wavelength of 570 nm. Figure 7-27 reveals that the reagent exhibits a high background absorption even at this wavelength, which suggests that a membrane reactor should be used. Arsenazo I does not form complexes with alkali metals, thus ensuring the selectivity for alkaline-earth metals that is required for the above-mentioned matrices. Arsenazo I may also be utilized as a derivatization reagent for the analysis of lanthanides. Figure 7-28 shows an example chromatogram, which was obtained by Wang et al. [47] using a silica-based anion exchanger with 2-methyllactic acid as an eluant.

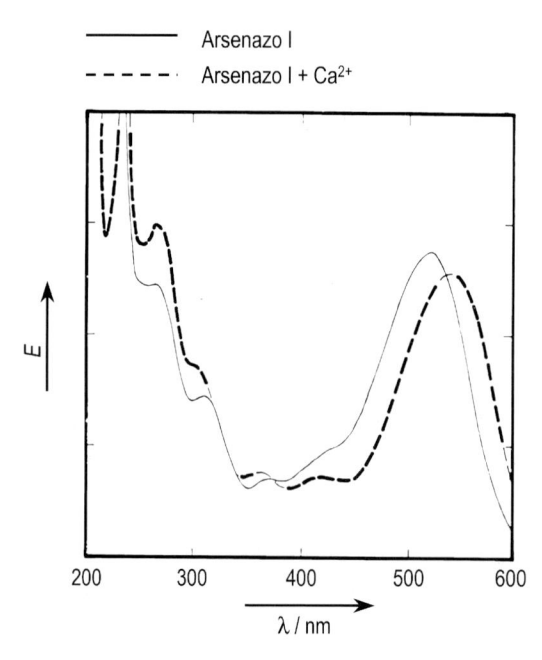

Fig. 7-27. Absorption spectra of Arsenazo I and its corresponding calcium complex.

Arsenazo III [Bis-(2-arsono-benzeneazo)-2,7-chromotropic acid] can also be used for the detection of lanthanides; it has the advantage of lower detection limits than PAR.

Arsenazo III

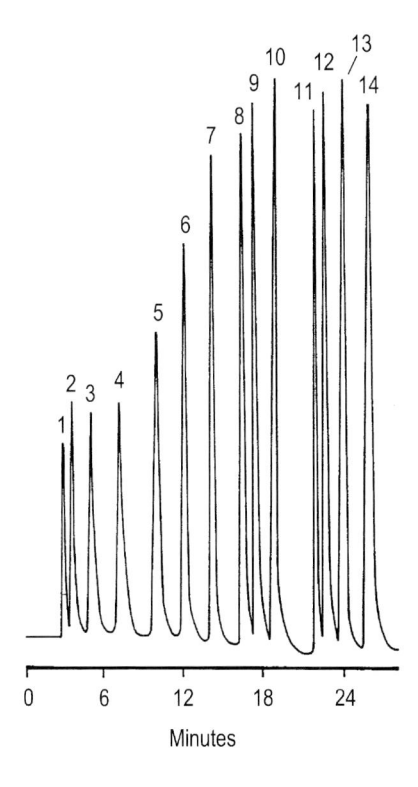

Fig. 7-28. Detection of lanthanides by post-column derivatization with Arsenazo I. — Separator column: Nucleosil 10 SA; eluant: 2-methyllactic acid; gradient: linear, 0.01 mol/L in 30 min to 0.04 mol/L; detection: photometry at 600 nm after reaction with Arsenazo I; reagent composition: 0.13 mmol/L Arsenazo I in 3 mol/L NH_4OH; solute concentrations: 10 mg/L each of Lu^{3+} (1), Yb^{3+} (2), Tm^{3+} (3), Er^{3+} (4), Ho^{3+} (5), Dy^{3+} (6), Tb^{3+} (7), Gd^{3+} (8), Eu^{3+} (9), Sm^{3+} (10), Nd^{3+} (11), Pr^{3+} (12), Ce^{3+} (13), and La^{3+} (14); (taken from [47]).

Cassidy et al. [48] used this reagent for the determination of lanthanides in thorium dioxide/uranium dioxide fuel rods. Arsenazo I and PAR are not suitable for this analysis, because both have to be used under strongly alkaline conditions. However, thorium hydrolyzes at alkaline pH, so that thorium-containing solutions cannot be injected directly under these conditions. In addition, the hydrolysis product is not soluble in water. Maximum sensitivity is obtained at pH 2.5, but the sensitivity obtained with a purely aqueous solution with the optimum concentration of $c = 0.15$ mol/L (pH 4.6) is sufficient. The measuring wavelength is 653 nm. A sample chromatogram was shown in Fig. 4-58 in Section 4.5.3.

Several derivatization techniques have been developed for trace analysis of bromate in drinking water (see Section 9.1), because concentrations below 5 µg/L cannot be determined by suppressed conductivity detection without pre-concentration [49]. Even at lowest µg/L levels, bromate has been classified by the World Health Organization (WHO) and the American Environmental Protection Agency (EPA) to be carcinogenic [50]. Lowering the detection limit by enlarging the injection volume compromises chromatographic efficiency and is, therefore, ruled out. However, sample pre-concentration, a method published by Joyce et al. [51] in 1994, is associated with enormous effort in sample preparation, because the bromate determination is interfered with by high amounts of chloride, sulfate, and carbonate in the sample. Thus, these anions have to be partly removed by using appropriate sample preparation cartridges prior to the pre-concentration step. With hyphenated techniques such as IC-MS via an electrospray interface [52] and IC/ICP-MS [53] (see also Section 7.4) detection limits in the sub-µg/L range are possible, but this kind of instrumentation is not at everyone's disposal. IC/ICP-MS is highly selective due to mass-selective detection of bromate, but even this method is not free of interferences (e.g., tribromoacetic acid).

Regarding selectivity and sensitivity, Weinberg et al. [54] presented an interesting alternative by introducing the post-column derivatization of bromate with bromide under acidic conditions to form tribromide, Br_3^-, which can be detected photometrically at 267 nm. The reaction scheme is described by Eqs. (198) and (199):

$$BrO_3^- + 5Br^- + 6H^+ \rightarrow 3Br_2 + 3H_2O \qquad (198)$$

$$Br_2 + Br^- \rightarrow Br_3^- \qquad (199)$$

In a temperature-depending reaction, bromate reacts with bromide under acidic conditions to form bromine, which further reacts with an excess of bromide to form tribromide. The indirect determination of bromate via the formation of tribromide is fascinating; the extinction coefficient for tribromide of $\varepsilon = 40{,}900$ L mol^{-1} cm^{-1} at $\lambda = 267$ nm is quite large. Moreover, this method can be used for other oxyhalides such as iodate and chlorite. Because the reaction of bromate to tribromide is nothing more than a decrease in the degree of oxidation from $+5$ to $-1/3$, the reaction requires a reducing agent that is compatible with all other reaction components. A combination of HBr and HNO_2 has proved to be suitable for this reaction. Both acids can be generated *in situ* by utilizing a membrane suppressor with cation exchange groups. For this, a mixture of 0.5 mol/L sodium bromide and 0.29 mol/L sodium nitrite is pumped through the suppressor, which is continuously regenerated with sulfuric acid of $c = 0.75$ mol/L. HBr and HNO_2 are generated *in situ* by exchanging sodium ions for hydronium ions then pumped via a manifold into a knitted reaction coil which is heated to 60 °C. Higher temperatures cannot be applied due to the formation of bubbles. As an example, Fig. 7-29 shows the chromatogram of a 1-µg/L bromate standard which was obtained by injecting 500 µL standard solution onto an IonPac AS9-HC separator. As can be seen from this chromatogram, a minimum detection limit of 0.5 µg/L bromate can be achieved without any problem when injecting such large volumes.

Recently, Wagner et al. [55] developed another derivatization technique for the analysis of bromate, which can be directly combined with the American EPA Method 300.1 [56]. In Part A, Method 300.1 describes the separation of the seven standard anions bromide, chloride, fluoride, nitrite, nitrate, phosphate, and sulfate, and in Part B the separation of bromate, bromide, chloride and chlorite on an IonPac AS9-HC high-capacity anion exchanger. The technique developed by

Wagner et al. is based on the reaction of bromate with o-dianisidine (ODA) in an acidic solution [57] with subsequent photometry of the reaction product at 540 nm.

o-Dianisidine

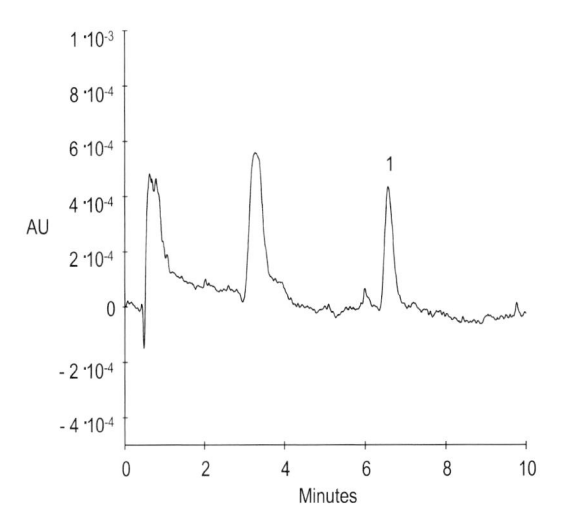

Fig. 7-29. Trace analysis of bromate by indirect determination via tribromide in form of a post-column derivatization. — Separator column: IonPac AS9-HC; eluant: 9 mmol/L Na_2CO_3; flow rate: 1 mL/min; detection: photometry at 267 nm after derivatization with $NaBr/NaNO_2$; reagent flow rate: 0.5 mL/min; reaction temperature: 60 °C; suppressor: ASRS-II; regenerant: 0.75 mol/L H_2SO_4; injection volume: 500 µL; solute concentration: 1 µg/L bromate (1).

The respective reagent is prepared as follows:

Put 300 mL de-ionized water into a 500-mL volumetric flask and add 40 mL of freshly distilled 70% nitric acid. 2.5 g potassium bromide (reagent grade) are added to this solution. 250 mg o-dianisidine are dissolved in 10 mL methanol (HPLC grade), which is then added to the nitric acid KBr solution. Finally, the volumetric flask is filled up to the meniscus with de-ionized water. The reagent is stable for at least a month when kept at room temperature.

As in the reaction yielding tribromide described above, the reaction coil is heated to 60 °C to speed up the derivatization reaction. The authors determined the minimum detection limit for bromate according to EPA standards [58] to be 0.5 µg/L. A representative chromatogram is shown in Fig. 7-30. The post-column derivatization with o-dianisidine is only interfered with by large amounts of chlorite (>200 mg/L) which also reacts with o-dianisidine and elutes ahead of bromate. Chlorite interferences are only possible, however, if drinking water is disinfected with a mixture of chlorine dioxide and ozone.

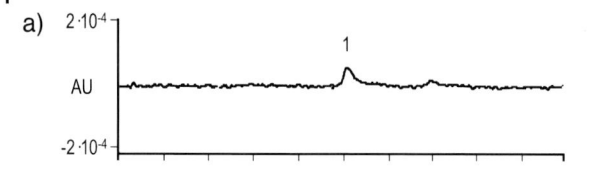

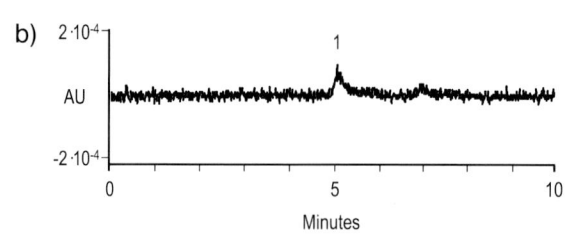

Fig. 7-30. Analysis of a bromate standard via derivatization with o-dianisidine (ODA). — Separator column: IonPac AS9-HC; eluant: 9 mmol/L Na_2CO_3; flow rate: 1.3 mL/min; detection: photometry at 450 nm after derivatization with ODA; reagent composition: see text; reagent flow rate: 0.7 mL/min; reaction temperature: 60 °C; injection volume: 220 µL; solute concentration: 0.5 µg/L bromate, (A) with electronic smoothing (Olympic, 25 points, 5 s and 1 repetition), (B) without electronic smoothing.

A very sensitive and selective derivatization technique for the determination of iodide and iodate in aqueous samples was described by von Gunten et al. [59] in their recent publication. After separation on an IonPac AS11 anion exchanger, iodide is determined photometrically as IBr_2^- which is formed in a bromide-containing eluant by adding a basic hypobromite solution with subsequent acidification:

$$I^- + OBr^- + Br^- + 2H^+ \rightarrow IBr_2^- + H_2O \qquad (200)$$

As with many trihalides, IBr_2^- is also characterized by a strong UV absorption (absorption maximum: 253 nm; $\varepsilon = 57{,}200$ L mol^{-1} cm^{-1}). The measuring wavelength is 249 nm, because the absorption difference between IBr_2^- and I_3^- is maximal at this wavelength. The instrumental setup illustrated in Fig. 7-31 is relatively simple. Acidification of both eluant and derivatization reagent is carried out with a membrane suppressor, which is positioned between the reaction coil and the UV detector cell. The suppressor is continuously regenerated with dilute sulfuric acid.

The formation of the interhalide IBr_2^- according to Eq. (200) can be expressed as a series of the following reactions:

$$I^- + OBr^- \rightarrow Br^- + OI^- \qquad (201)$$

$$OI^- + H^+ \rightarrow HOI \qquad (202)$$

$$HOI + H^+ + Br^- \rightarrow IBr + H_2O \qquad (203)$$

$$IBr + Br^- \rightarrow Br_2^- \qquad (204)$$

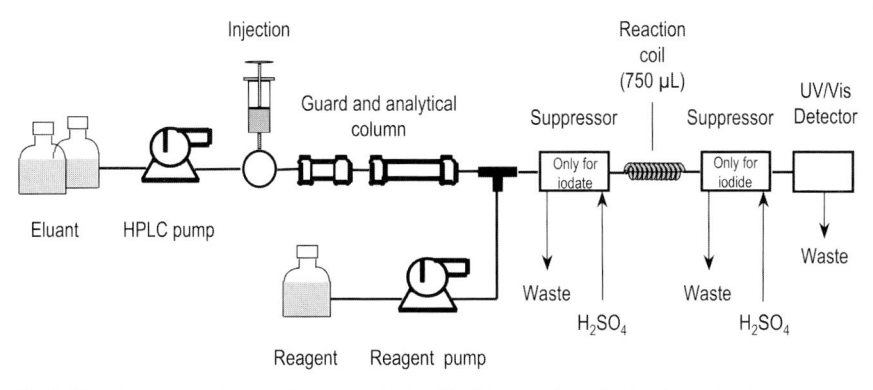

Fig. 7-31. Instrumental setup for the analysis of iodide or iodate via the determination of IBr_2^- and I_3^-.

The derivatization reagent consisting of a mixture of 5 µmol/L Br_2 and 2 mmol/L NaOH with the active species OBr^- and Br^- has to be adjusted at alkaline pH to avoid disproportionation of hypobromite to bromide and bromate. Only at low pH are the equilibria of the reactions in Eqs. (202) and (203) on the right side. Therefore, the basic derivatization reagent has to be acidified after passing through the reaction coil.

After its separation on an IonPac AS9(-SC) anion exchanger, iodate is determined photometrically as I_3^-, which is formed by adding an iodide solution with subsequent acidification:

$$IO_3^- + 8I^- + 6H^+ \rightarrow 3I_3^- + 3H_2O \tag{205}$$

The measuring wavelength of 288 nm is also the absorption maximum of I_3^- with $\varepsilon = 38{,}200$ L mol^{-1} cm^{-1} (I_3^- has a second absorption maximum at 351 nm with $\varepsilon = 25{,}700$ L mol^{-1} cm^{-1}). As can be seen from Fig. 7-31, the suppressor for eluant acidification is positioned between manifold and reaction coil. Direct addition of acid to the derivatization reagent (250 mmol/L KI) is not possible, because the interhalide would be formed via oxidation with oxygen present in solution. The resulting yellow color would greatly increase background absorption and, consequently, baseline noise.

The formation of the interhalide I_3^- according Eq. (205) can also be expressed as a series of the following reactions:

$$IO_3^- + 2I^- \rightarrow 3HOI \tag{206}$$

$$HOI + I^- + H^+ \rightarrow I_2 + H_2O \tag{207}$$

$$I_2 + I^- \rightarrow I_3 \tag{208}$$

The rate-determining step is the reaction in Eq. (206), which is again fast only at low pH. Thus, a suppressor is required for eluant acidification.

The reactions of iodide and iodate described by Eqs. (200) and (205) are quantitative and not interfered by other inorganic water constituents, even if those are present in the lower mg/L range. Using this technique, the minimum detection limits are 0.1 µg/L for both compounds. Von Gunten et al. successfully applied this technique for iodide speciation in various mineral water samples. Exemplified with a Swiss mineral water, Figures 7-32 and 7-33 show the analyses of iodide and iodate, respectively.

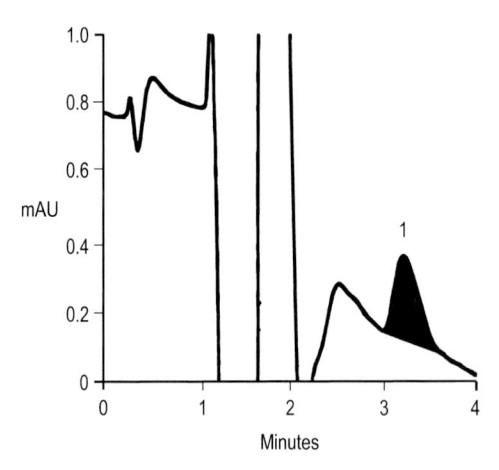

Fig. 7-32. Trace analysis of iodide in mineral water via indirect determination of IBr_2^- by means of post-column derivatization. – Separator column: IonPac AS11; eluant: 60 mmol/L NaBr + 1 mmol/L NaOH; flow rate: 1.5 mL/min; detection: photometry at 249 nm after derivatization with 5 µmol/L Br_2 + 2 mmol/L NaOH; reagent flow rate: 0.5 mL/min; suppressor: ASRS-II; regenerant: 0.45 mol/L H_2SO_4; regenerant flow rate: 3 mL/min; injection volume: 500 µL mineral water (Henniez, Switzerland); solute concentration: 0.4 µg/L iodide (1) besides 10 mg/L chloride, 18 mg/L nitrate, 13 mg/L sulfate, and 394 mg/L bicarbonate; (taken from [59]).

An extremely selective reagent for chromium(VI) is 1,5-diphenylcarbazide which forms an analytically useful inner complex with dichromate under acidic conditions.

1,5-Diphenylcarbazide

In this reaction, diphenylcarbazide is oxidized to diphenylcarbazone, while at the same time chromium(VI) is reduced to chromium(III), which forms a chelate complex with diphenylcarbazide. Like the metal-PAR complexes, it is red-purple colored and, thus, may be photometrically detected at 520 nm. As already

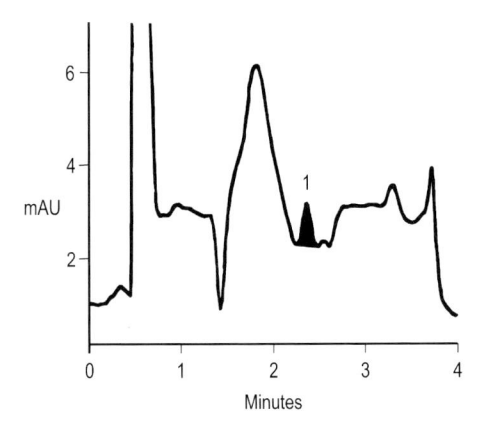

Fig. 7-33. Trace analysis of iodate in mineral water via indirect determination of I_3^- by means of post-column derivatization. − Separator column: IonPac AS9(-SC); eluant: 40 mmol/L B(OH)$_3$ + 20 mmol/L NaOH; flow rate: 1.5 mL/min; detection: photometry at 288 nm after derivatization with 250 mmol/L KI; reagent flow rate: 0.5 mL/min; suppressor: ASRS-II; regenerant: 0.45 mol/L H$_2$SO$_4$; regenerant flow rate: 3 mL/min; injection volume: 500 µL mineral water (Henniez, Switzerland); solute concentration: 0.9 µg/L iodate (1) besides 10 mg/L chloride, 18 mg/L nitrate, 13 mg/L sulfate, and 394 mg/L bicarbonate; (taken from [59]).

mentioned, chromate separation is performed by means of anion exchange chromatography. The high selectivity of this technique is revealed in Fig. 7-34, which shows the analysis of chromium(VI) in a flue gas scrubber solution. Despite high contents of chloride and sulfate, chromium(VI) can be determined in this sample without any interferences.

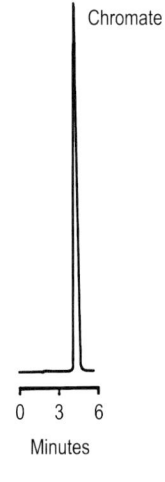

Fig. 7-34. Analysis of chromium(VI) in a flue gas scrubber solution. − Separator column: IonPac CS5; eluant: 2 mmol/L pyridine-2,6-dicarboxylic acid + 2 mmol/L Na$_2$HPO$_4$ + 10 mmol/L NaI + 50 mmol/L NH$_4$OAc; flow rate: 1 mL/min; detection: photometry at 520 nm after reaction with 1,5-DPC; reagent composition: 2 mmol/L 1,5-diphenylcarbazide + 0.5 mol/L H$_2$SO$_4$ − methanol (90:10 v/v); reagent flow rate: 0.5 mL/min; injection: 50 µL sample (1:10 diluted).

Polyvalent anions such as polyphosphates, polyphosphonates, and sequestering agents are analyzed by anion exchange chromatography with post-column derivatization using ferric nitrate in acidic solution as a reagent [60]. The reaction of these compounds with iron(III) causes a bathochromic shift of the background absorbance up to the wavelength range between 310 and 330 nm. Because other inorganic anions such as chloride, sulfate, etc. also cause such a shift under these conditions, the detection of complexing agents via derivatization with iron(III) is only possible in matrices with relatively low electrolyte content. Occasionally, the selectivity of the method is not sufficient for the analysis of complexing agents in detergents, cleansing agents, and wastewaters resulting from their production and use. These matrices partially exhibit very high concentration differences between complexing agents and electrolytes.

Phosphorus-containing complexing agents, in contrast, can be determined very selectively. The method is based on the selective determination of polyphosphonates developed by Baba et al. [61-63]. It was modified appropriately and successfully applied to the analysis of polyphosphates and polyphosphonates in detergents and cleansing agents by Vaeth et al. [64]. Using nitric acid or peroxodisulfate at 105 °C, polyphosphates and polyphosphonates are hydrolyzed to orthophosphate, which is then reacted in a second step with a molybdate/vanadate reagent to the known phosphorvanadatomolybdic acid. Its absorption is measured at 410 nm. Polyphosphate hydrolysis is quantitative under these conditions; the degradation of phosphonates reaches about 90% to 100% and, therefore, must be checked for each particular phosphonate under the given instrumental parameters. In comparison with the derivatization with iron(III), it is simpler to obtain chromatograms with this phosphorus-specific detection for such complex matrices as detergents, because only phosphorus-containing compounds are detected. Quantitative analysis is also quite simple. Calibration may be carried out with orthophosphate because the peak areas are directly proportional to the phosphorus content of the compound. In addition, phosphorus-specific detection allows the use of KCl/EDTA-mixtures as eluants instead of nitric acid. This results in shorter conditioning times for the separator column and rules out a potential hydrolysis of polyphosphates during the separation.

The derivatization with sodium molybdate can also be utilized for the analysis of water-soluble silicate. Its separation can be achieved with both ion-exchange- and ion-exclusion chromatography. With sodium molybdate, silicate also forms the yellow heteropolyacid $H_4[Si(Mo_3O_{10})_4]$ and exhibits an absorption maximum at 410 nm.

7.2.1.3 Indirect UV Detection

UV detection may also be performed indirectly. This method is called indirect photometric chromatography (IPC). Introduced independently by Small et al. [65] and by Cochrane and Hillman [66] in 1982, a UV-absorbing eluant is utilized for the determination of UV-transparent ions. In case of anion analysis, the anion exchanger being used is equilibrated with a UV-active eluant Na^+E^-. According to Fig. 7-35 (A), a constant signal at the detector outlet is observed at a constant flow rate and an appropriate absorption wavelength. If a sample Na^+S^-

is injected, the solute anion S^- is retained under suitable chromatographic conditions and elutes with a certain retention time. The nearly Gaussian-shaped elution profile is shown in Fig. 7-35 (B). Owing to the principle of electroneutrality and the ion-exchange equilibrium, the appearance of S^- is accompanied by a change in total absorption, while both the total anion concentration (E^- and S^-) and the concentration of counter ions (Na^+) remain constant. In this way, the concentration of S^- is determined indirectly by measuring the decreasing UV absorption of E^-.

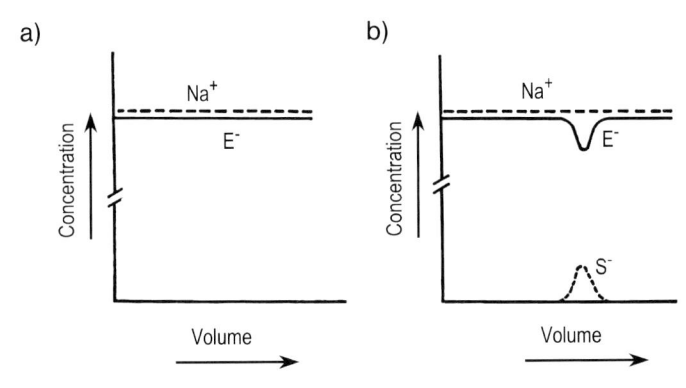

Fig. 7-35. Principle of the indirect photometric detection.

For the analysis of inorganic anions via indirect UV detection, aromatic carboxylic acid eluants are normally used, as they are in non-suppressed conductivity detection, because they possess a correspondingly high light absorptivity. Small et al. illustrated this with the separation of various inorganic anions as shown in Fig. 7-36; the separation was performed on a latexed anion exchanger using a sodium phthalate eluant. The anions shown in Fig. 7-36 exhibit no intrinsic absorption at the measuring wavelength of 285 nm.

When selecting suitable eluants for indirect photometric detection, the chemical and photometric properties of the eluant ions have to be taken into consideration. Knowing the absorption properties of the organic acid and the dependence of its degree of dissociation on the mobile phase pH is essential for a successful application of this method. These criteria were investigated by Small et al. [65] for a series of eluant ions. Phthalate, o-sulfobenzoate, iodide, and trimesate proved to be particularly suited for inorganic anion analysis (see also Section 3.7.3). Figure 7-37 reveals that indirect photometric detection also allows the detection of carbonate. Owing to its high pK value this ion cannot be determined together with strong mineral acids via suppressed conductivity detection, unless hydroxide eluants are used. The hydroxide eluants are converted to water in the suppressor, so the resulting background conductivity is almost zero.

The precise measurement of eluant ion absorption is also extremely important in indirect photometric detection. It is known from classical spectrophotometry [67] that the photometric error is only small in an absorbance range between 0.2

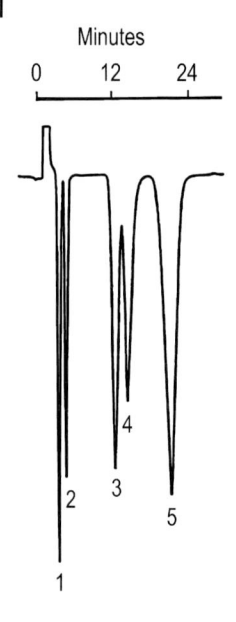

Fig. 7-36. Indirect photometric detection of various inorganic anions. — Separator column: 250 mm × 4 mm i.d. SAR-40-0.6; eluant: 1 mmol/L sodium phthalate (pH 7 to 8); flow rate: 2 mL/min; detection: UV (285 nm, indirect); injection volume: 20 µL; solute concentrations: 106 mg/L chloride (1), 138 mg/L nitrite (2), 400 mg/L bromide (3), 310 mg/L nitrate (4), and 480 mg/L sulfate (5); (taken from [65]).

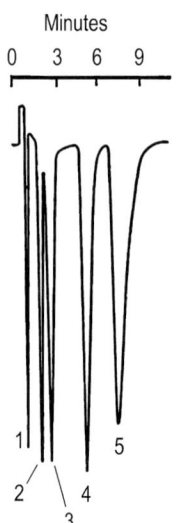

Fig. 7-37. Indirect photometric detection of weak and strong inorganic acids. — Separator column: see Fig. 7-36; eluant: 1 mmol/L sodium phthalate + 1 mmol/L H_3BO_3 (pH 10); flow rate: 5 mL/min; detection: UV (285 nm, indirect); injection volume: 20 µL; solute concentrations: 90 mg/L carbonate (1), 70 mg/L chloride (2), 190 mg/L orthophosphate (3), 250 mg/L azide (4), and 500 mg/L nitrate (5); (taken from [65]).

and 0.8. Therefore, the eluant concentration chosen must have a background absorption that lies within this range. However, because of the ion-exchange capacity of the separator column and the required elution power, the eluant ion concentration must be carefully selected. The background absorption must be adjusted accordingly by choosing a suitable measuring wavelength. Indirect photometric detection is compatible with separator columns of different ion-exchange capacities and with eluants of high and low concentrations. This detection method may be adapted to the chromatographic conditions.

The measuring wavelength directly affects detection sensitivity. If the molar extinction coefficient of the eluant ion increases due to changes in the measuring wavelength, the detection sensitivity is enhanced, as the absorbance difference becomes larger at constant eluant ion concentration. The highest sensitivity is obtained in the wavelength region of the absorption maximum of the eluant ion.

The principle of indirect photometric detection can also be applied to ion-pair chromatography provided the counter ion of the ion-pair reagent exhibits the respective absorption properties. For example, N-methyloctylammonium-p-toluenesulfonate was utilized as an ion-pair reagent to analyze chloride and sulfate [68] upon application of indirect photometric detection. The corresponding chromatogram is shown in Fig. 7-38.

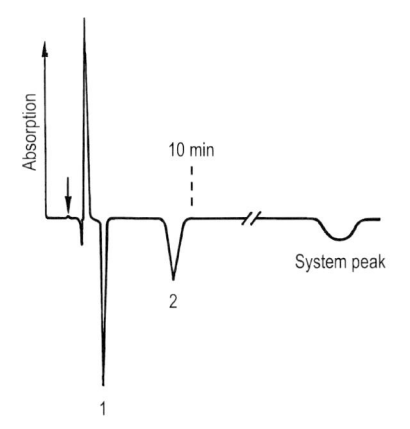

Fig. 7-38. Ion-pair chromatographic separation of chloride and sulfate with indirect photometric detection. — Separator column: 200 mm × 4.6 mm i.d. Zorbax C8; eluant: 1 mmol/L N-methyloctylammonium-p-toluenesulfonate; flow rate: 1.6 mL/min; injection volume: 20 µL; solute concentrations: 35 mg/L chloride (1) and 96 mg/L sulfate (2); (taken from [68]).

In connection with indirect photometric detection, a signal is observed in anion separations via ion-exchange- and ion-pair chromatography that cannot be attributed to any solute ion [68-70]. This signal, usually referred to as the *system peak*, changes its size and position depending on the given chromatographic conditions. It represents a potential interference for solute ions eluting in this range. A detailed explanation of this phenomenon is found in Section 3.7.3.

Indirect photometric detection is also utilized for cation determinations. Figure 7-39 illustrates the separation of sodium, ammonium, and potassium with a copper sulfate eluant obtained by Small et al. [65]. Aromatic bases as cationic analogues to aromatic acids used in anion analysis are inappropriate as eluants for this detection method. As monovalent cations, they are already eluted by oxonium ions. Thus, a significant change in absorption is only observed when the protonated base cation itself contributes to the ion-exchange process.

In summary, it must be pointed out that indirect photometric detection is characterized by a higher sensitivity than non-suppressed conductivity detection, while it is markedly less sensitive than suppressed conductivity detection. The use of indirect photometric detection is preferred for the determination of major

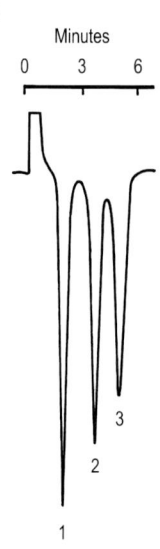

Fig. 7-39. Indirect photometric detection of sodium, ammonium, and potassium. — Separator column: Dowex 50; eluant: 5 mmol/L copper sulfate; flow rate: 0.7 mL/min; detection: UV (252 nm; indirect); injection volume: 20 µL; solute concentrations: 230 mg/L sodium (1), 180 mg/L ammonium (2), and 391 mg/L potassium (3); (taken from [65]).

components when it is desirable to avoid an extensive dilution of the sample. With direct sample injection, typical detection limits for inorganic anions are in the medium to high µg/L range.

7.2.2
Fluorescence Detection

In ion chromatography, fluorescence detection is mainly utilized in combination with post-column derivatization, because inorganic anions and cations, with the exception of the uranyl cation, UO_2^{2+}, do not exhibit any fluorescence. Fluorescence results from the excitation of molecules via absorption of electromagnetic radiation; it is the emission of fluorescence radiation when the excited system returns to the energetic ground level. The emitted wavelength is characteristic to the kind of molecule while the intensity is proportional to concentration.

The best known and most widely used fluorescence method was developed by Roth and Hampai [71] for the detection of primary amino acids, and was described in Section 5.9.2. It is based mainly on the reaction of α-amino acids with o-phthaldialdehyde (OPA) and 2-mercaptoethanol to yield an intensively blue-fluorescing complex. Even at room temperature, the reaction occurs within seconds. At an excitation wavelength of 340 nm and an emission wavelength of 455 nm, this method has a detection limit in the low pmol range. The derivatization with o-phthaldialdehyde is applicable to all compounds carrying a primary amino group. This includes the ammonium ion, primary amines, polyamines, and peptides.

The only drawback of the fluorescence method developed by Roth and Hampai is the relatively small stability of the N-substituted 1-alkylthioisoindoles that are formed from α-amino acids after reaction with OPA. This severely limits the applicability of OPA for the pre-column derivatization of primary amines.

N-substituted alkylthioisoindole

Much higher stabilities and partly higher fluorescence yields were obtained by Stobough et al. [72]. Naphthaline-2,3-dialdehyde (NDA), which reacts in the presence of cyanide ions with primary amines to N-substituted 1-cyanobenz[f]-isoindoles (CBI), was used as a reagent:

NDA → CBI derivative

The excitation spectrum of these derivatives shows maxima at 246 nm and 420 nm; the emission is measured at a wavelength of 490 nm. At an excitation wavelength of 246 nm, detection limits are obtained in the medium to low fmol range. However, secondary amines can be subjected to the two derivatization methods described above only after oxidation.

The phosphorus-specific detection demonstrated in Section 7.2.1.2 using the analysis of polyphosphates and polyphosphonates is a characteristic example of a two-step post-column derivatization that is now possible. The analysis of a secondary amine via a two-step derivatization and fluorescence detection is exemplified with an herbicide – the glyphosate [N-(methylphosphono)-glycine].

Glyphosate

After its separation on a 8-μm cation exchanger (Pickering Laboratories, Mountain View, CA, USA) it is oxidized to glycine with hypochlorite at 36 °C and subsequently reacted with OPA and N,N-dimethyl-2-mercaptoethylamine-hydrochloride (Thiofluor) to yield a strongly fluorescing isoindole. In comparison with the traditionally used 2-mercaptoethanol, Thiofluor offers the advantages of being completely inodorous and more stable in solution. The instrumental setup for this post-column derivatization consisting of two reagent pumps, two reaction coils (one has to be heated), and a column oven is schematically depicted in Fig. 7-40. If the backpressure of the analytical column falls below 3.4 MPa, the reagent pumps are automatically switched off. This safety switch prevents the derivatization reagent from diffusing back onto the analytical column. As seen in the corresponding chromatogram in Fig. 7-41, even the main glyphosate metabolite – aminomethanesulfonic acid (AMPA) – may be detected in the same run. After AMPA is eluted, the separator column is regenerated with 5 mmol/L KOH for two minutes and then conditioned with the eluant for seven minutes, before the next sample can be injected. Based on an injection volume of 50 µL, the minimum detection limit for glyphosate is 1.8 µg/L. The glyphosate method is linear in the investigated concentration range between 0.05 mg/L and 10 mg/L [73]

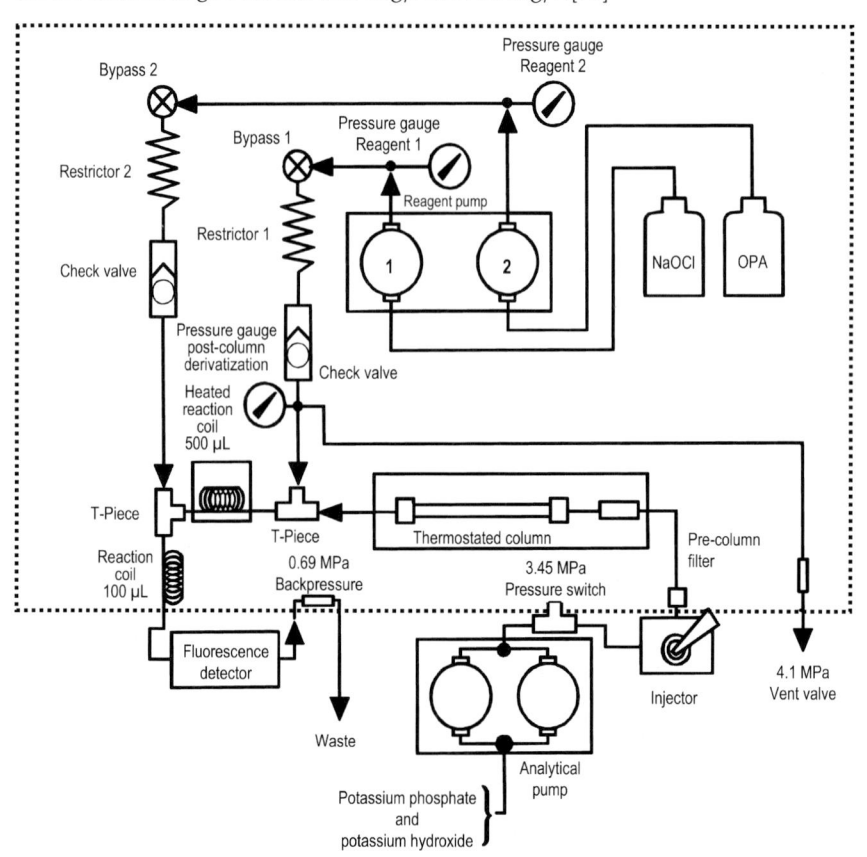

Fig. 7-40. Schematics of the two-step derivatization for the analysis of glyphosate.

Another derivatization technique for forming fluorophores has been described by Lee and Fields [74]. They reacted oxidizable inorganic anions such as nitrite, thiosulfate, and iodide with cerium(IV), thereby forming fluorescing cerium(III) according to Eqs. (209), (210), and (211):

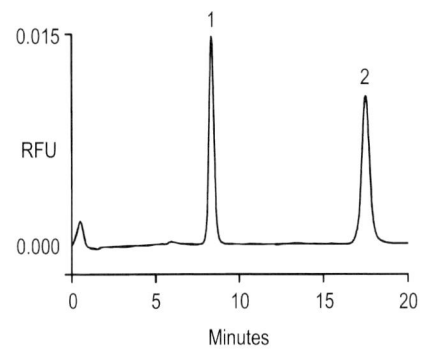

Fig. 7-41. Two-step derivatization of glyphosate with subsequent fluorescence detection. — Separator column: 150 mm × 4 mm i.d. glyphosate column (Pickering Laboratories); column temperature: 55 °C; eluant: 5 mmol/L potassium phosphate (pH 2); regenerant: 5 mmol/L KOH; flow rate: 0.4 mL/min; detection: fluorescence after oxidation with NaOCl and reaction with Thiofluor; reagent 1: 100 µL 5 % hypochlorite solution per bottle of hypochlorite diluent (Pickering); reagent 2: 100 mg OPA + 2 g Thiofluor + 10 mL methanol per bottle of OPA diluent (Pickering); injection volume: 100 µL; solute concentrations: 2.5 mg/L each of aminomethanephosphonic acid (1) and glyphosate (2).

$$HNO_2 + 2Ce^{4+} + H_2O \rightarrow NO_3^- + 2Ce^{3+} + 3H^+ \quad (209)$$

$$2S_2O_3^{2-} + 2Ce^{4+} \rightarrow S_4O_6^{2-} + 2Ce^{3+} \quad (210)$$

$$2I^- + 2Ce^{4+} \rightarrow I_2 + 2Ce^{3+} \quad (211)$$

Cerium(III) may be detected at an excitation wavelength of 247 nm and an emission wavelength of 350 nm. To stabilize the cerium(IV) reagent, it is prepared in 0.5 mol/L sulfuric acid. The addition of sodium bismutate serves to oxidize possible cerium(III) traces in the reagent, which keep the background fluorescence as low as possible.

For the reaction of the column effluent with the cerium(IV) reagent, Fields et al. used a solid-bed reactor with a volume of 2.8 mL instead of a simple reaction coil. This relatively large volume is necessary to ensure the required reaction time of at least two minutes for the oxidation of nitrite ions with cerium(IV). While the reaction of nitrite ions with cerium(IV) is comparatively slow, the maximum fluorescence yield with iodide is obtained in less than ten seconds. On the other hand, the reaction kinetics with thiosulfate appears to be completely different. As seen in the diagram in Fig. 7-42, this reaction is also characterized by a fast rise of the fluorescence yield within a short time and it increases even further because the reaction product from Eq. (210), tetrathionate, also reacts slowly with cerium(IV).

The choice of the eluant is decisively important for achieving maximum sensitivity. According to Fields et al., for the ion-exchange chromatographic separation of the three investigated anions in a single run, potassium hydrogenphthalate or succinic acid may be used as an eluant. Succinic acid exhibits lower background fluorescence. Figure 7-43 shows the chromatogram of a 1.2-nmol standard of these three anions, obtained on Vydac 302 IC with a potassium hydrogenphthalate eluant. The detection limit obtained with this method is in the low µg/L range for all three investigated anions. Even though it does not differ considerably from that of direct UV detection, the specificity of this method is significantly better, especially for samples having a high sodium chloride content.

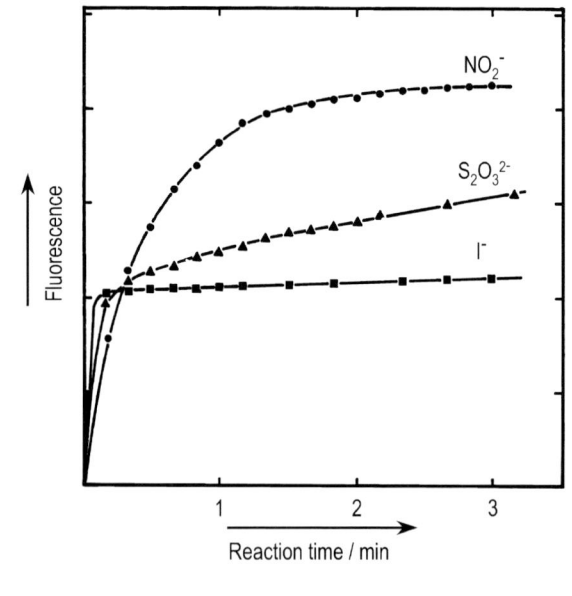

Fig. 7-42. Time dependences of the fluorescence yield for the reactions of nitrite, thiosulfate, and iodide with cerium(IV); (taken from [74]).

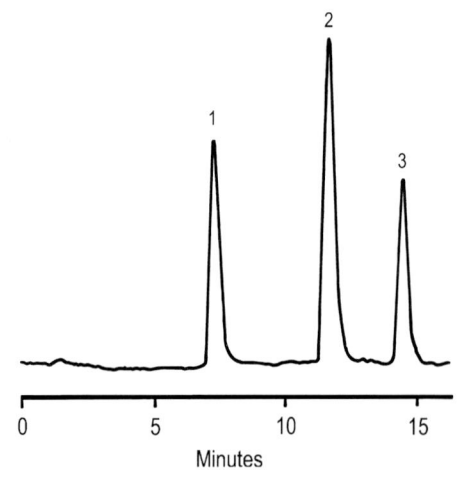

Fig. 7-43. Analysis of nitrite, thiosulfate, and iodide utilizing fluorescence detection after derivatization with cerium(IV). — Separator column: Vydac 302 IC; eluant: 1 mmol/L KHP + 10 mmol/L Na_2SO_4, pH 5.5 with $Na_2B_4O_7$; flow rate: 1 mL/min; detection: fluorescence after reaction with cerium(IV); injection volume: 100 µL; solute concentrations: 0.5 mg/L nitrite (1), 1.1 mg/L thiosulfate (2), and 1.5 mg/L iodide (3); (taken from [74]).

Another important application of the redox system cerium(III)/cerium(IV) is indirect fluorescence chromatography (IFC) of alkali- and alkaline-earth metals. As with indirect photometric detection (IPC), an eluant with a high background fluorescence is utilized which is lowered during the solute ion elution. Hence, negative signals are registered for analyte species. While it has been known for some time that an aqueous cerium(III) solution exhibits fluorescence [75, 76] and, as described above, is used analytically [74], only the works of Danielson et al. [77, 78] confirmed that strongly diluted cerium(III) solutions can also be employed to elute alkali metals and ammonium. The cerium(III) concentration required for an almost baseline-resolved separation of monovalent cations is only 10^{-5} mol/L. Figure 7-44 shows such a chromatogram, with a total analysis time of less than eight minutes. Lithium cannot be determined under these chromatographic conditions, because it elutes within the system void. The detection limits obtained for monovalent cations (3 µg/L for sodium and 100 µg/L for cesium) are comparable to the limits of other detection methods such as indirect photometric detection and conductivity detection. Again, in samples with complex matrices, the higher specificity of IFC is advantageous. This also applies to the indirect fluorescence detection of alkaline-earth metals that was described by Bächmann et al. [79]. Because a higher cerium(III) concentration is necessary to elute divalent cations, a significantly higher background fluorescence results that adversely affects baseline quality.

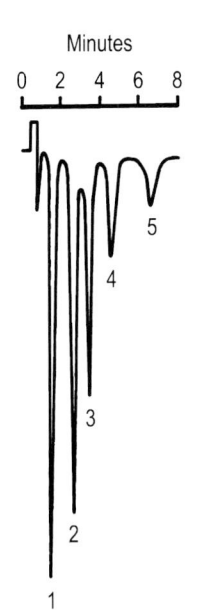

Fig. 7-44. Analysis of monovalent cations with indirect fluorescence detection. — Separator column: 100 mm × 3.2 mm i.d. ION-210; eluant: 10^{-5} mol/L cerium(III) sulfate; flow rate: 1 mL/min; detection: indirect fluorescence; injection volume: 20 µL; solute concentrations: 0.16 mg/L sodium (1), 0.15 mg/L ammonium (2), 0.21 mg/L potassium (3), 0.71 mg/L rubidium (4), and 1.2 mg/L cesium (5); (taken from [77]).

Indirect fluorescence detection also applies to anion analysis [80,81]. Mho and Yeung [80] utilized salicylate as fluorescent eluant ions. Owing to their structural similarities to phthalate and benzoate, these ions exhibit good elution properties. Salicylate shows a strong absorption at 325 nm and a fairly efficient emission at 420 nm. An HeCd laser with double-beam optics that emits polarized UV light with a wavelength of 325 nm at a power of about 7 mW was used as an excitation source. However, with regard to sensitivity, the method does not differ from the above-described cation analysis technique using cerium(III); i.e., the detection limits obtained for inorganic anions such as iodate and chloride correspond to those of indirect photometric detection and conductivity detection. This can be explained by the fact that the eluant ion concentration (sodium salicylate) of about $2.2 \cdot 10^{-4}$ mol/L necessary to elute solute ions is usually too high. The full potential of this analytical method will only be exploited when it becomes possible to manufacture separator columns with extremely low ion-exchange capacities that allow the use of very dilute eluants.

Nimura and Kinoshita [82] developed a derivatization technique for the analysis of fatty acids that is based on 9-anthryldiazomethane (ADAM). The latter is very stable in solution and reacts with fatty acids at room temperature without a catalyst to yield strongly fluorescent esters, which can be chromatographed on an ODS phase with acetonitrile/water mixtures. The derivatization of fatty acid is carried out with a 0.1% methanolic solution of 9-anthryldiazomethane, which is prepared via oxidation of 9-anthraldehyde-hydrazone following a procedure by Nakaya et al. [83]. The fluorescence- and excitation spectra of fatty acid-methylanthracene esters exhibit maxima at 412 nm and 365 nm, respectively. As an example, Fig. 7-45 shows the separation of ADAM derivatives — a mixture of saturated and unsaturated long-chain fatty acids — which can be detected down to the lowest pmol range.

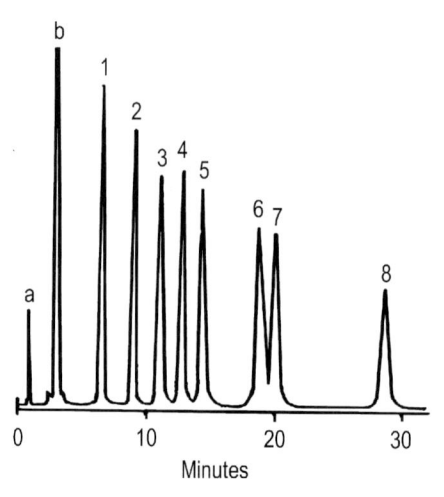

Fig. 7-45. Separation of ADAM derivatives of a mixture of long-chain fatty acids. — Separator column: 250 mm × 4 mm i.d. Lichrosorb RP8 (5-µm); eluent: acetonitrile-water (90:10 v/v); flow rate: 1.1 mL/min; detection: fluorescence after derivatization with ADAM; excitation wavelength: 365 nm; emission wavelength: 412 nm; injection volume: 10 µL; solute concentrations: (a) degradation product of the derivatization reagent, (b) not reacted derivatization reagent, 30 to 80 ng each of the derivatives (1) C10:0, (2) C12:0, (3) C18:3, (4) C14:0, (5) C18:2, (6) C16:0, (7) C18:1, and (8) C18:0 (fatty acids are characterized by two numbers: the first one is the number of C-atoms, the second one is the number of double bonds); (taken from [82]).

7.3
Other Detection Methods

Apart from the detection methods described thus far, refractive index detection is sometimes mentioned in the literature, although this detection method is very insensitive and, above all, unspecific. In contrast to other detection methods, refractive index detection is subject to much fewer restrictions. The refractive index of the eluant has no influence, because the measurement is usually carried out by comparing the refractive index of the column effluent with that of the pure eluant present in the reference cell. Thus, the sensitivity of this detection method solely depends on the detector performance. Commercially available detectors are usually designed for the detection of higher concentrations. For the separation of inorganic anions and their detection via changes in the refractive index, organic acids such as phthalic acid [69, 70, 84], salicylic acid [84], p-hydroxybenzoic acid [84], and o-sulfobenzoic acid [84] are utilized as eluants. Figure 7-46 illustrates such a chromatogram. The only advantage of refractive index detection is the ability to use comparatively concentrated eluants, which allow the use of high-capacity ion exchangers.

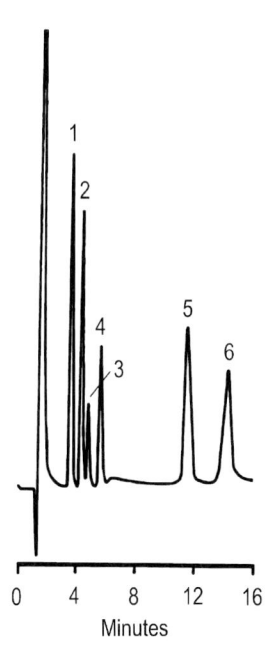

Fig. 7-46. Refractive index detection of various inorganic anions. — Separator column: 250 mm × 4.6 mm i.d. Vydac 302 IC; eluant: 6 mmol/L sodium hydrogenphthalate, pH 4.0; flow rate: 2 mL/min; injection volume: 100 µL; solute concentrations: 2.5 to 7.5 mg/L each of chloride (1), nitrite (2), bromide (3), nitrate (4), iodide (5), and sulfate (6); (taken from [85]).

The combination of an ion chromatographic separation with a radioactivity monitor (type LB505, Berthold, Wildbad, Germany) for the analysis of radio-strontium was described by Stadlbauer et al. [86]. The objective of their study was the development of a method for the simple separation of the fission products Sr-90 and Sr-89 from other radionuclides such as barium (Ba-133) which, like Cs-137, is a companion of Sr-90/Sr-89 and is formed with a similar yield (about 6%) in a nuclear reactor. Stadlbauer et al. separated these ions on a surface-sulfonated cation exchanger that was connected to a scintillation detector cell. Figure 7-47 shows the chromatogram of a Sr-90 standard that was obtained with this setup. The peak volume in this chromatogram corresponds to 76.4 Bq of Sr-90. Although online detection of radiostrontium is only feasible at medium to high activity concentrations, this method provides advantages for process control and self-monitoring of nuclear-medical- and nuclear power plants. The fractionated collection of the column effluent makes it possible to obtain Sr-90 in pure form from samples that have a high calcium excess or a complex radioactive contamination by various nuclides. It may be determined offline via a methane flow counter.

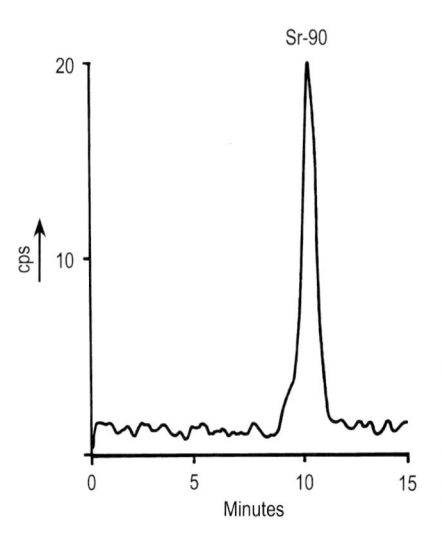

Fig. 7-47. Analysis of Sr-90 utilizing a radioactivity monitor. — Separator column: IonPac CS2; eluant: 30 mol/L HCl + 2 mmol/L histidine hydrochloride; flow rate: 1.5 mL/min; detection: scintillation measurement for Sr-90; injection volume: 50 µL; solute concentration: 76.4 Bq Sr-90 with inactive $SrCl_2$ as a carrier.

7.4 Hyphenated Techniques

In recent years, the coupling of ion chromatography with element-specific detection methods has increasingly gained importance. Element-specific detection is carried out with atomic spectrometric techniques including atomic absorption spectrometry (AAS), atomic emission spectrometry (ICP-OES) as well as the coupling between ICP and mass spectrometry (ICP-MS).

The coupling of an atomic absorption spectrometer with an ion chromatograph is relatively straightforward. It only requires a capillary that connects the separator column outlet with the nebulizer of the AAS instrument [87]. When choosing an eluant for analyte separation, it is important to prevent a high background signal and a quick sooting of the burner by the mobile phase. As early as 1980, Woolson and Aharonson [88] utilized this method for the determination of arsenic(III) and arsenic(V).

7.4.1
IC-ICP Coupling

Today, ICP-OES [89] is one of the most important methods for trace metal analysis; in combination with mass spectrometry [90] it is even used for ultra-trace metal analysis. The advantage of coupling those techniques with ion chromatography includes the ability to separate and detect metals with different oxidation states. The analytical interest in chemical speciation is based on the fact that the oxidation state of an element determines toxicity, environmental behavior, and biological effects. Chromium(VI), for example, is highly toxic even in very small amounts, whereas chromium(III) compounds are essential for lipid- and carbohydrate metabolism. On the other hand, heavy and transition metals often have to be determined in samples with very complex matrices such as body fluids, in which the concentration ratio between transition metals and alkali- and alkaline-earth metals is usually $1:10^6$ or even greater. Thus, considerable interferences in atomic spectrometric trace metal analysis may result. When modifying the sample to be analyzed by ion chromatographic techniques, the advantages of both methods are combined, so that many of these problems are solved. The performance of ICP-OES/ICP-MS is generally better than that of individual methods.

In ICP-OES, the analytical signal is generated via the atomic emission process. If a sample enters the inductively coupled plasma − a strongly ionized gas with temperatures between 6,000 K and 10,000 K − the solvent evaporates at first. Small particles are left behind, which also evaporate and partly dissociate into atoms and ions. During relaxation of these atoms and ions to a lower energy level, every element emits radiation of a characteristic wavelength, while the intensity of this radiation is proportional to the element concentration. ICP-OES spectrometers are offered in two versions. In a so-called simultaneous spectrometer, a holographic grating splits the polychromatic radiation coming from the plasma into distinct wavelengths or emission lines, which are detected by photomultiplier tubes. Spectral resolution is between 0.1 nm and 0.5 nm. Simultaneous spectrometers allow the simultaneous analysis of up to 60 elements in the mid to low µg/L range; reproducible signals are also achieved with transient samples. Sequential spectrometers are less suitable for coupling with ion chromatographs because they require a steady-state concentration of sample over a time period of 10 s, which is never the case in chromatography.

ICP-OES is not free of interferences. Physical interferences, for example, are caused by viscous samples or by samples with high amounts of dissolved solids such as sea water, brines, or urine, which clog the nebulizer and lead to flow variations. Although interferences of this kind can be minimized by diluting the sample, the sensitivity of the technique suffers depending on the dilution factor. Chemical interferences result from changes in the efficiency of evaporation, atomization and ionisation, which are caused by high concentrations of easily ionised elements such as sodium and potassium. Again, diluting the sample is the only solution to the problem, even in light of the negative consequences. Last but not least, there are spectral interferences in ICP-OES, which result from the overlapping of emission lines. Spectral interferences can be mathematically corrected by analyzing a known concentration of the interfering element and determining the ratio of the emission intensity for the element of interest. Even though it is possible to compensate for all these interferences, coupling with ion chromatography helps to minimize them and expands the application range of ICP spectrometers.

A characteristic example of the performance of IC/ICP-OES is the speciation of chromium [91, 92]. The common ion chromatographic method is based on the transformation of chromium(VI) into 1,5-diphenylcarbazone complexes following separation on an anion exchanger. Those complexes can be determined photometrically at 520 nm (see Section 7.2.1). For the separation of chromium(III)/chromium(VI) a mixed-bed ion exchanger is used as a stationary phase, on which chromium(III) is eluted as a cation and chromium(VI) as an anion. Detection is carried out via post-chromatographic oxidation of chromium(III) to chromium(VI) with peroxodisulfate and catalytic amounts of silver nitrate at elevated temperature (see Section 4.5.3). Chromium(VI) can then be detected directly by measuring its light absorption at 365 nm or, more specifically, by post-column derivatization with 1,5-DPC. Because element-specific detection with a simultaneous spectrometer only requires a few seconds, the time aspect has to be considered with respect to the preceding chromatographic separation. A very fast separation of chromium(III)/chromium(VI) is obtained on a 50 mm × 4 mm i.d. latexed anion exchanger, which is typically used as a guard column. However, its separation power is more than sufficient for separating these two species. Oxy metal anions are retained on this stationary phase and, thus, separated from metal cations, which elute as one peak in the system void. In this way it is possible to simultaneously determine metal cations and oxy metal anions. The separation of chromium(III)/chromium(VI) in Fig. 7-48 was obtained with 0.1 mol/L nitric acid as an eluant. Acidic eluants have proved to be especially suitable, because the concentration ratio between chromium(III) and chromium(VI) stays constant over several days when the samples to be analyzed are acidified. Thus, dilution is the only sample preparation required when using a nitric acid eluant. In addition, the solutes are present in nitric acid after their chromatographic separation, which is of great advantage for the subsequent atomic spectrometric detection. Stannate, vanadate, and molybdate can be separated and detected under similar chromatographic conditions [93].

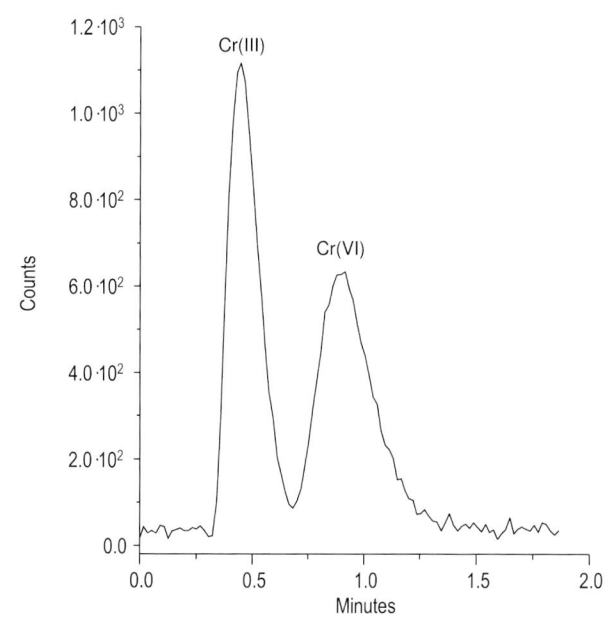

Fig. 7-48. Separation of chromium(III)/chromium(VI) with element-specific detection. — Separator column: 50 mm × 4 mm i.d. OmniPac PAX-100; eluant: 0.1 mol/L HNO_3, pH 2 with NH_4OH; flow rate: 2 mL/min; detection: ICP-OES (Spectroflame Modula, Spectro, Kleve, Germany); measuring wavelength: 267.716 nm; injection volume: 250 µL; solute concentrations: 0.1 mg/L each of chromium(III) and chromium(VI); (taken from [91]).

Quantitation is carried out based on peak heights or peak areas. The statistical data for this method are summarized in Table 7-5. As can be seen from this data, reproducibility of quantitation depends on the kind of data evaluation. The minimum detection limits which are calculated for both chromium species prove that this method is suitable for determining small amounts of chromium(III) and chromium(VI).

Significantly lower detection limits can be achieved with ICP-MS as a detection method. Early results obtained with this hyphenated technique were based on a concentric nebulization, which led to relatively broad peaks due to band broadening inside the nebulizer [94]. Powell et al. [93] applied the method of direct injection (DIN, **D**irect **I**njection **N**ebulization) for chromium speciation in environmental samples. Using this method, band broadening of the steady-state signals is reduced to a minimum due to the absence of a nebulizer [95]. The sample is introduced via an injection valve that is connected to a 10-µL injection loop. Via a switching valve, the sample is directed into the plasma either through a micro column for chromium(III)/chromium(VI) separation, or through a by-pass loop for determining total chromium. The plasma chamber and switching valve are connected with a 1.5 mm i.d. capillary. Powell et al. used a Sciex Elan

Table 7-5. Statistical data for the analysis of chromium(III)/chromium(VI) with element-specific detection.

	Cr(III)		Cr(VI)	
	Peak height	Peak area	Peak height	Peak area
Correlation coefficient of the linear calibration $c = 0.02 - 1$ mg/L	>0.9999	>0.99999	>0.999	>0.99999
Reproducibility [%] ($n = 10$; $c = 0.1$ mg/L)	1.2	0.7	1.6	0.7
Rel. standard deviation of retention times [%]		<0.1		<0.1
Rel. standard deviation of the linear calibration function [%]	0.32	0.26	0.85	0.4
Minimum detection limit [mg/L]*)	0.015	0.012	0.047	0.035

*) Determination was carried out following DIN 32645.

5000 ICP-MS spectrometer with a resolution of 0.8 Da. The sample introduction system of this instrument was replaced by the DIN system. Chromium determination was carried out via m/z 52, because Cr^{52} produces larger signals than Cr^{53}. Quantitative results are obtained by peak area evaluation with an algorithm in the instrument software. With an optimized system, minimum detection limits for total chromium, chromium(III), and chromium(VI) are 30, 60, and 180 ng/L, respectively. In comparison to the photometric method after derivatization with 1,5-DPC, those values are lower by one order of magnitude. The only spectral interference for Cr^{52} in aqueous solution is $Ar^{36}O^{16}$, which is responsible for a slightly increased background signal from a blank solution. In addition, Roehl et al. [96] reported matrix interferences, if the sample to be analyzed contains large amounts of carbon, sulfur, and chloride. In this case, Cr^{52} is interfered by $Ar^{40}C^{12}$ and $S^{36}O^{16}$, Cr^{53} by $Cl^{37}O^{16}$. Powell et al. successfully demonstrated the power of IC/ICP-MS for chromium speciation in industrial wastewaters.

Another important example for the benefit of IC/ICP-OES or IC/ICP-MS is the speciation of arsenic. For this purpose, Urasa et al. [97] coupled an ion chromatograph to a DCP atomic emission spectrometer[2] by connecting the separator column and nebulizer with a capillary that was slightly modified [98]. A general problem of DC plasma detection is the relatively poor sensitivity, which is attributed to both the inadequate nebulization efficiency and the dilution of the sample during the chromatographic process. In comparison with direct DCP-AES, the dilution effect alone accounts for a reduction in signal intensity by 50% to 75%. Urasa et al. compensated for this by injecting large sample volumes

2) DCP: Direct Current Plasma

(up to 1000 µL). Alternatively, they investigated the applicability of concentrator columns for improving the sensitivity. It turned out that 10 µg/L arsenic(V) may be correctly determined by concentrating a sample volume of 100 mL. This represents a significant improvement over direct DCP-AES, which is less sensitive by two orders of magnitude.

Heitkemper et al. [99] applied element-specific detection for the simultaneous determination of arsenic(III)/arsenic(V) in food additives. The separation shown in Fig. 7-49 was obtained on a latexed anion exchanger with a NaOH eluant.

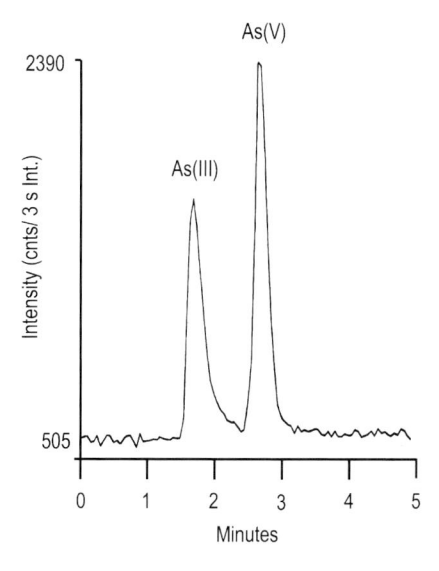

Fig. 7-49. Separation of arsenic(III)/arsenic(V) with element-specific detection. — Separator column: IonPac AS4A (-SC); eluant: 40 mmol/L NaOH; flow rate: 1 mL/min; suppressor: AMMS-II; regenerant: 12.5 mmol/L H_2SO_4; detection: ICP-OES; measuring wavelength: 193.7 nm; injection volume: 100 µL; solute concentrations: 5 mg/L each of arsenic(III) and arsenic(V); (taken from [91]).

Because the human body metabolizes inorganic arsenic by reductive methylation, all the reaction products have to be analyzed. ICP-MS is the only detection method that can achieve the required minimum detection limits for arsenic (low µg/L range) in body fluids. However, in body fluids such as urine with its high chloride content, interferences are observed due to the formation of Ar^+Cl^-, which has a mass-to-charge ratio (m/z 75) equal to that of arsenic. When coupling the ICP-MS with a high-capacity anion exchanger, this interference can be eliminated, because chloride is well separated from the arsenic compounds of interest. Figure 7-50 shows the analysis of a human urine sample (chromatogram A) containing a relatively high concentration of 400 µg/L arsenate, which the patient incorporated orally by drinking contaminated well water. Despite the high chloride concentration, neither overloading effects nor other interferences are observed. The various arsenic compounds can be separated to baseline on IonPac AS7 with a $NaOH/Na_2CO_3$ gradient. Arsenite cannot be detected under these conditions, because it is oxidized to arsenate in the mobile phase at high pH. Chromatogram B in Fig. 7-50 reveals that after one week the amount of excreted arsenic has decreased to 1/50 of the original value. The minimum detection limits for methylarsonic acid, dimethylarsinic acid, and arsenobetain reported in literature are between 2 µg/L and 5 µg/L (based on an injection volume of 20 µL) [100].

$$H_3C-\underset{\underset{OH}{|}}{\overset{\overset{O}{||}}{As}}-OH \qquad H_3C-\underset{\underset{OH}{|}}{\overset{\overset{O}{||}}{As}}-CH_3 \qquad H_3C-\underset{\underset{CH_3}{|}}{\overset{\overset{CH_3}{|}}{\overset{+}{As}}}-CH_2COO^-$$

Methylarsonic acid　　Dimethylarsinic acid　　　Arsenobetain

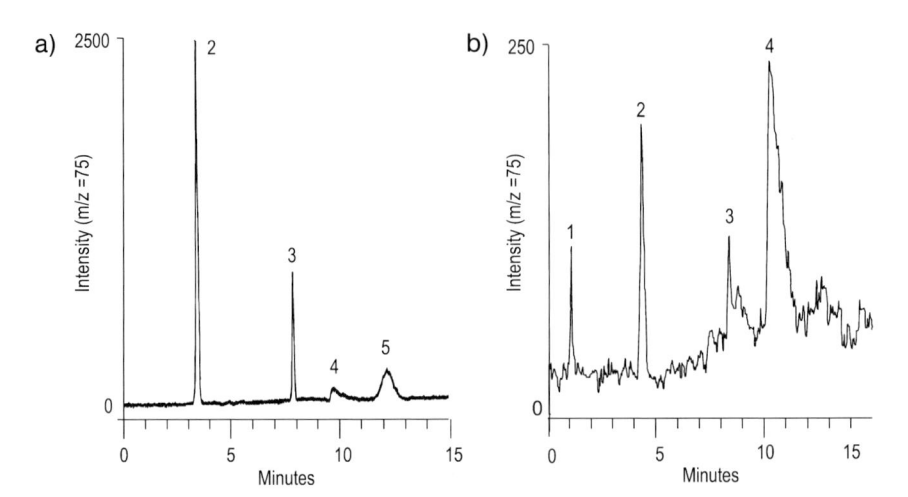

Fig. 7-50. Simultaneous determination of various arsenic species in human urine with element-specific detection. — Separator column: IonPac AS7; eluant: NaOH/Na$_2$CO$_3$ gradient; flow rate: 1 mL/min; injection volume: 100 µL; suppressor: AMMS-II; regenerant: 12.5 mmol/L H$_2$SO$_4$; detection: ICP-OES; measuring wavelength: 193.7 nm; analytes: (1) arsenobetain, (2) methylarsonic acid, (3) dimethylarsinic acid, (4) chloride, and (5) arsenate; (A) after oral incorporation of a contaminated well water, (B) after one week; (taken from [91]).

In this conjunction, the integration of a hydride generator in front of the sample introduction system of the ICP-MS described by Roehl et al. [93] and Schlegel et al. [101] should also be mentioned. It increases sensitivity for various arsenic- and selenium compounds by a factor of 2 to 5. It is true that in hydride formation with the subsequent separation of gas and liquid, temperature has to be strictly controlled for reproducibility reason, but detection is much more specific because molecular interferences do not occur. Mattusch and Wennrich [102] also separated anionic, neutral, and cationic arsenic compounds such as arsenite, dimethylarsenic acid, arsenobetain, and arsenocholine on IonPac AS7, but used a very steep nitric acid gradient. Arsenobetain and arsenocholine are retained, above all, by hydrophobic interactions with the stationary phase.

IC-ICP can be applied for the speciation of a number of other compounds. An interesting application of DC plasma detection is the method developed by Biggs et al. [103] for the determination of polyphosphates, which were separated on a neutral PS/DVB-based polymer phase by means of ion-pair chromatography. The separation of the phosphate oligomers P_1 to P_{12} in neutralized polyphosphoric acid was carried out with tetraethylammonium nitrate as an ion-pair reagent and by applying a potassium nitrate gradient. In contrast to expectations, the rise of the salt concentration during the gradient run did not cause a baseline drift nor did it affect the response factor. A representative chromatogram of a polyphosphoric acid sample neutralized with tetraethylammonium hydroxide is shown in Fig. 7-51. The detection limit of this method is 0.2 µg P.

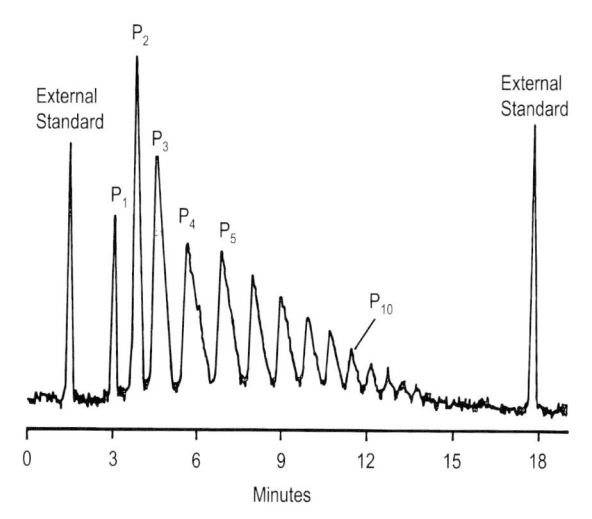

Fig. 7-51. Analysis of polyphosphoric acid upon application of DC plasma detection. — Separator column: 150 mm × 4.1 mm i.d. PRP-1; eluant: (A) 10 mmol/L tetraethylammonium nitrate (pH 9), (B) 10 mmol/L tetraethylammonium nitrate + 0.1 mol/L KNO_3 (pH 9); gradient: linear, 1% B to 40% B in 18 min; flow rate: 1.2 mL/min; detection: DC plasma; emission wavelength: 214.9 nm; injection: 20 µL of a solution with ca. 60 µg P; external standard: 8.24 µg P as orthophosphate; (taken from [103]).

7.4.2
IC-MS Coupling

The second important hyphenation is the coupling of an ion chromatograph to a mass spectrometer; this combination provides the analyst with information on analyte structure and molecular weight. While mass-selective detection in gas chromatography became routine years ago, the coupling of a liquid chromatograph to a mass spectrometer is problematic, because the relatively large amounts of liquid mobile phase are not compatible with the high vacuum in the ion source of a mass spectrometer. To solve these incompatibility problems, various types of LC-MS interfaces have been developed, which are briefly introduced in the following discussion. The most important characteristic of a LC-MS interface is the transfer of analyte molecules from the separator column into the high vacuum of a mass spectrometer. This means that an analyte molecule dissolved in the mobile phase has to make the transition to an analyte molecule isolated in the gas phase. To achieve high sensitivity, as many analyte molecules as possible have to be transferred, while the mobile phase constituents have to be largely removed. The maximum amount of liquid that can be introduced into the high vacuum of a mass spectrometer without increasing pressure in the ion source depends solely on the capacity of the vacuum pumps. Because the feed rate of modern turbomolecular pumps is between 0.3 and 0.7 m^3/s, the intake of water is limited to 1.8 µL/min. Conventional separator columns in ion chromatography are operated at 1 mL/min, so only 0.2 % of that amount of liquid can be directly introduced into the vacuum system. These numbers impressively demonstrate the dimension of this incompatibility problem. It can be solved by removing a large amount of solvent before the analytes enter the high vacuum. Split techniques as well as microbore columns with a small inner diameter of 2 mm and low flow rates of 0.25 mL/min also help to significantly decrease the amount of liquid. Combinations of all these measures can also be applied.

An additional problem are the electrolytes that are used for separating ionic species on ion exchangers. Such separations cannot be easily performed without electrolytes. Therefore, volatile eluants such as ammonium sulfate have to be used, which can be pumped out of the ion source. A second solution to this problem are continuously regenerated suppressor systems (see Section 3.6 and 4.3), with which eluants can be desalted prior to entering the MS interface. It is true that the use of a suppressor system limits the choice of eluant [104]. Meanwhile, however, so many ion exchangers for anion- and cation chromatography are available that anions and cations can be eluted efficiently and with high resolution using NaOH or mineral acids eluants. Because these ion exchangers are usually also solvent compatible, even surface-active anions can be eluted by adding organic solvents to the mobile phase. As early as 1990, Simson et al. [105] successfully employed a micromembrane suppressor for desalting eluants. They placed it between the anion exchanger and the thermospray interface and determined a residual sodium concentration of 13-30 µmol/L when pumping an eluant concentration of 10–100 mmol/L. Conboy et al. [106] also utilized suppressors for their investigations on quaternary ammonium compounds and organic sulfonates and sulfates, placing them between separator column and interface.

Particle-Beam Interface
More than 25 years of research in the field of LC-MS has resulted in the development of a number of different interfaces, some of which are commercially available today. Widely distributed interfaces include particle-beam-, thermospray- and APCI interfaces (APCI, Atmospheric-Pressure Chemical Ionization), which are briefly described here together with some characteristic applications. In a particle-beam interface [107, 108], the column effluent is fed either pneumatically or by thermospray nebulization into a desolvation chamber. The chamber is close to atmospheric pressure and connected to a jet separator, which separates the high molecular weight analytes from the low molecular weight eluant molecules. The analyte molecules are then transferred as small particles at almost the speed of sound into a conventional EI/CI ion source, where they disintegrate in evaporative collisions with a heated target. The released molecules are ionized by electron impact (EI) or chemical ionization (CI). Review articles on this subject were published by Creaser and Stygall [109] and Capiello [110]. The schematic diagram of a particle-beam interface (HP 59980A) is illustrated in Fig. 7-52.

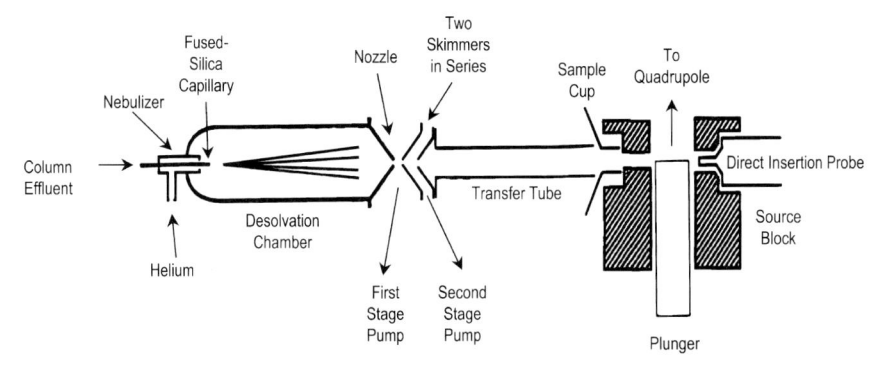

Fig. 7-52. Schematic diagram of a particle-beam interface (HP 59980A).

Most commercial systems utilize a pneumatic, concentric nebulizer that typically consists of a 100-µm fused-silica capillary, through which the column effluent is directed. A helium flow (1-3 L/min) circumvents the end of this capillary. After nebulization, the solvent is evaporated from the droplets in the externally heated desolvation chamber. Temperature is generally kept at about 50-70 °C. As a result of the high pumping efficiency in the jet separator, the pressure in the desolvation chamber generally is sub-ambient, i.e. between 20 and 30 kPa. The design of the jet separator is based on the development by Winkler et al. [111] (Fig. 7-53). By means of two mechanical pumps, the pressure of 25 kPa in the desolvation chamber is reduced to 10 kPa in the first pumping region, to 30 Pa in the second pumping region, and to $2 \cdot 10^{-3}$ Pa in the ion source. The submicron particles formed during the desolvation process strike the walls of the ion source, which is kept at about 250 °C, and evaporate. The released molecules are ionized by EI or CI. Other ionization techniques such as FAB (Fast Atom Bombardment) [112] and laser desorption ionization [113] in combination with a particle-beam interface are also described in the literature. A special feature of the setup illustrated in Fig. 7-52 is the two sample introduction systems (particle-beam interface and direct insertion probe) at two opposite sides of the ion source. Upon switching from EI to CI operation mode, a "Plunger assembly" in the ion source is advanced towards the quadrupole mass analyzer, changing the effective ionization chamber from an open configuration to a more enclosed one. LC-MS investigations with a particle-beam interface are usually carried out with quadrupole mass spectrometers. In addition, Finnigan MAT and Micromass also offer commercial solutions based on sectorfield instruments.

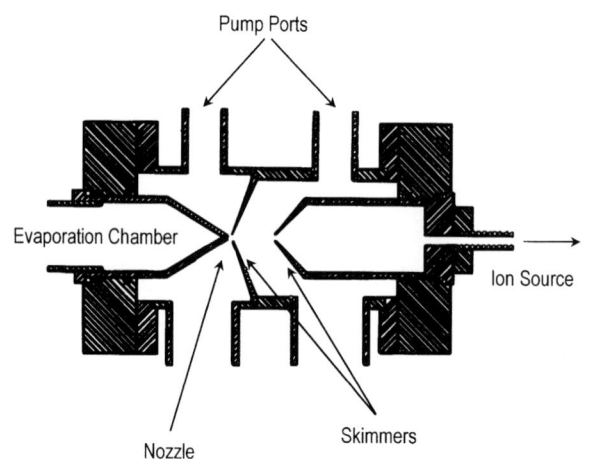

Fig. 7-53. Schematic diagram of a jet separator according to Winkler et al. [111].

The most important characteristic of a particle-beam interface is the possibility to obtain EI spectra, which can be evaluated according to known rules and compared to those of spectra libraries. For obtaining a meaningful spectrum of a substance, an absolute amount of 10 to 100 ng is required. In 90 % of all applications EI ionization is applied; in 20 % of the applications EI ionization is combined with chemical ionization in the positive ion mode. All other CI applications are carried out in the negative ion mode.

A characteristic example for a successful coupling of an ion chromatograph to a mass spectrometer via a particle-beam interface is described by Hsu [114] who analyzed aromatic sulfonic acids in leachates. Hsu used an OmniPac PAX-500 multimode phase in the microbore format to separate the sulfonic acids. The compounds of interest were eluted with a mixture of NaOH and acetonitrile under gradient conditions. This eluant mixture is very well suited for the desalting process via a membrane-based suppressor system. To minimize extra-column band broadening, the suppressor was also used in the microbore format. Prior to assembling the suppressor, the membranes were cleaned with organic solvents to remove potential organic contaminants resulting from the manufacturing process. Membrane-based suppressor systems exhibit a limited pressure stability of 0.7 MPa, which is a disadvantage. However, the pressure stability is sufficient enough to withstand the backpressure of a particle-beam interface. As an example, Fig. 7-54 shows the chromatogram of a leachate extract utilizing UV detection at 254 nm (A) and PBMS in the EI mode (B). Sample preparation was carried out following the method by Kim [115]. In the first step, the leachate is lyophilized to remove water and volatile organic compounds. The residuum is dissolved in methanol. Then acetone is slowly added to this solution to precipitate inorganic salts. The supernatant solution is then filtrated and concentrated with a rotary evaporator. The methanolic extract can be injected directly. As can be revealed from Fig. 7-54, three significant peaks could be detected in such an extract. In good agreement with an earlier publication [115], the mass spectra of these compounds are identical and correspond to p-chlorobenzenesulfonic acid (PCBSA) and its o- and m-isomers.

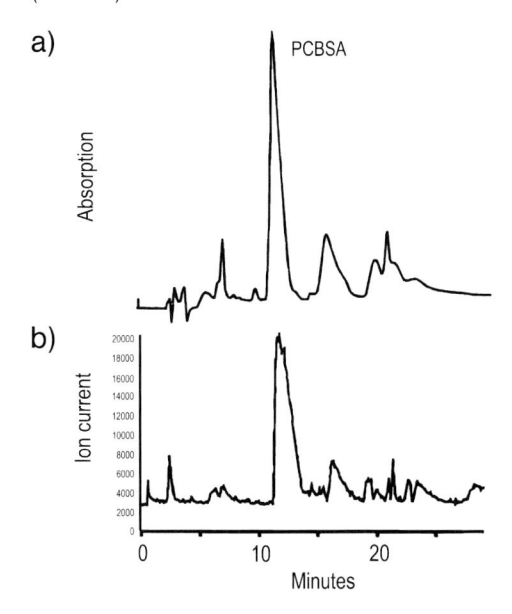

Fig. 7-54. Separation of a leachate extract on a multimode phase utilizing UV detection (A) and particle-beam mass spectrometry (B) in the EI mode. — Separator column: OmniPac PAX-500 (250 mm × 2 mm i.d.); flow rate: 0.25 mL/min; eluant: NaOH — acetonitrile; gradient: linear, 9 mmol/L NaOH—ACN (90:10 v/v) to 50 mmol/L NaOH—ACN (80:20 v/v) in 8 min, then to 80 mmol/L NaOH—ACN (73:27 v/v) in 5 min, then to 120 mmol/L NaOH—ACN (60:40 v/v) in 5 min; detection: (a) UV (254 nm), (b) PBMS, EI mode; injection volume: 5 µL; (taken from [114]).

The corresponding mass spectrum of the main peak is shown in Fig. 7-55 (top) and compared to that of a respective standard (bottom). In both cases, the background signal has been subtracted. Smaller signals such as m/z 104 and m/z 186 correspond to p-xylenesulfonic acid, m/z 94 and m/z 158 to benzenesulfonic acid, which are both present as impurities. Although the application of a particle-beam interface is limited to analytes with a significant vapour pressure in the

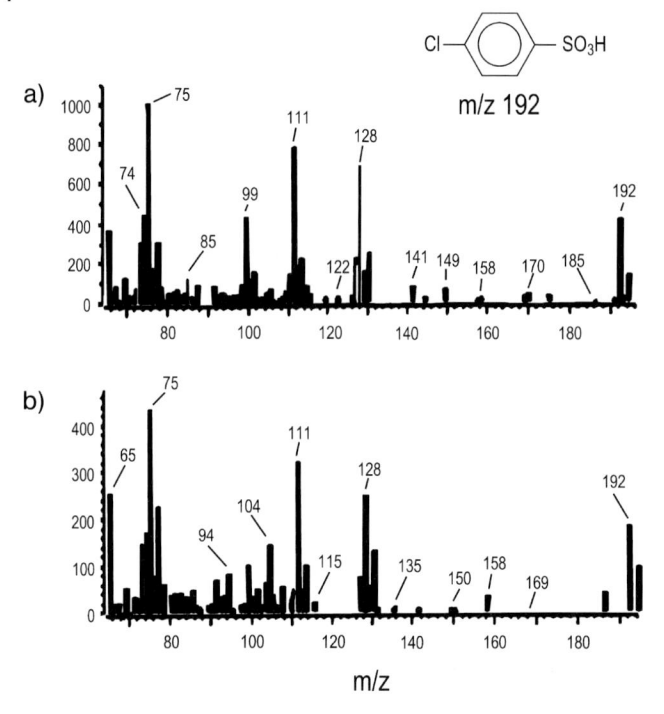

Fig. 7-55. Comparison of the mass spectra of p-chlorobenzenesulfonic acid in a standard (a) and a leachate extract (b) utilizing PBMS in the EI mode (taken from [114]).

ion source, it is still significant in relation to other LC-MS interfaces. The major reason for that is the ability to generate EI mass spectra online. Another limitation in the applicability of PBMS is the fact that it can only be applied to molecules < 1000 Da. The lack of sensitivity in the lower ng range can be compensated, for example, by pre-concentration techniques [116], which are essential for a more widespread distribution in environmental analysis.

Thermospray Interface

In the technique known as "thermospray" [108, 117, 118] volatile electrolytes dissolved in the eluant are nebulized out of a heated vaporizer tube. As a result of heating, the liquid is nebulized and partially vaporized. The capillary is heated under controlled conditions to avoid complete evaporation of the liquid in the capillary. Unvaporized solvent and sample are carried into the ion source as micro droplets or particles in a supersonic jet of vapour. By applying efficient pumping directly at the ion source, up to 2 mL/min of aqueous solvents can be introduced into the MS vacuum system. The ionization of volatile analytes takes place by means of ion-molecule reactions in the gas phase; non-volatile samples may be ionized by direct ion evaporation processes from highly charged droplets or particles. The CI reagent gas can be made either in a conventional way using energetic electrons from a filament or discharge electrode, or in a process called thermospray ionization, where the volatile buffer dissolved in the eluant is involved. Figure 7-56 shows the schematic diagram of a thermospray interface with an electrically heated capillary, in which the column effluent is nebulized and partially evaporated, forming an ultrasonic jet of vapour containing a mist of fine droplets or particles. As the droplets travel at high velocity through the heated ion source, they continue to vaporize due to the rapid heat input from the surrounding hot vapour. A portion of the vapour and ions produced in the ion source escapes into the vacuum system of the mass spectrometer through the entrance cone, and the remainder of the excessive vapour is pumped away by a mechanical vacuum pump. The heart of the thermospray interface is the electrically heated stainless steel tube that has an internal diameter of 0.1 to 0.15 mm. The vapour produced by vaporizing 1 mL/min of solvent is about one hundred times the amount that can be accommodated by a typical mass spectrometer equipped for chemical ionization. The thermospray ion source allows high mass flows to be vaporized by using a very tight ion chamber similar to that used in chemical ionization but with a mechanical vacuum pump (about 300 L/min) attached directly to the source chamber through a port opposite the vaporizer. The source block is strongly heated; temperature is monitored with a sensor. The vaporizer tip is surrounded by a separately heated block that allows the region cooled by the rapid adiabatic expansion to be properly heated without overheating the portions of the ion source further downstream. Thoriated iridium filaments are used instead of tungsten or rhenium filaments to achieve a satisfactory operating life under conditions of high pressures of water and solvent vapours. But even with thoriated iridium filaments, it is difficult to maintain satisfactory emission at high flows of water; the discharge electrode overcomes that problem.

A characteristic feature of thermospray ionization is the presence of ammonium acetate or another volatile electrolyte that contributes to the ionization process. For a predominant number of compounds the mass spectra obtained with thermospray can be interpreted based on the knowledge about ion-molecule reactions observed in the CI mode. The major differences are that the pressure in a thermospray source may be somewhat higher (typically 2-20 Torr) compared with typical CI and the gas throughput is very much higher. In conventional CI, the gas flows involved are typically in the range of 1-20 mL/min, while in thermospray, the vapour flow is in the range of 0.5−2 L/min. In a conventional CI source, ionization is initiated by high-energy electrons. If water vapour is used as the CI reagent gas, the initial process is:

$$H_2O + e^- \rightarrow H_2O^+ + 2e^- \tag{212}$$

The water molecular ions react on virtually every collision to produce the hydronium ion by the proton transfer reaction:

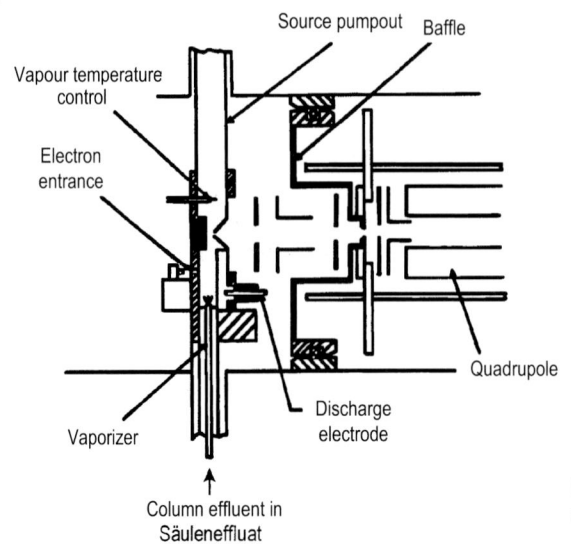

Fig. 7-56. Schematic design of a thermospray interface.

$$H_2O^+ + H_2O \rightarrow H_3O^+ + OH \tag{213}$$

The hydronium ion does not react further with water except to form clusters by the reaction:

$$H_3O^+ + 2H_2O \rightarrow H_3O^+(H_2O) + H_2O \tag{214}$$

Clustering may continue by a similar set of reactions:

$$H_3O^+(H_2O)_n + 2H_2O \rightarrow H_3O^+(H_2O)_{n+1} + H_2O \tag{215}$$

with $n = 0, 1, 2, ...$

The distribution of hydrated hydronium ions observed in the CI mass spectrum is determined by the equilibrium constants for the reactions described with Eq. (215) and the temperature and pressure in the ion source. Hydronium ions do not react further with water vapour, but if samples with higher proton affinity than water are present, they will react according to proton transfer reactions such as:

$$H_3O^+ + M \rightarrow MH^+ + H_2O \tag{216}$$

Polar samples of lower proton affinity may react by displacing a water molecule from a cluster, for example:

$$H_3O^+(H_2O) + M \rightarrow H_3O^+(M) + H_2O \tag{217}$$

Samples that are both less basic and less polar than water will not react and consequently will not be detected in the positive-ion CI spectrum when water is the reagent. One solution to this kind of problem is to employ negative ions. In the negative-ion mode, hydroxide ions can be produced that react with molecules more acidic than water vapour to form the deprotonated anion. If an electron beam or discharge is used to produce the initiate ionization, an excess of thermal electrons is also formed in the high pressure source. These may attach to molecules of high electron affinity to form M^- ions. Thus, both $(M-H)^-$- and M^- ions may be observed in the CI spectrum, depending on the relative acidity and electron affinity of the sample. If the sample has low electron affinity and is less acidic and less polar than water vapour, it will not be detected in the negative-ion mode.

In most respects, the thermospray ion source behaves as a conventional CI source. The only major difference is the relatively high gas flow involved in thermospray at conventional liquid chromatography flow rates. When an electron-emitting filament or discharge is used to initiate ionization, the processes occurring in the thermospray ion source appear to be identical to those occurring in a conventional CI source. In chemical ionization, fragment ions may be observed in addition to the usual protonated and deprotonated molecular ions. The only fragments observed are those that are unreactive with the gas in the ion source. The extent of fragmentation is determined by the internal energy of the ions at formation. More acidic reagents and higher temperatures tend to induce more fragmentation in the positive-ion mode, while less acidic reagents tend to produce more fragmentation in the negative-ion mode. In addition, a thermally labile molecule may fragment prior to ionization, with the fragments then being ionized, for example, by proton transfer. It is often difficult to determine whether fragmentation or ionization occurred first under CI conditions because the products are very often the same.

HPLC methods based on standard bore columns can be combined with thermospray without any modifications. Thermospray works best at LC flow rates between 0.5 and 1.5 mL/min. Flow rates outside that range are also possible but compromise sensitivity. Volatile salts and compounds such as ammonium acetate, ammonium formate, ammonium alkylsulfonates, trifluoroacetic acid, and tetrabutylammonium hydroxide are suitable as mobile phases. Phosphates and alkali salts in the mobile phase should be avoided in any case.

Biochemistry has many uses for LC-MS. While biochemical macromolecules such as nucleosides and nucleotides [119], peptides and proteins [120–122] are already investigated by LC-MS quite extensively, very little has been published about sugar oligomers. This might be due to the incompatibility of NaOH or NaOH/NaOAc mobile phases used for separating carbohydrates by anion exchange chromatography with online mass-selective detection. The publication by Simpson et al. [105] cited above can be regarded as a breakthrough on this issue. Simpson et al. managed to desalt NaOH/NaOAc eluants with a membrane suppressor prior to entering the thermospray interface. Simpson et al. investigated mono- and di-amino sugars for the first time in this way. However, the small number of protonated dimeric molecular ions in the corresponding mass spectra indicates thermal degradation in the thermospray interface. Niessen et al. [123]

attributed this thermal degradation of sugar oligomers to the presence of ammonium acetate in the eluant; this causes a thermally induced ammonolysis of oligosaccharides into their monomers. If sodium acetate is used instead of ammonium acetate, intact sodium-containing molecular ions are obtained for maltodextrins in the positive-ion mode after desalting with two membrane suppressors in series. Niessen et al. [123, 124] applied this system to the analysis of oligosaccharides after enzymatic degradation of plant cell wall polysaccharides using a Finnigan MAT TSQ-70 tandem mass spectrometer. Because the membrane suppressor used for desalting has a limited pressure stability of about 0.7 MPa, and because the thermospray interface requires a solvent pressure of about 4 MPa, a so-called booster pump was installed between the suppressor and the thermospray interface. The booster pump solves the pressure incompatibility by increasing the flow rate to the more favourable 2 mL/min. In the manifold, 1 mL/min of a dilute aqueous sodium acetate solution was added to achieve ionization. The ion-exchange capacity of two membrane suppressors in series is sufficient to exchange sodium ions up to a maximum concentration of $c = 0.4$ mol/L for hydronium ions. This represents a kind of limitation as the elution of higher oligosaccharides requires sodium acetate concentrations up to 0.5 mol/L. Thus, when rinsing the separator column with higher sodium acetate concentrations, the separator column and interface have to be disconnected.

In gradient elution with sodium acetate, a background signal is observed that can be attributed to the formation of sodium acetate cluster ions. They are distributed throughout the spectrum (m/z = 187, 269, 351 and subsequent values at 82 amu increments), corresponding to the general formula $[(NaOOCCH_3)_n + Na]^+$. In general, some contamination of the ion source through the presence of too high sodium acetate concentrations in the mobile phase was not found to be detrimental to sensitivity. During operation, the ion source was kept at 350 °C, which is assumed to reduce the contamination by sodium acetate.

As a result of the presence of sodium acetate cluster ions, in general no peaks are observed in the total-ion-current chromatograms. This means that the actual analyte signals must be searched for, using the known and/or expected m/z values for the oligosaccharides. Because fragmentation does not occur, the oligosaccharide of interest can be identified by a molecular-mass-related peak in the mass chromatogram. The mass spectra of oligosaccharides obtained under these conditions contain strong peaks due to the sodiated molecule, i.e. $[M_r + Na]^+$ at m/z = M_r + 23. As an example, Fig. 7-57(A) shows the mass spectrum of sucrose. Especially with the smaller oligomers, a signal is also observed at m/z = M_r + 45, though in lower abundance, which can be attributed to $[M_r - H + 2Na]^+$. This peak is assumed to result from the competition between hydronium- and sodium ions in the neutralization of the sugar anions in the suppressor. For the oligosaccharides with higher DP values (DP > 4), doubly charged disodium ions are observed, i.e. $[M_r + 2Na]^{2+}$ at m/z = $(M_r + 46)/2$. Triply charged ions are not observed. Uronic acids were also analyzed by Niessen et al. Their mass spectra contain signals such as $[M_r + Na]^+$ at m/z = M_r + 23, $[M_r - H + 2Na]^+$ at m/z = M_r + 45 as well as a sodiated fragment peak caused by the loss of water from the acid, $[M_r + Na - H_2O]$ at m/z = M_r + 5. In the mass spectrum of digalacturonic acid (M_r = 370), shown in Fig. 7-57(B), these peaks are found at m/z = 393, 415, and 375; a peak due to the sodiated disodium salt at m/z = 437 is also observed.

The expected m/z ratios for sodiated sugar oligomers can be calculated using the values in Table 7-6. For example, an oligosaccharide with the composition $(Glc)_4(Xyl)_2(Fuc)$ is expected to give a peak at m/z = 203 + 3 × 162 + 2 × 132 + 146 = 1099, while the expected doubly charged ion is detected at m/z = (1099 + 23)/2 = 561.

LC/MS data for α-1,4-glucans up to DP10 based on a reversed-phase separation using a dilute aqueous sodium acetate mobile phase ($c = 10^{-4}$ mol/L) were also reported by Niessen et al. [125]. With HPAEC-MS α-1,4-glucans could be detected up to DP6. For higher oligomers, high sodium acetate concentrations are required that can no longer be sufficiently removed by the membrane suppressor. In the HPAEC-MS analysis of an arabinan digest, α-1,5-arabinose oligomers up to DP9 could be detected, while oligomers up to DP15 were readily detectable by pulsed amperometric detection. The total-ion-current chromatogram of an MID run of the arabinan digest is shown in Fig. 7-58. As with other oligosaccharide samples, a dramatic decrease in response is observed at higher DP values. However, the peaks of interest are readily detected by mass spectrometry. Other homologous series of oligosaccharides such as β-1,4-galactans (up to DP6) and β-1,4-xylans (up to DP15) could also be detected by HPAEC-MS.

With the introduction of CarboPac PA-100 (see Section 3.10), HPAEC-MS analysis of oligosaccharides could be improved significantly. CarboPac PA-100 allows the elution of oligosaccharides at a relatively low sodium acetate concentration, so that HPAEC-MS is applicable for sugar oligomers with a higher degree of polymerization. With α-1,4-glucans, for example, oligomers up to DP6 can be detected with CarboPac PA1, while oligomers up to DP12 are detected when separated on CarboPac PA-100. Thus, the limiting factor for the detection of higher DP values is the concentration of these oligomers in the sample to be analyzed. In such samples, oligomers with DP > 10 often contribute less than 1% in the weight distribution. In practice, less than 0.2 nmol of DP10 is injected.

The application of the negative-ion mode is also promising, because better sensitivities are obtained for higher oligosaccharides. Niessen et al. [126] tested the negative-ion mode for α-1,4-glucose- and arabinogalactose oligomers. The mass spectra obtained under these conditions are surprisingly complex. For smaller α-1,4-glucose oligomers up to DP4, primarily three ions are detected: $[M - H]^-$ at m/z = M_r - 1, $[M + OAc]^-$ at m/z = M_r + 59, and $[M + HSO_4]^-$ at m/z = M_r + 97. At DP < 4 the acetate- and hydrogensulfate adducts are most abundant, whereas for DP5 to DP7 the deprotonated molecule is most abundant, and at DP > 7 the doubly charged ions are most abundant. The most abundant

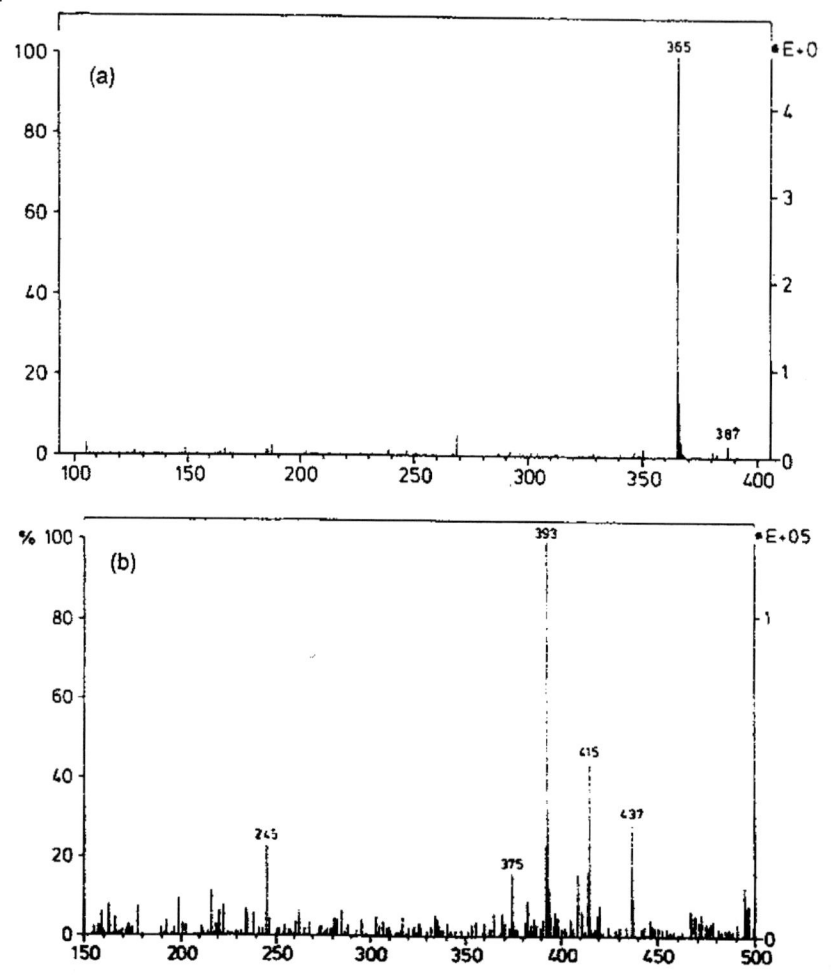

Fig. 7-57. Mass spectra of (A) sucrose and (B) digalacturonic acid with HPAEC-MS. — Separator column: CarboPac PA1; eluant: 0.1 mol/L NaOH — NaOAc; gradient: linear, 0.1 mol/L NaOAc to 0.3 mol/L NaOAc in 20 min; flow rate: 1 mL/min; detection: thermospray MS (Finnigan MAT TSQ-70) in the positive-ion mode and MID mode, scan range m/z = 150 to 1500 in 3 s; injection volume: 25 µL; (taken from [123]).

doubly charged peak is due to $[M + SO_4]^{2-}$, and doubly deprotonated ions and acetate- and hydrogensulfate adducts are also observed. In general, the formation of adduct ions is unfavourable to sensitivity, because the ion intensity is spread over a number of peaks instead of being concentrated in only one peak. However, when summing the peak areas of the most important peaks, it appears that a much better response is achieved in the negative-ion mode than in the positive-ion mode for most DP values, especially for the higher values. Negative-ion

Table 7-6. m/z values and m/z increments for sodium-containing oligosaccharides.

Sugar	m/z	Increment (1+)	Increment (2+)	Example
Aldopentose	173	132	66	Ribose, xylose, arabinose
Deoxyaldohexose	187	146	73	Rhamnose, fucose
Aldohexose	203	162	81	Glucose, mannose, galactose
Ketohexose	203	162	81	Fructose
Hexuronic acids	217, 239	176	88	Glucuronic acid, galacturonic acid
4-O-Methyl-hexuronic acids	231, 253	190	95	4-O-Methylglucuronic acid

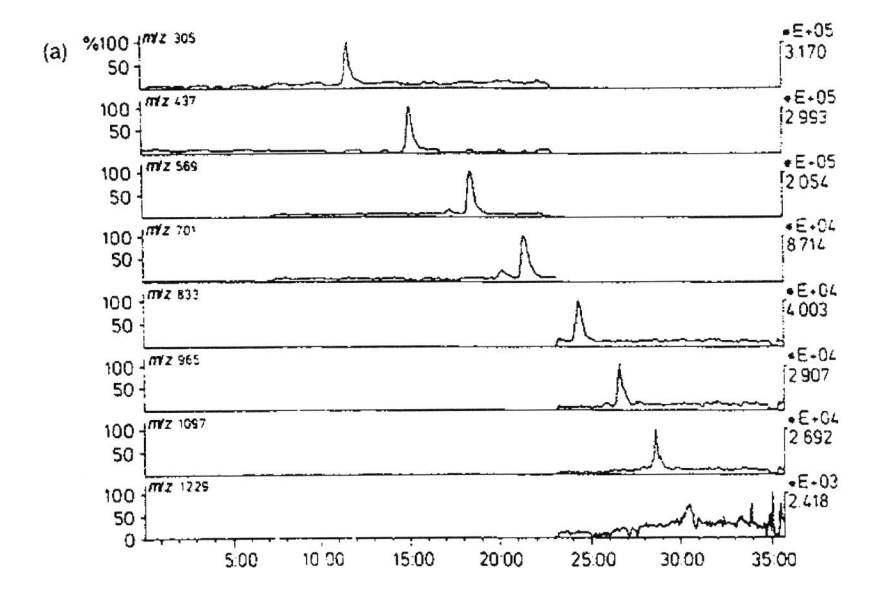

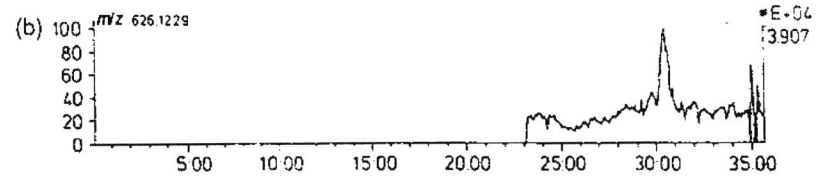

Fig. 7-58. HPAEC-MS analysis of an arabinan digest. (A) Mass chromatogram of singly charged ions from DP2 (m/z = 305) to DP9 (m/z = 1229) and (B) mass chromatogram of singly and doubly charged ions for DP9 (m/z = 626 und 1229). − Separator column: CarboPac PA1; eluant: 0.1 mol/L NaOH−NaOAc; gradient: linear, 0.1 mol/L NaOH isocratically for 5 min, then to 0.1 mol/L NaOH + 0.5 mol/L NaOAc in 35 min; flow rate: 1 mL/min; detection: thermospray MS (Finnigan MAT TSQ-70) in the MID mode, m/z = 173, 305, 437, 569, and 701 for 23 min, then m/z = 428, 494, 560, 626, 692, 833, 965, 1097, 1229, and 1361 for 12 min, 5 s per scan; injection volume: 25 μL; (taken from [124]).

detection may be especially helpful in more advanced studies directed at structure elucidation, because deprotonated molecules are more readily fragmented than sodiated molecules observed in the positive-ion mode.

This brings up the interesting topic of the origin of the various ions detected. The observation of deprotonated ions and acetate adducts is not very surprising. As a result of the removal of sodium ions by the suppressor, the solvent entering the mass spectrometer is acetic acid with concentrations that follow the sodium acetate gradient. The observation of hydrogensulfate- and sulfate adducts indicates that the sulfuric acid that is used for suppressor regeneration has leaked through the membrane. Such a diffusion through the membrane, which only occurs if the suppressor is chemically regenerated with sulfuric acid, significantly influences ionization conditions in the thermospray source. It might also prove to be useful in the negative-ion detection of other compounds.

With the advent of new, more versatile and more robust LC-MS interfaces based on atmospheric-pressure ionization, the use of thermospray will rapidly diminish. However, the above-mentioned applications in sugar analysis as well as a large number of other applications in organic ion analysis such as phenoxyacetic acids [128], quaternary ammonium pesticides [129, 130], anionic surfactants [131, 132], sulfated azo dyes [133, 134], sulfonamides [135], and β-lactam antibiotics [136, 137] have proved the potential of this coupling technique in qualitative and quantitative analysis.

Electrospray Interface

Electrospray ionization (ESI) is an atmospheric-pressure ionization technique (API). In an electrospray interface, the column effluent is nebulized into an atmospheric-pressure ion source. The nebulization is due to the application of an electric field that results from a potential difference (up to 5 kV) between the narrow-bore spray capillary (needle) and a surrounding counter electrode. The solvent emerging from that needle breaks into fine threads that subsequently disintegrate in small droplets. With progressive solvent evaporation, the ions inside the droplets, the charge of which is opposite to that of the applied field, migrate to the surface. During this process, the charge density at the surface increases so much that those droplets undergo electrohydrodynamic Raleigh instabilities, which results in a number of much smaller droplets with a relatively large number of charges. In a very complex process [138] that is not yet completely understood, repetitive Coulombic droplet explosion may take place, leading to ions in the gas phase. The emitted ions are then directed into the quadrupole mass analyzer by electrostatic lenses. A detailed monography about this subject was published by Cole [139].

The electric field required to generate aerosol largely depends on the surface tension of the solvent. If the composition of the column effluent remains constant, nebulization is not a problem. However, in the case of gradient elution, where eluant composition and, consequently, surface tension are changing with time, nebulization can be a problem. A rather elegant solution to this problem is the pneumatically assisted electrospray ionization ("IonSpray") developed by Perkin-Elmer Sciex, in which a sheath gas assists in the formation of droplets. A schematic diagram of this ionization process is shown in Fig. 7-59.

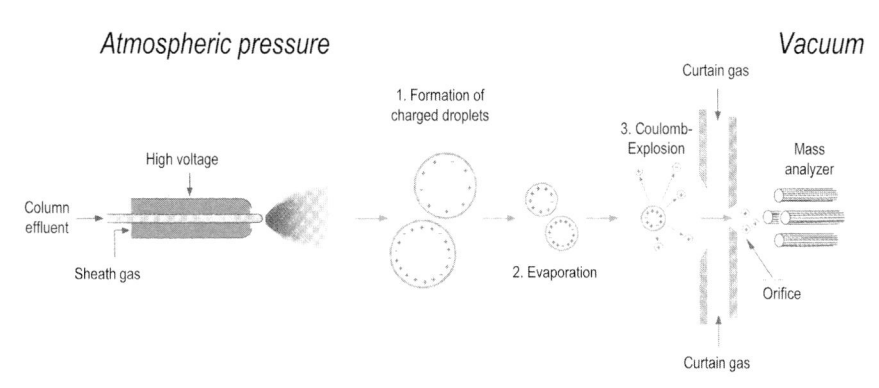

Fig. 7-59. Schematic diagram of the ionization process in electrospray ionization ("IonSpray", Perkin-Elmer Sciex).

ESI is especially appropriate for LC flow rates in the lower µL/min range. When a liquid chromatograph utilizing conventional separator columns with 4.6 mm internal diameters is coupled to ESI, the column effluent has to be split with a ratio of 0.01 prior to entering the interface. Higher flow rates up to 500 µL/min are compatible with thermally assisted electrospray [140], which heats the spray capillary to 150–240 °C. Ultrasonic nebulizers [141] in combination with a sheath gas for drying and focussing are also suited for higher flow rates.

In an interface for chemical ionization at atmospheric pressure (APCI), aerosol formation occurs in a heated tube with nitrogen as a reagent gas; the solvent is completely evaporated in the tube. The gas-vapour mixture is then directed into the atmospheric-pressure ion source, in which chemical ionization is initiated by electrons that are generated at the corona discharge needle. The solvent vapour acts as a reagent gas. The charge is transferred to the analyte molecules by ion-molecule reactions (cluster formation and declustering of ions). Subsequently, the ions generated are sampled into the high vacuum of a mass spectrometer for mass analysis. Figure 7-60 schematically depicts the APCI ionization process.

Both ionization modes can be operated in the positive- and negative-ion mode to obtain either protonated or deprotonated molecular ions as well as sodium-, potassium-, ammonium-, formate-, or acetate adducts. Because atmospheric-pressure ionization is a "soft" ionization technique (i.e. molecular ions are predominantly formed and fragmentation is very rare) the mass spectra of single components do not exhibit many signals. Under suitable MS conditions (voltage

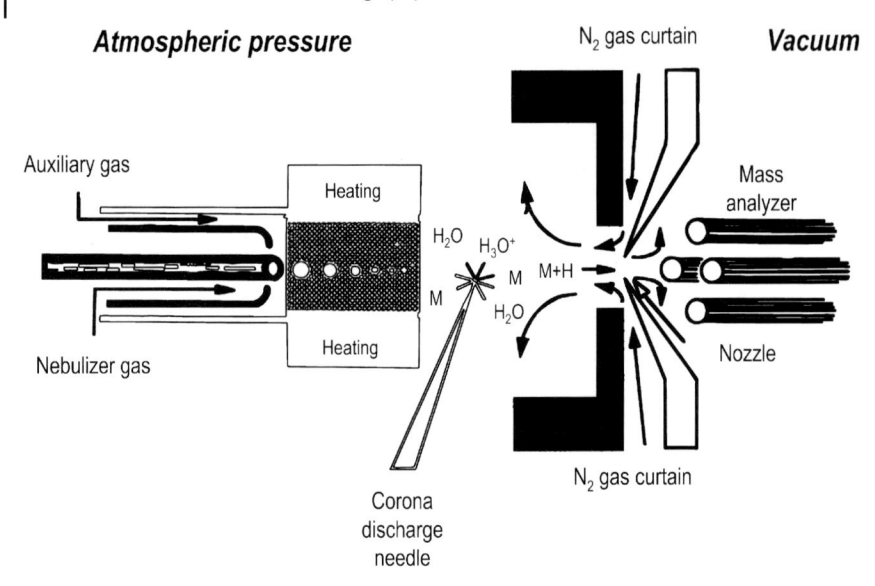

Fig. 7-60. Schematic diagram of the ionization process in APCI ("Heated Nebulizer", Perkin-Elmer Sciex) [141].

increase at the electrostatic lenses) or in tandem mass spectrometry, fragmentations can be induced for a more detailed structural elucidation. Such fragmentations, however, differ from those observed in electron impact ionization (EI). Because spectral libraries for API-MS do not exist at present, the ability to identify unknowns is rather limited.

Although ESI-MS/MS was originally developed for the analysis of high molecular weight organic compounds, it can also be applied to the analysis of strongly polar and ionic compounds. Charles et al. [143] applied this technique for the first time to the trace analysis of bromate in water at sub-µg/L levels. The ion chromatographic separation was carried out on an IonPac AG9-SC guard column. The optimal flow rate for ESI-MS/MS detection of 50 µL/min was obtained by a 1:20 split utilizing a T-piece of very low dead volume. However, the traditional carbonate/bicarbonate eluant for this column is not compatible with ESI-MS/MS detection, because bicarbonate forms a non-volatile residue during the evaporation process. This residue would precipitate around the entrance cone and eventually block it. Also, sensitivity is compromised when using a carbonate/bicarbonate eluant compared with pure de-ionized water. One of the two ways to reduce the electrolyte content in the column effluent prior to entering the electrospray interface is the use of a suppressor system. Alternatively, an electrolyte has to be selected that is compatible with ESI and capable of eluting bromate. For example, ammonium sulfate can be used; it elutes bromate at a concentration of $c = 27.5$ mg/L. If the ammonium sulfate solution is prepared with 90% (v/v) methanol instead of pure de-ionized water, nebulization and subsequent evaporation is supported which, in turn, increases sensitivity by a factor of 10. Because the stationary phase used for the separation is solvent-compatible, such an eluant can be used without any problem. In the course of sample preparation, anions that have a negative influence on separation or detection have to be removed. These anions include bicarbonate, sulfate, and chloride. The negative effect of bicarbonate on ESI-MS/MS detection was mentioned above. However, it can easily be removed by using a SPE cartridge filled with a cation exchanger in the hydrogen form (OnGuard-H, Dionex Corporation). As a result of this cation exchange process, carbon dioxide is formed due to the sample acidification. The carbon dioxide can be removed by sparging the sample with helium. Sulfate at concentrations $c > 20$ mg/L also interferes with the analysis. If sulfate is present in the sample at higher concentrations, SPE cartridges such as OnGuard-Ba (cation exchanger in the barium form) are used for precipitating sulfate. Higher concentrations of chloride are precipitated with the cation exchanger in the silver form (OnGuard-Ag).

In the mass spectrum, bromate is detected at m/z = 127 and 129, which reflects the natural occurence of the two evenly distributed bromine isotopes ^{79}Br and ^{81}Br. After fragmentation of both ions in the second quadrupole (Q2), the product ions can be analyzed in the third quadrupole (Q3). In the course of fragmentation, bromate can loose one, two, or three oxygen atoms to yield BrO_2^-, BrO^-, and Br^-, respectively, as the products. Due to the isotope distribution, six transitions generally result that can be registered in the multiple-reaction-monitoring mode (MRM): 127/111, 127/95, 127/79, 129/113, 129/97, and 129/81. Because a high background signal is observed after the elution of bromate, which is attributed to the formation of HSO_4^- (m/z = 97), only the first four transitions are selected to obtain a perfectly resolved chromatographic signal such as that of a 1-µg/L bromate standard in Fig. 7-61. With this technique, Charles et al. calculated a minimum detection limit for bromate of 0.1 µg/L; the standard deviation for $n = 7$ is 1.9% at this concentration level.

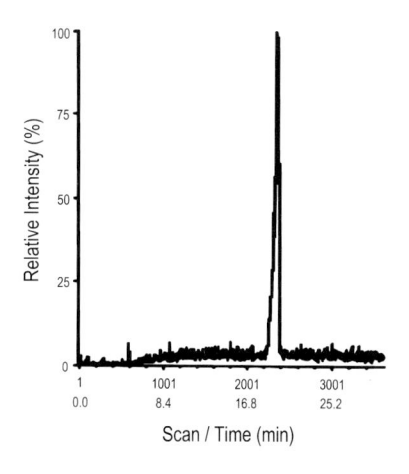

Fig. 7-61. Chromatogram of a 1-µg/L bromate standard with ESI-MS/MS detection. — Separator column: IonPac AG9-SC; eluant: 27.5 mg/L ammonium sulfate — methanol (10:90 v/v); flow rate: 1 mL/min; detection: ESI-MS/MS in the negative MRM mode (127/111, 127/95, 127/79, and 129/113); sample volume: 5 mL; sample: 1 µg/L bromate standard in de-ionized water; (taken from [143]).

In a recently published paper, Charles and Pépin [144] extended the ESI-MS/MS technique developed for trace analysis of bromate to the analysis of other oxyhalides such as chlorite, chlorate, and iodate that can occur when drinking water is ozonated (see Section 9-1). Because chlorine also occurs in two isotopic forms, oxychlorides are detected in the negative-ion mode at two different m/z values: m/z = 67 and 69 for ClO_2^-, and m/z = 83 and 85 for ClO_3^-. Iodate is detected at m/z = 175. The fragmentation of these ions also results from the successive loss of oxygen atoms. The eluant mixture consisting of ammonium sulfate and methanol developed for the analysis of bromate cannot be used for the analysis of oxyhalides, because it is contaminated with chlorate. For this reason, the authors used ammonium nitrate as an eluting agent. Because the affinity of nitrate towards the stationary phase is lower than that of sulfate, its concentration has to be increased accordingly. Optimal separations between iodate and chlorite, the elution order of which is reversed due to the high methanol content in the mobile phase [145], are obtained with 65 mg/L ammonium nitrate in the mobile phase. As an example, Fig. 7-62 shows the chromatogram of a drinking water sample, in which 1.95 µg/L chlorate could be detected. Detection was carried out following the transitions $^{35}ClO_3^-/^{35}ClO_2^-$ (83/67) and $^{37}ClO_3^-/^{37}ClO_2^-$ (85/69). With this technique, the minimum detection limits for iodate, chlorite, and chlorate are 0.5, 1.0, and 0.05 µg/L, respectively.

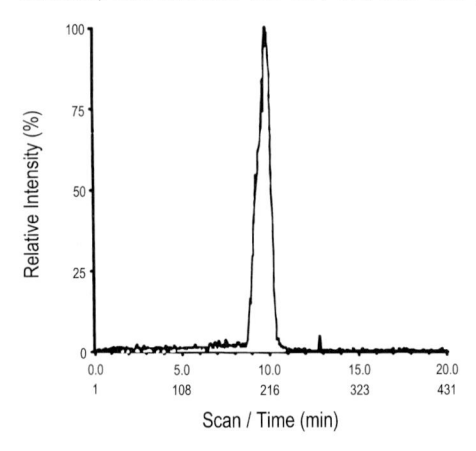

Fig. 7-62. Trace analysis of chlorate in drinking water with ESI⁻MS/MS detection. – Separator column: IonPac AG9-SC; eluant: 65 mg/L ammonium nitrate – methanol (10:90 v/v); flow rate: 1 mL/min; detection: ESI⁻MS/MS in the negative MRM mode (83/67, 85/69); sample volume: 5 mL; sample: drinking water with 1.95 µg/L chlorate as well as 11.9 mg/L chloride, 6.7 mg/L nitrate, 19.3 mg/L sulfate, 129.4 mg/L bicarbonate, 13.7 mg/L sodium, 3.7 mg/L potassium, 9.4 mg/L magnesium, and 25.8 mg/L calcium; (taken from [144]).

Finally, the IC-MS method that Mohsin [146] developed for the analysis of organophosphorous and organosulfur compounds in an insecticide such as monomethylphosphate, monomethylsulfate, and dimethylphosphate should be reviewed. All three compounds were separated on an IonPac AS11 anion exchanger using a NaOH/methanol mixture as an eluant. Mohsin desalted the eluant with an Alltech 1000HP suppressor, which contains two packed bed suppressor columns. One of the suppressor columns is used for suppression, while the other one is electrochemically regenerated at the same time. The suppressor effluent is split 99:1 and mixed with acetonitrile/water (90:10 v/v) containing 0.5% ammonium hydroxide. Detection is carried out in the negative-ion mode.

Ammonium hydroxide increases the pH of the mobile phase entering the electrospray interface and, thus, supports negative ion formation. The two compounds monomethylphosphate and monomethylsulfate, identified in the insecticide by electrospray MS, exhibit the same mass-to-charge ratio of m/z = 111, but differ in their mass spectra due to differences in the fragmentation and isotope distribution of the [M-H]$^-$ ions.

At present, electrospray interfacing is the most widely used technique for introducing liquid into a mass spectrometer. In addition, electrospray ionization is a very efficient ionization technique that significantly extended the analytical potential of mass spectrometry. Today, both electrospray ionization and chemical ionization at atmospheric pressure (APCI) are standard methods in LC-MS. Although the majority of current publications deals with the characterization of biologically relevant macromolecules, the few examples introduced above underline the significance of electrospray interfaces for the sensitive detection and structural elucidation of ionic compounds. The dominating role of electrospray is attributed, above all, to its ease-of-use, high sensitivity, reliability, and robustness. The future will show whether new interface techniques will supersede electrospray in its current dominant role for LC-MS.

Index

Numbers in front of the page numbers refer to Volume 1 or 2, respectively.

a

Accuracy 2/551
Acebutolol 1/449
Acesulfam K 2/752
Acetaldehyde 2/622
Acetate 1/44, 1/168 f, 1/199
Acetic acid 1/74, 1/362, 2/627, 2/650, 2/701, 2/714, 2/716, 2/786
N-Acetylchitobiose 1/241
Acetylcholine 1/302, 1/420, 1/451, 2/742
N-Acetylgalactosamine 1/241
N-Acetylglucosamine 1/238
N-Acetylmannosamine 1/251
N-Acetylneuraminic acid 1/238, 1/242, 2/799
Acid blue 40 1/459
Acid blue 113 1/459
Acid red 114 1/459
cis,trans-Aconitate 1/73
Acridine orange 1/459
Acrylic acid 1/363, 1/373
Activity coefficient 1/28, 1/466, 1/467
Additives
 – inorganic 1/406
 – organic 1/100, 2/690 ff, 2/731
Adenosine 1/178
 – diphosphate 1/179, 2/794
 – monophosphate 1/179, 2/794
 – triphosphate 1/179, 2/794
Adipic acid 1/372
Adsorption process, non-ionic 1/3, 1/359
Aerosol 2/622
Alanine 1/258, 1/387
Alcohol 1/364, 1/379, 1/485, 2/774
Aldehyde 1/379, 1/485
Alizarin red S 1/459
Alkali metal 1/280, 1/286, 1/299, 1/318, 1/518, 2/628, 2/718, 2/780, 2/787

Alkaline-earth metal 1/281, 1/286, 1/299, 1/303, 1/322, 1/501, 1/518, 2/628, 2/718, 2/742, 2/780, 2/787, 2/819
Alkaloid 1/402, 1/418, 1/455
Alkanesulfonate 1/404, 1/424
Alkanolamine 1/418, 1/491
Alkenesulfonate 1/424
Alkylamine 1/302, 1/320, 1/493
Alkylbenzene sulfonate 1/424, 1/430, 2/709
Alkylbenzyl-dimethyl-ammonium 2/706
Alkyl-dimethyl-benzylammonium chloride 1/435
Alkylether sulfate 2/709
Alkyl sulfate 1/425
Alkyl sulfonate 1/425, 2/706
Alkyltrimethylammonium 1/433
Allobarbital 1/442, 1/453
Allylsulfonic acid 2/689
Aluminum 1/341, 1/501, 2/649, 2/789
Amidosulfonic acid 1/170, 2/647
Amine 1/418, 1/487, 1/491, 1/514
 – aliphatic 1/288, 1/324, 2/759
 – aromatic 1/356, 1/418
 – biogenic 1/295, 1/302, 1/346, 1/451, 2/742, 2/747
 – catechol- 1/421, 1/451, 1/455, 2/796
 – di- 1/295, 1/345
 – hydrophobic 1/295
 – poly- 1/344 ff, 1/514
 – polyvalent 1/295
Amino acid 1/257 ff, 1/382 ff, 1/487, 1/493, 1/514
 – carbohydrate interference 1/259
 – detection 1/389 ff
 – dissociation behavior 1/99, 1/384
 – elution order 1/258
 – O-phosphorylated 1/264

– physiological 1/388
– sample preparation 1/391 f
p-Aminobenzoic acid 2/770
2-[(Aminocarbonyl)oxy]-N,N,N-trimethyl-1-
  propanaminium chloride
  (Bethanechol) 2/766
7-Aminocephalosporic acid 1/458
5-Amino-2,3-dihydro-1,4-phthalazindione
  (Luminol) 2/632
2-(2-Aminoethyl)pyridine 1/357
p-Aminohippuric acid 2/782
3-Amino-1-hydroxypropylidene-1,1-
  bisphosphonate (Pamidronate) 2/765
Aminomethanephosphonic acid 1/517
Aminomethanesulfonic acid 1/516
Aminopolycarboxylic acid 2/63, 1/181, 1/443
Aminopolyphosphonic acid 1/63, 1/181,
  1/183, 1/184
– di-N-oxide 1/185
– mono-N-oxide 1/184
2-Aminopyridine 1/357
Amino sugar 1/206, 1/214 f, 1/243
– N-acetylated 1/214
– de-N-acetylated 1/243
AminoTrap column 1/215, 1/243
Aminotris-(methylenephosphonic acid)
  (ATMP) 1/191
Ammonium 1/283, 1/293, 1/315, 1/318,
  1/514, 2/598, 2/609, 2/615, 2/621, 2/634,
  2/690, 2/734, 2/815
Ammonium compound, quaternary 1/355,
  1/418, 1/433
Amobarbital 1/442, 1/453
Amperometry
– integrated 1/391, 1/487 f
– pulsed 1/480 ff
Ampicillin 1/487, 2/735
Amylamine 2/759
Amylopectin 1/235
Amylose 1/234
Analysis
– air 2/675
– – clean room 2/675
– of chemicals 1/378 ff, 2/648 ff,
– function 2/560
– online 2/650, 2/652
– quantitative 2/549 ff
– time 1/7
Aniline 1/356

Anion
– halide 1/121 f, 1/133
– inorganic
– – survey 1/121 f
– – UV measuring wavelengths 1/498
– non-metal 1/125
– non-polarizable 1/94
– organic 1/168 ff
– oxyhalide 1/121
– peroxohydroxide 2/701
– polarizable 1/61, 1/66, 1/72, 1/79,
  1/94, 1/122, 1/159 ff, 1/413, 2/601
– polyvalent 1/63, 1/163, 1/181 ff, 1/509
– standard 1/39
– – limit of determination 2/628
– surface-active 1/72, 1/424 ff
– surface-inactive 1/409 ff
Anion exchanger
– high capacity 1/35
– – cellulose 1/35
– – polymer-based 1/35
– – silica-based 1/35
– low capacity
– – latexed 1/54 ff
– – polymethacrylate 1/47 ff
– – polyvinyl 1/54
– – PS/DVB 1/35 ff
– – silica-based 1/81 ff
Anion-trap column (ATC) 1/196
Anomer 1/224
Anserine 1/388
9-Anthryldiazomethane (ADAM) 1/520
Antibody, monoclonal 1/240, 1/269, 1/271
AOX/AOS 2/624
Application
– building material industry 1/302
– chemical industry 2/814 f
– clinical chemistry 2/781 ff
– electroplating industry 2/678 ff
– environmental analysis 2/588 ff
– food and beverage industry 2/711 ff
– household product industry 2/697 ff
– mineralogy 2/815
– petrochemical industry 1/302, 2/803 ff
– pharmaceutical industry 2/756 ff
– power plant chemistry 1/342, 2/626 ff
– pulp & paper industry 2/810 ff
– semiconductor industry 1/336, 1/370,
  2/651 ff
Approximation test, acc. to Mandel 2/565

Aprobarbital 1/442
Arabinose 1/213, 2/726, 2/750, 2/774
Arabitol 1/211, 2/774
Arginine 1/258, 1/262, 1/387
Arogenic acid 1/267
Arsenate 1/125 f, 2/622, 2/649
Arsenazo I
   see [o-(1,8-dihydroxy-3,6-disulfo-2-naphthylazo)benzenearsonic acid]
Arsenazo III
   see bis-(2-arsono-benzeneazo)-2,7-chromotropic acid
Arsenite 1/124, 1/370, 1/475
Arsenobetain 1/528
Arsenocholine 1/528
Arylalkylamine 1/418
Aryl sulfonate 1/427
Ascorbic acid 2/714, 2/722, 2/741, 2/769
Asparagine 1/258, 1/387
Aspargic acid 1/258
Aspartame 2/731
Aspirin 1/454
Asymmetry factor 1/14
Atropine 1/402, 1/421
Audit Trail 2/556
AutoNeutralization 2/639, 2/648, 2/667, 2/833
Azide 1/125, 1/152, 1/512, 1/787

b

Band broadening
   see peak broadening
Barbital 1/423, 1/442, 1/453
Barbiturate 1/418, 1/423, 1/441, 1/450
Barbituric acid 1/423, 1/453
Barium 1/283, 1/287, 1/305
Bath, galvanic
  – chromic acid 2/682, 2/690
  – copper, electroless 2/687
  – copper pyrophosphate 2/679, 2/689
  – copper sulfate, electrolytic 2/679
  – gold 2/686
  – nickel, electroless 2/680
  – nickelborohydride 2/693
  – nickel/iron 2/686, 2/692
  – nickel sulfamate 2/682

  – nickel/zinc 2/682
  – tin/lead 2/673
  – zinc 2/688
Benzalkonium chloride 2/761
Benzene 1/452
1,2-Benzenedicarboxylate 1/450
1,3-Benzenedicarboxylate 1/450
1,4-Benzenedicarboxylate 1/450
m-Benzene disulfonate 1/451
o-Benzene disulfonate 1/451
Benzenehexacarboxylate
   see phytic acid
Benzenepentacarboxylate 1/450
Benzene sulfonate 1/427
   1,2,4,5-Benzenetetracarboxylate
   (Pyromellitate) 1/448, 1/450
1,2,3-Benzenetricarboxylate 1/450
1,2,4-Benzenetricarboxylate 1/450
1,3,5-Benzenetricarboxylate 1/448, 1/450
1,3,5-Benzenetrisulfonate 1/451
Benzoate 1/410, 1/448, 2/731, 2/746, 2/762
Benzylalcohol 1/452
4-Benzylpyridine 1/357
Beryllium 2/615
Bethanechol chloride
   see 2-[(aminocarbonyl)oxy]-N,N,N-trimethyl-1-propanaminium chloride 2/766
Beverage 2/713 ff
2,2'-Bipyridine 1/357
Bis-(2-arsono-benzeneazo)-2,7-chromotropic acid (Arsenazo III) 1/502
Blank value 2/562, 2/627
Bleaching activator 2/701
Bleaching agent 2/697
Boltzmann's law of energy distribution 1/467
Borate 1/153, 1/366, 1/369, 2/633, 2/648, 2/686
BorateTrap column 1/218
Borophosphorosilicate glass film 1/131, 2/672
Brightener 2/694
Brine 2/817
Bromate 1/40, 2/65, 2/67, 1/136, 1/503, 1/544, 2/591 ff
Bromide 1/130, 1/411, 1/498, 2/588, 2/624, 2/781
p-Bromobenzoate 1/449
Bromocresol purple 1/459
Builder 2/697 f
Butabarbital 1/442, 1/450

4-Butene-18-crown-6   1/88
2,3-Butanediol   2/773
1,4-Butanedisulfonic acid   2/757
Butanesulfonic acid   1/428
n-Butanol   1/381
iso-Butylamine   1/353
n-Butylamine   1/353
sec.-Butylamine   1/353
tert.-Butylamine   1/353, 2/759
tert.-Butyl-ethane-1,2,2-trisphosphonic
   acid   1/189
n-Butyric acid   1/74, 1/362, 1/443

C

Cadaverine   1/344, 2/747
Cadmium   1/331, 1/335, 2/613
Caffeic acid   1/10
Caffeine   1/455, 1/457, 2/731
Caffeoylquinic acid   2/752
Calcium   1/283, 2/598, 2/758
Calibration   2/560 ff
  – area normalization   2/569
  – basic   2/560, 2/565
  – external standard   2/571
  – function   2/560 ff
  – internal standard   2/570
  – standard addition   2/572 f
Capacity factor   1/17, 1/29, 1/30, 1/319
Capacity ratio   1/21, 1/155
Capillary electrophoresis   1/3
Capric acid   1/443
Caproic acid   1/362, 1/443
Caprylic acid   1/443
Carbohydrate   1/62, 1/205 ff, 1/364,
   1/485, 2/736, 2/790
  – acidity   1/246
  – phosphoric acid ester   1/224
Carbonate   1/365, 1/369, 1/511, 2/591,
   2/633, 2/686, 2/698, 2/701
Carbonylphosphonate   2/764
Carboxylic acid
  – aliphatic   1/168, 1/448, 2/615
  – aromatic   1/448
  – di-   2/615
  – hydroxy-   1/168, 2/614
  – keto-   2/615
Carboxypeptidase B   1/270
Carnithine   1/388
Carprofen   1/454
Carrez precipitation   2/740

Casein   2/738
Cation
  – exchange process   1/279, 1/301
  – exchanger   1/280 ff
  – – latexed   1/298 ff
  – – silica-based   1/303 ff
  – – solvent influence   1/284
  – – surface-sulfonated   1/280
  – – weak acid   1/282, 1/324
  – surface-active   1/433 ff
  – surface-inactive   1/418 ff
  – simultaneous analysis   1/282 ff, 1/304,
     1/323, 2/598
Cavity effect   1/122, 1/143
Cefadroxil   1/458
Cefazolin   1/458
Cefotaxim   1/458
Cell
  – constant   1/469
  – pulsed amperometric   1/496
Cellobiose   1/221, 1/227, 2/774
Cellotriose   2/812
Cellulose   1/235
Cement analysis   2/815
Cephalexin   1/458
Cephaloridine   1/458
Cephalosporin   1/453, 1/458
Cephalosporin C   1/458
Cephalotin   1/458
Cephapirine   1/487, 2/736
Cerium   1/344
Cerium(III)/Cerium(IV)   1/518
Cesium   1/287
Cetylpyridinium
   see hexadecylpyridinium
Chlorate   1/40, 2/65, 1/409, 1/546, 2/602 f,
   2/811
Chloride   1/59 ff, 2/588, 2/624, 2/811, 2/816
Chlorine dioxide   2/602
Chlorite   1/65, 1/73, 1/504, 1/546, 2/596,
   2/602 f, 2/811
p-Chlorobenzenesulfonate   1/449, 1/533
Chlorocholine   1/420
Chlorogenic acid   1/10, 2/752
Chloromethyl methylether   1/37
Choline   1/302, 1/419, 1/451, 2/742, 2/767
Chromate   1/159, 1/336, 1/508
Chromatogram   1/13
Chromatography
  – affinity   1/267

- gas   1/1
- high performance liquid   1/1
- ion   1/1
-   - multi-dimensional   1/5, 1/444
-   - online   2/650, 2/652
-   - two-dimensional   1/369
- ion-exchange   1/2, 1/3
-   - anion exchange   1/27 ff
-   - cation exchange   1/279 ff
- ion-exclusion   1/4, 1/359 ff
- ion-pair   1/4, 1/393 ff
-   - retention model   1/394 ff
- liquid   1/1
- paper   1/1
- reversed-phase   1/4, 1/403
- thin layer   1/1

Chromazurol S   1/459, 2/618

Chromium
- chromium(III)   1/336
- chromium(III)/(VI) ratio   2/683
- chromium(IV)   1/508, 2/645
- simultaneous analysis   1/336 f, 1/524

α-Chymotrypsinogen   1/269

Cinchonin   1/457

Citrazinic acid
  see 2,6-dihydroxyisonicotinic acid

Citric acid   1/76, 1/174, 1/363, 1/413, 1/448, 2/704, 2/714, 2/731, 2/738, 2/762, 2/786

Citrus juice   2/714

Clodronate   2/763

Cluster, molybdenum-sulfur   1/418

Cobalt   1/306, 1/331, 1/334, 2/613, 2/631

Cocaine   1/421

Codeine   1/421

Colchicine   1/457

Collagen hydrolysate   1/266

Column
- body, material   1/5
- coefficient   1/21
- dead time   1/155
- efficiency   1/17
- length   1/17
- maintenance   2/835
- poison   2/614, 2/835
- temperature   1/208, 1/291, 1/400, 1/408

Complex
- aluminum halide   1/98
- cryptand-cation   1/93
- formation coefficient   1/327
- Gd-DTPA   1/414
- indicator   1/330
- iron cyanide   1/407, 1/415
- lanthanide-PDCA   1/343
- metal-chloro   1/163, 1/498
- metal-cyano   1/128, 1/498
- metal-DTPA   1/414
- metal-EDTA   1/163, 1/414, 2/687
- metal-oxalate   1/333
-   - stability constant   1/335
- metal-PDCA   1/333
-   - stability constant   1/335
- molydenum disulfido   1/419
- multimetal   1/416
- stability
-   - kinetic   1/330
-   - thermodynamic   1/329
- Zn-EDTA   1/500

Complexing agent   1/327, 1/510, 2/641

Complexon   1/509

Computer, personal   2/555

Concentrator column   2/628
- IonPac AG5   2/653
- MetPac CC-1   2/594, 2/819, 2/821
- TAC-1   2/653
- TAC-LP1   2/678
- TBC-1   1/368, 2/633
- TCC-LP1   2/677

Conditioning agent   2/634

Conductivity
- background   1/101, 1/103, 1/110, 1/469
- coefficient   1/469
- electrical   1/462
- equivalent   1/462, 1/464
-   - table   1/466
- of electrolyte solution   1/462 ff

Confidence interval   2/559, 2/564

Control card   2/578 ff
- blank value   2/580
- $\bar{x}$-R-Combination   2/583
- Cusum   2/583
- differences   2/583
- mean value   2/580
- quality   2/578
- recovery   2/580
- Shewhart   2/579
- Span   2/582

Copper   1/332, 1/334, 2/613, 2/789

Correlation coefficient   2/562

Cotrell equation 1/478
Coulomb explosion 1/542
Coupling
– HPICE/HPIC 1/359, 1/375 ff
– IC/AAS 1/522
– IC/DCP-AES 1/526
– IC/ICP 1/523 ff
– IC/ICP-MS 1/525
– IC/ICP-OES 1/125, 1/526
– IC/MS 1/504, 1/530 ff, 2/607
o-Cresol 1/441
Cross-linking 1/35
12-Crown-4 1/87
15-Crown-5 1/87
18-Crown-6
  (1,4,7,10,13,16-Hexaoxa-
  cyclooctadecane) 1/86
Crown ether 1/87
Cryptand 1/92, 1/179
– binding constant 1/93
Cumene sulfonate 1/424, 1/427, 2/703
Current, electrical 1/463
– charging 1/478
– diffusion 1/476
– Faraday 1/478
Cutter agents 2/741
Cyanate 1/124, 2/833
Cyanic acid 2/622
Cyanide 1/63, 1/126, 1/370, 1/475, 1/479, 2/810
– easily releasable 1/129
– free 1/128 f
– reaction to cyanate 2/833
– total 1/130
Cyanocobalamine 2/770
p-Cyanophenol 1/29, 1/62
Cyclamate 2/752
Cyclic voltammetric stripping (CVS) 2/692
Cyclitol 1/210
Cyclodextrin 1/232
Cyclohexylamine 1/350
Cystathionine 2/795
Cysteic acid 1/391
Cysteine 1/387, 1/392, 2/795
Cysteinylglycine 2/795
Cystine 1/258, 1/392
Cytidine 1/178
– 5'-monophosphate 1/181
Cytochrome c 1/269, 1/272

d

Dansyl chloride 1/383
Dead time 1/13
Debye-Hückel theory 1/466 f
1,10-Decanediamine 1/354
Decanesulfonic acid 1/428
Decyl-2.2.2 1/93
Decyl-dimethyl-benzylammonium 1/435
Decyl sulfate 1/427
Decyltrimethylammonium 1/355, 1/434
Degree
– of cross-linking 1/361
– – effective 1/144, 1/448
– of freedom 2/559
Deoxyadenosine-5'-monophosphate 1/181
Deoxycytidine-5'-monophosphate 1/181
2-Deoxyglucose 1/211, 1/217, 1/224, 1/242
2'-Deoxyguanosine 1/178
– 5'-monophosphate 1/181
2'-Deoxy-D-ribose 1/212
Depolarisator 1/475
Deproteination 1/391, 2/736, 2/758, 2/783, 2/790 f
Derivatization
– post-column 1/182, 1/499 ff
– – two-step 1/510, 1/515
– pre-column 1/383
Desalter, membrane 1/252
Detection 1/461 ff
– amperometric 1/461, 1/473 ff
– – integrated 1/487 f
– – pulsed 1/480 ff, 2/696
– – with fixed working potential 1/478
– chemiluminescence 2/632
– choice 1/9
– conductivity 1/461 ff
– – application form 1/469 ff
– – direct 1/469
– – indirect 1/102, 1/471
– criterium 2/574
– element-specific 1/125, 1/524
– fluorescence 1/461, 1/514 ff
– – indirect 1/518
– phosphorus-specific 1/165, 1/190, 1/510, 2/643, 2/699
– photometric 1/411, 1/461, 1/498 ff
– – after derivatization 1/499 ff
– – direct 1/498 f
– – indirect 1/510 ff
– refractive index 1/461, 1/521

- radioactivity 1/521
Detector
- amperometric 1/6
- - pulsed amperometric 1/496
- - cell 1/496
- choice 1/9
- conductivity 1/6
- fluorescence 1/6
- performance criteria 1/6
- szintillation 1/522
- UV/Vis 1/6
Dextran 1/232
- hydrolysate 1/232
Di-N-acetylchitobiose 1/241
Dialkyldimethylammonium 1/433
Dialkyl sulfosuccinate 1/431
Dialysis 2/733, 2/832
- Donnan 2/832
- electro 2/832
- passive 2/832
1,2-Diaminopropane 1/347
o-Dianisidine (ODA) 1/505, 2/597
Dibenzo-18-crown-6 1/87, 1/88
Dibromoacetic acid 1/171
Dibromomonochloroacetic acid 1/170, 1/171
Dibutylphosphate 1/176
Dicaffeoylquinic acid 2/752
Dicarboxylic acid 2/615
Dichloroacetic acid 1/171, 1/413
Dichlorobenzyl-alkyl-dimethylammonium 1/436
Dichloromethylenebisphosphonic acid (Clodronate) 2/763
Dichloromonobromoacetic acid 1/171
2,6-Dichloro-4-nitroaniline 1/356
2,4-Dichloro-3-nitrophenol 1/441
2,4-Dichlorophenoxyacetic acid (2,4-D) 1/176
Diethanolamine 1/296, 1/302
Diethylamine 1/287, 1/321, 1/325
2-Diethylaminoethanol 1/290, 1/350
N,N'-Diethylaniline 1/356
Diethylene glycol 1/383
Diethylenetriamine 1/348
Diethylenetriaminepentaacetic acid (DTPA) 1/183
Diethylenetriaminepentamethylenephosphonic acid (DTPP, DEQUEST 2060) 1/183
N,N-Diethylethylenediamine 1/347

Diethyltolueneamide 1/452
Diffusion
- current 1/476
- Eddy 1/19
- lateral 1/19, 1/20
- longitudinal 1/19, 1/20
Diflunisal 1/454
Digalacturonic acid
- mass spectrum 1/539
Digestion
- bomb 2/828
- combustion 2/827
- dry ashing 2/827
- fusion 2/831
- oxygen 2/830
- Schöninger 2/768
- wet 2/827
- Wickbold 2/828
Dihydrolipoic acid 2/770
4,5-Dihydroxy-1,3-benzenedisulfonic acid-disodium salt (Tiron) 1/501
3,4-Dihydroxybenzylamine 1/422, 2/797
o-(1,8-Dihydroxy-3,6-disulfo-2-naphthylazo)benzenearsonic acid (Arsenazo I) 1/501
2,6-Dihydroxyisonicotinic acid (Citrazinic acid) 1/175
D-Dihydroxyphenylglycine 1/458
Diisobutyl-[2-(2-phenoxyethoxy)ethyl]-dimethylbenzyl-ammonium chloride (Hyamine 1622) 1/436, 2/698
Dimethylamine 1/289, 1/321
3-Dimethylaminopropylamine 1/349
2-Dimethylaminopyridine 1/357
N,N'-Dimethylaniline 1/356
Dimethylarsinic acid 1/528
N,N-Dimethyl-2-mercaptoethylamine-hydrochloride (Thiofluor) 1/515
2,4-Dimethylphenol 1/441
Dimethylphosphate 1/546
1,2-Dimethylpropylamine 1/351
Dinaphthal-18-crown-6 1/87
3,4-Dinitrobenzoate 1/449
Diphenhydramine 1/402
1,5-Diphenylcarbazide (DPC) 1/508
Diphenylhydantoin 1/453
N,N'-Diphenylthiourea 2/693
Diphosphonic acid
- geminal 1/185
- vicinal 1/185

1,1-Diphosphonopropane-2,3-dicarboxylic
    acid (DPD) 1/191
Di-n-propylamine 1/351
Disaccharide 1/226
  − non-reducing 1/226
  − reducing 1/226
Distribution
  − coefficient, Nernst 1/17, 1/30, 1/32,
    1/38, 1/326
  − equilibrium 1/17
Disulfate 1/414
Disulfite 1/414
5,5′-Dithiobis(2-nitrobenzoic acid)
    (DTNB) 1/390
Dithiomolybdate 1/418
Dithionate 1/414, 2/646
Diuretic 1/455
Divinylbenzene (DVB) 1/35
1,12-Dodecanediamine 1/354
Dodecyl-dimethyl-benzylammonium 1/435
Dodecyl ether sulfate 1/429
Dodecylpyridinium chloride 1/437
Dodecyl sulfate 1/425, 1/427, 2/692,
    2/698, 2/710
Dodecyl sulfonate 1/425
Dodecyltrimethylammonium 1/355, 1/434
Donnan effect 1/104
Donor atom 1/86
Dopamine 1/422, 2/797
Double layer, electrical 1/398
Dulcitol 1/209
Dye
  − azo 1/451, 1/542
  − reactive 2/814
Dysprosium 1/344

e

Electrode
  − counter 1/474
  − recession 1/484
  − reference 1/474
  − types 1/475
  − working 1/474
Electron affinity 1/537
Electronic signature 2/556
Electrophoretic effect 1/468
Electro-transfer 2/798
Eluant
  − amino acid 1/99
  − aminoalkylsulfonic acid 1/100

− ammonium sulfate/sulfuric acid 1/341,
    1/501
− benzoic acid 1/145, 1/149, 1/366,
    1/470
− borate/gluconate 1/47, 1/51, 1/145
− boric acid 1/49
− carbonate hydroxide
    see sodium carbonate/sodium
    hydroxide
− cerium(III) nitrate 1/309
− choice 1/124, 1/145
− citrate buffer 1/386
− citric acid 1/147, 1/151, 1/326
− citric acid/tartaric acid 1/332
− concentration and pH value 1/137 ff,
    1/156 ff
− copper sulfate 1/513
− p-cyanophenol 1/100
− ethylenediamine/oxalic acid 1/309
− ethylenediamine/tartaric acid 1/281,
    1/309, 1/330
− flow rate 1/123
− for anion exchange chromatography
    1/98 ff
− for cation exchange chromatography
    1/308 ff
− for ion-exclusion chromatography
    1/365
− formic acid 1/304
− for non-suppressed systems 1/100
− fumaric acid 1/151
− heptafluoropropanoic acid
    (perfluorobutyric acid) 1/366
− Hi-Phi buffer 1/386
− hydrochloric acid 1/280, 1/308, 1/365
− hydrochloric acid/2,3-diaminopropionic
    acid 1/282, 1/300, 1/309
− hydroxide/benzoate 1/51
− p-hydroxybenzoic acid 1/49, 1/51,
    1/146, 1/521
− α-hydroxyisobutyric acid (HIBA)
    1/331, 1/343
− iodide 1/511
− isonicotinic acid 1/309
− lithium hydroxide 1/95
− lithium sulfate/sulfuric acid 1/491
− methanesulfonic acid 1/283, 1/308,
    1/310
− methanesulfonic acid/mannitol 1/368
− 2-methyllactic acid 1/502

- nicotinic acid   1/151, 2/787
- nitric acid   1/182, 1/280, 1/308, 1/370
- nitric acid/dipicolinic acid   2/618
- octanesulfonic acid   1/366
- octanesulfonic acid/mannitol   1/366
- oxalic acid   1/49, 1/326, 1/333
- oxalic acid/citric acid   1/332
- oxalic acid/diglycolic acid   1/343
- oxalic acid/ethylenediamine   1/309
- perchloric acid   1/381
- perfluorobutyric acid
    see heptafluoropropanoic acid
- perfluoroheptanoic acid
    see tridecafluoroheptanoic acid
- m-phenylenediamine-dihydrochloride
    1/282, 1/309
- phthalic acid   1/49, 1/51, 1/83, 1/102, 1/470, 1/511, 1/521, 2/787
- picolinic acid   1/298
- potassium chloride/EDTA   1/165, 1/348
- potassium hydrogenphthalate   1/145, 1/162
- potassium hydroxide   1/47, 1/94, 1/102, 1/196 f, 1/470
- pyridine-2,6-dicarboxylic acid   1/326, 1/333
- pyridine-2,6-dicarboxylic acid/oxalic acid   1/310
- salicylic acid   1/151, 1/521
- silver nitrate   1/309
- sodium benzoate   1/102
- sodium carbonate   1/66 f, 1/126, 1/162
- sodium carbonate/sodium bicarbonate   1/39, 1/51, 1/57, 1/99
- sodium carbonate/sodium dihydrogenborate/ethylenediamine   1/126
- sodium carbonate/sodium hydroxide   1/58
- sodium hydroxide   1/69, 1/77, 1/94, 1/100, 1/195
- – addition of barium acetate   1/221
- – addition of zinc acetate   1/228
- – preparation   1/196, 1/220
- sodium hydroxide/sodium acetate   1/232
- sodium p-hydroxybenzoate   1/39
- sodium nitrate   1/176
- sodium perchlorate   1/98
- sodium phenolate   1/99
- sodium phthalate   1/511
- sodium sulfate/sulfuric acid   1/493
- sodium tetraborate   1/100, 1/135, 1/199
- succinic acid   1/151
- o-sulfobenzoate   1/101, 1/511, 1/521
- sulfuric acid   1/288, 1/308, 1/365
- tartaric acid   1/49, 1/84, 1/296, 1/305, 1/326
- tartaric acid/oxalic acid   1/306
- tartaric acid/pyridine-2,6-dicarboxylic acid   1/304, 1/310, 1/323
- tetrabutylammonium salicylate   1/411
- tridecafluoroheptanoic acid (perfluoroheptanoic acid)   1/366
- trifluoroacetic acid   1/290
- trimesic acid   1/102, 1/151, 1/511
- tyrosine   1/100, 1/133
- vanillic acid/N-methyldiethanolamine   1/51
- water   1/90, 1/307, 1/365, 1/381
Eluant generator   1/196 f, 1/206, 1/223, 1/295, 2/591, 2/627, 2/750, 2/814
Endcapping   1/89
EndoF2   2/801
EndoH   2/801
Endoplasmatic reticulum   1/239
Endothall
    see 7-oxabicyclo[2,2,1]heptane-2,3-dicarboxylic acid
Enthalpy
- hydration   1/122
- sorption   1/33
- – free   1/31
Entropy
- configuration   1/34
- mixing   1/34
- sorption   1/32 ff
Ephedrine   1/421, 1/453
Epimerisation   1/251
Epinephrine   1/421, 2/797
Erbium   1/344
Error
- α-   2/574
- β-   2/574
- random   2/557
- statistical   2/558
- systematic   2/557
Erythritol   1/211, 2/774

Erythropoietin (rEPO) 1/240, 1/255
Etching solution 2/670
Ethane-1,2-bis(P-methyl-phosphinic acid) 1/186
1,2-Ethanediphosphonic acid 1/186
1,2-Ethanedisulfonic acid 2/684
Ethane-1,1,2,2-tetrakis(P-methyl-phospinic acid) 1/186
Ethane-1,2,2-tris(P-methyl-phosphinic acid) 1/186
Ethanol 1/365, 1/381, 2/722, 2/773
Ethanolamine 1/306, 1/419, 1/491, 2/634, 2/807
Ethylamine 1/320, 1/324, 1/419
2-Ethylaminoethanol 1/290
Ethylenediamine 1/345
Ethylenediaminediacetic acid (EDDA) 1/183, 1/443
Ethylenediaminetetraacetic acid (EDTA) 1/183
Ethylenediaminetetramethylenephosphonic acid (EDTP, DEQUEST 2041) 1/183
Ethylenediaminetriacetic acid (EDTriA) 1/183, 1/443
Ethylene glycol
  see monoethylene glycol
Europium 1/344
Exclusion
  − Donnan 1/4, 1/359
  − steric 1/4, 1/359
  − volume 1/360
Extraction cartridge 2/599, 2/751, 2/826

f

Faraday constant 1/477
Fatty acid 1/520
  − long-chain 1/365, 1/441
  − short-chain 1/76, 1/80, 1/362
Fatty alcohol ether sulfate 1/424, 1/428, 2/709
Fatty alcohol polyglycolethersulfosuccinate 1/432
Fatty alcohol sulfate 1/427
Fenbufen 1/454
Fermentation 2/772 ff
Ferulic acid 1/10
Feruloylquinic acid 2/752
Fetuin, bovine 1/243
Fibrinogen 1/243
Filler 2/697, 2/702

Flue gas
  − desulfurization 2/644
  − denitrification 2/647
  − scrubber solution 2/644
Fluorenylmethyloxycarbonyl chloride (FMOC) 1/383
Fluorescein 1/459
Fluoride 1/39, 1/44, 1/73, 1/76, 1/78, 1/133 ff, 1/369, 2/589
Formaldehyde 1/381, 2/622, 2/680
Formamidinium 1/320
Formate 1/44
Formic acid 1/74, 1/168, 1/362, 2/628, 2/650, 2/688, 2/786
Fronting effect
  see leading effect
Fructose 1/213, 2/718
D-Fructose-1,6-diphosphate 1/226
β-D-Fructose-2,6-diphosphate 1/226
D-Fructose-1-phosphate 1/226
D-Fructose-6-phosphate 1/226
Fruit juice analysis 2/714 ff
F-test 2/566
Fucose 1/212, 1/238, 2/750
Fucosidase 2/801
Fumaric acid 1/72, 1/172, 1/363, 2/716, 2/786

g

Gadolinium 1/344
β-1,4-Galactan 1/539
Galactinose 1/230
Galactitol 2/774
D-Galactosamine 1/214
α-D-Galactosamine-1-phosphate 1/226
Galactose 1/213, 1/239, 2/750
α-D-Galactose-1-phosphate 1/226
α-D-Galactose-6-phosphate 1/226
β-Galactosidase 2/801
Galacturonic acid 1/76, 1/374, 2/714
Gallium 1/336
Gentiobiose 1/227
Gentisate 1/449
α-1,6-Glucan 1/232
Gluconic acid 2/688, 2/704, 2/775, 2/812
α-D-Glucopyranosido-1,6-mannitol 1/228, 2/754
α-D-Glucopyranosido-1,6-sorbitol 1/228, 2/754
D-Glucosamine 1/214

α-D-Glucosamine-1-phosphate   1/226
α-D-Glucosamine-6-phosphate   1/226
Glucose   1/213, 1/219, 1/224, 2/718
  – oxidation mechanism   1/481
α-D-Glucose-1,6-diphosphate   1/226
α-D-Glucose-1-phosphate   1/226
β-D-Glucose-1-phosphate   1/226
D-Glucose-6-phosphate   1/226
Glucuronic acid   1/373, 2/812
α-D-Glucuronic acid-1-phosphate   1/226
Glutamic acid   1/258, 1/387
Glutamine   1/258, 1/387
Glutaric acid   1/74, 1/363
Glutaric dialdehyde   1/381
Glutathione   2/795
Glycerol   1/210, 1/381 f, 2/707, 2/722, 2/762
α-/β-Glycerophosphate   1/177
Glycine   1/258, 1/387
Glycolic acid   1/44, 1/76, 1/168, 1/199, 1/363, 2/627, 2/650
Glycolipid   1/237
N-Glycolylneuraminic acid   1/242, 2/799
N-Glycopeptidase F   1/246, 1/256, 2/801
Glycoprotein   1/236
  – antifreeze   1/240
  – hydrolysate   1/242
Glycosylation   1/251
Glycosyltransferase   1/239
Glyoxal   1/381
Glyphosate
  see [N-(methylphosphono)glycine]
Gold(I)/(III)   1/416, 2/686
Golgi apparatus   1/239
Gradient
  – aminoalkylsulfonic acid   1/201
  – capacity   1/92, 1/194
  – carbonate   1/106
  – composition   1/106, 1/112, 1/194
  – concentration   1/112, 1/194, 1/283, 1/312
  – – optimization   1/201 f
  – p-cyanophenol   1/201
  – 3-(N-cyclohexylamino)-1-propanesulfonic acid (CAPS)   1/201
  – inverse   1/213
  – pH   1/194
  – profile   1/233
  – step   1/191, 1/283, 1/325
  – taurine   1/201

  – tetraborate   1/44, 1/199
Gradient elution   1/67, 1/191
  – choice of eluants   1/194 f
  – isoconductive technique   1/202 ff
  – of amines   1/292, 1/349
  – of anilines   1/356
  – of carbohydrates   1/214
  – of ethanolamines   1/352
  – of ethylamines   1/352
  – of inorganic anions   2/627
  – of inorganic and organic anions   1/69, 1/72, 1/192, 2/627
  – of inorganic and organic cations   1/349 ff
  – of inositol phosphates   2/792
  – of lanthanides   1/343
  – of methylamines   1/351
  – of organic acids   2/715
  – of polyphosphates   1/195
  – of pyridines   1/357
  – of water-soluble vitamins   2/769
  – theoretical aspects   1/192 ff
Grafting   1/42 f
Griess method   2/785
Growth hormone, human   1/271
Guanidinium   1/320
Guanosin   1/178
  – 5'-monophosphate   1/181
Guanyl urea   1/321
Gypsum analysis   2/817

h

Hafnium   1/340
Haloacetic acid   1/170, 2/606
Heat of sorption   1/33
1,7-Heptanediamine   1/354
Heptanesulfonic acid   1/428
Heptyltriethylammonium   1/355
Hexacyanoferrate(II)/(III)   1/407, 1/415
Hexadecylpyridinium   1/436
Hexadecyl sulfate   1/427
Hexadecyltrimethylammonium   1/355, 1/434
Hexafluorosilicate   2/682
Hexametaphosphate   2/739
Hexamethylenediamine-tetramethylene-phosphonic acid (DEQUEST 2051)   1/184
1,6-Hexanediamine   1/354
Hexanesulfonic acid   1/428

1,4,7,10,13,16-Hexaoxacyclooctadecane
    (18-crown-6)   1/86
Hexasaccharide   1/226
Hexitol   1/209
Hexobarbital   1/423
Hexosaminidase   2/801
Hippuric acid   1/171, 2/786
Hirudine   1/271
Histamine   1/347, 2/748
Histidine   1/258, 1/387, 2/748
Holmium   1/344
Homocysteine   2/794
Homocystine   1/388, 1/391
Homovanillic acid   1/422
Hull probe   2/692
Humic acid   2/599
Hyamine 1622
    see diisobutyl-[2-(2-phenoxy-ethoxy)ethyl]-
    dimethylbenzyl-ammonium chloride
Hydration
    – primary   1/143
    – secondary   1/143
Hydrazine   1/475, 2/680
Hydrazinium compounds   1/439
Hydrazinolysis   1/245
Hydrogen
    – bonding   1/363, 1/404
    – chloride   2/624
    – cyanide   2/622
Hydrophobicity   1/398
Hydrotrope   1/424
Hydroxyalkane sulfonate   1/426
p-Hydroxybenzene sulfonate   1/449, 1/451
p-Hydroxybenzoate   1/449
Hydroxycarboxylic acid, aliphatic   1/363, 2/614
1-Hydroxyethane-1,1-diphosphonic acid
    (HEDP, DEQUEST 2010)   1/186
Hydroxyethyl-ethylenediaminetriacetic
    acid (HEDTA)   1/183
5-Hydroxy-3-indolylacetic acid   1/422, 2/797
Hydroxyisobutyric acid   1/363
α-Hydroxyisocaproic acid   1/170
Hydroxylamine   1/319, 2/705
Hydroxylaminedisulfonic acid (HADS)
    2/645, 2/647
Hydroxylaminetrisulfonic acid   2/647
Hydroxylysine   1/386, 1/388
Hydroxymethanesulfonic acid   1/130
4-Hydroxymethylbenzo-18-crown-6   1/88

Hydroxymethylene cation   1/38
D-Hydroxyphenylglycine   1/458
Hydroxyproline   1/266, 1/386
2-Hydroxypropyltrimethyl ammonium
    (2-HPTA)   2/766
Hypobromite   2/591
Hypophosphite   1/126, 2/680

i

Ibuprofen   1/454
2-Imidazolidinthion   2/693
Imino acid   1/390
Iminodiacetic acid   1/443
Iminodisulfonic acid   2/647
Immunoglobulin   1/240, 1/243, 1/269, 2/797
Indigocarmine   1/459
Indomethacin   1/454
Injection volume   1/5
Injector, loop   1/5
Inosin   1/178
Inositol   1/210, 2/720
    – 1,4-diphosphate   2/793
    – 4,5-diphosphate   2/793
    – monophosphate   1/210, 2/791
    – 1,2,5,6-tetraphosphate   2/793
    – 1,3,4,5-tetraphosphate   2/793
    – 1,4,5-triphosphate   2/791, 2/793
    – 1,5,6-triphosphate   2/793
Integrator, digital   1/6, 2/554
Interaction
    – adsorptive   1/398
    – Coulomb   1/398, 1/464
    – electrostatic   1/56, 1/298, 1/398, 481
    – interionic   1/466 ff
    – ion-dipole   1/122, 1/143
    – ion-water molecule   1/122
    – ion-molecule   1/467
    – non-ionic   1/29, 1/213
    – π-π   1/29, 1/176, 1/187, 1/364, 1/373, 1/410, 1/439
    – sorption   1/29
    – van-der-Waals   1/56, 1/276, 1/298, 1/398
Interface
    – electrospray   1/504, 1/542 ff, 2/607
    – particle beam   1/531 ff
    – thermospray   1/251, 1/534 ff
    – user   2/555
Interleukin-1   1/271

Inulin 1/234, 2/736
Inverted sugar 2/720, 2/754
Iodate 1/136, 1/504, 1/506, 1/546, 2/596
Iodide 1/61, 1/74, 1/133, 1/159, 1/413, 1/479, 1/498, 1/506, 1/516, 2/608, 2/621, 2/624, 2/733, 2/745, 2/818
Ion
 – cloud 1/466
 – exclusion process 1/359 f
 – interaction model 1/397
 – lipophilic 1/393, 1/400
 – mobility 1/462
 – pair 1/397, 1/457, 1/464
 – polarizable 1/3
 – radius 1/122
 – reflux 2/836
 – strength 1/89, 1/97, 1/466
 – suppression mode 1/441
 – surface-active 1/424 ff
 – surface-inactive 1/401, 1/409 ff
 – velocity 1/462
Ion chromatograph
 – computer controlled 2/555
 – process 2/651
 – schematics 1/5
Ion chromatography
 – advantage 1/7
 – method 1/3
Ion-exchange
 – capacity 1/34, 1/38, 1/57
 – function 1/3
 – model 1/394
 – process 1/3, 1/27, 1/279
Ionization
 – atmospheric pressure (API) 1/542
 – chemical (CI) 1/531
 – electron impact (EI) 1/531
 – electrospray 1/542
 – fast atom bombardment (FAB) 1/532
 – ion spray 1/543
 – thermospray 1/535
Ion-pair
 – formation 1/394
 – reagent 1/401
Iron 1/332, 1/334, 2/789
Isoascorbic acid 2/769
Isobutyric acid 1/374
Isocitric acid 1/76, 1/174, 1/448, 2/714
Isoelectric focussing (IEF) 1/255
Isoelectric point 1/99

Isoleucine 1/258, 1/387
Isomaltose 1/227, 2/754
Isomaltulose 1/228, 2/754
Isopropylethylphosphonic acid 1/73
Isopropylmethylphosphonic acid (IMPA) 2/621
Itaconic acid 1/373

j

Jet-Separator 1/531

k

1-Kestose 1/230, 2/738
6-Kestose 1/230
Ketocarboxylic acid 2/615
α-Ketoglutaric acid 1/364
α-Ketoisocaproic acid 1/170
α-Ketoisovaleric acid 1/376
Ketomalonate 1/73
Knox plot 1/23, 1/397
Kohlrausch square root law 1/464
Kraft process 2/810

l

Lactic acid 1/74, 1/168, 1/363, 1/376, 2/615, 2/682, 2/714, 2/716, 2/738, 2/786
Lactose 1/221, 1/227, 2/736, 2/745
Lactulose 1/227, 2/790
Lanthanide 1/325, 1/336, 1/342 ff, 1/502
Lanthanum 1/343
Latex
 – anion exchanger 1/54 ff
 – – overview 1/57 ff
 – cation exchanger 1/298 ff
 – particle 1/55, 1/299
Lauric acid 1/443
Lauryl sulfate
  see dodecyl sulfate
Lead 1/331, 1/333
Leading effect 1/15, 1/44
Leucine 1/258, 1/387
Leucrose 1/228
Ligand 1/336
 – choice 1/329
 – concentration
 – – complexing 1/329
 – – effective 1/328
 – – free 1/329
 – – total 1/329

– exchange 1/336
Limit of coverage 2/574
Limit of detection 2/574 f
Limit of determination 2/574, 2/577
 – for standard anions 2/588, 2/628
 – for standard cations 2/628
 – for transition metals 2/653
Lincomycin 1/489, 2/735
Lipoic acid 1/487, 2/770
Lithium 1/287, 2/619, 2/651
Loading capacity 1/60, 2/64
Lobry-de-Bruyn-von-Ekenstein rearrangement 1/208, 1/250
Low method 2/720
Luminol
 see 5-amino-2,3-dihydro-1,4-phthalazindione
Lutetium 1/343
Lysine 1/258, 1/387
Lysozyme 1/269

m

Magnesium 1/283, 2/598, 2/645
Malachite green 1/459
Maleic acid 1/73, 1/172, 1/372, 2/786
Malic acid 1/72, 1/174, 1/363, 2/714
Malonic acid 1/72, 1/73, 1/172, 1/174, 1/372, 2/786
Maltitol 2/762
Maltodecaose 2/725
Maltodextrin 2/745
Maltoheptaose 1/230, 2/725
Maltohexaose 1/230, 2/725
Maltononaose 2/725
Maltooctaose 2/725
Maltopentaose 1/230, 2/725
Maltose 1/206, 1/227, 2/722
 – oligomers 1/230
Maltotetraose 1/230, 2/722
Maltotriose 1/230, 2/722
Mandelic acid 1/171, 1/372, 830
Manganese 1/284, 1/324, 1/331, 2/619, 2/631, 2/778
Mannitol 1/209, 2/720, 2/754, 2/762, 2/790
Mannose 1/210, 1/218, 1/239, 2/750
Mass transfer
 – effect 1/20
 – resistance to 1/19 f, 1/22
Matrix elimination, inline 1/378, 2/639, 2/641, 2/659, 2/670

Mean, arithmetic 2/558
Measuring cell, amperometric 1/474, 1/496
Mecoprop
 see 2-(2-methyl-4-chlorphenoxy)-propionic acid
Median 2/551
Melibiose 1/227
Melizitose 2/755
Mellitate 1/448
Membrane
 – desalting 1/252, 1/537
 – Donnan 1/360
 – polyvinylenedifluoride (PVDF) 1/257, 2/798
 – reactor 1/325, 1/501
Mephobarbital 1/442, 1/453
Metanephrine 2/797
(4-Methacryloylamino)-benzo-15-crown-5 1/88
Methanedisulfonic acid 2/684
Methanesulfonic acid 1/170, 1/428, 2/682
Methanol 1/365, 1/381, 2/773
Metharbital 1/453
Methionine 1/258, 1/387, 2/794
Methohexital 1/453
Methylamine 1/320, 1/350
N-Methylaniline 1/356
Methylarsonic acid 1/528
3-Methylbutanol-1 1/381
2-(2-Methyl-4-chlorphenoxy)propionic acid (Mecoprop) 1/176
N-Methyldiethanolamine 1/302, 2/809
Methylenebisphosphonic acid 2/764
Methylene blue 1/459
4,4'-Methylenedianiline 1/356
(1S,2R)-(+)-N-Methylephedrine 1/453
3-O-Methylglucose 2/790
Methyl green 1/459
o-/p-Methylhippuric acid 1/171
N-Methyloctylammonium-p-toluene sulfonate 1/513
Methyl orange 1/451
Methylose 1/212
Methylphosphonic acid (MPA) 2/621
N-(Methylphosphono)glycine (Glyphosate) 1/515
2-Methylpropanol-1 1/381
Methyl red 1/451
Methyl sulfate 2/756
Methylsulfonate 1/73

Micro-extraction technique 2/657
Modifyer
 – inorganic 1/406
 – organic 1/400, 1/403 f
Molybdate 1/159, 1/524
Molybdenum sulfur cluster 1/418
Moment
 – central 2/551
 – first 2/551
 – second 2/551
 – zero 2/551
Monobromoacetic acid 1/171
Monobromomonochloroacetic acid 1/171
Monobutyl phosphate 1/176
Monochloroacetic acid 1/171
Monochloromethylenebisphosphonic acid 2/764
Monoethanolamine 1/289, 1/291, 1/296, 1/302
Monoethylamine 1/321, 1/325
Monoethylene glycol 1/381, 1/383
Monofluoro phosphate 2/671, 2/672, 2/710
Monoisopropyl sulfate 2/612
Monomethylamine 1/289, 1/321
Monomethyl phosphate 1/546
Monomethyl sulfate 1/546
Monosaccharides 1/206, 1/212 ff, 2/799
 – Alditol 1/244
Morphine 1/421, 1/457
Morpholine 1/288, 1/349, 2/634
Mucin 1/240
Myoglobin 1/269
Myristic acid 1/443

n

Naphthaline-2,3-dialdehyde (NDA) 1/515
Naphthaline-2-sulfonic acid 1/403
α-Naphthol 1/441
1,3,6-Naphtholenetrisulfonate 1/451
β-Naphthol orange 1/451
Naproxen 1/454
Negative ion mode 1/532, 1/537, 1/539, 1/543
Neodymium 1/344
Neo-kestose 1/230
Nernst equation 1/476
Neuraminic acid 1/274
Neuraminidase 1/246, 2/801
Neurotensin 1/273
Neutralization 2/825

Nickel 1/306, 1/332, 1/334, 2/613
Nicotine 1/457
Nicotinic acid 2/770
Nicotinic acid amide 2/770
Nile blue 1/459
Ninhydrin 1/389
Nitrate 1/65, 1/409, 1/411, 1/498, 2/588, 2/609, 2/621, 2/679, 2/733, 2/739
Nitrilotriacetic acid (NTA) 1/183, 2/701
Nitrilotri(methylenephosphonic acid) (NTP, DEQUEST 2000) 1/183
Nitrilotrisulfonic acid 2/647
Nitrite 1/475, 1/479, 1/498, 1/516, 2/588, 2/609, 2/611, 2/621, 2/733, 2/739, 2/784
2-Nitroaniline 1/356
4-Nitroaniline 1/356
Nitrogen oxide 2/622
1,9-Nonanediamine 1/354
Norepinephrine 1/422, 2/797
Norleucine 1/258
Normetanephrine 2/797
Nucleic acid 1/275 ff
Nucleoside 1/177
Nucleotide 1/177
 – phosphates 1/178
Nystose 1/230, 2/738

o

Obstruction factor 1/20
1,8-Octanediamine 1/354
Octanesulfonic acid 1/428
Octyl sulfate 1/427
Ohm's law 1/462
Olefin sulfonate 1/426
Oligonucleotide 1/275
 – secondary structures 1/276
 – use 1/275
Oligosaccharide 1/226 ff, 1/237
 – derived from glycoproteins 1/236 ff
 – – analysis 2/797 ff
 – – complex type 1/238
 – – data bank 1/254
 – – hybride type 1/238
 – – mannose type 1/238
 – – retention behavior 1/247
 – – separation and detection 1/246 ff
 – – sialylated 1/249
 – – structural analysis 1/240 ff
 – – structural isomer 1/246
 – – structure 1/238

– mass spectra 1/538
Orange I 1/451, 1/459
Orange II 1/451, 1/459
Orange G 1/459
Ornithine 1/258, 1/388
Orthophosphate 1/126, 1/195, 1/371, 2/588, 2/591, 707, 2/830
Orthophosphite 1/126, 1/195, 2/673
Orthosilicate 1/153, 1/371, 1/510, 2/591, 2/633, 2/653, 2/698
Osmate 2/604
Ostwald dilution law 1/150
Outlier 2/567
Outlier test 2/567 ff
 – acc. to Grubbs 2/568
 – acc. to Nalimov 2/567
Ovalbumine 1/240, 1/274
 – phosphorylation pattern 1/275
Overvoltage 1/497
 – passage 1/497
 – concentration 1/497
7-Oxabicyclo[2,2,1]-heptane-2,3-dicarboxylic acid (Endothall) 1/176
Oxalacetic acid 1/364
Oxalic acid 1/173, 2/674, 2/782, 2/810
Oxamic acid 1/170
Oxyhalide 1/78 f, 1/121
Ozonation, of drinking water 2/591

P

Palatinitol 1/228, 2/754
Palatinose
 see isomaltulose
Palmitic acid 1/443
Pamidronate
 see 3-amino-1-hydroxypropylidene-1,1-bisphosphonate
Papain 1/269
Papaverin 1/402, 1/424, 1/457
PAR
 see 4-(2-pyridylazo)resorcinol
Parameter
 – experimental retention-determining 1/400 ff
 – – of non-suppressed systems 1/145 ff
 – – of suppressed systems 1/123 ff
 – information 2/550
Peak 1/13
 – area 2/551, 2/552
 – asymmetry 1/14

– broadening 1/17
– Gauß curve 1/13, 1/15, 1/18
– height 2/551, 2/552
– form 1/14
– system 1/51, 1/84, 1/147, 1/153 ff, 1/513
– variance 1/18
– width 1/19
Pectin hydrolysate 1/214
D-Penicillamine 1/391
D-Penicillaminesulfonic acid 1/391
Penicillin V 2/771
n-Pentanol 1/381
Pentasaccharide 1/226
Pentitol 1/210
Perborate 1/184, 2/701
Perchlorate 1/74, 1/79, 1/94, 1/413, 2/599
Peroxohydroxide anion 2/669, 2/701
Peroxide bleach 2/701
Peroxoborate 2/701
Peroxodisulfate 1/149, 1/414
Peroxomonosulfate 1/149
Phase
 – aluminum oxide 1/95 ff
 – aminopropyl 1/411, 1/498
 – cyanopropyl 1/411, 1/498
 – crown ether 1/86 ff
 – – polyamide 1/88
 – – synthesis 1/88 f
 – cryptand 1/92 ff
 – multimode 1/445
 – octadecyl 1/4, 1/10
 – volume ratio 1/17, 1/31
Phenobarbital 1/423, 1/442, 1/450
Phenol 1/441, 1/475
Phenoxyacetic acid 1/542, 2/771
Phenoxycarboxylic acid 1/175, 1/542
Phenylalanine 1/258, 1/387
1-Phenyl-ethane-1,2-diphosphonic acid 1/186
1-Phenyl-ethane-1,2,2-triphosphonic acid 1/186
1-Phenyl-ethane-1,2,2-tris(P-methylphosphinic acid) 1/186
1-Phenyl-ethene-1-phosphonic acid 1/186
$trans$-1-Phenyl-ethene-2-phosphonic acid 1/186
Phenylglyoxylic acid 2/783
Phenylisothiocyanate (PITC) 1/383
Phenylphosphonate 1/449
Phenylthiohydantoine (PTH) 1/383

Phosphate
  see orthophosphate
Phosphite
  see orthophosphite
Phosphonium compound 1/438
2-Phosphonobutane-1,2,4-tricarboxylic
  acid (PBTC)1/185
Phosphonopropanetricarboxylic acid 1/189
Phosphorothioate 1/275
Photolysis 2/729, 2/827
Phthalate 1/73, 1/448
o-Phthaldialdehyde (OPA) 1/383, 1/390, 1/514
Phytic acid 1/448, 1/450, 2/743
4-Picoline 1/356
Piperazine 2/809
pK value
  – acetic acid 1/372
  – acrylic acid 1/372
  – adipic acid 1/372
  – aliphatic dicarboxylic acids 1/173
  – ammonium 1/317
  – aromatic monocarboxylic acids 1/172
  – arsenite 1/125
  – azide 2/787
  – barbital 1/423
  – barbituric acid 1/423
  – benzoic acid 1/150, 1/172
  – boric acid 1/368
  – *iso*-butyric acid 1/372
  – *n*-butyric acid 1/372
  – caffeine 2/731
  – carbohydrates 1/62, 1/205
  – carbonic acid 1/138
  – citric acid 1/372
  – cyanate 2/833
  – cyanide 2/833
  – 2,3-diaminopropionic acid (DAP) 1/282, 1/309
  – diethylamine 1/317
  – formic acid 1/372
  – fumaric acid 1/372
  – glycolic acid 1/372
  – hippuric acid 1/172
  – hydroxide substituted dicarboxylic acids 1/174
  – *p*-hydroxybenzoic acid 1/158
  – lactic acid 1/372
  – maleic acid 1/372
  – malic acid 1/174
  – malonic acid 1/173, 1/372
  – mandelic acid 1/172, 1/372
  – monoethylamine 1/317
  – monomethylamine 1/317
  – morpholine 1/317
  – nicotinic acid 1/151
  – orthophosphoric acid 1/138
  – oxalic acid 1/173, 1/372
  – phthalic acid 1/158
  – poly(butadiene-maleic acid) 1/304
  – propionic acid 1/372
  – pyruvic acid 1/372
  – saccharin 772
  – salicylic acid 1/151
  – succinic acid 1/151, 1/173, 1/372
  – tartaric acid 1/174, 1/372
  – triethanolamine 1/317
  – trimethylamine 1/317
Plasma etching 2/654
Plate
  – hight 1/17, 1/19, 1/24
  – – effective 1/18
  – – reduced 1/23
  – – theoretical 1/17, 1/23, 1/24
  – number 1/17, 1/19
  – – effective 1/18
Polyalcohol, cyclic
  see cyclitol
Poly(butadiene-maleic acid) (PBDMA) 1/282, 1/304, 1/323
Polydeoxyadenosine 1/276
Polydeoxyguanosine 1/277
Polyether
  – bicyclic 1/179
  – cyclic 1/86
Polyfructan 1/234
Polyphosphate 1/80, 1/164, 1/509, 1/529, 2/738
Polyphosphinic acid 1/186
  – retention behavior 1/187 f
  – stereoisomer 1/188
  – structural isomer 1/187
Polyphosphonic acid 1/186, 1/509, 2/641, 2/698
  – diastereomer 1/189
  – retention behavior 1/187 f
  – rotational isomer 1/188
  – stereoisomer 1/188
  – structural isomer 1/188
Polysaccharide 1/230 ff

Poly(thiometalate)   1/416
Polythionate   1/414, 2/646
Positive ion mode   1/532, 1/536, 1/543
Post-column
  − addition, of NaOH   1/210, 1/213
  − derivatization   1/182
Potassium   1/283, 2/598, 2/645
Potential
  − Galvani   1/497
  − − equilibrium   1/497
  − limit   1/496
  − standard   1/476
  − working   1/476
Praseodymium   1/344
Precision   2/551, 2/566
Pre-concentration
  − technique   2/628
  − via chelation   2/821
Preservative   2/746
Proline   1/258, 1/387
1,2-Propanediamine   1/353
Propanesulfonic acid   1/428
1,2,3-Propanetricarboxylate   1/448
2-Propanol   1/381
Propionaldehyde   1/486
Propionic acid   1/74, 1/362
n-Propylamine   1/353
Propylene glycol   1/382, 2/762
Protein   1/267 ff
  − deamidation   1/271
  − hydrolysis   1/392
  − microheterogeneous   1/273
  − recombinant   1/271
  − therapeutic   1/270
Proteoglycan   1/237
(1R,2R)-(-)-Pseudoephedrine   1/453
(1S,2S)-(+)-Pseudoephedrine   1/453
Pullulan   1/236
Pullulanase   1/236
Pulse sequence
  − for amino acids   1/494
  − with three potentials   1/483
  − with four potentials   1/484
  − multicyclic   1/487 f, 775
Pump, analytical   1/5
Purine base   1/178
Putrescine   1/344, 2/747
Pyridine   1/357
Pyridoxine   2/770
4-(2-Pyridylazo)resorcinol (PAR)   1/499

Pyrimidine base   1/178
Pyrogallic acid   1/441
Pyroglutamate aminopeptidase   1/272
Pyromellitate
  see 1,2,4,5-benzenetetracarboxylate
Pyrophosphate   1/163, 1/195, 2/769, 2/830
Pyrophosphoric acid   1/284
Pyruvic acid   1/73, 1/364, 1/376, 2/738, 2/786

q

Qualification
  − installation (IQ)   2/556
  − operation (OQ)   2/556
  − performance (PQ)   2/556
  − vendor (VQ)   2/556
Quantity
  − chromatographic   1/13
  − statistical   2/557
  − thermodynamic   1/30
Quinic acid   1/73, 1/74, 2/715
Quinine   1/457

r

Radiostrontium analysis   1/521
Raffinose   1/206, 1/228
Reagent
  − 9-anthryldiazomethane (ADAM)   1/520
  − Arsenazo I   1/501
  − Arsenazo III   1/340, 1/502
  − cerium(IV)   1/516
  − delivery   1/500
  − o-dianisidine (ODA)   1/505
  − 1,5-diphenylcarbazide (DPC)   1/508
  − ion-pair   1/4
  − iron(III) nitrate   1/510
  − Luminol   2/632
  − naphthaline-2,3-dialdehyde (NDA)   1/515
  − PAR   1/339, 1/499
  − PAR/ZnEDTA   1/332
  − o-phthaldialdehyde (OPA)   1/383, 1/514
  − sodium molybdate   1/510
  − Tiron   1/341, 1/501
Recovery   2/572
Reddening agent   2/741
Regenerant
  − delivery   1/113 f
Regeneration

- continuous 1/108, 1/110, 1/113, 1/115, 1/311, 1/313, 1/367, 1/399 f
- periodic 1/104, 1/311, 1/399

Regression
- coefficient 2/561
- linear 2/561
- weighed 2/567

Relaxation
- constant 1/469
- effect 1/468
- time 1/468

Residual 2/566
- analysis 2/566
- standard deviation 2/563, 2/565

Resin
- divinylbenzene 1/393
- Dowex 1x10 1/310
- ethylvinylbenzene/divinylbenzene 1/35, 1/61, 1/282 f
- functionalization 1/36, 1/37 f
- – amination 1/37
- – chloromethylation 1/37
- gel-type 1/35
- hydroxyethylmethacrylate 1/49
- iminodiacetic acid 2/821
- ionic form 1/36
- laurylmethacrylate 1/83
- macroreticular 1/35, 1/37
- mesoporous 1/37
- microporous 1/35
- polyamide crown ether 1/88
- poly(benzo-15-crown-5) 1/307
- polymethacrylate 1/35, 1/47 f, 1/296
- polyvinyl 1/35, 1/47 f, 1/296
- polyvinylpyrrolidone 2/599
- porosity 1/35
- shrinking process 1/36
- swelling process 1/36
- styrene/divinylbenzene 1/35, 1/280
- XAD-1 1/38

Resistivity 1/462
Resolution 1/15
Resorcinol 1/441
Result 2/559
Retention
- temperatur dependance 1/32
- time
- – gross 1/13
- – solute 1/13

Rhamnose 1/213, 2/726, 2/750, 2/774
Rhodamine B 1/459
Riboflavin 2/770
Ribonuclease B 1/240, 1/269
- crystal structure 1/272
D-Ribose 1/221, 2/750
α-D-Ribose-1-phosphate 1/226
Rubidium 1/287
Ruhemann's purple 1/389

S

Saccharin 2/692, 2/731, 2/752, 2/762
Sarcosine 1/388
Salicylic acid 2/688
Saliva analysis 2/781
Samarium 1/344
Sample
- loading capacity 1/65
- preparation 2/550, 2/823 ff
- – for amino acid analysis 1/391 f
- – on-column 1/449
- storage 2/550
Sampling 2/549
Scatter 2/559
Schiff base 1/205
Schöninger flask 2/769
SDS Polyacrylamide gel electrophoresis (SDS-PAGE) 1/255
Secobarbital 1/442, 1/453
Selectivity 1/8, 1/16
- coefficient 1/28, 1/38, 1/326, 1/472
- of latexed anion exchangers, overview 1/57 ff
- solvent influence 1/143
Selenate 1/126
Selenite 1/73, 1/126
Selenometalate 1/416
Sensitivity 1/473, 2/560
- of the detection system 2/560
- of the method 2/561, 2/588
Separator column
- Aminex 50W-X4 1/365
- Aminex HPX-85H 1/379
- Aminex HPX 87H 2/723
- AminoPac NA-1 1/386
- AminoPac PA-1 1/385
- AminoPac PA-10 1/257, 1/385
- AminoPac PC-1 1/385
- AN1 1/41
- AN2 1/41
- AN300 1/41

- BTC 2710   1/385
- CarboPac MA1   1/210, 1/245, 2/762
- CarboPac PA1   2/62, 1/129, 1/136, 1/169
- CarboPac PA10   1/212, 1/242
- CarboPac PA20   1/212, 1/219
- CarboPac PA-100   1/233, 1/539
- DNAPac PA-100   1/273, 1/276
- efficiency   1/17 ff
- ExcelPak ICS A23   1/40
- Fast-Sep Anion   1/64
- Fast-Sep Cation   1/299
- Hypersil 5 MOS   1/429, 2/710
- IC Pak A   1/47, 1/146
- Inertsil ODS II   1/438
- ION-100/110   1/52
- ION-200/210   1/296, 1/519
- ION-300   1/364
- IonPac AS1   1/57
- IonPac AS2   1/58
- IonPac AS3   1/58, 1/126
- IonPac AS4   1/60
- IonPac AS4A   1/60
- IonPac AS4A-SC   1/61
- IonPac AS5   1/61, 1/160
- IonPac AS5A   1/69
- IonPac AS6
   see CarboPac PA1
- IonPac AS7   1/63, 1/163, 1/182, 1/527
- IonPac AS9   1/65
- IonPac AS9-SC   1/66, 1/162, 2/593
- IonPac AS9-HC   1/66, 1/504, 2/609
- IonPac AS10   1/77, 1/125, 2/641
- IonPac AS11   1/70, 1/164, 1/196, 2/715
- IonPac AS11-HC   1/73
- IonPac AS12A   1/78, 1/126, 1/169, 2/812
- IonPac AS14   1/42
- IonPac AS14A   1/45
- IonPac AS15   1/45, 1/199
- IonPac AS15A   1/46
- IonPac AS16   1/79, 2/601
- IonPac AS17   1/80, 2/590
- IonPac Cryptand A1   1/93
- IonPac CS1   1/280
- IonPac CS2   1/281, 1/341
- IonPac CS3   1/298, 1/341, 1/345
- IonPac CS5   1/62, 1/332
- IonPac CS5A   1/338, 2/620
- IonPac CS10   1/283, 1/300, 1/324
- IonPac CS11   1/302
- IonPac CS12   1/282, 2/599
- IonPac CS12A   1/285
- IonPac CS14   1/288, 1/346, 1/493
- IonPac CS15   1/291, 1/351, 2/634,
- IonPac CS16   1/292, 2/806
- IonPac CS17   1/295, 1/346
- IonPac ICE-AS1   1/361
- IonPac ICE-AS6   1/361, 2/615
- IonPac ICE-Borate   1/369
- IonPac NS1   1/446
- LCA A01   1/39
- LCA A03   1/48
- LCA A04   1/39
- LCA K01   1/281
- LCA K02   1/296, 1/332
- length   1/123
- LiChrosil IC CA   1/306
- LiChrosorb RP18   1/411
- maintenance   2/835 f
- MCI Gel SCA04   1/51
- MCI Gel SCK01   1/281
- Metrosep Anion Dual 1   1/49, 1/163
- Metrosep Anion Dual 2   1/49
- Metrosep Anion SUPP 1   2/663
- Metrosep Anion SUPP 4   1/54
- Metrosep Anion SUPP 5   1/54
- Metrosep Cation 1-2   1/304, 2/599
- Metrosep Organic Acids   2/689
- Mikropac MCH10   1/443
- Novosep A-1   1/107
- Nucleosil 5 SA   1/303
- Nucleosil 10 SA   1/503
- Nucleosil 10 Anion   1/86
- Nucleosil 10 C_8 1/431
- OmniPac PAX-100   1/445
- OmniPac PAX-500   1/445, 2/554, 2/750
- OmniPac PCX-100   1/303, 1/445, 2/768
- OmniPac PCX-500   1/445, 2/794
- ORH 801   1/364
- Polygosil-60-D-10 CN   1/412
- Polyspher CHCA   2/726
- Polyspher IC AN-1   1/49
- Polyspher OA-HY   2/615
- ProPac PA1   1/271
- ProPac SAX-10   1/274, 2/738
- ProPac SCX-10   1/269
- ProPac WCX-10   1/269
- PRP-1   1/401
- PRP-X100   1/39

- PRP-X200   1/281
- PRP-X300   1/364
- Rezex RPM Monosaccharide   1/211
- SAR-40-0.6   1/157
- Shimpack IC-A1   1/48, 1/149
- Shimpack IC-C1   1/281, 1/322
- Shimpack IE   2/716
- Spherisorb A5Y   1/98
- Spherisorb ODS 2   2/769
- stability   1/9
- Star Ion A300 IC Anion   1/41
- Star Ion A300 HC   1/42
- Supersep Anion   1/480
- TSK Gel 620 SA   1/471
- TSK Gel IC Cation   1/281
- TSK Gel IC Cation SW   1/303
- TSK Gel IC-PW   1/47
- TSK Gel IC-SW   1/84
- Universal Anion   1/50
- Universal Cation   1/307
- Vydac 300 IC 405   1/83, 1/163
- Vydac 302 IC 4.6   1/83
- Vydac 400 IC 405   1/303, 1/320
- Vydac C8 (208TP5451)   1/490
- Waters C18 Radial-Pak   1/412
- Wescan 269-001   1/84
- Wescan 269-029   1/147
- Wescan 269-031   1/152
- Zorbax-NH$_2$ 1/413

Serine   1/258, 1/387
Serotonin   1/422, 2/797
Serum analysis   2/783
Shikimic acid   1/373
Sialic acid
   see N-acetylneuraminic acid
Silicate
   see orthosilicate
Silvex
   see 2-(2,4,5-trichlorophenoxy)-propionic acid
Sodium   1/283, 2/598, 2/635, 2/645
Soil analysis   2/619
Sorbic acid   2/76, 1/442, 2/746
Sorbitol   1/209, 1/211, 2/718, 2/726, 2/754, 2/762
Speciation   1/523, 2/630
- arsenic   1/526
- chromium   1/524, 1/526
Spermidine   1/344, 2/747
Spermine   1/344

Stachyose   1/206, 1/228
Standard
- addition   2/572 f
- external   2/571
- internal   2/570
Standard deviation   2/558
- method   2/563, 2/565
- – relative   2/563, 2/565
- residual   2/563, 2/565
- theoretical   2/558
Stannate   1/524
Stearic acid   1/442
Strontium   1/287, 1/502, 2/619
Strychnine   1/457
Student factor   2/559
Succinic acid   1/74, 1/172, 1/363, 2/615, 2/688
Succinylcholine   2/767
Sucrose   1/226, 2/718, 2/745
- mass spectrum   1/538
Sugar
- alcohol   1/209 ff
- substitute   2/753
Sulfadiazin   1/457
Sulfadimethoxin   1/457
Sulfamate   2/682
Sulfamerazin   1/457
Sulfamethazin   1/457
Sulfanilamide   1/457
Sulfanilic acid   1/455
Sulfate   1/65, 2/588, 2/644, 2/702, 2/771
Sulfathiazole   1/457
Sulfide   1/63, 1/126, 1/475, 1/479, 2/810
- free   1/127
Sulfisoxazole   1/457
Sulfite   1/65, 1/130, 1/370, 2/722, 2/728, 2/761
- purity determination   1/131
- stabilization   1/130
α-Sulfofatty acid methyl ester   1/424
Sulfonamide   1/455, 1/542
Sulfonium compound   1/437
5-Sulfosalicylate   1/451
Sulfosuccinic acid ester   1/431
Sulfur dioxide   2/622
Sulfur-nitrogen compound   2/645, 2/647
Supporting electrolyte   1/102, 1/475
Suppressor
- capacity   1/112

− column   1/2, 1/103 ff, 1/310, 1/366, 1/399
− − regeneration   1/104, 1/311, 1/399
− DS-Plus™   1/105
− ERIS   1/105
− hollow fiber membrane
− − for anion exchangechromatography   1/108 ff
− − for ion-exclusionchromatography   1/367
− − for ion-pair chromatography   1/399
− − for cation exchange chromatography   1/311
− − regeneration   1/110, 1/311, 1/367, 1/399
− − schematics   1/109
− micromembrane
− − DCR™ mode   1/115
− − for anion exchange chromatography   1/111 ff
− − for ion-exclusion chromatography   1/367
− − for ion-pair chromatography   1/400
− − for cation exchange chromatography   1/312
− − regeneration   1/113, 1/115, 1/313, 1/367, 1/400
− − schematics   1/111
− monolithic
− − for anion exchange chromatography   1/120
− − for cation exchange chromatography   1/318
− − schematic   1/120
− MSM   1/104
− reaction   1/103, 1/366
− self-regenerating
− − converter mode   1/315 f
− − for anion exchange chromatography   1/115 ff
− − for cation exchange chromatography   1/313 f
− − operation modes   1/118 f, 1/314
− system   1/103
− void volume   1/104, 1/106, 1/109, 1/111
Surfactant   2/698 ff
− anionic   1/542, 2/698, 2/709
− cationic   2/698
Sweetener   1/228, 2/752

t

Tailing effect   1/14, 1/44
Tartaric acid   1/72, 1/172, 1/363, 2/688, 2/714
Tartronic acid   1/174
Taurine   1/258, 1/388
Terbium   1/344
Terephthalate   1/449
Tetraacetyl-ethylenediamine (TAED)   1/184, 2/701
Tetrabutylammonium   1/355, 1/421, 1/434
Tetradecylpyridinium   1/436
Tetradecyl sulfate   1/427
Tetraethylammonium   1/355
Tetraethylenepentamine   1/348
Tetrafluoroborate   1/131, 2/672, 2/833
Tetraheptylammonium   1/355
Tetrahexylammonium   1/355
Tetrametaphosphate   1/166, 2/740
Tetramethylammonium   1/355, 1/420
Tetrapentylammonium   1/355, 1/434
Tetrapolyphosphate   1/166 f, 1/195, 2/740
Tetrapropylammonium   1/355, 1/420, 1/434
Tetrasaccharide   1/226
Tetrathionate   1/414
Theobromine   1/455
Theophyllin   1/455
Thiamine   2/770
Thiamylal   1/453
Thiocyanate   1/61, 1/74, 1/159
Thiofluor
   see N,N-dimethyl-2-mercaptoethylamine-hydrochloride
Thioglycolic acid   1/408
Thiometalate   1/416
Thiomolybdate   1/417
Thiosulfate   1/61, 1/74, 1/159, 1/475, 1/479, 1/516
Thorium   1/339, 1/503
Threonine   1/258, 1/387
Thulium   1/344
Thymidine   1/178
− 5'-monophosphate   1/181
Thymol   1/441
Thymol blue   1/459
Tiron
   see 4,5-dihydroxy-1,3-benzene-disulfonic acid-disodium salt
Tissue plasminogen activator (tPA)   1/240
Tolmetin   1/454

Toluene sulfonate 1/424, 1/427, 1/451, 2/703
p-Toluenesulfonic acid 1/403
3-Toluidine 1/356
Transferrin, human serum 1/239, 1/254, 1/273
Transglucosidase 1/232
Transition metal 1/325 ff, 2/630, 2/690, 2/778, 2/821
 – limit of detection 2/653
 – separation 1/326 ff
Transport number 1/465
Trehalose 1/226, 1/228, 2/726, 2/774
Triangulation 2/551, 2/553
Tribromide 1/504, 2/596
Tribromoacetic acid 1/170 f
Tributylmethylammonium 1/355, 1/434
Tricarballylate 1/73
Trichloroacetic acid 1/171
2,4,5-Trichlorophenoxyacetic acid (2,4,5-T) 1/176
2-(2,4,5-Trichlorphenoxy)propionic acid (Silvex) 1/176
Triethanolamine 1/296, 1/302, 2/704
Triethylamine 1/321, 1/325
Triethylene glycol 1/383
Triethylenetetramine 1/348
Trifluoroacetate 1/76, 2/756
Triiodide 1/507, 2/596
Trimesate 1/449
Trimetaphosphate 2/740, 2/830
Trimethoprim 1/457
Trimethylamine 1/289, 1/321
Triphenylarsonium compound 1/439
Triphenyl-mono(β-jonylidene-ethylene)-phosphonium chloride 1/438
Triphenylphosphonium compound 1/439
Tripolyphosphate 1/163, 1/195, 2/698, 2/708, 2/740, 2/769
Trisaccharide 1/226
Tropaeolin O 1/459
Tryptophane 1/265, 1/388, 1/392
t-Test 2/569
Tungstate 1/73, 1/159
Turanose 1/227
Two-phase titration 1/436, 2/698
Tyrosine 1/258, 1/387

u

Ultrafiltration 1/392
Ultracentrifugation 1/392

Uranium 1/336, 1/339, 1/503
Uranyl cation 1/340, 1/514
Uridine 1/178
 – 5′-monophosphate 1/181
Urine analysis 2/782

v

Valency 1/122, 371
Valeric acid 1/362
Validation 2/556
Valine 1/258, 1/260, 1/387
Vanadate 1/524
Vanadium 1/336
van Deemter
 – curves 1/24 f
 – equation 1/19, 1/22
 – theory 1/19 ff, 1/24
van't Hoff plot 1/32
Variance
 – homogeneity 2/561, 2/566
 – inhomogeneity 2/561, 2/566
4-Vinylbenzo-18-crown-6 1/88
Vinylsulfonic acid 2/689
Vitamin, water-soluble 1/455, 2/769
V-Mask 2/585
Voltammetry 1/475
 – basics 1/475 ff
 – cyclic 1/481
 – hydrodynamic 1/475
 – pulsed 1/475
Voltammogram 1/476
 – cyclo
 – – of formaldehyde 1/486
 – – of glucose 1/481
 – – of propionaldehyde 1/486
Volume
 – breakthrough 2/634
 – dead 1/17
 – exclusion 1/360
 – totally permeated 1/360

w

Water analysis
 – conditioned water 2/634
 – cooling water 1/294, 2/640
 – drinking water 2/589
 – feed water 2/627
 – formation water 2/803

- ground water 2/598
- ice 2/598
- landfill leachate 1/532, 2/614
- rain water 2/598
- sea water 2/615, 2/786
- seepage water 2/614
- snow 2/598
- surface water 2/608
- swimming pool water 2/598
- ultra-pure water 2/626, 2/631, 2/651
- wastewater 2/609

Weight distribution coefficient 1/29
Western transfer 1/257
Working range 2/561

x

Xanthine 1/455
β-1,4-Xylan 1/539
Xylene sulfonate 1/424, 1/427, 1/533
Xylitol 1/211, 2/720
Xylose 1/212, 1/218, 1/239, 2/750

y

Ytterbium 1/343

z

Zeolith A 2/698
Zinc 1/306, 1/331, 1/334, 2/613, 2/789
Zirkonium 1/340
ZnEDTA 1/332, 1/500